1500 1000 500
25 0-21 0-29 0-25

Plasma Physics for Nuclear Fusion

Kenro Miyamoto

Plasma Physics for Nuclear Fusion

Revised edition

The MIT Press
Cambridge, Massachusetts, and London, England

PLASMA PHYSICS FOR NUCLEAR FUSION by Kenro Miyamoto
Originally published in Japanese by Iwanami Shoten, Publishers, Tokyo, 1976

This book was set in Lasercomp Times Roman by Asco Trade Typesetting Limited, Hong Kong, and printed and bound by Halliday Lithograph Corporation in the United States of America.

Library of Congress Cataloging-in-Publication Data

Miyamoto, Kenrō, 1931–
[Kakuyūgō no tame no purazuma butsuri. English]
Plasma physics for nuclear fusion/Kenro Miyamoto.—Rev. ed.

p. cm.
Translation of: Kakuyūgō no tame no purazuma butsuri.
Includes bibliographies and index.
ISBN 0-262-13237-0 0-262-63117-2 (pbk.)
1. Plasma (Ionized gases) 2. Controlled fusion. I. Title.
QC718.M5913 1989 88-12879
621.48′4—dc 19 CIP

Contents

Preface to the Second Edition

My purpose in preparing this edition was to update the text to include recent results on the confinement and heating of high-temperature plasmas. The first edition was published in 1976 by Iwanami Shoten, and the English version was published in 1980 by The MIT Press; a Chinese translation was published in 1981 by Science Publishing Company, Beijing. Since then, remarkable developments have occurred in nuclear fusion research. Now intensive experiments with large tokamaks aim at scientifically demonstrating fusion reactors, and several alternative approaches to tokamaks are being explored. There have also been great advances in understanding the physics of wave heating. I have tried to present these recent developments in a concise way in this edition, especially through a complete rewriting of chapter 16.

Part III presents the analyses of wave phenomena and velocity-space instabilities, using the kinetic model. This method appears to be quite different from the magnetohydrodynamic method given in pt. II, but note that both methods are based upon the same Boltzmann equation. The methods are based upon different assumptions; the two will be seen to be complementary rather than competitive.

Chapter 10 describes waves in a cold plasma. The cold-plasma approximation is simple, while still retaining the important wave properties. In ch. 11 electrostatic waves in a hot plasma are treated. The electric field of the waves can be expressed by the scalar potential, which simplifies the analysis. Electrostatic waves play an important role in low-beta plasmas. Velocity space instabilities are described in ch. 12. Investigation of various magnetohydrodynamic and velocity-space instabilities is the most important subject of plasma confinement. The analysis of general electromagnetic waves in a hot plasma is introduced in ch. 13.

Part IV includes discussion of *heating*, *diagnostics*, and *confinement*. These topics are most important in plasma experiments aiming at nuclear fusion. Various kinds of heating mechanisms and plasma diagnostics are described in chs. 14 and 15, respectively. The discussions given in these chapters are based on the related theory as developed in the previous chapters. In ch. 16, confinement experiments are reviewed and various types of confinement devices are explained.

Sections which may be skipped without loss of continuity are marked by asterisks in the text, as well as in the table of contents. This does not mean, however, that the marked sections are not important. They are marked to serve the readers who want to read through this book as quickly as possible. References to the literature are limited due to the introductory nature of this book. The mks rationalized system of units is used throughout unless otherwise specified.

I express my sincere gratitude to Professor Koji Husimi, former director of the Institute of Plasma Physics, Nagoya University, for his continuous encouragement in preparing this book. It is a pleasure to acknowledge the many stimulating discussions with my colleagues in the Institute of Plasma Physics.

Preface to the First Edition

My primary objective in writing this book has been to provide a basic text for students beginning the study of plasma physics as it relates to controlled nuclear fusion. Prerequisites for understanding the material in this book are courses in dynamics, electromagnetic theory, and mathematical analysis; the present work is therefore appropriate for use in graduate or senior undergraduate courses. I also expect that it will be a useful reference for scientists and engineers working in the relevant fields. Work on plasma confinement and heating continues to progress steadily toward the goal of a controlled nuclear fusion reaction. Accordingly, I have here attempted to present a comprehensive and unified description of plasma physics, including the results of recent research.

The 16 chapters of this book can be divided into four parts, which are: I, Fundamentals (chs. 1–5); II, The magnetohydrodynamic description of a plasma (chs. 6–9); III, Kinetic descriptions of waves and instabilities (chs. 10–13); IV, Heating, diagnostics, and confinement (chs. 14–16).

In ch. 1 Lawson's condition is introduced and the concept of the fusion reactor is presented. The properties of several magnetic configurations used to confine the plasma are explained in ch. 2. The behavior of charged particles moving in a magnetic field is studied in ch. 3, where the orbit of a particle is analyzed by applying the quite general and fairly accurate drift approximation. In ch. 4, Coulomb collisions between charged particles are described, and the changes of momentum and energy due to Coulomb collisions are obtained. Then, in ch. 5, the velocity distribution function is introduced and the Boltzmann equation is formulated, from which we derive the rate of change of the distribution function in phase space. This equation is the starting point of plasma kinetic theory. As an example, application to the diffusion of a plasma in velocity space is treated for the case of a mirror magnetic field.

A plasma behaves as a fluid under some conditions. In ch. 6, the magnetohydrodynamic (MHD) equations are derived by taking appropriate averages of the Boltzmann equation. The MHD equations describe the "fluid model" of a plasma in a magnetic field. In ch. 7 the properties of the equilibrium plasma in tokamak, stellerator, and mirror configurations are described. Even if the plasma satisfies the equilibrium condition, it diffuses along or across the magnetic field. Diffusion due to binary collisions (classical or neoclassical diffusion) are considered in ch. 8. Pseudoclassical and anomalous diffusion due to the fluctuations within the plasma are also examined. In ch. 9 the characteristics of magnetohydrodynamic instabilities are studied and stability conditions are discussed.

Part III presents the analyses of wave phenomena and velocity-space instabilities, using the kinetic model. This method appears to be quite different from the magnetohydrodynamic method given in pt. II, but note that both methods are based upon the same Boltzmann equation. The methods are based upon different assumptions; the two will be seen to be complementary rather than competitive.

Chapter 10 describes waves in a cold plasma. The cold-plasma approximation is simple, while still retaining the important wave properties. In ch. 11 electrostatic waves in a hot plasma are treated. The electric field of the waves can be expressed by the scalar potential, which simplifies the analysis. Electrostatic waves play an important role in low-beta plasmas. Velocity space instabilities are described in ch. 12. Investigation of various magnetohydrodynamic and velocity-space instabilities is the most important subject of plasma confinement. The analysis of general electromagnetic waves in a hot plasma is introduced in ch. 13.

Part IV includes discussion of *heating*, *diagnostics*, and *confinement*. These topics are most important in plasma experiments aiming at nuclear fusion. Various kinds of heating mechanisms and plasma diagnostics are described in chs. 14 and 15, respectively. The discussions given in these chapters are based on the related theory as developed in the previous chapters. In ch. 16, confinement experiments are reviewed and various types of confinement devices are explained.

Sections which may be skipped without loss of continuity are marked by asterisks in the text, as well as in the table of contents. This does not mean, however, that the marked sections are not important. They are marked to serve the readers who want to read through this book as quickly as possible. References to the literature are limited due to the introductory nature of this book. The mks rationalized system of units is used throughout unless otherwise specified.

I express my sincere gratitude to Professor Koji Husimi, former director of the Institute of Plasma Physics, Nagoya University, for his continuous encouragement in preparing this book. It is a pleasure to acknowledge the many stimulating discussions with my colleagues in the Institute of Plasma Physics.

March 1979

I FUNDAMENTALS

1 Nuclear Fusion

1.1 Energy Demand and Resources

As a consequence of the development of civilization, the worldwide demand for energy has been increasing rapidly. The energy consumption of mankind is estimated or predicted as follows:

1850	0.3–0.5 Q/century
1851–1950	4 Q/century
1951–2000	≈ 15 Q/half-century

The consumption of energy has been growing rapidly since the nineteenth century, with an estimated rate of about 0.25 Q/year in 1975. 1 Q is 10^{18} Btu (British thermal units) or 1.05×10^{21} Joules.

The estimated world reserves of fossil fuels—petroleum, coal, and other—are about 80 Q. If current trends continue, these resources will be almost depleted within a century. Thus nuclear energy, already applied on a commercial scale, is recognized as the most important near-term replacement for fossil fuels.

Fission reactors, which utilize the energy released in fission chain reactions, are already used as practical power plants. Most currently operational power reactors rely on fission of U^{235}, triggered by thermal neutrons. The percentage of this isotope contained in natural uranium, however, is only 0.7%. Based upon the world reserves of uranium, the total amount of energy obtainable from nonbreeding fission reactors is estimated to be about 3 Q, which is far from equal to the long-term world energy demand.

The nuclides U^{238} and Th^{232} may be converted into new fissionable nuclear fuel by the absorption of fast neutrons (> 1 MeV) as follows:

$$ {}_{92}U^{238} + n \rightarrow {}_{92}U^{239} \rightarrow {}_{93}Np^{239} \rightarrow {}_{94}Pu^{239} $$
$$ \searrow e \qquad \searrow e $$

(neutron absorption) (β decay)

$$ \nearrow e \qquad \nearrow e $$
$$ {}_{90}T^{232} + n \rightarrow {}_{90}Th^{233} \rightarrow {}_{91}Pa^{233} \rightarrow {}_{92}U^{233} $$

Fast breeder reactors use the fission of U^{235} not only as the energy source but also as the source of fast neutrons, which collide with U^{238} or Th^{232} to produce the new nuclear fuels Pu^{239} and U^{233} as the results of reactions given above.[1]* Intensive development efforts have been and still are

* References will be found at the end of each chapter.

expended to make this type of reactor technically and economically feasible. If fast breeder reactors become available, the fission-reactor contribution to the energy reserves will rise to an estimated 350 Q. Even so, if a time span of several hundred years or more is considered, this will still not be a sufficient reserve. Furthermore—and this is perhaps the most important constraint—there seems to be no safe way, at present or in the foreseeable future, to dispose of the vast quantities of radioactive waste associated with a 300–400 Q fission program. Nor does there seem to be any way, in such a large program, to protect against proliferation of and trafficking in plutonium.

1.2 Fusion Reactions

A possible alternative and successor to the fast breeder reactor is a system in which useable energy results from the fusion of light nuclides such as deuterium, tritium, helium-3, and lithium. Deuterium exists abundantly in nature; for example, it comprises 0.015 atom percent of the hydrogen in sea water. This amount of deuterium could supply enough energy for 4×10^9 Q. The deuterium is not maldistributed. The fusion process itself does not leave long-lived radioactive products, and the problem of radioactive-waste disposal is much less serious than that for fission reactors. However, this immense reserve of nuclear energy is not yet usable; in fact controlled fusion is still in the stage of basic research, although fusion energy was released in an uncontrolled, explosive manner by the hydrogen bomb in 1951.

Nuclear reactions of interest for fusion reactors are as follows:

(1) $D + D \rightarrow T(1.01\ \text{MeV}) + p(3.03\ \text{MeV})$
(2) $D + D \rightarrow He^3(0.82\ \text{MeV}) + n(2.45\ \text{MeV})$
(3) $D + T \rightarrow He^4(3.52\ \text{MeV}) + n(14.06\ \text{MeV})$
(4) $D + He^3 \rightarrow He^4(3.67\ \text{MeV}) + p(14.67\ \text{MeV})$
(5) $Li^6 + n \rightarrow T + He^4 + 4.8\ \text{MeV}$
(6) $Li^7 + n \rightarrow T + He^4 + n - 2.5\ \text{MeV}$

A binding energy per nucleon is smaller in very light or very heavy nuclides and largest in the nuclides with atomic mass numbers around 60. Therefore, large amount of the energy can be released when the light nuclides are fused. The cross sections of the reactions (1) + (2), (3), and (4) are shown in figs. 1.1 and 1.2.[2]

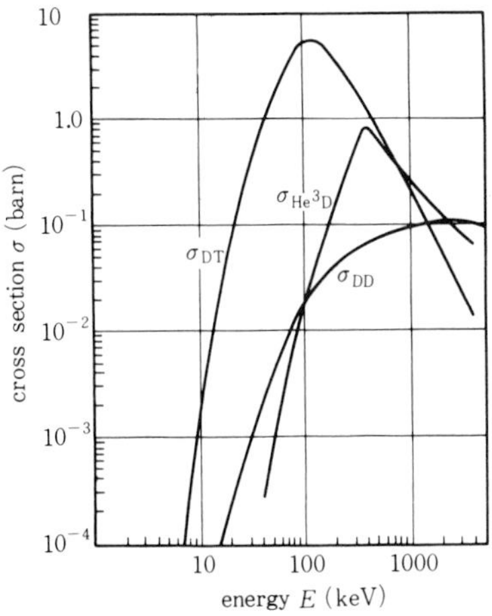

Fig. 1.1
Fusion cross sections plotted against the kinetic energy of relative motion. σ_{DD} is the sum of the cross sections of reactions (1) and (2). One barn is 10^{-24} cm^2.

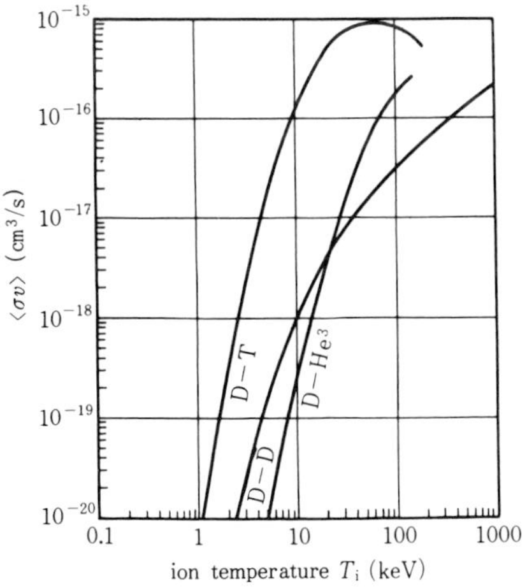

Fig. 1.2
The expected value of $\langle\sigma v\rangle$ when the ions obey the Maxwell distribution at the temperature T_i. $\langle\sigma v\rangle_{DD}$ is the sum of the cross sections of reactions (1) and (2).

The D-T reaction has the largest cross section of those shown, and the amount of released energy is also largest. Taking account of the difficulties of plasma confinement, the D-T reaction should be the one undertaken in the first development stages, although D-D reactors should be aimed at ultimately. For a D-T reactor, a lithium blanket surrounding the plasma moderates the fast neutrons, converting their kinetic energy to heat. Furthermore, the lithium blanket breeds tritium due to the reaction (5) and (6). Figure 1.3 shows an example of electric power plant based upon such a controlled fusion reactor. Lithium at high temperature comes out from the blanket, gives up its heat to liquid potassium in a heat exchanger; a potassium turbine generates the electric power. The potassium-turbine exhaust, itself at a high temperature, is circulated in a second heat exchanger to generate steam for a steam-turbine generator. The technologies concerning potassium heat exchangers and turbines are the same as those required for similar components used in fission breeder reactors. Abundances of Li^6 and Li^7 are 7.5% and 92.5% respectively. Reactions between lithium and neutrons are endothermic (-1.97 MeV) on the average. Since the D-T reaction releases 17.58 MeV, a lithium atom may be considered as a 15.6 MeV energy source. The world reserve of lithium is estimated to be $(8\text{–}9) \times 10^6$ tons; this is equivalent to about 1700 Q. It should be noted that the lithium content of the seas, of total volume of 1.37×10^9 km^3, is 0.17 g/m^3 (while the uranium content is 0.003 g/m^3), and that it is possible to extract this lithium from the sea water. In the next section we begin our study of the fusion reactor core with a discussion of necessary conditions and plasma parameters.

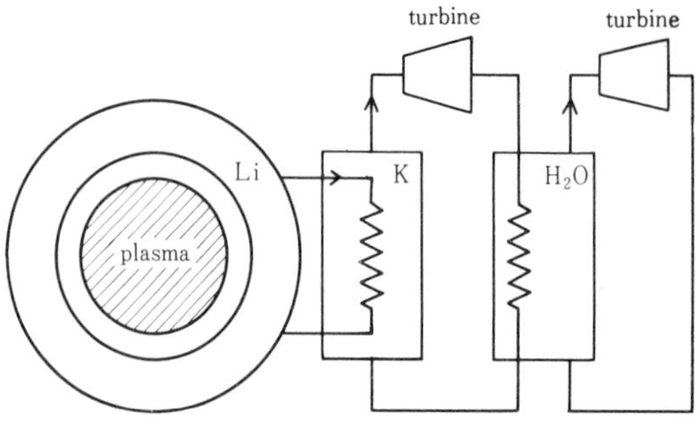

Fig. 1.3
An electric power plant based upon a fusion reactor.

1.3 Lawson's Condition and the Ignition Condition

As the temperature of a material is raised, its state changes from solid to liquid and then to gas. If the temperature is elevated further, an appreciable number of the gas atoms are ionized. The word "plasma" is used in physics to designate the high-temperature gaseous state in which the charge numbers of ions and electrons are almost the same and charge neutrality is satisfied over a physical dimension larger than a Debye length (see section 4.7). Plasmas are found in fluorescent and neon lamps. Examples of plasmas in nature are the aurora, the ionosphere, the Van Allen belt, and the corona of the sun.

A plasma for a fusion reactor should have very high temperature, so that the ions have velocities large enough to overcome their mutual Coulomb repulsion, so that collision and fusion can take place. The energy released must be larger than that necessary to maintain the hot plasma.

Let us denote the plasma density by n and the plasma temperature (temperatures of ions and electrons are assumed to be the same) by T; then the thermal energy per unit volume is $3nT$. The thermal energy decreases by heat conduction and also by particle losses. When the notation P_L denotes the energy loss of the plasma per unit volume and per unit time (power loss per unit volume), the loss time τ is defined by the equation,

$$P_L \equiv \frac{3nT}{\tau}. \tag{1.1}$$

We may also call this time τ the *energy confinement time*, as we can consider that the plasma is confined during the period τ.

When an electron in the plasma collide with an ion, the electron radiates, which is an additional loss mechanism. This phenomena will be discussed in detail in sec. 15.2. The power loss P_b per unit volume due to this electron bremsstrahlung is

$$P_b = 1.5 \times 10^{-38} Z^2 n^2 \left(\frac{T}{e}\right)^{1/2} \text{W/m}^3, \tag{1.2}$$

where Z is the ratio of the ion charge to electron charge. The mks system is used throughout this book unless otherwise mentioned and the temperature T means $T \equiv \kappa T'$, T' (K) being the absolute temperature and κ the Boltzmann constant ($\kappa = 1.3804 \times 10^{-23}$ J/K). Therefore the unit of T is the Joule (J). In plasma physics, one electron volt (eV) is frequently used as a unit of energy. This is the energy necessary to move an electron, charge $e = 1.6021 \times 10^{-19}$ Coulomb, against a potential difference of 1 volt; the conversion factor is 1 eV = 1.6021×10^{-19} J. 1 eV corresponds to 1.16×10^4 K ($=e/\kappa$). The unit of density n is m^{-3}.

The output power of the D-T reaction per unit volume of plasma, with 50% deuterium and 50% tritium, is

$$P_{\mathrm{T}} = \frac{n^2}{4}\langle\sigma v\rangle Q_{\mathrm{T}}, \tag{1.3}$$

where σ is the cross section of the reaction, v is the relative velocity of the deuterium and tritium, and $\langle\ \rangle$ indicates the average. Q_{T} is 22.4 MeV, the sum of 17.58 MeV and 4.8 MeV. The first of these two quantities is the sum of the energies of He^4 and neutron due to the D-T reaction; the second is the energy released by the Li^6-n reaction (1 MeV = 10^6 eV).

Lawson derived his condition for the energy balance in the fusion reactor taking the efficiency η of thermal-to-electric energy conversion and heating efficiency into consideration. That is, the energy necessary to maintain the required condition should be less than η times the energy released from the plasma including the thermonuclear fusion energy. The efficiency of the thermal-to-electric energy conversion is of the order of 0.4–0.5 and the efficiency of plasma heating by electric energy is expected to be 0.5–0.9, so that η is of the order of 1/3. Then *Lawson's condition*[3] is expressed by

$$\begin{aligned} P_{\mathrm{b}} + P_{\mathrm{L}} &= \eta(P_{\mathrm{T}} + P_{\mathrm{b}} + P_{\mathrm{L}}) \\ \alpha n^2 T^{1/2} + \frac{3nT}{\tau} &= \frac{\eta}{1-\eta}\frac{1}{4}n^2\langle\sigma v\rangle Q_{\mathrm{T}} \\ n\tau &= \frac{3T}{\dfrac{\eta}{1-\eta}\dfrac{Q_{\mathrm{T}}}{4}\langle\sigma v\rangle - \alpha T^{1/2}}. \end{aligned} \tag{1.4}$$

Since $\langle\sigma v\rangle$ is a function of the temperature T only, the last relation of (1.4) may be plotted on an $n\tau$, T diagram (fig. 1.4). When $T \approx 20$ keV, $\langle\sigma v\rangle \approx 5 \times 10^{-22}$ m^3/s and $n\tau$ must be larger than 3×10^{19} $m^{-3}s$.

Next the *ignition condition* will be considered. If α particles He^{++} which come out of the D-T reaction are confined to the plasma and thus heat it, one can expect to compensate for the energy losses of the plasma. The heating power P_α of the α particles per unit volume is

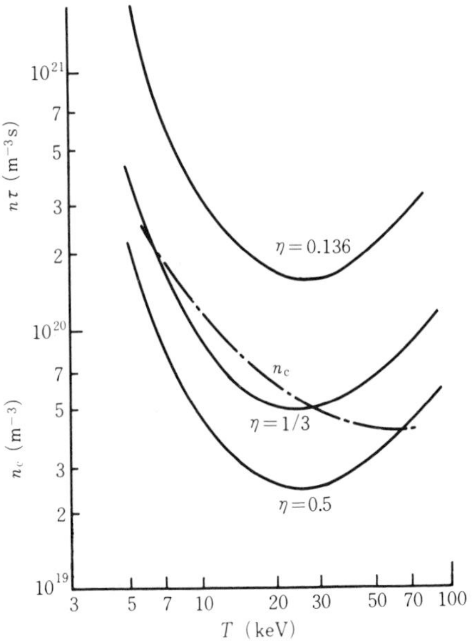

Fig. 1.4
Lawson's condition on an $n\tau$, T diagram for the efficiencies $\eta = 1/3$ and the ignition condition ($\eta = 0.136$). The critical condition is $P_b + P_L = P_T$, which corresponds to $\eta = 0.5$. The upper limit n_c of the density is also shown under the conditions of $a_p = 2$ m, $a_w = 2.5$ m, $p_w = 1 \times 10^6$ W/m^2. The meanings of a_p, a_w, and p_w are explained in sec. 1.4.

$$P_\alpha = \frac{n^2}{4} \langle \sigma v \rangle Q_\alpha, \tag{1.5}$$

where Q_α is the energy (3.52 MeV) of an α particle produced in the D-T reaction. The condition that α particle heating balances the energy loss is $P_b + P_L = P_\alpha$, and this condition is equivalent to Lawson's condition with $\eta = 0.136$.

1.4 Outline of a Fusion Reactor

In order to satisfy Lawson's condition, many problems remain to be solved concerning plasma confinement and heating. However, recent experimental developments on the confinement of hot plasma have encouraged scientists to start investigating the technological problems of nuclear fusion reactors. In this section, these technological problems will be

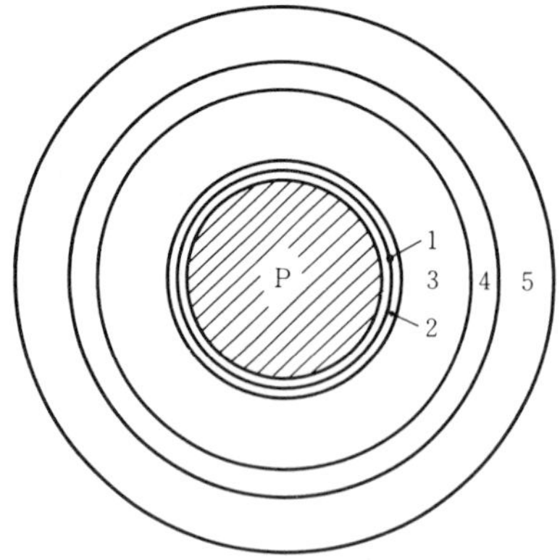

Fig. 1.5
Cross section of a fusion reactor. P, plasma; 1, vacuum wall; 2, wall cooling region; 3, lithium blanket; 4, shields; and 5, superconducting coils.

discussed briefly. The structure of a fusion reactor, as presently conceived, is shown in fig. 1.5. It consists of a vacuum wall (1), a wall cooling region (2), a liquid lithium blanket (3), which serves as a neutron moderator and a heat carrier (1.5–2 m thick), neutron and x-ray shields (4) (0.5–0.7 m thick), and superconducting coils and mechanical supports (5) (1 m thick). Therefore the thickness from the vacuum wall to the outside of coils is estimated to be 3–4 m. As the vacuum wall suffers from the heavy exposure to intensive neutrons and x-rays, the permissible amount of the output power through the unit area of the wall is considered to be less than about $1 \times 10^6\ \mathrm{W/m^2}$ (This limit may in future be revised, depending upon the results of engineering studies.) Let the radius of the plasma be denoted by a_p and the wall radius by a_w; then the output power p_w per unit area of the vacuum wall is

$$p_\mathrm{w} = \frac{\pi a_\mathrm{p}^2}{2\pi a_\mathrm{w}} \frac{n^2}{4} \langle \sigma v \rangle Q_\mathrm{T} < 1 \times 10^6\ \mathrm{W/m^2}. \tag{1.6}$$

The output power P_T per unit length of plasma is $P_\mathrm{T} = 2\pi a_\mathrm{w} p_\mathrm{w} = \pi a_\mathrm{p}^2 (n^2/4) \langle \sigma v \rangle Q_\mathrm{T}$ W/m. The solution for n yields

$$n_\mathrm{c} = \left(\frac{a_\mathrm{w}}{a_\mathrm{p}^2} p_\mathrm{w} \left(\frac{8}{Q_\mathrm{T}} \right) \right)^{1/2} \frac{1}{\langle \sigma v \rangle^{1/2}}. \tag{1.7}$$

This equation gives the upper limit of the plasma density.[4] Figure 1.4 shows the critical density $n_\mathrm{c}(T)$ for $a_\mathrm{p} = 2$ m, $a_\mathrm{w} = 2.5$ m and $p_\mathrm{w} = 1 \times 10^6\ \mathrm{W/m^2}$.

In this case, the total output power P_T per unit length of plasma is $P_T < 1.6 \times 10^7$ W/m. When the plasma temperature is 15 keV, the density limitation is $n_c < 10^{20}\ \mathrm{m}^{-3}$, while Lawson's condition requires $n\tau > 0.6 \times 10^{20}\ \mathrm{m}^{-3}/\mathrm{s}$ ($n \approx 0.6 \times 10^{20}\ \mathrm{m}^{-3}$, $\tau \approx 1$ s). Investigations of technological problems, cost effectiveness, and the optimal design of fusion reactors are progressing steadily in cooperation with the advancing research in plasma confinement and heating.[5]

References

1. D. Jakeman: Physics of Nuclear Reactors, University of London Press, London, 1966

2. W. R. Arnold, J. A. Phillips, G. A. Sawyer, E. J. Stovall, and J. L. Tuck: Phys. Rev. **93**, 483(1954)
J. L. Tuck: Nucl. Fusion **1**, 201(1961)
C. F. Wandel, T. Hesselberg Jensen, and O. Kofoed-Hansen: Nuclear Instr. and Methods **4**, 249(1959)
S. Glasstone and R. H. Lovberg: Controlled Thermonuclear Reactions, Van Nostrand, Princeton, New Jersey, 1960

3. J. D. Lawson: Proc. Phys. Soc. (London) **B70**, 6(1957)

4. R. Carruthers: Engineering Parameters of a Fusion Reactor, Nuclear Fusion Reactor Conference, Culham 1969

5. International Tokamak Reactor. Zero Phase, 1980; Phase One, 1982; Phase Two A, Part I, 1984; Phase Two A, Part II, 1986. International Atomic Energy Agency, Vienna.

2 Magnetic-Field Configurations

At the present time, the use of an appropriate magnetic field is regarded as the most promising method for the confinement of a hot plasma. The most desirable configuration of the magnetic field cannot be discussed without the detailed study of plasmas stability, which will be presented in subsequent chapters. In this introductory chapter, however, the properties of magnetic fields, especially of steady state fields, will be analyzed, and then several typical examples—as mirror, cusp, tokamak, and stellarator fields—will be introduced and their characteristics discussed.

2.1 Maxwell Equations[1]

Let us denote the *electric intensity*, the *magnetic induction*, the *electric displacement* and the *magnetic intensity by* $\boldsymbol{E}$, $\boldsymbol{B}$, $\boldsymbol{D}$, and $\boldsymbol{H}$ respectively. If the *charge density* and the *current density* are denoted by ρ and $\boldsymbol{j}$, Maxwell's equations are

$$\nabla \times \boldsymbol{E} + \frac{\partial \boldsymbol{B}}{\partial t} = 0 \tag{2.1}$$

$$\nabla \times \boldsymbol{H} - \frac{\partial \boldsymbol{D}}{\partial t} = \boldsymbol{j} \tag{2.2}$$

$$\nabla \cdot \boldsymbol{B} = 0 \tag{2.3}$$

$$\nabla \cdot \boldsymbol{D} = \rho. \tag{2.4}$$

As is clear from eqs. (2.2) and (2.4), ρ and $\boldsymbol{j}$ satisfy the relation

$$\nabla \cdot \boldsymbol{j} + \frac{\partial \rho}{\partial t} = 0. \tag{2.5a}$$

From eq. (2.3), the vector $\boldsymbol{B}$ can be expressed by the rotation of the vector $\boldsymbol{A}$:

$$\boldsymbol{B} = \nabla \times \boldsymbol{A}. \tag{2.6}$$

$\boldsymbol{A}$ is called the *vector potential.* If eq. (2.6) is substituted into (2.1), we obtain

$$\nabla \times \left(\boldsymbol{E} + \frac{\partial \boldsymbol{A}}{\partial t}\right) = 0. \tag{2.7}$$

The quantity in parenthesis can be expressed by a *scalar potential* ϕ and

$$E = -\nabla\phi - \frac{\partial A}{\partial t}. \tag{2.8a}$$

Since any other set of A' and ϕ',

$$A' = A - \nabla\psi \tag{2.9}$$

$$\phi' = \phi + \frac{\partial\psi}{\partial t}, \tag{2.10}$$

can also satisfy eqs. (2.6) and (2.8), ϕ and A are not uniquely determined.

When the medium is uniform and isotropic, B and D are expressed by

$$D = \varepsilon E, \qquad B = \mu H; \tag{2.11}$$

ε and μ are called *dielectric constant* and *permeability* respectively. The value of ε_0 and μ_0 in vacuum are

$$\varepsilon_0 = \frac{10^7}{4\pi c^2} \mathrm{C}^2 \cdot \mathrm{s}^2/\mathrm{kg} \cdot \mathrm{m}^3 = 8.854 \times 10^{-12} \mathrm{F/m} \tag{2.12}$$

$$\mu_0 = 4\pi \times 10^{-7} \mathrm{kg} \cdot \mathrm{m/C}^2 = 1.257 \times 10^{-6} \mathrm{H/m}, \tag{2.13}$$

where c is the velocity of light in vacuum. There is an Ohm's law relation between j and E; i.e.,

$$E = \eta j \tag{2.14}$$

$$j = \sigma E. \tag{2.15}$$

Here η is the *specific resistivity* and σ is the *specific conductivity*. When the medium is anisotropic, ε, μ and η, σ must in general be expressed in tensor form. By means of (2.11), eqs. (2.2) and (2.4) may be reduced to

$$\nabla \times \nabla \times A + \mu\varepsilon\nabla\frac{\partial\phi}{\partial t} + \mu\varepsilon\frac{\partial^2 A}{\partial t^2} = \mu j \tag{2.16}$$

$$\nabla^2\phi + \nabla\frac{\partial A}{\partial t} = -\frac{1}{\varepsilon}\rho. \tag{2.17}$$

As ϕ and A are arbitrary, by eqs. (2.9) and (2.10), we impose the supplementary condition

$$\nabla \cdot A + \mu\varepsilon\frac{\partial\phi}{\partial t} = 0. \tag{2.18a}$$

Then eqs. (2.16) and (2.17) reduce to the wave equations

$$\nabla^2\phi - \mu\varepsilon\frac{\partial^2\phi}{\partial t^2} = -\frac{1}{\varepsilon}\rho \tag{2.19a}$$

$$\nabla^2\boldsymbol{A} - \mu\varepsilon\frac{\partial^2\boldsymbol{A}}{\partial t^2} = -\mu\boldsymbol{j}. \tag{2.20a}$$

The propagation velocity of the electromagnetic field is $1/(\mu\varepsilon)^{1/2}$. The difference of the scalar products of eq. (2.1) with $\boldsymbol{H}$ and eq. (2.2) with $\boldsymbol{E}$ gives a statement of the conservation of energy:

$$\nabla(\boldsymbol{E}\times\boldsymbol{H}) + \frac{\partial}{\partial t}\left(\frac{\boldsymbol{E}\cdot\boldsymbol{D}}{2} + \frac{\boldsymbol{H}\cdot\boldsymbol{B}}{2}\right) + \boldsymbol{E}\cdot\boldsymbol{j} = 0. \tag{2.21}$$

The quantity

$$\boldsymbol{S} = \boldsymbol{E}\times\boldsymbol{H} \tag{2.22}$$

is called the *Poynting vector*; it gives the flow of electromagnetic energy per unit area per unit time. The second term in (2.21) is the time derivative of the electromagnetic-energy density, and the third term gives the Joule heating. When the medium is isotropic, the sum of the vector products of eqs. (2.1) with $\varepsilon\boldsymbol{E}$ and eq. (2.2) with $\boldsymbol{B}$ is

$$\varepsilon(\nabla\times\boldsymbol{E})\times\boldsymbol{E} + \frac{1}{\mu}(\nabla\times\boldsymbol{B})\times\boldsymbol{B} = \boldsymbol{j}\times\boldsymbol{B} + \varepsilon\mu\frac{\partial}{\partial t}\boldsymbol{S}. \tag{2.23}$$

This can be rewritten as the equation of conservation of momentum:

$$\rho\boldsymbol{E} + \boldsymbol{j}\times\boldsymbol{B} = \sum_{\alpha=1}^{3}\sum_{\beta=1}^{3}\boldsymbol{i}_\alpha\frac{\partial T_{\alpha\beta}}{\partial x_\beta} - \varepsilon\mu\frac{\partial}{\partial t}\boldsymbol{S} \tag{2.24}$$

where $T_{\alpha\beta}$ is the *electromagnetic stress tensor*, which is of the form

$$T_{\alpha\beta} = \left(E_\alpha D_\beta - \frac{1}{2}(\boldsymbol{E}\cdot\boldsymbol{D})\delta_{\alpha\beta}\right) + \left(B_\alpha H_\beta - \frac{1}{2}(\boldsymbol{B}\cdot\boldsymbol{H})\delta_{\alpha\beta}\right). \tag{2.25}$$

When $\boldsymbol{E} = 0$ and $\boldsymbol{B} = (0, 0, B)$, the stress tensor $T_{\alpha\beta}$ is

$$T_{\alpha\beta} = \begin{bmatrix} -B^2/2\mu & 0 & 0 \\ 0 & -B^2/2\mu & 0 \\ 0 & 0 & B^2/2\mu \end{bmatrix}.$$

The stress is the tension along the line of magnetic force; it manifests itself as a pressure perpendicular to the direction of the magnetic field.

When the fields do not change with time, the field equations reduce to

$$\boldsymbol{E} = -\nabla\phi \tag{2.8b}$$

$$\boldsymbol{B} = \nabla \times \boldsymbol{A} \tag{2.6}$$

$$\nabla^2\phi = -\frac{1}{\varepsilon}\rho \tag{2.19b}$$

$$\nabla^2\boldsymbol{A} = -\mu\boldsymbol{j} \tag{2.20b}$$

$$\nabla\cdot\boldsymbol{A} = 0 \tag{2.18b}$$

$$\nabla\cdot\boldsymbol{j} = 0. \tag{2.5b}$$

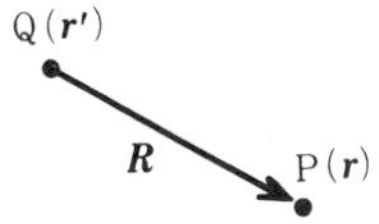

Fig. 2.1
Observation point P and the location Q of a charge or current.

The scalar and vector potentials ϕ and $\boldsymbol{A}$ at an observation point P (given by the position vector $\boldsymbol{r}$) are expressed in terms of the charge and current densities at the point Q (given by $\boldsymbol{r}'$) by

$$\phi(\boldsymbol{r}) = \frac{1}{4\pi\varepsilon}\int_{\mathrm{v}} \frac{\rho(\boldsymbol{r}')}{R}\,\mathrm{d}\boldsymbol{r}' \tag{2.26}$$

$$\boldsymbol{A}(\boldsymbol{r}) = \frac{\mu}{4\pi}\int_{\mathrm{v}} \frac{\boldsymbol{j}(\boldsymbol{r}')}{R}\,\mathrm{d}\boldsymbol{r}', \tag{2.27}$$

where $\boldsymbol{R} \equiv \boldsymbol{r} - \boldsymbol{r}'$ and R is the absolute magnitude $|\boldsymbol{R}|$ of $\boldsymbol{R}$, i.e.,

$$R = ((x - x')^2 + (y - y')^2 + (z - z')^2)^{1/2}. \tag{2.28}$$

The notation $\mathrm{d}\boldsymbol{r}'$ means $\mathrm{d}x'\,\mathrm{d}y'\,\mathrm{d}z'$. Since the fields $\boldsymbol{E}$ and $\boldsymbol{B}$ are expressed by eqs. (2.8b) and (2.6), we find

$$\boldsymbol{E} = \frac{1}{4\pi\varepsilon}\int_{\mathrm{v}} \frac{\boldsymbol{R}\rho}{R^3}\,\mathrm{d}\boldsymbol{r}' \tag{2.29}$$

$$\boldsymbol{B} = \frac{\mu}{4\pi} \int_{\mathrm{V}} \frac{\boldsymbol{j} \times \boldsymbol{R}}{R^3} \mathrm{d}\boldsymbol{r}'. \tag{2.30}$$

When the current distribution is given by a current I flowing in closed loops, magnetic intensity is described by *Biot-Savart equation*

$$\boldsymbol{H} = \frac{\boldsymbol{B}}{\mu} = \frac{I}{4\pi} \int_{\mathrm{C}} \frac{\boldsymbol{s} \times \boldsymbol{n}}{R^2} \mathrm{d}s \tag{2.31}$$

where $\boldsymbol{s}$ and $\boldsymbol{n}$ are the unit vectors in the directions of $\boldsymbol{j}$ and $\boldsymbol{R}$, respectively.

2.2 Magnetic Surfaces

2.2a Exact Solutions of Magnetic Surfaces

A *line of magnetic force* satisfies the equations

$$\frac{\mathrm{d}x}{B_x} = \frac{\mathrm{d}y}{B_y} = \frac{\mathrm{d}z}{B_z}. \tag{2.32}$$

If length along a line of magnetic force is denoted by l, and the absolute magnitude of $\boldsymbol{B}$ by B, then eq. (2.32) is equal to $\mathrm{d}l/B$. The *magnetic surface* $\psi(\boldsymbol{r}) = \text{const.}$ is such that all lines of magnetic force lie upon on that surface which satisfies the condition

$$(\nabla\psi(\boldsymbol{r}))\cdot\boldsymbol{B} = 0. \tag{2.33}$$

($\nabla\psi$ is normal to the magnetic surface.) When a magnetic field has some symmetry, the exact solution of the magnetic surface can easily be derived. When a field is invariant under a shift along some given direction, the field is said to possess translational symmetry. A field with axial symmetry is invariant under rotation around some axis. When the field is invariant under movement along a helical path, we say that the field has helical symmetry. In these cases, the magnetic surfaces can be expressed in term of the vector potential. In terms of the cylindrical coordinates r, θ, z, the magnetic field $\boldsymbol{B}$ is given by

$$B_r = \frac{1}{r}\frac{\partial A_z}{\partial \theta} - \frac{\partial A_\theta}{\partial z} \qquad B_\theta = \frac{\partial A_r}{\partial z} - \frac{\partial A_z}{\partial r} \qquad B_z = \frac{1}{r}\frac{\partial}{\partial r}(rA_\theta) - \frac{1}{r}\frac{\partial A_r}{\partial \theta}. \tag{2.34}$$

$\boldsymbol{B}$ and $\boldsymbol{A}$ are independent of z in the case of translational symmetry. In the case of axial symmetry, they are independent of θ. For helical symmetry, they are dependent only on r and $\theta - \alpha z$. Therefore

$$A_z(r, \theta) = \text{const.} \tag{2.35}$$

$$rA_\theta(r, z) = \text{const.} \tag{2.36}$$

$$A_z(r, \theta - \alpha z) + \alpha r A_\theta(r, \theta - \alpha z) = \text{const.} \tag{2.37}$$

can satisfy the condition (2.33); thus eqs. (2.35)–(2.37) are equations of magnetic surfaces.[2]

2.2b Approximate Solutions of Average Magnetic Surfaces

Although the exact solution of the magnetic surface cannot be obtained except special cases, the average magnetic surface may be derived in many cases.* Let us consider the case in which the magnetic field has a large component in the z direction, the components in the r and θ directions being small; and let there be periodicity in the z direction. The magnetic field $\boldsymbol{B}(r, \theta, z)$ is then expressed by

$$\boldsymbol{B}(r, \theta, z) = \boldsymbol{B}_0(r, \theta) + \boldsymbol{b}(r, \theta, z), \tag{2.38}$$

where $\boldsymbol{B}_0$ has a z component only, and where $|\boldsymbol{b}|$ is much smaller than $|\boldsymbol{B}_0|$ ($|\boldsymbol{b}| \ll |\boldsymbol{B}_0|$). Let L stand for the repetition length (pitch) of $\boldsymbol{B}$. Then the lines of magnetic force are given by

$$\frac{\mathrm{d}r}{\mathrm{d}z} = \frac{b_r}{B_0 + b_z} \approx \frac{b_r}{B_0} - \frac{b_r b_z}{{B_0}^2} \tag{2.39}$$

$$\frac{\mathrm{d}\theta}{\mathrm{d}z} = \frac{1}{r}\frac{b_\theta}{B_0 + b_z} \approx \frac{b_\theta}{rB_0} - \frac{b_\theta b_z}{r{B_0}^2}. \tag{2.40}$$

Denote the average coordinates of the lines of magnetic force by $\bar{r}$, $\bar{\theta}$, $\bar{z}$; then the average coordinates satisfy

$$\frac{\mathrm{d}\bar{r}}{\mathrm{d}z} = \frac{\bar{b}_r}{B_0} - \overline{\frac{b_r b_z}{{B_0}^2}} + \overline{\frac{\partial}{\partial \bar{r}}\left(\frac{b_r}{B_0}\right)\cdot\frac{\langle b_r\rangle}{B_0}} + \overline{\frac{\partial}{\partial \bar{\theta}}\left(\frac{b_r}{B_0}\right)\cdot\frac{\langle b_\theta\rangle}{\bar{r}B_0}} \tag{2.41}$$

$$\frac{\mathrm{d}\bar{\theta}}{\mathrm{d}z} = \frac{\bar{b}_\theta}{\bar{r}B_0} - \overline{\frac{b_\theta b_z}{\bar{r}{B_0}^2}} + \overline{\frac{\partial}{\partial \bar{r}}\left(\frac{b_\theta}{\bar{r}B_0}\right)\frac{\langle b_r\rangle}{B_0}} + \overline{\frac{\partial}{\partial \bar{\theta}}\left(\frac{b_\theta}{\bar{r}B_0}\right)\frac{\langle b_\theta\rangle}{\bar{r}B_0}}. \tag{2.42}$$

(The averaging method used to derive these equations will be described in sec. 2.2c.) The overbar $\bar{\ }$ and bracket $\langle\ \ \rangle$ notations mean

* In the case of a toroidal stellarator, the helical symmetry is lost and the magnetic surfaces do not necessarily exist.

$$\bar{a} \equiv \frac{1}{L}\int_0^L a\,\mathrm{d}z \qquad \langle a\rangle \equiv \int_0^z \tilde{a}\,\mathrm{d}z - \overline{\int_0^z \tilde{a}\,\mathrm{d}z} \qquad (\tilde{a} \equiv a - \bar{a}).$$

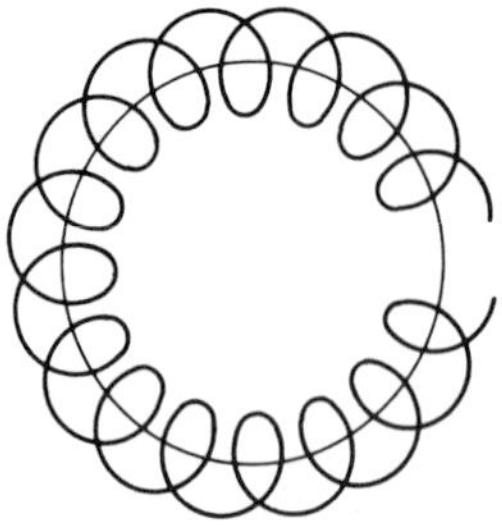

Fig. 2.2
Average magnetic surface.

On the other hand, $\nabla\cdot\boldsymbol{B} = 0$ reduces to

$$\nabla\cdot\boldsymbol{B} = \frac{1}{r}\frac{\partial}{\partial r}(rb_r) + \frac{1}{r}\frac{\partial b_\theta}{\partial\theta} + \frac{\partial b_z}{\partial z} = 0. \tag{2.43}$$

Using the relations $\partial\langle a\rangle/\partial z = \tilde{a}$, $\overline{\langle a\rangle b} + \overline{a\langle b\rangle} = 0$, and $\overline{\langle a\rangle a} = 0$, we find

$$\overline{\frac{\partial}{\partial\bar{r}}\left(\frac{b_r}{B_0}\right)\frac{\langle b_r\rangle}{B_0}} = \overline{\frac{1}{B_0}\frac{\partial b_r}{\partial\bar{r}}\frac{\langle b_r\rangle}{B_0}} + \overline{\frac{\partial}{\partial\bar{r}}\left(\frac{1}{B_0}\right)\frac{b_r\langle b_r\rangle}{B_0}}$$

$$= -\frac{1}{B_0^{\,2}}\overline{\left(\frac{b_r}{r} + \frac{1}{r}\frac{\partial b_\theta}{\partial\theta} + \frac{\partial b_z}{\partial z}\right)\langle b_r\rangle}$$

$$= -\frac{1}{B_0^{\,2}}\frac{1}{\bar{r}}\overline{\frac{\partial b_\theta}{\partial\theta}\langle b_r\rangle} + \frac{1}{B_0^{\,2}}\overline{b_r b_z},$$

$$\overline{\frac{\partial}{\partial\bar{\theta}}\left(\frac{b_r}{B_0}\right)\frac{\langle b_\theta\rangle}{\bar{r}B_0}} = -\overline{\frac{\partial}{\partial\bar{\theta}}\left(\frac{\langle b_r\rangle}{B_0}\right)b_\theta\frac{1}{\bar{r}B_0}}.$$

Then eq. (2.41) is reduced to

$$\frac{\mathrm{d}\bar{r}}{\mathrm{d}z} = \frac{\bar{b}_r}{B_0} - \frac{1}{\bar{r}B_0}\frac{\partial}{\partial\bar{\theta}}\overline{\left(\frac{\langle b_r\rangle b_\theta}{B_0}\right)}. \tag{2.44}$$

Similarly, eq. (2.42) becomes

$$\frac{\mathrm{d}\bar{\theta}}{\mathrm{d}z} = \frac{\bar{b}_\theta}{\bar{r}B_0} + \frac{1}{\bar{r}B_0}\frac{\partial}{\partial\bar{r}}\overline{\left(\frac{\langle b_r\rangle b_\theta}{B_0}\right)}. \tag{2.45}$$

By use of the vector potential $\boldsymbol{A}$, we can express the components $\bar{b}_r$ and $\bar{b}_\theta$ by $\bar{b}_r = (1/r)(\partial \bar{A}_z/\partial\theta)$ and $\bar{b}_\theta = \partial\bar{A}_z/\partial r$, respectively. Finally, the relations (2.44) and (2.45) can be written as

$$\frac{\mathrm{d}\bar{r}}{\mathrm{d}z} = \frac{1}{\bar{r}B_0}\frac{\partial\psi_{\mathrm{a}}}{\partial\bar{\theta}} \tag{2.46}$$

$$\frac{\mathrm{d}\bar{\theta}}{\mathrm{d}z} = \frac{-1}{\bar{r}B_0}\frac{\partial\psi_{\mathrm{a}}}{\partial\bar{r}}, \tag{2.47}$$

where

$$\psi_{\mathrm{a}}(\bar{r}, \bar{\theta}) = \bar{A}_z + \frac{\overline{b_r\langle b_\theta\rangle}}{B_0} = \bar{A}_z - \frac{\overline{\langle b_r\rangle b_\theta}}{B_0}. \tag{2.48}$$

Equation (2.48) gives the average magnetic surface[2] to an accuracy of first order in $\varepsilon = L/\mathscr{L}$, the ratio of the pitch L to the characteristic length $\mathscr{L}$ along z by which $\bar{\theta}$ changes by 1 radian (a few times the circumference of torus). The equations of the lines of magnetic force, to first order in ε, are (see sec. 2.2c)

$$r = \bar{r} + \frac{\langle b_r\rangle}{B_0} \tag{2.49}$$

$$\theta = \bar{\theta} + \frac{\langle b_\theta\rangle}{rB_0}. \tag{2.50}$$

In the case of a toroidal field, only the θ component of the magnetic field is large and $\boldsymbol{B}$ is given by

$$\boldsymbol{B}(r, \theta, z) = \boldsymbol{B}_0(r, z) + \boldsymbol{b}(r, \theta, z),$$

where $\boldsymbol{B}_0$ has a θ component only. In this case the average magnetic surfaces are derived by means similar to the foregoing, to yield

$$\frac{\mathrm{d}\bar{r}}{\mathrm{d}\bar{\theta}} = -\frac{1}{B_0}\frac{\partial\psi_{\mathrm{a}}}{\partial\bar{z}} \tag{2.51}$$

$$\frac{\mathrm{d}\bar{z}}{\mathrm{d}\bar{\theta}} = \frac{1}{B_0}\frac{\partial\psi_{\mathrm{a}}}{\partial\bar{r}} \tag{2.52}$$

$$\psi_{\mathrm{a}}(\bar{r}, \bar{z}) = \overline{rA_\theta} + \frac{\bar{r}^2}{B_0}\overline{b_z\langle b_r\rangle}. \tag{2.53}$$

2.2c The Averaging Method for a System of First-Order Differential Equations

Let us assume that ε is a small parameter and that f_k are periodic functions of θ, with period 2π, in the system of first-order differential equations

$$\frac{dx_k}{dt} = f_k(x_i, t, \theta) \qquad \theta = \frac{t}{\varepsilon}. \tag{2.54}$$

To solve eqs. (2.54), assume that x_k can be expanded as

$$x_k = \xi_k + \varepsilon g_{1k}(\xi_i, t, \theta) + \varepsilon^2 g_{2k}(\xi_i, t, \theta) + \cdots \tag{2.55}$$

$$\frac{d\xi_k}{dt} = \varphi_{0k}(\xi_i, t) + \varepsilon\varphi_{1k}(\xi_i, t) + \varepsilon^2\varphi_{2k}(\xi_i, t) + \cdots. \tag{2.56}$$

Here the ξ_k do not depend on θ; they are considered to be averages of x_k. If we substitute eqs. (2.55) and (2.56) into (2.54), we obtain

$$\varphi_{0k} + \frac{\partial g_{1k}}{\partial \theta} = f_k \tag{2.57}$$

$$\varphi_{1k} + \sum_i \frac{\partial g_{1k}}{\partial \xi_i}\varphi_{0i} + \frac{\partial g_{1k}}{\partial t} + \frac{\partial g_{2k}}{\partial \theta} = \sum_i \frac{\partial f_k}{\partial \xi_i} g_{1i}. \tag{2.58}$$

$\cdots.$

Define the following quantities:

$$\bar{f} \equiv \frac{1}{2\pi}\int_0^{2\pi} f d\theta \tag{2.59}$$

$$\tilde{f} \equiv f - \bar{f} \tag{2.60}$$

$$\langle f \rangle \equiv \int_0^{\theta} \tilde{f} d\theta - \overline{\int_0^{\theta} \tilde{f} d\theta}. \tag{2.61}$$

As the number of the g_{ik} and φ_{ik} is larger than the number of equations in (2.57) and (2.58), one must impose some appropriate restrictions. In this case, these are that the average of g_{ik} is zero. Then, using the definitions (2.59), (2.60), and (2.61), eqs. (2.57) and (2.58) are reduced to

$$\varphi_{0k} = \bar{f}_k \qquad g_{1k} = \langle f_k \rangle \tag{2.62}$$

$$\varphi_{1k} = \overline{\sum_i \frac{\partial f_k}{\partial \xi_i} g_{1i}} \qquad g_{2k} = \sum_i \left\langle \frac{\partial f_k}{\partial \xi_i} g_{1i} \right\rangle - \sum_i \frac{\partial \langle g_{1k} \rangle}{\partial \xi_i} \varphi_{0i} - \frac{\partial \langle g_{1k} \rangle}{\partial t} \tag{2.63}$$

$\cdots,$

and finally we find

$$\frac{\mathrm{d}\xi_k}{\mathrm{d}t} = \bar{f}_k + \varepsilon \overline{\sum_i \frac{\partial f_k}{\partial \xi_i} \langle f_i \rangle} + \cdots \tag{2.64}$$

$$x_k = \xi_k + \varepsilon \langle f_k \rangle + \cdots. \tag{2.65}$$

2.3 Configurations of Various Magnetic Fields

Several field configurations are introduced in this section. Some properties of mirror, cusp, multipole, spherator, levitron, tokamak, and stellarator fields are described.

2.3a Magnetic Field of Ring Currents and Straight Currents (Spherator, Levitron, Tokamak, and Multipole Fields)

Coils producing circular, or ring, currents are the most popular amongst plasma experimenters. The coil in fig. 2.3 lies in the x, y plane; its position is given by $z = 0$ and $r = a$ in cylindrical coordinates. When a current I flows in the ring coil, the vector potential has only a θ component A_θ, given by

$$A_\theta = \frac{\mu}{\pi k} I \left(\frac{a}{r}\right)^{1/2} \left(\left(1 - \frac{1}{2}k^2\right) K(k) - E(k)\right) \tag{2.66}$$

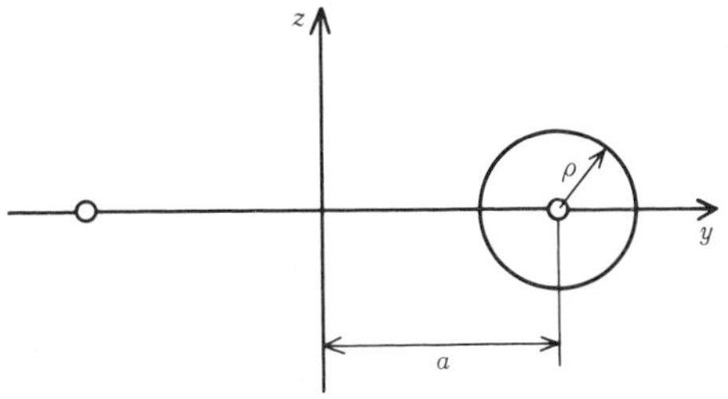

Fig. 2.3
Ring current.

$$k^2 = \frac{4ar}{(a+r)^2 + z^2}, \tag{2.67}$$

where K and E are complete elliptic integrals of the first and second kind, respectively. These integrals may be written in the following form for the vicinity of the ring, where k is nearly equal to 1:

$$E(k) = 1 + \frac{1}{2}\left(\ln\frac{4}{(1-k^2)^{1/2}} - \frac{1}{2}\right)(1-k^2) + \cdots \tag{2.68}$$

$$K(k) = \ln\frac{4}{(1-k^2)^{1/2}} + \frac{1}{4}\left(\ln\frac{4}{(1-k^2)^{1/2}} - 1\right)(1-k^2) + \cdots; \tag{2.69}$$

the magnetic surface is given by

$$rA_\theta = \frac{\mu a I}{2\pi}\left(\ln\frac{8a}{\rho} - 2\right) = \text{const.} \tag{2.70}$$
$$\rho^2 = (r-a)^2 + z^2.$$

Far from the ring, the magnetic field becomes dipole, with dipole moment $M = I\pi a^2$. When $r^2 + z^2 \gg a^2$ and k is small, E and K can be expressed by

$$E = \frac{\pi}{2}\left(1 - \frac{k^2}{4} - \frac{3}{64}k^4 - \cdots\right) \tag{2.71}$$

$$K = \frac{\pi}{2}\left(1 + \frac{k^2}{4} + \frac{9}{64}k^4 + \cdots\right), \tag{2.72}$$

and the magnetic surface by

$$rA_\theta = \frac{\mu M}{4\pi}\frac{r^2}{(r^2+z^2)^{3/2}}. \tag{2.73}$$

Near the z axis, i.e., for $r \ll a$, the vector potential is given by

$$rA_\theta = \frac{\mu M}{4\pi}\frac{r^2}{((a+r)^2+z^2)^{3/2}}. \tag{2.74}$$

Since the magnetic field (B_r, B_θ, B_z) is given by

$$B_r = -\frac{\partial A_\theta}{\partial z} \qquad B_\theta = 0 \qquad B_z = \frac{1}{r}\frac{\partial}{\partial r}(rA_\theta), \tag{2.75}$$

these components can be expressed as

$$B_r = \frac{\mu I}{2\pi} \frac{z}{r((a+r)^2 + z^2)^{1/2}} \left(-K + \frac{a^2 + r^2 + z^2}{(a-r)^2 + z^2} E \right) \tag{2.76}$$

$$B_z = \frac{\mu I}{2\pi} \frac{1}{((a+r)^2 + z^2)^{1/2}} \left(K + \frac{a^2 - r^2 - z^2}{(a-r)^2 + z^2} E \right), \tag{2.77}$$

where we have used

$$\frac{\partial K}{\partial k} = \frac{1}{k}\left(\frac{E}{1-k^2} - K\right) \qquad \frac{\partial E}{\partial k} = \frac{1}{k}(E - K). \tag{2.78}$$

When a uniform field $B_\perp$ in the z direction is superposed on the ring field I, a vector potential

$$A_{\theta\perp} = \frac{1}{2} B_\perp r \tag{2.79}$$

must be added to the ring field potential.

When a constant current I_t flows along the z axis, the vector potential of the field has a z component only:

$$A_z = -\frac{\mu}{2\pi} I_t \ln r. \tag{2.80}$$

The magnetic field is given by

$$B_\theta = \frac{\mu I_t}{2\pi r} \qquad B_r = B_z = 0. \tag{2.81}$$

This is a simple toroidal field.

The magnetic surfaces $r(A_\theta + A_{\theta\perp}) = \text{const.}$ of the combined ring current and the uniform field $B_\perp$ are shown in fig. 2.4. Even if the current I_t flows along the axis, only the θ component of the field is superposed; the

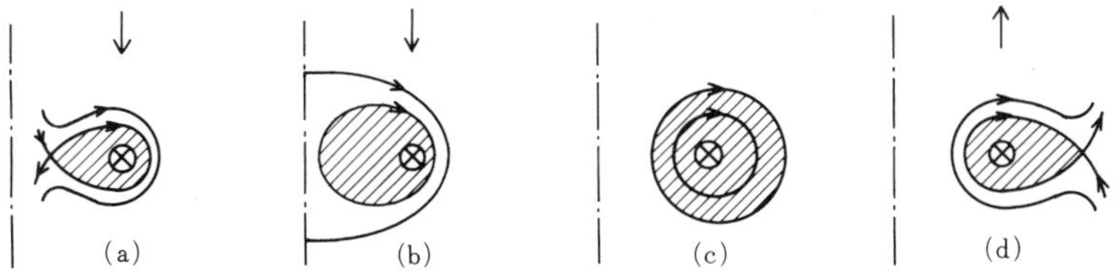

Fig. 2.4
Magnetic surfaces produced by combined I_t, I, and $B_\perp$ fields. The arrows show the direction of $B_\perp$.

shape of the magnetic surface is not affected. By means of a change of the ratios between I_t, I, and $B_\perp$, the configurations of spherator[3] (fig. 2.4b) and levitron[4] (fig. 2.4a) are obtained. When the circular ring current is replaced by a plasma current, a tokamak[5] field is obtained. The tokamak configuration will be discussed in secs. 7.7 and 16.2 in more detail.

The $2n$th *multipole field* is produced by a set of $2n$ parallel currents in the z direction, n being of magnitude of I at the positions $r = a, \theta = (2l/n)\pi$, $l = 0, \ldots, n-1$, and n being of magnitude $-I$ at $r = a, \theta = ((2l+1)/n)\pi$, $l = 0, \ldots, n-1$. A hexapole field is shown in fig. 2.5.

The vector potential A_z of a multipole field is

$$A_z = -\frac{\mu I}{4\pi} \ln\left(\frac{(r^{2n} + a^{2n}) - 2r^n a^n \cos n\theta}{(r^{2n} + a^{2n}) + 2r^n a^n \cos n\theta}\right), \tag{2.82}$$

and the field components are

$$B_r = -\frac{\mu I}{\pi} n r^{n-1} \frac{a^n \sin n\theta (a^{2n} + r^{2n})}{(r^{2n} + a^{2n})^2 - (2r^n a^n \cos n\theta)^2} \tag{2.83}$$

$$B_\theta = -\frac{\mu I}{\pi} n r^{n-1} \frac{a^n \cos n\theta (a^{2n} - r^{2n})}{(r^{2n} + a^{2n})^2 - (2r^n a^n \cos n\theta)^2}. \tag{2.84}$$

For $r < a$, these become

$$B_r \approx -\left(\frac{n\mu I}{\pi a}\right)\left(\frac{r}{a}\right)^{n-1} \sin n\theta \tag{2.85}$$

$$B_\theta \approx -\left(\frac{n\mu I}{\pi a}\right)\left(\frac{r}{a}\right)^{n-1} \cos n\theta \tag{2.86}$$

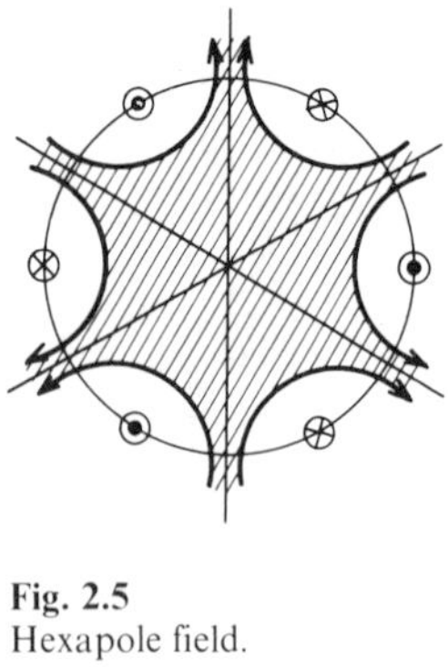

Fig. 2.5
Hexapole field.

and the magnitude is $B = (B_r^2 + B_\theta^2)^{1/2} \approx \left|\frac{n\mu I}{\pi a}\right|\left(\frac{r}{a}\right)^{n-1}$, so that the magnitude B increases with r.

A $2n$th multipole field is also produced by a set of n parallel currents I all of the same direction, giving

$$A_z = -\frac{\mu I}{4\pi}\ln\left((r^{2n} + a^{2n}) - 2r^n a^n \cos n\theta\right) \tag{2.87}$$

$$B_r = -\frac{\mu I}{2\pi} n r^{n-1} \frac{a^n \sin n\theta}{r^{2n} + a^{2n} - 2r^n a^n \cos n\theta} \tag{2.88}$$

$$B_\theta = -\frac{\mu I}{2\pi} n r^{n-1} \frac{a^n \cos n\theta - r^n}{r^{2n} + a^{2n} - 2r^n a^n \cos n\theta}. \tag{2.89}$$

For example, we may consider the toroidal multipole field shown in fig. 2.6. This configuration satisfies the stabilizing condition for plasma confinement. Ohkawa and Kerst[6] have demonstrated experimentally that a plasma can be stably confined in a toroidal multipole field.

2.3b Mirror Field (Minimum-B Field), Cusp Field

A periodic field along the z axis with axial symmetry about that axis is called a *bumpy field*. The vector potential of a bumpy field is

$$A_\theta = \frac{\mu J k}{2\pi}\left(\frac{r}{2} + 2a \sum_{n=1}^{\infty} \alpha_n K_1(nka) I_1(nkr) \cos nkz\right) \qquad r < a \tag{2.90}$$

$$A_\theta = \frac{\mu J k}{2\pi} 2a \sum_{n=1}^{\infty} \alpha_n I_1(nka) K_1(nkr) \cos nkz \qquad r > a. \tag{2.91}$$

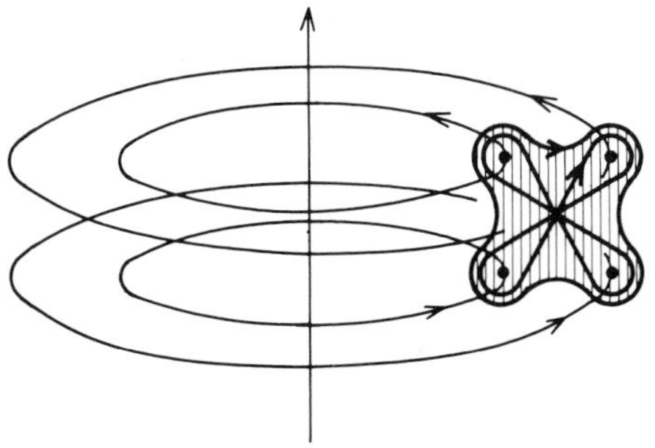

Fig. 2.6
Toroidal octopole field.

As the rotation of the magnetic field is zero in the current-free region, the field can be expressed by a scalar potential $\phi_B(B = \nabla\phi_B)$, i.e.,

$$\phi_B = \frac{Jk}{2\pi}\left(z + 2a \sum_{n=1}^{\infty} \alpha_n K_1(nka) I_0(nkr) \sin nkz\right) \qquad r < a \tag{2.92}$$

$$\phi_B = -\frac{Jk}{2\pi}\left(2a \sum_{n=1}^{\infty} \alpha_n I_1(nka) K_0(nkr) \sin nkz\right) \qquad r > a, \tag{2.93}$$

where $K_n(x)$ and $I_n(x)$ are the modified Bessel functions of the nth order. If the pitch of the field is L, then k is $k = 2\pi/L$. The magnetic surfaces are given by $rA_\theta = \text{const}$. If the magnitude of the field on the z axis is given by $b(z) \equiv \phi'(z)$, then the vector potential is

$$A_\theta = \frac{r}{2}b(z) - \frac{r^3}{16}b''(z) + \cdots = J_1\left(r\frac{\mathrm{d}}{\mathrm{d}z}\right)\phi(z). \tag{2.94}$$

Let us consider the simple case,

$$A_\theta = B_0\left(\frac{r}{2} - \frac{b}{k}I_1(kr)\cos kz\right) \tag{2.95}$$

$$\phi_B = B_0\left(z - \frac{b}{k}I_0(kr)\sin kz\right). \tag{2.96}$$

Where $kr \ll 1$, it follows that

$$\frac{B_z}{B_0} = 1 - bI_0(kr)\cos kz \approx 1 - b\left(1 + \frac{1}{8}(kr)^2\right)\cos kz \tag{2.97}$$

$$\frac{B_r}{B_0} = -bI_1(kr)\sin kz \approx -\frac{b}{4}(kr)\sin kz. \tag{2.98}$$

With the assumption $b > 0$, the absolute value of the field is $B \approx |B_z| \approx |B_0|(1 - b)$ at $z = 0$ and $B \approx |B_0|(1 + b)$ at $z = \pm\pi/k = \pm L/2$. Let us consider only the region $|z| < L/2$. The magnitude of the field increases as $z \to L/2$ or $z \to -L/2$ along the lines of force; the magnitude is small around $z = 0$. Charged particles are reflected at $\pm L/2$ under conditions which will be described in ch. 3. Such a magnetic field is called a *mirror field* (fig. 2.7). The ratio of the magnitude of the field at the ends against the magnitude at the center is called the *mirror ratio*. The larger the mirror ratio, the more effectively will particles be reflected at the ends and thus be trapped in the mirror field. Equations (2.97) and (2.98) analytically

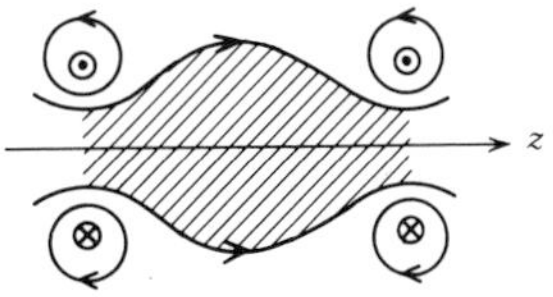

Fig. 2.7
Mirror field.

express a mirror field of mirror ratio $R = (1 + b)/(1 - b)$ in the region $|z| < L/2$.*

Although the magnitude of the field expressed by eqs. (2.97) and (2.98) increases toward the both ends along the lines of force, the magnitude decreases in the outward radial direction $[B = |B_z| \approx |B_0|(1 - b\cos kz - b(kr)^2(\cos kz)/8)]$. When a multipole field is superposed on the mirror field, then the magnitude of the field becomes

$$\frac{B}{|B_0|} = 1 - b\cos kz - \frac{(kr)^2}{8} b\cos kz + \frac{1}{2B_0{}^2}\left(\frac{n\mu I}{\pi a}\right)^2 \left(\frac{r}{a}\right)^{2(n-1)}$$

$$\approx (1 - b) + \frac{b}{2}(kz)^2 + \frac{1}{2B_0{}^2}\left(\frac{n\mu I}{\pi a}\right)^2 \left(\left(\frac{r}{a}\right)^{2n-2} - \beta\left(\frac{r}{a}\right)^2\right) \tag{2.99}$$

$$\beta = \frac{(ka)^2 b B_0{}^2}{4}\left(\frac{\pi a}{n\mu I}\right)^2. \tag{2.100}$$

The magnitude of the combined field increases outward in all direction if the multipole field is larger than some value. The property that the field increases outward in all directions is one of the important conditions that must be met if we are to confine the plasma in a stable way. As Ioffe firstly demonstrated that such a system confines the plasma in stable way experimentally, this field is called *Ioffe field.*[7] As the magnitude of the magnetic field is minimum in the confining region of plasma, the Ioffe field is also called *minimum-B field*. A mirror field without a superposed multipole field (eqs. (2.95) and (2.96)) is called a *simple mirror field*. Variations of minimum-B fields are shown in fig. 2.8a–d. Experiments with the baseball-seam coil (the American expression) or tennis-ball-seam coil (the English expression) have been carried out[8] (fig. 2.8d). The Yin-Yang coil,[9] producing larger mirror ratios, (fig. 2.8e), has also been studied.

*Equations (2.97) and (2.98) express a bumpy field which is periodic extension of the central mirror field. They analytically express only that part of the bumpy field that lies within the region $|z| < L/2$.

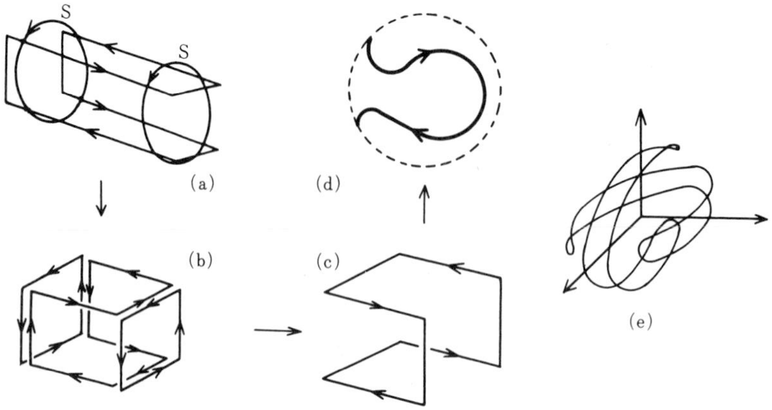

Fig. 2.8
Ioffe-field coils. (a) Ring coils produce a simple mirror, and four straight bars produce the quadrupole field. (b), (c) Modifications of this array. (d) Baseball coil. (e) Yin-Yang coil.

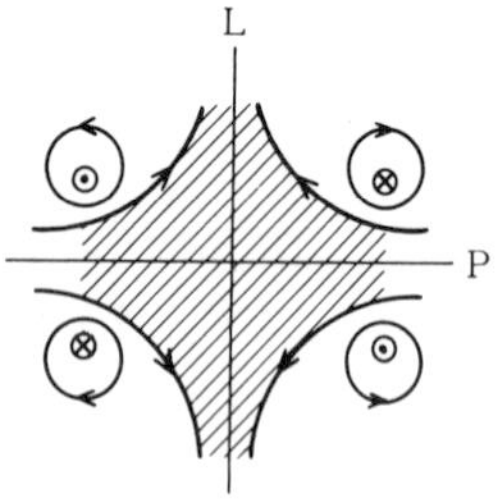

Fig. 2.9
Cusp field. *P*, point cusp; *L*, line cusp.

A *cusp field* is produced by two ring currents of opposed sense, as is shown in fig. 2.9. The cusp field is expressed by

$$A_\theta = arz, \tag{2.101}$$

$$\left.\begin{aligned} B_r &= -ar \\ B_z &= 2az. \end{aligned}\right\} \tag{2.102}$$

The lines of magnetic force are given by $rA_\theta = ar^2z = \text{const}$. The magnitude of the magnetic field is $B^2 = a^2(r^2 + 4z^2)$; the field satisfies the minimum-B condition. But the magnitude of B is zero at the origin ($r = 0$, $z = 0$); the disadvantage of this is that the resultant magnetic moment is not adiabatically invariant, as will be discussed in ch. 3.

2.3c Stellarator Field

Let us consider a magnetic field of helical symmetry. By means of cylindrical coordinates r, θ, z, we can express the field in terms of r and $\varphi \equiv \theta - \delta\alpha z$. (fig. 2.10) Here we assume that $\alpha > 0$ and $\delta = \pm 1$. A magnetic field in a current-free region ($j = 0$) can be expressed by a scalar potential ϕ_B, satisfying $\Delta\phi_B = 0$, and we can write

$$\phi_B = B_0 z + \frac{1}{\alpha}\sum_{l=1}^{\infty} b_l I_l(lar)\sin(l\varphi) \tag{2.103}$$

$$\varphi \equiv \theta - \delta\alpha z. \tag{2.104}$$

The field components B_r, B_θ, B_z of $B = \nabla\phi_B$ are given by

$$\left.\begin{aligned} B_r &= \sum_{l=1}^{\infty} l b_l I_l'(l\alpha r)\sin(l\varphi) \\ B_\theta &= \sum_{l=1}^{\infty}\left(\frac{1}{\alpha r}\right) l b_l I_l(l\alpha r)\cos(l\varphi) \\ B_z &= B_0 - \delta\sum_{l=1}^{\infty} l b_l I_l(l\alpha r)\cos(l\varphi). \end{aligned}\right\} \tag{2.105}$$

The vector potential corresponding to this field has components

$$\left.\begin{aligned} A_r &= -\frac{\delta}{\alpha^2 r}\sum_{l=1}^{\infty} b_l I_l(l\alpha r)\sin(l\varphi) \\ A_\theta &= \frac{B_0}{2} r - \frac{\delta}{\alpha}\sum_{l=1}^{\infty} b_l I_l'(l\alpha r)\cos(l\varphi) \\ A_z &= 0. \end{aligned}\right\} \tag{2.106}$$

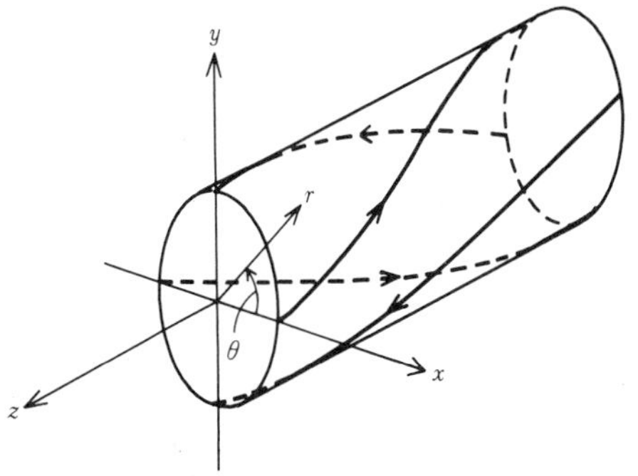

Fig. 2.10
Current coils of stellarator field.

Using these, we can write

$$\left.\begin{aligned} B_r &= -\frac{\partial A_\theta}{\partial z} \\ B_\theta &= \frac{\partial A_r}{\partial z} \\ B_z &= \frac{1}{r}\frac{\partial (rA_\theta)}{\partial r} - \frac{1}{r}\frac{\partial A_r}{\partial \theta}. \end{aligned}\right\} \tag{2.107}$$

The magnetic surface $\psi = A_z + \delta\alpha r A_\theta = \delta\alpha r A_\theta = \text{const.}$ is given by

$$\psi(r, \varphi) = B_0 \frac{\delta\alpha r^2}{2} - r \sum_{l=1}^{\infty} b_l I_l'(lar)\cos(l\varphi) = \text{const.} \tag{2.108}$$

Let the magnetic fluxes in the z and θ directions inside the magnetic surface be denoted by Φ and X (X is the integral over the pitch along z, i.e., over $2\pi/\alpha$); then these may be expressed by

$$\Phi = \int_0^{2\pi}\int_0^{r(\varphi)} B_z(r, \varphi) r dr d\theta \tag{2.109}$$

$$X = \int_0^{2\pi/\alpha}\int_0^{r(\varphi)} B_\theta(r, \varphi) dr dz = \frac{1}{\alpha}\int_0^{2\pi}\int_0^{r(\varphi)} B_\theta(r, \varphi) dr d\theta. \tag{2.110}$$

As the relation $\alpha r B_z - \delta B_\theta = \alpha\partial(rA_\theta)/\partial r = \delta^{-1}\partial\psi/\partial r$ holds, we find that

$$\Phi - \delta X = 2\pi\psi/(\delta\alpha). \tag{2.111}$$

Let us consider only one harmonic component of the field. The scalar potential and the magnetic surface are expressed by

$$\phi_B = B_0 z + \frac{b}{\alpha} I_l(l\alpha r)\sin(l\theta - \delta l\alpha z) \tag{2.112}$$

$$\psi = \frac{B_0}{2\delta\alpha}\left((\alpha r)^2 - \frac{2\delta(\alpha r)b}{B_0} I_l'(l\alpha r)\cos(l\theta - \delta l\alpha z)\right) = \frac{B_0}{2\delta\alpha}(\alpha r_0)^2. \tag{2.113}$$

The singular points (r_s, θ_s) in the $z = 0$ plane are given by

$$\frac{\partial\psi}{\partial r} = 0 \qquad \frac{\partial\psi}{\partial\theta} = 0. \tag{2.114}$$

Since the modified Bessel function $I_l(x)$ satisfies

$$I_l''(x) + \frac{1}{x}I_l'(x) - \left(1 + \frac{l^2}{x^2}\right)I_l = 0, \tag{2.115}$$

the singular points are given by

$$\sin(l\theta_s) = 0 \tag{2.116}$$

$$\alpha r\left(1 - \frac{\delta bl}{B_0}\left(1 + \frac{1}{(\alpha r_s)^2}\right) I_l(l\alpha r_s)\cos(l\theta_s)\right) = 0, \tag{2.117}$$

or

$$\left.\begin{aligned} \theta_s &= 2\pi(j-1)/l \qquad && \delta b/B_0 > 0 \\ &= 2\pi\left(j - \frac{1}{2}\right)\frac{1}{l} && \delta b/B_0 < 0 \end{aligned}\right\} \quad j = 1, \ldots, l \tag{2.118}$$

$$\left|\frac{\delta bl}{B_0}\right| = \frac{1}{(1 + (\alpha r_s)^{-2}) I_l(l\alpha r_s)}. \tag{2.119}$$

The magnetic surfaces for $l = 1$, $l = 2$, and $l = 3$ are shown in fig. 2.11. The *hyperbolic* singular points are called *separatrix points*, and the surfaces crossing these points are called *separatrices*. When $x \ll 1$, the modified Bessel function is

$$I_l(x) \approx \frac{1}{l!}\left(\frac{x}{2}\right)^l.$$

The magnetic surface in the region $\alpha r \ll 1$ is expressed by

$$(\alpha r)^2 - \frac{\delta b(1/2)^{l-1}}{B_0(l-1)!}(\alpha r)^l \sin l(\theta - \delta\alpha z) = \text{const.} \tag{2.120}$$

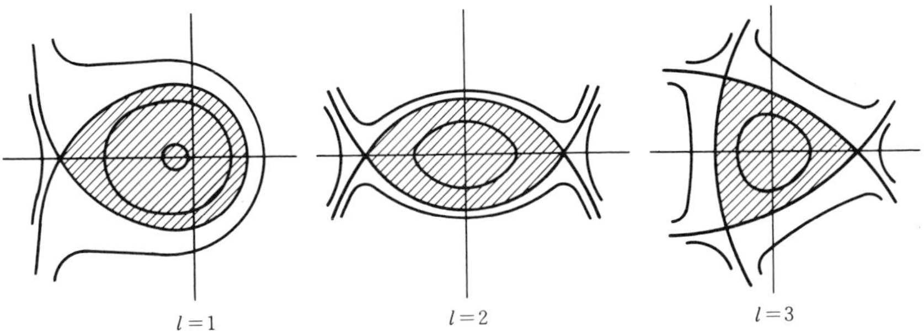

Fig. 2.11
Magnetic surfaces, showing separatrix points and separatrices, of the stellarator field.

The magnitude B is

$$\left(\frac{B}{B_0}\right)^2 = 1 - 2\frac{\delta l b}{B_0} I_l \cos(l\varphi) + \left(\frac{lb}{B_0}\right)^2 \left(I_l^2\left(1 + \frac{1}{(\alpha r)^2}\right)\cos^2(l\varphi) + (I_l')^2 \sin^2(l\varphi)\right).$$

The magnitude B at the separatrix (r_s, θ_s) is

$$\left(\frac{B}{B_0}\right)^2 = 1 - \frac{(\alpha r)^2}{1 + (\alpha r)^2},$$

and B at the point $(r_s, \theta_s + \pi/l)$ is

$$\left(\frac{B}{B_0}\right)^2 = 1 + \frac{(\alpha r)^2}{1 + (\alpha r)^2}.$$

We have discussed the exact magnetic surface in the foregoing. When the radial coordinate r is sensibly smaller than r_s (say $r < r_s/2$), the approximate magnetic surface deduced by the averaging method is useful. Using eqs. (2.46) and (2.47), we obtain the average magnetic surface (overbars on r and θ are omitted here for simplicity)

$$\frac{dr}{dz} = \frac{1}{B_0}\frac{1}{r}\frac{\partial \psi_a}{\partial \theta} \qquad \frac{d\theta}{dz} = \frac{-1}{rB_0}\frac{\partial \psi_a}{\partial r} \tag{2.121}$$

$$\psi_a = \frac{-1}{B_0}\overline{\langle b_r \rangle b_\theta} = \frac{-\delta b^2 l}{2B_0 \alpha}\frac{1}{\alpha r} I_l(l\alpha r) I_l'(l\alpha r). \tag{2.122}$$

The equation for the average line of magnetic force is

$$r = r_0 \tag{2.123}$$

$$\theta = \delta\left(\frac{b}{B}\right)^2 \frac{l^3}{2r}\left(\frac{d}{dx}\left(\frac{I_l I_l'}{x}\right)_{x = l\alpha r}\right) z + \theta_0 = \gamma z + \theta_0. \tag{2.124}$$

The more accurate equation of the line of magnetic force, derived from eqs. (2.49) and (2.50), is

$$r = r_0 + \frac{\delta}{\alpha}\frac{b}{B_0} I_l' \cos l(\theta - \delta \alpha z) \tag{2.125}$$

$$\theta = \gamma z + \theta_0 - \delta \frac{b}{B_0}\left(\frac{1}{\alpha r}\right)^2 I_l \sin l(\theta - \delta \alpha z). \tag{2.126}$$

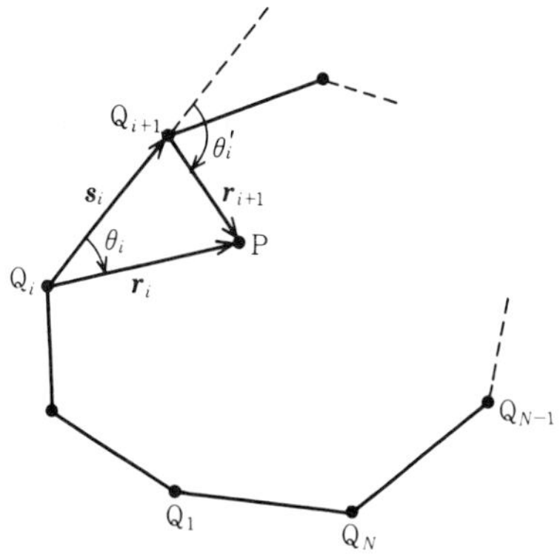

Fig. 2.12
A closed filamentary current loop consisting of straight segments.

As $I_l' \approx I_l/\alpha r$ for $\alpha r \ll 1$, the line of magnetic force is the superposition of the average line (eqs. (2.123) and (2.124)) and of the gyration of radius $bI_l'/(\alpha B_0)$ (see fig. 2.2).

Thus we have studied the properties of the helical symmetric magnetic field. Since the most famous device generating this type of field is the stellarator,[10] this field configuration is called the *stellarator field.*

2.3d Relations between Current Configuration and Magnetic Field

When a current distribution $\boldsymbol{j}(\boldsymbol{r})$ is given in a homogeneous medium, the magnetic field (eq. (2.30)) is

$$\boldsymbol{B}(\boldsymbol{r}) = \frac{\mu}{4\pi}\int \frac{\boldsymbol{j}(\boldsymbol{r}') \times (\boldsymbol{r}-\boldsymbol{r}')}{|\boldsymbol{r}-\boldsymbol{r}'|^3}\,\mathrm{d}\boldsymbol{r}'. \tag{2.127}$$

Where we have closed filamentary currents, then the Biot-Savart equation (2.31) can be used. Where a closed current loop is composed of straight segments $Q_1, Q_2, \ldots, Q_N$, the magnetic field is given by

$$\boldsymbol{B}(\boldsymbol{r}) = \sum_{i=1}^{N} \Delta\boldsymbol{B}_i(\boldsymbol{r}) \tag{2.128}$$

$$\Delta\boldsymbol{B}_i = \frac{\mu I}{4\pi}\int\frac{\sin\theta}{R^2}\,\mathrm{d}l\,\frac{\boldsymbol{s}_i\times\boldsymbol{r}_i}{|\boldsymbol{s}_i\times\boldsymbol{r}_i|}$$

$$= \frac{\mu I}{4\pi r_i \sin\theta_i}(\cos\theta_i - \cos\theta_i')\frac{\boldsymbol{s}_i\times\boldsymbol{r}_i}{|\boldsymbol{s}_i\times\boldsymbol{r}_i|}. \tag{2.129}$$

The vector $\boldsymbol{s}_i$ is the vector $\overrightarrow{Q_iQ_{i+1}}$ taking one to Q_{i+1} from Q_i, as shown in fig. 2.12. The vector $\boldsymbol{r}_i = \overrightarrow{Q_iP}$ is the distance to the observation point P from the infinitesimal current element at Q_i, and θ_i is the angle between $\boldsymbol{s}_i$ and $\boldsymbol{r}_i$. Equations (2.128) and (2.129) can be conveniently used to calculate the magnetic field.

The Biot-Savart law is generalized to a magnetic field in a finite region V surrounded by a closed surface S as follows:[1]

$$\boldsymbol{B}(\boldsymbol{r}) = \frac{\mu}{4\pi}\int_V \boldsymbol{j}(\boldsymbol{r}') \times \nabla'\left(\frac{1}{R}\right) \mathrm{d}\boldsymbol{r}' - \frac{1}{4\pi}\int_S (\boldsymbol{n} \times \boldsymbol{B}) \times \nabla'\left(\frac{1}{R}\right) \mathrm{d}S - \frac{1}{4\pi}\int_S (\boldsymbol{n}\cdot\boldsymbol{B})\nabla'\left(\frac{1}{R}\right) \mathrm{d}S, \tag{2.130}$$

where ∇' is the gradient along $\boldsymbol{r}'$ and $\boldsymbol{n}$ is the (outward) unit normal vector of S. The contribution of the region outsides is accounted for by the surface current $\mu\boldsymbol{K} = -\boldsymbol{n} \times \boldsymbol{B}$ on S and the magnetic charge $-(\boldsymbol{n}\cdot\boldsymbol{B})$ on S. When S is chosen to be a magnetic surface and the current density in V is zero, then one has $(\boldsymbol{n}\cdot\boldsymbol{B}) = 0$ and $\boldsymbol{j} = 0$. Then eq. (2.130) is reduced to

$$\boldsymbol{B}(\boldsymbol{r}) = \frac{\mu}{4\pi}\int_{S_m} \boldsymbol{K} \times \nabla'\left(\frac{1}{R}\right) \mathrm{d}S. \tag{2.131}$$

Sometimes it is necessary to find a current distribution that would produce a given magnetic field. This problem is difficult to solve in general; however it may be possible to find the current distributions in the following case. Let the magnetic field inside the boundary surface S be B_{in} and that outside be B_{out}. Both B_{in} and B_{out} must satisfy the boundary conditions at S, which are

$$(\boldsymbol{B}_{\text{in}} - \boldsymbol{B}_{\text{out}})\cdot\boldsymbol{n} = 0 \tag{2.132}$$

$$(\boldsymbol{B}_{\text{in}} - \boldsymbol{B}_{\text{out}}) \times \boldsymbol{n} = \mu\boldsymbol{K}. \tag{2.133}$$

Equation (2.132) gives the relation between $\boldsymbol{B}_{\text{in}}$ and $\boldsymbol{B}_{\text{out}}$ and eq. (2.133) gives the surface current $\boldsymbol{K}$.

Let us consider as an example the stellarator field. In the region $r < a$, the magnetic field $\boldsymbol{B}_{\text{in}}$ is given by (2.105). In the region $r > a$, the magnetic field $\boldsymbol{B}_{\text{out}}$ must be finite even at $r = \infty$, so that the expression for $\boldsymbol{B}_{\text{out}}$ is given by replacing b_l and $I_l(x)$ of eq. (2.105) by c_l and $K_l(x)$, respectively, where $K_l(x)$ is the lth modified Bessel function of the second kind. The condition (2.132) yields the relation between c_l and b_l. Equation (2.133) gives the surface current distribution

$$K_\theta = \frac{\delta b \cos l(\theta - \delta\alpha z)}{\mu(\alpha a)K_l'} \tag{2.134}$$

$$K_z = \frac{b \cos l(\theta - \delta\alpha z)}{\mu(\alpha a)^2 K_l'}. \tag{2.135}$$

This current distribution corresponds to the fundamental harmonic component of the line current I shown in fig. 2.10. The value of I is given by

$$\frac{2\mu I}{\pi a B_0} = \frac{bl}{B_0}\frac{1}{(l\alpha a)^2 K_l'}(1 + (\alpha a)^2)^{1/2}. \tag{2.136}$$

When current does not flow except on the boundary surface S, the magnetic fields are expressed by scaler potentials. In this case the boundary conditions (2.132) and (2.133) are reduced to

$$\nabla(\phi_{\text{in}} - \phi_{\text{out}})\cdot \boldsymbol{n} = 0$$

$$\nabla(\phi_{\text{in}} - \phi_{\text{out}}) \times \boldsymbol{n} = \mu K,$$

where ϕ_{in} and ϕ_{out} are solutions of $\nabla^2\phi = 0$.

2.4 Rotational Transform Angle, Shear, and Magnetic Well Depth of Toroidal Systems

The lines of magnetic force of a toroidal field encircle the major axis of the torus infinitely many times while they also encircle an inner toroidal volume V_{i}. The lines do not cross each other inside V_{i}. For the axisymmetric torus described in sec. 2.3a or the linear stellarator described in sec. 2.3c, the boundary surface S between V_{i} and the "external" volume V_{e} is the separatrix. However, the separatrix is not always clearly defined. In toroidal stellarator fields, there is a thin region outside V_{i}, in which the lines of magnetic force circle the major axis many times before they cross the wall of the discharge tube.

A line of magnetic force at the center of a toroidal region V_{i}, which comes back to the same position after one turn around the major axis of the torus, is called a *magnetic axis* (or, sometimes, a *minor axis*).

2.4a Rotational Transform Angle

Let us consider a plane P normal to the magnetic axis O in fig. 2.13. A line of force starting from a point on P circles the major axis of the torus and comes back to cross the plane P. The cross point rotates around O by an angle ι in P. The angle ι varies for different lines of force. When a line of force circles the major axis N times, the cross point rotates around O in the amount $\Sigma_{k=1}^{N} \iota_k$. The average value of ι is defined by

$$\iota = \lim_{N\to\infty} \frac{\sum_{k=1}^{N} \iota_k}{N}, \tag{2.137}$$

and this average angle is called the *rotational transform angle*.

Let the magnetic flux in the φ direction within the magnetic surface ψ be $\Phi(\psi)$ and the magnetic flux in the θ direction be $X(\psi)$. Then the rotational transform angle ι is given by

$$\iota = 2\pi \frac{\mathrm{d}X}{\mathrm{d}\Phi}. \tag{2.138}$$

We define the *safety factor* q_s by

$$q_s \equiv \frac{2\pi}{\iota}. \tag{2.139}$$

The reason that we call this quantity the safety factor will be given in ch. 9.

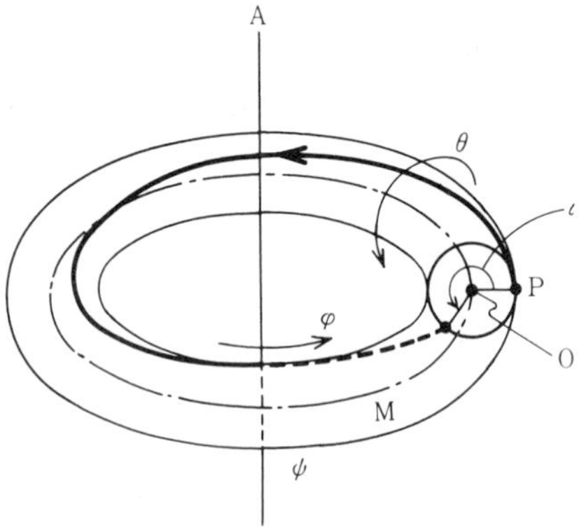

Fig. 2.13
The major axis A and the magnetic axis M of a toroidal field.

2.4b Shear

The rotational transform angle ι is, in general, a function of the magnetic surface ψ and of the radial coordinate r. In general, the lines of magnetic force are sheared with a change of the radial coordinate, as is shown in fig. 2.14. To indicate the amount of shear, the *shear parameter* Θ is defined:

$$\Theta = \frac{r^2 \mathrm{d}\iota/\mathrm{d}r}{2\pi R}. \tag{2.140}$$

Here $2\pi R$ is the circumference taken along the magnetic axis of the torus. When the torus is circular, R is called the major radius and the plasma radius a is called the minor radius. The ratio of R to a is known as the *aspect ratio*:

$$A = \frac{R}{a}, \tag{2.141}$$

while $\varepsilon = A^{-1} = a/R$ is the inverse aspect ratio. It will be shown that shear has an important effect upon the stability of plasma confinement schemes.

2.4c Magnetic Well Depth

Let the volume inside a magnetic surface ψ be V and the magnetic flux in the toroidal direction φ inside the magnetic surface ψ be Φ. We define the *specific volume* U by

$$U = \frac{\mathrm{d}V}{\mathrm{d}\Phi}. \tag{2.142}$$

If the unit vector of the magnetic field $\boldsymbol{B}$ is denoted by $\boldsymbol{b}$ and the normal unit vector of the infinitesimal cross-sectional area $\mathrm{d}S_i$ is denoted by $\boldsymbol{n}$, then we have

$$\mathrm{d}V = \int\left(\sum_i (\boldsymbol{b}\cdot\boldsymbol{n})_i \mathrm{d}S_i\right)\mathrm{d}l \qquad \mathrm{d}\Phi = \sum_i (\boldsymbol{b}\cdot\boldsymbol{n})_i B_i \mathrm{d}S_i. \tag{2.143}$$

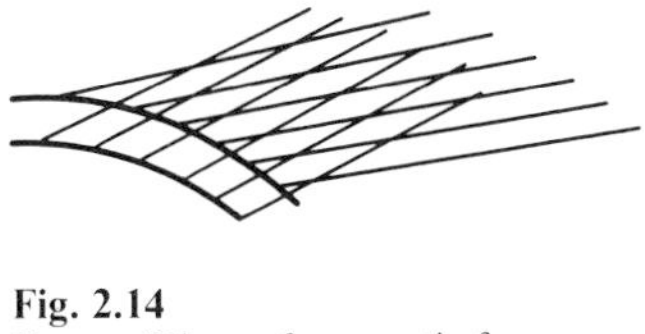

Fig. 2.14
Shear of lines of magnetic force.

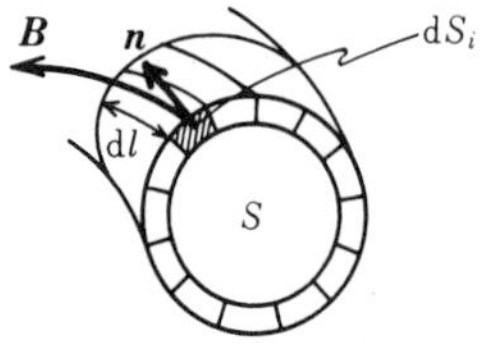

Fig. 2.15
Specific volume of a toroidal field.

When the lines of magnetic force close upon a single circuit of the torus, the specific volume U is

$$U = \frac{\oint \left(\sum_i (\boldsymbol{b} \cdot \boldsymbol{n})_i \mathrm{d}S_i \right) \mathrm{d}l}{\sum_i (\boldsymbol{b} \cdot \boldsymbol{n})_i B_i \mathrm{d}S_i} = \frac{\sum_i (\boldsymbol{b} \cdot \boldsymbol{n})_i B_i \mathrm{d}S_i \oint \frac{\mathrm{d}l}{B_i}}{\sum_i (\boldsymbol{b} \cdot \boldsymbol{n})_i B_i \mathrm{d}S_i}. \tag{2.144}$$

As the integral over l is carried out along a small tube of the magnetic field, $\sum_i (\boldsymbol{b} \cdot \boldsymbol{n})_i \mathrm{d}S_i B_i$ is independent of l (conservation of magnetic flux). As will be explained in ch. 7, $\oint \mathrm{d}l/B_i$ on the same magnetic surface is constant, so that U is reduced to

$$U = \oint \frac{\mathrm{d}l}{B}. \tag{2.145a}$$

When the lines of magnetic force close at N circuits, U is

$$U = \frac{1}{N} \int_N \frac{\mathrm{d}l}{B}. \tag{2.145b}$$

When the lines of magnetic force are *not* closed, U is given by

$$U = \lim_{N \to \infty} \frac{1}{N} \int_N \frac{\mathrm{d}l}{B}. \tag{2.145c}$$

Therefore, U may be considered to be an average of $1/B$. When U decreases outward, it means that the magnitude B of the magnetic field increases outward in an average sense, so that the plasma region is the so-called *average minimum-B* region. That $\bar{B}$ be a minimum in the plasma region is also an important condition for stability (see sec. 9.1). When the value of U on the magnetic axis and on the outermost magnetic surface are U_0 and U_a respectively, we define a *magnetic well depth* $-\Delta U/U$ as

$$-\frac{\Delta U}{U}=\frac{U_0-U_a}{U_0}. \tag{2.146}$$

2.4d Examples: Tokamak and Stellarator

Let us study the rotational transform angle, shear, and magnetic well depth in simple cases. In the analysis of a toroidal system, it is useful to first consider the corresponding linear system, as the latter retains many properties (such as the rotational transform angle and shear) of the former. When the linear system is treated by use of cylindrical coordinates, periodicity in both z and θ is required. Some integral multiple of the linear pitch (along z) corresponds to the circumference of the toroid. As simple examples we consider here the linear magnetic field and the current distribution $j_z(r)$ as shown in fig. 2.16 (tokamak) and the linear stellarator field.

(i) Rotational Transform Angle and Shear The magnetic field of the linear tokamak is given by

$$\begin{aligned} B_r &= 0 \\ B_\theta &= \frac{\mu_0 2\pi\int_0^r j_z r\mathrm{d}r}{2\pi r}=\frac{\mu_0}{r}\int_0^r j_z(r)r\mathrm{d}r \\ B_z &= B_0. \end{aligned} \tag{2.147}$$

The rotational transform angle is

$$\iota=\frac{\mathrm{d}\theta}{\mathrm{d}z}\times 2\pi R=2\pi\frac{R}{r}\frac{B_\theta}{B_0}=2\pi\frac{R}{r^2}\frac{\mu_0}{B_0}\int_0^r j_z(r)r\mathrm{d}r. \tag{2.148}$$

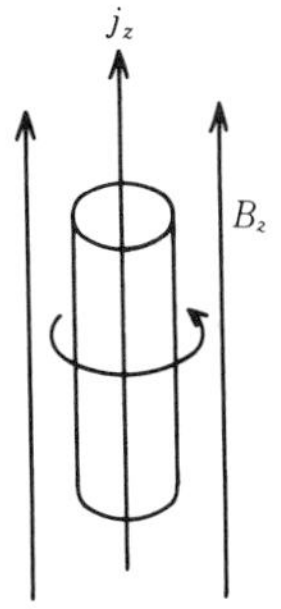

Fig. 2.16
Linear magnetic field superposed upon the current $j_z(r)$ (tokamak).

The shear parameter is

$$\Theta = \frac{r^2 \mathrm{d}\iota/\mathrm{d}r}{2\pi R} = \frac{\mu_0 r^2}{B_0}\frac{\mathrm{d}}{\mathrm{d}r}\left(\frac{1}{r^2}\int_0^r j_z(r) r \mathrm{d}r\right)$$
$$= \frac{\mu_0}{B_0}\left(r j_z(r) - \frac{2}{r}\int_0^r j_z(r) r \mathrm{d}r\right). \tag{2.149}$$

When the current distribution is uniform and $j_z = I/\pi a^2$, ι and Θ are

$$\iota = \frac{R}{a}\frac{\mu_0 I}{a B_0} \qquad \Theta = 0. \tag{2.150}$$

On the other hand, eqs. (2.147) and (2.148) together yield

$$\frac{\mathrm{d}X}{\mathrm{d}\Phi} = \frac{2\pi R B_\theta}{2\pi r B_0} = \frac{\iota}{2\pi}, \tag{2.151}$$

which satisfies eq. (2.138).

In the linear stellarator configuration, the magnetic field (eqs. (2.105)) is given by

$$\left.\begin{aligned} B_r &= l b I_l'(l\alpha r)\sin l(\theta - \delta\alpha z) \\ B_\theta &= l b(\alpha r)^{-1} I_l(l\alpha r)\cos l(\theta - \delta\alpha z) \\ B_z &= B_0 - \delta l b I_l(l\alpha r)\cos l(\theta - \delta\alpha z). \end{aligned}\right\} \tag{2.152}$$

The average line of magnetic force is expressed by eqs. (2.123) and (2.124): and ι is

$$\iota = \delta\pi\left(\frac{b}{B}\right)^2 \frac{R l^3}{r}\frac{\mathrm{d}}{\mathrm{d}x}\left(\frac{I_l I_l'}{x}\right)\Bigg|_{x = l\alpha r}. \tag{2.153}$$

If we take $l\alpha r \ll 1$, then $I_l(x) = (x/2)^l/l!$ and ι is

$$\iota = \delta 2\pi\left(\frac{b}{B}\right)^2\left(\frac{1}{2^l l!}\right)^2 l^5 (l-1)\alpha R (l\alpha r)^{2(l-2)} \qquad l \geq 2. \tag{2.154}$$

The parameter b is related to the parameters of the current configuration by eq. (2.136). As is clear from the foregoing equations, ι for a $l = 2$ stellarator field with $l\alpha r \ll 1$ is uniform and the shear is zero. When $l = 3$, the shear is strong but $\iota = 0$ at the magnetic axis (i.e., for $r = 0$). For toroids, the pseudotoroidal coordinates R, φ, z; r, θ shown in fig. 2.17(a) are often used. We have

$$R = R_0 + r\cos\theta \qquad z = r\sin\theta. \tag{2.155}$$

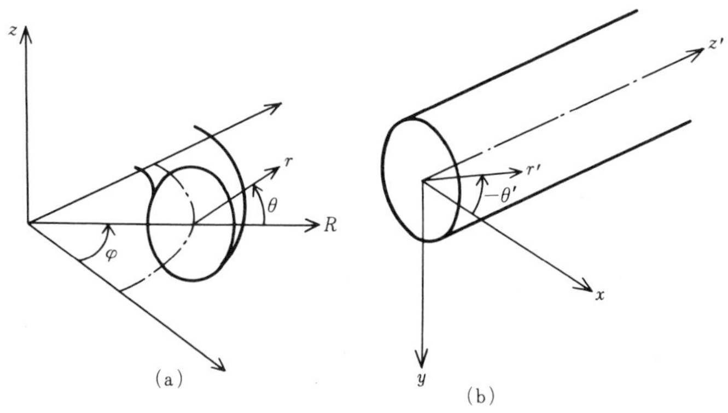

Fig. 2.17
Pseudotoroidal coordinates (a) and the corresponding cylindrical coordinates (b).

The cylindrical coordinates r', θ', z' for the corresponding linear configuration (fig. 2.17b) are related to the pseudotoroidal coordinates by

$$r \leftrightarrow r' \qquad \theta \leftrightarrow -\theta' \qquad \varphi \leftrightarrow \frac{z'}{R_0}.$$

Note particularly the relation $\theta \leftrightarrow -\theta'$.

(ii) Magnetic Well Depth Let us consider a toroidal magnetic field whose magnetic surfaces are concentric and of circular cross section, with the rotational transform angle $\iota(r)$. By the use of pseudotoroidal coordinates, we see that the average lines of magnetic force are given by

$$\frac{\partial r}{\partial (R_0 \varphi)} = \frac{-1}{B_0}\frac{1}{r}\frac{\partial \psi_a}{\partial \theta} \qquad \frac{\partial \theta}{\partial (R_0 \varphi)} = +\frac{1}{B_0}\frac{1}{r}\frac{\partial \psi_a}{\partial r}, \tag{2.156}$$

where $\psi_a = \psi_0(r)$ is the average magnetic surface and the rotational transform angle $\iota(r)$ is

$$\frac{\iota(r)}{2\pi} = \frac{1}{B_0}\frac{R_0}{r}\frac{d\psi_0(r)}{dr}. \tag{2.157}$$

When a uniform vertical field $B_\perp$ is superposed,[11] the resultant magnetic force line is obtained by noting that ψ_a in eq. (2.156) is given by

$$\psi_a = \psi_0(r) + B_\perp r \cos\theta. \tag{2.158}$$

When the second term in the right-hand side of eq. (2.158) is much smaller than the first term, then ψ_a is given by

$$\psi_a \approx \psi_0(\rho) \tag{2.159}$$

where

$$\left.\begin{aligned} \rho &= r - \Delta(r)\cos\theta \\ \Delta(r) &= \frac{-B_\perp r}{\partial\psi_0/\partial r} = -R_0\frac{B_\perp}{B_0}\frac{2\pi}{\iota(r)}. \end{aligned}\right\} \tag{2.160}$$

The new magnetic surface is still a circular cylinder, but the center is shifted by an amount $\Delta(r)$. The magnetic axis is shifted from the center to x so that there is satisfied

$$\frac{\partial\psi_0(x)}{\partial x} + B_\perp = 0, \tag{2.161}$$

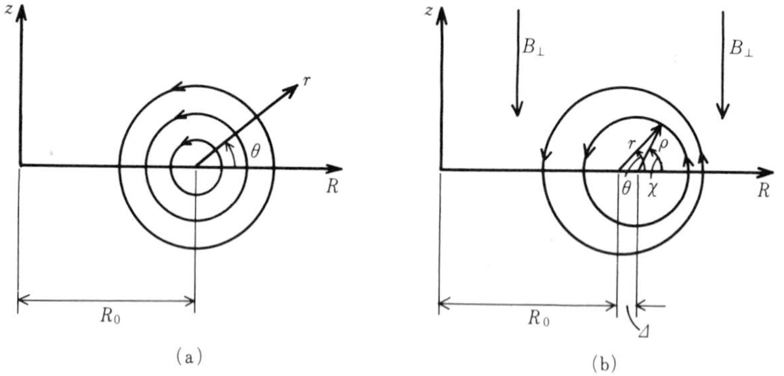

Fig. 2.18
Superposition of a vertical magnetic field.

i.e., the axis shifts by an amount $\Delta_0 = \Delta(\Delta_0)$. Let us estimate the specific volume $U = \mathrm{d}V/\mathrm{d}\Phi = \oint \mathrm{d}l/B$ in the case where B_r, $B_\theta \ll B_\varphi$, with $B_\varphi = B_0R_0/R$. The specific volume U on the magnetic axis is

$$U_0 \approx 2\pi(R_0 + \Delta_0)\frac{(R_0 + \Delta_0)}{B_0R_0} \approx \frac{2\pi R_0}{B_0}\left(1 + 2\frac{\Delta_0}{R_0}\right). \tag{2.162}$$

The magnetic surface $\psi_0(\rho) = \text{const.}$ is the torus of major radius $R = R_0 + \Delta(r)$, and the specific volume is

$$U = \frac{\mathrm{d}V}{\mathrm{d}\Phi} \approx \frac{2\pi(R_0 + \Delta(\rho))2\pi\rho}{\dfrac{B_0R_0}{(R_0 + \Delta(\rho))}2\pi\rho} \approx \frac{2\pi R_0}{B_0}\left(1 + 2\frac{\Delta(\rho)}{R_0}\right), \tag{2.163}$$

so that the magnetic well depth is

$$\frac{U_0 - U}{U_0} \approx 2\frac{\Delta_0 - \Delta(\rho)}{R_0}. \tag{2.164}$$

When the rotational transform angles at the magnetic axis and on the outermost surface are different, the average minimum-B field can be realized by superposition of the appropriate vertical field $B_\perp$ as is clear from eqs. (2.160) and (2.164). When the position of the magnetic axis is shifted outward as is shown in fig. 2.18b, the field is an average minimum-B field. When the system is axisymmetric, the system is average minimum-B if the separatrix point is located in the inner side near the major axis of the toroid (see fig. 2.4a).

References

1. J. A. Stratton: Electromagnetic Theory, McGraw-Hill, New York, 1941
W. R. Smythe: Static and Dynamic Electricity, McGraw-Hill, New York, 1941

2. A. I. Morozov and L. S. Solovev: Rev. of Plasma Phys. **2**, 1 (ed. by M. A. Leontovich) Consultants Bureau, New York, 1966

3. S. Yoshikawa, M. Barrault, W. Harris, D. Meade., R. Palladino, and S. von Goeler: Plasma Phys. and Controlled Nucl. Fusion Research **1**, 403 (Conf. Proceedings, Novosibirsk 1968) (IAEA, Vienna, 1969)
S. Yoshikawa: Nucl. Fusion **13**, 433(1973)

4. D. H. Birdsall, R. J. Briggs, S. A. Colgate, H. P. Furth, and C. W. Hartman: Plasma Phys. and Controlled Nucl. Fusion Research **2**, 291 (Conf. Proceedings, Culham 1965) (IAEA, Vienna, 1966)

5. L. A. Artsimovich, V. V. Afrosimov, I. P. Glaskovskii, M. P. Petrov, and V. S. Strelkov: ibid., 595
L. A. Artsimovich, V. A. Vershkov, A. V. Glukhov, E. P. Gorbunov, V. S. Zaveryaev, S. E. Lysenko, S. V. Mirnov, I. B. Semenov, and V. A. Strelkov: Plasma Phys. and Controlled Nucl. Fusion Research **1**, 443 (Conf. Proceedings, Novosirbirsk 1968) (IAEA, Vienna, 1969). English translation in Nucl. Fusion, suppl., p. 41 (1972)

6. T. Ohkawa and D. W. Kerst: Phys. Rev. Lett. **7**, 41(1961)
T. Ohkawa, A. A. Schupp, M. Yoshikawa, and H. G. Voorhies: Plasma Phys. and Controlled Nucl. Fusion Research **2**, 531 (Conf. Proceedings, Culham 1965) (IAEA, Vienna, 1966)

7. Yu. B. Gott, M. S. Ioffe, and V. G. Telkovsky: Plasma Phys. and Controlled Nucl. Fusion Research (Conf. Proceedings, Salzburg 1961). Also in Nucl. Fusion, pt. 3, 1045 (1962)

8. C. C. Damn, J. H. Foote, A. H. Futch Jr., R. K. Goodman, F. J. Gordon, A. L. Hunt, R. G. Mallon, K. G. Moses, J. E. Osher, R. F. Post, and J. F. Steinhaus: Plasma Phys. and Controlled Nucl. Fusion Research **2**, 253 (Conf. Proceedings, Novosibirsk 1968) (IAEA, Vienna, 1969)

9. R. W. Moir and R. F. Post: Nucl. Fusion **9**, 253(1969)

10. L. Spitzer, Jr: Phys. of Fluids **1**, 253(1958)
A. S. Bishop and E. Hinnov: Plasma Phys. and Controlled Nuclear Fusion Research. **2**, 673 (Conf. Proceedings, Culham 1965) (IAEA, Vienna, 1966)

11. A. Gibson, J. Hugill, G. W. Reid, R. A. Rowe, and B. C. Sanders: Plasma Phys. and Controlled Nucl. Fusion Research **1**, 465 (Conf. Proceedings, Novosibirsk 1968) (IAEA, Vienna, 1969)
J. B. Taylor: Phys. of Fluids **8**, 1203(1965)

3 Motion of a Single Particle

In this chapter, the motion of individual charged particles in a given magnetic field is studied in detail. There are a large number of charged particles in a plasma, thus their movements do affect the magnetic field. But this effect is neglected in this chapter. It is justified when β, which is the ratio of the plasma pressure $p = nT$ to the pressure of the magnetic field $B^2/2\mu$, is small. In other words, the analytical method given here will be sufficiently accurate to predict the behavior of the particles in low-β plasmas. The influence of Coulomb collisions, which is the subject of the next chapter, is also ignored. Therefore, here each particle is assumed to have a long mean free path.

In general, exact solutions for the equation of motion of a single particle are very difficult to obtain except for some simple cases (sec. 3.2). When the Larmor radius, which will be defined in sec 3.3, is much smaller than any characteristic length of the magnetic field (for example, the radius of curvature of the lines of magnetic force), the movement of a particle may be accurately analyzed by means of the drift approximation (sec. 3.3). The drift approximation is used to study the behavior of particles trapped in a mirror field or confined in a toroidal field (sec. 3.4). The banana orbit of a particle in a tokamak or stellarator field is described in sec. 3.5. The influence of high-frequency electric fields on particle motion is considered in sec. 3.6.

3.1 Equation of Motion of a Charged Particle

The equation of motion of a particle of rest mass m_0 and charge q in an electromagnetic field $\boldsymbol{E}$, $\boldsymbol{B}$ is

$$\frac{\mathrm{d}}{\mathrm{d}t} m\boldsymbol{v} = q(\boldsymbol{E} + \boldsymbol{v} \times \boldsymbol{B}) \qquad m = \frac{m_0}{(1 - v^2/c^2)^{1/2}}, \tag{3.1}$$

where c is velocity of light. When generalized coordinates q_i are used, the Lagrangian formulation is more convenient:

$$\frac{\mathrm{d}}{\mathrm{d}t}\frac{\partial L}{\partial \dot{q}_i} - \frac{\partial L}{\partial q_i} = 0. \tag{3.2}$$

In the relativistic case, $L(q_i, \dot{q}_i)$ is given by

$$L = -m_0 c^2 \left(1 - \frac{v^2}{c^2}\right)^{1/2} + q\boldsymbol{v} \cdot \boldsymbol{A} - q\phi \tag{3.3}$$

while in the nonrelativistic case, $L(q_i, \dot{q}_i)$ is given by

$$L = \frac{mv^2}{2} + q\boldsymbol{v}\cdot\boldsymbol{A} - q\phi. \tag{3.4}$$

The equations of motion in the Hamiltonian form use the generalized momentum, defined by

$$p_i = \partial L/\partial \dot{q}_i, \tag{3.5}$$

as the independent variables; the Hamiltonian equations are

$$\left.\begin{aligned} \frac{\mathrm{d}p_i}{\mathrm{d}t} &= -\frac{\partial \mathscr{H}}{\partial q_i} \\ \frac{\mathrm{d}q_i}{\mathrm{d}t} &= -\frac{\partial \mathscr{H}}{\partial p_i}. \end{aligned}\right\} \tag{3.6}$$

Here the Hamiltonian $\mathscr{H}$ is given by

$$\mathscr{H} = -L + \sum_i p_i\dot{q}_i. \tag{3.7}$$

The relativistic Hamiltonian is

$$\mathscr{H} = (m_0{}^2c^4 + c^2(\boldsymbol{p} - q\boldsymbol{A})^2)^{1/2} + q\phi \tag{3.8}$$

and the nonrelativistic Hamiltonian is

$$\mathscr{H} = \frac{1}{2m}(\boldsymbol{p} - q\boldsymbol{A})^2 + q\phi \tag{3.9}$$

where $\boldsymbol{p} = (p_1, p_2, p_3)$ is the momentum vector.

When the electromagnetic field has translational symmetry, the z component of the momentum is constant as is clear from the Langrangian equation,

$$p_z = \frac{\partial L}{\partial \dot{z}} = m\dot{z} + qA_z = \text{const.} \tag{3.10}$$

In the axisymmetric system, we have

$$p_\theta = \frac{\partial L}{\partial \dot{\theta}} = mr^2\dot{\theta} + qrA_\theta = \text{const.} \tag{3.11}$$

When the system has helical symmetry, i.e., when the field depends on only r and $\theta - \alpha z$, the following relation holds:

$$\frac{\partial L}{\partial z} + \alpha \frac{\partial L}{\partial \theta} = 0;$$

and

$$\frac{\partial L}{\partial \dot{z}} + \alpha \frac{\partial L}{\partial \dot{\theta}} = m(\dot{z} + \alpha r^2 \dot{\theta}) + q(A_z + \alpha r A_\theta) = \text{const.} \tag{3.12}$$

If the Hamiltonian does not depend explicitly on the time t, then

$$\mathscr{H} = \text{const.} \tag{3.13}$$

is one integral of the Hamiltonian equations. This integral expresses the conservation of energy, which may be written as

$$mc^2 + q\phi = W \tag{3.14}$$

$$\frac{mv^2}{2} + q\phi = W \tag{3.15}$$

for the relativistic and nonrelativistic cases, respectively.

3.2 Particle Orbits in Axially Symmetric Systems

3.2a Relation between Particle Orbits and Magnetic Surfaces in Toroidal Fields

The coordinates r^*, θ^*, z^* on a magnetic surface of an axisymmetric field satisfy

$$r^* A_\theta(r^*, z^*) = c_m. \tag{3.16}$$

The orbit (r, θ, z) of a charged particle in the field (eq. (3.11)) is given by

$$rA_\theta(r, z) + \frac{m}{q} r^2 \dot{\theta} = \frac{p_\theta}{q}. \tag{3.17}$$

If c_m is chosen to be $c_m = p_\theta/q$, the relation between the magnetic surface and the particle orbit is reduced to

$$rA_\theta(r, z) - r^* A_\theta(r^*, z^*) = -\frac{m}{q} r^2 \dot{\theta}. \tag{3.18}$$

The distance δ (fig. 3.1) between the surface and the orbit is given by

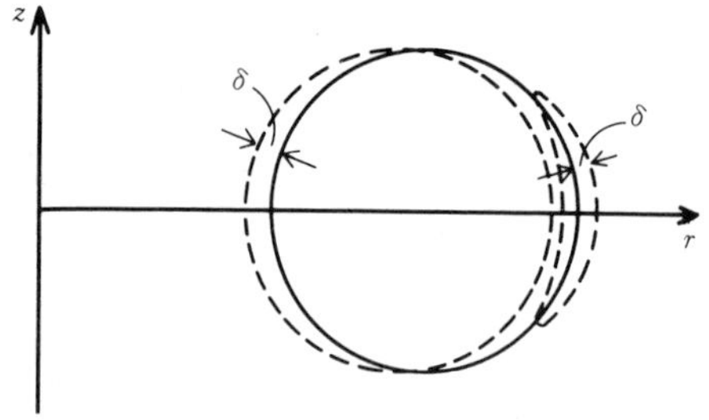

Fig. 3.1
Magnetic surface (solid line) and particle orbit (dotted line).

$$|\nabla(rA_\theta)|\delta \approx \left|\frac{m}{q} r\dot{\theta}\right| r, \tag{3.19}$$

where δ is small compared to r. As is clear from eq. (2.34), one has $|\nabla(rA_\theta)| = r(B_r^2 + B_z^2)^{1/2}$; thus

$$\delta \approx \left|\frac{m}{q}\frac{r\dot{\theta}}{B_p}\right| \equiv \rho_{B,p}, \tag{3.20}$$

where $B_p^2 \equiv B_r^2 + B_z^2$. $\rho_{B,p}$ is equal to the Larmor radius corresponding to the magnetic field B_p and the tangential velocity $r\dot{\theta}$ (see sec. 3.3). From eq. (3.15), the relation $|r\dot{\theta}| < (2(w - q\phi)/m)^{1/2}$ is derived, so that the orbit of a charged particle cannot deviate from the magnetic surface by more than the Larmor radius corresponding to B_p. Therefore, particle orbits will be limited to a finite region if the magnetic surface is itself of finite extent. This suggests that the magnitude of the poloidal components r, z (fig. 3.1) of the magnetic field play important roles in toroidal confinement.

3.2b Cusp Field

For simplicity, here the electrostatic potential ϕ is assumed to be zero. Then the cusp field is expressed by

$$B_r = -ar \qquad B_\theta = 0 \qquad B_z = 2az \tag{3.21}$$

$$A_r = 0 \qquad A_\theta = arz \qquad A_z = 0. \tag{3.22}$$

Equations (3.11) and (3.15) are reduced to

$$\frac{m}{2}(\dot{r}^2 + \dot{z}^2) + \frac{(p_\theta - qar^2 z)^2}{2mr^2} = W \tag{3.23}$$

$$mr\dot{\theta} = \frac{p_\theta}{r} - qazr. \tag{3.24}$$

The *Störmer potential* is defined by

$$X = \frac{1}{2m}\left(\frac{p_\theta}{r} - q(arz)\right)^2. \tag{3.25}$$

Then the region containing orbits of the particles of energy $mv^2/2$ is limited by

$$X = \frac{1}{2m}\left(\frac{p_\theta}{r} - qarz\right)^2 < \frac{m}{2}v^2. \tag{3.26}$$

This is called the *Störmer region* (fig. 3.2). When $p_\theta \cdot qaz > 0$, X has a minimum value $X_{\min} = 0$; this occurs at $r = (p_\theta/qaz)^{1/2}$, and $r\dot{\theta}$ becomes zero at this point. When $p_\theta \cdot qaz < 0$, then $X_{\min} = (2m^2)^{-1}(-p_\theta \cdot qaz)^{1/2}$ and $mr\dot{\theta} = \pm(-p_\theta qaz)^{1/2}$. The sign $\pm$ depends on the sign of $p_\theta \gtrless 0$. Projections of the orbits onto the r, θ plane are shown in fig. 3.3 for both cases, $p_\theta \cdot qaz \gtrless 0$.

3.2c Mirror Field

The vector potential of the mirror field, in which the magnitude of the magnetic field on the axis is $b = b(z)$, is given by

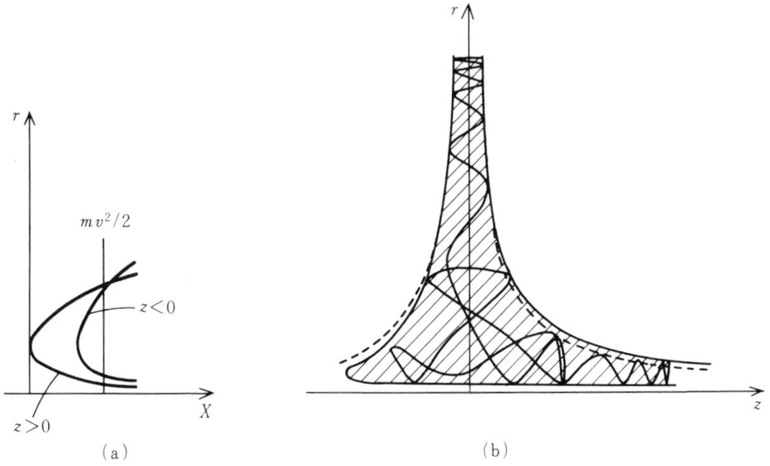

Fig. 3.2
(a) Störmer potential X and (b) Störmer region. The dotted lines in (b) are lines of magnetic force. This is the case $aqp_\theta > 0$. After Schuurman and de Kluiver.[1]

$$A_\theta(r, z) = \frac{r}{2}b(z) - \frac{r^3}{16}b''(z) + \cdots = \left(\frac{J_1(r\mathrm{d}/\mathrm{d}z)}{\mathrm{d}/\mathrm{d}z}\right)b(z). \tag{3.27}$$

The potential X corresponding to the Störmer potential for the cusp field is

$$X = \frac{1}{2m}\left(\frac{p_\theta}{r} - qA_\theta(r, z)\right)^2. \tag{3.28}$$

The singular points of X are determined by $\partial X/\partial r = 0$, $\partial X/\partial z = 0$. When $b(z)$ is a minimum b_0 at $z = 0$ and is maximum b_L at $z = \pm L$, then the positions of the singular points are ($z_0 = 0$, $r_0{}^2 = -2p_\theta/qb_0$), ($z_L = \pm L$, $r_L{}^2 = -2p_\theta/qb_L$): The potential X has a minimum at the first set of points and exhibits saddle points at the latter set (fig. 3.4).

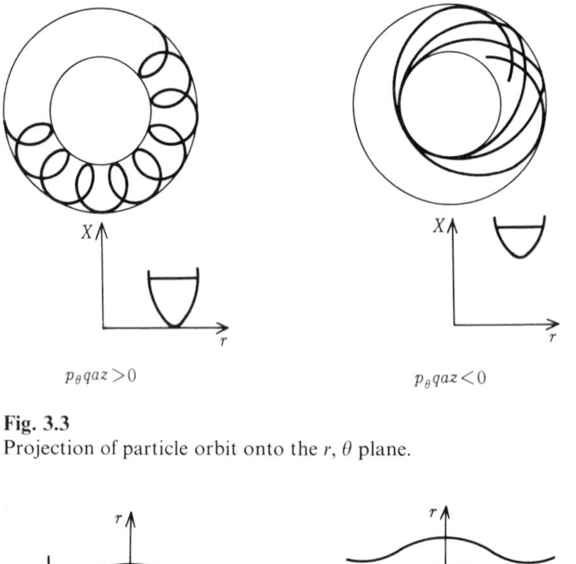

Fig. 3.3
Projection of particle orbit onto the r, θ plane.

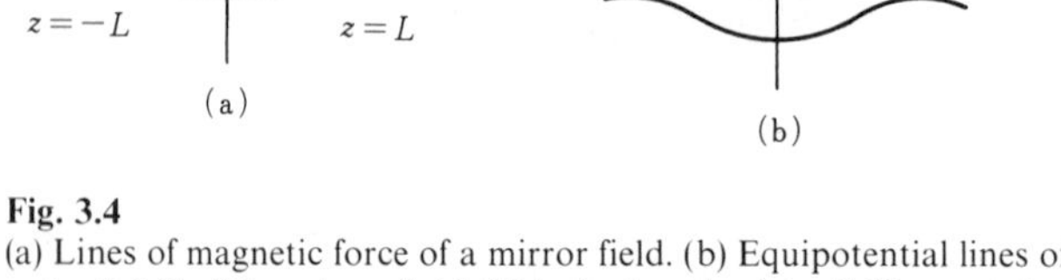

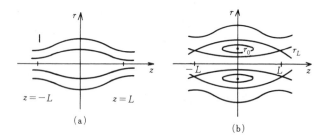

Fig. 3.4
(a) Lines of magnetic force of a mirror field. (b) Equipotential lines of the equivalent potential X of the mirror field. $2L$ is the length of the field. $z = \pm L$, $r = r_L$ are the positions of the saddle points.

Unless the condition

$$\frac{p_\theta}{qb_0} < 0 \tag{3.29}$$

is satisfied, X does not have minimum points and particles are not confined. A sufficient condition for particle confinement is

$$\frac{mv^2}{2} < \frac{1}{2m}\left(\frac{p_\theta}{r} - qA_\theta\right)_L^2 = \frac{1}{2m}\left(\frac{p_\theta}{r_L} - qA_\theta(r_L, z_L)\right)^2, \tag{3.30}$$

where (r_L, z_L) is the saddle point. When the approximate vector potential $A_\theta = (r/2)b(z)$ is used, then one has $r_L{}^2 = -2p_\theta/qb_L$ and the condition (3.30) reduces to

$$-2qb_L p_\theta > m^2v^2.$$

Since $mr\dot{\theta} = p_\theta/r - qA_\theta$, $r_0{}^2 = -2p_\theta/qb_0$, then p_θ is

$$p_\theta = \frac{1}{2}m(r_0\dot{\theta}_0)r_0 = \frac{mv_{\theta 0}}{2}\left(-\frac{2p_\theta}{qb_0}\right)^{1/2},$$

and one has

$$p_\theta = \frac{-m^2 v_{\theta 0}{}^2}{2qb_0}.$$

Consequently the *condition of mirror confinement* is given by[2]

$$\frac{v_{\theta 0}{}^2}{v^2} > \frac{b_0}{b_L}. \tag{3.31}$$

This condition is more strict than the following condition (3.32) of mirror confinement deduced from the drift approximation, which will be discussed in the following section:

$$\frac{v_\perp{}^2}{v^2} = \frac{v_{\theta 0}{}^2 + v_{r0}{}^2}{v^2} > \frac{b_0}{b_L}. \tag{3.32}$$

3.3 The Drift Approximation

3.3a Equation of Motion of the Guiding Center

The equation of motion of a charged particle in an electric field $\boldsymbol{E}$, magnetic field $\boldsymbol{B}$, and gravitational field $\boldsymbol{g}$ is

$$m\frac{\mathrm{d}\boldsymbol{v}}{\mathrm{d}t} = q\boldsymbol{E} + m\boldsymbol{g} + q(\boldsymbol{v} \times \boldsymbol{B}).$$

When these fields are static and uniform, the equation of motion can be obtained in a simple form if the following velocities are introduced:

$$\boldsymbol{v} = \boldsymbol{u}_{\mathrm{E}} + \boldsymbol{u}_{\mathrm{G}} + \boldsymbol{u} \tag{3.33}$$

$$\boldsymbol{u}_{\mathrm{E}} \equiv \frac{\boldsymbol{E} \times \boldsymbol{b}}{B} \tag{3.34}$$

$$\boldsymbol{u}_{\mathrm{G}} \equiv \frac{(m/q)}{B}\boldsymbol{g} \times \boldsymbol{b} = -\frac{1}{\Omega}(\boldsymbol{g} \times \boldsymbol{b}). \tag{3.35}$$

Then

$$\frac{\mathrm{d}\boldsymbol{u}_{\perp}}{\mathrm{d}t} = \Omega(\boldsymbol{b} \times \boldsymbol{u}_{\perp}) \tag{3.36}$$

$$\frac{\mathrm{d}\boldsymbol{u}_{\parallel}}{\mathrm{d}t} = q\boldsymbol{E}_{\parallel} + m\boldsymbol{g}_{\parallel}, \tag{3.37}$$

where the *cyclotron frequency* Ω is defined by

$$\Omega \equiv \frac{-qB}{m}. \tag{3.38}$$

$\boldsymbol{b}$ is the unit vector of $\boldsymbol{B}$. $\boldsymbol{u}_{\mathrm{E}}$ is the drift velocity due to the crossed electric and magnetic fields and does not depend on the sign of the charge: electrons and ions drift in the same direction (eq. (3.34) and fig. 3.5). $\boldsymbol{u}_{\mathrm{G}}$ is the drift velocity due to the crossed gravitational and magnetic fields. Electrons and ions drift in opposite directions in this case (eq. (3.35) and fig. 3.5). The parallel and perpendicular components of the velocity due to the magnetic field alone are denoted by $\boldsymbol{u}_{\parallel}$, $\boldsymbol{u}_{\perp}$:

$$\boldsymbol{u}_{\parallel} = (\boldsymbol{u} \cdot \boldsymbol{b})\boldsymbol{b} = u_{\parallel}\boldsymbol{b} \tag{3.39}$$

$$\boldsymbol{u}_{\perp} = \boldsymbol{u} - \boldsymbol{u}_{\parallel} = \boldsymbol{b} \times (\boldsymbol{u} \times \boldsymbol{b}). \tag{3.40}$$

The orbit of a charged particle is given by the solution of eqs. (3.36) and (3.37) as follows:

$$\boldsymbol{r} = \rho(t) + \boldsymbol{R}(t) \tag{3.41}$$

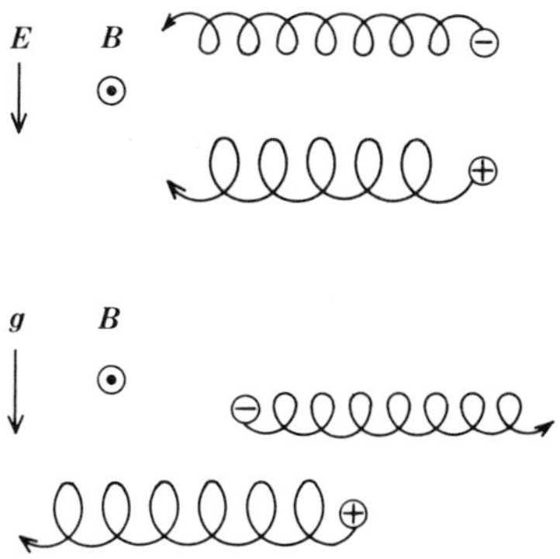

Fig. 3.5
Drift motions produced by crossing the electric field $\boldsymbol{E}$ and gravitational field $\boldsymbol{g}$, respectively, with $\boldsymbol{B}$.

$$\boldsymbol{R}(t) = (\boldsymbol{u}_{\mathrm{E}} + \boldsymbol{u}_{\mathrm{G}})t + \boldsymbol{R}_0 + \left(v_{\parallel,0}t + \frac{1}{2}\left(\frac{qE_{\parallel}}{m} + g_{\parallel}\right)t^2\right)\boldsymbol{b} \tag{3.42}$$

$$\rho(t) = \rho_B(\cos \Omega t \boldsymbol{e}_1 + \sin \Omega t \boldsymbol{e}_2), \tag{3.43}$$

where $\boldsymbol{e}_1, \boldsymbol{e}_2, \boldsymbol{b}$ are mutually orthogonal, satisfying the relation $\boldsymbol{b} = (\boldsymbol{e}_1 \times \boldsymbol{e}_2)$. The *Larmor radius* ρ_B is defined by

$$\rho_B = \frac{u_\perp}{|\Omega|}. \tag{3.44}$$

Consequently the orbit of a charged particle in static and uniform fields is a superposition of the gyration expressed by eq. (3.43), due to the magnetic field, and the drift motion due to the electric and gravitational fields. The magnitudes of $\boldsymbol{u}_{\mathrm{E}}$ and $\boldsymbol{u}_{\mathrm{G}}$ are inversely proportional to the magnitude of the magnetic field. The cyclotron frequency Ω of the gyration is proportional to B, and the Larmor radius is inversely proportional to B. Accordingly, the larger the magnetic field, the smaller the Larmor radius and the more rapid is the gyration, while the center of gyration drifts across the magnetic field more slowly. This drifting center of gyration is called the *guiding center*.

Let us next consider the case where the electric and magnetic fields change gradually in space and time. Denoting characteristic length and time for these changes by L and T respectively, we assume that

$$\frac{\rho_B}{L} \approx O(\varepsilon) \ll 1 \qquad \frac{1}{T\Omega} \approx O(\varepsilon) \ll 1, \tag{3.45}$$

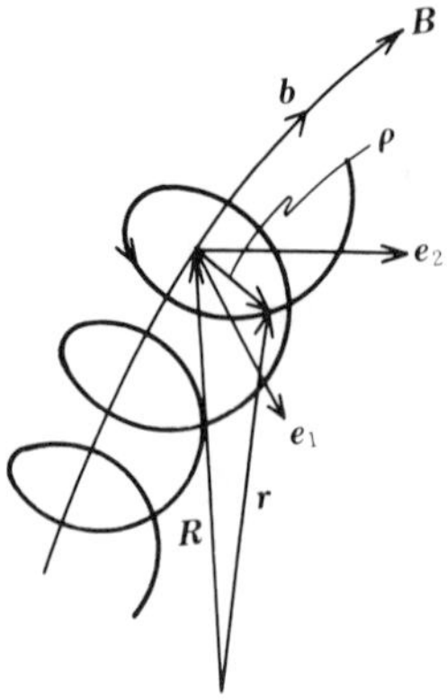

Fig. 3.6
The guiding center, for $q < 0$.

where $O(\varepsilon)$ means "to the first order of the small quantity ε."
Introducing the quantities

$$\theta \equiv \int \Omega \mathrm{d}t \tag{3.46}$$

$$\rho \equiv \rho_B(\cos\theta \boldsymbol{e}_1 + \sin\theta \boldsymbol{e}_2), \tag{3.47}$$

we define the coordinates of the guiding center $\boldsymbol{R}$ by

$$\boldsymbol{r} \equiv \boldsymbol{R} + \rho \tag{3.48}$$

where $\Omega = (-e/m)B$, $\boldsymbol{e}_1$, $\boldsymbol{e}_2$, $\rho_B = u_\perp/|\Omega|$, and $\boldsymbol{u}_\perp = \boldsymbol{v}_\perp - \boldsymbol{u}_{\mathrm{E}} - \boldsymbol{u}_{\mathrm{G}}$ are to be taken at $\boldsymbol{R}$, not at $\boldsymbol{r}$ (fig. 3.6). To examine the time variation of $\boldsymbol{R}$, eq. (3.48) is substituted in the equation of motion. Expanding $\boldsymbol{B}$, $\boldsymbol{E}$, and $\boldsymbol{g}$ in the vicinity of $\boldsymbol{R}$,

$$\boldsymbol{B}(\boldsymbol{r}) = \boldsymbol{B}(\boldsymbol{R}) + (\rho\cdot\nabla)\boldsymbol{B}(\boldsymbol{R}) + O(\varepsilon^2)\boldsymbol{B}, \tag{3.49}$$

we find

$$\begin{aligned}(\ddot{\boldsymbol{R}} + \ddot{\rho}) = {}& \frac{1}{m}(q\boldsymbol{E}(\boldsymbol{R}) + m\boldsymbol{g}(\boldsymbol{R}) + (\rho\cdot\nabla)(q\boldsymbol{E}(\boldsymbol{R}) + m\boldsymbol{g}(\boldsymbol{R}))\\ &+ \frac{q}{m}[\dot{\boldsymbol{R}} \times \boldsymbol{B}(\boldsymbol{R}) + \dot{\rho} \times \boldsymbol{B}(\boldsymbol{R}) + \dot{\boldsymbol{R}} \times (\rho\cdot\nabla)\boldsymbol{B}(\boldsymbol{R}) + \dot{\rho}\\ &\times (\rho\cdot\nabla)\boldsymbol{B}(\boldsymbol{R})] + O(\varepsilon^2)v\Omega,\end{aligned} \tag{3.50}$$

where the dots mean time derivative. $\mathrm{d}\rho/\mathrm{d}t$ is expressed by

$$\dot{\rho} = \Omega\rho_B(-\sin\theta \boldsymbol{e}_1 + \cos\theta \boldsymbol{e}_2) + \frac{\mathrm{d}}{\mathrm{d}t}(\rho_B \boldsymbol{e}_1)\cos\theta + \frac{\mathrm{d}}{\mathrm{d}t}(\rho_B \boldsymbol{e}_2)\sin\theta. \tag{3.51}$$

Substitute Eq. (3.51) into Eq. (3.50) and take time average over a gyration period $2\pi/|\Omega|$. When the time average $\int_0^{2\pi} f\,\mathrm{d}\theta/2\pi$ is denoted by $\langle f\rangle$, then $\langle\boldsymbol{\rho}\rangle = O(\varepsilon^2)L$, $\langle\ddot{\boldsymbol{\rho}} - (q/m)(\dot{\boldsymbol{\rho}} \times \boldsymbol{B})\rangle = O(\varepsilon^2)v\Omega$. So that $\ddot{\boldsymbol{R}}$ is

$$\ddot{\boldsymbol{R}} = \frac{1}{m}(q\boldsymbol{E}(\boldsymbol{R}) + m\boldsymbol{g}(\boldsymbol{R})) + \frac{q}{m}\dot{\boldsymbol{R}} \times \boldsymbol{B}(\boldsymbol{R})$$
$$+ \frac{q}{m}\langle\dot{\rho} \times (\rho\cdot\nabla)\boldsymbol{B}(\boldsymbol{R})\rangle + O(\varepsilon^2)v\Omega. \tag{3.52}$$

The third term on the right is reduced to $(q/m)(\Omega\rho_B{}^2/2)\nabla B(\boldsymbol{R})$ with accuracy of $O(\varepsilon^2)$. With use of $\nabla\boldsymbol{B} = 0$ and $\nabla \approx \boldsymbol{e}_1(\boldsymbol{e}_1\cdot\nabla) + \boldsymbol{e}_2(\boldsymbol{e}_2\cdot\nabla) + \boldsymbol{e}_3(\boldsymbol{e}_3\cdot\nabla)$, eq. (3.52) is converted to

$$m\ddot{\boldsymbol{R}} = (q\boldsymbol{E}(\boldsymbol{R}) + m\boldsymbol{g}(\boldsymbol{R})) + q\dot{\boldsymbol{R}} \times \boldsymbol{B}(\boldsymbol{R}) + \left(\frac{q\Omega}{2}\rho_B{}^2\right)\nabla B(\boldsymbol{R}) + O(\varepsilon^2)mv\Omega. \tag{3.53}$$

The third term $q(\Omega/2\pi)\pi\rho_B{}^2$ is the magnetic moment μ_m of the current loop of the gyration with current $I = -q\Omega/2\pi$ and encircling area $\pi\rho_B{}^2$; i.e.,

$$\mu_{\mathrm{m}} = IS = -\frac{q\Omega}{2}\rho_B{}^2 = \frac{mu_\perp{}^2/2}{B} \approx \frac{mv_\perp{}^2/2}{B}. \tag{3.54}$$

It will be shown in sec. 3.3c that the magnetic moment is adiabatically invariant.

The time variation of the parallel component of the velocity $v_{\mathrm{g}\parallel} \equiv \dot{\boldsymbol{R}}\cdot\boldsymbol{b}$ is given by

$$\frac{\mathrm{d}}{\mathrm{d}t}(\dot{\boldsymbol{R}}\cdot\boldsymbol{b}) = \ddot{\boldsymbol{R}}\cdot\boldsymbol{b} + \dot{\boldsymbol{R}}\cdot\dot{\boldsymbol{b}}.$$

Since $\boldsymbol{b}\cdot\boldsymbol{b} = 1$, then $\boldsymbol{b}\cdot\dot{\boldsymbol{b}} = 0$. As the magnitude of $\dot{\boldsymbol{R}}_\perp$ is $|\dot{\boldsymbol{R}}_\perp| = O(\varepsilon)v$, as is shown in eq. (3.56), the time variation of $v_{\mathrm{g}\parallel}$ is expressed by

$$m\frac{\mathrm{d}}{\mathrm{d}t}(v_{\mathrm{g}\|}) = qE_{\|} + mg_{\|} - \mu_{\mathrm{m}}\frac{\partial B}{\partial l}, \tag{3.55}$$

where l is the length along the line of magnetic force.

3.3b Drift Velocity

The vector product of eq. (3.53) and $\boldsymbol{b}(R)$ gives the perpendicular component of guiding center's velocity $\boldsymbol{v}_{\mathrm{g}\perp} \equiv \dot{\boldsymbol{R}}_{\perp}$,

$$\boldsymbol{v}_{\mathrm{g}\perp} = \frac{1}{B}(\boldsymbol{E}\times\boldsymbol{b}) + \frac{m}{qB}(\boldsymbol{g}\times\boldsymbol{b}) - \frac{\mu_{\mathrm{m}}}{qB}(\nabla B\times\boldsymbol{b}) - \frac{m}{qB}(\dot{\boldsymbol{v}}_{\mathrm{g}}\times\boldsymbol{b}) + O(\varepsilon^2)v. \tag{3.56}$$

The necessary accuracy of the quantity $\dot{\boldsymbol{v}}_{\mathrm{g}}$ in the fourth term is only the zeroth order and $\dot{\boldsymbol{v}}_{\mathrm{g}} = (\mathrm{d}/\mathrm{d}t)(\boldsymbol{v}_{\mathrm{g}\perp} + \boldsymbol{v}_{\mathrm{g}\|})$ is

$$\dot{\boldsymbol{v}}_{\mathrm{g}} \approx \frac{\mathrm{d}}{\mathrm{d}t}(v_{\mathrm{g}\|}\boldsymbol{b}) = \frac{\mathrm{d}v_{\mathrm{g}\|}}{\mathrm{d}t}\boldsymbol{b} + v_{\mathrm{g}\|}\left(\frac{\partial\boldsymbol{b}}{\partial t} + v_{\mathrm{g}\|}\frac{\partial\boldsymbol{b}}{\partial l}\right) + O(\varepsilon)\frac{v}{T}. \tag{3.57}$$

Since $\nabla\times\boldsymbol{E} = -\partial\boldsymbol{B}/\partial t$, then $|\partial\boldsymbol{B}/\partial t| \approx O(\varepsilon)$, so that $|\partial\boldsymbol{b}/\partial t| \approx O(\varepsilon)$ is the first-order quantity. Since $\partial/\partial l$ is the differentiation along the line of magnetic force, we find

$$\frac{\partial\boldsymbol{b}}{\partial l} = -\frac{\boldsymbol{n}}{\boldsymbol{R}}. \tag{3.58}$$

R is the radius of curvature of the line of magnetic force and $\boldsymbol{n}$ is the unit vector from the center O of curvature to a point P on the line of magnetic force.

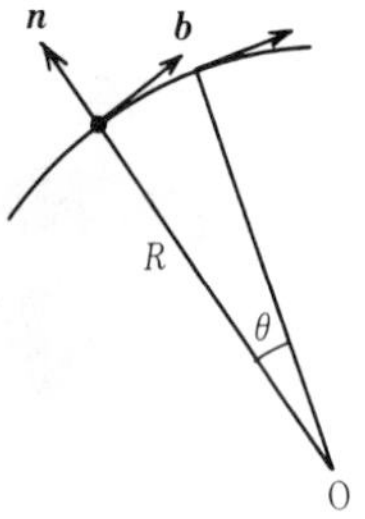

Fig. 3.7
Radius of curvature of a line of magnetic force.

The parallel component $v_{g\parallel} = \boldsymbol{v}_g \cdot \boldsymbol{b}(\boldsymbol{R})$ of the guiding center's velocity is nearly the same as the parallel component $v_\parallel = \boldsymbol{v} \cdot \boldsymbol{b}(\boldsymbol{r})$ of the particle velocity:

$$v_{g\parallel} = v_\parallel + O(\varepsilon)v. \tag{3.59}$$

Substitution of eq. (3.57) into eq. (3.56) gives

$$\boldsymbol{v}_{g\perp} = \boldsymbol{u}_E + \boldsymbol{u}_G + \frac{\boldsymbol{b}}{\Omega} \times \left(-\frac{\mu_m}{m} \nabla B - v_\parallel^2 \frac{\partial \boldsymbol{b}}{\partial l} \right) + O(\varepsilon^2)v. \tag{3.60}$$

The third term is the magnetic gradient drift due to the acceleration $-(\mu_m/m)\nabla B$ and the fourth term is the magnetic curvature drift due to the centrifugal force. (The fourth term can be expressed as $v_\parallel^2 \boldsymbol{n}/\boldsymbol{R}$ by virtue of eq. (3.58). The centrifugal force arises when the charged particle runs along the curved line of magnetic force.) By means of the vector formula

$$\frac{1}{2}\nabla B^2 = (\boldsymbol{B} \cdot \nabla)\boldsymbol{B} + \boldsymbol{B} \times (\nabla \times \boldsymbol{B}) = (\boldsymbol{B} \cdot \nabla)\boldsymbol{B} - \mu_0(\boldsymbol{j} \times \boldsymbol{B}), \tag{3.61}$$

we find

$$\nabla B \approx (b \cdot \nabla)\boldsymbol{B} = \boldsymbol{b}(\boldsymbol{b} \cdot \nabla)B + B(\boldsymbol{b} \cdot \nabla)\boldsymbol{b} = \frac{\partial B}{\partial l}\boldsymbol{b} - \frac{B}{R}\boldsymbol{n}. \tag{3.62}$$

Here we assume that $\nabla \times \boldsymbol{B} = \mu_0 \boldsymbol{j}$ is negligible ($\boldsymbol{j} \times \boldsymbol{B} = \nabla p$, $p \ll B^2/2\mu$). Then the third and fourth terms of eq. (3.60) are expressed by

$$\frac{\boldsymbol{b}}{\Omega} \times \left(\frac{-v_\perp^2}{2B} \nabla B - v_\parallel^2 \frac{\partial \boldsymbol{b}}{\partial l} \right) = \frac{(v_\perp^2/2 + v_\parallel^2)}{R} \frac{\boldsymbol{b} \times \boldsymbol{n}}{\Omega}. \tag{3.63}$$

Equation (3.60) is finally reduced to ($\mathrm{v}_\perp = u_\perp(1 + O(\varepsilon))$)

$$\begin{aligned} \boldsymbol{v}_{g\perp} &= \frac{\boldsymbol{b}}{\Omega} \times \left(\frac{q}{m}\boldsymbol{E} + \boldsymbol{g} - \frac{v_\perp^2/2}{B} \nabla B + v_\parallel^2 \frac{\boldsymbol{n}}{R} \right) + O(\varepsilon^2)v \\ &= \frac{\boldsymbol{E} \times \boldsymbol{b}}{B} + \frac{\boldsymbol{b}}{\Omega} \times \left(\boldsymbol{g} + \frac{v_\perp^2/2 + v_\parallel^2}{R}\boldsymbol{n} \right) + O(\varepsilon^2)v, \end{aligned} \tag{3.64}$$

where $\Omega = -qB/m$, $\mu_m = mv_\perp^2/2B$. The quantities $\boldsymbol{E}$, $\boldsymbol{B}$, $\boldsymbol{g}$, $\boldsymbol{n}$ and Ω, R are taken at the guiding center while $v_\parallel$, $v_\perp$ are components of the particle velocity.

When the electric field is very strong, another term

$$\dot{\boldsymbol{v}}_{\mathrm{gE}} = \frac{\mathrm{d}\boldsymbol{u}_{\mathrm{E}}}{\mathrm{d}t} + v_{\mathrm{g}\parallel}\left(\frac{\partial \boldsymbol{b}}{\partial t} + (\boldsymbol{u}_{\mathrm{E}}\cdot\nabla)\boldsymbol{b}\right) \tag{3.65}$$

must be added in eq. (3.57) for the expression of $\dot{\boldsymbol{v}}_{\mathrm{g}}$. $(\partial/\partial t + \boldsymbol{u}_{\mathrm{E}}\cdot\nabla)\boldsymbol{b}$ is the total time derivative in the coordinates system drifting with the velocity $\boldsymbol{u}_{\mathrm{E}}$, and the magnitude of $\boldsymbol{b}$ is $|\boldsymbol{b}| = 1$. Accordingly the terms in parentheses in the second term of eq. (3.65) may be expressed by means of a cross product of $\boldsymbol{b}$, i.e., as a rotation $\boldsymbol{\omega}$ given by

$$\boldsymbol{\omega} = \boldsymbol{b} \times \left(\frac{\partial}{\partial t} + \boldsymbol{u}_{\mathrm{E}}\cdot\nabla\right)\boldsymbol{b}. \tag{3.66}$$

Then the second term of eq. (3.65) is equal to $-v_{\mathrm{g}\parallel}\boldsymbol{b}\times\boldsymbol{\omega}$. The term due to the electric field,

$$\frac{\boldsymbol{b}}{\Omega} \times \left(-\frac{\mathrm{d}\boldsymbol{u}_{\mathrm{E}}}{\mathrm{d}t} + \boldsymbol{v}_{\mathrm{g}\parallel} \times \boldsymbol{\omega}\right), \tag{3.67}$$

must be added to eq. (3.64). The first term of eq. (3.67) is the inertial drift of acceleration $-\mathrm{d}\boldsymbol{u}_{\mathrm{E}}/\mathrm{d}t$ and the second term is the Coriolis drift.

In this section, the expression for v_{g} has been derived by a simple method. More systematic, but also more complicated, methods have been described.[2-4]

3.3c Adiabatic Invariant

When a system is periodic in time, the action integral, in terms of the canonical variables p, q, is

$$J = \oint p\mathrm{d}q; \tag{3.68}$$

this is an adiabatic invariant, as is well known.

(i) Magnetic Moment When the action integral is taken for a gyration of velocity $v_\perp$ and Larmor radius ρ_B, eq. (3.68) may be written

$$J_\perp = \oint mv_\perp\rho_B\mathrm{d}\varphi = 2\pi m\frac{v_\perp^2}{\Omega} = \frac{-4\pi m}{q}\frac{mv_\perp^2/2}{B}$$

$$= \frac{-4\pi m}{q}\mu_{\mathrm{m}}. \tag{3.69}$$

The *magnetic moment* μ_m is defined by eq. (3.54). Let us check the invariance of μ_m in the presence of a slowly changing magnetic field. Then an electric field is induced and the equation of motion is

$$\frac{\mathrm{d}}{\mathrm{d}t}\left(\frac{m v_\perp^2}{2}\right) = q\boldsymbol{v}_\perp \cdot \boldsymbol{E}_\perp. \tag{3.70}$$

The change of the kinetic energy ΔW per gyration is

$$\begin{aligned}\Delta W_\perp &= q\int(v_\perp \cdot \boldsymbol{E})\mathrm{d}t = \mp q\oint \boldsymbol{E}\cdot \mathrm{d}s = \mp q\int(\nabla \times \boldsymbol{E})\cdot \boldsymbol{n}\mathrm{d}S \\ &= \mp q\int \frac{\partial \boldsymbol{B}}{\partial t}\cdot \boldsymbol{n}\mathrm{d}S = \mp q\pi\rho_B{}^2\frac{\partial B}{\partial t} = \frac{2\pi}{|\Omega|}\frac{\partial B}{\partial t}\frac{|q\Omega|\rho_B{}^2}{2} \\ &= \Delta B\frac{W_\perp}{B} + O(\varepsilon^2)W_\perp.\end{aligned} \tag{3.71}$$

Therefore μ_m is invariant* to $O(\varepsilon^2)$. When the magnetic field increases slowly ($|\partial \boldsymbol{B}/\partial t| \ll |\Omega B|$), the kinetic energy $W_\perp \propto B$ is increased and the particle is heated.

(ii) Longitudinal Adiabatic Invariant As will be described in sec. 3.4b, a charged particle trapped in a mirror field moves back and forth along the line of magnetic force. The action integral of this motion

$$J_\parallel = m\oint v_\parallel \mathrm{d}l \tag{3.72}$$

is another adiabatic invariant. $J_\parallel$ is called the longitudinal adiabatic invariant. $v_\parallel$ and $\mathrm{d}l$ are expressed by

$$v_\parallel = (\boldsymbol{v}\cdot \boldsymbol{b}) \qquad \mathrm{d}l = (\mathrm{d}\boldsymbol{r}\cdot \boldsymbol{b}). \tag{3.73}$$

When the length between the mirror points is denoted by l and the average longitudinal velocity is denoted by $\langle v \rangle$, $J_\parallel$ is given by $J_\parallel = 2m\langle v_\parallel \rangle l$. As one makes the mirror length l shorter, $\langle v_\parallel \rangle$ increases (for $J_\parallel$ is conserved), and the particles are accelerated. This phenomenon is called *Fermi acceleration*.

* When all forces except the electromagnetic force are zero, μ_m is constant to $O(\varepsilon^3)$[4].

3.4 Orbit of the Guiding Center

3.4a Guiding Center and Magnetic Surface

When $\boldsymbol{g}$ is zero, the velocity of the guiding center is

$$\begin{aligned}\boldsymbol{v}_{\mathrm{g}} &= v_{\|}\boldsymbol{b} + \frac{1}{B}(\boldsymbol{E}\times\boldsymbol{b}) + \frac{mv_{\perp}^{2}/2}{qB^{2}}(\boldsymbol{b}\times\nabla B)\\ &\quad+\frac{mv_{\|}^{2}}{qB^{2}}(\boldsymbol{b}\times(\boldsymbol{b}\cdot\nabla)\boldsymbol{B}) + O(\varepsilon^{2})v,\end{aligned} \tag{3.74}$$

and

$$\mu_{\mathrm{m}} = \frac{mv_{\perp}^{2}/2}{B} = \mathrm{const.}\,(1 + O(\varepsilon^{2})). \tag{3.75}$$

When the electromagnetic field is static, the electric field $\boldsymbol{E}$ can be expressed in terms of the potential $\boldsymbol{E} = -\nabla\phi$ and the conservation of energy,

$$\frac{m}{2}(v_{\|}^{2} + v_{\perp}^{2}) + q\phi = W(1 + O(\varepsilon^{2})), \tag{3.76}$$

holds. Then $v_{\|}$ is expressed by

$$\begin{aligned} v_{\|} &= \pm\left(\frac{2}{m}(W - q\phi) - v_{\perp}^{2}\right)^{1/2}\\ &= \pm\left(\frac{2}{m}(W - q\phi - \mu_{\mathrm{m}}B)\right)^{1/2}. \end{aligned} \tag{3.77}$$

Noting that $v_{\|}$ is a function of the coordinates (eq. (3.77)), we can write $\nabla\times(mv_{\|}\boldsymbol{b})$ as

$$\nabla\times(mv_{\|}\boldsymbol{b}) = mv_{\|}\nabla\times\boldsymbol{b} + \nabla(mv_{\|})\times\boldsymbol{b}. \tag{3.78}$$

As the relation $\nabla(\boldsymbol{b}\cdot\boldsymbol{b}) = 2(\boldsymbol{b}\cdot\nabla)\boldsymbol{b} + 2\boldsymbol{b}\times(\nabla\times\boldsymbol{b}) = 0$ holds (app. 1), eq. (3.74) for $\boldsymbol{v}_{\mathrm{g}}$ is reduced to

$$\boldsymbol{v}_{\mathrm{g}} = \boldsymbol{b}\left(v_{\|} - \frac{mv_{\|}^{2}}{qB}(\boldsymbol{b}\cdot\nabla\times\boldsymbol{b})\right) + \frac{v_{\|}}{qB}\nabla\times(mv_{\|}\boldsymbol{b}). \tag{3.79}$$

When $\boldsymbol{b}\cdot\nabla\times\boldsymbol{b} = 0$ (this is the case when $\boldsymbol{b}\cdot\boldsymbol{j} = \boldsymbol{b}\cdot\nabla\times\boldsymbol{B}/\mu_{0} = 0$), eq. (3.79) becomes

$$\frac{\mathrm{d}\boldsymbol{r}_{\mathrm{g}}}{\mathrm{d}t} = \frac{v_{\|}}{B}\nabla\times\left(\boldsymbol{A} + \frac{mv_{\|}}{qB}\boldsymbol{B}\right), \tag{3.80}$$

where $\boldsymbol{A}$ is the vector potential of the magnetic field $\boldsymbol{B}$. When the new vector potential

$$\boldsymbol{A}^* = \boldsymbol{A} + \frac{mv_{\parallel}}{qB}\boldsymbol{B} \tag{3.81}$$

is introduced, the orbit of drift motion is equal to the line of magnetic force of the magnetic field $\boldsymbol{B}^* = \nabla \times \boldsymbol{A}^*$. By reasoning analogous to that in sec. 2.2a, the orbit surface of drift motion is given by[2]

$$A_z^* = \text{const.} \tag{3.82}$$

for translationally symmetric systems, by

$$rA_\theta^* = \text{const.} \tag{3.83}$$

for axisymmetric systems, and by

$$A_z^* + \alpha r A_\theta^* = \text{const.} \tag{3.84}$$

for helically symmetric systems, respectively.

3.4b Trapping in Mirror Fields

The component of the velocity parallel to the magnetic field is given by

$$v_{\parallel} = \pm\left(\frac{2}{m}(W - q\phi - \mu_{\mathrm{m}}B)\right)^{1/2} = \pm\left(v^2 - \frac{2}{m}\mu_{\mathrm{m}}B\right)^{1/2}. \tag{3.85}$$

Denote the maximum value of the mirror field by B_{M}. The value $v_{\parallel}$ of the particle with

$$v^2 < \frac{2}{m}\mu_{\mathrm{m}}B_{\mathrm{M}} \tag{3.86}$$

becomes imaginary, so that the particle must be reflected by the mirror field (see fig. 3.8a). Writing $v_{\perp}/v = \sin\alpha$ and using the relation

$$\frac{2}{m}\mu_{\mathrm{m}} = \frac{{v_{\perp}}^2}{B},$$

we can deduce that the *condition of mirror confinement* is

$$\frac{{v_{\perp}}^2}{v^2} > \sin^2\alpha_{\mathrm{cr}} \equiv \frac{B}{B_{\mathrm{M}}}. \tag{3.87}$$

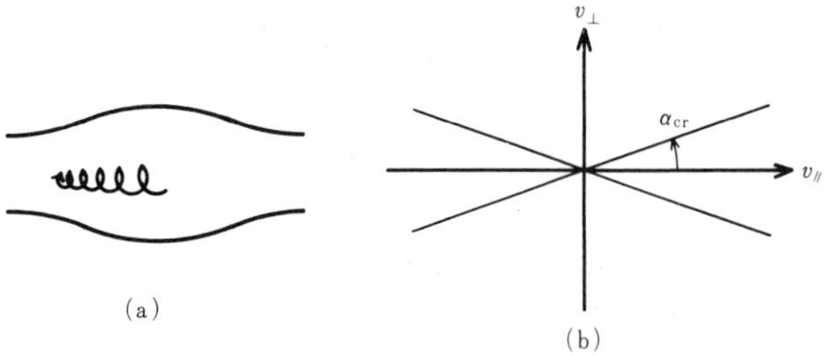

Fig. 3.8
Trapping in a mirror field.

When a particle has larger value of α than the critical value α_{cr} defined by eq. (3.87) in the velocity space $(v_\parallel, v_\perp)$, the particle is trapped. Particles in the conical region $\alpha < \alpha_{cr}$ escape from the mirror field; we call this region in velocity space the *loss cone* and the critical angle α_{cr} is called the *loss cone angle.*

3.4c Orbit of the Guiding Center in Toroidal Fields (Banana Orbit)

A simple toroidal field is expressed by

$$B_r = 0 \qquad B_\varphi = B_0 \frac{R_0}{R} \qquad B_z = 0 \tag{3.88}$$

in terms of cylindrical coordinates (R, φ, z). The drift velocity of the guiding center is

$$\boldsymbol{v}_g = v_\parallel{}^* \boldsymbol{e}_\varphi + \frac{m}{2qB_\varphi R}(2v_\parallel{}^2 + v_\perp{}^2)\boldsymbol{e}_z \tag{3.89}$$

$$v_{dr} \equiv \frac{m}{2qB_\varphi R}(2v_\parallel{}^2 + v_\perp{}^2) = \frac{m}{2qB_0 R_0}(2v_\parallel{}^2 + v_\perp{}^2) \tag{3.90}$$

$$v_\parallel = (\boldsymbol{v}\cdot\boldsymbol{b}) \qquad v_\parallel{}^* = (\boldsymbol{v}\cdot\boldsymbol{e}_\varphi),$$

where $\boldsymbol{e}_\varphi$ and $\boldsymbol{e}_z$ are the unit vectors in the φ and z directions, respectively. Ions and electrons drift in opposite direction along z. This drift is called *toroidal drift*. As a result of the resultant charge polarization, an electric field $\boldsymbol{E}$ is induced and both ions and electrons drift outward $(\boldsymbol{E} \times \boldsymbol{B})$. Consequently, a simple toroidal field cannot confine a plasma unless the polarization field is cancelled out by an appropriate method (fig. 3.9).

If lines of magnetic force connect the upper and lower regions, the polarization can be short-circuited, as the charged particles can move freely along the lines of force. When some rotation is introduced in the particle orbit around a magnetic axis **O** (fig. 3.10), the orbit projected onto the R, z plane is expressed by

$$\frac{dx}{dt} = -\omega z \qquad \frac{dz}{dt} = \omega x + v_{dr}. \tag{3.91}$$

If v_{dr} is nearly constant, the orbit is

$$\left(x + \frac{v_{dr}}{\omega}\right)^2 + z^2 = \text{const.} \tag{3.92}$$

Accordingly, the orbit of the guiding center deviates from the magnetic surface only by an amount $\Delta = -v_{dr}/\omega$. If the rotational transform angle

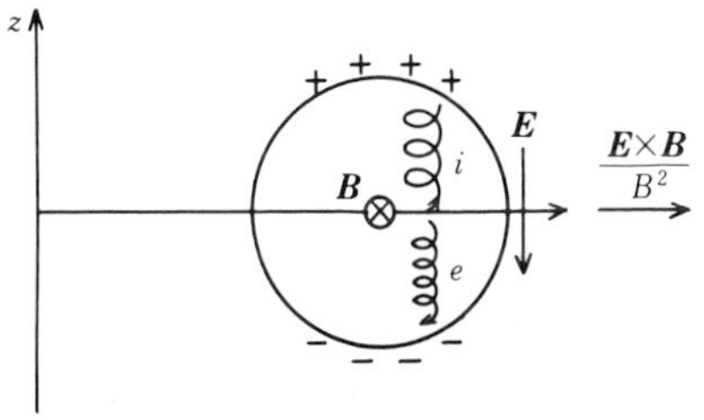

Fig. 3.9
Toroidal drift.

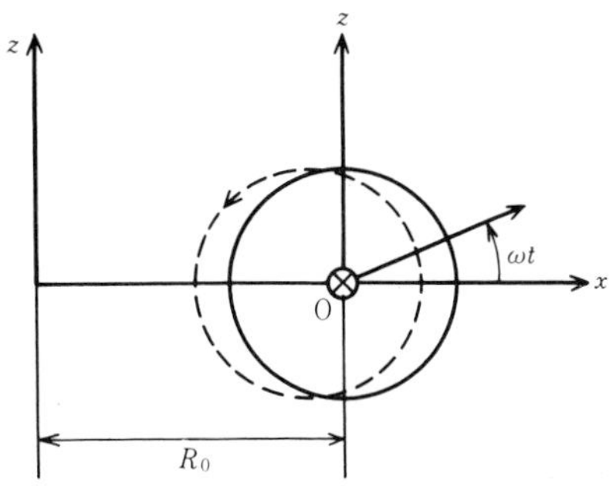

Fig. 3.10
Compensation for toroidal drift by the introduction of rotation.

ι is introduced into the toroidal field, ω is given by $\omega = \iota v_{\parallel}^*/2\pi R_0$ and the sign of ω follows the sign of $v_{\parallel}^* = (\boldsymbol{v}\cdot\boldsymbol{e}_{\varphi})$. The amount of deviation is[5]

$$\Delta = -v_{\mathrm{dr}}\frac{2\pi R_0}{\iota v_{\parallel}^*} = -\frac{m}{qB_0}\frac{v_{\parallel}^2 + v_{\perp}^2/2}{v_{\parallel}^*}\frac{2\pi}{\iota} = \mp\beta\rho_B\frac{2\pi}{\iota}, \tag{3.93}$$

where

$$\rho_B = \frac{m(v_{\parallel}^2 + v_{\perp}^2)^{1/2}}{qB} \qquad \beta = \frac{1 + (v_{\perp}/v_{\parallel})^2/2}{(1 + (v_{\perp}/v_{\parallel})^2)^{1/2}}. \tag{3.94}$$

Particles that satisfy the foregoing are called *transit particles* or *untrapped particles*; they can go around the torus without being trapped. Let us next consider there particles which cannot go around the torus. In this case the r, θ coordinates of fig. 3.11 are used instead of the x, z coordinates of fig. 3.10. The magnitude B of the toroidal field is expressed by

$$B \approx |B_0|(1 - \varepsilon\cos\theta) = |B_0|(1 - \varepsilon\cos\kappa l) \qquad \varepsilon = \frac{r}{R},$$

where l is taken as the length measured along the line of magnetic force. For $(v_{\perp}/v_{\parallel})^2 > 1/\varepsilon$, the particles are trapped outside in the weak region of the magnetic field. These particles are called *trapped particles.* Let us consider the orbit of the guiding center of a trapped particle. Due to the toroidal drift, the radial velocity is

$$\dot{r} = \frac{m}{qB_0}\frac{v_{\perp}^2}{2R}\sin\theta. \tag{3.95}$$

Here we have assumed $v_{\parallel}^2 \ll v_{\perp}^2$. The equation of motion along the parallel component is

$$m\frac{\mathrm{d}v_{\parallel}^*}{\mathrm{d}t} = qE_{\parallel} - \mu_{\mathrm{m}}\frac{\partial B}{\partial l}. \tag{3.96}$$

When $E_{\parallel} \approx 0$, eq. (3.96) is

$$\dot{v}_{\parallel}^* = -\frac{\mu_{\mathrm{m}}}{m}\frac{\partial B}{\partial l} \approx -\frac{v_{\perp}^2}{2}\varepsilon\kappa\sin\theta = -\frac{v_{\perp}^2}{2}\frac{B_\theta}{RB_0}\sin\theta. \tag{3.97}$$

(As $r\theta/l \approx B_\theta/B_0$, κ is given by $\kappa = B_\theta/(B_0 r)$.) The integral of eqs. (3.95) and (3.97) is reduced to

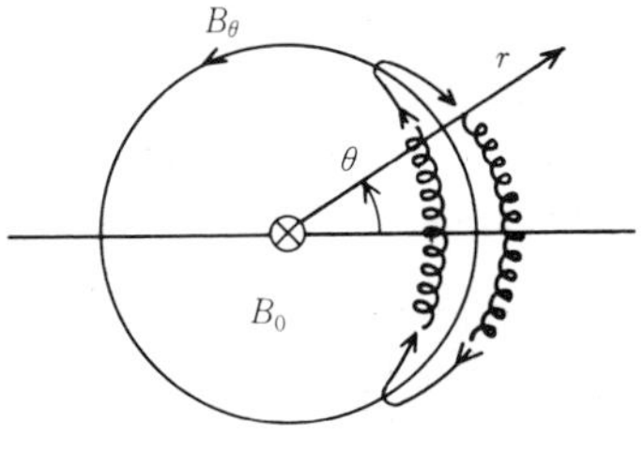

Fig. 3.11
Banana orbit of an ion.

$$r - r_0 = -\frac{m}{qB_\theta} v_\parallel^*, \tag{3.98}$$

where $r = r_0$ is the turning point at which $v_\parallel^* = 0$. As the quantity $v_\parallel^*$ can take either sign, eq. (3.98) expresses an orbit of *banana* shape as is shown in fig. 3.11. The width of the banana is given by eq. (3.98); the width is approximately the same as the Larmor radius $\rho_{B\mathrm{p}}$ corresponding to the poloidal magnetic field B_p and the velocity $v_\parallel$.

When the magnetic field is axisymmetric, the drift surface can be solved exactly by means of eq. (3.83), to give

$$R\left(A_\varphi + \frac{mv_\parallel}{qB} B_\varphi\right) = \text{const.} \tag{3.99}$$

The deviation δr of the drift surface from the magnetic surface $RA_\varphi =$ const. is given by

$$\delta r = -\frac{B_\varphi m(\pm v_\parallel^*)}{B \quad qB_\theta} \approx -\frac{mv_\parallel^*}{qB_\theta} \qquad (\text{for } B_\varphi \gtrless 0), \tag{3.100}$$

with a discussion similar to that of sec. 3.2. Equation (3.100) gives the same result as eq. (3.48). ($v_\parallel^*$ in eq. (3.98) is $(\boldsymbol{v} \cdot \boldsymbol{e}_\varphi)$ and $v_\parallel$ in eq. (3.99) is $(\boldsymbol{v} \cdot \boldsymbol{b})$.) These derivations of orbits of guiding centers are very important for the analyses of plasma diffusion and confinement times of toroidal systems, all of which will be discussed in ch. 8.

3.5 Drift of the Banana Center

3.5a Longitudinal Adiabatic Invariant

Particles with large values of $(v_\perp/v_\parallel)^2 > 1/\varepsilon$ are trapped in the weak region of a toroidal magnetic field and the guiding centers of the trapped particles assume banana orbits. When the size of a banana orbit is small compared to the dimensions of the plasma, the guiding center of the banana itself can be considered. The drift motion of the banana center is a characteristic of the motion of particles in the toroidal field, and this motion may be described by means of the longitudinal adiabatic invariant $J = \oint mv_{\parallel} \mathrm{d}l$ along the banana orbit.[6] Let us use generalized curvilinear coordinates (u^1, u^2, u^3) to study the properties of the longitudinal adiabatic invariant. We let the u^3 axis align with the line of the magnetic field. When the vectors $\boldsymbol{a}_i \equiv \partial \boldsymbol{r}/\partial u^i$ are defined, any vector $\boldsymbol{F}$ is described by $\boldsymbol{F} = \Sigma_i f^i \boldsymbol{a}_i$. The f^i are the contravariant components of $\boldsymbol{F}$. Next we define the vectors $\boldsymbol{a}^j$, which satisfy the relations $\boldsymbol{a}_i \cdot \boldsymbol{a}^j = \delta_{ij}$ (see app. 2). The vector $\boldsymbol{F}$ is also described by $\boldsymbol{F} = \Sigma_j f_j \boldsymbol{a}^j$, where the f_j are the covariant components of $\boldsymbol{F}$. Then f^i and f_j are given by $f^i = \boldsymbol{F} \cdot \boldsymbol{a}^i$, $f_j = \boldsymbol{F} \cdot \boldsymbol{a}_j$ respectively. Define $g_{ij} \equiv (\boldsymbol{a}_i \cdot \boldsymbol{a}_j)$ and $g^{ij} \equiv (\boldsymbol{a}^i \cdot \boldsymbol{a}^j)$; then the relation of f_i and f^i is $f_i = \Sigma_j g_{ij} f^i$ and $f^i = \Sigma_j g^{ij} f_j$. The unit vector $\boldsymbol{b}$ of the magnetic field $\boldsymbol{B}$ has contravariant components $b^1 = 0$, $b^2 = 0$, $b^3 = \pm (g_{33})^{-1/2}$ as $\boldsymbol{b} = b^3 \boldsymbol{a}_3$, $(\boldsymbol{a}_3 \cdot \boldsymbol{a}_3) = g_{33}$ (fig. 3.12). The equation of motion of the guiding center of the particle is given by eq. (3.80). With use of rotation operator $\nabla \times$ in these general curvilinear coordinates (app. 1), eq. (3.80) may be written as

$$\frac{\mathrm{d}u^1}{\mathrm{d}t} = \frac{v_{\parallel}}{qBg^{1/2}} \left(\frac{\partial}{\partial u^2}(mv_{\parallel} b_3) - \frac{\partial}{\partial u^3}(mv_{\parallel} b_2) \right) \tag{3.101}$$

$$\frac{\mathrm{d}u^2}{\mathrm{d}t} = \frac{v_{\parallel}}{qBg^{1/2}} \left(\frac{\partial}{\partial u^3}(mv_{\parallel} b_1) - \frac{\partial}{\partial u^1}(mv_{\parallel} b_3) \right) \tag{3.102}$$

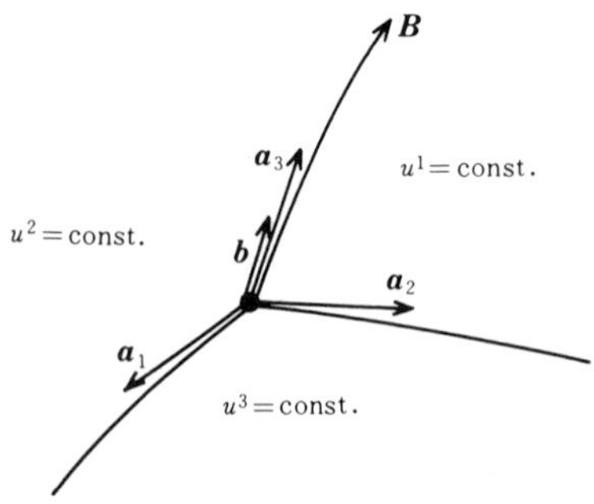

Fig. 3.12
Curvilinear coordinates.

$$\frac{du^3}{dt} = \left(v_{\|} - \frac{mv_{\|}{}^2}{qB}(\boldsymbol{b}\cdot\nabla\times\boldsymbol{b})\right)b^3$$

$$+ \frac{v_{\|}}{qBg^{1/2}}\left(\frac{\partial}{\partial u^1}(mv_{\|}b_2) - \frac{\partial}{\partial u^2}(mv_{\|}b_1)\right). \tag{3.103}$$

$\nabla\boldsymbol{B} = 0$ may be written in the form

$$\frac{\partial}{\partial u^3}\left(g^{1/2}\frac{B}{(g_{33})^{1/2}}\right) = 0. \tag{3.104}$$

As $dl = \pm du^3 a_3 = \pm du^3(g_{33})^{1/2}$, then $v_{\|}dt = dl = \pm du^3(g_{33})^{1/2}$. The integrals of eqs. (3.101), (3.102) over the time t yield

$$\Delta u^1 = \pm\frac{1}{qB}\left(\frac{g_{33}}{g}\right)^{1/2}\int\left(\frac{\partial}{\partial u^2}(mv_{\|}b_3) - \frac{\partial}{\partial u^3}(mv_{\|}b_2)\right)du^3 \tag{3.105}$$

$$\Delta u^2 = \pm\frac{1}{qB}\left(\frac{g_{33}}{g}\right)^{1/2}\int\left(\frac{\partial}{\partial u^3}(mv_{\|}b_1) - \frac{\partial}{\partial u^1}(mv_{\|}b_3)\right)du^3. \tag{3.106}$$

Accordingly, the amount of displacement of the banana center over a single period is given by

$$\Delta u^1 = \pm\frac{1}{qB}\left(\frac{g_{33}}{g}\right)^{1/2}\frac{\partial J}{\partial u^2} \tag{3.107a}$$

$$\Delta u^2 = \mp\frac{1}{qB}\left(\frac{g_{33}}{g}\right)^{1/2}\frac{\partial J}{\partial u^1} \tag{3.108a}$$

$$J = \oint mv_{\|}b_3 du^3 = \oint mv_{\|}\, dl. \tag{3.109}$$

(As $f_j = \Sigma_i g_{ji} f^i$, then $b_3 = g_{33}b^3 = (g_{33})^{1/2}$, $(g_{33})^{1/2}du^3 = dl$.) The total differential of J is

$$\Delta J = \frac{\partial J}{\partial u^2}\Delta u^2 + \frac{\partial J}{\partial u^1}\Delta u^1 = qB\left(\frac{g}{g_{33}}\right)^{1/2}(\Delta u^1\Delta u^2 - \Delta u^2\Delta u^1) = 0, \tag{3.110}$$

and the adiabatic invariant $J =$ const. is therefore reduced. The period τ of the banana orbit is given by

$$\tau = \oint\frac{dl}{v_{\|}} = m\frac{\partial}{\partial W}\oint v_{\|}\,dl$$
$$= \frac{\partial J}{\partial W},$$

where $v_{\parallel} = \pm(2/m)^{1/2}(W - \mu_m B - q\phi)^{1/2}$ and W is the total energy of the particle. Consequently the drift velocity of the banana center is expressed by

$$\frac{du^1}{dt} \approx \frac{\Delta u^1}{\tau} = \frac{\pm 1}{qB}\left(\frac{g_{33}}{g}\right)^{1/2} \frac{\partial J/\partial u^2}{\partial J/\partial W} \tag{3.107b}$$

$$\frac{du^2}{dt} \approx \frac{\Delta u^2}{\tau} = \frac{\mp 1}{qB}\left(\frac{g_{33}}{g}\right)^{1/2} \frac{\partial J/\partial u^1}{\partial J/\partial W}. \tag{3.108b}$$

Let us consider an axially symmetric torus as an example. The magnetic field is given by

$$\boldsymbol{B} = B_0\left(1 - \frac{r}{R_0}\cos\theta\right)\boldsymbol{e}_\varphi + B_\theta \boldsymbol{e}_\theta, \tag{3.111}$$

where $\boldsymbol{e}_\varphi$ is the unit vector along the magnetic axis and $\boldsymbol{e}_\theta$ the unit vector in the θ direction. We assume here that $|B_\theta| \ll |B_0|$. When the energy and the magnetic moment of the particle are denoted by W and μ_m respectively, the longitudinal adiabatic invariant is

$$\begin{aligned} J &= m \oint v_{\parallel}\, dl \\ &= m \oint \left(\frac{2}{m}(W - \mu_m B)\right)^{1/2} dl \\ &= m \oint \left(\frac{2}{m}(W - \mu_m |B_0|(1 - \varepsilon\cos\theta))\right)^{1/2} dl \\ &= (2m\varepsilon\mu_m|B_0|)^{1/2} \oint (2\kappa^2 + \cos\theta - 1)^{1/2}\, dl \\ &= 16(m\varepsilon\mu_m|B_0|)^{1/2} R_0 |q_s| \int_0^{\theta_0/2} \left(\kappa^2 - \sin^2\frac{\theta}{2}\right)^{1/2} d\left(\frac{\theta}{2}\right) \\ &= 8 \cdot 2^{1/2} m v_\perp \varepsilon^{1/2} R_0 |q_s| (E(\kappa) - (1 - \kappa^2)K(\kappa)). \end{aligned} \tag{3.112}$$

Here the relation $|r d\theta/dl| \approx |B_\theta/B_0|$ has been used. The quantities κ^2, q_s, and ε are

$$\kappa^2 = \frac{W - \mu_m|B_0|(1 - \varepsilon)}{2\varepsilon\mu_m|B_0|} \approx \frac{R}{2r}\frac{v_{\parallel}^2}{v_\perp^2} < 1 \tag{3.113}$$

$$q_s \equiv \frac{r}{R_0}\frac{B_0}{B_\theta} \tag{3.114}$$

$$\varepsilon \equiv \frac{r}{R_0}. \tag{3.115}$$

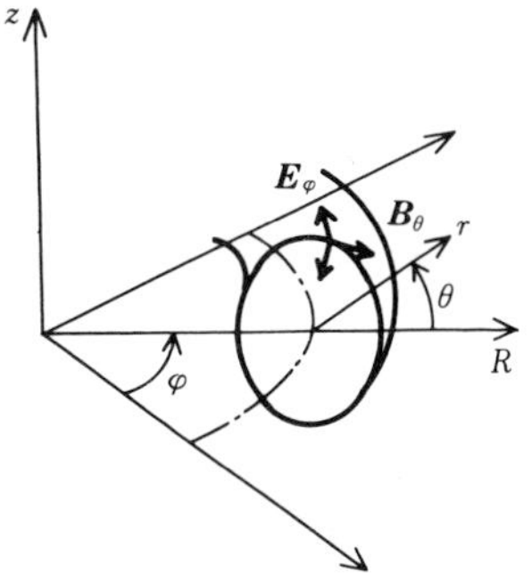

Fig. 3.13
Coordinate system for a torus.

The quantity κ must satisfy $\kappa^2 < 1$ for particles to be trapped. The equations of the lines of magnetic force is $r\mathrm{d}\theta/R\mathrm{d}\varphi \approx B_\theta/B_0$ and $\mathrm{d}\theta/\mathrm{d}\varphi \approx 1/q_\mathrm{s}$; i.e.,

$$\theta - \frac{\varphi}{q_\mathrm{s}} = \text{const.} \tag{3.116}$$

When u^1 and u^2 are $u^1 = \theta - \varphi/q_\mathrm{s}$ and $u^2 = r$, the displacement of the banana center over a single period (eqs. 3.107a, b) is given by eqs. (3.107) and (3.108) as follows:

$$\left.\begin{aligned} \Delta u^1 &= \frac{1}{qB_0 r}\frac{\partial}{\partial r}J \\ \Delta r &= \frac{-1}{qB_0 r}\frac{\partial}{\partial u^1}J = 0. \end{aligned}\right\} \tag{3.117}$$

Since $\partial v_\parallel/\partial W = 1/mv_\parallel$, the period τ of the banana motion is given by

$$\tau = \oint \frac{\mathrm{d}l}{v_\parallel} = m\frac{\partial}{\partial W}\oint v_\parallel \mathrm{d}l = \frac{\partial J}{\partial W} = 4\left(\frac{m}{\varepsilon\mu_\mathrm{m}|B_0|}\right)^{1/2} R|q_\mathrm{s}|K(\kappa)$$

$$= 4{\cdot}2^{1/2}\frac{R|q_\mathrm{s}|}{v_\perp \varepsilon^{1/2}}K(\kappa). \tag{3.118}$$

The drift velocity in the u^1 direction is

$$\frac{\Delta u^1}{\tau} = \frac{1}{qB_0 r}\frac{\partial J/\partial r}{\partial J/\partial W} = \frac{m v_\perp{}^2 \varepsilon}{qB_0 r^2}\left(\frac{E}{K} - \frac{1}{2}\right). \tag{3.119}$$

As $\Delta u^1 = \Delta(\theta - \varphi/q_s) = -\Delta R\varphi/Rq_s$, it follows that

$$\frac{\Delta R\varphi}{\tau} = -\frac{Rq_s}{qB_0 r}\frac{\partial J/\partial r}{\partial J/\partial W} = \frac{-1}{qB_\theta}\frac{\partial J/\partial r}{\partial J/\partial W} = \frac{-m v_\perp{}^2}{qB_\theta R}\left(\frac{E}{K} - \frac{1}{2}\right). \tag{3.120}$$

Here $K(\kappa^2)$ and $E(\kappa^2)$ are complete elliptic integrals of the first and second kind respectively; i.e.,

$$K(\kappa) = \int_0^{\pi/2} \frac{dx}{(1 - \kappa^2 \sin^2 x)^{1/2}}$$

$$E(\kappa) = \int_0^{\pi/2} (1 - \kappa^2 \sin^2 x)^{1/2}\, dx$$

$$\frac{dE}{d(\kappa^2)} = \frac{1}{2\kappa^2}(E - K)$$

$$\frac{dK}{d(\kappa^2)} = -\frac{1}{2\kappa^2}\left(K - \frac{E}{1-\kappa^2}\right).$$

The banana drifts along the magnetic axis with the velocity given by eq. (3.120) in the case of an axially symmetric torus (see fig. 3.14).

*3.5b Orbit of the Banana Center in a Stellarator Field[6,7]

The magnetic field of a linear stellarator (eq. (2.105)) is given as follows:

$$B_r = B_0\left(\frac{lR}{ka}\right)\varepsilon_h{}^0\left(\frac{r}{a}\right)^{l-1}\sin\left(\delta l\theta' - \frac{nz}{R}\right)$$

$$B_\theta{}' = \delta B_0\left(\frac{lR}{ka}\right)\varepsilon_h{}^0\left(\frac{r}{a}\right)^{l-1}\cos\left(\delta l\theta' - \frac{nz}{R}\right)$$

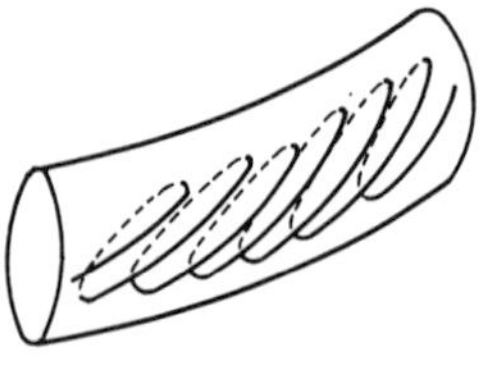

Fig. 3.14
Drift of a banana in an axisymmetric torus.

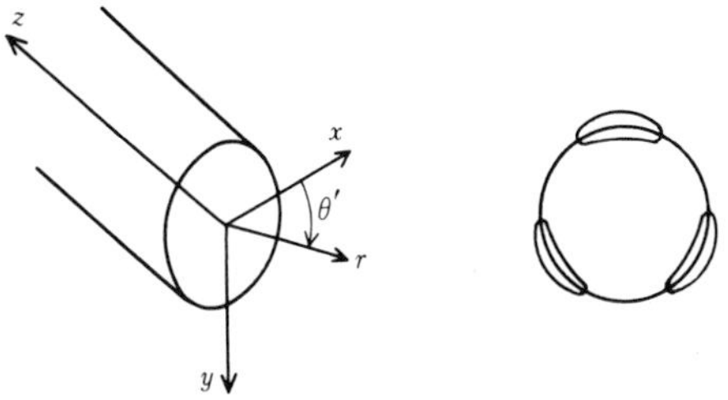

Fig. 3.15
Cylindrical coordinates and banana orbits in a linear stellarator ($l = 3$).

$$B_z = B_0\left(1 - \varepsilon_h{}^0\left(\frac{r}{a}\right)^l \cos\left(\delta l\theta' - \frac{nz}{R}\right)\right),$$

where $\delta = \pm 1$, and l, n are positive integers. The condition $l\alpha r \ll 1$ is assumed here. The parameter $\varepsilon_h{}^0$ expresses the ratio of the magnitude of the helical to the uniform component. We can assume $\varepsilon_h{}^0 > 0$ without loss of generality. r, θ', z are the cylindrical coordinates shown in fig. 3.15. Because of the helical symmetry, the orbit of the guiding center is given by $A_z{}^* + (k/\delta l)(r/R)A_{\theta'}{}^* = \text{const}$.

The longitudinal adiabatic invariant J and the period τ of the banana orbit are given by*

$$J = \oint mv_\parallel \mathrm{d}l = 8{\cdot}2^{1/2} mv_\perp(\varepsilon_h)^{1/2}\frac{R}{n}(E(\kappa) - (1 - \kappa^2)K(\kappa)) \tag{3.121}$$

$$\tau = \oint \frac{\mathrm{d}l}{v_\parallel} = 4{\cdot}2^{1/2}\frac{R}{nv_\perp(\varepsilon_h)^{1/2}}K(\kappa) \tag{3.122}$$

$$\kappa^2 = \frac{W - \mu_m|B_0|(1 - \varepsilon_h)}{2\varepsilon_h\mu_m B} \approx \frac{1}{2\varepsilon_h}\frac{v_\parallel{}^2}{v_\perp{}^2} \tag{3.123}$$

$$\varepsilon_h = \varepsilon_h{}^0\left(\frac{r}{a}\right)^l. \tag{3.124}$$

Here we use $2\pi/\iota$, where ι is the rotational transform angle, instead of q_s of eq. (3.116) for the line of magnetic force ($\mathrm{d}\theta/\mathrm{d}\varphi = \iota/2\pi$). The drift of the banana center is described by eqs. (3.117), (3.118), and (3.119), and may be written

* The cylindrical (r, θ', z) coordinates correspond to the pseudotoroidal coordinates (r, $-\theta$, $R\varphi$).

$$\left.\begin{aligned}\frac{\mathrm{d}\theta}{\mathrm{d}t} &= \frac{\Delta\theta}{\tau} = \frac{mv_\perp{}^2\varepsilon_{\mathrm{h}}l}{qB_0r^2}\left(\frac{E}{K}-\frac{1}{2}\right) = \frac{T}{qB_0r^2}2\varepsilon_{\mathrm{h}}l\left(\frac{E}{K}-\frac{1}{2}\right)\\ \frac{\mathrm{d}r}{\mathrm{d}t} &= \frac{\Delta r}{\tau} \approx 0.\end{aligned}\right\} \tag{3.125}$$

Here we assume $\Delta u^1 = \Delta(\theta - (\iota/2\pi)\varphi) \approx \Delta\theta$. Accordingly the banana in the linear stellarator rotates in the θ direction with an angular velocity given by eq. (3.125).

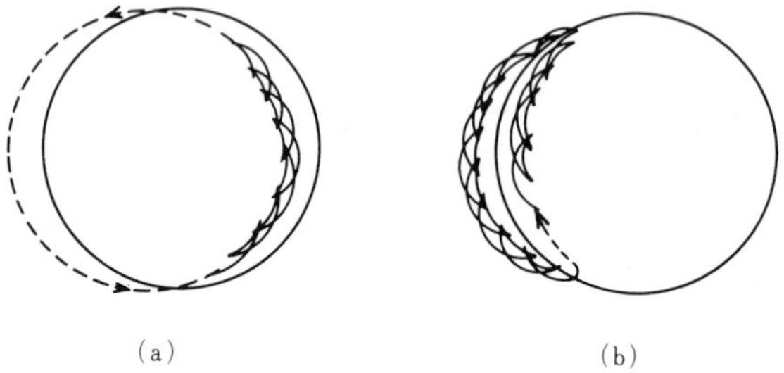

Fig. 3.16
Untrapped banana (a), and trapped banana (b).

When the linear stellarator is curved to form the toroidal stellarator, the magnitude of the magnetic field is $B \approx |B_z| = |B_0|(1 - \varepsilon_{\mathrm{t}}\cos\theta - \varepsilon_{\mathrm{h}}\cos(-\delta l\theta - \mathrm{n}\varphi))$. The term $-\varepsilon_{\mathrm{t}}B_0\cos\theta(\varepsilon_{\mathrm{t}} = r/R)$ is the toroidal term, and $\delta = \pm 1$. Then the longitudinal adiabatic invariant is given by

$$\begin{aligned} J(r, \theta, W, \mu_{\mathrm{m}}) &\\ &= m\oint v_\parallel \mathrm{d}l\\ &= (2m)^{1/2}\oint[W - q\phi - \mu_{\mathrm{m}}|B_0|(1-\varepsilon_{\mathrm{t}}\cos\theta - \varepsilon_{\mathrm{h}}\cos(-\delta l\theta - n\varphi))]^{1/2}\mathrm{d}l\\ &= 16Rn^{-1}(m\mu_{\mathrm{m}}|B_0|\varepsilon_{\mathrm{h}}(r))^{1/2}(E(\kappa) - (1-\kappa^2)K(\kappa)), \end{aligned} \tag{3.126}$$

where

$$\kappa^2(r, \theta, W, \mu_{\mathrm{m}}) = \frac{W - \mu_{\mathrm{m}}|B_0|(1 - \varepsilon_{\mathrm{t}}(r)\cos\theta - \varepsilon_{\mathrm{h}}(r)) - q\phi(r)}{2\mu_{\mathrm{m}}|B_0|\varepsilon_{\mathrm{h}}(r)}, \tag{3.127}$$

with $\phi(r)$ the electrostatic potential of the plasma. By using eqs. (3.117),

(3.118), and (3.119), we obtain for the equation of the orbit of the banana center (r, θ)

$$\left.\begin{aligned} \frac{d\theta}{dt} &= \frac{1}{qB_0 r}\frac{\partial J/\partial r}{\partial J/\partial W} \\ \frac{dr}{dt} &= \frac{-1}{qB_0 r}\frac{\partial J/\partial \theta}{\partial J/\partial W}. \end{aligned}\right\} \tag{3.128}$$

Since

$$\frac{\partial \kappa^2}{\partial r} = -\frac{l}{r}\kappa^2 + \frac{\varepsilon_t \cos\theta}{2\varepsilon_h r} + \frac{l}{2r} - \frac{q\partial\phi/\partial r}{2\varepsilon_h \mu_m |B_0|},$$

we can reduce eq. (3.128) to[8]

$$\frac{d\theta}{dt} = \frac{V_\perp}{r}\cos\theta + \omega_h + \omega_E \tag{3.129}$$

$$\frac{dr}{dt} = V_\perp \sin\theta, \tag{3.130}$$

where

$$\left.\begin{aligned} \frac{V_\perp}{r} &= \varepsilon_t \frac{T}{qB_0 r^2} = -\varepsilon_t \omega_0 \\ \omega_h &= -2l\left(\frac{E}{K} - \frac{1}{2}\right)\varepsilon_h \omega_0 \\ \omega_E &= \frac{E_r}{B_0 r} \\ T &= \frac{1}{2}m v_\perp{}^2 \qquad \omega_0 = \frac{-T}{qB_0 r^2}. \end{aligned}\right\} \tag{3.131}$$

The first term of eq. (3.129) is the term due to the toroidal drift, the second term is the angular velocity of the banana center due to the stellarator field, and the third term is the angular velocity of the $\boldsymbol{E} \times \boldsymbol{B}$ rotation due to the radial electric field. The ratio of these terms is approximately $\varepsilon_t : \varepsilon_h : \alpha = qE_r r/T$. The angular velocity ω_0 defined in eq. (3.131) is roughly equal to the drift frequency which will be introduced in sec. 12.1.

Unless $d\theta/dt \propto \partial J/\partial r = 0$, the banana can circulate in the θ direction. This type of banana is called an *untrapped banana* or a *transit banana* (fig. 3.16a). On the contrary, if $\partial J/\partial r = 0$ at some point, it follows that $d\theta/dt = 0$. In this case, the banana center is trapped (fig. 3.16b). This type of banana is called a *trapped banana*. The orbit of this banana center is also of banana shape, so that the trapped banana is also called a *super-banana*.

Let us consider the case where the ω_h is dominant in eq. (3.129) compared with the toroidal-drift term and ω_E (i.e., $\varepsilon_h \gg \varepsilon_t, \alpha$). The displacement of the untrapped banana orbit from the magnetic surface is

$$\Delta = -\frac{V_\perp}{\omega_h}, \tag{3.132}$$

as is clear from eqs. (3.129) and (3.130). However the superbanana orbit is quite different from the untrapped banana orbit. As is clear from eq. (3.125) or (3.131), we find $\omega_h = 0$ at κ^2 which satisfies $E/K = 0.5$, i.e., $\kappa^2 = 0.826$ ($E = 1.66$, $K = 2.32$). This relation determines the position $r = r_0$ at which $d\theta/dt = 0$ for $\varepsilon_h \gg \varepsilon_t$. Expanding ω_h around $r = r_0$, we have

$$\begin{aligned}\frac{d\theta}{dt} &= -2\varepsilon_h l\omega_0 \left(\frac{d}{d(\kappa^2)}\left(\frac{E}{K}\right)\right)_{\kappa=\kappa_0} (\kappa^2(r) - \kappa_0{}^2) \\ &= 1.74\varepsilon_h l\omega_0 \frac{l}{r_0}\left(\frac{1}{2} - \kappa_0{}^2\right)(r - r_0),\end{aligned} \tag{3.133}$$

where

$$\left.\frac{d}{d(\kappa^2)}\frac{E}{K}\right|_{\kappa^2 = \kappa_0{}^2} = -0.87.$$

On the other hand, dr/dt is given by

$$\frac{dr}{dt} = -r\varepsilon_t\omega_0 \sin\theta, \tag{3.134}$$

so that eqs. (3.133) and (3.134) are

$$\frac{1}{\omega_0 r_0}\frac{d(r - r_0)}{dt} = -\varepsilon_t \sin\theta \tag{3.135}$$

$$\frac{1}{\omega_0}\frac{d\theta}{dt} = -0.57\varepsilon_h \frac{l^2}{r_0}(r - r_0). \tag{3.136}$$

The solution is

$$\frac{(r - r_0)^2}{r_0{}^2} = \frac{3.5}{l^2}\frac{\varepsilon_t}{\varepsilon_h}(\cos\theta_0 - \cos\theta). \tag{3.137}$$

θ_0 is the value of θ at the turning point. The period T_s in which the banana center circulates the superbanana orbit can be estimated by means of eqs. (3.136) and (3.137):

$$T_s|\omega_0| = \frac{4\cdot 2^{1/2}\times 0.94}{l(\varepsilon_h\varepsilon_t)^{1/2}}\int_{\theta_0/2}^{\pi/2}\frac{d(\theta/2)}{\left(\sin^2\dfrac{\theta}{2}-\sin^2\dfrac{\theta_0}{2}\right)^{1/2}},$$

i.e.,

$$T_s = \frac{1}{|\omega_0|}\frac{5.3}{l(\varepsilon_h\varepsilon_t)^{1/2}}K\left(\cos\frac{\theta_0}{2}\right). \tag{3.138}$$

The size of the superbanana orbit can be estimated from eq. (3.137), and is

$$\Delta r \approx \left(\frac{\varepsilon_t}{\varepsilon_h}\right)^{1/2} r_0. \tag{3.139}$$

The size is independent of the magnitude of the magnetic field. In other words, the superbanana does not become small even if the magnetic field is strong.

We have discussed the orbit of the banana center under the assumption that the toroidal effect is very small ($\varepsilon_h \gg \varepsilon_t$, α). When the radial electric field is dominant ($\alpha \gg \varepsilon_t$, ε_h), the banana is untrapped and the orbit deviates from the magnetic field by the amount $\Delta = -V_\perp/\omega_E$. When toroidal drift is dominant ($\varepsilon_t \gg \varepsilon_h$, α), the banana drifts in such a way that it eventually escapes from the plasma region.

Banana motion plays an important role in the studies of diffusion of toroidal plasmas, which will be discussed in ch. 8.

3.5c Effect of a Longitudinal Electric Field on Banana Motion

In the tokamak configuration, a toroidal electric field is applied in order to induce the plasma current. The guiding center of a particle drifts by $\boldsymbol{E}\times\boldsymbol{B}/B^2$ but the banana center moves in different way. The toroidal electric field is described by

$$E_\varphi = -\frac{\partial A_\varphi}{\partial t} \tag{3.140}$$

in R, φ, z coordinates. Since angular momentum is conserved, we can write

$$R(mR\dot{\varphi} + qA_\varphi) = \text{const.} \tag{3.141}$$

Taking the average of eq. (3.141) over a Larmor period, and using the relation

$$\langle R\dot{\varphi}\rangle = \frac{B_\varphi}{B}v_\parallel, \tag{3.142}$$

we find

$$R\left(\frac{mv_\parallel B_\varphi}{B} + qA_\varphi\right) = \text{const.} \tag{3.143}$$

For particles in banana motion, $v_\parallel$ becomes $v_\parallel = 0$ at the turning points of the banana orbit. The displacement of a turning point per period Δt is obtained from

$$0 = \Delta(RA_\varphi(R, z)) = \Delta r\frac{\partial}{\partial r}RA_\varphi + \Delta t\frac{\partial}{\partial t}RA_\varphi, \tag{3.144}$$

where r is the radial coordinate of the magnetic surface (fig. 3.17). The differentiations of RA_φ with respect to φ and θ are zero, since $RA_\varphi = \text{const.}$ is the magnetic surface. By means of the relation

$$\left(\frac{1}{R}\right)\frac{\partial}{\partial r}(RA_\varphi) = B_\theta,$$

we obtain the drift velocity of the banana center:[9]

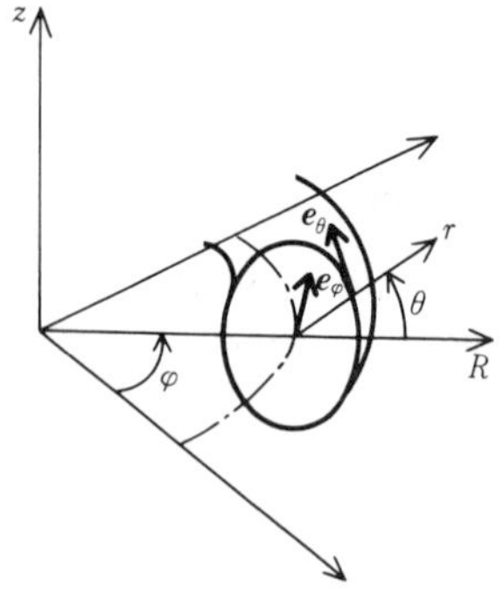

Fig. 3.17
Coordinate system for a torus.

$$\frac{\Delta r}{\Delta t} = \frac{E_\varphi}{B_\theta}. \tag{3.145}$$

The sign of B_θ, which is produced by the current induced by the toroidal electric field E_φ, is opposite to the sign of E_φ, as is clear from fig. 3.17. In other words, $\Delta r/\Delta t$ is negative and the banana center moves inward. Since $\Delta r/\Delta t = E_\varphi/B_\theta$, the velocity is faster than the drift velocity $E_\varphi B_\theta/B^2$ of the guiding center of the particle in the ratio of $(B/B_\theta)^2$. This phenomenon is called Ware's pinch effect.

3.6 Motion of Particles in High-Frequency Electric Fields

When an electric field of angular frequency ω is applied to a charged particle, the motion of the particle consists of a forced oscillation of frequency ω and the Larmor motion of cyclotron frequency Ω. But we can expect that the guiding center of the particle moves slowly, and will see that a high-frequency electric field acts as an effective potential for the motion of the guiding center. When the frequency ω approaches the cyclotron frequency, a strong resonance acceleration of the particle occurs. The equation of motion of the charged particle is

$$\begin{aligned}\frac{\mathrm{d}\boldsymbol{v}}{\mathrm{d}t} &= \boldsymbol{F}\sin\omega t + \frac{e}{m}\boldsymbol{v}\times\boldsymbol{B}\\ &= \boldsymbol{F}\sin\omega t - \Omega\boldsymbol{v}\times\boldsymbol{b}\end{aligned} \tag{3.146}$$

$$\frac{\mathrm{d}\boldsymbol{r}}{\mathrm{d}t} = \boldsymbol{v}, \tag{3.147}$$

where $\boldsymbol{b} = \boldsymbol{B}/B$, $\Omega = -qB/m$. $\boldsymbol{F}$ and $\boldsymbol{B}$ are assumed to be constant in time.

First let us consider the simple case that $\boldsymbol{F}$ and $\boldsymbol{B}$ are uniform. The differentiation of eq. (3.146) with respect to t gives

$$\begin{aligned}\frac{\mathrm{d}^2\boldsymbol{v}}{\mathrm{d}t^2} &= \omega\boldsymbol{F}\cos\omega t - \Omega\frac{\mathrm{d}\boldsymbol{v}}{\mathrm{d}t}\times\boldsymbol{b}\\ &= \omega\boldsymbol{F}\cos\omega t - \Omega\boldsymbol{F}\times\boldsymbol{b}\sin\omega t - \Omega^2(\boldsymbol{v}-(\boldsymbol{v}\cdot\boldsymbol{b})\boldsymbol{b}),\end{aligned}$$

i.e.,

$$\frac{\mathrm{d}^2\boldsymbol{v}_\perp}{\mathrm{d}t^2} + \Omega^2\boldsymbol{v}_\perp = \omega\boldsymbol{F}_\perp\cos\omega t - \Omega\boldsymbol{F}\times\boldsymbol{b}\sin\omega t$$

$$\frac{\mathrm{d}^2\boldsymbol{v}_\parallel}{\mathrm{d}t^2} = \omega\boldsymbol{F}_\parallel\cos\omega t.$$

The solution is

$$\boldsymbol{v}_\omega = \frac{\omega \boldsymbol{F}_\perp \cos \omega t - \Omega \boldsymbol{F} \times \boldsymbol{b} \sin \omega t}{\Omega^2 - \omega^2} - \frac{\boldsymbol{F}_\parallel}{\omega} \cos \omega t \tag{3.148}$$

$$\boldsymbol{r}_\omega = \frac{\omega \boldsymbol{F}_\perp \sin \omega t + \Omega (\boldsymbol{F} \times \boldsymbol{b}) \cos \omega t}{\omega(\Omega^2 - \omega^2)} - \frac{\boldsymbol{F}_\parallel}{\omega^2} \sin \omega t. \tag{3.149}$$

Next we consider the case that $\boldsymbol{F}$ and $\boldsymbol{B}$ vary gradually in space. Let us introduce the orthogonal unit vectors $\boldsymbol{e}_1, \boldsymbol{e}_2$ which are also orthogonal to $\boldsymbol{b}$ and satisfy $\boldsymbol{e}_1 \times \boldsymbol{e}_2 = \boldsymbol{b}$. Let us transform $\boldsymbol{v}$, $\boldsymbol{r}$ to (u, w, α), $\boldsymbol{x}$ by the following relations:

$$\boldsymbol{v} = u\boldsymbol{b} + w(\boldsymbol{e}_1 \cos \alpha + \boldsymbol{e}_2 \sin \alpha) + \boldsymbol{v}_w \tag{3.150}$$

$$\boldsymbol{r} = \boldsymbol{x} + \frac{w}{\Omega}(\boldsymbol{e}_1 \sin \alpha - \boldsymbol{e}_2 \cos \alpha) + \boldsymbol{r}_w, \tag{3.151}$$

where $\boldsymbol{v}_w$ and $\boldsymbol{r}_w$ are the same as in eqs. (3.148) and (3.149). Here $\boldsymbol{F}$ and $\boldsymbol{B}$ are no longer uniform. The quantities in the right-hand sides of eqs. (3.150) and (3.151) are to be taken at the guiding center $\boldsymbol{x}$. The substitution of eqs. (3.150) and (3.151) into eqs. (3.146) and (3.147) yields the differential equation of (u, w, α) and $\boldsymbol{x}$, i.e.,

$$\begin{aligned}\frac{\mathrm{d}\boldsymbol{x}}{\mathrm{d}t} &= u\boldsymbol{b} + w(\boldsymbol{e}_1 \cos \alpha + \boldsymbol{e}_2 \sin \alpha)\left(1 - \frac{1}{\Omega}\frac{\mathrm{d}\alpha}{\mathrm{d}t}\right) \\ &\quad - \frac{\mathrm{d}}{\mathrm{d}t}\left(\frac{\boldsymbol{F}_\perp}{\Omega^2 - \omega^2}\right) \sin \omega t - \frac{\mathrm{d}}{\mathrm{d}t}\left(\frac{\Omega(\boldsymbol{F} \times \boldsymbol{b})}{\omega(\Omega^2 - \omega^2)}\right) \cos \omega t \\ &\quad + \frac{\mathrm{d}}{\mathrm{d}t}\left(\frac{\boldsymbol{F}_\parallel}{\omega^2}\right) \sin \omega t - \frac{\mathrm{d}}{\mathrm{d}t}\left(\frac{w}{\Omega}\boldsymbol{e}_1\right) \sin \alpha + \frac{\mathrm{d}}{\mathrm{d}t}\left(\frac{w}{\Omega}\boldsymbol{e}_2\right) \cos \alpha \end{aligned} \tag{3.152}$$

$$\begin{aligned}&\frac{\mathrm{d}u}{\mathrm{d}t}\boldsymbol{b} + u\frac{\mathrm{d}\boldsymbol{b}}{\mathrm{d}t} + \frac{\mathrm{d}w}{\mathrm{d}t}(\boldsymbol{e}_1 \cos \alpha + \boldsymbol{e}_2 \sin \alpha) + w(-\boldsymbol{e}_1 \sin \alpha + \boldsymbol{e}_2 \cos \alpha)\frac{\mathrm{d}\alpha}{\mathrm{d}t} \\ &\quad + w\left(\cos \alpha \frac{\mathrm{d}\boldsymbol{e}_1}{\mathrm{d}t} + \sin \alpha \frac{\mathrm{d}\boldsymbol{e}_2}{\mathrm{d}t}\right) - \omega^2 \boldsymbol{r}_w + \frac{\mathrm{d}}{\mathrm{d}t}\left(\frac{\omega \boldsymbol{F}_\perp}{\Omega^2 - \omega^2}\right) \cos \omega t \\ &\quad - \frac{\mathrm{d}}{\mathrm{d}t}\left(\frac{\Omega \boldsymbol{F} \times \boldsymbol{b}}{\Omega^2 - \omega^2}\right) \sin \omega t - \frac{1}{\omega}\frac{\mathrm{d}\boldsymbol{F}_\parallel}{\mathrm{d}t} \cos \omega t \\ &= \boldsymbol{F}^* \sin \omega t - \Omega^*(u\boldsymbol{b} + w(\boldsymbol{e}_1 \cos \alpha + \boldsymbol{e}_2 \sin \alpha) + \boldsymbol{v}_w) \times \boldsymbol{b}^*, \end{aligned} \tag{3.153}$$

where $\boldsymbol{F}^* = \boldsymbol{F}(r)$, $\Omega^* = \Omega(\boldsymbol{r})$, and $\boldsymbol{b}^* = \boldsymbol{b}(\boldsymbol{r})$. It is clear that $-\omega^2 \boldsymbol{r}_\omega = \boldsymbol{F}\sin\omega t - \Omega\boldsymbol{v}_\omega \times \boldsymbol{b}$. Scalar products of eq. (3.153) with $\boldsymbol{b}$, $\boldsymbol{e}_1\cos\alpha + \boldsymbol{e}_2\sin\alpha$, and $-\boldsymbol{e}_1\sin\alpha + \boldsymbol{e}_2\cos\alpha$ yield the expressions for $\mathrm{d}u/\mathrm{d}t$, $\mathrm{d}w/\mathrm{d}t$, and $\mathrm{d}\alpha/\mathrm{d}t$, respectively. By means of $\boldsymbol{b}\mathrm{d}\boldsymbol{b}/\mathrm{d}t = (1/2)\mathrm{d}\boldsymbol{b}^2/\mathrm{d}t = 0$, we find

$$\frac{\mathrm{d}u}{\mathrm{d}t} = \boldsymbol{b}\cdot\Bigg(-w\left(\cos\alpha\frac{\mathrm{d}\boldsymbol{e}_1}{\mathrm{d}t} + \sin\alpha\frac{\mathrm{d}\boldsymbol{e}_2}{\mathrm{d}t}\right) - w(\boldsymbol{e}_1\cos\alpha + \boldsymbol{e}_2\sin\alpha)\times(\Omega^*\boldsymbol{b}^* - \Omega\boldsymbol{b}) + (\boldsymbol{F}^* - \boldsymbol{F})\sin\omega t - \boldsymbol{v}_\omega\times(\Omega^*\boldsymbol{b}^* - \Omega\boldsymbol{b}) - \frac{\mathrm{d}}{\mathrm{d}t}\boldsymbol{A}\Bigg) \tag{3.154}$$

$$\frac{\mathrm{d}w}{\mathrm{d}t} = (\boldsymbol{e}_1\cos\alpha + \boldsymbol{e}_2\sin\alpha)\cdot\Bigg(-u\frac{\mathrm{d}\boldsymbol{b}}{\mathrm{d}t} - w\left(\cos\alpha\frac{\mathrm{d}\boldsymbol{e}_1}{\mathrm{d}t} + \sin\alpha\frac{\mathrm{d}\boldsymbol{e}_2}{\mathrm{d}t}\right) + (\boldsymbol{F}^* - \boldsymbol{F})\sin\omega t - (u\boldsymbol{b} + \boldsymbol{v}_\omega)\times(\Omega^*\boldsymbol{b}^* - \Omega\boldsymbol{b}) - \frac{\mathrm{d}\boldsymbol{A}}{\mathrm{d}t}\Bigg) \tag{3.155}$$

$$w\frac{\mathrm{d}\alpha}{\mathrm{d}t} = (-\boldsymbol{e}_1\sin\alpha + \boldsymbol{e}_2\cos\alpha)\cdot\Bigg(-u\frac{\mathrm{d}\boldsymbol{b}}{\mathrm{d}t} - w\left(\cos\alpha\frac{\mathrm{d}\boldsymbol{e}_1}{\mathrm{d}t} + \sin\alpha\frac{\mathrm{d}\boldsymbol{e}_2}{\mathrm{d}t}\right) + (\boldsymbol{F}^* - \boldsymbol{F})\sin\omega t - \boldsymbol{v}_\omega\times(\Omega^*\boldsymbol{b}^* - \Omega\boldsymbol{b}) - \Omega^*(u\boldsymbol{b} + w(\boldsymbol{e}_1\cos\alpha + \boldsymbol{e}_2\sin\alpha))\times\boldsymbol{b}^* - \frac{\mathrm{d}}{\mathrm{d}t}\boldsymbol{A}\Bigg), \tag{3.156}$$

where

$$\frac{\mathrm{d}\boldsymbol{A}}{\mathrm{d}t} \equiv \frac{\mathrm{d}}{\mathrm{d}t}\left(\frac{\omega\boldsymbol{F}_\perp}{\Omega^2 - \omega^2}\right)\cos\omega t - \frac{\mathrm{d}}{\mathrm{d}t}\left(\frac{\Omega\boldsymbol{F}\times\boldsymbol{b}}{\Omega^2 - \omega^2}\right)\sin\omega t - \frac{1}{\omega}\frac{\mathrm{d}\boldsymbol{F}_\parallel}{\mathrm{d}t}\cos\omega t. \tag{3.157}$$

The time averages of eqs. (3.152), (3.154), (3.155), and (3.156) reduce to[10]

$$\frac{\mathrm{d}u}{\mathrm{d}t} = \frac{1}{4}\boldsymbol{b}\cdot\nabla\left(\frac{F_\perp{}^2}{\Omega^2 - \omega^2} - \frac{F_\parallel{}^2}{\omega^2}\right) - \frac{1}{2}\frac{w^2}{\Omega}\boldsymbol{b}\cdot\nabla\Omega - \frac{1}{2}\frac{\boldsymbol{b}\cdot(\boldsymbol{F}\times(\nabla\times\boldsymbol{F}))}{\Omega^2 - \omega^2} \tag{3.158}$$

$$\frac{\mathrm{d}w}{\mathrm{d}t} = \frac{-wu}{2}\nabla\boldsymbol{b} \tag{3.159}$$

$$\frac{\mathrm{d}\alpha}{\mathrm{d}t} = \Omega + \frac{u}{2}[\boldsymbol{e}_1\cdot(\boldsymbol{b}\cdot\nabla)\boldsymbol{e}_2 - \boldsymbol{e}_2\cdot(\boldsymbol{b}\cdot\nabla)\boldsymbol{e}_1 + \boldsymbol{e}_1\cdot(\boldsymbol{e}_2\cdot\nabla)\boldsymbol{b} - \boldsymbol{e}_2\cdot(\boldsymbol{e}_1\cdot\nabla)\boldsymbol{b}] \tag{3.160}$$

$$\frac{dx}{dt} = u\boldsymbol{b}. \tag{3.161}$$

(The derivation of these equations will be described at the end of this section.) From $\nabla \cdot \boldsymbol{B} = 0$ we have the relation $\nabla \cdot \boldsymbol{b} = -(\nabla \Omega/\Omega) \cdot \boldsymbol{b}$. Since $d/dt = (dx/dt) \cdot \nabla \approx u\boldsymbol{b} \cdot \nabla$, it follows that

$$\frac{dw}{dt} = \frac{wu}{2} \frac{\boldsymbol{b}}{\Omega} \cdot \nabla \Omega = \frac{w}{2} \frac{1}{\Omega} \frac{d}{dt} \Omega$$

and

$$\frac{w^2}{2\Omega} \equiv \frac{\mu_m}{-q} = \text{const.} \tag{3.162}$$

This expresses the adiabatic invariance of the magnetic moment. For

$$\nabla \times \boldsymbol{F} = 0, \tag{3.163}$$

[eq. (3.159)] $\times\, w$ + [eq. (3.158)] $\times\, u$ yields

$$\frac{1}{2} \frac{d}{dt} (u^2 + w^2) = \frac{u}{4} (\boldsymbol{b} \cdot \nabla) \left(\frac{F_\perp{}^2}{\Omega^2 - \omega^2} - \frac{F_\parallel{}^2}{\omega^2} \right) = \frac{1}{4} \frac{d}{dt} \left(\frac{F_\perp{}^2}{\Omega^2 - \omega^2} - \frac{F_\parallel{}^2}{\omega^2} \right);$$

i.e.,

$$\frac{1}{2}(u^2 + w^2) + \frac{1}{4} \left(\frac{F_\perp{}^2}{\omega^2 - \Omega^2} + \frac{F_\parallel{}^2}{\omega^2} \right) = \text{const.} \tag{3.164}$$

This expresses the conservation of energy. The second term of eq. (3.164) is the effective potential due to the high-frequency electric field. The equation of motion along the line of magnetic force, deduced from eqs. (3.158) and (3.162), is

$$m \frac{du}{dt} = -\boldsymbol{b} \cdot \nabla \left(\frac{m}{4} \left(\frac{F_\perp{}^2}{\omega^2 - \Omega^2} + \frac{F_\parallel{}^2}{\omega^2} \right) + \mu_m B \right). \tag{3.165}$$

The first term in the right-hand side of eq. (3.165) becomes very large at $\omega \approx \Omega$. When there is a nonzero gradient of B, a force in the direction of $-\nabla B$ is induced. This phenomena can be applied to particle acceleration or confinement.[11,12] This acceleration will occur even in the presence of a nonzero electric-field gradient. (In other words, acceleration does not occur when the magnetic and the electric fields are uniform.)

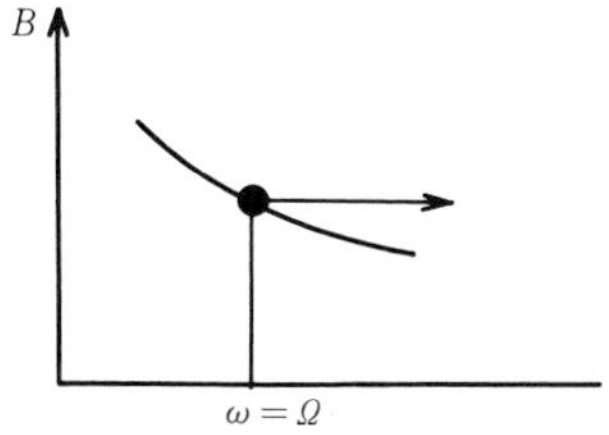

Fig. 3.18
Particle acceleration $-\nabla B$.

The calculation of the time averages of eqs. (3.152)–(3.156) is not difficult, but complicated. We use the relations $\mathrm{d}/\mathrm{d}t = \partial/\partial t + (\mathrm{d}\boldsymbol{x}/\mathrm{d}t)\nabla$, $\langle \sin \omega t\rangle = \langle \cos \omega t\rangle = 0$, $\langle \sin \alpha\rangle = \langle \cos \alpha\rangle = 0$, $\langle \sin^2 \omega t\rangle = \langle \cos^2 \omega t\rangle = \langle \sin^2 \alpha\rangle = \langle \cos^2 \alpha\rangle = 1/2$, $\langle \sin \omega t \cdot \cos \omega t\rangle = \langle \sin \alpha \cdot \cos \alpha\rangle = \langle \sin \omega t \cdot \sin \alpha\rangle = \ldots = 0$, and

$$\Omega^* \boldsymbol{b}^* - \Omega \boldsymbol{b} \approx \left(\left(\frac{w}{\Omega}(\boldsymbol{e}_1 \sin \alpha - \boldsymbol{e}_2 \cos \alpha) + \boldsymbol{r}_\omega\right) \cdot \nabla\right) \Omega \boldsymbol{b}.$$

A similar relation holds for $\boldsymbol{F}^* - \boldsymbol{F}$. We assume that F, Ω, ω are large quantities, and neglect terms $O(1/\Omega)$. Then $\mathrm{d}/\mathrm{d}t = (\mathrm{d}\boldsymbol{x}/\mathrm{d}t)\cdot\nabla \approx u(\boldsymbol{b}\cdot\nabla)$. The time average of eq. (3.154) is

$$\begin{aligned}
\frac{\mathrm{d}u}{\mathrm{d}t} &= \boldsymbol{b}\cdot\Big\langle -w(\boldsymbol{e}_1 \cos \alpha + \boldsymbol{e}_2 \sin \alpha) \times \left(\frac{w}{\Omega}(\boldsymbol{e}_1 \sin \alpha - \boldsymbol{e}_2 \cos \alpha)\cdot\nabla\right)\Omega \boldsymbol{b} \\
&\qquad + (\boldsymbol{r}_\omega \cdot \nabla)\boldsymbol{F} \sin \omega t - \boldsymbol{v}_\omega \times (\boldsymbol{r}_\omega \cdot \nabla)\Omega \boldsymbol{b}\Big\rangle \\
&= \frac{\boldsymbol{b}}{2}\cdot\Bigg[-\frac{\omega \boldsymbol{F}_\perp}{\Omega^2 - \omega^2} \times \left(\frac{\Omega(\boldsymbol{F}\times\boldsymbol{b})}{\omega(\Omega^2 - \omega^2)}\cdot\nabla\right)\Omega \boldsymbol{b} \\
&\qquad + \frac{\Omega(\boldsymbol{F}\times\boldsymbol{b})}{\Omega^2 - \omega^2} \times \left(\frac{\boldsymbol{F}_\perp \cdot \nabla}{\Omega^2 - \omega^2}\right)\Omega \boldsymbol{b} - \frac{\Omega(\boldsymbol{F}\times\boldsymbol{b})}{\Omega^2 - \omega^2} \times \left(\frac{\boldsymbol{F}_\parallel \cdot \nabla}{\omega^2}\right)\Omega \boldsymbol{b} \\
&\qquad + \left(\left(\frac{\boldsymbol{F}_\perp}{\Omega^2 - \omega^2} - \frac{\boldsymbol{F}_\parallel}{\omega^2}\right)\cdot\nabla\right)\boldsymbol{F} + \frac{w^2}{\Omega}\boldsymbol{e}_1 \times (\boldsymbol{e}_2 \cdot \nabla)\Omega \boldsymbol{b} \\
&\qquad - \frac{w^2}{\Omega}\boldsymbol{e}_2 \times (\boldsymbol{e}_1 \cdot \nabla)\Omega \boldsymbol{b}\Bigg]
\end{aligned}$$

$$= \frac{\boldsymbol{b}}{2}\cdot\left[-\frac{\Omega^2}{(\Omega^2-\omega^2)^2}\boldsymbol{F}_\perp \times ((\boldsymbol{F}\times\boldsymbol{b})\cdot\nabla)\boldsymbol{b} + \frac{\Omega^2(\boldsymbol{F}\times\boldsymbol{b})\times(\boldsymbol{F}_\perp\cdot\nabla)\boldsymbol{b}}{(\Omega^2-\omega^2)^2} \right.$$

$$- \frac{\Omega^2 F_\parallel(\boldsymbol{F}\times\boldsymbol{b})\times(\boldsymbol{b}\cdot\nabla)\boldsymbol{b}}{\omega^2(\Omega^2-\omega^2)} + \frac{(\boldsymbol{F}_\perp\cdot\nabla)\boldsymbol{F}}{\Omega^2-\omega^2} - \frac{1}{\omega^2}(\boldsymbol{F}_\parallel\cdot\nabla)\boldsymbol{F}$$

$$\left. + w^2(\boldsymbol{e}_1\times(\boldsymbol{e}_2\cdot\nabla)\boldsymbol{b} - \boldsymbol{e}_2\times(\boldsymbol{e}_1\cdot\nabla)\boldsymbol{b}) \right]$$

$$= \frac{\Omega^2}{2(\Omega^2-\omega^2)^2}[-(\boldsymbol{b}\times\boldsymbol{F}_\perp)\cdot((\boldsymbol{F}_\perp\times\boldsymbol{b})\cdot\nabla)\boldsymbol{b} + \boldsymbol{F}_\perp\cdot(\boldsymbol{F}_\perp\cdot\nabla)\boldsymbol{b}]$$

$$- \frac{\Omega^2 F_\parallel}{2\omega^2(\Omega^2-\omega^2)}\boldsymbol{F}_\perp\cdot(\boldsymbol{b}\cdot\nabla)\boldsymbol{b} + \frac{\boldsymbol{b}\cdot(\boldsymbol{F}_\perp\cdot\nabla)\boldsymbol{F}}{2(\Omega^2-\omega^2)} - \frac{F_\parallel \boldsymbol{b}\cdot(\boldsymbol{b}\cdot\nabla)\boldsymbol{F}}{2\omega^2}$$

$$+ \frac{w^2}{2}[\boldsymbol{e}_2\cdot(\boldsymbol{e}_2\cdot\nabla)\boldsymbol{b} + \boldsymbol{e}_1\cdot(\boldsymbol{e}_1\cdot\nabla)\boldsymbol{b} + \boldsymbol{b}\cdot(\boldsymbol{b}\cdot\nabla)\boldsymbol{b}].$$

Here we have used $\nabla \boldsymbol{b}^2 = 2\boldsymbol{b}\cdot\nabla\boldsymbol{b} = 0$. From the vector formulas for $\nabla(\boldsymbol{A}\cdot\boldsymbol{C})$ and $\nabla\times(\boldsymbol{A}\times\boldsymbol{C})$ (see app. 1) we have

$$(\boldsymbol{C}\cdot\nabla)\boldsymbol{A} = \frac{1}{2}(\nabla(\boldsymbol{A}\cdot\boldsymbol{C}) + \nabla\times(\boldsymbol{A}\times\boldsymbol{C}) - \boldsymbol{A}\times(\nabla\times\boldsymbol{C}) - \boldsymbol{C}\times(\nabla\times\boldsymbol{A})$$

$$- \boldsymbol{A}(\nabla\boldsymbol{C}) + \boldsymbol{C}(\nabla\boldsymbol{A}))\,.$$

By similar manipulations, $\nabla\boldsymbol{F}^2/2$ is reduced to $\nabla\boldsymbol{F}^2/2 = \boldsymbol{F}\times(\nabla\times\boldsymbol{F}) + (\boldsymbol{F}\cdot\nabla)\boldsymbol{F}$. The substitution of $\boldsymbol{A}\to\boldsymbol{b}$, $\boldsymbol{C}\to(\boldsymbol{F}_\perp\times\boldsymbol{b})$ and $\boldsymbol{A}\to\boldsymbol{b}$, $\boldsymbol{C}\to\boldsymbol{F}_\perp$ into the formula for $(\boldsymbol{C}\cdot\nabla)\boldsymbol{A}$ yields

$$\frac{\mathrm{d}u}{\mathrm{d}t} = \frac{\Omega^2}{4(\Omega^2-\omega^2)^2}\Big(-(\boldsymbol{b}\times\boldsymbol{F}_\perp)\cdot[\nabla\times\boldsymbol{F}_\perp - \boldsymbol{b}\times(\nabla\times(\boldsymbol{F}_\perp\times\boldsymbol{b}))$$

$$+ (\boldsymbol{F}_\perp\times\boldsymbol{b})(\nabla\boldsymbol{b})] + \boldsymbol{F}_\perp\cdot[\nabla\times(\boldsymbol{b}\times\boldsymbol{F}_\perp) - \boldsymbol{b}\times(\nabla\times\boldsymbol{F}_\perp) + \boldsymbol{F}_\perp(\nabla\boldsymbol{b})]\Big)$$

$$- \frac{\Omega^2 F_\parallel}{2\omega^2(\Omega^2-\omega^2)}\boldsymbol{F}_\perp\cdot(-\boldsymbol{b}\times(\nabla\times\boldsymbol{b}))$$

$$+ \frac{\boldsymbol{b}}{2(\Omega^2-\omega^2)}\left(\frac{1}{2}\nabla\boldsymbol{F}^2 - \boldsymbol{F}\times(\nabla\times\boldsymbol{F})\right) - \frac{F_\parallel \boldsymbol{b}\cdot((\boldsymbol{b}\cdot\nabla)\boldsymbol{F})}{2(\Omega^2-\omega^2)}$$

$$- \frac{F_\parallel \boldsymbol{b}\cdot((\boldsymbol{b}\cdot\nabla)\boldsymbol{F})}{2\omega^2} + \frac{w^2}{2}\nabla\boldsymbol{b}$$

$$= \frac{\Omega^2}{2(\Omega^2-\omega^2)^2}\boldsymbol{F}_\perp{}^2\cdot\nabla\boldsymbol{b} + \frac{\Omega^2 F_\parallel}{2\omega^2(\Omega^2-\omega^2)}\boldsymbol{F}_\perp\cdot(\boldsymbol{b}\times(\nabla\times\boldsymbol{b}))$$

$$+ \frac{\boldsymbol{b}}{4(\Omega^2-\omega^2)}\cdot\nabla(F_\perp{}^2+F_\parallel{}^2) - \frac{\boldsymbol{b}}{2(\Omega^2-\omega^2)}\cdot(\boldsymbol{F}\times(\nabla\times\boldsymbol{F}))$$

$$- \frac{\Omega^2 F_\parallel \boldsymbol{b}\cdot((\boldsymbol{b}\cdot\nabla)\boldsymbol{F})}{2\omega^2(\Omega^2-\omega^2)} + \frac{w^2}{2}\nabla\boldsymbol{b}$$

$$= \frac{\Omega^2 F_\perp{}^2(\nabla\boldsymbol{b})}{2(\Omega^2-\omega^2)^2} + \frac{\boldsymbol{b}\cdot\nabla(F_\perp{}^2+F_\parallel{}^2)}{4(\Omega^2-\omega^2)} - \frac{\boldsymbol{b}}{2(\Omega^2-\omega^2)}\cdot(\boldsymbol{F}\times(\nabla\times\boldsymbol{F}))$$

$$- \frac{\Omega^2 F_\parallel(\boldsymbol{b}\cdot\nabla)(\boldsymbol{F}\cdot\boldsymbol{b})}{2\omega^2(\Omega^2-\omega^2)} + \frac{w^2}{2}(\nabla\boldsymbol{b}).$$

Here we have used the vector formula for $\nabla(\boldsymbol{F}\cdot\boldsymbol{b})$ and $\boldsymbol{b}\cdot((\boldsymbol{F}\cdot\nabla)\boldsymbol{b}) = (1/2)(\boldsymbol{F}\cdot\nabla)(\boldsymbol{b}\cdot\boldsymbol{b}) = 0$. By means of $\nabla\boldsymbol{b} = \nabla(\boldsymbol{B}/\boldsymbol{B}) = -\boldsymbol{b}\nabla\Omega/\Omega$, $\mathrm{d}u/\mathrm{d}t$ is reduced to

$$\frac{\mathrm{d}u}{\mathrm{d}t} = \frac{1}{4}\boldsymbol{b}\cdot\nabla\left(\frac{F_\perp{}^2}{\Omega^2-\omega^2} - \frac{F_\parallel{}^2}{\omega^2}\right) - \frac{w^2}{2\Omega}\boldsymbol{b}\cdot\nabla\Omega - \frac{\boldsymbol{b}\cdot(\boldsymbol{F}\times(\nabla\times\boldsymbol{F}))}{2(\Omega^2-\omega^2)}.$$

Similarly we have $\mathrm{d}\boldsymbol{x}/\mathrm{d}t = u\boldsymbol{b}$. The expression for $\mathrm{d}w/\mathrm{d}t$ is

$$\frac{\mathrm{d}w}{\mathrm{d}t} = \left\langle(\boldsymbol{e}_1\cos\alpha+\boldsymbol{e}_2\sin\alpha)\cdot\left[-w(u\cos\alpha\cdot(\boldsymbol{b}\cdot\nabla)\boldsymbol{e}_1 - u\sin\alpha\cdot(\boldsymbol{b}\cdot\nabla)\boldsymbol{e}_2)\right.\right.$$

$$\left.\left. - u\boldsymbol{b}\times\left(\frac{w}{\Omega}(\boldsymbol{e}_1\sin\alpha-\boldsymbol{e}_2\cos\alpha)\cdot\nabla\right)\Omega\boldsymbol{b}\right]\right\rangle$$

$$= -\frac{wu}{2}[\boldsymbol{e}_1\cdot(\boldsymbol{b}\cdot\nabla)\boldsymbol{e}_1 + \boldsymbol{e}_2\cdot(\boldsymbol{b}\cdot\nabla)\boldsymbol{e}_2 + \boldsymbol{e}_1\cdot(\boldsymbol{e}_1\cdot\nabla)\boldsymbol{b} + \boldsymbol{e}_2\cdot(\boldsymbol{e}_2\cdot\nabla)\boldsymbol{b}]$$

$$= -\frac{wu}{2}[\boldsymbol{e}_1\cdot(\boldsymbol{e}_1\cdot\nabla)\boldsymbol{b} + \boldsymbol{e}_2\cdot(\boldsymbol{e}_2\cdot\nabla)\boldsymbol{b}] = -\frac{wu}{2}(\nabla\boldsymbol{b} - \boldsymbol{b}\cdot(\boldsymbol{b}\cdot\nabla)\boldsymbol{b})$$

$$= -\frac{wu}{2}\nabla\boldsymbol{b}.$$

The expression for $\mathrm{d}\alpha/\mathrm{d}t$ can be obtained in a similar manner.

References

1. W. Schuurman and H. de Kluiver: Plasma Phys. **7**, 245(1965)

2. A. I. Morozov and L. S. Solovév: Rev. of Plasma Phys. **2**, 201 (ed. by M. A. Leontovich) Consultants Bureau, New York, 1966

3. D. V. Sivukhin: ibid. **1**, 1 (ed. by M. A. Leontovich), Consultants Bureau, New York 1966

4. N. N. Bogolyubov and Yu. A. Mitropolyskii: Asimptoticheskie Metody v Teori Nelineinykh Kolebanii, Gosudarstvennde Izdatel'stvo Fiziko-Matematiteskoi Literatury, Moscow, 1963. English translation: Asymptotic Methods in The Theory of Nonlinear Oscillations, Hindustan Publication Corp., New Delhi, 1961

5. K. Miyamoto: Phys. of Fluids **14**, 722(1971)

6. H. P. Furth and M. N. Rosenbluth: Plasma Phys. and Controlled Nucl. Fusion Research **1**, 821 (Conf. Proceedings, Novosibirsk 1968) (IAEA, Vienna, 1969)

7. A. A. Galeev, R. Z. Sagdeev, H. P. Furth, and M. N. Rosenbluth: Phys. Rev. Lett. **22**, 511(1969)

8. K. Miyamoto: Phys. of Fluid **17**, 1476 (1974)

9. A. A. Ware: Phys. Rev. Lett. **25**, 15(1970)

10. T. Watanabe: lecture note in February 1973, Inst. of Plasma Phys. Nagoya University.

11. R. Bardet, T. Consoli, R. Geller: Nucl. Fusion **5**, 7(1965)

12. S. Miyake, T. Sato, K. Takayama, T. Watari, S. Hiroe, T. Watanabe, and K. Husimi: J. Phys. Soc. Japan **31**, 265(1971)
T. Watari, S. Hiroe, T. Watanabe, T. Hatori, T. Sato, K. Takayama, and K. Husimi: Proc. 5th European Conf. on Controlled Fusion and Plasma Physics, p. 104, Grenoble, 1971

4 Coulomb Collision

The motions of charged particles were analyzed in the previous chapter without considering the effects of collisions between particles. In this chapter, phenomena associated with electron-electron (e-e), ion-ion (i-i) and electron-ion (e-i) collisions will be discussed. Since the range of the Coulomb force acting among charged particles is relatively large, the Coulomb collision is considered to be a long-range interaction. The energy and momentum changes in binary collisions (sec. 4.1), the energy exchange between a plasma and test particles (in other words, the heating of plasmas by test particles (sec. 4.2)), the temperature relaxation of a two-component plasma (sec. 4.3), and the momentum variation of particles passing through a plasma (sec. 4.4) are described. Runaway electrons (sec. 4.5) and the electric conductivity of plasmas (sec. 4.6) are also discussed.

We first study the interaction of two particles and calculate the Coulomb scattering cross section (Rutherford cross section). Let the impact parameter be denoted by b, and let the deflection angle in the center-of-mass system by χ, as shown in fig. 4.1. The magnitude of the relative velocity of the two particles remains unchanged in a Coulomb collision. The scattering $\sigma(\chi)\mathrm{d}\Omega$ into the infinitesimal solid angle $\mathrm{d}\Omega$ at the deflection angle χ is obtained as follows:

$$2\pi b\,\mathrm{d}b = \sigma(\chi)\mathrm{d}\Omega \qquad (4.1)$$

$$\mathrm{d}\Omega = 2\pi \sin\chi\,\mathrm{d}\chi$$

$$\sigma(\chi) = \frac{b}{\sin\chi}\frac{\mathrm{d}b}{\mathrm{d}\chi}. \qquad (4.2)$$

When a particle of mass m and charge q collides with another particle of mass m^* and charge q^*, the relation between b and χ is

$$\cot(\chi/2) = \frac{4\pi\varepsilon_0 m_{\mathrm{r}} b u^2}{qq^*}, \qquad (4.3)$$

where ε_0 is the dielectric constant in vacuum and m_{r} is the reduced mass,

$$m_{\mathrm{r}} = \frac{mm^*}{m + m^*}. \qquad (4.4)$$

Therefore the scattering cross section is given by

$$\sigma(\chi) = \frac{(qq^*)^2}{(8\pi\varepsilon_0 m_{\mathrm{r}} u^2 \sin^2(\chi/2))^2}. \qquad (4.5)$$

Sivukhin's review paper[1] gives an excellent account of Coulomb scattering.

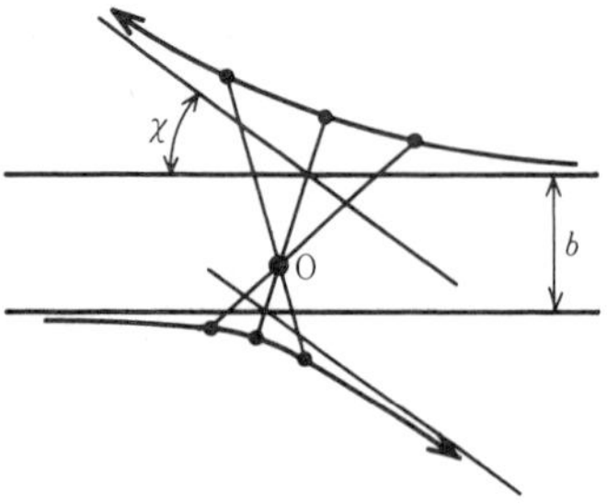

Fig. 4.1
Orbits of two colliding particles in center-of-mass coordinates.

4.1 Change of Energy and Momentum

Suppose that the velocities of two charged particles before collision are $\boldsymbol{v}$, $\boldsymbol{v}^*$ and the velocities after the collision are $\boldsymbol{v} + \delta\boldsymbol{v}$, $\boldsymbol{v}^* + \delta\boldsymbol{v}^*$, respectively. The conservation of momentum is

$$m\delta\boldsymbol{v} + m^*\delta\boldsymbol{v}^* = 0. \tag{4.6}$$

The relative velocity of the two particle is $\boldsymbol{u} = \boldsymbol{v} - \boldsymbol{v}^*$. Accordingly, $\delta\boldsymbol{v}$ and $\delta\boldsymbol{v}^*$ are expressed in terms of $\delta\boldsymbol{u} = \delta\boldsymbol{v} - \delta\boldsymbol{v}^*$ as follows:

$$\left.\begin{aligned}\delta\boldsymbol{v} &= \frac{m^*}{m + m^*}\delta\boldsymbol{u} = \frac{m_{\mathrm{r}}}{m}\delta\boldsymbol{u}\\ \delta\boldsymbol{v}^* &= \frac{-m}{m + m^*}\delta\boldsymbol{u} = -\frac{m_{\mathrm{r}}}{m^*}\delta\boldsymbol{u}.\end{aligned}\right\} \tag{4.7}$$

By using the momenta $\boldsymbol{p} = m\boldsymbol{v}$ and $\boldsymbol{p}^* = m^*\boldsymbol{v}^*$, eq. (4.7) may be written as

$$\delta\boldsymbol{p} = -\delta\boldsymbol{p}^* = m_{\mathrm{r}}\delta\boldsymbol{u}. \tag{4.8}$$

The change of energy of a particle is given by

$$\delta\varepsilon = \frac{m}{2}(\boldsymbol{v} + \delta\boldsymbol{v})^2 - \frac{m}{2}\boldsymbol{v}^2 = m\boldsymbol{V}\cdot\delta\boldsymbol{v} + m_{\mathrm{r}}\boldsymbol{u}\cdot\delta\boldsymbol{v} + \frac{m}{2}(\delta\boldsymbol{v})^2,$$

where

$$\boldsymbol{V} = \frac{m\boldsymbol{v} + m^*\boldsymbol{v}^*}{m + m^*} \tag{4.9}$$

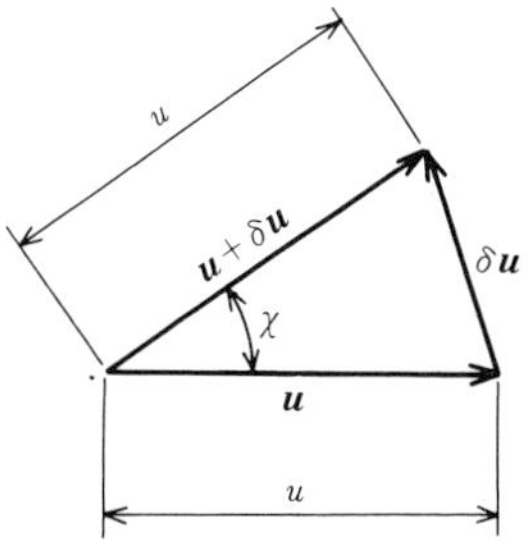

Fig. 4.2
Deflection angle χ and change $\delta\boldsymbol{u}$ of the relative velocity vector.

is the velocity of the center of mass, which is unchanged by the collision ($\delta \boldsymbol{V} = 0$). As the equation of the conservation of energy is

$$(m + m^*)\frac{V^2}{2} + \frac{m_r}{2}\boldsymbol{u}^2 = (m + m^*)\frac{V^2}{2} + \frac{m_r}{2}(\boldsymbol{u} + \delta\boldsymbol{u})^2, \tag{4.10}$$

we find that

$$\delta(\boldsymbol{u})^2 = 2\boldsymbol{u}\cdot\delta\boldsymbol{u} + (\delta\boldsymbol{u})^2 = 0. \tag{4.11}$$

From eqs. (4.7) and (4.11), the changes $\delta\varepsilon$ and $\delta\varepsilon^*$ are given by

$$\left.\begin{aligned} \delta\varepsilon &= m(\boldsymbol{V}\cdot\delta\boldsymbol{v}) = m_r(\boldsymbol{V}\cdot\delta\boldsymbol{u}) \\ \delta\varepsilon^* &= m^*(\boldsymbol{V}\cdot\delta\boldsymbol{v}^*) = -m_r(\boldsymbol{V}\cdot\delta\boldsymbol{u}). \end{aligned}\right\} \tag{4.12}$$

In terms of the deflection angle χ as indicated in fig. 4.2, $\delta\boldsymbol{u}$ may be expressed by

$$\begin{aligned} \delta\boldsymbol{u} &= u\sin\chi\,\boldsymbol{n} - (1 - \cos\chi)\boldsymbol{u} \\ &= u\sin\chi\,\boldsymbol{n} - 2\sin^2\frac{\chi}{2}\boldsymbol{u}, \end{aligned} \tag{4.13}$$

where $\boldsymbol{n}$ is the unit vector normal to the vector $\boldsymbol{u}$ in the plane determined by $\boldsymbol{u}$ and $\delta\boldsymbol{u}$. The changes of energy and momentum are expressed by

$$\delta\varepsilon = -\,\delta\varepsilon^* = m_r u\sin\chi(\boldsymbol{n}\cdot\boldsymbol{V}) - 2m_r\sin^2\frac{\chi}{2}(\boldsymbol{V}\cdot\boldsymbol{u}) \tag{4.14}$$

$$\delta\boldsymbol{p} = -\,\delta\boldsymbol{p}^* = m_r u\sin\chi\,\boldsymbol{n} - 2m_r\boldsymbol{u}\sin^2\frac{\chi}{2}. \tag{4.15}$$

Using eqs. (4.8) and (4.11), we can write

$$\delta p_{\parallel} = m_r \frac{\boldsymbol{u}\cdot\delta\boldsymbol{u}}{u} = -\frac{m_r}{2u}(\delta\boldsymbol{u})^2 = -\frac{(\delta\boldsymbol{p})^2}{2m_r u}. \tag{4.16}$$

Suppose that the *field particles* of mass m^*, charge q^*, and the velocity $\boldsymbol{v}^*$ are distributed uniformly with the density n^*. When a *test particle* of mass m and charge q runs among the field particles, the energy and the momentum of the test particle change by the amounts $\delta\varepsilon$ and $\delta\boldsymbol{p}$ per collision. Accordingly the average change of energy and momentum in time $\mathrm{d}t$ are given by $n^*u\mathrm{d}t\int\delta\varepsilon\sigma(\chi,u)\,\mathrm{d}\Omega$ and $n^*u\mathrm{d}t\int\delta\boldsymbol{p}\sigma(\chi,u)\,\mathrm{d}\Omega$, respectively. Then the rates of change of energy and momentum of the test particle are expressed by

$$\left\langle\frac{\mathrm{d}\varepsilon}{\mathrm{d}t}\right\rangle = n^*u\int\delta\varepsilon\sigma(\chi,u)\mathrm{d}\Omega \tag{4.17 a}$$

$$\left\langle\frac{\mathrm{d}\boldsymbol{p}}{\mathrm{d}t}\right\rangle = n^*u\int\delta\boldsymbol{p}\sigma(\chi,u)\mathrm{d}\Omega. \tag{4.18 a}$$

Substitution of eqs. (4.14) and (4.15) into eqs. (4.17a) and (4.18a) yields

$$\left\langle\frac{\mathrm{d}\varepsilon}{\mathrm{d}t}\right\rangle = -4\pi m_r n^* u(\boldsymbol{V}\cdot\boldsymbol{u})\int_0^\pi \sin^2\frac{\chi}{2}\sigma(\chi,u)\sin\chi\,\mathrm{d}\chi \tag{4.17 b}$$

$$\left\langle\frac{\mathrm{d}\boldsymbol{p}}{\mathrm{d}t}\right\rangle = -4\pi m_r n^* u\boldsymbol{u}\int_0^\pi \sin^2\frac{\chi}{2}\sigma(\chi,u)\sin\chi\,\mathrm{d}\chi, \tag{4.18 b}$$

since the averages of $\sin\chi\,\boldsymbol{n}$ and $u\sin\chi(\boldsymbol{V}\cdot\boldsymbol{n})$ are zero.

When the expression (4.5) for $\sigma(\chi,u)$ is used, the integral $\int_0^\pi\cot(\chi/2)\mathrm{d}\chi$ diverges. However, charged particles of opposite sign tend to cancel the Couloumb field of any given charged particle, so that the electric field outside the Debye radius λ_{d} becomes zero as will be further explained in sec. 4.7. Therefore there is a lower limit $\chi_{\min}$ of the integrals of (4.17b) and (4.18b), and the integral is

$$\ln\Lambda \equiv \frac{1}{2}\int_{\chi_{\min}}^{\pi}\cot\frac{\chi}{2}\mathrm{d}\chi = \ln\frac{1}{\sin\dfrac{\chi_{\min}}{2}} \tag{4.19}$$

which we call the *Couloumb logarithm*. With the Coulomb logarithm, eqs. (4.17b) and (4.18b) are

$$\left\langle \frac{d\varepsilon}{dt} \right\rangle = -\frac{n^*(qq^*)^2 \ln \Lambda}{4\pi\varepsilon_0{}^2 m_r u^3}(\boldsymbol{V}\cdot\boldsymbol{u}) \tag{4.20}$$

$$\left\langle \frac{d\boldsymbol{p}}{dt} \right\rangle = -\frac{n^*(qq^*)^2 \ln \Lambda}{4\pi\varepsilon_0{}^2 m_r u^3}\boldsymbol{u}. \tag{4.21}$$

When the field particles are distributed in velocity space so that the number of field particles in the small region $d\boldsymbol{v}^* \equiv dv_x{}^* dv_y{}^* dv_z{}^*$ around $\boldsymbol{v}^*$ is $f^*(\boldsymbol{v}^*)d\boldsymbol{v}^*$, the rates of change of energy and momentum of the test particle are

$$\left\langle \frac{d\varepsilon}{dt} \right\rangle = -\frac{q^2}{4\pi\varepsilon_0{}^2}\sum{}^* \int \frac{q^{*2}\ln \Lambda}{m_r u^3}(\boldsymbol{V}\cdot\boldsymbol{u}) f^*(\boldsymbol{v}^*) d\boldsymbol{v}^* \tag{4.22}$$

$$\left\langle \frac{d\boldsymbol{p}}{dt} \right\rangle = -\frac{q^2}{4\pi\varepsilon_0{}^2}\sum{}^* \int \frac{q^{*2}\ln \Lambda}{m_r u^3}\boldsymbol{u} f^*(\boldsymbol{v}^*) d\boldsymbol{v}^*. \tag{4.23}$$

The summation is to be taken over all kinds of field particles (electron and ions). As the dependence of $\ln\Lambda$ on u is weak, $\ln\Lambda$ may be taken outside the integrals, to give

$$\left\langle \frac{d\varepsilon}{dt} \right\rangle = -\sum{}^* \ln \Lambda \left(\frac{\boldsymbol{v}\cdot\boldsymbol{E}_v}{m_r} - \frac{\varphi_v}{m} \right) \tag{4.24}$$

$$\left\langle \frac{d\boldsymbol{p}}{dt} \right\rangle = -\sum{}^* \ln \Lambda \frac{\boldsymbol{E}_v}{m_r}, \tag{4.25}$$

where

$$\boldsymbol{E}_v = \frac{1}{4\pi\varepsilon_0}\int \frac{\boldsymbol{u}}{u^3}\rho_v(\boldsymbol{v}^*) d\boldsymbol{v}^* \tag{4.26}$$

$$\varphi_v = \frac{1}{4\pi\varepsilon_0}\int \frac{1}{u}\rho_v(\boldsymbol{v}^*) d\boldsymbol{v}^* \tag{4.27}$$

$$\rho_v \equiv \frac{(qq^*)^2}{\varepsilon_0} f^*(\boldsymbol{v}). \tag{4.28}$$

The expressions (4.26) and (4.27) are those of the electric field and the electrostatic potential of the charge distribution ρ_v.

4.2 Heating of a Plasma by a Particle Beam

When an electron beam or an ion beam is injected into a plasma, energy is exchanged between the beam and the plasma. When the velocity distribution function of the plasma is isotropic, i.e., expressible by $f^*(\boldsymbol{v}^*) = f^*(v^*)$ (v^* being the absolute magnitude of $\boldsymbol{v}^*$), $\boldsymbol{E}_\mathrm{v}$ and φ_v can be easily obtained as the electric field and the electrostatic potential of a spherically symmetric charge density (fig. 4.3):

$$\boldsymbol{E}_\mathrm{v}(\boldsymbol{v}) = \frac{1}{4\pi\varepsilon_0}\frac{\boldsymbol{v}}{v^3}\int_{v^*<v} \rho_\mathrm{v} \mathrm{d}\boldsymbol{v}^* \tag{4.29}$$

$$\varphi_\mathrm{v}(\boldsymbol{v}) = \frac{1}{4\pi\varepsilon_0}\left(\frac{1}{v}\int_{v^*<v} \rho_\mathrm{v} \mathrm{d}\boldsymbol{v}^* + \int_{v^*>v} \frac{\rho_\mathrm{v}}{v^*}\mathrm{d}\boldsymbol{v}^*\right). \tag{4.30}$$

When the distribution function is Maxwellian,

$$\left.\begin{aligned} f^*(v^*) &= n^*\left(\frac{b^*}{\pi^{1/2}}\right)^3 \exp(-b^{*2}v^{*2}) \\ b^* &= \left(\frac{m^*}{2T^*}\right)^{1/2}, \end{aligned}\right\} \tag{4.31}$$

$\boldsymbol{E}_\mathrm{v}$ and φ_v are given by

$$\boldsymbol{E}_\mathrm{v}(\boldsymbol{v}) = \frac{(qq^*)^2 n^*}{4\pi\varepsilon_0{}^2 v^3}\Phi_1(b^*v)\boldsymbol{v} \tag{4.32}$$

$$\varphi_\mathrm{v}(\boldsymbol{v}) = \frac{(qq^*)^2 n^*}{4\pi\varepsilon_0{}^2 v}\Phi(b^*v), \tag{4.33}$$

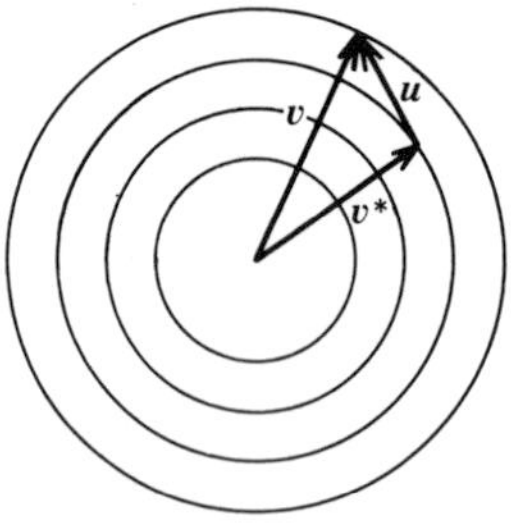

Fig. 4.3
Relationship of $\boldsymbol{v}$, $\boldsymbol{v}^*$, and $\boldsymbol{u}$.

where

$$\Phi(x) = \frac{2}{\pi^{1/2}} \int_0^x \exp(-\xi^2)\,\mathrm{d}\xi \tag{4.34}$$

$$\Phi_1(x) = \Phi(x) - \frac{2x}{\pi^{1/2}} \exp(-x^2) = \Phi(x) - x\frac{\mathrm{d}\Phi}{\mathrm{d}x}. \tag{4.35}$$

When $x > 2$, it can be noted that $\Phi(x) \approx \Phi_1(x) \approx 1$. Substitution of eqs. (4.32) and (4.33) into eq. (4.24) yields

$$\left\langle \frac{\mathrm{d}\varepsilon}{\mathrm{d}t} \right\rangle = -\frac{q^2}{4\pi\varepsilon_0{}^2 v} \sum{}^* \ln \Lambda n^* \frac{q^{*2}}{m^*} \left(\Phi(b^* v) - \left(1 + \frac{m^*}{m}\right) \frac{2b^* v}{\pi^{1/2}} \exp(-b^{*2} v^2) \right). \tag{4.36}$$

Setting $x = b^* v$, $\beta = m^*/m$, and introducing the function

$$F(x, \beta) = \Phi(x) - \frac{2x}{\pi^{1/2}}(1 + \beta)\exp(-x^2), \tag{4.37}$$

we find

$$\left\langle \frac{\mathrm{d}\varepsilon}{\mathrm{d}t} \right\rangle = -\sum{}^* \frac{\ln \Lambda}{4\pi\varepsilon_0{}^2 m^*} (qq^*)^2 n^* b^* \frac{F(x, \beta)}{x}. \tag{4.38}$$

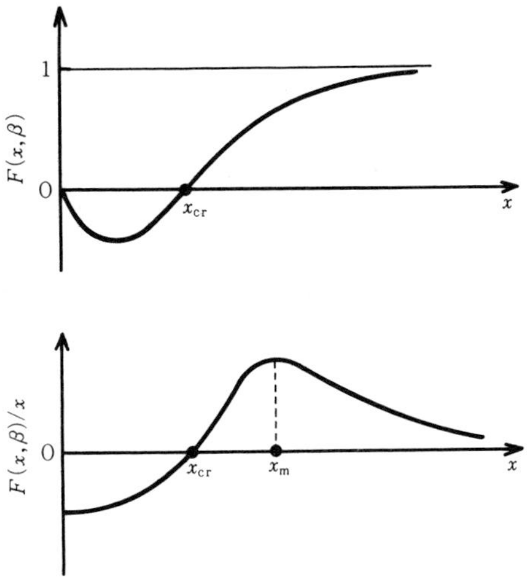

Fig. 4.4
Graphs of $F(x, \beta)$ and $F(x, \beta)/x$ vs. x.

$F(x, \beta)$ is minimum at $x = (\beta/2(1+\beta))^{1/2}$. The value of x which satisfies $F(x, \beta) = 0$ is denoted by $x = x_{cr}$ (fig. 4.4). When $0 < x < x_{cr}$, then $F < 0$. This means that the beam gains energy from the plasma. Denote the value of x at which $F(x, \beta)/x$ is a maximum by x_m; this is the point where the rate of transfer of energy from the beam to the plasma is most effective. Values of x_{cr} and x_m vs. $\beta = m^*/m$ are listed in table 4.1.

Table 4.1
Values of x_{cr}, x_m vs. $\beta = m^*/m$.

$\beta = m^*/m$	0	0.1	1	10	10^2	10^3	10^4
x_{cr}	0	0.377	0.990	1.76	2.37	2.84	3.24
x_m	1.52	1.57	1.85	2.42	2.94	3.37	3.74

When an electron beam runs through a plasma, the beam heats the plasma electrons and ions (the latter of charge $Z_i e$) in different ways. Let us compare the heating rate of both species of the plasma. If the energy of the electron beam $\varepsilon_{be} = m_e v_{be}^2/2$ is larger than the electron temperature of the plasma ($\varepsilon_{be} > 2T_e$, say), the functions $\Phi(b^* v)$ and $\Phi_1(b^* v)$ defined by eqs. (4.34) and (4.35) are nearly equal to 1 and the rate of change of the beam energy is

$$\left\langle \frac{d\varepsilon_{be}}{dt} \right\rangle \approx -\frac{e^4 \ln \Lambda}{4\pi\varepsilon_0^2} \left(\frac{n_i Z_i^2}{m_i} + \frac{n_e}{m_e} \right) \frac{1}{v_{be}}. \tag{4.39}$$

Therefore the heating rate of the plasma electrons is $m_i/(m_e Z)$ time as large as the heating rate of the ions.

Let us consider the case where an ion beam (charge Ze) is injected into the plasma. When the velocity v_{bi} of beam ions is larger than the thermal velocity of the plasma electrons, the situation is the same as in the foregoing case. When v_{bi} is smaller than the thermal velocity of the plasma electrons, the heating rate is given by

$$\left\langle \frac{d\varepsilon_{bi}}{dt} \right\rangle = -\frac{Z^2 e^4 \ln \Lambda n_e}{4\pi\varepsilon_0^2 v_{bi} m_e} \left[\sum_i \frac{m_e n_i Z_1^2}{m_i n_e} \left(\Phi(y) - \left(1 + \frac{m_i}{m_{bi}}\right) \frac{2y}{\pi^{1/2}} \exp(-y^2) \right) + \Phi(x) - \left(1 + \frac{m_e}{m_{bi}}\right) \frac{2x}{\pi^{1/2}} \exp(-x^2) \right], \tag{4.40}$$

where

$$y = b_i{}^* v_{bi} = \left(\frac{m_i}{2T_i}\right)^{1/2} \times v_{bi} = \left(\frac{m_i}{m_{bi}}\right)^{1/2} \times (\varepsilon_{bi}/T_i)^{1/2}$$

$$x = b_e{}^* v_{bi} = \left(\frac{m_e}{2T_e}\right)^{1/2} \times v_{bi} = \left(\frac{m_e}{m_{bi}}\right)^{1/2} \times (\varepsilon_{bi}/T_e)^{1/2} \qquad y = \left(\frac{m_i\, T_e}{m_e\, T_i}\right)^{1/2} x.$$

When $y > 2$, $x < 0.3$, i.e., $0.5v_{te} > v_{bi} > 3v_{ti}$ (v_{te} and v_{ti} are electron and ion thermal velocity, respectively), then the rate of change of the beam energy is

$$\left\langle \frac{d\varepsilon_{bi}}{dt} \right\rangle \approx -\frac{Z^2 e^4 \ln \Lambda n_e}{4\pi\varepsilon_0{}^2 m_e v_{bi}} \left(\sum_i \frac{m_e}{m_i} \frac{n_i Z_i{}^2}{n_e} + \frac{4}{3\pi^{1/2}} \left(\frac{m_e \varepsilon_{bi}}{m_{bi} T_e}\right)^{3/2} \right). \tag{4.41}$$

The energy of that ion beam which heats the plasma electrons and plasma ions at an equal rate is derived from eq. (4.41), and is[2]

$$\varepsilon_{cr} = 15T_e \left(A_b{}^{3/2} \sum \frac{n_i Z_i{}^2}{n_e} \frac{1}{A_i} \right)^{2/3}, \tag{4.42}$$

where A_b, A_i are the atomic weights of the beam ion and plasma ion, respectively. $(\ )^{2/3}$ is of the order of 1. The slowing down time τ of the beam is

$$\tau = -\int_0^{\varepsilon} \frac{d\varepsilon_{bi}}{d\varepsilon_{bi}/dt} = \frac{\tau_\varepsilon{}^{ei}}{1.5} \ln \left(1 + \left(\frac{\varepsilon}{\varepsilon_{cr}}\right)^{3/2} \right), \tag{4.43}$$

where

$$\tau_\varepsilon{}^{ei} = 3 \times 10^{14} \frac{A_b (T_e/e)^{3/2}}{Z^2 n_e \ln \Lambda} \qquad \text{(s)}. \tag{4.44}$$

$\tau_\varepsilon{}^{ei}$ is the relaxation time of for energy exchange between electrons and ions, which will be explained in the next section.

4.3 Temperature Relaxation Time of a Two-Component Plasma

From eq. (4.36) we can obtain the energy relaxation between the beam particles (m, q) with a specified velocity and the plasma particles (m^*, q^*), the latter taken to be of Maxwellian velocity distribution at the temperature T^*. In this section we treat the temperature relaxation between two components of the plasma with different temperatures T and T^*. We may consider the component at T to be an ensemble of beams with different velocities, with a weighing function (velocity distribution function) $f(v) =$

$(b/\pi^{1/2})^3 \exp(-b^2v^2)$ $(b^2 = m/2T)$. If we take the average of eq. (4.36) with $f(v)$, the rate of energy transfer $Q = \langle\langle d\varepsilon/dt \rangle\rangle$ from the component at T to the component at T^* can be deduced. In the averaging process, we utilize following relations:

$$\left\langle \frac{\Phi(b^*v)}{v} \right\rangle \equiv \frac{4b^3}{\pi^{1/2}} \int_0^\infty \Phi(b^*v) \exp(-b^2v^2)\, v\mathrm{d}v$$

$$= \frac{8b^3}{\pi b^{*2}} \int_0^\infty \exp\left(-\left(\frac{b}{b^*}x\right)^2\right) \cdot x \left(\int_0^x \exp(-\xi^2)\mathrm{d}\xi \right) \mathrm{d}x$$

$$= \frac{8b^3}{\pi b^{*2}} \int_0^\infty \exp(-\xi^2)\mathrm{d}\xi \int_\varepsilon^\infty \exp\left(-\left(\frac{b}{b^*}x\right)^2\right) x\mathrm{d}x$$

$$= \frac{2}{\pi^{1/2}} \frac{bb^*}{(b^2 + b^{*2})^{1/2}}$$

$$\langle \exp(-(b^*v)^2) \rangle = \left(\frac{b}{\pi^{1/2}}\right)^3 \int_0^\infty \exp(-(b^2 + b^{*2})v^2) \cdot 4\pi v^2\, \mathrm{d}v$$

$$= \frac{b^3}{(b^2 + b^{*2})^{3/2}}.$$

With $Q = \langle\langle d\varepsilon/dt \rangle\rangle$ thus calculated, we can define the *temperature relaxation time* (also called the *energy relaxation time*) τ_ε by the relation

$$Q \equiv \frac{(3/2)(T - T^*)}{\tau_\varepsilon}; \tag{4.45}$$

thus τ_ε is given by

$$\tau_\varepsilon = \frac{(2\pi)^{1/2}\, 3\pi\varepsilon_0^2\, mm^*}{n^* \ln \Lambda (qq^*)^2} \left(\frac{T}{m} + \frac{T^*}{m^*} \right)^{3/2}. \tag{4.46}$$

Denote electron-electron, ion-ion, and electron-ion temperature relaxation times by $\tau_\varepsilon^{\mathrm{ee}}$, $\tau_\varepsilon^{\mathrm{ii}}$, and $\tau_\varepsilon^{\mathrm{ei}}$, respectively; then

$$\tau_\varepsilon^{\mathrm{ee}} \approx \frac{(2\pi)^{1/2}\, 6\pi\varepsilon_0^2\, m_{\mathrm{e}}^{1/2}\, T_{\mathrm{e}}^{3/2}}{n_{\mathrm{e}} e^4 \ln \Lambda} \tag{4.47}$$

$$\tau_\varepsilon^{\mathrm{ii}} \approx \frac{(2\pi)^{1/2}\, 6\pi\varepsilon_0^2\, m_{\mathrm{i}}^{1/2}\, T_{\mathrm{i}}^{3/2}}{Z^4 n_{\mathrm{i}} e^4 \ln \Lambda} \tag{4.48}$$

$$\tau_\varepsilon^{\mathrm{ei}} \approx \frac{(2\pi)^{1/2}\, 3\pi\varepsilon_0^2 m_{\mathrm{i}} m_{\mathrm{e}}^{-1/2}\, T_{\mathrm{e}}^{3/2}}{Z^2 n_{\mathrm{i}} e^4 \ln \Lambda}. \tag{4.49}$$

The quantities $-e$ and Ze are the charges of the electrons and the ions, respectively ($n_e = Zn_i$). Ratios are

$$\tau_\varepsilon{}^{ee} : \tau_\varepsilon{}^{ii} : \tau_\varepsilon{}^{ei} \approx 1 : \frac{1}{Z^3}\left(\frac{m_i}{m_e}\right)^{1/2}\left(\frac{T_i}{T_e}\right)^{3/2} : \frac{1}{(2Z)}\frac{m_i}{m_e}.$$

It should be noted that the e-i temperature relaxation time is much longer than the others.

4.4 Momentum Relaxation Times

The momentum change $\delta \boldsymbol{p}$ of a test particle is expressed by eq. (4.25). When the field particles are distributed isotropically, $\boldsymbol{E}_v$ has the same direction as $\boldsymbol{v}$ and eq. (4.25) gives the rate of change of momentum parallel to the velocity $\boldsymbol{v}$ of the test particle. The rate of change of momentum perpendicular to the velocity $\boldsymbol{v}$ is also an important quantity. Let us consider the square of the magnitude of the momentum i.e.,

$$\frac{\mathrm{d}p^2}{\mathrm{d}t} = \frac{\mathrm{d}p_\parallel{}^2}{\mathrm{d}t} + \frac{\mathrm{d}p_\perp{}^2}{\mathrm{d}t} \tag{4.50}$$

$$\frac{\mathrm{d}p^2}{\mathrm{d}t} = 2m\frac{\mathrm{d}\varepsilon}{\mathrm{d}t} \tag{4.51}$$

$$\frac{\mathrm{d}p_\parallel{}^2}{\mathrm{d}t} = 2p_\parallel\frac{\mathrm{d}p_\parallel}{\mathrm{d}t} = 2\boldsymbol{p}\frac{\mathrm{d}\boldsymbol{p}}{\mathrm{d}t}. \tag{4.52}$$

Take the average over the test particles with the distribution function $f^*(v^*)$ (definition of the distribution function is given in ch. 5), and utilize eqs. (4.50) and (4.52); then $\langle \mathrm{d}p_\perp{}^2/\mathrm{d}t\rangle$ is reduced to

$$\left\langle\frac{\mathrm{d}p_\perp{}^2}{\mathrm{d}t}\right\rangle = 2m\left\langle\frac{\mathrm{d}\varepsilon}{\mathrm{d}t}\right\rangle - 2\boldsymbol{p}\left\langle\frac{\mathrm{d}\boldsymbol{p}}{\mathrm{d}t}\right\rangle. \tag{4.53}$$

From the relations (4.24) and (4.25) and eqs. (4.32) and (4.33), the rate of change of the momentum is given by

$$\left\langle\frac{\mathrm{d}\boldsymbol{p}}{\mathrm{d}t}\right\rangle = -\frac{q^2\boldsymbol{v}}{4\pi\varepsilon_0{}^2v^3}\sum{}^*\frac{\ln \Lambda q^{*2}}{m_r}n^*\,\Phi_1(b^*v) \tag{4.54}$$

$$\left\langle\frac{\mathrm{d}p_\perp{}^2}{\mathrm{d}t}\right\rangle = \frac{2q^2}{4\pi\varepsilon_0{}^2v}\sum{}^*\ln \Lambda n^*q^{*2}\,\Phi(b^*v). \tag{4.55}$$

Define the *momentum relaxation times* $\tau_{\parallel}$ and $\tau_{\perp}$ by

$$\left\langle \frac{\mathrm{d}p_{\parallel}}{\mathrm{d}t} \right\rangle \equiv -\frac{p}{\tau_{\parallel}} \tag{4.56}$$

$$\left\langle \frac{\mathrm{d}p_{\perp}{}^2}{\mathrm{d}t} \right\rangle \equiv \frac{p^2}{\tau_{\perp}}; \tag{4.57}$$

then these are given by

$$\tau_{\parallel} = \frac{4\pi\varepsilon_0{}^2 m v^3}{q^2 \Sigma^* \ln \Lambda q^{*2} n^* \Phi_1(b^* v)/m_{\mathrm{r}}} \tag{4.58}$$

$$\tau_{\perp} = \frac{2\pi\varepsilon_0{}^2 m^2 v^3}{q^2 \Sigma^* \ln \Lambda q^{*2} n^* \Phi(b^* v)}. \tag{4.59}$$

$\tau_{\parallel}$ is the slowing down time of the velocity parallel to $\boldsymbol{v}$ and $\tau_{\perp}$ is the diffusion time in the direction perpendicular to $\boldsymbol{v}$ in velocity space. $\tau_{\parallel}$ and $\tau_{\perp}$ are also called *collision times.*

When the test particle is an electron and the plasma consists of electrons and ions of charge Ze, the e-e and e-i collision times are

$$\tau_{\mathrm{e}} \equiv \tau_{\parallel}{}^{\mathrm{ee}} = \tau_{\perp}{}^{\mathrm{ee}} = \frac{Z}{2}\tau_{\parallel}{}^{\mathrm{ei}} = Z\tau_{\perp}{}^{\mathrm{ei}} = \frac{2\pi\varepsilon_0{}^2 m_{\mathrm{e}}{}^2 v_{\mathrm{e}}{}^3}{n_{\mathrm{e}} e^4 \ln \Lambda}, \tag{4.60}$$

because $\Phi \approx \Phi_1 \approx 1$ for $b^* v > 1$.

Taking the average value of v_{e} from $m_{\mathrm{e}} v_{\mathrm{e}}{}^2/2 = (3/2) T_{\mathrm{e}}$, we find the collision time τ_{e} to be given by

$$\tau_{\mathrm{e}} = \frac{3^{1/2} 6\pi\varepsilon_0{}^2 m_{\mathrm{e}}{}^{1/2} T_{\mathrm{e}}{}^{3/2}}{n_{\mathrm{e}} e^4 \ln \Lambda}. \tag{4.61}$$

This value is within 20% of Spitzer's result,[3]

$$\tau_{\mathrm{e}} = \frac{25.8\pi^{1/2} \varepsilon_0{}^2 m_{\mathrm{e}}{}^{1/2} T_{\mathrm{e}}{}^{3/2}}{n_{\mathrm{e}} e^4 \ln \Lambda}. \tag{4.62}$$

The i-i collision time is similarily obtained ($n_{\mathrm{e}} = Z n_{\mathrm{i}}$):

$$\tau_{\mathrm{i}} = \frac{3^{1/2} 6\pi\varepsilon_0{}^2 m_{\mathrm{i}}{}^{1/2} T_{\mathrm{i}}{}^{3/2}}{Z^4 n_{\mathrm{i}} e^4 \ln \Lambda}. \tag{4.63}$$

When the test particle is an ion and the field particles are plasma electrons, $x = b^* v$ is small and the relations $\Phi(x) \approx 2x/\pi^{1/2}$ and $\Phi_1(x) \approx 4x^3/(3\pi^{1/2})$ can be utilized. The result is

$$\tau_{ie} \equiv \tau_{\|}{}^{ie} \approx \tau_{\perp}{}^{ie} \approx \frac{(2\pi)^{1/2} 3\pi\varepsilon_0{}^2 m_e{}^{-1/2} m_i T_e{}^{3/2}}{Z^2 n_e e^4 \ln \Lambda}. \tag{4.64}$$

From eqs. (4.61), (4.63), and (4.64), the ratios of the three collision times are

$$\tau_e : \tau_i : \tau_{ie} \approx 1 : \frac{1}{Z^3}\left(\frac{m_i}{m_e}\right)^{1/2}\left(\frac{T_i}{T_e}\right)^{3/2} : \frac{m_i}{Z^2 m_e}. \tag{4.65}$$

These collision times correspond to the energy relaxation times $\tau_\varepsilon{}^{ee}$, $\tau_\varepsilon{}^{ii}$, and $\tau_\varepsilon{}^{ei}$, and the values are of the same order, respectively.

The mean free paths of the electrons and ions are given by $\lambda_e = (3T_e/m_e)^{1/2}\,\tau_e, \lambda_i = (3T_i/m_i)^{1/2}\,\tau_i$, respectively:

$$\lambda_e = \frac{25\pi\varepsilon_0{}^2 T_e^2}{n_e e^4 \ln \Lambda} \tag{4.66}$$

$$\lambda_i = \frac{25\pi\varepsilon_0{}^2 T_i^2}{Z^4 n_i e^4 \ln \Lambda} \tag{4.67}$$

When $T_e = T_i/Z$, then $\lambda_e = \lambda_i$.

4.5 Runaway Electron

When a uniform electric field $\boldsymbol{E}$ is applied to a plasma, the motion of a test particle is

$$\left\langle \frac{d\boldsymbol{p}}{dt} \right\rangle = -\left(\frac{q^2}{4\pi\varepsilon_0{}^2 v^3} \sum{}^* \frac{\ln \Lambda q^{*2}}{m_r} n^* \,\Phi(b^* v) \right)\boldsymbol{v} + q\boldsymbol{E}. \tag{4.68}$$

The first term of the right-hand side, the deceleration term, decreases as v increases; the deceleration term becomes smaller than the acceleration term $q\boldsymbol{E}$ at a critical value v_{cr}. When $v > v_{cr}$, the test particle is accelerated. The deceleration term becomes smaller (proportional to v^{-3}) and the velocity starts to increase without limit. Such an electron is called a *runaway electron*. When $x > 2$, then $\Phi \approx 1$ and eq. (4.68) is reduced to

$$\left\langle \frac{d\boldsymbol{p}}{dt} \right\rangle = -\frac{e^4 \ln \Lambda n_e}{2\pi\varepsilon_0{}^2 m_e}\frac{\boldsymbol{v}}{v^3} - e\boldsymbol{E}. \tag{4.69}$$

Accordingly the critical velocity is obtained by making the right-hand side of eq. (4.69) zero, i.e.,

$$\frac{m_e v_{cr}^2}{2e} = \frac{e^2 n \ln \Lambda}{4\pi\varepsilon_0{}^2 \boldsymbol{E}}. \tag{4.70}$$

Taking $\ln \Lambda = 20$, we find that the critical energy of the runaway electron is

$$\frac{m_e v_{cr}^2}{2e} = 5 \times 10^{-16} \frac{n}{E} \qquad \text{(mks units).} \tag{4.71}$$

When $n \approx 10^{20}/\text{m}^3$, $E = 10$ V/m, electrons with energy larger than 5 keV become runaway electrons. Even in the presence of a confining magnetic field, the runaway condition obtains whenever the electric field has a component parallel to the magnetic field.

4.6 Electric Resistivity

When an electric field is applied to a plasma, a current flows

$$\boldsymbol{j} = n_e e(\bar{\boldsymbol{v}}_1 - \bar{\boldsymbol{v}}_e). \tag{4.72}$$

by definition. But the plasma exhibits resistivity as an effect of collisions. If the specific resistivity is denoted by η, then the electric field is $\boldsymbol{E} = \eta \boldsymbol{j}$. The increase of momentum of the plasma electrons per unit time and unit volume due to the electric field is

$$-en_e \boldsymbol{E} = -n_e e\eta \boldsymbol{j}. \tag{4.73}$$

The increase of any single electron's momentum by a collision with an ion is $m_e(\boldsymbol{v}_e - \boldsymbol{v}_i)$ as is clear from eq. (4.8), so that the increase of momentum of the plasma electrons per unit time and unit volume is given by

$$-n_e m_e(\boldsymbol{v}_e - \boldsymbol{v}_i)\nu_{ei\parallel} = \frac{m_e \boldsymbol{j} \nu_{e\parallel}}{e}, \tag{4.74}$$

where $\nu_{ei\parallel}$ is the collision frequency and $\nu_{e\parallel} = 1/\tau_{\parallel}{}^{ei} = Z/2\tau_e$. As the sum of eqs. (4.73) and (4.74) should be zero, the specific resistivity $\eta_{\parallel}$ is

$$\eta_{\parallel} = \frac{m_e \nu_{ei\parallel}}{n_e e^2} = \frac{(m_e)^{1/2} Ze^2 \ln \Lambda}{3^{1/2}\, 12\pi\varepsilon_0{}^2\, T_e{}^{3/2}}$$

$$= 2.3 \times 10^{-9} \frac{Z \ln \Lambda}{T_e{}^{*3/2}} \qquad (\Omega\text{-m}). \tag{4.75}$$

T^* is the electron temperature expressed in keV. Spitzer's formula of $\eta_{\parallel}$ is

$$\eta_{\parallel} = 1.65 \times 10^{-9} \frac{Z \ln \Lambda}{T_{\mathrm{e}}^{*3/2}} \qquad (\Omega\text{-m}) \tag{4.76}$$

The specific resistivity of a plasma with $T_{\mathrm{e}} = 1$ keV is slightly larger than the specific resistivity of copper at 20°C, 1.8×10^{-8} Ω-m. A similar analysis can be applied when collisions with neutral particles must also be taken into account.

4.7 Debye Shielding and the Coulomb Logarithm

Assume that there is small perturbation of the electron density in a uniform neutral plasma. When the electrons have a Boltzmann distribution, the density n_{e} is

$$n_{\mathrm{e}} = n_0 \exp \frac{e\phi}{T_{\mathrm{e}}} \approx n_0 \left(1 + \frac{e\phi}{T_{\mathrm{e}}}\right). \tag{4.77}$$

Poisson's equation is

$$-\varepsilon_0 \nabla^2 \phi = -e(n_{\mathrm{e}} - n_0),$$

i.e.,

$$\nabla^2 \phi - \phi/\lambda_{\mathrm{d}}^2 = 0, \tag{4.78}$$

where

$$\lambda_{\mathrm{d}} = (\varepsilon_0 T_{\mathrm{e}}/n_{\mathrm{e}} e^2)^{1/2} = 7.45 \times 10^3 \left(\frac{T_{\mathrm{e}}}{e n_{\mathrm{e}}}\right)^{1/2} \qquad (\mathrm{m}) \tag{4.79}$$

(T_{e}/e in eV; n_{e} in m^{-3}, λ_{d} in m). A spherically symmetric solution of eq. (4.78) is

$$\phi = \alpha \frac{\exp(-r/\lambda_{\mathrm{d}})}{r}. \tag{4.80}$$

It is clear from eq. (4.80) that the electrostatic perturbation is shielded out to a distance λ_{d}. This distance is called the *Debye length* or *Debye radius.* The number n_{λ} of electrons included in the sphere of radius λ_{d} is given by

$$n_{\lambda} = \frac{4\pi}{3} \frac{(\varepsilon_0 T/e^2)^{3/2}}{n^{1/2}}. \tag{4.81}$$

If the average distance between the particles is taken as $(4\pi/3)r_0{}^3 \equiv 1/n$, the number n_λ is

$$n_\lambda = (\lambda_d/r_0)^3. \tag{4.82}$$

For most plasmas, any characteristic dimension a_p, is much larger than the Debye length,

$$a_p \gg \lambda_d.$$

If $\lambda_d > a_p$ (this would be a plasma of very low density), electrostatic interactions between the charged particle can be neglected and the system becomes only a group of individual charged particles. This is no longer considered a plasma. When the density becomes very high, $\lambda_d < r_0$, quantum statistics should be introduced.

Because of Debye shielding, the lower limit $\chi_{\min}$ of the scattering angle for the Coulomb logarithm can be obtained by substitution of λ_d for the impact parameter b in eq. (4.3):

$$\cot\frac{\chi_{\min}}{2} = 4\pi\varepsilon_0 m_r \lambda_d \frac{u^2}{qq^*}. \tag{4.83}$$

The value of b where $\chi = \pi/2$ is

$$b_I = \frac{qq^*}{4\pi\varepsilon_0 m_r u^2}, \tag{4.84}$$

so that $\cot(\chi_{\min}/2) = \lambda_d/b_I$ and the Coulomb logarithm is

$$\ln\Lambda = -\ln\left(\sin\frac{\chi_{\min}}{2}\right) = \ln\frac{(\lambda_d{}^2 + b_I{}^2)^{1/2}}{b_I} \approx \ln\frac{\lambda_d}{b_I}$$

$$= \ln\left(\frac{4\pi\varepsilon_0 m_r}{qq^*}\left(\frac{\varepsilon_0 T_e}{n_e e^2}\right)^{1/2} u^2\right). \tag{4.85}$$

As the dependence of $\ln\Lambda$ on u is logarithmic, we may neglect the distribution of u. Substitution of the average value $m_r u^2/2 = 3T_e/2$ into eq. (4.85) yields

$$\ln\Lambda \approx 7 + 2.3\log_{10}\left(\left(\frac{T_e}{e}\right)^{3/2}\bigg/\left(\frac{n_e}{10^{20}}\right)^{1/2}\right), \tag{4.86}$$

where T_e/e is in eV and n_e is in m^{-3}. The values of the Coulomb logarithm for laboratory plasmas are mostly about 10 to 20. The classical analysis can be applied only for the case that the impact parameter (4.84) is larger than the De Broglie wavelength $\lambda = h/m_e u$. A discussion of the Coulomb logarithm including quantum effects is contained in ref. 1.

References

1. D. V. Sivukhin: Rev. of Plasma Phys. **4** (ed. by M. A. Leontovich), Consultants Bureau, New York, 1966
2. T. H. Stix: Plasma Phys. **14**, 367(1972)
3. L. Spitzer Jr.: Physics of Fully Ionized Gases, Wiley Interscience, New York, 1962

5 Velocity-Space Distribution Function and Fundamental Plasma Equation

A plasma consists of many ions and electrons, but the individual behavior of each particle can hardly be observed. What can be observed instead are statistical averages. In order to describe the properties of a plasma, it is necessary to define a distribution function that indicates particle density in the phase space whose ordinates are the particle positions and velocities. The distribution function may or may not be stationary with respect to time. In sec. 5.1, the equation governing the distribution function $f(q_i, p_i, t)$ is derived by means of Liouville's theorem. Boltzmann's equation for the distribution function $f(r, v, t)$ is formulated in sec. 5.2. When the collision term is neglected, Boltzmann's equation is called Vlasov's equation. In sec. 5.3, the effects of collisions are discussed. In secs. 5.4 and 5.5, the diffusion of particles in velocity space is studied for the mirror field and then the mirror loss is estimated.

5.1 Phase Space and Distribution Function

A particle can be specified by its coordinates x, y, z, velocity v_x, v_y, v_z, and time t. More generally, the particle can be specified by *canonical variables* $q_1, q_2, q_3, p_1, p_2, p_3$ and t in phase space. When canonical variables are used, an infinitesimal volume in phase space $\Delta = \delta q_1\, \delta q_2\, \delta q_3\, \delta p_1\, \delta p_2\, \delta p_3$ is conserved (Liouville's theorem). The motion of a particle in phase space is described by Hamilton's equations

$$\frac{\mathrm{d}q_i}{\mathrm{d}t} = \frac{\partial H(q_i, p_i, t)}{\partial p_i} \qquad \frac{\mathrm{d}p_i}{\mathrm{d}t} = -\frac{\partial H(q_i, p_i, t)}{\partial q_i}. \tag{5.1}$$

The variation over time of Δ is given by

$$\frac{\mathrm{d}\Delta}{\mathrm{d}t} = \left(\frac{\mathrm{d}(\delta q_1)}{\mathrm{d}t}\delta p_1 + \frac{\mathrm{d}(\delta p_1)}{\mathrm{d}t}\delta q_1\right)\delta q_2\,\delta p_2\,\delta q_3\,\delta p_3 + \dots$$
$$+ \left(\frac{\mathrm{d}(\delta q_3)}{\mathrm{d}t}\delta p_3 + \frac{\mathrm{d}(\delta p_3)}{\mathrm{d}t}\delta q_3\right)\delta q_1\,\delta p_1\,\delta q_2\,\delta p_2, \tag{5.2}$$

$$\frac{\mathrm{d}}{\mathrm{d}t}(\delta q_i) = \delta\left(\frac{\partial H}{\partial p_i}\right) = \frac{\partial^2 H}{\partial p_i \partial q_i}\delta q_i$$

$$\frac{\mathrm{d}}{\mathrm{d}t}(\delta p_i) = -\delta\left(\frac{\partial H}{\partial q_i}\right) = -\frac{\partial^2 H}{\partial q_i \partial p_i}\delta p_i$$

$$\frac{\mathrm{d}\Delta}{\mathrm{d}t} = \sum_i \left(\frac{\partial^2 H}{\partial p_i \partial q_i} - \frac{\partial^2 H}{\partial q_i \partial p_i}\right)\Delta = 0. \tag{5.3}$$

This is a proof of Liouville's theorem (see fig. 5.1).

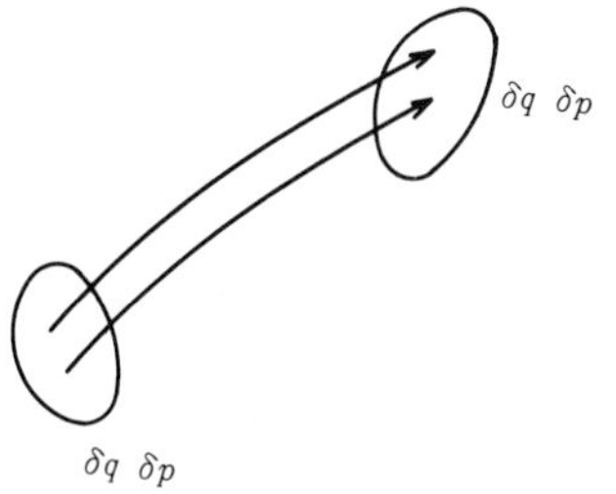

Fig. 5.1
Movement of particles in phase space.

Let the number of particles in a small volume of phase space be

$$f(q_i, p_i, t)\delta\boldsymbol{q}\delta\boldsymbol{p}, \tag{5.4}$$

where $\delta\boldsymbol{q} = \delta q_1\,\delta q_2\,\delta q_3$, $\delta\boldsymbol{p} = \delta p_1\,\delta p_2\,\delta p_3$, and $f(q_i, p_i, t)$ is the *distribution function in phase space.* If the particles move according to the equation of motion and are not scattered by collisions, the small volume in phase space is conserved. As the particle number is conserved, the distribution function is also constant, i.e.,

$$\frac{\mathrm{d}f}{\mathrm{d}t} = 0 \tag{5.5}$$

$$\frac{\partial f}{\partial t} + \sum_i \left(\frac{\mathrm{d}q_i}{\mathrm{d}t}\frac{\partial f}{\partial q_i} + \frac{\mathrm{d}p_i}{\mathrm{d}t}\frac{\partial f}{\partial p_i}\right) = 0 \tag{5.6a}$$

$$\frac{\partial f}{\partial t} + \sum_i \left(\frac{\partial H}{\partial p_i}\frac{\partial f}{\partial q_i} - \frac{\partial H}{\partial q_i}\frac{\partial f}{\partial p_i}\right) = 0. \tag{5.6b}$$

With use of Poisson's brackets, eq. (5.6b) can be written as

$$\frac{\partial f}{\partial t} + \{fH\} = 0. \tag{5.6c}$$

When the velocity $\boldsymbol{V}$ in phase space is defined by $\boldsymbol{V} = \left(\frac{\mathrm{d}q_i}{\mathrm{d}t}, \frac{\mathrm{d}p_i}{\mathrm{d}t}\right)$, it follows that

$$\mathrm{div}\ \boldsymbol{V} = \sum_i \frac{\partial}{\partial q_i}\frac{\mathrm{d}q_i}{\mathrm{d}t} + \sum_i \frac{\partial}{\partial p_i}\frac{\mathrm{d}p_i}{\mathrm{d}t} = 0$$

and eq. (5.6a) can be described by

$$\frac{\partial f}{\partial t} + f \operatorname{div} \boldsymbol{V} + \boldsymbol{V}\cdot \operatorname{grad} f = \frac{\partial f}{\partial t} + \operatorname{div} \boldsymbol{V}f = 0. \tag{5.6d}$$

A stationary distribution function ($\partial f/\partial t = 0$) must satisfy

$$\frac{\partial H}{\partial p_i}\frac{\partial f}{\partial q_i} - \frac{\partial H}{\partial q_i}\frac{\partial f}{\partial p_i} = 0,$$

i.e.,

$$f = f(H(p, q)).$$

In the foregoing discussion we did not take collisions into account. If we denote the variation of f due to the collisions by $(\delta f/\delta t)_{\text{coll}}$, eq. (5.6a) becomes

$$\frac{\partial f}{\partial t} + \sum_i \left(\frac{\mathrm{d}q_i}{\mathrm{d}t}\frac{\partial f}{\partial q_i} + \frac{\mathrm{d}p_i}{\mathrm{d}t}\frac{\partial f}{\partial p_i}\right) = \left(\frac{\delta f}{\delta t}\right)_{\text{coll}}. \tag{5.6e}$$

The collision term will be explained in more detail in sec. 5.3.

5.2 Boltzmann's and Vlasov's Equations

Let us use the space and velocity-space coordinates $x_1, x_2, x_3, v_1, v_2, v_3$ instead of the canonical coordinates. The Hamiltonian is

$$H = c({m_0}^2c^2 + (\boldsymbol{p} - q\boldsymbol{A})^2)^{1/2} + q\phi \tag{5.7}$$

$$p_i = m_0\gamma v_i + qA_i \qquad \gamma \equiv (1 - v^2/c^2)^{-1/2}, \tag{5.8}$$

and $\mathrm{d}x_i/\mathrm{d}t$ and $\mathrm{d}p_i/\mathrm{d}t$ are given by

$$\left.\begin{aligned} \frac{\mathrm{d}x_i}{\mathrm{d}t} &= \frac{\partial H}{\partial p_i} = v_i \\ \frac{\mathrm{d}p_i}{\mathrm{d}t} &= -\frac{\partial H}{\partial x_i} = \frac{q}{m_0\gamma}\sum_k (p_k - qA_k)\frac{\partial A_k}{\partial x_i} - q\frac{\partial \phi}{\partial x_i} \\ &= q\sum_k v_k \frac{\partial A_k}{\partial x_i} - q\frac{\partial \phi}{\partial x_i}. \end{aligned}\right\} \tag{5.9}$$

Consequently eq. (5.6a) becomes

$$\frac{\partial f}{\partial t} + \sum_i v_i \frac{\partial}{\partial x_i} f + \sum_i \left(q \sum_k v_k \frac{\partial A_k}{\partial x_i} - q \frac{\partial \phi}{\partial x_i} \right) \frac{\partial f}{\partial p_i} = \left(\frac{\delta f}{\delta t} \right)_{\text{coll}}. \tag{5.10}$$

$(\delta f/\delta t)_{\text{coll}}$ is the collision term. By using the relations

$$\sum_k v_k \frac{\partial A_k}{\partial x_i} = (\boldsymbol{v} \cdot \nabla) A_i + (\boldsymbol{v} \times (\nabla \times \boldsymbol{A}))_i, \tag{5.11}$$

$$\left.\begin{aligned} \frac{\partial}{\partial p_i} f(v_k(p_j, x_j, t), x_k, t) &= \sum_k \frac{\partial f}{\partial v_k} \frac{\partial v_k}{\partial p_i} \\ \frac{\partial}{\partial x_i} f(v_k(p_j, x_j, t), x_k, t) &= \sum_k \frac{\partial f}{\partial v_k} \frac{\partial v_k}{\partial x_i} + \frac{\partial f}{\partial x_i} \\ \frac{\partial}{\partial t} f(v_k(p_j, x_j, t), x_k, t) &= \sum_k \frac{\partial f}{\partial v_k} \frac{\partial v_k}{\partial t} + \frac{\partial f}{\partial t}, \end{aligned}\right\} \tag{5.12}$$

and

$$\left.\begin{aligned} \frac{\partial v_k(p_j, x_j, t)}{\partial p_i} &= \frac{1}{m_0 \gamma} \left(\delta_{ik} - \frac{v_i v_k}{c^2} \right) \\ \frac{\partial v_k(p_j, x_j, t)}{\partial x_i} &= \frac{-q}{m_0 \gamma} \sum_j \frac{\partial A_j}{\partial x_i} \left(\delta_{jk} + \frac{v_j v_k}{c^2} \right) \\ \frac{\partial v_k(p_j, x_j, t)}{\partial t} &= \frac{-q}{m_0 \gamma} \sum_j \frac{\partial A_j}{\partial t} \left(\delta_{jk} - \frac{v_j v_k}{c^2} \right), \end{aligned}\right\} \tag{5.13}$$

eq. (5.10) is reduced to [1]

$$\frac{\partial f}{\partial t} + \sum_k v_k \frac{\partial f}{\partial x_k} + \sum_{k,i} \frac{F_i}{m_0 \gamma} \left(\delta_{ik} - \frac{v_i v_k}{c^2} \right) \frac{\partial f}{\partial v_k} = \left(\frac{\partial f}{\partial t} \right)_{\text{coll}} \tag{5.14a}$$

$$\left.\begin{aligned} F_i &= -q \left(\frac{\partial \phi}{\partial x_i} + \frac{\partial A_i}{\partial t} \right) + q(\boldsymbol{v} \times \boldsymbol{B})_i \\ \boldsymbol{F} &= q(\boldsymbol{E} + \boldsymbol{v} \times \boldsymbol{B}). \end{aligned}\right\} \tag{5.15}$$

In the nonrelativistic case ($\gamma \to 1$, $v^2/c^2 \to 0$) we find

$$\frac{\partial f}{\partial t} + \boldsymbol{v} \cdot \nabla_{\boldsymbol{r}} f + \frac{\boldsymbol{F}}{m} \nabla_{\boldsymbol{v}} f = \left(\frac{\delta f}{\partial t} \right)_{\text{coll}}, \tag{5.14b}$$

where $\nabla_{\boldsymbol{r}} \equiv (\partial/\partial x, \partial/\partial y, \partial/\partial z)$, $\nabla_{\boldsymbol{v}} \equiv (\partial/\partial v_x, \partial/\partial v_y, \partial/\partial v_z)$. Equation (5.14a) or (5.14b) is Boltzmann's equation.

When the plasma is rarefied, the collision term $(\delta f/\delta t)_{\text{coll}}$ may be neglected. However, the interactions of the charged particles are still included through the internal electric and magnetic field which are calculated from the charge and current densities by means of Maxwell's equations. The charge and current densities are expressed by the distribution functions for the electron and the ion. Then

$$\left.\begin{aligned}
&\frac{\partial f}{\partial t} + \boldsymbol{v}\cdot\nabla_r f + \frac{q}{m}(\boldsymbol{E} + \boldsymbol{v}\times\boldsymbol{B})\cdot\nabla_v f = 0\\
&\nabla\cdot\boldsymbol{E} = \frac{1}{\varepsilon_0}\sum q\int f\,\mathrm{d}\boldsymbol{v}\\
&\frac{1}{\mu_0}\nabla\times\boldsymbol{B} = \frac{\partial\boldsymbol{D}}{\partial t} + \sum q\int \boldsymbol{v}f\,\mathrm{d}\boldsymbol{v}\\
&\nabla\times\boldsymbol{E} = -\frac{\partial\boldsymbol{B}}{\partial t}\\
&\nabla\cdot\boldsymbol{B} = 0.
\end{aligned}\right\} \tag{5.16}$$

The first of these equations is the *collisionless Boltzmann equation* or *Vlasov's equation*. Boltzmann's or Vlasov's equation or the Fokker-Planck equation to be described in sec. 5.3 is the fundamental equation of plasma kinematics.

Next let us consider Boltzmann's equation in the drift approximation. The drift approximation to a particle orbit was described in ch. 3; for this case the particle can be specified in the phase space by the coordinates of the guiding center (x, y, z) and the magnitude of parallel component $p_\parallel = mv_\parallel$ and perpendicular component $p_\perp = mv_\perp$ of the momentum. In this case

$$p_\perp \mathrm{d}x\,\mathrm{d}y\,\mathrm{d}z\,\mathrm{d}p_\parallel\,\mathrm{d}p_\perp$$

can be proved to be invariant.[2] When the distribution function is denoted by $f(x, y, z, p_\parallel, p_\perp, t)$, Boltzmann's equation in the drift approximation is reduced to

$$\frac{\partial f}{\partial t} + \dot{\boldsymbol{r}}\cdot\nabla_r f + \dot{p}_\parallel\frac{\partial f}{\partial p_\parallel} + \dot{p}_\perp\frac{\partial f}{\partial p_\perp} = \left(\frac{\delta f}{\delta t}\right)_{\text{coll}}.$$

If the magnetic moment μ_{m} is used instead of $p_\perp$ $(p_\perp{}^2 = 2mB(\boldsymbol{r})\mu_{\text{m}})$, the volume element is

$$p_\perp \mathrm{d}x\mathrm{d}y\mathrm{d}z\mathrm{d}p_\parallel \mathrm{d}p_\perp = mB\mathrm{d}x\mathrm{d}y\mathrm{d}z\mathrm{d}p_\parallel \mathrm{d}\mu_\mathrm{m}.$$

Because $\dot{\mu}_\mathrm{m} = 0$, Boltzmann's equation, in terms of $f(x, y, z, p_\parallel, \mu_\mathrm{m}, t)$ in the drift approximation, becomes

$$\frac{\partial f}{\partial t} + \dot{\boldsymbol{r}}\nabla_{\boldsymbol{r}} f + \dot{p}_\parallel \frac{\partial f}{\partial p_\parallel} = \left(\frac{\delta f}{\delta t}\right)_{\mathrm{coll}}.$$

5.3 Collision Terms

*5.3a Boltzmann's Collision Integral[3]

Let us consider the collision of particles P and P_1 as is shown in fig. 5.2. The velocities of the two particles before the collision are $\boldsymbol{v}$, $\boldsymbol{v}_1$ and the velocities after the collision are $\boldsymbol{v}'$, $\boldsymbol{v}_1'$. The prime $'$ indicates quantities taken after the collision. It is assumed that the interaction is symmetric, so that the reverse process is possible (see fig. 5.3). It is also assumed that the collision is local in space and short in time.

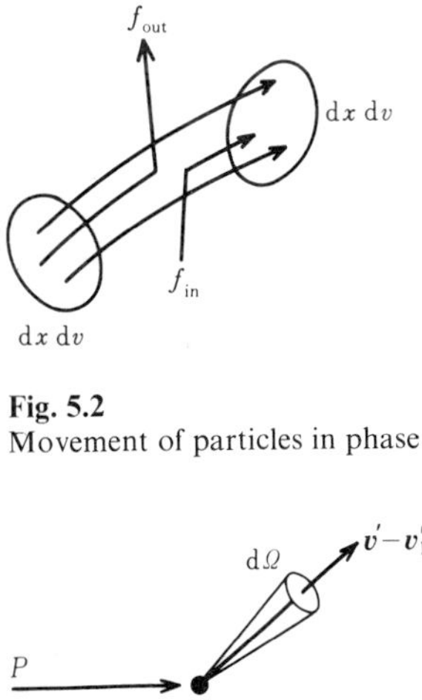

Fig. 5.2
Movement of particles in phase space and the effect of a collision.

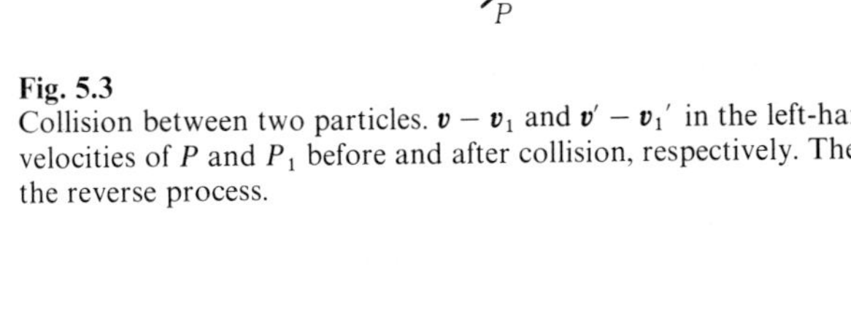

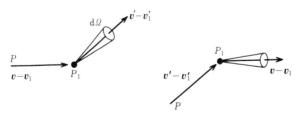

Fig. 5.3
Collision between two particles. $\boldsymbol{v} - \boldsymbol{v}_1$ and $\boldsymbol{v}' - \boldsymbol{v}_1'$ in the left-hand side are relative velocities of P and P_1 before and after collision, respectively. The figure on the right shows the reverse process.

When the particle P within $(\boldsymbol{v}, \boldsymbol{v} + \delta\boldsymbol{v})$ collides with the P_1 particles distributed with the density $f_1 \, d\boldsymbol{v}_1$, the scattering probability into an infinitesimal solid angle $d\Omega$ per unit time is

$$\sigma_c d\Omega |\boldsymbol{v} - \boldsymbol{v}_1| f_1 d\boldsymbol{v}_1,$$

where $\sigma_c d\Omega$ gives the scattering into $d\Omega$. Consequently the density of particles which escape per unit time from the region $(\boldsymbol{v}, \boldsymbol{v} + \delta\boldsymbol{v})$ by virtue of collision is

$$\left(\frac{\delta f}{\delta t}\right)_{\text{out}} d\boldsymbol{v} = \int d\boldsymbol{v} f \sigma_c d\Omega |\boldsymbol{v} - \boldsymbol{v}_1| f_1 d\boldsymbol{v}_1. \tag{5.17}$$

Next we must take account of the entering of the particle P into the region $(\boldsymbol{v}, \boldsymbol{v} + \delta\boldsymbol{v})$ by virtue of the collision. When P within the region $(\boldsymbol{v}', \boldsymbol{v} + \delta\boldsymbol{v}')$ collides with P_1 particles within the region $(\boldsymbol{v}_1', \boldsymbol{v}_1' + \delta\boldsymbol{v}')$, the density of particles which enter into the region $(\boldsymbol{v}, \boldsymbol{v} + \delta\boldsymbol{v})$ is

$$\left(\frac{\delta f}{\delta t}\right)_{\text{in}} d\boldsymbol{v} = \int d\boldsymbol{v}' f' \cdot \sigma_c' d\Omega' |\boldsymbol{v}' - \boldsymbol{v}_1'| f_1' d\boldsymbol{v}_1'. \tag{5.18}$$

Since the collision is symmetric, we may set

$$\sigma_c d\Omega = \sigma_c' d\Omega'.$$

From Liouville's theorem, it follows that

$$d\boldsymbol{v} d\boldsymbol{v}_1 = d\boldsymbol{v}' d\boldsymbol{v}_1'.$$

For elastic collision, conservation of energy and momentum yield $|\boldsymbol{v} - \boldsymbol{v}_1| = |\boldsymbol{v}' - \boldsymbol{v}_1'| \equiv g$ so that,

$$\left(\frac{\delta f}{\delta t}\right)_{\text{coll}} = \left(\frac{\delta f}{\delta t}\right)_{\text{in}} - \left(\frac{\delta f}{\delta t}\right)_{\text{out}} = \int (f' f_1' - f f_1) g \sigma_c d\Omega d\boldsymbol{v}_1 \tag{5.19}$$

Equation (5.19) is called *Boltzmann's collision integral.* In the process of the derivation of Boltzmann's collision integral, it was assumed that the collision is local in space and is short in time and the scattering cross section is small for small deflection angle. Accordingly this integral can be applied to the collision between charged and neutral particles, but it cannot be applied to the case of Coulomb collision.

5.3b Fokker-Planck Collision Term

In the case of Coulomb collision, scattering into small angles has a large cross section, and a test particle interacts with many field particles at the same time, since the Coulomb force is a long-range interaction. Consequently it is appropriate to treat Coulomb collision statistically. Assume that the velocity $\boldsymbol{v}$ of a particle is changed to $\boldsymbol{v} + \Delta\boldsymbol{v}$ after the time Δt by Coulomb collisions; denote the probability of this process by $W(\boldsymbol{v}, \Delta\boldsymbol{v})$. Then the distribution function $f(\boldsymbol{r}, \boldsymbol{v}, t)$ satisfies

$$f(\boldsymbol{r}, \boldsymbol{v}, t + \Delta t) = \int f(\boldsymbol{r}, \boldsymbol{v} - \Delta\boldsymbol{v}, t) W(\boldsymbol{v} - \Delta\boldsymbol{v}, \Delta\boldsymbol{v}) \mathrm{d}(\Delta\boldsymbol{v}). \tag{5.20}$$

In this process the state at $t + \Delta t$ depends only on the state at t. Such a process (i.e., one independent of the history of the process) is called a Markoff process. The change of the distribution function by virtue of Coulomb collision is

$$\left(\frac{\delta f}{\delta t}\right)_{\mathrm{coll}} \Delta t = f(\boldsymbol{r}, \boldsymbol{v}, t + \Delta t) - f(\boldsymbol{r}, \boldsymbol{v}, t). \tag{5.21}$$

Taylor expansion of the integrand of eq. (5.20) gives

$$f(\boldsymbol{r}, \boldsymbol{v} - \Delta\boldsymbol{v}, t) W(\boldsymbol{v} - \Delta\boldsymbol{v}, \Delta\boldsymbol{v})$$

$$= f(\boldsymbol{r}, \boldsymbol{v}, t) W(\boldsymbol{v}, \Delta\boldsymbol{v}) - \sum_r \frac{\partial (fW)}{\partial v_r} \Delta v_r + \sum_{rs} \frac{1}{2} \frac{\partial^2 (fW)}{\partial v_r \partial v_s} \Delta v_r \Delta v_s + \cdots. \tag{5.22}$$

From the definition of $W(\boldsymbol{v}, \Delta\boldsymbol{v})$, the integral of W is

$$\int W \mathrm{d}(\Delta\boldsymbol{v}) = 1.$$

Introducing the quantities

$$\int W \Delta\boldsymbol{v} \mathrm{d}(\Delta\boldsymbol{v}) = \langle \Delta\boldsymbol{v} \rangle_t \Delta t$$

$$\int W \Delta v_r \Delta v_s \mathrm{d}(\Delta\boldsymbol{v}) = \langle \Delta v_r \Delta v_s \rangle_t \Delta t$$

and substituting eq. (5.22) into eq. (5.20), we find

$$\left(\frac{\delta f}{\delta t}\right)_{\text{coll}} = -\nabla_v(\langle \Delta \boldsymbol{v}\rangle_t f) + \sum \frac{1}{2}\frac{\partial^2}{\partial v_r \partial v_s}(\langle \Delta v_r \Delta v_s\rangle_t f). \tag{5.24}$$

This term is called the *Fokker-Planck collision term* and $\langle \Delta \boldsymbol{v}\rangle_t$, $\langle \Delta \boldsymbol{v}_r \Delta \boldsymbol{v}_s\rangle_t$ are called *Fokker-Planck coefficients.* It is easy to show that $\int W \Delta v_r \Delta v_s$ $\mathrm{d}(\Delta \boldsymbol{v})$ is proportional to Δt. Δv_r is considered to be the total sum of $\Delta v_r{}^i$ which is, the change of $v_r{}^i$ due to the ith collision during Δt, i.e., $\Delta \boldsymbol{v}_r = \Sigma_i \Delta v_r{}^i$, so that $\Delta v_r \Delta v_s = \Sigma_i \Sigma_j \Delta v_r{}^i \Delta v_s{}^j$. When the collisions are statistically independent, the statistical average of $\Delta v_r{}^i \Delta v_s{}^j (i \neq j)$ is zero, or

$$\int W \Delta v_r \Delta v_s \mathrm{d}(\Delta \boldsymbol{v}) = \sum_i \int W \Delta v_r{}^i \Delta v_s{}^i \mathrm{d}(\Delta \boldsymbol{v}).$$

This expression is proportional to Δt. When the Fokker-Planck collision term (5.24) is used in eq. (5.15), this equation is called the *Fokker-Planck equation.*

5.4 Diffusion in Velocity Space for a Mirror (Mirror Loss)

The Fokker-Planck equation can be expressed in the form[4]

$$\frac{\partial f}{\partial t} + \boldsymbol{v}\cdot\nabla_r f + \frac{\boldsymbol{F}}{m}\cdot\nabla_v f + \nabla_v\cdot \boldsymbol{J} = 0, \tag{5.25}$$

where

$$J_i = A_i f - \sum_j D_{ij}\frac{\partial f}{\partial v_j} \tag{5.26}$$

$$A_i = \langle \Delta v_i\rangle_t - \frac{1}{2}\sum_j \frac{\partial}{\partial v_j}\langle \Delta v_i \Delta v_j\rangle_t \tag{5.27}$$

$$D_{ij} = \frac{1}{2}\langle \Delta v_i \cdot \Delta v_j\rangle_t. \tag{5.28}$$

The tensor D is called the *diffusion tensor* in velocity space and A is called the coefficient of dynamic friction. For convenience we consider the components of $\boldsymbol{J}$ parallel and perpendicular to the velocity $\boldsymbol{v}$ of the test particle. When the distribution function of field particles is isotropic. we find

$$\left.\begin{aligned} \boldsymbol{J}_{\parallel} &= -D_{\parallel}\nabla_{\parallel}f + \boldsymbol{A}f \\ \boldsymbol{J}_{\perp} &= -D_{\perp}\nabla_{\perp}f. \end{aligned}\right\} \tag{5.29}$$

$\boldsymbol{A}$ is parallel to $\boldsymbol{v}$ and the diffusion tensor becomes diagonal. When the distribution function is Maxwellian, A and D will be given by (see sec. 5.5)

$$\begin{aligned} mvD_{\parallel} &= -T^*A \\ D_{\parallel} &= \frac{(qq^*)^2 n^* \ln \Lambda}{8\pi\varepsilon_0{}^2 vm^2}\frac{\Phi_1(b^*v)}{b^{*2}v^2} \\ D_{\perp} &= \frac{(qq^*)^2 n^* \ln \Lambda}{8\pi\varepsilon_0{}^2 vm^2}\left(\Phi(b^*v) - \frac{\Phi_1(b^*v)}{2b^{*2}v^2}\right). \end{aligned} \tag{5.30}$$

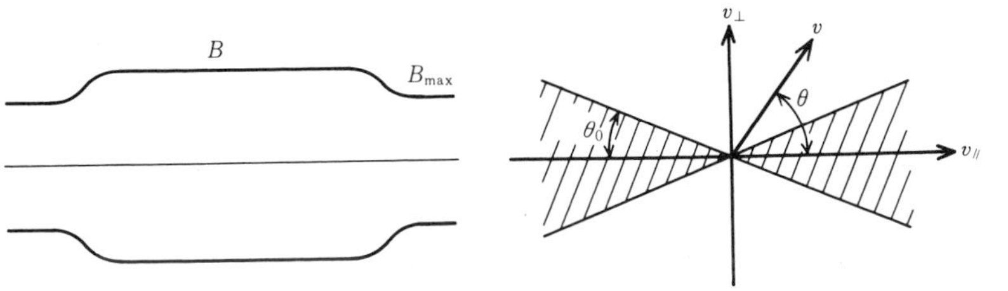

Fig. 5.4
Mirror field and loss cone.

Let us consider a mirror field as is shown in fig. 5.4; the loss-cone angle is

$$\sin\theta_0 = \left(\frac{B}{B_{\max}}\right)^{1/2} = \left(\frac{1}{R}\right)^{1/2}, \tag{5.31}$$

as was discussed in ch. 3. When a charged particle enters the loss cone $\theta < \theta_0$ in the velocity space (v, θ, φ) shown in fig. 5.4, the particle escapes from the mirror field. One can consider the diffusion process into the loss cone due to the Coulomb collision as a loss mechanism for the mirror. We do not consider here other loss mechanisms due to instabilities of the plasma. For the analysis of diffusion loss in velocity space, one must solve the Fokker-Planck equation (5.25). The term q of plasma production or injection is introduced to compensate for plasma loss, so that the problem may be treated as a stationary one. Then eq. (5.25) is reduced to

$$\frac{\partial f}{\partial t} + \boldsymbol{v}\nabla_r f + \frac{\boldsymbol{F}}{m}\nabla_v f + \nabla_v \boldsymbol{J} = q(\boldsymbol{r}, \boldsymbol{v}). \tag{5.32}$$

We utilize spherical coordinates v, θ, φ in velocity space. It is assumed that the plasma is uniform with respect to the coordinates $\boldsymbol{r}$, and that the plasma is composed of electrons and one kind of ion. The scattering of ions by electrons is neglected. Consequently, we have $\partial f/\partial t = 0, \boldsymbol{v}\cdot\nabla_{\boldsymbol{r}} f = 0$. It is assumed that the distribution function depends on v and θ and not on φ. It follows that $q(\boldsymbol{v}\times\boldsymbol{B})\cdot\nabla_v f = 0$. When the electric field is neglected, eq. (5.32) is reduced to

$$v\frac{\partial}{\partial\theta}(\sin\theta\,\boldsymbol{J}_\theta) + \sin\theta\frac{\partial}{\partial v}(v^2\boldsymbol{J}_v) = qv^2\sin\theta. \tag{5.33}$$

The following boundary conditions must be satisfied:

$$f = 0 \quad \text{at} \quad \theta = \theta_0$$

$$\frac{\partial f}{\partial\theta} = 0 \quad \text{at} \quad \theta = \frac{\pi}{2}.$$

Expressions for D_{ij}, A_i are necessary to solve eq. (5.33). However it is necessary to know the distribution function in order to obtain these expressions. So, as an approximation, we adopt the expression (5.30) for D_{ij}, A_i, assuming a homogeneous Maxwellian distribution. Then eq. (5.33) is reduced to

$$D_\perp\frac{\partial}{\partial\theta}\left(\sin\theta\frac{\partial f}{\partial\theta}\right) + \sin\theta\frac{\partial}{\partial v}\left(v^2\left(D_\parallel\frac{\partial f}{\partial v} - Af\right)\right) = -qv^2\sin\theta. \tag{5.34}$$

Choosing the distribution function to be

$$f = \Theta(\theta)\exp\left(-\frac{mv^2}{2T}\right), \tag{5.35}$$

we find that

$$D_\parallel\frac{\partial f}{\partial v} - Af = -\left(\frac{mv}{T}D_\parallel + A\right)f = 0,$$

from eq. (5.30). Finally, the Fokker-Planck equation is simplified to

$$\frac{\partial}{\partial\theta}\left(\sin\theta\frac{\partial f}{\partial\theta}\right) = -\frac{qv^2}{D_\perp}\sin\theta. \tag{5.36}$$

When the source term $q(v, \theta)$ is independent of θ, the solution satisfying the boundary condition is

$$f = C \ln\left(\frac{\sin\theta}{\sin\theta_0}\right) \exp\left(-\frac{mv^2}{2T}\right), \tag{5.37a}$$

where

$$\frac{q(v)v^2}{D_\perp(v)} = C \exp\left(-\frac{mv^2}{2T}\right), \tag{5.38a}$$

and C is a constant. The ion density n is

$$\int f \, \mathrm{d}\boldsymbol{v} = n,$$

and the constant C is given by

$$C\left(2\pi\frac{T}{m}\right)^{3/2}\left(\ln\left(\cot\frac{\theta_0}{2}\right) - \cos\theta_0\right) = n. \tag{5.39a}$$

In the stationary state, the ion loss I_{L} is the same as the plasma production rate, so that

$$I_{\mathrm{L}} = \int q \mathrm{d}\boldsymbol{v}. \tag{5.40}$$

The integral region is $\theta_0 < \theta < \pi - \theta_0$.

When we define the confinement time τ of a mirror field by

$$I_{\mathrm{L}} = \frac{n}{\tau}, \tag{5.41}$$

the confinement time τ is then simply n/I_{L}. With the constant C given by eq. (5.39a), the source term q is determined from eq. (5.38a) ($D_\perp$ is given by eq. (5.30)). I_{L} is given by eq. (5.40) and τ is obtained by eq. (5.41), i.e.,

$$\begin{aligned} I_{\mathrm{L}} &= 4\pi C \cos\theta_0 \int_0^\infty D_\perp(v) \exp\left(-\frac{mv^2}{2T}\right) \mathrm{d}v \\ &= \frac{\ln\Lambda\, e^4 n^2}{(2\pi)^{1/2}\, 8\pi\varepsilon_0{}^2 m^{1/2} T^{3/2}} \frac{(3\ln(2^{1/2}+1) - 2^{1/2})}{\left(\ln\left(\cot\frac{\theta_0}{2}\right) - \cos\theta_0\right)} \cos\theta_0 \end{aligned} \tag{5.42a}$$

$$\tau = 1.57 \frac{\ln\left(\cot\frac{\theta_0}{2}\right) - \cos\theta_0}{\cos\theta_0} \tau_i. \tag{5.43a}$$

The derivation of eq. (5.42a) is described in sec. 5.5. When the mirror ratio R is large,

$$\tau \approx (0.785 \ln R - 0.5)\tau_i, \tag{5.44a}$$

where τ_i is the ion-ion collision time given in sec. 4.4.

When the source term q is expressed by $q = q_0(v)\varphi(\theta)$, the distribution function is

$$f = C \int_{\theta_0}^{\theta} \frac{d\theta'}{\sin\theta'} \int_{\theta'}^{\pi/2} \varphi(\theta'') \sin\theta'' \, d\theta'' \cdot \exp\left(-\frac{mv^2}{2T}\right), \tag{5.37b}$$

where

$$\frac{q_0 v^2}{D_\perp(v)} = C \exp\left(-\frac{mv^2}{2T}\right). \tag{5.38b}$$

The function $\varphi(\theta)$ is normalized to satisfy

$$\int_{\theta_0}^{\pi/2} \varphi(\theta) \sin\theta \, d\theta = \cos\theta_0,$$

so that the case $\varphi(\theta) = 1$ is reduced to the foregoing result.

When the ions are injected in the perpendicular direction, $\varphi(\theta)$ is

$$\varphi(\theta) = \cos\theta_0 \delta\left(\theta - \frac{\pi}{2}\right).$$

In a similar manner, we can derive the confinement time and other parameters as follows:

$$f = C \cos\theta_0 \ln\frac{\tan(\theta/2)}{\tan(\theta_0/2)} \exp\left(-\frac{mv^2}{2T}\right) \tag{5.37c}$$

$$C\left(2\pi\frac{T}{m}\right)^{3/2} \ln\frac{1}{\sin\theta_0} = n \tag{5.39b}$$

$$I_L = 4\pi C \cos\theta_0 \int_0^\infty D_\perp(v) \exp\left(-\frac{mv^2}{2T}\right) dv$$

$$= \frac{\ln \Lambda\, e^4 n^2 (3\ln(2^{1/2}+1) - 2^{1/2})}{(2\pi)^{1/2}\, 8\pi\varepsilon_0{}^2 m^{1/2} T^{3/2} \cdot \ln \dfrac{1}{\sin\theta_0}} \cos\theta_0 \tag{5.42b}$$

$$\tau = 1.57 \times \frac{\ln\left(\dfrac{1}{\sin\theta_0}\right)}{\cos\theta_0}\tau_{\mathrm{i}} \tag{5.43b}$$

$$\approx 0.78(\ln R)\tau_{\mathrm{i}}. \tag{5.44b}$$

The confinement time τ increases only as $\ln R$.[5] It should be noted that the increase depends only upon the ion temperature ($\tau \propto \tau_{\mathrm{i}}$), not on the magnitude of B or the plasma size.

5.5 The Diffusion Tensor and the Coefficient of Dynamic Friction

The expressions (5.30) for the diffusion tensor and the coefficient of dynamic friction were introduced in sec. 5.4. Let us derive eq. (5.30) in this section. The change $\delta p_{\parallel}$ of the momentum due to collision is given by eq. (4.16) as follows:

$$\delta p_{\parallel} = -2m_{\mathrm{r}} u \sin^2(\chi/2). \tag{5.45}$$

Taking the z axis in the direction of $\boldsymbol{u}$, $\langle \Delta p_{\parallel} \rangle = \langle \Delta p_z \rangle$ is given by

$$\Delta t \langle \Delta p_z \rangle_t \equiv \int W \Delta p_{\parallel} \mathrm{d}(\Delta v) = f^*(\boldsymbol{v}^*) u \mathrm{d}\boldsymbol{v}^* \Delta t \int \delta p_z \sigma(\chi, u) \mathrm{d}\Omega, \tag{5.46}$$

and

$$\langle \Delta p_z \rangle_t = -\frac{(qq^*)^2}{4\pi\varepsilon_0{}^2 m_{\mathrm{r}} u^2} \ln \Lambda f^*(\boldsymbol{v}^*) \mathrm{d}\boldsymbol{v}^*. \tag{5.47}$$

From eq. (4.16),

$$\delta p_{\parallel} = -\frac{(\delta \boldsymbol{p})^2}{2m_{\mathrm{r}} u}, \tag{5.48}$$

we find

$$\langle \Delta p_x \cdot \Delta p_x \rangle_t = \langle \Delta p_y \cdot \Delta p_y \rangle_t = \frac{1}{2}\langle \Delta p_{\perp} \cdot \Delta p_{\perp} \rangle_t = -m_{\mathrm{r}} u \langle \Delta p_{\parallel} \rangle_t$$

$$= \frac{(qq^*)^2}{4\pi\varepsilon_0{}^2 u} \ln \Lambda f^*(\boldsymbol{v}^*) \mathrm{d}\boldsymbol{v}^*, \tag{5.49}$$

where we used the approximation $\langle \Delta p_z \cdot \Delta p_z \rangle = 0$.

When one expresses these relations in generalized coordinates, $\langle \Delta p_i \cdot \Delta p_j \rangle_t$ and $\langle \Delta p_i \rangle_t$ are given by

$$\langle \Delta p_i \cdot \Delta p_j \rangle_t = \frac{(qq^*)^2}{4\pi\varepsilon_0{}^2} \ln \Lambda u_{ij} f^*(\boldsymbol{v}^*) \mathrm{d}\boldsymbol{v}^* \tag{5.50}$$

$$\langle \Delta p_i \rangle_t = -\frac{(qq^*)^2}{4\pi\varepsilon_0{}^2} \frac{\ln \Lambda \, u_i}{m_r u^3} f^*(\boldsymbol{v}^*) \mathrm{d}\boldsymbol{v}^*, \tag{5.51}$$

where

$$u_{ij} = \frac{u^2 \delta_{ij} - u_i u_j}{u^3} = \frac{\partial^2 u}{\partial u_i \partial u_j}. \tag{5.52}$$

Accordingly D_{ij} and A_i are given by

$$\begin{aligned} D_{ij} &= \frac{1}{2} \langle \Delta v_i \Delta v_j \rangle_t = \sum \frac{1}{2m^2} \langle \Delta p_i \Delta p_j \rangle_t \\ &= \sum{}^* \frac{(qq^*)^2 \ln \Lambda}{8\pi\varepsilon_0{}^2 m^2} \int u_{ij} f^*(\boldsymbol{v}^*) \mathrm{d}\boldsymbol{v}^* \end{aligned} \tag{5.53}$$

$$\begin{aligned} A_i &= \langle \Delta v_i \rangle_t - \frac{1}{2} \sum_j \frac{\partial}{\partial v_j} \langle \Delta v_i \Delta v_j \rangle_t \\ &= \frac{1}{m} \sum \langle \Delta p_i \rangle_t - \frac{1}{2m^2} \sum_j \frac{\partial}{\partial v_j} \langle \Delta p_i \Delta p_j \rangle_t \\ &= -\sum{}^* \frac{(qq^*)^2 \ln \Lambda}{8\pi\varepsilon_0{}^2 m} \int \left(\frac{2u_i}{m_r u^3} + \sum_j \frac{1}{m} \frac{\partial u_{ij}}{\partial v_j} \right) f^*(\boldsymbol{v}^*) \mathrm{d}\boldsymbol{v}^*. \end{aligned} \tag{5.54}$$

Since

$$\sum_j \frac{\partial u_{ij}}{\partial v_j} = \sum_j \frac{\partial u_{ij}}{\partial u_j} = \sum_j \frac{\partial^3 u}{\partial u_j \partial u_i \partial u_j} = -2\frac{u_i}{u^3}$$

$$\frac{2u_i}{m_r u^3} - \frac{2}{m} \frac{u_i}{u^3} = \frac{2u_i}{u^3 m^*},$$

A_i is

$$A_i = -\sum{}^* \frac{(qq^*)^2 \ln \Lambda}{4\pi\varepsilon_0{}^2 m} \int \frac{u_i}{m^* u^3} f^*(\boldsymbol{v}^*)\mathrm{d}\boldsymbol{v}^*. \tag{5.55}$$

Define $\boldsymbol{u}^*$ by $\boldsymbol{u}^* = -\boldsymbol{u}$. Since

$$\sum_j \frac{\partial u_{ij}}{\partial v_j{}^*} = -2\frac{u_i{}^*}{u^3} = 2\frac{u_i}{u^3},$$

A_i is reduced to

$$A_i = -\sum{}^* \frac{(qq^*)^2 \ln \Lambda}{8\pi\varepsilon_0{}^2 m} \sum_j \int \frac{1}{m^*} \frac{\partial u_{ij}}{\partial v_j{}^*} f^*(\boldsymbol{v}^*)\mathrm{d}\boldsymbol{v}^*$$

$$= \sum{}^* \frac{(qq^*)^2 \ln \Lambda}{8\pi\varepsilon_0{}^2 m} \sum_j \int u_{ij} \frac{1}{m^*} \frac{\partial f^*(\boldsymbol{v}^*)}{\partial v_j{}^*} \mathrm{d}\boldsymbol{v}^*. \tag{5.56}$$

Consequently,

$$J_i = A_i f - \sum_j D_{ij} \frac{\partial f}{\partial v_j}$$

$$= \sum{}^* \frac{(qq^*)^2 \ln \Lambda}{8\pi\varepsilon_0{}^2 m} \int \sum_j \left(\frac{f}{m^*} \frac{\partial f^*}{\partial v_j{}^*} - \frac{f^*}{m} \frac{\partial f}{\partial v_j} \right) u_{ij} \mathrm{d}\boldsymbol{v}^*. \tag{5.57}$$

This is called the *Landau integral.*[6] From eqs. (5.55) and (5.53), we find

$$\boldsymbol{A} = -\sum{}^* \frac{(qq^*)^2 \ln \Lambda}{4\pi\varepsilon_0{}^2 m^* m} \int \frac{\boldsymbol{u}}{u^3} f^* \mathrm{d}\boldsymbol{v}^* \tag{5.58}$$

$$D_{ij} = \sum{}^* \frac{(qq^*)^2 \ln \Lambda}{8\pi\varepsilon_0{}^2 m^2} \frac{\partial^2}{\partial u_i \partial u_j} \int u f^* \mathrm{d}\boldsymbol{v}^*, \tag{5.59}$$

i.e.,

$$\boldsymbol{A} = -\sum{}^* \frac{\ln \Lambda}{mm^*} \boldsymbol{E}_{\mathrm{v}} \tag{5.60}$$

$$D_{ij} = \sum{}^* \frac{\ln \Lambda}{m^2} \frac{\partial^2 \psi(\boldsymbol{v})}{\partial u_j \partial u_j}, \tag{5.61}$$

where

$$\boldsymbol{E}_{\mathrm{v}} = \frac{1}{4\pi\varepsilon_0} \int \frac{\boldsymbol{u}}{u^3} \rho_{\mathrm{v}}(\boldsymbol{v}^*)\mathrm{d}\boldsymbol{v}^* \tag{5.62}$$

$$\psi(\boldsymbol{v}) = \frac{1}{8\pi\varepsilon_0} \int u\rho_v(\boldsymbol{v}^*)\mathrm{d}\boldsymbol{v}^* \tag{5.63}$$

$$\rho_v = \frac{(qq^*)^2}{\varepsilon_0} f^*(\boldsymbol{v}^*). \tag{5.64}$$

When the distribution function $f^*(\boldsymbol{v}^*)$ is Maxwellian, $n^*(b^*/\pi^{1/2})^3 \exp(-b^{*2}v^2)$, eq. (4.32) gives

$$A = -\sum{}^* \frac{(qq^*)^2 n^* \ln \Lambda}{4\pi\varepsilon_0{}^2 mm^* v^3} \Phi_1(b^*v)\boldsymbol{v}. \tag{5.65}$$

The function $\psi(\boldsymbol{v})$ is given by

$$\psi(\boldsymbol{v}) = \frac{(qq^*)^2 n^*}{8\pi\varepsilon_0{}^2} \int_0^\infty v^{*2}\mathrm{d}v^* \left(\frac{b^*}{\pi^{1/2}}\right)^3 \exp(-b^{*2}v^2)\cdot 2\pi \int_0^\pi \boldsymbol{u}\sin\theta\mathrm{d}\theta. \tag{5.66}$$

It follows from $\boldsymbol{u}^2 = v^2 + v^{*2} - 2vv^*\cos\theta$, $u\mathrm{d}u = vv^*\sin\theta\mathrm{d}\theta$ that

$$\int_0^\pi u\sin\theta\,\mathrm{d}\theta = \begin{cases} 2\left(v + \dfrac{1}{3}\dfrac{v^{*2}}{v}\right) & v^* < v \\ 2\left(v^* + \dfrac{1}{3}\dfrac{v^2}{v^*}\right) & v^* > v, \end{cases}$$

and $\psi(\boldsymbol{v})$ is reduced to

$$\psi(\boldsymbol{v}) = \frac{(qq^*)^2 n^* v}{8\pi\varepsilon_0{}^2}\left(\left(1 + \frac{1}{2b^{*2}v^2}\right)\Phi(b^*v) + \frac{1}{\pi^{1/2}b^*v}\exp(-b^{*2}v^2)\right). \tag{5.67}$$

Thus eq. (5.30) is obtained:

$$D_\parallel = \sum{}^* \frac{(qq^*)^2 n^* \ln \Lambda}{8\pi\varepsilon_0{}^2 vm^2} \frac{\Phi_1(b^*v)}{b^{*2}v^2} \tag{5.68}$$

$$D_\perp = \sum{}^* \frac{(qq^*)^2 n^* \ln \Lambda}{8\pi\varepsilon_0{}^2 vm^2}\left(\Phi(b^*v) - \frac{\Phi_1(b^*v)}{2b^{*2}v^2}\right) \tag{5.69}$$

$$mvD_\parallel = -T^*A. \tag{5.70}$$

Next let us derive the integrals (5.42a) and (5.42b). This problem is the same as that of proving

$$\int_0^\infty D_\perp \exp\left(-\frac{mv^2}{2T}\right) dv = \frac{\ln \Lambda (qq^*)^2 n^*}{8\pi\varepsilon_0{}^2 m^2}\left(\frac{3\ln(2^{1/2}+1) - 2^{1/2}}{2}\right),$$

i.e.,

$$\int_0^\infty \frac{1}{x}\left(\Phi(x) - \frac{\Phi_1(x)}{2x^2}\right)\exp(-x^2)dx = \left(\frac{3\ln(2^{1/2}+1) - 2^{1/2}}{2}\right). \tag{5.71}$$

Define the integral

$$I(\alpha) \equiv \int_0^\infty \frac{\Phi(x)}{x}\exp(-\alpha x^2)dx,$$

and differentiate:

$$\begin{aligned}\frac{dI(\alpha)}{dx} &= -\int_0^\infty x\Phi(x)\exp(-\alpha x^2)dx \\ &= -\frac{2}{\pi^{1/2}}\int_0^\infty x\exp(-\alpha x^2)dx \int_0^x \exp(-y^2)dy \\ &= -\frac{2}{\pi^{1/2}}\int_0^\infty \exp(-y^2)dy \int_y^\infty x\exp(-\alpha x^2)dx \\ &= -\frac{1}{2\alpha(1+\alpha)^{1/2}}.\end{aligned}$$

Since $I(\infty) = 0$, $I(\alpha)$ is

$$I(\alpha) = -\int_\infty^\alpha (2x)^{-1}(1+x)^{-1/2}dx = \ln\left(\frac{(1+\alpha)^{1/2}+1}{\alpha^{1/2}}\right),$$

and we find

$$\int_0^\infty \frac{\Phi(x)}{x}\exp(-x^2)dx = I(x) = \ln(2^{1/2}+1).$$

Since $\Phi_1(x) = \Phi - x\,\Phi'(x)$ and $\Phi'(x) = (2/\pi^{1/2})\exp(-x^2)$, we find

$$\begin{aligned}&\int_0^\infty \frac{\Phi_1(x)}{2x^3}\exp(-x^2)dx \\ &= -\frac{\pi^{1/2}}{8}\int_0^\infty (\Phi(x) - x\,\Phi'(x))\,\Phi'(x)\,d\frac{1}{x^2}\end{aligned}$$

$$= \frac{\pi^{1/2}}{8}\int_0^\infty \frac{\Phi\Phi'' - 2x\Phi'\Phi''}{x^2}\mathrm{d}x$$

$$= -\frac{1}{2}\int_0^\infty \frac{\Phi}{x}\exp(-x^2)\mathrm{d}c + \frac{2}{\pi^{1/2}}\int_0^\infty \exp(-2x^2)\mathrm{d}x$$

$$= -\frac{1}{2}\ln(2^{1/2}+1) + \frac{1}{2^{1/2}}.$$

Thus eq. (5.71) has been derived.

References

1. J. G. Linhart: Plasma Physics, North-Holland, Amsterdam, 1961

2. D. V. Sivukhin: Rev. of Plasma Phys. **1** (ed. by M. A. Leontovich), Consultants Bureau, New York, 1966

3. D. C. Montgomery and D. A. Tidman: Plasma Kinetic Theory, McGraw-Hill, 1964

4. D. V. Sivukhin: Rev. of Plasma Phys. **4** (ed. by M. A. Leontovich), Consultants Bureau, New York, 1966

5. D. Judd, W. M. MacDonald, and M. N. Rosebluth: US Report WASH-289, Feb. 1955

6. L. D. Landau: Zh. Exp. Teor. Fyz. 7, 203(1937)

II THE MAGNETOHYDRODYNAMIC DESCRIPTION OF A PLASMA

6 Magnetohydrodynamic Equations

In this chapter the fundamental equations governing the relations among statistical averages of mass density, current density, charge density, flow velocity, pressure, etc., will be obtained by manipulating the Boltzmann, Vlasov, and Fokker-Planck equations, which were derived in the previous chapter. The averages are functions of time t and the position coordinates. Their mutual relations are obtained by considering appropriate averages of Boltzmann's equation in velocity space. In sec. 6.1, the magnetohydrodynamic equations of motion are derived for a plasma composed of ions and electrons. In sec. 6.2, a plasma is viewed as a single-component fluid and the corresponding equation of motion and Ohm's law are introduced. Momentum and energy transport due to collision are described in sec. 6.3. Then in sec. 6.4, a collisionless plasma is considered and the anisotropic pressure tensor and the CGL equation are introduced. Finally, the magnetohydrodynamic equations are discussed for a simplified case in sec. 6.5.

6.1 Derivation of the Macroscopic Equations

Multiply a function $g(\boldsymbol{r}, \boldsymbol{v}, t)$ of the variables $\boldsymbol{r}, \boldsymbol{v}, t$ by Boltzmann's equation

$$\frac{\partial f}{\partial t} + \boldsymbol{v}\cdot\nabla_{\boldsymbol{r}} f + \frac{\boldsymbol{F}}{m}\nabla_{\boldsymbol{v}} f = \left(\frac{\delta f}{\delta t}\right)_{\text{coll}} \tag{6.1}$$

and integrate over $\boldsymbol{v}$. When $g = 1, g = m\boldsymbol{v}$, or $g = mv^2/2$, equations in terms of the density, the momentum, or the energy, respectively, can be obtained. Averages of g weighed by the distribution function are denoted by $\langle g\rangle$, *i.e.*,

$$\langle g(\boldsymbol{r}, t)\rangle = \frac{\displaystyle\int g(\boldsymbol{r}, \boldsymbol{v}, t) f(\boldsymbol{r}, \boldsymbol{v}, t)\mathrm{d}\boldsymbol{v}}{\displaystyle\int f(\boldsymbol{r}, \boldsymbol{v}, t)\mathrm{d}\boldsymbol{v}}. \tag{6.2}$$

Since density is expressed by

$$n(\boldsymbol{r}, t) = \int f(\boldsymbol{r}, \boldsymbol{v}, t)\mathrm{d}\boldsymbol{v}, \tag{6.3}$$

we find that

$$n(\boldsymbol{r}, t)\langle g(\boldsymbol{r}, t)\rangle = \int g(\boldsymbol{r}, \boldsymbol{v}, t) f(\boldsymbol{r}, \boldsymbol{v}, t)\mathrm{d}\boldsymbol{v}. \tag{6.4}$$

By integrating by parts, we obtain

$$\int g\frac{\partial f}{\partial t}\,\mathrm{d}\boldsymbol{v} = \frac{\partial}{\partial t}(n\langle g\rangle) - n\frac{\partial}{\partial t}\langle g\rangle$$

$$\int g v_i \frac{\partial f}{\partial x_i}\,\mathrm{d}\boldsymbol{v} = \frac{\partial(n\langle v_i g\rangle)}{\partial x_i} - n\frac{\partial}{\partial x_i}\langle v_i g\rangle$$

$$\int g\frac{F_i}{m}\frac{\partial f}{\partial v_i}\,\mathrm{d}\boldsymbol{v} = -\frac{n}{m}\left\langle\frac{\partial}{\partial v_i} g F_i\right\rangle.$$

For the force $\boldsymbol{F} = q\boldsymbol{E} + q(\boldsymbol{v}\times\boldsymbol{B})$, the relation

$$\frac{\partial F_i}{\partial v_i} = 0$$

holds, and

$$\int g\frac{F_i}{m}\frac{\partial f}{\partial v_i}\,\mathrm{d}\boldsymbol{v} = -\frac{n}{m}\left\langle F_i\frac{\partial g}{\partial v_i}\right\rangle.$$

Consequently we find, in terms of averages,[1] that

$$\frac{\partial}{\partial t}(n\langle g\rangle) - n\left\langle\frac{\partial g}{\partial t}\right\rangle + \nabla_r n\langle \boldsymbol{v}g\rangle - n\langle\nabla_r \boldsymbol{v}g\rangle - \frac{n}{m}\langle\boldsymbol{F}\nabla_v g\rangle$$
$$= \int g\left(\frac{\delta f}{\delta t}\right)_{\mathrm{coll}}\mathrm{d}\boldsymbol{v}. \tag{6.5}$$

Equation (6.5), with $g = 1$, is

$$\frac{\partial n(\boldsymbol{r},t)}{\partial t} + \nabla_r(n(\boldsymbol{r},t)\langle\boldsymbol{v}(\boldsymbol{r},t)\rangle) = \int\left(\frac{\delta f}{\delta t}\right)_{\mathrm{coll}}\mathrm{d}\boldsymbol{v}. \tag{6.6}$$

This is the equation of continuity. When the effects of ionization and recombination are neglected, the collision term is zero.

Equation (6.5) for $g = m\boldsymbol{v}$ yields the equation of motion:

$$\frac{\partial}{\partial t}(mn(\boldsymbol{r},t)\langle\boldsymbol{v}\rangle) + \sum_j\frac{\partial}{\partial x_j}n(\boldsymbol{r},t)m\langle v_j\boldsymbol{v}\rangle - n(\boldsymbol{r},t)\langle\boldsymbol{F}(\boldsymbol{r},t)\rangle$$
$$= \int m\boldsymbol{v}\left(\frac{\delta f}{\delta t}\right)_{\mathrm{coll}}\mathrm{d}\boldsymbol{v}. \tag{6.7}$$

Let us define the velocity of random motion $\boldsymbol{v}_{\mathrm{r}}$ by

$$\boldsymbol{v} \equiv \langle \boldsymbol{v}(\boldsymbol{r}, t) \rangle + \boldsymbol{v}_\mathrm{r}. \tag{6.8}$$

By definition of $\boldsymbol{v}_\mathrm{r}$, its average is zero:

$$\langle \boldsymbol{v}_\mathrm{r} \rangle = 0. \tag{6.9}$$

Since

$$\langle v_i v_j \rangle = \langle v_i \rangle \langle v_j \rangle + \langle v_{\mathrm{r}i} v_{\mathrm{r}j} \rangle, \tag{6.10}$$

we find

$$\begin{aligned} m \sum_j \frac{\partial}{\partial x_j} (n \langle v_i v_j \rangle) &= m \sum_j \frac{\partial}{\partial x_j} (n \langle v_i \rangle \langle v_j \rangle) + m \sum_j \frac{\partial}{\partial x_j} (n \langle v_{\mathrm{r}i} v_{\mathrm{r}j} \rangle) \\ &= mn \sum_j \langle v_j \rangle \frac{\partial}{\partial x_j} \langle v_i \rangle + m \langle v_i \rangle \sum_j \frac{\partial}{\partial x_j} (n \langle v_j \rangle) \\ &\quad + m \sum_j \frac{\partial}{\partial x_j} (n \langle v_{\mathrm{r}i} v_{\mathrm{r}j} \rangle). \end{aligned} \tag{6.11}$$

Multiplication of the equation of continuity (6.6) by $m\boldsymbol{v}$ gives

$$m \frac{\partial}{\partial t} (n \langle \boldsymbol{v} \rangle) = mn \frac{\partial}{\partial t} \langle \boldsymbol{v} \rangle - m \langle \boldsymbol{v} \rangle \sum_j \frac{\partial}{\partial x_j} (n \langle v_j \rangle) + m \langle \boldsymbol{v} \rangle \int \left(\frac{\delta f}{\delta t} \right)_{\mathrm{coll}} \mathrm{d}\boldsymbol{v}. \tag{6.12}$$

Define the *pressure tensor* by

$$mn \langle v_{\mathrm{r}i} v_{\mathrm{r}j} \rangle \equiv P_{ij}; \tag{6.13}$$

then the equation of motion is

$$\begin{aligned} mn(\boldsymbol{r}, t) \left(\frac{\partial}{\partial t} + \langle \boldsymbol{v} \rangle \nabla \right) \langle \boldsymbol{v} \rangle &= n(\boldsymbol{r}, t) \langle \boldsymbol{F} \rangle - \sum_j \frac{\partial}{\partial x_j} P_{ij} + \int m\boldsymbol{v} \left(\frac{\delta f}{\delta t} \right)_{\mathrm{coll}} \mathrm{d}\boldsymbol{v} \\ &\quad - m \langle \boldsymbol{v} \rangle \int \left(\frac{\delta f}{\delta t} \right)_{\mathrm{coll}} \mathrm{d}\boldsymbol{v}. \end{aligned} \tag{6.14}$$

As $\boldsymbol{v} = \langle \boldsymbol{v} \rangle + \boldsymbol{v}_\mathrm{r}$, the collision term $\boldsymbol{R}$ is

$$\boldsymbol{R} = \int m\boldsymbol{v}_\mathrm{r} \left(\frac{\delta f}{\delta t} \right)_{\mathrm{coll}} \mathrm{d}\boldsymbol{v}. \tag{6.15}$$

When the distribution function is isotropic, the pressure tensor is

$$P_{ij} = nm\langle v_{ri}{}^2\rangle \delta_{ij} = nm\frac{\langle v_r{}^2\rangle}{3}\delta_{ij} = p\delta_{ij}, \tag{6.16}$$

and $p = nT$ for a Maxwellian distribution.

We introduce the tensor

$$\Pi_{ij} = nm\langle v_{ri}v_{rj} - (\langle v_r{}^2\rangle/3)\delta_{ij}\rangle; \tag{6.17}$$

then the pressure tensor may be expressed by

$$P_{ij} = p_i\delta_{ij} + \Pi_{ij}. \tag{6.18}$$

The tensor Π_{ij} is zero for an isotropic distribution function, and a nonzero Π_{ij} indicates anisotropic properties of the distribution function. We have the relations $\Pi_{ij} = \Pi_{ji}$ and $\Sigma_i \Pi_{ii} = 0$. Consequently, the equation of motion is

$$mn(\boldsymbol{r}, t)\frac{\mathrm{d}}{\mathrm{d}t}\langle \boldsymbol{v}\rangle = n(\boldsymbol{r}, t)\langle \boldsymbol{F}\rangle - \nabla p - \sum_j \frac{\partial \Pi_{ij}}{\partial x_j} + \boldsymbol{R}. \tag{6.19}$$

Here $\boldsymbol{R}$ measures the change of momentum due to collision with particles of the other kind.

When $g = mv^2/2$, then $\boldsymbol{v} \times \boldsymbol{B}\cdot\nabla_v v^2 = 0$. The energy-transport equation is reduced to

$$\frac{\partial}{\partial t}\left(\frac{nm}{2}\langle v^2\rangle\right) + \nabla_{\boldsymbol{r}}\left(\frac{n}{2}m\langle \boldsymbol{v}v^2\rangle\right) = qn\boldsymbol{E}\cdot\langle \boldsymbol{v}\rangle + \int \frac{mv^2}{2}\left(\frac{\delta f}{\delta t}\right)_{\mathrm{coll}} \mathrm{d}\boldsymbol{v}. \tag{6.20}$$

Using the average velocity $\langle \boldsymbol{v}\rangle$, we obtain

$$\frac{nm}{2}\langle v^2\rangle = \frac{nm}{2}(\langle \boldsymbol{v}\rangle)^2 + \frac{3}{2}p$$

$$\begin{aligned}\langle \boldsymbol{v}v^2\rangle = {} & (\langle \boldsymbol{v}\rangle)^2\langle \boldsymbol{v}\rangle + 2\langle(\langle \boldsymbol{v}\rangle\cdot\boldsymbol{v}_r)\rangle\langle \boldsymbol{v}\rangle + \langle v_r{}^2\rangle\langle \boldsymbol{v}\rangle \\ & + \langle \boldsymbol{v}\rangle^2\langle \boldsymbol{v}_r\rangle + 2\langle(\langle \boldsymbol{v}\rangle\cdot\boldsymbol{v}_r)\boldsymbol{v}_r\rangle + \langle v_r{}^2\boldsymbol{v}_r\rangle.\end{aligned}$$

The 2nd and 4th terms are zero, and the 5th term is

$$\begin{aligned}\sum_i \langle v_i\rangle\langle v_{ri}v_{rj}\rangle &= \sum_i \langle v_i\rangle\frac{P_{ij}}{nm} = \frac{1}{nm}\sum_i \langle v_i\rangle(p\delta_{ij} + \Pi_{ij}) \\ &= \frac{1}{nm}p\langle \boldsymbol{v}\rangle + \frac{1}{nm}\sum_i \langle v_i\rangle\Pi_{ij};\end{aligned}$$

and

$$\langle \boldsymbol{v}v^2\rangle = \left(\langle \boldsymbol{v}\rangle^2 + 5\frac{p}{nm}\right)\langle \boldsymbol{v}\rangle + 2\sum_i \langle v_i\rangle \frac{\Pi_{ij}}{nm} + \langle v_r{}^2 \boldsymbol{v}_r\rangle.$$

Consequently eq. (6.20) is reduced to

$$\frac{\partial}{\partial t}\left(\frac{nm}{2}\langle \boldsymbol{v}\rangle^2 + \frac{3}{2}p\right) + \nabla\left(\left(\frac{nm}{2}\langle \boldsymbol{v}\rangle^2 + \frac{5}{2}p\right)\langle \boldsymbol{v}\rangle + \sum_i \Pi_{ij}\langle v_i\rangle + \boldsymbol{q}\right)$$

$$= qn\boldsymbol{E}\cdot\langle \boldsymbol{v}\rangle + \boldsymbol{R}\cdot\langle \boldsymbol{v}\rangle + Q \tag{6.21}$$

$$\boldsymbol{q}(\boldsymbol{r}, t) = \int \frac{m}{2} v_r{}^2 \boldsymbol{v}_r f(\boldsymbol{r}, \boldsymbol{v}, t)\,\mathrm{d}\boldsymbol{v} \tag{6.22}$$

$$Q(\boldsymbol{r}, t) = \int \frac{mv_r{}^2}{2}\left(\frac{\delta f}{\delta t}\right)\mathrm{d}\boldsymbol{v}. \tag{6.23}$$

Q is the heat generated by collision and $\boldsymbol{q}$ is the energy-flux density due to random motion. The left-hand side of eq. (6.21) is the total time derivative of the kinetic energy $nm\langle \boldsymbol{v}\rangle^2/2$ and the internal energy $(3/2)p$, the work done by the total pressure $\Sigma_j(\partial/\partial x_j)\,\Sigma_i p_{ij}\langle v_i\rangle$, and the term $\boldsymbol{q}$. By the use of the equation of motion (6.19) and the equation of continuity (6.6), the kinetic-energy term in eq. (6.21) can be eliminated and we find

$$\frac{3}{2}\frac{\partial p}{\partial t} + \nabla\left(\frac{3}{2}p\langle \boldsymbol{v}\rangle\right) + nT\nabla\langle \boldsymbol{v}\rangle + \sum_{ij}\Pi_{ij}\frac{\partial}{\partial x_j}\langle v_i\rangle + \nabla\boldsymbol{q} = Q. \tag{6.24}$$

The relation $p = nT$ and the equation of continuity yield

$$\frac{3}{2}\frac{\partial nT}{\partial t} + \nabla\left(\frac{3}{2}nT\langle \boldsymbol{v}\rangle\right) = \frac{3}{2}n\frac{\mathrm{d}T}{\mathrm{d}t}.$$

By setting $s \equiv \ln(T^{3/2}/n) = \ln(p^{3/2}/n^{5/2})$, we may write eq. (6.24) as

$$Tn\frac{\mathrm{d}s}{\mathrm{d}t} = T\left(\frac{\partial ns}{\partial t} + \nabla(ns\langle \boldsymbol{v}\rangle)\right) = -\nabla\boldsymbol{q} - \sum_{ij}\Pi_{ij}\frac{\partial\langle v_i\rangle}{\partial x_j} + Q. \tag{6.25}$$

This is the equation of entropy.

Let us summarize the equation of continuity, the equation of motion, and the thermal-energy transport equation for electrons and ions. With the notation $\langle \boldsymbol{v}_e\rangle = \boldsymbol{V}_e$ and $\langle \boldsymbol{v}_i\rangle = \boldsymbol{V}_i$, we have

$$\frac{\partial n_{\mathrm{e}}}{\partial t} + \nabla (n_{\mathrm{e}} \boldsymbol{V}_{\mathrm{e}}) = 0 \tag{6.26}$$

$$\frac{\partial n_{\mathrm{i}}}{\partial t} + \nabla (n_{\mathrm{i}} \boldsymbol{V}_{\mathrm{i}}) = 0 \tag{6.27}$$

$$m_{\mathrm{e}} n_{\mathrm{e}} \frac{\mathrm{d}_{\mathrm{e}} \boldsymbol{V}_{\mathrm{e}}}{\mathrm{d}t} = -\nabla p_{\mathrm{e}} - \sum_{\beta} \frac{\partial}{\partial x_{\beta}} \Pi_{\alpha\beta}{}^{\mathrm{e}} - e n_{\mathrm{e}} (\boldsymbol{E} + \boldsymbol{V}_{\mathrm{e}} \times \boldsymbol{B}) + \boldsymbol{R} \tag{6.28}$$

$$m_{\mathrm{i}} n_{\mathrm{i}} \frac{\mathrm{d}_{\mathrm{i}} \boldsymbol{V}_{\mathrm{i}}}{\mathrm{d}t} = -\nabla p_{\mathrm{i}} - \sum_{\beta} \frac{\partial}{\partial x_{\beta}} \Pi_{\alpha\beta}{}^{\mathrm{i}} + e Z n_{\mathrm{i}} (\boldsymbol{E} + \boldsymbol{V}_{\mathrm{i}} \times \boldsymbol{B}) - \boldsymbol{R} \tag{6.29}$$

$$\frac{3}{2} n_{\mathrm{e}} \frac{\mathrm{d}_{\mathrm{e}} T_{\mathrm{e}}}{\mathrm{d}t} + p_{\mathrm{e}} \nabla \boldsymbol{V}_{\mathrm{e}} = -\nabla \boldsymbol{q}_{\mathrm{e}} - \sum_{\alpha\beta} \Pi_{\alpha\beta}{}^{\mathrm{e}} \frac{\partial V_{\mathrm{e}\alpha}}{\partial x_{\beta}} + Q_{\mathrm{e}} \tag{6.30}$$

$$\frac{3}{2} n_{\mathrm{i}} \frac{\mathrm{d}_{\mathrm{i}} T_{\mathrm{i}}}{\mathrm{d}t} + p_{\mathrm{i}} \nabla \boldsymbol{V}_{\mathrm{i}} = -\nabla \boldsymbol{q}_{\mathrm{i}} - \sum_{\alpha\beta} \Pi_{\alpha\beta}{}^{\mathrm{i}} \frac{V_{\mathrm{i}\alpha}}{\partial x_{\beta}} + Q_{\mathrm{i}}, \tag{6.31}$$

where

$$p_{\mathrm{e}} = n_{\mathrm{e}} T_{\mathrm{e}} \qquad p_{\mathrm{i}} = n_{\mathrm{i}} T_{\mathrm{i}} \tag{6.32}$$

$$\frac{\mathrm{d}_{\mathrm{e}}}{\mathrm{d}t} = \frac{\partial}{\partial t} + (\boldsymbol{V}_{\mathrm{e}} \cdot \nabla) \qquad \frac{\mathrm{d}_{\mathrm{i}}}{\mathrm{d}t} = \frac{\partial}{\partial t} + (\boldsymbol{V}_{\mathrm{i}} \cdot \nabla). \tag{6.33}$$

6.2 Magnetohydrodynamic Equations for a Plasma

Macroscopic equations for an electron fluid and an ion fluid were introduced in the previous section. In the present section we will treat the plasma itself as a single fluid. Define the total mass density ρ_{m}, the average flow velocity $\boldsymbol{V}$, the charge density ρ, and the current density $\boldsymbol{j}$ as follows:

$$\rho_{\mathrm{m}} = n_{\mathrm{e}} m_{\mathrm{e}} + n_{\mathrm{i}} m_{\mathrm{i}} \tag{6.34a}$$

$$\boldsymbol{V} = \frac{n_{\mathrm{e}} m_{\mathrm{e}} \boldsymbol{V}_{\mathrm{e}} + n_{\mathrm{i}} m_{\mathrm{i}} \boldsymbol{V}_{\mathrm{i}}}{\rho_{\mathrm{m}}} \tag{6.35a}$$

$$\rho = -e n_{\mathrm{e}} + Z n_{\mathrm{i}} e \tag{6.36a}$$

$$\boldsymbol{j} = -e n_{\mathrm{e}} \left(\boldsymbol{V}_{\mathrm{e}} - \frac{Z n_{\mathrm{i}}}{n_{\mathrm{e}}} \boldsymbol{V}_{\mathrm{i}} \right). \tag{6.37a}$$

From eqs. (6.26) and (6.27), it follows that

$$\frac{\partial \rho_m}{\partial t} + \nabla(\rho_m \boldsymbol{V}) = 0 \tag{6.38}$$

$$\frac{\partial \rho}{\partial t} + \nabla \boldsymbol{j} = 0. \tag{6.39}$$

The addition of eqs. (6.28) and (6.29) yields

$$\rho_m \frac{\partial \boldsymbol{V}}{\partial t} + (m_e n_e (\boldsymbol{V}_e \cdot \nabla)\boldsymbol{V}_e + m_i n_i (\boldsymbol{V}_i \cdot \nabla)\boldsymbol{V}_i)$$

$$= -\sum_\beta \frac{\partial}{\partial x_\beta} P_{\alpha\beta} + \rho \boldsymbol{E} + (\boldsymbol{j} \times \boldsymbol{B}) - \rho_m \nabla \phi_g, \tag{6.40}$$

where $P_{\alpha\beta} = P_{\alpha\beta}{}^{i} + P_{\alpha\beta}{}^{e}$. Note that the gravitational field $\nabla\phi_g$ has been included. Quasi neutrality of the plasma allows us to write $n_e \approx Zn_i$. Since even the smallest difference in these densities will result in an observable electric field, the net electric charge density is nonzero ($\rho = -e\Delta n_e \equiv -e(n_e - Zn_i) \neq 0$), although $\Delta n_e \ll n_e$. Due to the large mass ratio $m_i/m_e \gg 1$, eqs. (6.34a)–(6.37a) may be expressed as

$$\rho_m = n_i m_i \left(1 + \frac{m_e}{m_i} Z\right) \tag{6.34b}$$

$$\boldsymbol{V} = \boldsymbol{V}_i + \frac{m_e Z}{m_i}(\boldsymbol{V}_e - \boldsymbol{V}_i) \tag{6.35b}$$

$$\rho = -e\Delta n_e \tag{6.36b}$$

$$\boldsymbol{j} = -en_e(\boldsymbol{V}_e - \boldsymbol{V}_i). \tag{6.37b}$$

The second term on the left-hand side of eq. (6.40) may be set equal to $(\boldsymbol{v} \cdot \nabla)\boldsymbol{v}$ to an accuracy of m_e/m_i.

Combining eq. (6.28) with (6.37b) ($\boldsymbol{v}_e = \boldsymbol{v}_i - \boldsymbol{j}/en_e \approx \boldsymbol{v} - \boldsymbol{j}/en_e$), we find

$$\boldsymbol{E} + \left(\boldsymbol{V} - \frac{\boldsymbol{j}}{en_e}\right) \times \boldsymbol{B} + \frac{1}{en_e}\sum_\beta \frac{\partial}{\partial x_\beta} P_{\alpha\beta}{}^{e} - \frac{\boldsymbol{R}}{en_e}$$

$$= \frac{m_e}{e^2 n_e}\frac{\partial}{\partial t}\boldsymbol{j} - \frac{m_e}{e}\frac{\partial \boldsymbol{V}_i}{\partial t},$$

where $(\boldsymbol{v}_e \cdot \nabla)\boldsymbol{v}_e$ is neglected. When $|\partial \boldsymbol{v}_i/\partial t| \ll |\partial \boldsymbol{v}_e/\partial t|$, the 2nd term of the right-hand side of the foregoing equation may be neglected. The collision term $\boldsymbol{R}$ can be expressed by

$$\boldsymbol{R} = \int m_e \boldsymbol{v}_r \left(\frac{\delta f}{\delta t}\right)_{\text{coll}} \mathrm{d}\boldsymbol{v} = \int m\boldsymbol{v}\left(\frac{\delta f}{\delta t}\right)_{\text{coll}} \mathrm{d}\boldsymbol{v} = -m_e n_e \nu_{ei}(\boldsymbol{V}_e - \boldsymbol{V}_i) = n_e e\eta\boldsymbol{j},$$

where η is the specific resistivity (see sec. 4.6).

The equation of motion and Ohm's law are

$$\rho_m \frac{\mathrm{d}\boldsymbol{V}}{\mathrm{d}t} = -\sum_\beta \frac{\partial}{\partial x_\beta} P_{\alpha\beta} + \rho\boldsymbol{E} + (\boldsymbol{j} \times \boldsymbol{B}) - \rho_m \nabla\phi_g \tag{6.42a}$$

$$\boldsymbol{E} + \left(\boldsymbol{V} - \frac{\boldsymbol{j}}{en_e}\right) \times \boldsymbol{B} + \frac{1}{en_e}\sum_\beta \frac{\partial}{\partial x_\beta} P_{\alpha\beta}{}^{e} - \eta\boldsymbol{j} = \frac{m_e}{e^2 n_e}\frac{\partial \boldsymbol{j}}{\partial t}, \tag{6.43a}$$

respectively. The equation of continuity and Maxwell's equations are

$$\frac{\partial \rho_m}{\partial t} + \nabla\cdot(\rho_m \boldsymbol{V}) = 0 \tag{6.44}$$

$$\frac{\partial \rho}{\partial t} + \nabla\cdot\boldsymbol{j} = 0 \tag{6.45}$$

$$\nabla \times \boldsymbol{E} = -\frac{\partial \boldsymbol{B}}{\partial t} \tag{6.46}$$

$$\frac{1}{\mu_0}\nabla \times \boldsymbol{B} = \boldsymbol{j} + \varepsilon_0 \frac{\partial \boldsymbol{E}}{\partial t} \tag{6.47}$$

$$\varepsilon_0 \nabla\cdot\boldsymbol{E} = \rho \tag{6.48}$$

$$\nabla\cdot\boldsymbol{B} = 0. \tag{6.49}$$

Since the ratio of the 1st term and the 2nd term of the right-hand side of eq. (6.47) is $(\eta\varepsilon_0\omega)^{-1} \approx ne^2/(m_e\nu_e\varepsilon_0\omega) \approx \Pi_e{}^2/\omega\nu_e$ (where $\Pi_e{}^2 \equiv ne^2/m_e\varepsilon_0$, $\partial/\partial t \approx \omega$), the 2nd term can be neglected for the low frequency case. When the distribution function is isotropic ($1/\omega > \tau_e, \tau_i$ i.e., $v_T{}^{e,i}/\omega > \lambda_{e,i}$) and the gravitational field can be neglected, eqs. (6.42a) and (6.43a) may be expressed in the form

$$\rho_m \frac{\mathrm{d}\boldsymbol{V}}{\mathrm{d}t} = -\nabla p + \rho\boldsymbol{E} + \boldsymbol{j} \times \boldsymbol{B} \tag{6.42b}$$

$$\boldsymbol{E} + \left(\boldsymbol{V} - \frac{\boldsymbol{j}}{en_e}\right) \times \boldsymbol{B} + \frac{1}{en_e}\nabla p_e - \eta\boldsymbol{j} = \frac{m_e}{e^2 n_e}\frac{\partial \boldsymbol{j}}{\partial t}. \tag{6.43b}$$

The right-hand side of eq. (6.43b) can be neglected for $\omega \ll \Omega_e$ (Ω_e is the electron cyclotron frequency). Equation (6.43b) is then

$$E + V \times B - \frac{1}{en_e}\nabla p_i - \eta j = \frac{\Delta n_e}{n_e} E + \frac{m_i}{Ze}\frac{dV}{dt} + \frac{m_e}{e}\frac{\partial(V_i - V_e)}{\partial t}.$$

The 1st term may be neglected, since $\Delta n_e/n_e \ll 1$. The 2nd term is of the order of ω/Ω_i times $|v \times B|$ and the 3rd term is of the order of ω/Ω_e. Consequently, in the low-frequency case ($\omega \ll \Omega_i$, i.e., $v_T{}^i/\omega \gg \rho_B{}^i$), eq. (6.43b) is reduced to

$$E + V \times B - \frac{1}{en_e}\nabla p_i = \eta j. \tag{6.43c}$$

From eq. (6.43c), the perpendicular component of the velocity is

$$V_\perp = \left(E - \frac{1}{en_e}\nabla p_i - \eta j\right) \times \frac{B}{B^2}.$$

In the formula for the velocity of the guiding center of a single charged particle, there is a B-gradient drift term (see ch. 3). However, such a term does not appear in the formula for $V_\perp$. On the other hand, the p-gradient drift that appears in the foregoing equation for $V_\perp$ does not appear in the formula for the velocity of the guiding center of a single particle. We must be careful not to confuse the relation between the flow velocity of the plasma fluid and the drift velocity of individual particles. The relationship of the two is discussed in Spitzer's text (ref. 3 to ch. 2).

When $|\nabla p_i/n| \approx |\nabla T_i| \ll |eE|$ or $\rho_B{}^i v_T{}^i/(a|V|) \ll 1$, the term in ∇p_i may be neglected and we find

$$E + V \times B = \eta j. \tag{6.43d}$$

When the collision frequency ν_{ei} is much smaller than Ω_e, ($\lambda_e \gg \rho_B{}^e$), we have the relation $|\eta j|/|V \times B| \approx \nu_{ei}/\Omega_e \approx \rho_B{}^e/\lambda_e$ and ηj can be neglected.

6.3 Collision Terms[2]

Let us consider two kind of particles a, b and neglect the effect of ionization and recombination. For elastic collisions, the laws of conservation of particle number, momentum, and energy are expressed by

$$\int \left(\frac{\delta f}{\delta t}\right)^{ab}_{coll} dv = 0 \tag{6.50}$$

$$\int m_{\mathrm{a}}\boldsymbol{v}\left(\frac{\delta f}{\delta t}\right)_{\mathrm{coll}}^{\mathrm{aa}} \mathrm{d}\boldsymbol{v} = 0 \tag{6.51}$$

$$\int \frac{m_{\mathrm{a}}\boldsymbol{v}^2}{2}\left(\frac{\delta f}{\delta t}\right)_{\mathrm{coll}}^{\mathrm{aa}} \mathrm{d}\boldsymbol{v} = 0 \tag{6.52}$$

$$\int m_{\mathrm{a}}\boldsymbol{v}\left(\frac{\delta f}{\delta t}\right)_{\mathrm{coll}}^{\mathrm{ab}} \mathrm{d}\boldsymbol{v} + \int m_{\mathrm{b}}\boldsymbol{v}\left(\frac{\delta f}{\delta t}\right)_{\mathrm{coll}}^{\mathrm{ba}} \mathrm{d}\boldsymbol{v} = 0 \tag{6.53}$$

$$\int \frac{m_{\mathrm{a}}\boldsymbol{v}^2}{2}\left(\frac{\delta f}{\delta t}\right)_{\mathrm{coll}}^{\mathrm{ab}} \mathrm{d}\boldsymbol{v} + \int \frac{m_{\mathrm{b}}\boldsymbol{v}^2}{2}\left(\frac{\delta f}{\delta t}\right)_{\mathrm{coll}}^{\mathrm{ba}} \mathrm{d}\boldsymbol{v} = 0. \tag{6.54}$$

From eqs. (6.15) and (6.23), it follows that

$$\boldsymbol{R} = \int m\boldsymbol{v}_{\mathrm{r}}\left(\frac{\delta f}{\delta t}\right)_{\mathrm{coll}} \mathrm{d}\boldsymbol{v} \tag{6.55}$$

$$Q = \int \frac{m\boldsymbol{v}_{\mathrm{r}}^{2}}{2}\left(\frac{\delta f}{\delta t}\right)_{\mathrm{coll}} \mathrm{d}\boldsymbol{v}; \tag{6.56}$$

and, from eqs. (6.53) and (6.54), we find that

$$\boldsymbol{R}_{\mathrm{ei}} = -\boldsymbol{R}_{\mathrm{ie}} \tag{6.57}$$

$$Q_{\mathrm{ei}} + Q_{\mathrm{ie}} = -(\boldsymbol{R}_{\mathrm{ei}}\cdot \boldsymbol{V}_{\mathrm{e}} + \boldsymbol{R}_{\mathrm{ie}}\cdot \boldsymbol{V}_{\mathrm{i}}) = -\boldsymbol{R}_{\mathrm{ei}}\cdot(\boldsymbol{V}_{\mathrm{e}} - \boldsymbol{V}_{\mathrm{i}}). \tag{6.58}$$

6.3a Transport of Momentum and Energy by Collision

The force $\boldsymbol{R}$ on electrons due to collision with ions, which has already appeared in eqs. (6.28) and (6.29), consists of a *friction force* $\boldsymbol{R}_{\mathrm{u}}$ and a *thermal force* $\boldsymbol{R}_{\mathrm{T}}$. When is $Z = 1$, $\boldsymbol{R}_{\mathrm{u}}$ and $\boldsymbol{R}_{\mathrm{T}}$ are given by

$$\boldsymbol{R}_{\mathrm{u}} = en_{\mathrm{e}}\left(\frac{\boldsymbol{j}_{\parallel}}{\sigma_{\parallel}} + \frac{\boldsymbol{j}_{\perp}}{\sigma_{\perp}}\right) \tag{6.59}$$

$$\boldsymbol{R}_{\mathrm{T}} = -0.71 n_{\mathrm{e}}\nabla_{\parallel} T_{\mathrm{e}} - \frac{3}{2}\frac{n_{\mathrm{e}}}{\Omega_{\mathrm{e}}\tau_{\mathrm{e}}}(\boldsymbol{b}\times\nabla T_{\mathrm{e}}), \tag{6.60}$$

where $\sigma_{\parallel}$, $\sigma_{\perp}$, the *specific conductivity* in the parallel and perpendicular directions, respectively, are given by

$$\sigma_{\perp} = \frac{n_{\mathrm{e}}e^2\tau_{\mathrm{e}}}{m_{\mathrm{e}}} \qquad \sigma_{\parallel} = \frac{2n_{\mathrm{e}}e^2\tau_{\mathrm{e}}}{m_{\mathrm{e}}}. \tag{6.61}$$

The heat flux $\boldsymbol{q}^{\mathrm{e}}$ due to the electrons is given by eq. (6.22); it consists of the term $\boldsymbol{q}_{\mathrm{u}}{}^{\mathrm{e}}$ carried by the particle flow and the term $\boldsymbol{q}_{\mathrm{T}}{}^{\mathrm{e}}$ due to heat conduction across the temperature gradient. These quantities are given by

$$\boldsymbol{q}_{\mathrm{u}}{}^{\mathrm{e}} = 0.71 n_{\mathrm{e}} T_{\mathrm{e}} \boldsymbol{u}_{\parallel} + \frac{3}{2}\frac{n_{\mathrm{e}} T_{\mathrm{e}}}{\Omega_{\mathrm{e}} \tau_{\mathrm{e}}}(\boldsymbol{b} \times \boldsymbol{u}) \tag{6.62}$$

$$\boldsymbol{q}_{\mathrm{T}}{}^{\mathrm{e}} = -\kappa_{\parallel}{}^{\mathrm{e}} \nabla_{\parallel} T_{\mathrm{e}} - \kappa_{\perp}{}^{\mathrm{e}} \nabla_{\perp} T_{\mathrm{e}} - \frac{5}{2}\frac{n_{\mathrm{e}} T_{\mathrm{e}}}{\Omega_{\mathrm{e}} m_{\mathrm{e}}}(\boldsymbol{b} \times \nabla T_{\mathrm{e}}), \tag{6.63}$$

where

$$\boldsymbol{u} = \boldsymbol{V}_{\mathrm{e}} - \boldsymbol{V}_{\mathrm{i}} = -\frac{\boldsymbol{j}}{e n_{\mathrm{e}}}$$

$$\kappa_{\parallel}{}^{\mathrm{e}} = 3.16 \frac{n_{\mathrm{e}} T_{\mathrm{e}} \tau_{\mathrm{e}}}{m_{\mathrm{e}}} \tag{6.64}$$

$$\kappa_{\perp}{}^{\mathrm{e}} = 4.66 \frac{n_{\mathrm{e}} T_{\mathrm{e}}}{m_{\mathrm{e}} \Omega_{\mathrm{e}}{}^2 \tau_{\mathrm{e}}}. \tag{6.65}$$

The heat flux $\boldsymbol{q}^{\mathrm{i}}$ due to the ions is given by

$$\boldsymbol{q}^{\mathrm{i}} = -\kappa_{\parallel}{}^{\mathrm{i}} \nabla_{\parallel} T_{\mathrm{i}} - \kappa_{\perp}{}^{\mathrm{i}} \nabla_{\perp} T_{\mathrm{i}} + \frac{5}{2}\frac{n_{\mathrm{i}} T_{\mathrm{i}}}{|\Omega_{\mathrm{i}}| m_{\mathrm{i}}}(\boldsymbol{b} \times \nabla T_{\mathrm{i}}) \tag{6.66}$$

$$\kappa_{\parallel}{}^{\mathrm{i}} = 3.9 \frac{n_{\mathrm{i}} T_{\mathrm{i}} \tau_{\mathrm{i}}}{m_{\mathrm{i}}} \tag{6.67}$$

$$\kappa_{\perp}{}^{\mathrm{i}} = 2 \frac{n_{\mathrm{i}} T_{\mathrm{i}}}{m_{\mathrm{i}} \Omega_{\mathrm{i}}{}^2 \tau_{\mathrm{i}}}. \tag{6.68}$$

κ is the heat conductivity; τ_{e} is given by eq. (4.61) or (4.62), and τ_{i} is given by eq. (4.63).

The change of the ion thermal energy density due to e-i collision is

$$Q_{\mathrm{i}} = \frac{(3/2)(T_{\mathrm{e}} - T_{\mathrm{i}})}{\tau_{\varepsilon}{}^{\mathrm{ei}}}, \tag{6.69}$$

where $\tau_{\varepsilon}{}^{\mathrm{ei}}$ is given by eq. (4.49). The change of the electron thermal energy density due to e-i collision is

$$Q_{\mathrm{e}} = -\boldsymbol{R} \cdot \boldsymbol{u} - Q_{\mathrm{i}} = \frac{j_{\parallel}{}^2}{\sigma_{\parallel}} + \frac{j_{\perp}{}^2}{\sigma_{\perp}} + \frac{1}{e n_{\mathrm{e}}} \boldsymbol{j} \cdot \boldsymbol{R}_{\mathrm{T}} - \frac{(3/2)(T_{\mathrm{e}} - T_{\mathrm{i}})}{\tau_{\varepsilon}{}^{\mathrm{ei}}}. \tag{6.70}$$

Let us discuss the terms $\boldsymbol{R}_{\mathrm{T}}$, $\boldsymbol{q}_{\mathrm{u}}{}^{\mathrm{e}}$, $\boldsymbol{q}_{\mathrm{T}}{}^{\mathrm{e}}$, and $\boldsymbol{q}^{\mathrm{i}}$ qualitatively. The meanings of the other terms $\boldsymbol{R}_{\mathrm{u}}$, Q_{i}, and Q_{e} are already clear. (A more formal discussion can be made by using the Landau integral (5.57). The reader should refer to ref. 2 for the formal treatment.)

(i) Thermal Force $\boldsymbol{R}_{\mathrm{T}}$ It is assumed that the average velocities of electrons and ions, V_{e} and V_{i}, are zero. Consider any given plane within the plasma. Then the number of electrons crossing the plane from the left-hand side is equal to the number of electrons crossing the plane from the right-hand side. The number crossing per unit time in either direction is given by $n_{\mathrm{e}}v_{\mathrm{e}}$. These electron fluxes experience the resistances $\boldsymbol{R}_{+}$ and $\boldsymbol{R}_{-}$, respectively, due to e-i collisions in the plane; the collision force is $m_{\mathrm{e}}n_{\mathrm{e}}v_{\mathrm{e}}/\tau_{\mathrm{e}}$. When the plasma is isotropic, $\boldsymbol{R}_{+}$ and $\boldsymbol{R}_{-}$ are equal and the total force across any chosen plane will be zero. However, when the distribution function of the electrons on the right-hand side of the plane is different from that of left-hand side, $\boldsymbol{R}_{+}$ and $\boldsymbol{R}_{-}$ are unequal, and there is a resultant force on the electrons.

Let the z axis be the direction of the lines of magnetic force and assume that there is a temperature gradient in the z direction. The difference of the temperature of electrons coming from $z < 0$ and the temperature of electrons from $z > 0$ is around $\lambda\partial T_{\mathrm{e}}/\partial z$, where λ is the electron mean free path. As the magnitudes of $\boldsymbol{R}_{+}$ and $\boldsymbol{R}_{-}$ are inversely proportional to the temperature, their difference is

$$R_{\mathrm{T}z} \approx -\frac{\lambda}{T_{\mathrm{e}}}\frac{\partial T_{\mathrm{e}}}{\partial z}\frac{m_{\mathrm{e}}n_{\mathrm{e}}v_{\mathrm{e}}}{\tau_{\mathrm{e}}} \approx -\frac{m_{\mathrm{e}}v_{\mathrm{e}}{}^{2}}{T_{\mathrm{e}}}n_{\mathrm{e}}\frac{\partial T_{\mathrm{e}}}{\partial z} \approx -n_{\mathrm{e}}\frac{\partial T_{\mathrm{e}}}{\partial z}.$$

This is the 1st term of eq. (6.60).

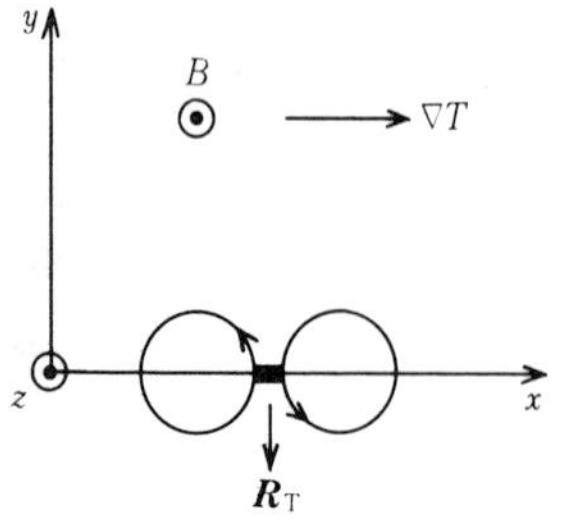

Fig. 6.1
Thermal force due to a temperature gradient perpendicular to the magnetic field.

When the temperature gradient is perpendicular to the magnetic field, we take the x axis to be the direction of ∇T. Electrons gyrate with a Larmor radius $\rho_B{}^{\mathrm{e}}$. Let us consider the movement of electrons in the y direction at $x = x_0$, as shown in fig. 6.1. The temperature of electrons coming from $x = x_0 + \rho_B{}^{\mathrm{e}}$ differs from the temperature of electrons from $x = x_0 - \rho_B{}^{\mathrm{e}}$ in the amount $\rho_B{}^{\mathrm{e}} \partial T_{\mathrm{e}}/\partial x$. The resultant thermal force in the y direction is

$$R_{\mathrm{T}y} \approx -\frac{\rho_B{}^{\mathrm{e}}}{T_{\mathrm{e}}} \frac{\partial T_{\mathrm{e}}}{\partial x} \frac{m_{\mathrm{e}} n_{\mathrm{e}} v_{\mathrm{e}}}{\tau_{\mathrm{e}}} \approx \frac{-n_{\mathrm{e}}}{\Omega_{\mathrm{e}} \tau_{\mathrm{e}}} \frac{\partial T_{\mathrm{e}}}{\partial x}. \tag{6.71}$$

This is the 2nd term of eq. (6.60).

(ii) Flux of Electron Thermal Energy As the collision time τ_{e} is proportional to v^3, most of the current parallel to the magnetic field is carried by the faster electrons. The current is given by $\boldsymbol{j} = -en_{\mathrm{e}}(\boldsymbol{V}_{\mathrm{e}} - \boldsymbol{V}_{\mathrm{i}}) = -en_{\mathrm{e}}\boldsymbol{u}$. Consider a coordinate system in which $\boldsymbol{V}_{\mathrm{e}} = 0$. Then the faster electrons move in the direction $\boldsymbol{u}$ and the slower electrons move in the direction $-\boldsymbol{u}$. The numbers of electrons which flow in the direction of $\boldsymbol{u}$ and $-\boldsymbol{u}$ are balanced, but the flow of the energy is of the order of $n_{\mathrm{e}} T_{\mathrm{e}} \boldsymbol{u}$. This is the 1st term of eq. (6.62).

Next, assume that the plasma flows in the x direction, perpendicular to the magnetic field, with the flow velocity u_x (fig. 6.2). The friction force on electrons due to e-i collisions decelerates the electrons in the region $y > 0$ and accelerates the electrons in the region $y < 0$ (fig. 6.2). (Note that the guiding center for the electrons is on the $y = 0$ plane.) Electrons at $x = x_0$, $y = 0$ coming from the upper region $y > 0$ are decelerated and those coming from the lower region $y < 0$ are accelerated; the energy

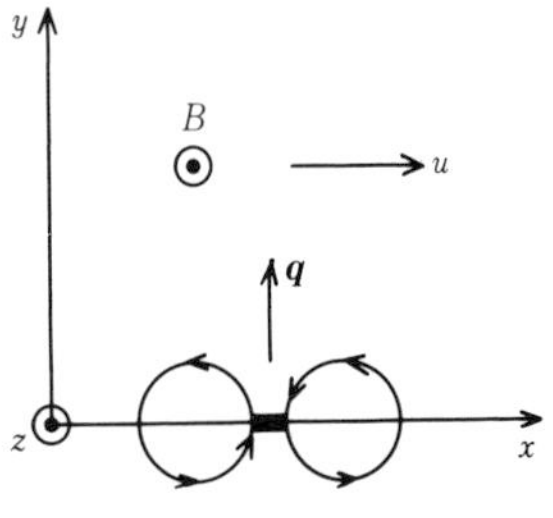

Fig. 6.2
Flux of thermal energy due to plasma flow perpendicular to the magnetic field.

difference is $(m_e u/\tau_e)\rho_B{}^e$. Consequently there is a net energy flow in the positive y direction, i.e.,

$$q_y = \frac{m_e u_x}{\tau_e}\rho_B{}^e n_e v_e \approx \frac{m_e u_x}{\tau_e}\frac{v_e{}^2}{\Omega_e}n_e \approx \frac{n_e T_e}{\Omega_e \tau_e}u_x. \tag{6.72}$$

This is the 2nd term of eq. (6.62).

Next, consider particles in random motion, with a step length Δx and a time τ between steps. The diffusion coefficient D and the particle flow I are given by

$$D = \frac{(\Delta x)^2}{\tau} \qquad I = -D\frac{\partial n}{\partial x} = -\frac{(\Delta x)^2}{\tau}\frac{\partial n}{\partial x}, \tag{6.73}$$

as is well known. The energy flow due to thermal diffusion is thus

$$q = -\frac{n(\Delta x)^2}{\tau}\frac{\partial T_e}{\partial x} = -\kappa\frac{\partial T_e}{\partial x} \qquad \kappa \approx \frac{n(\Delta x)^2}{\tau} \approx nD. \tag{6.74}$$

The step lengths $(\Delta x)_\parallel$, $(\Delta x)_\perp$ parallel and perpendicular to the magnetic field are $(\Delta x)_\parallel \approx \lambda \approx v\tau$, $(\Delta x)_\perp \approx v/\Omega$, respectively. Consequently, the thermal diffusion coefficients for both electrons and ions are given by

$$\kappa_\parallel \approx \frac{n\lambda^2}{\tau} \approx \frac{nT\tau}{m} \tag{6.75}$$

$$\kappa_\perp \approx \frac{n(v/\Omega)^2}{\tau} \approx \frac{nT}{m\Omega^2\tau}. \tag{6.76}$$

These are the 1st and 2nd terms of eqs. (6.63) and (6.66).

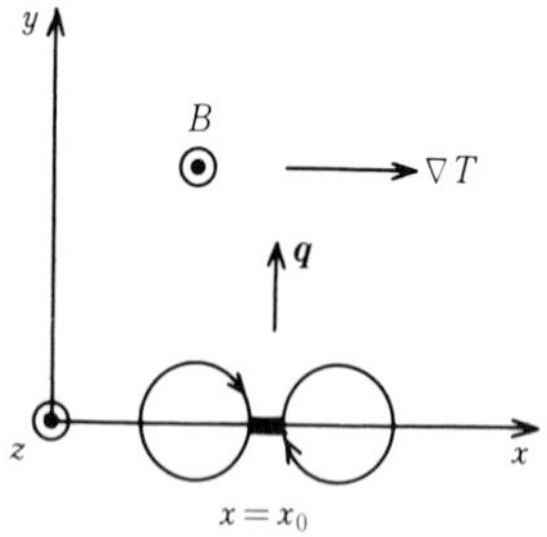

Fig. 6.3
Flux of thermal energy due to a temperature gradient perpendicular to the magnetic field.

Where there is a temperature gradient in the x direction, perpendicular to the magnetic field (fig. 6.3), the particles at $x = x_0$, $y = 0$ coming from $y < 0$ have larger energy than the particles coming from $y > 0$; the difference is $\rho_B \partial T/\partial x$ (fig. 6.3). Accordingly there is an energy flow in the positive y direction:

$$q_y = nv\rho_B \frac{\partial T}{\partial x} \approx \frac{\pm nT}{|\Omega|m} \frac{\partial T}{\partial x} . \tag{6.77}$$

This is the 3rd term of eqs. (6.63) and (6.66). The sign corresponds to the sign of the charge.

*6.3b Velocity Gradient and Viscosity

The anisotropy term $\Pi_{\alpha\beta}$ in the pressure tensor $P_{\alpha\beta}$

$$\Pi_{\alpha\beta} = nm(\langle v_{r\alpha} v_{r\beta} \rangle - \langle v_r^2 \rangle / 3\delta_{\alpha\beta}) \tag{6.78}$$

can be expressed in terms of a velocity gradient, or the viscosity. Where there is a flow-velocity gradient, anisotropy is introduced into the distribution function and the anisotropy term $\Pi_{\alpha\beta}$ appears. For example, $\Pi_{zx} = nm\langle v_{rz} v_{rx} \rangle$ expresses the flow of the z component of momentum nmv_{rz} in the x direction. Let us assume that V_z changes only in the x direction (fig. 6.4).

The flows of the momentum nmV_z from the region $x > x_0$ or from the region $x < x_0$ due to diffusion are $\frac{\Delta x}{\tau} nmV_z$. When the plasma is homogeneous, these flows cancel each other. However, if a gradient of the flow velocity exists, the transverse flows are unequal. Their difference is Π_{zx}, i.e.,

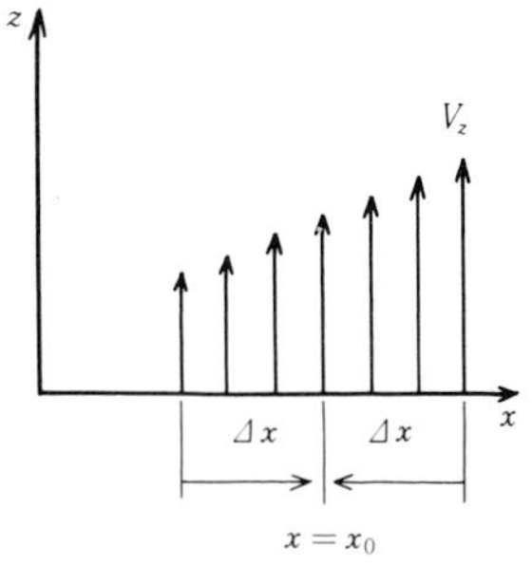

Fig. 6.4
Flow of momentum due to a velocity gradient.

$$-\Delta x\frac{\partial}{\partial x}\left(\frac{\Delta x}{\tau}nmV_z\right) = -\eta_2\frac{\partial V_z}{\partial x} = \Pi_{zx}$$

$$\eta_2 \approx \frac{(\Delta x)^2}{\tau}nm \approx nmD, \tag{6.79}$$

where η_2 is a coefficient of viscosity.

Generally, the velocity gradient is expressed by the rate of strain tensor

$$W_{\alpha\beta} = \frac{\partial V_\alpha}{\partial x_\beta} + \frac{\partial V_\beta}{\partial x_\alpha} - \frac{2}{3}\delta_{\alpha\beta}\nabla V. \tag{6.80}$$

When the magnetic field is strong ($\Omega\tau \gg 1$) and the field is directed along the z axis, $\Pi_{\alpha\beta}$ can be expressed by

$$\left.\begin{aligned}
\Pi_{zz} &= -\eta_0 W_{zz}\\
\Pi_{xx} &= -\eta_0\frac{1}{2}(W_{xx}+W_{yy}) - \eta_1\frac{1}{2}(W_{xx}-W_{yy}) - \eta_3 W_{xy}\\
\Pi_{yy} &= -\eta_0\frac{1}{2}(W_{xx}+W_{yy}) - \eta_1\frac{1}{2}(W_{yy}-W_{xx}) + \eta_3 W_{xy}\\
\Pi_{xy} &= \Pi_{yx} = -\eta_1 W_{xy} + \eta_3\frac{1}{2}(W_{xx}-W_{yy})\\
\Pi_{xz} &= \Pi_{zx} = -\eta_2 W_{xz} - \eta_4 W_{yz}\\
\Pi_{yz} &= \Pi_{zy} = -\eta_2 W_{yz} + \eta_4 W_{xz}.
\end{aligned}\right\} \tag{6.81}$$

The coefficients of viscosity for ions are

$$\eta_0{}^{\mathrm{i}} = 0.96 n_{\mathrm{i}} T_{\mathrm{i}} \tau_{\mathrm{i}} \tag{6.82}$$

$$n_1{}^{\mathrm{i}} = 0.3\frac{n_{\mathrm{i}}T_{\mathrm{i}}}{\Omega_{\mathrm{i}}{}^2\tau_{\mathrm{i}}} \qquad \eta_2{}^{\mathrm{i}} = 4\eta_1{}^{\mathrm{i}} \tag{6.83}$$

$$\eta_3{}^{\mathrm{i}} = 0.5\frac{n_{\mathrm{i}}T_{\mathrm{i}}}{|\Omega_{\mathrm{i}}|} \qquad \eta_4{}^{\mathrm{i}} = 2\eta_3{}^{\mathrm{i}} \tag{6.84}$$

and the coefficients of viscosity of electrons are (for $Z = 1$)

$$\eta_0{}^{\mathrm{e}} = 0.73 n_{\mathrm{e}} T_{\mathrm{e}} \tau_{\mathrm{e}} \tag{6.85}$$

$$\eta_1{}^{\mathrm{e}} = 0.51\frac{n_{\mathrm{e}}T_{\mathrm{e}}}{\Omega_{\mathrm{e}}{}^2\tau_{\mathrm{e}}} \qquad \eta_2{}^{\mathrm{e}} = 4\eta_1{}^{\mathrm{e}} \tag{6.86}$$

$$\eta_3{}^{e} = -\,0.5\frac{n_e T_e}{\Omega_e} \qquad \eta_4{}^{e} = 2\eta_3{}^{e}. \tag{6.87}$$

The values of these coefficients will be arrived at by means of the following qualitative discussion (fig. 6.5).

First, let the all velocity gradients except $\partial V_z/\partial_z$ be zero. Π_{zz} is the flow of the z component of momentum, i.e., the flow parallel to the magnetic field. By putting $\Delta x \approx \lambda$ in eq. (6.79), the coefficient of viscosity is

$$\eta_0 \approx nT\tau \tag{6.88}$$

and the stress due to the viscosity is

$$F_z = \tfrac{4}{3}(\partial/\partial z)\eta_0(\partial V_z/\partial z). \tag{6.89}$$

Next, let the all velocity gradients except $\partial V_z/\partial x$ be zero. The contribution of $\partial V_z/\partial x$ to Π_{zx} is obtained by substituting $\Delta x \approx \rho_B$ in eq. (6.79). We find that $\Pi_{zx} \approx -\eta_2 \partial V_z/\partial x$, and

$$\eta_2 \approx \frac{\eta_0}{(\Omega\tau)^2}. \tag{6.90}$$

When we consider the case where only the component of $\partial V_y/\partial x$ is nonzero, we find $\Pi_{xy} \approx -\eta_1 \partial V_y/\partial x$, and

$$\eta_1 \approx \eta_2. \tag{6.91}$$

Next, let us consider the contribution of $\partial V_y/\partial x \neq 0$ to Π_{yy} (fig. 6.6). Π_{yy} is the flow of the y component of momentum in the y direction. As is shown

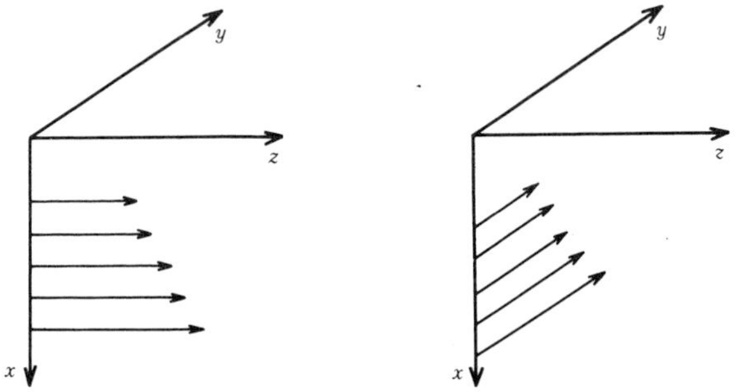

Fig. 6.5
Contribution of the velocity gradient $\partial V_z/\partial x$ to the component Π_{zx}, and contribution of the gradient $\partial V_y/\partial x$ to Π_{xy}.

in fig. 6.6, the difference of the y component of the momentum of particles at $x = x_0$ coming from the region $y > 0$ and that of particles coming from the region $y < 0$ is $\rho_B(\partial/\partial x)(mV_y)$. Consequently, we have

$$\Pi_{yy} = \pm nv\rho_B \frac{\partial}{\partial x}(mV_y) = \frac{\pm nT}{|\Omega|}\frac{\partial V_y}{\partial x} \approx \pm \frac{\eta_0}{|\Omega|\tau}\frac{\partial V_y}{\partial x}, \tag{6.92}$$

i.e.,

$$\eta_3 \approx \pm \frac{\eta_0}{|\Omega|\tau}. \tag{6.93}$$

The + or − sign is chosen according to the charge of the particles being considered.

The contribution of $\partial V_z/\partial x \neq 0$ to Π_{yz} (fig. 6.7) can be derived in a similar manner. As is clear from fig. 6.7, we find

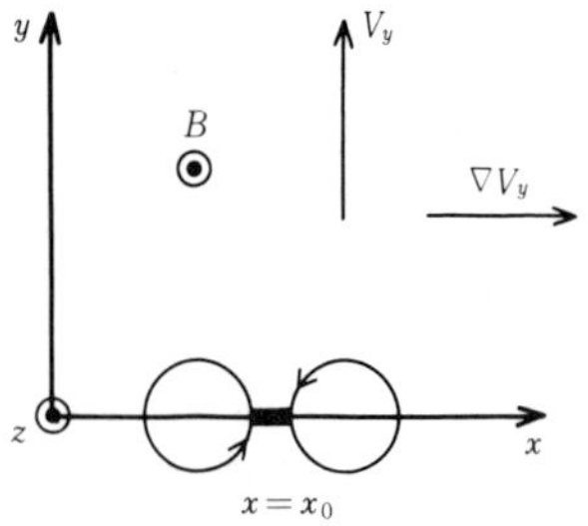

Fig. 6.6
Contribution of the velocity gradient $\partial V_y/\partial x$ to Π_{yy}.

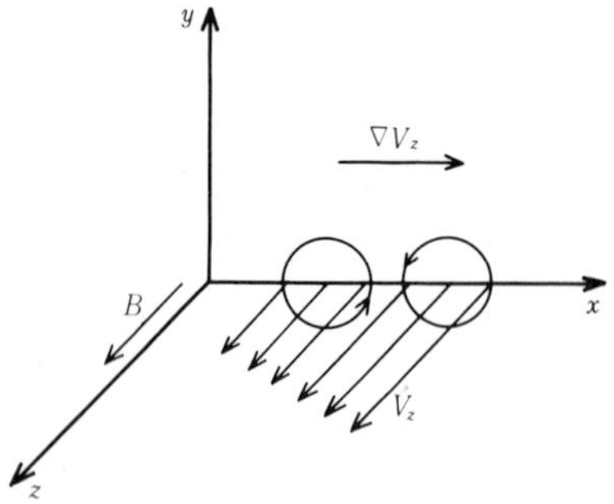

Fig. 6.7
Contribution of the velocity gradient $\partial V_z/\partial x$ to Π_{yz}.

$$II_{yz} = \pm nv\rho_B \frac{\partial}{\partial x}(mV_z) = \pm \frac{nT}{|\Omega|}\frac{\partial V_z}{\partial x} \approx \pm \frac{\eta_0}{|\Omega|\tau}\frac{\partial V_z}{\partial x}, \tag{6.94}$$

and

$$\eta_4 \approx \pm \frac{\eta_0}{|\Omega|\tau}. \tag{6.95}$$

6.4 The CGL Equations[3]

Macroscopic equations for a collisionless plasma have been derived by G. F. Chew, M. L. Goldberger, and F. E. Low. Boltzmann's equation for ions is

$$\frac{\partial f}{\partial t} + (\boldsymbol{v}\cdot\nabla_r)f + \frac{e}{m_i}(\boldsymbol{E} + \boldsymbol{v}\times\boldsymbol{B})\cdot\nabla_v f = 0. \tag{6.96}$$

When the ion Larmor frequency is larger than the other characteristic frequencies and the Larmor radius is smaller than the characteristic scale, the operator $(e/m^i)\,(\boldsymbol{E} + \boldsymbol{v}\times\boldsymbol{B})\cdot\nabla_v$ is larger than the others in eq. (6.96). Expanding the distribution function f

$$f = f_0 + f_1 + f_2 + \cdots, \tag{6.97}$$

we find

$$\frac{e}{m_i}(\boldsymbol{E} + \boldsymbol{v}\times\boldsymbol{B})\cdot\nabla_v f_0 = 0 \tag{6.98}$$

$$\frac{e}{m_i}(\boldsymbol{E} + \boldsymbol{v}\times\boldsymbol{B})\cdot\nabla_v f_1 + \left(\frac{\partial}{\partial t} + \boldsymbol{v}\cdot\nabla_r\right)f_0 = 0 \tag{6.99}$$

$\cdots$.

Set $\boldsymbol{u}_E = \boldsymbol{E}\times\boldsymbol{B}/B^2$ and assume that the electric field $\boldsymbol{E}$ and the magnetic field $\boldsymbol{B}$ are perpendicular to each other (If $E_\parallel \neq 0$, electrons are accelerated —since the plasma is collisionless—and $E_\parallel$ is short circuited and becomes zero). Then it follows that

$$\boldsymbol{E} = -\boldsymbol{u}_E \times \boldsymbol{B}, \tag{6.100}$$

and eq. (6.98) becomes

$$((\boldsymbol{v} - \boldsymbol{u}_E)\times\boldsymbol{B})\cdot\nabla_v f_0 = 0,$$

so that $\nabla_v f_0$, $(\boldsymbol{v} - \boldsymbol{u}_{\mathrm{E}})$, and $\boldsymbol{B}$ are seen to be coplaner. Therefore the general form of f_0 is

$$f_0((\boldsymbol{v} - \boldsymbol{u}_{\mathrm{E}})^2, (\boldsymbol{v} \cdot \boldsymbol{B}), \boldsymbol{r}, t). \tag{6.101}$$

The equation of continuity is

$$\frac{\partial n_0}{\partial t} + \nabla(n_0 \boldsymbol{V}) = 0, \tag{6.102}$$

where $\boldsymbol{V}$ is the average velocity of the ions. Multiply eq. (6.99) by $\boldsymbol{v}$ and integrate over $\boldsymbol{v}$. In a manner similar to that in sec. 6.2, we find

$$\rho_{\mathrm{m}} \frac{\mathrm{d}\boldsymbol{V}}{\mathrm{d}t} = -\sum_\beta \frac{\partial}{\partial x_\beta} P_{\alpha\beta} + \left(e \int (\boldsymbol{v} - \boldsymbol{V}) f_1 \mathrm{d}\boldsymbol{v} \right) \times \boldsymbol{B}. \tag{6.103}$$

Here eq. (6.100) and the relation

$$\boldsymbol{V}_\perp = \boldsymbol{u}_{\mathrm{E}}$$

have been used. (The relation $\boldsymbol{V}_\perp = \boldsymbol{u}_{\mathrm{E}}$ is easily obtained from eq. (6.101).) $P_{\alpha\beta}$ can be identified as the *anisotropic pressure tensor*

$$\overleftrightarrow{P} = p_\parallel \boldsymbol{bb} + p_\perp(\overleftrightarrow{\boldsymbol{I}} - \boldsymbol{bb}), \tag{6.104}$$

where $\boldsymbol{b}$ is the unit vector of $\boldsymbol{B}$. When the x_3 axis is chosen in the direction of $\boldsymbol{B}$, the tensor is

$$P_{ij} = \begin{pmatrix} p_\perp & 0 & 0 \\ 0 & p_\perp & 0 \\ 0 & 0 & p_\parallel \end{pmatrix}; \tag{6.105}$$

$p_\parallel$ and $p_\perp$ are the pressures parallel and perpendicular to the magnetic field, respectively. The temperatures parallel and perpendicular to the magnetic field can be defined by $p_\parallel = nT_\parallel$ and $p_\perp = nT_\perp$, respectively, n being the density.

When the distribution functions of electrons is denoted by f_{e}, Maxwell's equations are reduced to the following by the use of $\boldsymbol{E} = -\boldsymbol{u}_{\mathrm{E}} \times \boldsymbol{B}$:

$$-\nabla(\boldsymbol{u}_{\mathrm{E}} \times \boldsymbol{B}) = \frac{e}{\varepsilon_0} \int \mathrm{d}\boldsymbol{v}(f_0 + f_1 - f_{\mathrm{e}}) \tag{6.106}$$

$$\nabla \boldsymbol{B} = 0 \tag{6.107}$$

$$\frac{1}{\mu_0}\nabla \times \boldsymbol{B} = -\varepsilon_0 \frac{\partial}{\partial t}(\boldsymbol{u}_{\mathrm{E}} \times \boldsymbol{B}) + e\int \mathrm{d}\boldsymbol{v}\boldsymbol{v}(f_0 + f_1 - f_{\mathrm{e}}) \tag{6.108}$$

$$-\frac{\partial \boldsymbol{B}}{\partial t} = -\nabla \times (\boldsymbol{u}_{\mathrm{E}} \times \boldsymbol{B}). \tag{6.109}$$

Let us define the quantity $\boldsymbol{J}_1$ by

$$\begin{aligned} \boldsymbol{J}_1 &= e\int \mathrm{d}\boldsymbol{v}\boldsymbol{v}(f_0 + f_1 - f_{\mathrm{e}}) - e\int \mathrm{d}\boldsymbol{v}\,\boldsymbol{V}(f_0 + f_1 - f_{\mathrm{e}}) \\ &= e\int \mathrm{d}\boldsymbol{v}(\boldsymbol{v} - \boldsymbol{V})f_1 - e\int \mathrm{d}\boldsymbol{v}(\boldsymbol{v} - \boldsymbol{V})f_{\mathrm{e}}. \end{aligned} \tag{6.110}$$

Since the distribution function of electrons must also satisfy eq. (6.101), the perpendicular component of the average velocity of electrons is equal to $\boldsymbol{u}_{\mathrm{E}}$ and the term including f_{e} in eq. (6.110) is parallel to $\boldsymbol{B}$. Consequently the 2nd term in the right-hand side of eq. (6.103) is expressed by $\boldsymbol{J}_1 \times \boldsymbol{B}$. From eqs. (6.106), (6.108), and (6.110), we see that $\boldsymbol{J}_1$ can be expressed as

$$\boldsymbol{J}_1 = \frac{1}{\mu_0}\nabla \times \boldsymbol{B} + \varepsilon_0 \frac{\partial}{\partial t}(\boldsymbol{u}_{\mathrm{E}} \times \boldsymbol{B}) + \varepsilon_0 \boldsymbol{V}\nabla(\boldsymbol{u}_{\mathrm{E}} \times \boldsymbol{B}). \tag{6.111}$$

Finally, the following equations are deduced:

$$\begin{aligned} \rho_{\mathrm{m}}\frac{\mathrm{d}\boldsymbol{V}}{\mathrm{d}t} = &-\sum_{\beta}\frac{\partial}{\partial x_\beta}P_{\alpha\beta} + \frac{1}{\mu_0}(\nabla \times \boldsymbol{B}) \times \boldsymbol{B} + \varepsilon_0\left(\frac{\partial}{\partial t}(\boldsymbol{V} \times \boldsymbol{B})\right) \times \boldsymbol{B} \\ &+ \varepsilon_0(\boldsymbol{V} \times \boldsymbol{B})\nabla(\boldsymbol{V} \times \boldsymbol{B}) \end{aligned} \tag{6.112}$$

$$\frac{\partial \rho_{\mathrm{m}}}{\partial t} + \nabla(\rho_{\mathrm{m}}\boldsymbol{V}) = 0 \tag{6.113}$$

$$\frac{\partial \boldsymbol{B}}{\partial t} = \nabla \times (\boldsymbol{V} \times \boldsymbol{B}) \tag{6.114}$$

(since $\boldsymbol{V}_\perp = \boldsymbol{u}_{\mathrm{E}}$, $\boldsymbol{V} \times \boldsymbol{B} = \boldsymbol{u}_{\mathrm{E}} \times \boldsymbol{B}$).

These equations, relating the macroscopic quantities $\boldsymbol{V}$, ρ_{m}, and $\boldsymbol{B}$, are called the CGL equations. The equations of the pressure are necessary in addition to CGL equations. For this purpose we multiply eq. (6.99) by the tensor $(\boldsymbol{v} - \boldsymbol{V})(\boldsymbol{v} - \boldsymbol{V})$ and integrate; we find

$$\frac{\mathrm{d}p_\parallel}{\mathrm{d}t} = -p_\parallel \nabla \boldsymbol{V} - 2p_\parallel \boldsymbol{b}\cdot(\boldsymbol{b}\cdot\nabla)\boldsymbol{V} - \nabla(q_n + q_s)\boldsymbol{b} - 2\boldsymbol{b}\cdot\nabla q_s \tag{6.115}$$

$$\frac{\mathrm{d}p_\perp}{\mathrm{d}t} = -2p_\perp\nabla\cdot\boldsymbol{V} + p_\perp\boldsymbol{b}\cdot(\boldsymbol{b}\cdot\nabla)\boldsymbol{V} - \nabla\cdot q_s\boldsymbol{b} - q_s\nabla\cdot\boldsymbol{b}, \tag{6.116}$$

where q_n, q_s are the coefficients of the 3rd-order moment tensor

$$q_{ijk}{}^{(0)} = m_\mathrm{i}\int(v_i - V_i)(v_j - V_j)(v_k - V_k)f_0\,\mathrm{d}\boldsymbol{v}.$$

Since f_0 is given by eq. (6.101), $q_{ijk}{}^{(0)}$ is given by

$$q_{ijk}{}^{(0)} = q_n b_i b_j b_k + q_s(\delta_{ij}b_k + \delta_{ik}b_j + \delta_{jk}b_i). \tag{6.117}$$

However, eqs. (6.115) and (6.116) are not closed because q_n and q_s are still unknown quantities. To obtain expressions for q_n, q_s, we need the 4th-order moment tensor, and so on. If the 3rd-order moments q_n, q_s can be neglected, we can derive the pressure equations

$$\frac{\dot{p}_\parallel}{p_\parallel} = -\nabla\cdot\boldsymbol{V} - 2\boldsymbol{b}\cdot(\boldsymbol{b}\cdot\nabla)\boldsymbol{V} \tag{6.118}$$

$$\frac{\dot{p}_\perp}{p_\perp} = -2\nabla\cdot\boldsymbol{V} + \boldsymbol{b}\cdot(\boldsymbol{b}\cdot\nabla)\boldsymbol{V}. \tag{6.119}$$

As the following relations

$$\frac{\partial\boldsymbol{B}}{\partial t} = \nabla\times(\boldsymbol{V}\times\boldsymbol{B}) = -\boldsymbol{B}(\nabla\cdot\boldsymbol{V}) + (\boldsymbol{B}\cdot\nabla)\boldsymbol{V} - (\boldsymbol{V}\cdot\nabla)\boldsymbol{B}$$

$$(\boldsymbol{B}\cdot\nabla)\boldsymbol{V} = \dot{\boldsymbol{B}} + \boldsymbol{B}(\nabla\cdot\boldsymbol{V}) = \dot{\boldsymbol{B}} - \boldsymbol{B}\frac{\dot{\rho}_\mathrm{m}}{\rho_\mathrm{m}}$$

hold, we find

$$\frac{\mathrm{d}}{\mathrm{d}t}\left(\frac{p_\parallel B^2}{\rho_\mathrm{m}{}^3}\right) = 0 \qquad \frac{\mathrm{d}}{\mathrm{d}t}\left(\frac{p_\perp}{\rho_\mathrm{m}B}\right) = 0. \tag{6.120}$$

6.5 Simplified Magnetohydrodynamic Equations

The magnetohydrodynamic equations were derived in sec. 6.2. These equations can be simplified under the following assumptions. When the typical frequency is denoted by ω, the operator $\partial/\partial t$ is of the order of ω. For

$$\frac{1}{\omega} \gg \tau_e, \tau_i \qquad \frac{v_T{}^{e,i}}{\omega} \gg \lambda_{e,i}, \tag{6.121}$$

the distribution function can be considered to be isotropic. Under the conditions

$$\omega \ll \Omega_i \qquad \frac{v_T{}^i}{\omega} \gg \rho_B{}^i, \tag{6.122}$$

Ohm's law is reduced to $\boldsymbol{E} + \boldsymbol{V} \times \boldsymbol{B} - \nabla p_i/en = \eta \boldsymbol{j}$. If the term in ∇p_i can be neglected, the equation of motion, Ohm's law, and Maxwell's equations are

$$\boldsymbol{E} + \boldsymbol{V} \times \boldsymbol{B} = \frac{\boldsymbol{j}}{\sigma} \tag{6.123}$$

$$\rho_m \frac{d\boldsymbol{V}}{dt} = -\nabla p + \boldsymbol{j} \times \boldsymbol{B} \tag{6.124}$$

$$\nabla \times \boldsymbol{B} = \mu_0 \boldsymbol{j} \tag{6.125}$$

$$\nabla \times \boldsymbol{E} = -\frac{\partial \boldsymbol{B}}{\partial t}. \tag{6.126}$$

The displacement current is omitted in eq. (6.125) under the assumption that $\Pi_e{}^2/\omega\nu_e \gg 1$ ($\Pi_e{}^2 \equiv ne^2/(m_e\varepsilon_0)$). The equation of continuity and the equation of state are

$$\frac{d\rho_m}{dt} + \rho_m \nabla \boldsymbol{V} = 0 \tag{6.127}$$

$$\frac{d}{dt}(p\rho_m{}^{-r}) = 0 \quad \text{or} \quad \nabla \boldsymbol{V} = 0. \tag{6.128}$$

When $\nu_e \ll \Omega_e$, one may consider σ to be infinity. Substitution of eq. (6.123) into eq. (6.126) yields

$$\begin{aligned}\frac{\partial \boldsymbol{B}}{\partial t} &= \nabla \times (\boldsymbol{V} \times \boldsymbol{B}) - \frac{1}{\sigma}\nabla \times \boldsymbol{j} \\ &= \nabla \times (\boldsymbol{V} \times \boldsymbol{B}) + \frac{1}{\sigma\mu_0}\Delta \boldsymbol{B}.\end{aligned} \tag{6.129}$$

Here the relation $\nabla \times (\nabla \times \boldsymbol{B}) = -\Delta \boldsymbol{B}$ is used. $1/\sigma\mu_0$ is called the *magnetic viscosity*. The ratio R_m of the 1st term and the 2nd term is

$$\frac{|\nabla \times (\boldsymbol{V} \times \boldsymbol{B})|}{|\Delta \boldsymbol{B}/\sigma\mu_0|} \approx \frac{VB/L}{(B/L^2)(1/\sigma\mu_0)} = \sigma\mu_0 VL \equiv R_m. \tag{6.130}$$

We call R_m the *magnetic Reynolds number*. When $R_m \gg 1$, it can be shown that the lines of magnetic force are frozen in the plasma. When $R_m \ll 1$, the magnetic field changes according to the diffusion equation. Let the magnetic flux within the surface element ΔS be $\Delta\,\Phi$, and take the z axis in the B direction. Then $\Delta\,\Phi$ is

$$\Delta\,\Phi = \boldsymbol{B}\cdot\boldsymbol{n}\Delta S = B\Delta x\Delta y.$$

As the boundary of ΔS moves, the rate of change of ΔS is

$$\frac{d}{dt}(\Delta x) = \frac{d}{dt}(x + \Delta x - x) = V_x(x + \Delta x) - V_x(x) = \frac{\partial V_x}{\partial x}\Delta x$$

$$\frac{d}{dt}(\Delta S) = \left(\frac{\partial V_x}{\partial x} + \frac{\partial V_y}{\partial y}\right)\Delta x\Delta y.$$

The rate of change of the flux is

$$\frac{d}{dt}(\Delta\,\Phi) = \frac{dB}{dt}\Delta S + B\frac{d}{dt}(\Delta S)$$

$$= \left(\frac{d\boldsymbol{B}}{dt} + \boldsymbol{B}(\nabla\boldsymbol{V}) - (\boldsymbol{B}\cdot\nabla)\boldsymbol{V}\right)_z \Delta S. \tag{6.131}$$

By means of the relation $\nabla \times (\boldsymbol{V} \times \boldsymbol{B}) = -\boldsymbol{B}(\nabla\boldsymbol{V}) + (\boldsymbol{B}\cdot\nabla)\boldsymbol{V} - (\boldsymbol{V}\cdot\nabla)\boldsymbol{B}$, eq. (6.129) is reduced to

$$\frac{\partial \boldsymbol{B}}{\partial t} + (\boldsymbol{V}\cdot\nabla)\boldsymbol{B} + \boldsymbol{B}(\nabla\boldsymbol{V}) - (\boldsymbol{B}\cdot\nabla)\boldsymbol{V} = \frac{1}{\sigma\mu_0}\Delta\boldsymbol{B}. \tag{6.132}$$

When $R_m \to \infty$ or $\sigma \to \infty$ the rate of change of flux becomes zero, i.e., $d(\Delta\Phi)/dt \to 0$. This means that the magnetic flux is frozen in the plasma. Using eqs. (6.124) and (6.125), we can simplify the magnetohydrodynamic equations to

$$\rho_m \frac{d\boldsymbol{V}}{dt} = -\nabla p - \frac{1}{2\mu_0}\nabla B^2 + \frac{1}{\mu_0}(\boldsymbol{B}\cdot\nabla)\boldsymbol{B} \tag{6.133}$$

$$\frac{d\boldsymbol{B}}{dt} + \boldsymbol{B}(\nabla\boldsymbol{V}) - (\boldsymbol{B}\cdot\nabla)\boldsymbol{V} = \frac{1}{\sigma\mu_0}\Delta\boldsymbol{B} \tag{6.134}$$

$$\nabla\boldsymbol{B} = 0 \tag{6.135}$$

$$\frac{d}{dt}(p\rho_m{}^{-r}) = 0 \quad \text{or} \quad \nabla V = 0. \tag{6.136}$$

References

1. For example, D. J. Rose and M. Clark, Jr: Plasmas and Controlled Fusion, MIT Press, Cambridge, Mass., 1961

2. S. I. Braginskii: Rev. of Plasma Phys. **1**, 205 (ed. by M. A. Leontovich), Consultants Bureau, New York, 1966

3. G. F. Chew, M. L. Goldberger, and F. E. Low: Proc. Roy. Soc. A**236**, 112(1956)

7 Equilibrium

In order to maintain a hot plasma, we must keep it away from the vacuum-container wall. The most promising method for such confinement of a hot plasma is the use of appropriate magnetic fields. If plasmas are to be confined successfully, an equilibrium state must exist. In sec. 7.1, the conditions for plasma equilibrium are studied; magnetohydrodynamic phenomena such as adiabatic compression, shock, and diffusion are discussed in secs. 7.2 and 7.3. In sec. 7.4, equilibrium conditions of a toroidal plasma are developed. Equilibrium equations in the axisymmetric and helically symmetric systems are explained in secs. 7.5 and 7.6, respectively. Finally, in secs. 7.7 and 7.8, the foregoing formulations are applied to plasmas confined in tokamak and stellarator configurations.

7.1 Pressure Equilibrium

When a plasma is in the steady state, eq. (6.42b) yields the *equilibrium equation*

$$\nabla p = \boldsymbol{j} \times \boldsymbol{B}, \tag{7.1}$$

i.e., it is assumed that the electric-field term is negligible. In the stationary state, Maxwell's equations are

$$\nabla \times \boldsymbol{B} = \mu_0 \boldsymbol{j} \tag{7.2}$$

$$\nabla \boldsymbol{B} = 0 \tag{7.3}$$

$$\nabla \boldsymbol{j} = 0. \tag{7.4}$$

From the equilibrium equation (7.1), it follows that

$$\boldsymbol{B} \cdot \nabla p = 0 \tag{7.5}$$

$$\boldsymbol{j} \cdot \nabla p = 0. \tag{7.6}$$

Equation (7.5) indicates that $\boldsymbol{B}$ and ∇p are orthogonal, and the surfaces of constant pressure coincide with the magnetic surfaces. Equation (7.6) shows that the current-density vector $\boldsymbol{j}$ is everywhere parallel to the constant-pressure surfaces. Substitution of eq. (7.2) into eq. (7.1) yields

$$\nabla\left(p + \frac{B^2}{2\mu_0}\right) = (\boldsymbol{B} \cdot \nabla)\frac{\boldsymbol{B}}{\mu_0}. \tag{7.7a}$$

By using the vector relations

$$(\boldsymbol{B} \times (\nabla \times \boldsymbol{B}) + (\boldsymbol{B}\cdot\nabla)\boldsymbol{B} = \nabla \boldsymbol{B}^2/2)$$

$$(\boldsymbol{B}\cdot\nabla)\boldsymbol{B} = B^2[(\boldsymbol{b}\cdot\nabla)\boldsymbol{b} + \boldsymbol{b}((\boldsymbol{b}\cdot\nabla)B)/B] = \boldsymbol{B}^2(-\boldsymbol{n}/\boldsymbol{R} - (\nabla\boldsymbol{b})\boldsymbol{b})$$

(R is the radius of curvature of the line of magnetic force and $\boldsymbol{n}$ is the unit vector directed toward a point on the line of magnetic force from the center of curvature), we find that the right-hand side of eq. (7.7a) can be neglected when the radius of curvature is much larger than the length over which the magnitude B changes appreciably, i.e., the size of the plasma, and the variation of $\boldsymbol{B}$ along the line of magnetic force is much smaller than the variation of $\boldsymbol{B}$ in the perpendicular direction. Then eq. (7.7a) becomes

$$p + \frac{B^2}{2\mu_0} \approx \frac{B_0{}^2}{2\mu_0}. \tag{7.8a}$$

When the system is axially symmetric and $\partial/\partial z = 0$, eq. (7.7a) exactly reduces to

$$\frac{\partial}{\partial r}\left(p + \frac{B_z{}^2 + B_\theta{}^2}{2\mu_0}\right) = -\frac{B_\theta{}^2}{r\mu_0}. \tag{7.7b}$$

By integrating by parts we obtain

$$\left(p + \frac{B_z{}^2 + B_\theta{}^2}{2\mu_0}\right)_{r=a} = \frac{1}{\pi a^2}\int_0^a \left(p + \frac{B_z{}^2}{2\mu_0}\right) 2\pi r \mathrm{d}r,$$

i.e.,

$$\langle p \rangle + \frac{\langle B_z{}^2 \rangle}{2\mu_0} = p_a + \frac{B_z{}^2(a)}{2\mu_0} + \frac{B_\theta{}^2(a)}{2\mu_0}. \tag{7.8b}$$

$\langle \quad \rangle$ is here the volume average. As $B^2/2\mu_0$ is the pressure of the magnetic field, eq. (7.8b) is the equation of pressure. The ratio of plasma pressure to the pressure of the external magnetic field B_0

$$\beta = \frac{p}{B_0{}^2/2\mu_0} = \frac{n(T_\mathrm{e} + T_\mathrm{i})}{B_0{}^2/2\mu_0} \tag{7.9}$$

is called the *beta ratio* (fig. 7.1). For a confined plasma, β is always smaller than 1, and is used as a figure of merit of the confining magnetic field. The fact that the internal magnetic field is smaller than the external field indicates the *diamagnetism* of the plasma.

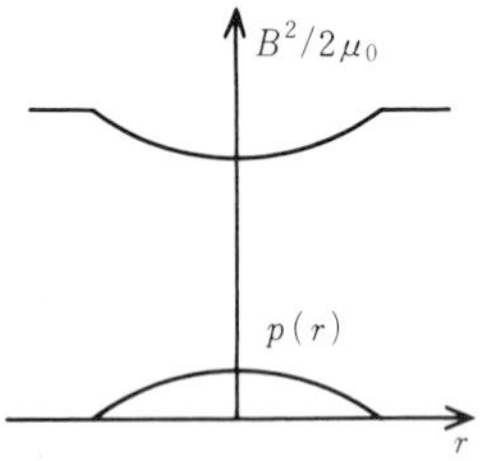

Fig. 7.1
Pressure of magnetic field and pressure of plasma.

7.2 The Hydrodynamic Description of a Plasma

7.2a Adiabatic Compression

As is clear from the pressure equation, the external magnetic field plays the role of a soft wall confining the plasma. By increasing the magnitude of the surrounding field, the plasma can be compressed. When the compression is slower than the energy-equipartition time, the compression is said to be three-dimensional. When the compression is perpendicular to the magnetic field, and the compression time is longer than the ion-cyclotron period but shorter than the relaxation time between the perpendicular and parallel temperatures, the compression is said to be two-dimensional. When the compression is parallel to the magnetic field, and the compression time is shorter than the relaxation time between the perpendicular and parallel temperatures but longer than the transit time of particles between mirror points, the compression is said to be one-dimensional. The relation of the plasma pressure p and the volume $\mathscr{V}$ in *adiabatic compression* is

$$p\mathscr{V}^{\gamma} = \text{const.} \tag{7.10}$$

The quantity γ is the ratio of specific heats and is given by

$$\gamma = \frac{2+\delta}{\delta}, \tag{7.11}$$

where δ is the number of degrees of freedom. The value of γ is 5/3 for $\delta = 3$, $\gamma = 2$ for $\delta = 2$, and $\gamma = 3$ for $\delta = 1$. If T is used instead of p, eq. (7.10) is

$$T\mathscr{V}^{\gamma-1} = \text{const.} \tag{7.12}$$

This equation shows that the plasma temperature can be raised by adiabatic compression. The smaller the number of degrees of freedom, the larger the increase of temperature.

7.2b Simple Shock

When a shock forms and travels across a plasma, there are discontinuities in the density, temperature, and flow velocity across the *shock front*. We assume that momentum and energy are not transported across the shock front. Here we do not try to study the internal structure of the shock. The relation the macroscopic quantities ahead of the shock front to those behind the front can be deduced by application of the laws of conservation of particles, momentum, and energy. Here we will treat *simple shock*, in which the plasma flows parallel to the magnetic-field lines and the interaction of the plasma and the magnetic field is negligible. Furthermore, it is assumed that the shock front does not move (in other words, we observe the shock in coordinates moving with the shock front). Let the mass density, the flow velocity, and the pressure ahead of the shock front be denoted by n, V, and p, respectively, and the quantities behind the front be denoted by n', V', and p' (fig. 7.2). Then the conservation laws are

$$\rho' V' = \rho V \tag{7.13}$$

$$\rho' V'^2 - \rho V^2 = p - p' \tag{7.14}$$

$$V'\left(\frac{\rho' V'^2}{2} + \frac{p'}{\gamma - 1}\right) - V\left(\frac{\rho V^2}{2} + \frac{p}{\gamma - 1}\right) = Vp - V'p'. \tag{7.15}$$

These are called the *Rankine-Hugoniot relations*. Introducing the velocity of sound c_s, we find

$$c_s{}^2 = \frac{\gamma p}{\rho} \qquad \frac{p}{\rho(\gamma - 1)} + \frac{p}{\rho} = \frac{c_s{}^2}{\gamma - 1}, \tag{7.16}$$

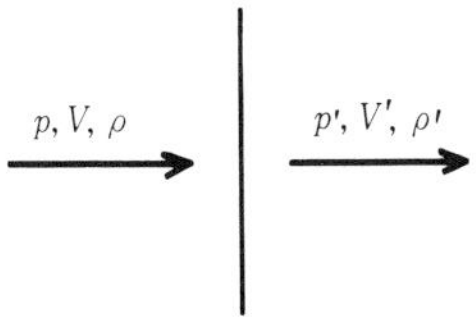

Fig. 7.2
Densities n, n', flow velocities V, V', and pressures p, p' ahead of and behind a shock front.

and the conservation laws become

$$\frac{\rho'}{\rho} = \frac{V}{V'} \tag{7.17}$$

$$\frac{\rho'}{\gamma} c_s'^2 + \rho' V'^2 = \frac{\rho}{\gamma} c_s^2 + \rho V^2 \tag{7.18}$$

$$\frac{V'^2}{2} + \frac{c_s'^2}{\gamma - 1} = \frac{V^2}{2} + \frac{c_s^2}{\gamma - 1}. \tag{7.19}$$

The Mach number, the ratio of the flow velocity to the velocity of sound, is

$$M = \frac{V}{c_s}. \tag{7.20}$$

A necessary condition for a shock to occur is $M > 1$. In terms of the Mach number, the Rankine-Hugoniot relations are written as

$$\frac{\rho'}{\rho} = \frac{1}{\alpha} \qquad \frac{V'}{V} = \alpha \tag{7.21}$$

$$\left(\frac{c_s'}{c_s}\right)^2 = 1 + \frac{\gamma - 1}{2}(1 - \alpha^2)M^2 \tag{7.22}$$

$$\alpha = \frac{2 + (\gamma - 1)M^2}{(\gamma + 1)M^2}. \tag{7.23}$$

As c_s is given by $c_s^2 = \gamma T/m_i$, the ratio $(c_s'/c_s)^2$ is equal to the ratio of the temperatures T'/T. When $\gamma = 5/3$, it follows that

$$\frac{\rho'}{\rho} = \frac{4}{1 + 3/M^2} \tag{7.24}$$

$$\frac{V'}{V} = \frac{1 + 3/M^2}{4} \tag{7.25}$$

$$\left(\frac{c_s'}{c_s}\right)^2 = 1 + \frac{5}{16}\left(1 - \frac{1}{M^2}\right)\left(1 + \frac{0.6}{M^2}\right)M^2. \tag{7.26}$$

These relations have been confirmed experimentally.[1] The temperature rise in the wake of a shock is high for large Mach number.

7.2c Plasma Flow

Let us study the variation of plasma parameters when a plasmoid runs along a uniform magnetic field in the z direction. When diffusion across the magnetic field can be neglected, the problem is one-dimensional. Let the line density, the flow velocity, and the temperature be $n(z,t)$, $V(z,t)$, and $T(z, t)$, respectively. Then the equation of continuity, the equation of motion, and the conservation of energy are described by

$$\frac{\partial n}{\partial t} + \frac{\partial}{\partial z}(nV) = 0 \tag{7.27}$$

$$nm_{\rm i}\left(\frac{\partial}{\partial t} + V\frac{\partial}{\partial z}\right)V = -\frac{\partial}{\partial z}(nT) \tag{7.28}$$

$$\frac{3}{2}\int_{-\infty}^{\infty} nT{\rm d}z + \frac{1}{2}\int_{-\infty}^{\infty} nm_{\rm i}V^2{\rm d}z = \text{const.} \tag{7.29}$$

(where $T = T_{\rm e} + T_{\rm i}$). Let us assume that the temperature is independent of z and that the density distribution is given by

$$n(z,t) = \frac{N_0}{\pi^{1/2}\lambda(t)}\exp\left(-\frac{(z - V_0 t)^2}{\lambda^2(t)}\right). \tag{7.30}$$

Then eqs. (7.27) and (7.28) are satisfied by

$$V = \frac{\dot{\lambda}}{\lambda}(z - V_0 t) + V_0 \tag{7.31}$$

$$T = \frac{m_{\rm i}}{2}\lambda\ddot{\lambda}. \tag{7.32}$$

($\dot{\lambda} \equiv {\rm d}\lambda/{\rm d}t$ and $\ddot{\lambda} = {\rm d}^2\lambda/{\rm d}t^2$). Equation (7.29) is reduced to

$$\frac{3N_0}{4}m_{\rm i}\lambda\ddot{\lambda} + \frac{N_0}{4}m_{\rm i}(\dot{\lambda})^2 + \frac{N_0}{2}m_{\rm i}V_0{}^2 = \frac{3}{2}N_0T_0 + \frac{N_0}{2}m_{\rm i}V_0{}^2. \tag{7.33}$$

Using the notation $\lambda_0 \equiv \lambda(0)$ and $t_0 \equiv (m_{\rm i}\lambda_0{}^2/6T_0)^{1/2}$, eq. (7.33) can be written as

$$3\lambda\frac{{\rm d}^2\lambda}{{\rm d}t^2} + \left(\frac{{\rm d}\lambda}{{\rm d}t}\right)^2 = \left(\frac{\lambda_0}{t_0}\right)^2, \tag{7.34}$$

and the solution is

$$\frac{t}{t_0} = \left(\left(\frac{\lambda}{\lambda_0}\right)^{2/3} - 1\right)^{1/2}\left(\left(\frac{\lambda}{\lambda_0}\right)^{2/3} + 2\right). \tag{7.35}$$

From eqs. (7.32), (7.34), and (7.31), we find[2]

$$\frac{T}{T_0} = 1 - \left(\frac{t_0/\lambda_0}{\mathrm{d}t/\mathrm{d}\lambda}\right)^2 = \left(\frac{\lambda_0}{\lambda}\right)^{2/3} \tag{7.36}$$

$$V = \frac{(z - V_0 t)}{t_0}\left(\frac{\lambda_0}{\lambda}\right)\left(1 - \left(\frac{\lambda_0}{\lambda}\right)^{2/3}\right)^{1/2} + V_0. \tag{7.37}$$

When $t/t_0 \gg 1$ (say $t > 4t_0$), the time variation of these parameters are given approximately by

$$\frac{\lambda}{\lambda_0} \approx \frac{t}{t_0} \qquad \frac{T}{T_0} \approx \left(\frac{t_0}{t}\right)^{2/3} \qquad V \approx \frac{z}{t}. \tag{7.38}$$

The width λ of the Gaussian distribution of the density is given by eqs. (7.35) and (7.38) and is proportional to t for large t. The temperature decreases according to eqs. (7.36) and (7.38). It is interesting to note that the flow velocity $V(z,t)$ of the plasma at z and t is given by z/t. These relations satisfy the relation of adiabatic expansion

$$\left(\frac{\partial}{\partial t} + V\frac{\partial}{\partial z}\right)(n^{1-\gamma}T) = 0, \tag{7.39}$$

where $\gamma = 5/3$.

These analytic equations are a special solution. Nevertheless, this solution expresses the general characteristics of the flow of the plasmoid and this simple analytic solution is very convenient to handle.[1]

7.3 Movement of Particles in a Plasma

7.3a Particle Movement in a Fully Ionized Plasma

When the distribution function of a plasma is isotropic and $\omega \ll \Omega_e$, the equation of motion and Ohm's law (eqs. (6.42b) and (6.43)) may be written as

$$\rho_m \frac{\mathrm{d}\boldsymbol{V}}{\mathrm{d}t} = -\nabla p + \rho \boldsymbol{E} + \boldsymbol{j} \times \boldsymbol{B} \tag{7.40}$$

$$E' \equiv E + V \times B + \frac{1}{en}\nabla p_e = \frac{1}{en}(j \times B) + \frac{j_{\parallel}}{\sigma_{\parallel}} + \frac{j_{\perp}}{\sigma_{\perp}}. \tag{7.41}$$

The electric conductivities are $\sigma_{\parallel} = 2n_e e^2/m_e \nu_e$, $\sigma_{\perp} = \sigma_{\parallel}/2$ (n being the density). From eq. (7.41), j is expressed in terms of E' as

$$j = \sigma_{\parallel} E_{\parallel}' + \frac{\sigma_{\perp}}{1 + (\Omega_e/\nu_e)^2}\left(E_{\perp}' + \frac{\Omega_e}{\nu_e}(b \times E')\right). \tag{7.42}$$

The relation for the component parallel to the magnetic field does not change from the field-free case, but the relation for the perpendicular component changes greatly. When $\Omega_e \gg \nu_e$, the perpendicular component of the current is nearly orthogonal to E' and the magnitude becomes very small (see fig. 7.3).

When $\omega \ll \Omega_i$, Ohm's law (eq. (6.43c)) is given by

$$E + V \times B - \frac{1}{en}\nabla p_i = \frac{1}{\sigma_{\parallel}} j_{\parallel} + \frac{1}{\sigma_{\perp}} j_{\perp}. \tag{7.43}$$

From this equation, we find

$$V_{\perp} = \frac{1}{B}\left(\left(E - \frac{1}{en}\nabla p_i\right) \times b\right) - \frac{1}{\sigma_{\perp} B^2} j_{\perp} \times B$$

$$= \frac{1}{B}\left(\left(E - \frac{1}{en}\nabla p_i\right) \times b\right) - \frac{m_e \nu_e}{ne^2 B^2}\nabla_{\perp} p, \tag{7.44}$$

i.e.

$$nV_{\perp} = \frac{1}{B}\left(\left(nE - \frac{T_i}{en}\nabla n\right) \times b\right) - (\rho_B{}^e)^2 \nu_e\left(1 + \frac{T_i}{T_e}\right)\nabla n, \tag{7.45}$$

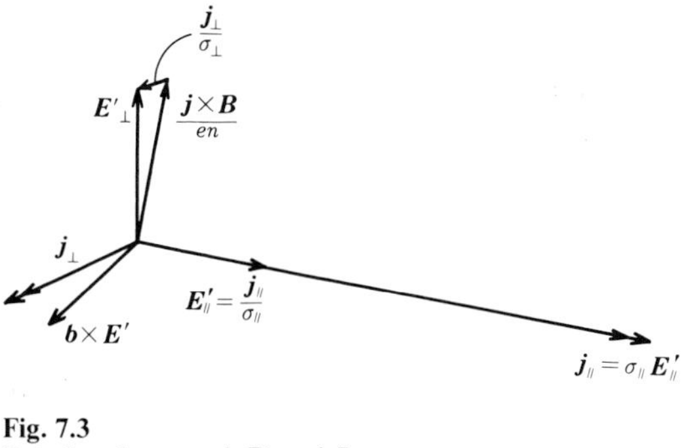

Fig. 7.3
Relations between j, E', and B.

where $\rho_B{}^{\rm e} = v_{\rm Te}/\Omega_{\rm e}$, $v_{\rm Te} = (T_{\rm e}/m_{\rm e})^{1/2}$. The second term of eq. (7.44) expresses the diffusion due to collisions, or the plasma resistivity. The first term is the drift due to the electric field $\boldsymbol{E}$ and the ion pressure. The parallel component of eq. (7.43) is

$$\boldsymbol{E}_{\parallel} - \frac{1}{en}\nabla_{\parallel} p_{\rm i} = \frac{1}{\sigma}\boldsymbol{j}_{\parallel}. \tag{7.46}$$

7.3b Particle Movement in a Magnetized Lorentzian Gas

Let us consider the *diffusion* and the *mobility* of charged particles in a gas which scatters from fixed centers. Such a system is called a *Lorentzian gas.* The equations of motion of ions and electrons are given (from eqs. (6.28) and (6.29)) as

$$mn\frac{\mathrm{d}\boldsymbol{V}}{\mathrm{d}t} = -\nabla p + qn(\boldsymbol{E} + \boldsymbol{V} \times \boldsymbol{B}) - mn\nu\boldsymbol{V}. \tag{7.47}$$

The difference from the case of a fully ionized plasma is in the collision term. The present collision term is applicable when the charged particles are scattered by heavy neutral atoms. When the cyclotron frequency is denoted by $\Omega = -qB/m$, the steady-state flow velocity $\boldsymbol{V}$ is expressed by

$$nV_\alpha = -\sum_{\beta=1}^{3}\frac{\partial}{\partial x_\beta}(D_{\alpha\beta}n) + \sum_{\beta=1}^{3}\mu_{\alpha\beta}nE_\beta, \tag{7.48}$$

where $D_{\alpha\beta}$ and $\mu_{\alpha\beta}$ are

$$[D_{\alpha\beta}] = \frac{T}{m}\begin{bmatrix} \dfrac{\nu}{\nu^2+\Omega^2} & \dfrac{-\Omega}{\nu^2+\Omega^2} & 0 \\[2ex] \dfrac{\Omega}{\nu^2+\Omega^2} & \dfrac{\nu}{\nu^2+\Omega^2} & 0 \\[2ex] 0 & 0 & \dfrac{1}{\nu} \end{bmatrix}$$

$$= \begin{bmatrix} D_\perp & -D_H & 0 \\ D_H & D_\perp & 0 \\ 0 & 0 & D_\parallel \end{bmatrix} \tag{7.49}$$

$$[\mu_{\alpha\beta}] = \frac{q}{T}[D_{\alpha\beta}]. \tag{7.50}$$

$D_{\alpha\beta}$ is called the *diffusion tensor* and $\mu_{\alpha\beta}$ is called the *mobility tensor*. Equation (7.50) is Einstein's relation, which is quite general. When $f(\boldsymbol{r}, \boldsymbol{v}, t)$ is the distribution function, Boltzmann's equation is

$$\frac{\partial f}{\partial t} + \boldsymbol{v}\cdot\nabla_r f + \frac{q}{m}(\boldsymbol{E} + \boldsymbol{v}\times\boldsymbol{B})\cdot\nabla_v f = \left(\frac{\delta f}{\delta t}\right)_{\text{coll}}. \tag{7.51}$$

Let the zeroth-order distribution function be $f_0(\boldsymbol{r}, \boldsymbol{v}) = f_0(\boldsymbol{r}, v_{\parallel}, v_{\perp})$ in the equilibrium state without collision. This function satisfies

$$(\boldsymbol{v}\times\boldsymbol{B})\nabla_v f_0 = 0. \tag{7.52}$$

It was assumed that $(\delta f_0/\delta t)_{\text{coll}} = 0$ and the electric field is a 1st-order quantity. Let the distribution function be $f = f_0 + f_1$ and linearize Boltzmann's equation with regard to f_1 as

$$\left(\boldsymbol{v}\cdot\nabla_r + \frac{q}{m}(\boldsymbol{v}\times\boldsymbol{B})\cdot\nabla_v + \left(\frac{\delta}{\delta t}\right)_{\text{coll}}\right)f_1 = -\left(\boldsymbol{v}\cdot\nabla_r f_0 + \frac{q}{m}\boldsymbol{E}\cdot\nabla_v f_0\right). \tag{7.53}$$

When f_0 is Maxwellian,

$$f_0(\boldsymbol{v}, \boldsymbol{r}) = n(\boldsymbol{r})\left(\frac{m}{2\pi T}\right)^{3/2}\exp\left(-\frac{mv^2}{2T}\right), \tag{7.54}$$

the right-hand side of eq. (7.53) is

$$-\left(\frac{1}{n}\nabla n - \frac{q}{T}\boldsymbol{E}\right)\cdot\boldsymbol{v}f_0. \tag{7.55}$$

As the operator on f_1 in the left-hand side of eq. (7.53) is linear, the solution is

$$f_1 = \boldsymbol{\Phi}(\boldsymbol{v})\cdot\left(\frac{1}{n}\nabla n - \frac{q}{T}\boldsymbol{E}\right). \tag{7.56}$$

The average velocity of the charged particles is

$$n\boldsymbol{V} = -D\nabla n + \mu n\boldsymbol{E}, \tag{7.57}$$

where

$$D = -\frac{1}{n}\int \boldsymbol{v}\boldsymbol{\Phi}\,\mathrm{d}\boldsymbol{v} \tag{7.58}$$

$$\mu = -\frac{q}{Tn}\int \boldsymbol{v}\Phi d\boldsymbol{v}. \tag{7.59}$$

Therefore we have Einstein's relation

$$\mu = \frac{q}{T}D. \tag{7.60}$$

We did not impose any restriction on the form of the collision term in the derivation of eq. (7.60), so that scattering of the particles by a random electric field may be included in the foregoing analysis.[3] However, the Maxwell distribution was assumed since the relation $\nabla_v f_0 = -(m\boldsymbol{v}/T)f_0$ was used.

The diagonal perpendicular component $D_\perp$ of the diffusion tensor is smaller than the parallel component $D_\parallel$ by the factor $[1 + (\Omega/\nu)^2]^{-1}$. When $\Omega \gg \nu$, then $D_\perp = \rho_B{}^2\nu$ (ρ_B is the Larmor radius), and $D_\parallel = T/(m\nu) = \lambda^2\nu$ ($v_T = (T/m)^{1/2}$, $\lambda = v_T/\nu$). Accordingly, the perpendicular diffusion coefficient $D_\perp$ for ions in a Lorentzian gas is larger than $D_\perp$ for electrons; but the parallel coefficient $D_\parallel$ for electrons is larger than that for ions.

When the magnetic field is in the z direction and the density gradient and the electric field are in the x direction, the diffusions of ions and electrons are given by

$$\left.\begin{aligned} nV_x{}^{\mathrm{i}} &= -D_\perp{}^{\mathrm{i}}\left(\frac{\partial n}{\partial x} - \frac{e}{T_{\mathrm{i}}}nE_x\right) \\ nV_x{}^{\mathrm{e}} &= -D_\perp{}^{\mathrm{e}}\left(\frac{\partial n}{\partial x} + \frac{e}{T_{\mathrm{e}}}nE_x\right). \end{aligned}\right\} \tag{7.61}$$

In the steady state, $V_x{}^{\mathrm{i}}$ must be equal to $V_x{}^{\mathrm{e}}$, so that

$$\frac{\partial n}{\partial x} = \frac{-\left(\dfrac{D_\perp{}^{\mathrm{e}}}{T_{\mathrm{e}}} + \dfrac{D_\perp{}^{\mathrm{i}}}{T_{\mathrm{i}}}\right)}{(D_\perp{}^{\mathrm{e}} - D_\perp{}^{\mathrm{i}})} enE_x, \tag{7.62}$$

i.e.,

$$\begin{aligned} nV_x &= nV_x{}^{\mathrm{i}} = nV_x{}^{\mathrm{e}} \\ &= -D_\perp{}^{\mathrm{i}}D_\perp{}^{\mathrm{e}}\frac{\left(\dfrac{1}{T_{\mathrm{e}}} + \dfrac{1}{T_{\mathrm{i}}}\right)}{\left(\dfrac{D_\perp{}^{\mathrm{e}}}{T_{\mathrm{e}}} + \dfrac{D_\perp{}^{\mathrm{i}}}{T_{\mathrm{i}}}\right)}\frac{\partial n}{\partial x}. \end{aligned} \tag{7.63}$$

When $D_\perp{}^{\rm i}/T_{\rm i} \gg D_\perp{}^{\rm e}/T_{\rm e}$, the particle flow is

$$nV_x = -D_\perp{}^{\rm e}\left(1 + \frac{T_{\rm i}}{T_{\rm e}}\right)\frac{\partial n}{\partial x}. \tag{7.64}$$

It is clear that the diffusion of a plasma is determined by the smaller of the ion and electron diffusion coefficients. Such diffusion is called *ambipolar diffusion.*

7.4 Relations of Plasma Equilibrium

7.4a Virial Theorem

By using the relation $\mu_0 \boldsymbol{j} = \nabla \times \boldsymbol{B}$, we may write the equation of equilibrium $\boldsymbol{j} \times \boldsymbol{B} = \nabla P$ in the form

$$\sum_i \frac{\partial}{\partial x_i} T_{ik} - \sum_i \frac{\partial}{\partial x_i} P_{ik} = 0, \tag{7.65}$$

where

$$T_{ik} = \frac{1}{\mu_0}\left(B_i B_k - \frac{1}{2}B^2\delta_{ik}\right). \tag{7.66}$$

This is called the *magnetic stress tensor* (see sec. 2.1). Since the relations

$$\sum_i \frac{\partial}{\partial x_i}(x_k(T_{ik} - P_{ik})) = \sum_i (T_{ik} - P_{ik})\delta_{ik} + x_k \sum_i \frac{\partial}{\partial x_i}(T_{ik} - P_{ik})$$

$$= \sum_i (T_{ik} - P_{ik})\delta_{ik} \tag{7.67}$$

hold, it follows that

$$\int\left(3p + \frac{B^2}{2\mu_0}\right)\mathrm{d}V = \oint\left(\left(p + \frac{B^2}{2\mu_0}\right)\boldsymbol{r}\,\mathrm{d}\boldsymbol{S} - \frac{(\boldsymbol{B}\cdot\boldsymbol{r})(\boldsymbol{B}\cdot\mathrm{d}\boldsymbol{S})}{\mu_0}\right) \tag{7.68}$$

for the scalar pressure p. This is called the *virial theorem.* When a plasma fills a finite region with $p = 0$ outside the region, and where no solid conductor carries the current anywhere inside or outside the plasma, the magnitude of the magnetic field is of the order of $1/r^3$, so the surface integral approaches zero as the plasma surface approaches infinity ($r \to \infty$). This contradicts the fact that the volume integral of eq. (7.68) is positive definite. In other words, a plasma of the finite extent cannot be in equilibrium unless there exist solid conductors to carry the current.

7.4b Anisotropic Pressure Equilibrium (Equilibrium of a Mirror Field)

The anisotropic pressure tensor (sec. 6.4) may be written as

$$P = p_\perp \boldsymbol{I} + (p_\parallel - p_\perp)\boldsymbol{bb}. \tag{7.69}$$

The right-hand side of the equilibrium condition $\boldsymbol{j} \times \boldsymbol{B} = \nabla P$ is given by

$$\begin{aligned}(\nabla P)_j &= \sum_k \frac{\partial P_{jk}}{\partial x_k} \\ &= \sum_k \frac{\partial p_\perp}{\partial x_k}\delta_{jk} + \sum_k \frac{\partial (p_\parallel - p_\perp)}{\partial x_k} b_j b_k + (p_\parallel - p_\perp)\sum_k \frac{\partial}{\partial x_k}(b_j b_k) \\ &= \frac{\partial p_\perp}{\partial x_j} + b_j \boldsymbol{b}\cdot\nabla(p_\parallel - p_\perp) + (p_\parallel - p_\perp)((\boldsymbol{b}\cdot\nabla)b_j + b_j(\nabla \boldsymbol{b})).\end{aligned} \tag{7.70}$$

The parallel component $(\nabla P)_\parallel$ is expressed by

$$(\nabla P)_\parallel = \frac{\partial}{\partial s} p_\perp + \frac{\partial}{\partial s}(p_\parallel - p_\perp) + (p_\parallel - p_\perp)\frac{-1}{B}\frac{\partial B}{\partial s} \tag{7.71}$$

(s is length measured along a line of the magnetic field). So the parallel component of the equilibrium condition is

$$\frac{\partial}{\partial s} p_\parallel + \frac{p_\perp - p_\parallel}{B}\frac{\partial B}{\partial s} = 0. \tag{7.72}$$

Since the perpendicular component $(\nabla P)_\perp$ is

$$(\nabla P)_\perp = \nabla_\perp p_\perp + (p_\parallel - p_\perp)(\boldsymbol{b}\cdot\nabla)\boldsymbol{b}, \tag{7.73}$$

the perpendicular component of the equilibrium condition is

$$\begin{aligned}\boldsymbol{j}_\perp &= \frac{\boldsymbol{b} \times (\nabla P)_\perp}{B} \\ &= \frac{\boldsymbol{b}}{B} \times (\nabla_\perp p_\perp + (p_\parallel - p_\perp)(\boldsymbol{b}\cdot\nabla)\boldsymbol{b}).\end{aligned} \tag{7.74}$$

Using the relations

$$(\boldsymbol{b}\cdot\nabla)\boldsymbol{b} = \nabla(\boldsymbol{b}^2)/2 + (\nabla \times \boldsymbol{b}) \times \boldsymbol{b} \qquad \nabla \times \boldsymbol{b} = \nabla \times \boldsymbol{B}/B + (\nabla \cdot B^{-1}) \times \boldsymbol{B},$$

we can write eq. (7.74) in the form

$$\boldsymbol{j}_\perp = \frac{\boldsymbol{b}}{B} \times \nabla_\perp p_\perp + \frac{p_\parallel - p_\perp}{B}\left(\boldsymbol{B} \times \frac{\nabla B}{B^2}\right) + \frac{(p_\parallel - p_\perp)}{B^2}\mu_0 \boldsymbol{j}_\perp, \tag{7.75}$$

i.e.,

$$\tau \boldsymbol{j}_\perp = \frac{\boldsymbol{b}}{B} \times \left(\nabla_\perp p_\perp + \frac{(p_\parallel - p_\perp)}{B} \nabla_\perp B \right) \tag{7.76}$$

$$\tau \equiv 1 + \frac{\mu_0 (p_\perp - p_\parallel)}{B^2}. \tag{7.77}$$

The divergence of $\tau \boldsymbol{j}_\perp$ is

$$\begin{aligned}
\nabla \cdot (\tau \boldsymbol{j}_\perp) &= \nabla \cdot \left(\frac{\boldsymbol{B}}{B^2} \times \nabla p_\perp \right) + \nabla \cdot \left(\frac{(p_\parallel - p_\perp)}{B^3} \boldsymbol{B} \times \nabla B \right) \\
&= \nabla p_\perp \cdot \left(\nabla \left(\frac{1}{B^2} \right) \times \boldsymbol{B} \right) + \frac{1}{B^2} \nabla p_\perp \cdot (\nabla \times \boldsymbol{B}) \\
&\quad + \nabla B \cdot \left(\nabla \left(\frac{p_\parallel - p_\perp}{B^3} \right) \times \boldsymbol{B} \right) + \frac{p_\parallel - p_\perp}{B^3} \nabla B \cdot (\nabla \times \boldsymbol{B}) \\
&= \nabla (p_\parallel + p_\perp) \cdot \frac{\boldsymbol{B} \times \nabla B}{B^3} + \left(\frac{1}{B^2} \nabla p_\perp + \frac{p_\parallel - p_\perp}{B^3} \nabla B \right) \cdot (\nabla \times \boldsymbol{B}) \\
&= \nabla (p_\parallel + p_\perp) \cdot \frac{\boldsymbol{B} \times \nabla B}{B^3} + \nabla_\parallel (p_\parallel + p_\perp) \frac{\nabla \times \boldsymbol{B}}{B^2}.
\end{aligned} \tag{7.78}$$

Here the relations $\boldsymbol{j} \cdot (\nabla P)_\perp = (\nabla P)_\perp \cdot (\nabla \times \boldsymbol{B}) = 0$ have been used. From $\nabla \cdot \boldsymbol{j} = 0$, the expression of $\nabla \cdot (\tau \boldsymbol{j}_\parallel)$ is given by

$$\begin{aligned}
\nabla \cdot (\tau \boldsymbol{j}_\parallel) &= -\nabla \cdot (\tau \boldsymbol{j}_\perp) + \boldsymbol{j} \cdot \nabla \left(\frac{\mu_0 (p_\perp - p_\parallel)}{B^2} \right) \\
&= -\nabla (p_\parallel + p_\perp) \frac{\boldsymbol{B} \times \nabla B}{B^3} - \nabla (p_\parallel + p_\perp) \cdot \frac{\mu_0 \boldsymbol{j}_\parallel}{B^2} + \nabla (p_\perp - p_\parallel) \cdot \frac{\mu_0 \boldsymbol{j}}{B^2} \\
&\quad - 2 (p_\perp - p_\parallel) \frac{\nabla B \cdot \mu_0 \boldsymbol{j}}{B^3} \\
&= -\nabla (p_\parallel + p_\perp) \cdot \frac{\boldsymbol{B} \times \nabla B}{B^3} + \left(\nabla (p_\perp - p_\parallel) - 2 (p_\perp - p_\parallel) \frac{\nabla B}{B} \right) \cdot \frac{\mu_0 \boldsymbol{j}_\perp}{B^2} \\
&= -\nabla (p_\parallel + p_\perp) \cdot \frac{\boldsymbol{B} \times \nabla B}{B^3} - \nabla (p_\parallel + p_\perp) \cdot \frac{\mu_0 \boldsymbol{j}_\perp}{B^2}.
\end{aligned} \tag{7.79}$$

Here eq. (7.72) and $(\nabla P)_\perp \cdot \boldsymbol{j}_\perp = 0$ have been used. Using

$$(\boldsymbol{b}\cdot\nabla)\boldsymbol{b}\times\boldsymbol{B} = B((\nabla\times\boldsymbol{b})\times\boldsymbol{b})\times\boldsymbol{b} = -B(\nabla\times\boldsymbol{b})_{\perp}$$
$$= -B(\nabla\times\boldsymbol{b} - \mu_0\boldsymbol{j}_{\parallel}/B),$$

we can reduce the expression for $\nabla(\tau\boldsymbol{j})$ to

$$\nabla(\tau\boldsymbol{j}_{\parallel}) = -\frac{\nabla(p_{\parallel}+p_{\perp})}{B}\cdot\left(\frac{\nabla\times\boldsymbol{B}}{\boldsymbol{B}} - \frac{\nabla B\times\boldsymbol{B}}{B^2} - \frac{\mu_0\boldsymbol{j}_{\parallel}}{B}\right)$$

$$= \frac{\nabla(p_{\parallel}+p_{\perp})}{B^2}\cdot((\boldsymbol{b}\cdot\nabla)\boldsymbol{b}\times\boldsymbol{B}). \tag{7.80}$$

By means of

$$B^2\tau\boldsymbol{B}\times(\boldsymbol{b}\cdot\nabla)\boldsymbol{b} = \left((\nabla\times\boldsymbol{B})_{\perp} + \left(\boldsymbol{B}\times\frac{\nabla B}{B}\right)\right)B^2\tau = \mu_0\boldsymbol{j}_{\perp}\tau B^2 + \tau\boldsymbol{B}\times\frac{\nabla B^2}{2}$$

$$= \mu_0\boldsymbol{B}\times\nabla_{\perp}\left(p_{\perp} + \frac{B^2}{2\mu_0}\right), \tag{7.81}$$

we reduce $\nabla(\tau\boldsymbol{j}_{\parallel})$ to

$$\nabla(\tau\boldsymbol{j}_{\parallel}) = \nabla(p_{\parallel}+p_{\perp})\cdot\left(\nabla\left(p_{\perp}+\frac{B^2}{2\mu_0}\right)\times\boldsymbol{B}\right)\frac{\mu_0}{\tau B^4}. \tag{7.82}$$

By writing $\tau\boldsymbol{j}_{\parallel} = \lambda\boldsymbol{B}$ and using $\nabla(\lambda\boldsymbol{B}) = \boldsymbol{B}\cdot\nabla\lambda$, we find

$$\frac{\mathrm{d}\lambda}{\mathrm{d}s} = \nabla(p_{\parallel}+p_{\perp})\cdot\left(\nabla\left(p_{\perp}+\frac{B^2}{2\mu_0}\right)\times\boldsymbol{B}\right)\frac{\mu_0}{\tau B^5}\Bigg). \tag{7.83}$$

When a magnetic-field line is closed or penetrates the plasma in the finite region, the integral of eq. (7.83) along the field line is zero, i.e.,

$$\oint\frac{\nabla(p_{\parallel}+p_{\perp})\cdot(\nabla(p_{\perp}+B^2/2\mu_0)\times\boldsymbol{B})\mu_0}{\tau B^5}\,\mathrm{d}s = 0. \tag{7.84}$$

Equations (7.72) and (7.84) are equilibrium conditions.[4] When the pressure is isotropic, the equilibrium conditions are

$$\frac{\partial p}{\partial s} = 0 \tag{7.85}$$

$$\oint\frac{\nabla p\cdot(\nabla(p+B^2/2\mu_0)\times\boldsymbol{B})}{B^5}\,\mathrm{d}s = 0. \tag{7.86}$$

When the beta ratio is small, eq. (7.84) is reduced to

$$\oint \frac{\nabla(p_{\parallel} + p_{\perp})\cdot(\nabla B \times \boldsymbol{B})}{B^4} \mathrm{d}s = 0. \tag{7.87}$$

The equations (7.78) and (7.80) for $\boldsymbol{j}_{\perp}$ and $\boldsymbol{j}_{\parallel}$ in terms of ∇p are used to reduce the plasma diffusion coefficient (see sec. 8.1).

Let us consider the equilibrium of a plasma confined in a mirror field as an example of anisotropic pressure. As the magnetic moment μ_{m} is conserved, one can specify the distribution function $\rho_{\mathrm{m}}(\mu_{\mathrm{m}}, W)$ of particles in terms of magnetic moment μ_{m} and the total energy W as follows:

$$p_{\perp} \propto \int \rho_{\mathrm{m}}(\mu_{\mathrm{m}}, W)\frac{\mu_{\mathrm{m}}B}{2}\mathrm{d}\mu_{\mathrm{m}}\mathrm{d}W \tag{7.88}$$

$$p_{\parallel} \propto \int \rho_{\mathrm{m}}(\mu_{\mathrm{m}}, W)(W - \mu_{\mathrm{m}}B)\mathrm{d}\mu_{\mathrm{m}}\mathrm{d}W. \tag{7.89}$$

ρ_{m} is proportional to the number of particles $f(\mu_{\mathrm{m}}, W, s)$ on a line of magnetic force, the density of the lines of magnetic force, and the time $\mathrm{d}t$ during which a particle stays in $\mathrm{d}s$, i.e., the time $\mathrm{d}t \propto \mathrm{d}s/(W - \mu_{\mathrm{m}}B)^{1/2}$. Consequently the perpendicular and parallel pressures are given by

$$p_{\perp} = \int f(\mu_{\mathrm{m}}, W, s)\frac{\mu_{\mathrm{m}}B^2}{2(W - \mu_{\mathrm{m}}B)^{1/2}}\mathrm{d}\mu_{\mathrm{m}}\mathrm{d}W \tag{7.90}$$

$$p_{\parallel} \infty \int f(\mu_{\mathrm{m}}, W, s)B(W - \mu_{\mathrm{m}}B)^{1/2}\mathrm{d}\mu_{\mathrm{m}}\mathrm{d}W. \tag{7.91}$$

These expressions satisfy eq. (7.72). As the drift velocity of the guiding center is

$$\boldsymbol{V}_{\mathrm{D}} = \frac{m}{e}\frac{\boldsymbol{B} \times \nabla B}{B^3}\left(\frac{1}{2}v_{\perp}{}^2 + v_{\parallel}{}^2\right), \tag{7.92}$$

the current $\boldsymbol{j}_{\mathrm{D}}$ induced by the drift is

$$\boldsymbol{j}_{\mathrm{D}} = \frac{(\boldsymbol{B} \times \nabla B)}{B^3}(p_{\parallel} + p_{\perp}) \tag{7.93}$$

$$\nabla \boldsymbol{j}_{\mathrm{D}} = \frac{\nabla(p_{\parallel} + p_{\perp})\cdot(\boldsymbol{B} \times \nabla B)}{B^3}. \tag{7.94}$$

Consequently the second equilibrium condition (7.84) is reduced to

$$\int \nabla \boldsymbol{j}_{\mathrm{D}}\frac{\mathrm{d}s}{B} = 0. \tag{7.95}$$

When the cross section of a small magnetic tube of flux $d\Phi$ is denoted by dA, *then* $d\Phi = B\,dA$ and eq. (7.95) becomes the volume integral

$$\int_{\text{flux tube}} \nabla \boldsymbol{j}_{\mathrm{D}} d\tau = 0. \tag{7.96}$$

When the system is an axisymmetric mirror, ∇p, ∇B, and $\boldsymbol{B}$ are coplaner and eq. (7.94) is zero; consequently the condition (7.84) is automatically satisfied.

7.4c Plasma Diffusion vs. Current Density and Resistivity

The steady-state equation of continuity and Ohm's law are

$$\nabla(p\boldsymbol{V}) = Q \tag{7.97}$$

$$\boldsymbol{E} + \boldsymbol{V} \times \boldsymbol{B} = \eta \boldsymbol{j} - \frac{T_{\mathrm{i}}}{e}\frac{\nabla p}{p}. \tag{7.98}$$

From $\nabla \times \boldsymbol{E} = 0$, $\boldsymbol{E}$ is expressed by $-\nabla\phi$; then eq. (7.98) is

$$\nabla f + \boldsymbol{V} \times \boldsymbol{B} = \eta \boldsymbol{j}, \tag{7.99}$$

where

$$f = -\phi + \frac{T_{\mathrm{i}}}{e}\ln p, \tag{7.100}$$

so that

$$\boldsymbol{B}\cdot\nabla f = \eta \boldsymbol{j}\cdot\boldsymbol{B} \tag{7.101}$$

$$\boldsymbol{V}_{\perp} = \frac{1}{B^2}(\nabla f - \eta \boldsymbol{j}) \times \boldsymbol{B}. \tag{7.102}$$

When $\boldsymbol{V}_{\parallel}$ is given by $\alpha\boldsymbol{B}$, then $\boldsymbol{V} = \boldsymbol{V}_{\perp} + \alpha\boldsymbol{B}$. From eq. (7.97) we find

$$\nabla(p\alpha\boldsymbol{B}) = Q + \nabla\left(\frac{p}{B^2}(\boldsymbol{B} \times \nabla f + \eta\nabla p)\right). \tag{7.103}$$

As the left-hand side of eq. (7.103) becomes $\boldsymbol{B}\cdot\nabla(\alpha p) = B\,d(\alpha p)/ds$, eq. (7.103) has the form of a *magnetic differential equation*

$$\boldsymbol{B}\cdot\nabla g = h. \tag{7.104}$$

When the line of magnetic force is closed, the loop integral is zero:

$$\oint \frac{h}{B} \mathrm{d}s = 0, \tag{7.105}$$

since the function g is a single-valued function. When the field lines are not closed, and form magnetic surfaces ergodically, we integrate eq. (7.104) in the shell bounded by two magnetic surfaces ψ and $\psi + \mathrm{d}\psi$ in which the plasma pressures are p and $p + \mathrm{d}p$, respectively. Since the left-hand side is equal to $\nabla(g\boldsymbol{B})$, the integral of the left-hand side is zero due to Gauss's theorem. Accordingly we find

$$\int_\psi h \frac{\mathrm{d}S}{|\nabla p|} = 0. \tag{7.106}$$

(The thickness $\mathrm{d}t$ of the shell is given by $|\nabla p| \mathrm{d}t = \mathrm{d}p = \text{const.}$) When this formula is applied to eqs. (7.101) and (7.103), we find

$$\int_\psi \frac{\mathrm{d}S}{|\nabla p|} (\boldsymbol{j} \cdot \boldsymbol{B}) = 0 \tag{7.107}$$

$$\int_\psi \frac{\mathrm{d}S}{|\nabla p|} \left(Q + \nabla \left(\frac{p}{B^2} (\boldsymbol{B} \times \nabla f + \eta \nabla p) \right) \right) = 0. \tag{7.108}$$

Multiply eq. (7.108) by $\mathrm{d}p$ and extend the integral region to the whole region inside a magnetic surface. By use of Gauss's theorem, we can reduce eq. (7.108) to

$$\int \mathrm{d}VQ + \int_\psi \mathrm{d}S \frac{\mp \nabla p}{|\nabla p|} \cdot (\boldsymbol{B} \times \nabla f + \eta \nabla p) \frac{p}{B^2} = 0, \tag{7.109}$$

where $\mp \nabla p / |\nabla p|$ is the outward unit vector normal to the magnetic surface (the pressure usually decreases outward, so that the minus sign is usual). Substitution of $\nabla p = \boldsymbol{j} \times \boldsymbol{B}$ into eq. (7.109) and expansion of the scalar product of two vectors which are expressed by vector products yields

$$\int \mathrm{d}VQ \mp \int_\psi \frac{\mathrm{d}S}{|\nabla p|} (-\boldsymbol{j} \cdot \nabla f + \eta \boldsymbol{j}^2) p = 0, \tag{7.110}$$

using eq. (7.101). The integral of the ∇f term can be reduced to the volume integral of the region bounded by two adjacent magnetic surfaces by using $\mathrm{d}t = \mp \mathrm{d}p / |\nabla p|$. The relation $\boldsymbol{j} \cdot \nabla f = \nabla (f \boldsymbol{j})$ reduces the volume integral to the surface integral. Since the relation $\nabla p \cdot \boldsymbol{j} = 0$ holds, the integral of the ∇f term is zero. Accordingly, we find

$$\int_V \mathrm{d}VQ = p\int_\psi \frac{\mathrm{d}S}{|\nabla p|}\eta \boldsymbol{j}^2. \tag{7.111}$$

From eq. (7.97), eq. (7.111) becomes[5]

$$\int_\psi \boldsymbol{V}\cdot\boldsymbol{n}\mathrm{d}S = \int_\psi \frac{\eta \boldsymbol{j}^2}{|\nabla p|}\mathrm{d}S. \tag{7.112}$$

The vector $\boldsymbol{n}$ is the outward unit normal vector of the magnetic surface. This relation is important since it allows us to estimate the amount of outward plasma flux across a magnetic surface from the specific resistivity and the plasma current.

7.5 Toroidal Equilibrium and Natural Coordinates

7.5a Surface Quantities[6]

From the equation of equilibrium $\nabla p = \boldsymbol{j} \times \boldsymbol{B}$, there follows

$$\boldsymbol{B}\cdot\nabla p = 0 \tag{7.113}$$

$$\boldsymbol{j}\cdot\nabla p = 0. \tag{7.114}$$

Then p is constant on a magnetic surface and the vector $\boldsymbol{j}$ is everywhere parallel to a magnetic surface. To study the characteristics of equilibrium in toroidal systems, the curvilinear coordinates $\mathrm{d}\boldsymbol{a}$, $\mathrm{d}\boldsymbol{s}$, and $\mathrm{d}\boldsymbol{\chi}$ are introduced as is shown in fig. 7.4. $\mathrm{d}\boldsymbol{a}$ is normal to the magnetic surface, i.e.,

$$\mathrm{d}\boldsymbol{a} = \frac{\nabla p}{|\nabla p|^2}\mathrm{d}p. \tag{7.115}$$

$\mathrm{d}\boldsymbol{s}$ is the coordinate along the magnetic axis and $\mathrm{d}\boldsymbol{\chi}$ is the poloidal coordinate. Let us consider the total plasma current I, of direction $\mathrm{d}\boldsymbol{S}_\chi \equiv$

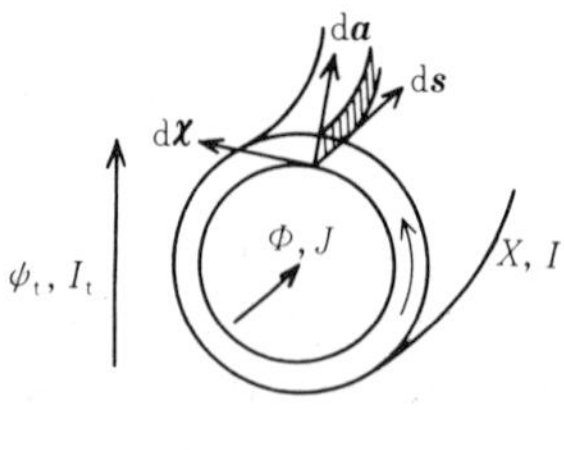

Fig. 7.4
Curvilinear coordinates for a toroidal system.

$\mathrm{d}\boldsymbol{a} \times \mathrm{d}\boldsymbol{s}$. For closed field lines, I is given by

$$I = \oint \boldsymbol{j} \cdot (\mathrm{d}\boldsymbol{a} \times \mathrm{d}\boldsymbol{s}), \tag{7.116}$$

while the plasma current is expressed by

$$\boldsymbol{j} = \frac{\boldsymbol{B} \times \nabla p}{B^2} + \lambda \boldsymbol{B}. \tag{7.117}$$

We find the relation

$$(\boldsymbol{B} \times \nabla p) \cdot (\mathrm{d}\boldsymbol{a} \times \mathrm{d}\boldsymbol{s}) = -(\boldsymbol{B} \cdot \mathrm{d}\boldsymbol{s})(\nabla p \cdot \mathrm{d}\boldsymbol{a}) = -\mathrm{d}p(\boldsymbol{B} \cdot \mathrm{d}\boldsymbol{s}) \tag{7.118}$$

by using $\boldsymbol{B} \cdot \mathrm{d}\boldsymbol{a} = 0$, $\nabla p \cdot \mathrm{d}\boldsymbol{s} = 0$. When $\mathrm{d}\boldsymbol{s}$ is taken along the field lines,

$$\mathrm{d}\boldsymbol{s} = \boldsymbol{b}\,\mathrm{d}l \tag{7.119}$$

and

$$I = -\mathrm{d}p \oint \frac{\mathrm{d}l}{B}. \tag{7.120}$$

The current I does not depend on what field line we choose, as long as the line belongs to the same magnetic surface. Consequently the specific volume

$$U = \oint \frac{\mathrm{d}l}{B} \tag{7.121}$$

is constant on a magnetic surface. (This result has already been utilized, in sec. 2.4.) Even for nonclosed field lines, one can define the specific volume by

$$U = \lim_{N \to \infty} \frac{1}{N} \int_N \frac{\mathrm{d}l}{B}. \tag{7.122}$$

As the equation of equilibrium is symmetric with respect to $\boldsymbol{B}$ and $\boldsymbol{j}$, we can consider the total flux X in the direction of $\mathrm{d}\boldsymbol{S}_\chi$. The flux is given by

$$X = \mathrm{d}p \oint \frac{\mathrm{d}l_j}{j}, \tag{7.123}$$

where the integration is carried out along the line of current. We can also define the quantity

$$U_j = \lim_{N \to \infty} \frac{1}{N} \int_N \frac{\mathrm{d}l_j}{j} \tag{7.124}$$

on the magnetic surface.

Let us next consider a closed loop in the χ (poloidal) direction and let the magnetic flux be Φ and the current be J through the cross section surrounded by the loop. When the vector potential is denoted by $\boldsymbol{A}$, we find from Stokes's theorem that

$$\oint \boldsymbol{B} \cdot \mathrm{d}\boldsymbol{\chi} = -\mu_0 J \tag{7.125}$$

$$\oint \boldsymbol{A} \cdot \mathrm{d}\boldsymbol{\chi} = -\Phi. \tag{7.126}$$

In a similar manner, we can consider a closed loop in the s (toroidal) direction, with magnetic flux Ψ_t and current I_t through the cross section surrounded by the toroidal loop. Then I_t and Ψ_t are given by

$$\oint \boldsymbol{B} \cdot \mathrm{d}\boldsymbol{s} = \mu_0 I_\mathrm{t} \tag{7.127}$$

$$\oint \boldsymbol{A} \cdot \mathrm{d}\boldsymbol{s} = \Psi_\mathrm{t}. \tag{7.128}$$

The quantities $J, \Phi; I, X; I_\mathrm{t}, \Psi_\mathrm{t}; P, U, U_j$ are functions of the magnetic surface Ψ only. These quantities are called *surface quantities.*

7.5b Equilibrium Equations

Let us express the equilibrium equation $\boldsymbol{j} \times \boldsymbol{B} = \nabla p$ in the coordinates described in the previous subsection. Define the two vectors

$$\delta \boldsymbol{S}_\chi \equiv \mathrm{d}\boldsymbol{a} \times \mathrm{d}\boldsymbol{s} \qquad \delta \boldsymbol{S}_s \equiv \mathrm{d}\boldsymbol{\chi} \times \mathrm{d}\boldsymbol{a}, \tag{7.129}$$

and take the vector product. From the vector-product formula (see app. 1), the vector product is

$$\delta \boldsymbol{S}_s \times \delta \boldsymbol{S}_\chi = \mathrm{d}\boldsymbol{a}(\mathrm{d}\boldsymbol{a} \cdot (\mathrm{d}\boldsymbol{s} \times \mathrm{d}\boldsymbol{\chi})) = \mathrm{d}\boldsymbol{a}\delta V. \tag{7.130}$$

As $\nabla p \cdot \mathrm{d}\boldsymbol{a} = \mathrm{d}p$, the scalar product of the equilibrium equation and eq. (7.130) is

$$\begin{aligned} \mathrm{d}p\delta V &= (\boldsymbol{j} \times \boldsymbol{B}) \cdot (\delta \boldsymbol{S}_s \times \delta \boldsymbol{S}_\chi) = (\boldsymbol{j} \cdot \delta \boldsymbol{S}_s)(\boldsymbol{B} \cdot \delta \boldsymbol{S}_\chi) - (\boldsymbol{j} \cdot \delta \boldsymbol{S}_\chi)(\boldsymbol{B} \cdot \delta \boldsymbol{S}_s) \\ &= \delta J \delta X - \delta I \delta \Phi. \end{aligned} \tag{7.131}$$

When eq. (7.131) is integrated over the volume region bounded by the two magnetic surfaces corresponding to p and $p + \mathrm{d}p$, the equilibrium equation becomes

$$\mathrm{d}p\mathrm{d}V = \mathrm{d}J\mathrm{d}X - \mathrm{d}I\mathrm{d}\Phi; \tag{7.132}$$

$\mathrm{d}J$ and $\mathrm{d}\Phi$ are the total current and the total flux bounded by the shell and lying in the s direction, while $\mathrm{d}I$ and $\mathrm{d}X$ are the total current and the total flux in the χ direction. Since $\mathrm{d}\boldsymbol{s}$ is taken along the magnetic-field lines, it follows that $\boldsymbol{B}\cdot\delta\boldsymbol{S}_\chi = 0$ and eq. (7.131) is

$$\mathrm{d}p\delta V = -\delta I\delta\Phi. \tag{7.133}$$

When eq. (7.133) is integrated along a line of magnetic force, we find

$$\mathrm{d}p\int \mathrm{d}\boldsymbol{s}\cdot\delta\boldsymbol{S}_s = -\int \delta I\delta\Phi.$$

As $\mathrm{d}\boldsymbol{s} = (\boldsymbol{B}/B)\mathrm{d}l$ and $\delta\Phi = \boldsymbol{B}\cdot\delta\boldsymbol{S}_s$ is constant along the field line, eq. (7.133) is

$$\mathrm{d}p\oint\frac{\mathrm{d}l}{B} = -\mathrm{d}I.$$

Consequently the specific volume is

$$U = \oint\frac{\mathrm{d}l}{B} = -\frac{\mathrm{d}I}{\mathrm{d}p} = \frac{\oint\delta V}{\delta\Phi} = \frac{\mathrm{d}V}{\mathrm{d}\Phi}, \tag{7.134}$$

where V is the volume enclosed by the magnetic surface.

*7.5c Natural Coordinates for Toroidal Systems[7,8]

Let us consider the coordinate system a, θ, φ as is shown in fig. 7.5. It is assumed that $a =$ const. is a magnetic surface, i.e., that $\boldsymbol{B}\cdot\nabla a = 0$. Any

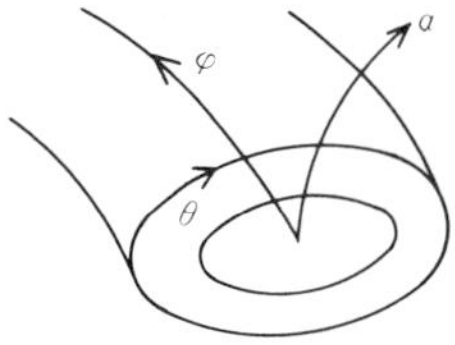

Fig. 7.5
Natural coordinates.

single-valued function on this surface must be periodic with respect to θ and φ, with periods 2π. Denoting $u^1 = a$, $u^2 = \theta$, $u^3 = \varphi$, we define the vectors

$$\boldsymbol{a}^j \equiv \nabla u^j$$

$$\boldsymbol{a}_j \equiv \frac{\partial \boldsymbol{r}}{\partial u^j}.$$

The contravariants $B^i = \boldsymbol{B}\cdot\boldsymbol{a}^i$ of the magnetic field $\boldsymbol{B}$ are

$$B^1 = \boldsymbol{B}\cdot\nabla a \qquad B^2 = \boldsymbol{B}\cdot\nabla\theta \qquad B^3 = \boldsymbol{B}\cdot\nabla\varphi, \tag{7.135}$$

and $\boldsymbol{B}$ is expressed by $\boldsymbol{B} = \Sigma B^i \boldsymbol{a}_i$ (see sec. 3.5a). Since the expression of $\nabla\boldsymbol{B} = 0$ in curvilinear coordinates is

$$\frac{\partial g^{1/2} B^2}{\partial\theta} + \frac{\partial g^{1/2} B^3}{\partial\varphi} = 0, \tag{7.136}$$

where g is the determinant $|g_{ij}|$ of the metric tensor $g_{ij} \equiv \boldsymbol{a}_i\cdot\boldsymbol{a}_j$, B^2 and B^3 are given by

$$\left.\begin{aligned} B^2 &= -\frac{1}{2\pi g^{1/2}}\frac{\partial\Theta}{\partial\varphi} \\ B^3 &= \frac{1}{2\pi g^{1/2}}\frac{\partial\Theta}{\partial\theta}. \end{aligned}\right\} \tag{7.137}$$

The magnetic-field lines are represented by

$$\frac{\mathrm{d}a}{B^1} = \frac{\mathrm{d}\theta}{B^2} = \frac{\mathrm{d}\varphi}{B^3}. \tag{7.138}$$

Consequently, the relation

$$\mathrm{d}\Theta = \frac{\partial\Theta}{\partial\theta}\mathrm{d}\theta + \frac{\partial\Theta}{\partial\varphi}\mathrm{d}\varphi = 0$$

holds along a line of magnetic force. As B^2 and B^3 are single-valued functions, the function Θ is of the form

$$\Theta(a,\theta,\varphi) = A(a)\theta + B(a)\varphi + \tilde{\Theta}(a,\theta,\varphi), \tag{7.139}$$

where $\tilde{\Theta}$ is a periodic function of θ and φ, with periods 2π. The magnetic fluxes in the $\nabla\theta$ and $\nabla\varphi$ directions within the region V bounded by the magnetic surfaces a and $a + \mathrm{d}a$ are

$$\mathrm{d}\Phi \equiv \frac{1}{2\pi}\int B^3 \mathrm{d}V = \frac{1}{2\pi}\int (\boldsymbol{B}\cdot\nabla\varphi)\mathrm{d}V = \frac{\mathrm{d}a}{2\pi}\int\int B^3 g^{1/2}\mathrm{d}\theta\mathrm{d}\varphi$$

$$= \frac{\mathrm{d}a}{4\pi^2}\int\int\frac{\partial\Theta}{\partial\theta}\mathrm{d}\theta\mathrm{d}\varphi = A(a)\mathrm{d}a \tag{7.140}$$

$$\mathrm{d}X \equiv \frac{1}{2\pi}\int B^2 \mathrm{d}V = \frac{1}{2\pi}\int (\boldsymbol{B}\cdot\nabla\theta)\mathrm{d}V = -B(a)\mathrm{d}a, \tag{7.141}$$

respectively. Because $\boldsymbol{B}\cdot\nabla\varphi = \nabla(\varphi\boldsymbol{B})$, the volume integral (7.140) may be written as a surface integral on the $\varphi = 0$ surface; so that eq. (7.140) gives the magnetic flux in the $\nabla\varphi$ direction. With the introduction of the variable $\tilde{\theta} \equiv \theta + \tilde{\Theta}/A$, eq. (7.139) becomes

$$\Theta(a, \tilde{\theta}, \varphi) = \Phi'(a)\tilde{\theta} - X'(a)\varphi.$$

The magnetic field is given by

$$\tilde{B}^1 = 0 \qquad \tilde{B}^2 = \frac{X'(a)}{2\pi g^{1/2}} \qquad \tilde{B}^3 = \frac{\Phi'(a)}{2\pi g^{1/2}} \tag{7.142}$$

in the new coordinates. Here the prime indicates differentiation with respect to a. Lines of magnetic force are expressed by

$$\Phi'(a)\tilde{\theta} - X'(a)\varphi = \text{const.} \tag{7.143}$$

in the new coordinates; in these coordinates, the field lines are straight. From now on, we rewrite $\tilde{\theta}$ as θ.

Next let us repeat a similar procedure for $\boldsymbol{j}$. From $\nabla\boldsymbol{j} = 0$, the contravariants of $\boldsymbol{j}$ are given by

$$\left.\begin{aligned} j^2 &= -\frac{1}{2\pi g^{1/2}}\frac{\partial J_{\mathrm{p}}}{\partial\varphi} \\ j^3 &= \frac{1}{2\pi g^{1/2}}\frac{\partial J_{\mathrm{p}}}{\partial\theta}, \end{aligned}\right\} \tag{7.144}$$

where J_{p} is expressed by

$$J_{\mathrm{p}} = J'(a)\theta - I'(a)\varphi + \nu(a, \theta, \varphi). \tag{7.145}$$

From the relations

$$\frac{1}{2\pi}\int j^3 \mathrm{d}V = \frac{\mathrm{d}a}{4\pi^2}\int\int\frac{\partial J_{\mathrm{p}}}{\partial\theta}\mathrm{d}\theta\mathrm{d}\varphi = J'(a)\mathrm{d}a \tag{7.146}$$

$$\frac{1}{2\pi}\int j^2 \mathrm{d}V = \frac{\mathrm{d}a}{4\pi^2}\int\int\left(-\frac{\partial J_{\mathrm{p}}}{\partial \varphi}\right)\mathrm{d}\theta\mathrm{d}\varphi = I'(a)\mathrm{d}a \tag{7.147}$$

it is clear that $\mathrm{d}J$ and $\mathrm{d}I$ are the total currents in the $\nabla\varphi$ and $\nabla\theta$ directions within the region between the magnetic surfaces a and $a + \mathrm{d}a$. Consequently the contravariants of J are given by

$$j^1 = 0 \qquad j^2 = \left(I'(a) - \frac{\partial v}{\partial \varphi}\right)\Big/ 2\pi g^{1/2} \qquad j^3 = \left(J'(a) + \frac{\partial v}{\partial \theta}\right)\Big/ 2\pi g^{1/2}. \tag{7.148}$$

Let us next discuss covariants of $\boldsymbol{B}$. From the relations $\nabla \times \boldsymbol{B} = \mu \boldsymbol{j}$, $j^1 = 0$, it follows that

$$\frac{1}{g^{1/2}}\left(\frac{\partial B_3}{\partial \theta} - \frac{\partial B_2}{\partial \varphi}\right) = 0 \qquad \frac{1}{g^{1/2}}\left(\frac{\partial B_1}{\partial \varphi} - \frac{\partial B_3}{\partial a}\right) = \mu_0 j^2$$

$$\frac{1}{g^{1/2}}\left(\frac{\partial B_2}{\partial \theta} - \frac{\partial B_1}{\partial \varphi}\right) = \mu_0 j^3.$$

Consequently, covariants B_2 and B_3 of $\boldsymbol{B}$ are

$$B_2 = \frac{1}{2\pi}\frac{\partial \phi}{\partial \theta} + A_2(a)$$

$$B_3 = \frac{1}{2\pi}\frac{\partial \phi}{\partial \varphi} + A_3(a).$$

B_1 is formally expressed by

$$B_1 = \frac{1}{2\pi}\frac{\partial \phi}{\partial a} + A_1(a, \theta, \varphi),$$

where $A_1(a, \theta, \varphi)$ is a periodic function of θ and φ with periods 2π. From the foregoing equations, $J(a)$ is

$$J(a) = \frac{1}{2\pi}\int j^3 g^{1/2}\mathrm{d}a\,\mathrm{d}\theta\,\mathrm{d}\varphi = \frac{1}{2\pi\mu_0}\int\left(\frac{\partial B_2}{\partial a} - \frac{\partial B_1}{\partial \theta}\right)\mathrm{d}a\,\mathrm{d}\theta\,\mathrm{d}\varphi = \frac{2\pi}{\mu_0}A_2(a),$$

and similarly $I(a)$ is

$$I(a) = \frac{1}{2\pi}\int j^2 g^{1/2}\mathrm{d}a\,\mathrm{d}\theta\,\mathrm{d}\varphi = -\frac{2\pi}{\mu_0}A_3(a).$$

Using $J(a)$ and $I(a)$, we may express the covariants of $\boldsymbol{B}$ by

$$\left.\begin{aligned} B_1 &= \left(\frac{\partial \phi}{\partial a} - \mu_0 v\right)\Big/ 2\pi \\ B_2 &= \left(\frac{\partial \phi}{\partial \theta} + \mu_0 J(a)\right)\Big/ 2\pi \\ B_3 &= \left(\frac{\partial \phi}{\partial \varphi} - \mu_0 I(a)\right)\Big/ 2\pi, \end{aligned}\right\} \tag{7.149}$$

where $v(a, \theta, \varphi)$ has already been given in eq. (7.145). Accordingly J and I are equal to

$$\left.\begin{aligned} \mu_0 J &= \int B_2 \mathrm{d}\theta \\ \mu_0 I &= -\int B_3 \mathrm{d}\varphi. \end{aligned}\right\} \tag{7.150}$$

When the equation of equilibrium is expressed in terms of the covariants and contravariants of $\boldsymbol{j}$ and $\boldsymbol{B}$, the relation $(\boldsymbol{j} \times \boldsymbol{B})_1 = g^{1/2}(j^2 B^3 - j^3 B^2)$ is reduced to

$$\begin{aligned} \boldsymbol{B} \cdot \nabla v &= \frac{1}{2\pi g^{1/2}}\left(X' \frac{\partial v}{\partial \theta} + \Phi' \frac{\partial v}{\partial \varphi}\right) \\ &= \frac{1}{2\pi g^{1/2}}((I'\Phi' - J'X') - 4\pi^2 g^{1/2} p') \equiv S. \end{aligned} \tag{7.151}$$

This is the magnetodifferential equation. The volume integral of S within a region surrounded by magnetic surfaces and the line integral of s along a closed line of magnetic force are both zero, i.e.,

$$\int_V S \mathrm{d}V = 0 \qquad \oint \frac{S}{B} \mathrm{d}l = 0. \tag{7.152}$$

If we take a region between any shell surrounded by the adjacent magnetic surfaces a and $a + \mathrm{d}a$, then $\delta V = \int g^{1/2} \mathrm{d}a \mathrm{d}\theta \mathrm{d}\varphi$ and $V' = \int g^{1/2} \mathrm{d}\theta \mathrm{d}\varphi$, so the equation of the equilibrium is

$$I'\Phi' - J'X' = p'V'. \tag{7.153}$$

Accordingly, eq. (7.151) is reduced to

$$X'\frac{\partial v}{\partial \theta} + \Phi'\frac{\partial v}{\partial \varphi} = p'(V' - 4\pi^2 g^{1/2}). \tag{7.154}$$

The curvilinear coordinates a, θ, φ in which the field lines are straight still have some arbitrariness. If we use a, $\bar{\theta}$, $\bar{\varphi}$ defined by

$$\bar{\theta} = \theta + X'(a)\frac{G(a, \theta, \varphi)}{2\pi} \tag{7.155}$$

$$\bar{\varphi} = \varphi + \Phi'(a)\frac{G(a, \theta, \varphi)}{2\pi}, \tag{7.156}$$

the lines of magnetic force are still expressed by

$$\Phi'\bar{\theta} - X'\bar{\varphi} = \text{const.} \tag{7.157}$$

The determinant $\bar{g}$ of the metric tensor in the new coordinates is

$$\frac{1}{\bar{g}^{1/2}} = \nabla a \cdot (\nabla\bar{\theta} \times \nabla\bar{\varphi}) = \frac{1}{g^{1/2}}\left(1 + \frac{X'}{2\pi}\frac{\partial G}{\partial \theta} + \frac{\Phi'}{2\pi}\frac{\partial G}{\partial \varphi}\right),$$

i.e.,

$$\frac{1}{\bar{g}^{1/2}} = \frac{1}{g^{1/2}} + \boldsymbol{B}\cdot\nabla G \tag{7.158}$$

(see app. 2). So we can choose the determinant of $(a, \bar{\theta}, \bar{\varphi})$ to be

$$\bar{g}^{1/2} = \frac{1}{4\pi^2}.$$

When we take the variable a as the volume V surrounded by a magnetic surface, eq. (7.154) is reduced to

$$X'\frac{\partial v}{\partial \bar{\theta}} + \Phi'\frac{\partial v}{\partial \bar{\varphi}} = 0 \tag{7.159}$$

in the coordinates $V, \bar{\theta}, \bar{\varphi}$, so that v can be chosen to be zero. Consequently, we have

$$B^1 = 0 \qquad B^2 = 2\pi\dot{X}(V) \qquad B^3 = 2\pi\dot{\Phi}(V) \tag{7.160}$$

$$j^1 = 0 \qquad j^2 = 2\pi\dot{I}(V) \qquad j^3 = 2\pi\dot{J}(V) \tag{7.161}$$

$$\dot{I}(V)\dot{\Phi}(V) - \dot{J}(V)\dot{X}(V) = \dot{p}(V). \tag{7.162}$$

(Here the dot indicates differentiation with respect to V.) This coordinate system, called the *natural system*, was introduced by Hamada.[8]

7.6 Axially Symmetric and Helically Symmetric Systems

7.6a Equilibrium Equation for Axially Symmetric Systems

Let us use cylindrical coordinates r, φ, z and denote the magnetic surface by ψ. Then ψ satisfies

$$\boldsymbol{B}\cdot\nabla\psi = 0 \tag{7.163}$$

$$\nabla\boldsymbol{B} = 0. \tag{7.164}$$

As the system is independent of φ, the foregoing equations may be written out as

$$B_r\frac{\partial\psi}{\partial r} + B_z\frac{\partial\psi}{\partial z} = 0$$

$$\frac{1}{r}\frac{\partial r B_r}{\partial r} + \frac{\partial B_z}{\partial z} = 0.$$

Consequently, one can set

$$rB_r = -\frac{\partial\psi}{\partial z} \qquad rB_z = \frac{\partial\psi}{\partial r}. \tag{7.165}$$

The relation $\nabla p\cdot\boldsymbol{B} = 0$ follows from the equilibrium equation and is expressed by

$$-\frac{\partial p}{\partial r}\frac{\partial\psi}{\partial z} + \frac{\partial p}{\partial z}\frac{\partial\psi}{\partial r} = 0.$$

Accordingly p is a function of ψ only, i.e.,

$$p = p(\psi). \tag{7.166}$$

Similarly, from $\nabla p\cdot\boldsymbol{j} = 0$ and $\mu_0\boldsymbol{j} = \nabla\times\boldsymbol{B}$, we may write

$$-\frac{\partial p}{\partial r}\frac{\partial r B_\varphi}{\partial z} + \frac{\partial p}{\partial z}\frac{\partial r B_\varphi}{\partial r} = 0;$$

and since rB_φ is a function of ψ only, then

$$rB_\varphi = \frac{\mu_0}{2\pi} I(\psi). \tag{7.167}$$

Inspecting the r component of $\boldsymbol{j} \times \boldsymbol{B} = \nabla p$ leads to the equation on ψ:[9]

$$L(\psi) + \mu_0 r^2 \frac{\partial p(\psi)}{\partial \psi} + \frac{\mu_0{}^2}{8\pi^2}\frac{\partial I^2(\psi)}{\partial \psi} = 0, \tag{7.168}$$

where

$$L(\psi) \equiv \left(r \frac{\partial}{\partial r}\frac{1}{r}\frac{\partial}{\partial r} + \frac{\partial^2}{\partial z^2}\right)\psi. \tag{7.169}$$

Equation (7.168) is the equilibrium equation for axially symmetric systems. The current density is expressed in terms of functions of the magnetic surface as

$$j_r = \frac{-1}{2\pi r}\frac{\partial I(\psi)}{\partial z} \qquad j_z = \frac{1}{2\pi r}\frac{\partial I(\psi)}{\partial r}$$

$$j_\varphi = \frac{-1}{\mu_0}\left(\frac{\partial}{\partial r}\frac{1}{r}\frac{\partial \psi}{\partial r} + \frac{\partial^2 \psi}{r \partial z^2}\right) = \frac{-L(\psi)}{\mu_0 r} = \frac{1}{\mu_0 r}\left(\mu_0 r^2 p' + \frac{\mu_0{}^2}{8\pi^2}(I^2)'\right) \tag{7.170}$$

or

$$\boldsymbol{j} = \frac{I'}{2\pi}\boldsymbol{B} + p' r \boldsymbol{e}_\varphi. \tag{7.171}$$

The functions $p(\psi)$ and $I^2(\psi)$ are arbitrary functions of ψ. However, for other than quadratic functions of ψ, eq. (7.168) becomes a nonlinear equation and is then difficult to treat analytically. Let us assume that p and I^2 are quadratic functions of ψ. The value ψ_s at the plasma boundary can be chosen to be zero ($\psi_s = 0$) without loss of generality. When the values at the boundary are $p = p_s$, $I^2 = I_s{}^2$ and the values at the magnetic axis are $\psi = \psi_0$, $p = p_0$, $I^2 = I_0{}^2$, then p and I^2 are expressed by [10]

$$\left.\begin{aligned} p(\psi) &= p_s + (p_0 - p_s)\frac{\psi^2}{\psi_0{}^2} \\ I^2(\psi) &= I_s{}^2 + (I_0{}^2 - I_s{}^2)\frac{\psi^2}{\psi_0{}^2}. \end{aligned}\right\} \tag{7.172}$$

In this case the pressure gradient and j_r, j_z, j_φ are zero at the plasma boundary $\psi = 0$. The equilibrium equation is then reduced to

$$L(\psi) + (\alpha r^2 + \beta)\psi = 0 \tag{7.173}$$

$$\alpha = \frac{2\mu_0(p_0 - p_s)}{\psi_0^2} \qquad \beta = \frac{\mu_0^2}{4\pi^2}\frac{(I_0^2 - I_s^2)}{\psi_0^2}. \tag{7.174}$$

The equations for $\boldsymbol{B}(r, z)$, $p(r, z)$, and $\boldsymbol{j}(r, z)$ are obtained from the solution of $\psi(r, z)$. So the problem of plasma equilibrium is here reduced to a boundary-value problem of a two-dimensional differential equation. From eqs. (7.172) and (7.174), the following relations are easily obtained:

$$\int_V \frac{\psi}{r^2}(\alpha r^2 + \beta)\psi \, \mathrm{d}V = 2\mu_0 \int_V (p - p_s)\mathrm{d}V + \frac{\mu_0^2}{4\pi^2}\int_V \frac{(I^2 - I_s^2)}{r^2}\mathrm{d}V$$

$$\int_V \frac{1}{r^2}\psi L(\psi)\mathrm{d}V = \int_S \frac{1}{r^2}\psi\nabla\psi \cdot \boldsymbol{n}\mathrm{d}S - \int_V \frac{1}{r^2}(\nabla\psi)^2\mathrm{d}V$$

$$= -\int_V (B_r^2 + B_z^2)\mathrm{d}V.$$

Then eq. (7.173) becomes

$$\int (p - p_s)\mathrm{d}V = \int \frac{1}{2\mu_0}(B_{\varphi v}^2 - B_\varphi^2 + (B_r^2 + B_z^2))\mathrm{d}V. \tag{7.175}$$

This is the equation of pressure balance ($B_{\varphi v}$ is the φ component of $\boldsymbol{B}$ outside the plasma (in vacuum)).

For problems of plasma equilibrium and stability in cases of circular or noncircular cross sections with general $p(\psi)$ and $I(\psi)$, one must solve the differential equation (7.168).[11]

7.6b Equilibrium Equations for Helically Symmetric Systems

In helically symmetric systems the physical quantities are dependent only on r and $\phi \equiv \varphi - hz$, so eqs. (7.163) and (7.164) are expressed by

$$B_r\frac{\partial\psi}{\partial r} + \left(B_\varphi\frac{1}{r} - B_z h\right)\frac{\partial\psi}{\partial\phi} = 0$$

$$\frac{\partial rB_r}{\partial r} + \frac{\partial}{\partial\phi}(B_\varphi - (hr)B_z) = 0.$$

Accordingly, we have

$$rB_r = \frac{\partial\psi}{\partial\phi} \tag{7.176}$$

$$B_\varphi - (hr)B_z = -\frac{\partial\psi}{\partial r}. \tag{7.177}$$

From $\nabla p \cdot \boldsymbol{B} = 0$, it follows that

$$p = p(\psi). \tag{7.178}$$

Writing

$$B_z + (hr)B_\varphi = B^*(r, \phi) \tag{7.179}$$

and using $\nabla p \cdot \boldsymbol{j} = 0$, we find

$$B^*(r, \phi) = B^*(\psi). \tag{7.180}$$

Consequently the expressions for $\boldsymbol{B}$, $\boldsymbol{j} = \nabla \times \boldsymbol{B}/\mu_0$, and the equilibrium equation are given by[12]

$$\left.\begin{aligned} B_r &= \frac{1}{r}\frac{\partial\psi}{\partial\phi} \\ B_\varphi &= \frac{-\dfrac{\partial\psi}{\partial r} + (hr)B^*}{1 + (hr)^2} \\ B_z &= \frac{(hr)\dfrac{\partial\psi}{\partial r} + B^*}{1 + (hr)^2} \end{aligned}\right\} \tag{7.181}$$

$$\left.\begin{aligned} \mu_0 j_r &= \frac{B^{*\prime}}{r}\frac{\partial\psi}{\partial\phi} \\ \mu_0 j &= -(hr)L_{\mathrm{h}}(\psi) - \frac{B^{*\prime}\dfrac{\partial\psi}{\partial r}}{1 + (hr)^2} + \frac{2h^2 r B^*}{(1 + (hr)^2)^2} \\ \mu_0 j_z &= -L_{\mathrm{h}}(\psi) + \frac{(hr)B^{*\prime}\dfrac{\partial\psi}{\partial r}}{1 + (hr)^2} + \frac{2hB^*}{(1 + (hr)^2)^2} \\ \mu_0(j_\varphi - hrj_z) &= -\frac{\partial B^*}{\partial r} \end{aligned}\right\} \tag{7.182a}$$

$$L_{\mathrm{h}}(\psi) + \frac{B^* B^{*\prime}}{1 + (hr)^2} - \frac{2hB^*}{(1 + (hr)^2)^2} = -\mu_0 p' \tag{7.183}$$

$$L_{\rm h}(\psi) \equiv \left(\frac{1}{r}\frac{\partial}{\partial r}\frac{r}{1+(hr)^2}\frac{\partial}{\partial r} + \frac{1}{r^2}\frac{\partial^2}{\partial \phi^2}\right)\psi. \tag{7.184}$$

Equation (7.183) is the equilibrium equation for helically symmetric systems. By means of eq. (7.183), eq. (7.182a) is reduced to

$$\mu_0 \boldsymbol{j} = (B^*)' \boldsymbol{B} + \mu_0 p'(hr\boldsymbol{e}_\varphi + \boldsymbol{e}_z). \tag{7.182b}$$

The first term is the parallel component to the magnetic field, i.e., $\boldsymbol{j}_{\parallel} \times \boldsymbol{B} = 0$; this is the so-called force-free term. In the case

$$B^*(\psi) = B_0 + \gamma\psi \qquad p(\psi) = p_0 + p'\psi,$$

eq. (7.183) is reduced to the linear differential equation

$$L_{\rm h}(\psi) + \frac{\gamma(B_0 + \gamma\psi)}{1+(hr)^2} - \frac{2h(B_0+\gamma\psi)}{(1+(hr)^2)^2} + \mu_0 p' = 0.$$

Write ψ as

$$\psi = \psi_0(r) + \sum_{n=1}^{\infty} \psi_n^{\rm c}(r)\cos n\phi + \sum_{n=1}^{\infty} \psi_n^{\rm s}(r)\sin n\phi;$$

then ψ_0 and ψ_n must satisfy

$$\left.\begin{aligned} &\frac{{\rm d}^2\psi_0}{{\rm d}r^2} + \frac{1-(hr)^2}{1+(hr)^2}\frac{1}{r}\frac{{\rm d}\psi_0}{{\rm d}r} + \gamma^2\psi_0 - \frac{2hr}{1+(hr)^2}\psi_0 + B_0\gamma - \frac{2hB_0}{1+(hr)^2} \\ &\qquad + \mu_0 p'(1+(hr)^2) = 0 \\ &\frac{{\rm d}^2\psi_n}{{\rm d}r^2} + \frac{1-(hr)^2}{1+(hr)^2}\frac{1}{r}\frac{{\rm d}\psi_n}{{\rm d}r} - \frac{n^2(1+(hr)^2)}{r^2}\psi_n + \gamma^2\psi_n - \frac{2hr}{1+(hr)^2}\psi_n = 0. \end{aligned}\right\} \tag{7.185}$$

The solutions of eq. (7.185) are[13]

$$\begin{aligned} \psi_0(r) = &-\frac{B_0}{\gamma} - \frac{\mu_0 p_0'}{\gamma^2}\left(1+(hr)^2+\frac{2h}{\gamma}\right) + a_0\left(\gamma r J_0'(\gamma r) - \frac{\gamma}{h}J_0(\gamma r)\right) \\ &+ b_0\left(\gamma r N_0'(\gamma r) - \frac{\gamma}{h}N_0(\gamma r)\right) \end{aligned}$$

$$\begin{aligned} \psi_n^{\rm c,s}(r) = a_n^{\rm c,s}\Big[&(n^2h^2-\gamma^2)^{1/2} r I_n'((n^2h^2-\gamma^2)^{1/2}r) \\ &-\frac{\gamma}{h}I_n((n^2h^2-\gamma^2)^{1/2}r)\Big] \end{aligned}$$

$$+b_n^{c,s}\left[(n^2h^2-\gamma^2)^{1/2} r K_n'((n^2h^2-\gamma^2)^{1/2} r) - \frac{\gamma}{h} K_n((n^2h^2-\gamma^2)^{1/2} r)\right]. \tag{7.186}$$

$J_0(z)$ and $N_0(z)$ are the zeroth-order Bessel functions of the 1st and 2nd kind and $I_n(z)$ and $K_n(z)$ are the nth-order modified Bessel functions of the 1st and 2nd kind, respectively.

7.7 Tokamak Equilibrium

7.7a Magnetic Surfaces in a Tokamak[14]

The equilibrium equation for an axisymmetric system (eq. (7.168)) is

$$\frac{\partial^2 \psi}{\partial r^2} - \frac{1}{r}\frac{\partial \psi}{\partial r} + \frac{\partial^2 \psi}{\partial z^2} = -\left(\mu_0 r^2 p'(\psi) + \frac{\mu_0{}^2}{8\pi^2}(I^2(\psi))'\right). \tag{7.187}$$

The right-hand side of the equation is zero outside the plasma region. Let us use toroidal coordinates b, ω, φ. The relations between these to cylindrical coordinates are

$$r = \frac{R_0 \sinh b}{\cosh b - \cos\omega} \tag{7.188}$$

$$z = \frac{R_0 \sin\omega}{\cosh b - \cos\omega}. \tag{7.189}$$

The curves $b = b_0$ are circles of radius $a = R_0/\sinh b_0$, centered at $r = R_0 \coth b_0$, $z = 0$.
When the magnetic-surface function ψ is replaced by F, according to

$$\psi = \frac{F(b,\omega)}{2^{1/2}(\cosh b - \cos\omega)^{1/2}}, \tag{7.190}$$

the function F satisfies

$$\frac{\partial^2 F}{\partial b^2} - \coth b \frac{\partial F}{\partial b} + \frac{\partial^2 F}{\partial \omega^2} + \frac{1}{4}F = 0 \tag{7.191}$$

outside the plasma region. When F is expanded,

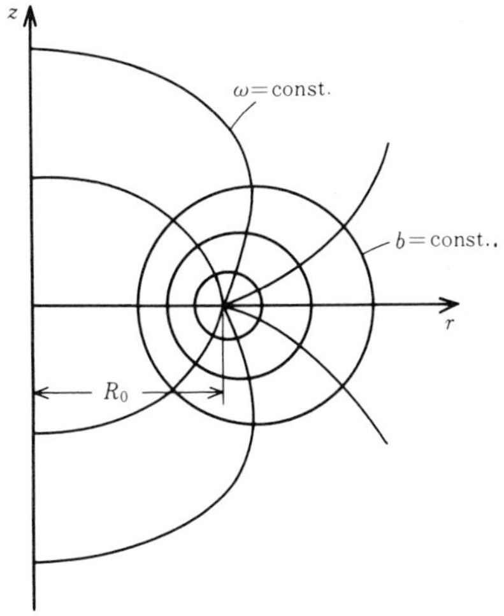

Fig. 7.6
Toroidal coordinates.

$$F = \Sigma g_n(b) \cos n\omega, \tag{7.192}$$

the coefficient g_n satisfies

$$\frac{d^2 g_n}{db^2} - \coth b \frac{dg_n}{db} - \left(n^2 - \frac{1}{4}\right) g_n = 0. \tag{7.193}$$

There are two independent solutions of eq. (7.193):

$$\left(n^2 - \frac{1}{4}\right) g_n = \sinh b \frac{d}{db} Q_{n-1/2}(\cosh b) \tag{7.194}$$

$$\left(n^2 - \frac{1}{4}\right) f_n = \sinh b \frac{d}{db} P_{n-1/2}(\cosh b). \tag{7.195}$$

$P_\nu(x)$ and $Q_\nu(x)$ are Legendre functions. If the ratio of the plasma radius to the major radius a/R_0 is small, i.e., when $e^{b_0} \gg 1$, then g_n and f_n are given by

$$g_0 = e^{b/2} \qquad g_1 = -\frac{1}{2} e^{-b/2}$$
$$f_0 = \frac{2}{\pi} e^{b/2}(b + \ln 4 - 2) \qquad f_1 = \frac{2}{3\pi} e^{3b/2}. \tag{7.196}$$

If we take terms up to $\cos\omega$, F and ψ are

$$F = c_0 g_0 + d_0 f_0 + 2(c_1 g_1 + d_1 f_1)\cos\omega \tag{7.197}$$

$$\psi = \frac{F}{2^{1/2}(\cosh b - \cos\omega)^{1/2}} \approx \mathrm{e}^{-b/2}(1 + \mathrm{e}^{-b}\cos\omega)F. \tag{7.198}$$

Use the coordinates ρ, ω' shown in fig. 7.7. These are related to the cylindrical and toroidal coordinates as follows:

$$\left.\begin{aligned} r &= R_0 + \rho\cos\omega' = \frac{R_0 \sinh b}{\cosh b - \cos\omega} \\ z &= \rho\sin\omega' = \frac{R_0 \sin\omega}{\cosh b - \cos\omega}. \end{aligned}\right\} \tag{7.199}$$

When b is large, the relations are

$$\omega' \approx \omega \qquad \frac{\rho}{2R_0} \approx \mathrm{e}^{-b}. \tag{7.200}$$

Accordingly the magnetic surface ψ is expressed by

$$\begin{aligned} \psi &= c_0 + \frac{2}{\pi}d_0(b + \ln 4 - 2) \\ &\quad + \left[\left(c_0 + \frac{2}{\pi}d_0(b + \ln 4 - 2)\right)\mathrm{e}^{-b} + \left(\frac{4}{3\pi}d_1\mathrm{e}^{b} - c_1\mathrm{e}^{-b}\right)\right]\cos\omega \\ &= d_0'\left(\ln\frac{8R}{\rho} - 2\right) + \left(\frac{d_0'}{2R}\left(\ln\frac{8R}{\rho} - 1\right)\rho + \frac{h_1}{\rho} + h_2\rho\right)\cos\omega. \end{aligned} \tag{7.201}$$

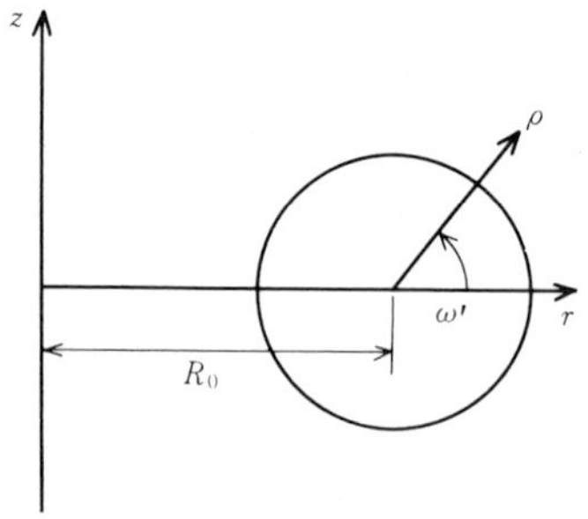

Fig. 7.7
The coordinates r, z and ρ, ω'.

In terms of ψ, the magnetic-field components are given by

$$rB_r = -\frac{\partial\psi}{\partial z} \qquad rB_z = \frac{\partial\psi}{\partial r} \tag{7.202}$$

$$rB_\rho = -\frac{\partial\psi}{\rho\partial\omega'} \qquad rB_{\omega'} = \frac{\partial\psi}{\partial\rho}. \tag{7.203}$$

From the relation

$$-\frac{d_0'}{\rho} = r\overline{B_{\omega'}} \approx R\frac{-\mu_0 I_p}{2\pi\rho},$$

the parameter d_0' can be taken as $d_0' = \mu_0 I_p R/2\pi$. Here I_p is the total plasma current in the φ direction. The expression of the magnetic surface is reduced to

$$\psi = \frac{\mu_0 I_p R}{2\pi}\left(\ln\frac{8R}{\rho} - 2\right) + \left(\frac{\mu_0 I_p}{4\pi}\left(\ln\frac{8R}{\rho} - 1\right)\rho + \frac{h_1}{\rho} + h_2\rho\right)\cos\omega', \tag{7.204}$$

where R_0 has been replaced by R. In the case of $a/R \ll 1$, the equation of pressure equilibrium (7.8b) is

$$\langle p\rangle - p_a = \frac{1}{2\mu_0}((B_{\varphi v}{}^2)_a + (B_r{}^2 + B_z{}^2)_a - \langle B_\varphi{}^2\rangle). \tag{7.205}$$

Here $\langle\ \ \rangle$ indicates the volume average and p_a is the plasma pressure at the plasma boundary. The value of $B_r{}^2 + B_z{}^2$ is nearly equal to $B_{\omega'}{}^2$. The ratio of $\langle p\rangle$ to $B_{\omega'}{}^2/2\mu_0$ is called the *poloidal beta ratio* β_p. When $p_a = 0$, β_p is

$$\beta_p = 1 + \frac{B_{\varphi v}{}^2 - \langle B_\varphi{}^2\rangle}{B_{\omega'}{}^2} \approx 1 + \frac{2B_{\varphi v}}{B_{\omega'}{}^2}\langle B_{\varphi v} - B_\varphi\rangle. \tag{7.206}$$

B_φ and $B_{\varphi v}$ are the toroidal magnetic fields in the plasma and the vacuum toroidal fields respectively. When B_φ is smaller than $B_{\varphi v}$, the plasma is diamagnetic, $\beta_p > 1$. When B_φ is larger than $B_{\varphi v}$, the plasma is *paramagnetic*, $\beta_p < 1$. When the plasma current flows along a line of magnetic force, the current produces the poloidal magnetic field $B_{\omega'}$ and a poloidal component of the plasma current appears and induces an additional toroidal magnetic field. This is the origin of the paramagnetism.

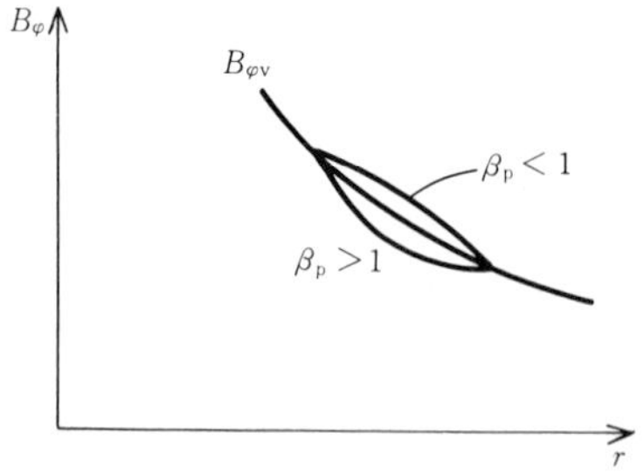

Fig. 7.8
Diamagnetism ($\beta_p > 1$) and paramagnetism ($\beta_p < 1$).

When the function (7.204) is used, the magnetic field is given by

$$\left.\begin{aligned} B_{\omega'} &= \frac{1}{r}\frac{\partial \psi}{\partial \rho} = \frac{-\mu_0 I_p}{2\pi\rho} + \left(\frac{\mu_0 I_p}{4\pi R}\ln\frac{8R}{\rho} + \frac{1}{R}\left(h_2 - \frac{h_1}{\rho^2}\right)\right)\cos\omega' \\ B_\rho &= -\frac{1}{r\rho}\frac{\partial \psi}{\partial \omega'} = \left(\frac{\mu_0 I_p}{4\pi R}\left(\ln\frac{8R}{\rho} - 1\right) + \frac{1}{R}\left(h_2 + \frac{h_1}{\rho^2}\right)\right)\sin\omega'. \end{aligned}\right\} \tag{7.207}$$

The cross section of the magnetic surface is

$$\psi(\rho, \omega') = \psi_0(\rho) + \psi_1 \cos\omega'.$$

When $\Delta = -\psi_1/\psi_0{}'$ is much smaller than ρ, the cross section is a circle whose center is displaced by an amount

$$\Delta(\rho) = \frac{\rho^2}{2R}\left(\ln\frac{8R}{\rho} - 1\right) + \frac{2\pi}{\mu_0 I_p R}(h_1 + h_2\rho^2) \tag{7.208}$$

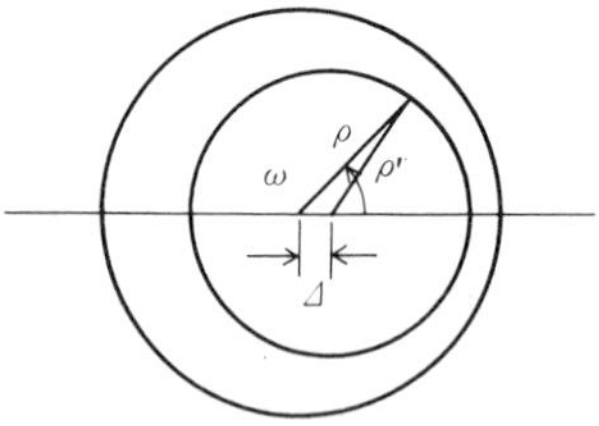

Fig. 7.9
Displacement of the plasma column.

Let us consider the parameters h_1 and h_2. As will be shown in sec. 7.7b, the poloidal component $B_{\omega'}$ of the magnetic field outside of the plasma must be

$$B_{\omega'}(a, \omega') = B_a\left(1 + \frac{a}{R}\Lambda \cos \omega'\right) \tag{7.209}$$

at equilibrium. a is the plasma radius and

$$\Lambda = \beta_p + \frac{l_i}{2} - 1, \tag{7.210}$$

and the poloidal beta ratio is

$$\beta_p = \frac{p}{(B_a^2/2\mu_0)} \tag{7.211}$$

and l_i is

$$l_i = \frac{\int B_{\omega'}^2 \rho \mathrm{d}\rho \mathrm{d}\omega'}{\pi a^2 B_a^2} \tag{7.212}$$

The parameters h_1 and h_2 are chosen to satisfy $B_\rho = 0$ and $B_{\omega'} = B_a \times (1 + (a/R)\Lambda \cos \omega')$ at the plasma boundary, i.e.,

$$h_1 = \frac{\mu_0 I_p}{4\pi} a^2\left(\Lambda + \frac{1}{2}\right) \qquad h_2 = -\frac{\mu_0 I_p}{4\pi}\left(\ln\frac{8R}{a} + \Lambda - \frac{1}{2}\right). \tag{7.213}$$

Accordingly ψ is given by

$$\psi = \frac{\mu_0 I_p R}{2\pi}\left(\ln\frac{8R}{\rho} - 2\right) - \frac{\mu_0 I_p}{4\pi}\left(\ln\frac{\rho}{a} + \left(\Lambda + \frac{1}{2}\right)\left(1 - \frac{a^2}{\rho^2}\right)\right)\rho \cos \omega'. \tag{7.214}$$

The term $h_2 \rho \cos \omega'$ in ψ (eq. 7.201) brings in the homogeneous vertical field

$$B_z = \frac{h_2}{R}, \tag{7.215}$$

which is to say that we must impose a vertical external field. When we write $\psi_e = h_2 \rho \cos \omega'$ so that ψ is the sum of two terms, $\psi = \psi_p + \psi_e$, ψ_e and ψ_p are expressed by

$$\psi_e = -\frac{\mu_0 I_p}{4\pi}\left(\ln\frac{8R}{a} + \Lambda - \frac{1}{2}\right)\rho\cos\omega' \tag{7.216}$$

$$\psi_p = \frac{\mu_0 I_p R}{2\pi}\left(\ln\frac{8R}{\rho} - 2\right) + \frac{\mu_0 I_p}{4\pi}\left(\left(\ln\frac{8R}{\rho} - 1\right)\rho + \frac{a^2}{\rho}\left(\Lambda + \frac{1}{2}\right)\right)\cos\omega'. \tag{7.217}$$

These formulas show that a uniform magnetic field in the z direction,

$$B_\perp = -\frac{\mu_0 I_p}{4\pi R}\left(\ln\frac{8R}{a} + \Lambda - \frac{1}{2}\right), \tag{7.218}$$

must be applied in order to maintain a toroidal plasma in equilibrium (fig. 7.10).

The amount of $B_\perp$ (eq. 7.218) for the equilibrium can be derived more intuitively. The *hoop force* by which the current ring of a plasma tends to expand is given by[14a]

$$F_h = \frac{\partial}{\partial R}\frac{L_p I_p^2}{2}\bigg|_{I_p = \text{const.}}, \tag{7.219}$$

where L_p is the self-inductance of the current ring:

$$L_p = \mu_0 R\left(\ln\frac{8R}{a} + \frac{l_i}{2} - 2\right). \tag{7.220}$$

Accordingly, the hoop force is

$$F_h = \frac{\mu_0 I_p^2}{2}\left(\ln\frac{8R}{a} + \frac{l_i}{2} - 1\right). \tag{7.221}$$

The outward force F_p exerted by the plasma pressure is

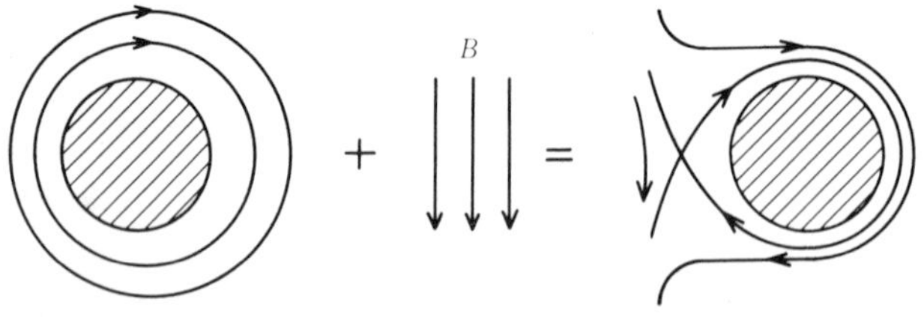

Fig. 7.10
Poloidal magnetic field due to the combined plasma current and vertical magnetic field.

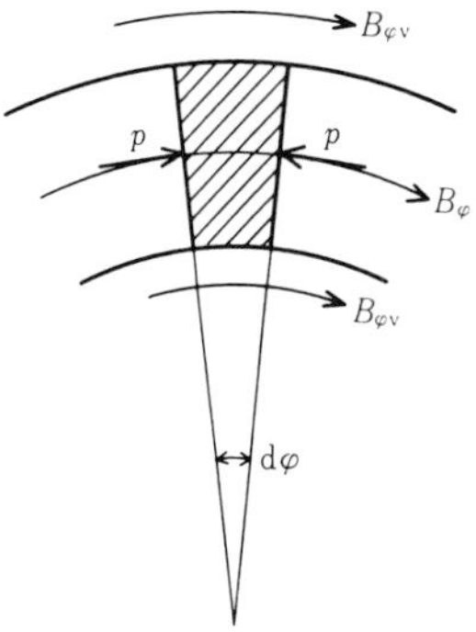

Fig. 7.11
Equilibrium of forces acting on a toroidal plasma.

$$\left.\begin{aligned} F_p \delta R &= \langle p \rangle \delta \mathscr{V} = \langle p \rangle 2\pi \delta (R\pi a^2) \\ F_p &= 2\pi^2 a^2 \langle p \rangle, \end{aligned}\right\} \tag{7.222}$$

where $\delta \mathscr{V}$ is the plasma volume (fig. 7.11).
The inward (contractive) force F_{B1} due to the tension of the toroidal field inside the plasma is

$$F_{B1} = -\frac{\langle B_\varphi{}^2 \rangle}{2\mu_0} 2\pi^2 a^2, \tag{7.223}$$

and the outward force F_{B2} by the pressure due to the external magnetic field $B_{\varphi v}$ is

$$F_{B2} = \frac{B_{\varphi v}{}^2}{2\mu_0} 2\pi^2 a^2. \tag{7.224}$$

The force F_J acting on the plasma due to the vertical field $B_\perp$ is

$$F_J = I_p B_\perp 2\pi R. \tag{7.225}$$

Balancing these forces gives

$$\frac{\mu_0 I_p{}^2}{2}\left(\ln\frac{8R}{a} + \frac{l_i}{2} - 1\right) + 2\pi^2 a^2 \left(\langle p \rangle + \frac{B_{\varphi v}{}^2}{2\mu_0} - \frac{\langle B_\varphi{}^2 \rangle}{2\mu_0}\right) + 2\pi R I_p B_\perp = 0,$$

and the amount of $B_\perp$ necessary is

$$B_\perp = \frac{-\mu_0 I_p}{4\pi R}\left(\ln\frac{8R}{a} + \frac{l_i}{2} - 1 + \beta_p - \frac{1}{2}\right). \tag{7.226}$$

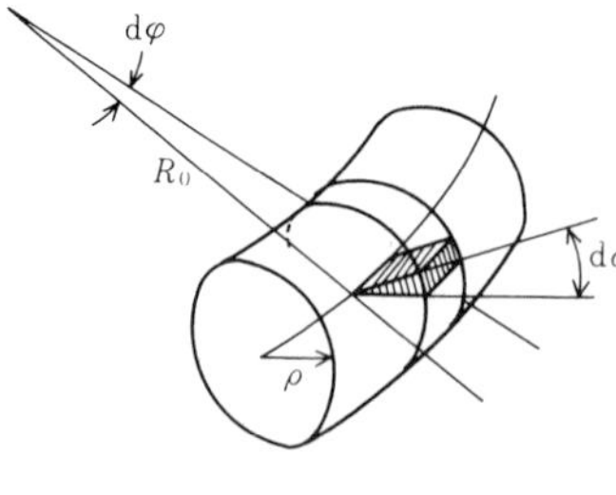

Fig. 7.12
Volume element of a toroidal plasma.

7.7b Poloidal Field for Tokamak Equilibrium[15]

The plasma pressure and the magnetic stress tensor are given by

$$T_{\alpha\beta} = \left(p + \frac{B^2}{2\mu_0}\right)\delta_{\alpha\beta} - \frac{B_\alpha B_\beta}{\mu_0}. \tag{7.227}$$

Let us consider a volume element bounded by $(\omega, \omega + \mathrm{d}\omega)$, $(\varphi, \varphi + \mathrm{d}\varphi)$, and $(0, a)$ as is shown in fig. 7.12. Denote the unit vectors in the directions r, z, φ and ρ, ω by $\boldsymbol{e}_r$, $\boldsymbol{e}_z$, $\boldsymbol{e}_\varphi$ and $\boldsymbol{e}_\rho$, $\boldsymbol{e}_\omega$, respectively. The relations between these are

$$\left.\begin{aligned} &\boldsymbol{e}_\rho = \boldsymbol{e}_r \cos\omega + \boldsymbol{e}_z \sin\omega \\ &\boldsymbol{e}_\omega = \boldsymbol{e}_z \cos\omega - \boldsymbol{e}_r \sin\omega \\ &\frac{\partial \boldsymbol{e}_\omega}{\partial \omega} = -\boldsymbol{e}_\rho \qquad \frac{\partial \boldsymbol{e}_\rho}{\partial \omega} = \boldsymbol{e}_\omega. \end{aligned}\right\} \tag{7.228}$$

(Here ω is the same as ω' of sec. 7.7a). Let $\mathrm{d}S_\rho$, $\mathrm{d}S_\omega$, $\mathrm{d}S_\varphi$ be surface-area elements with the normal vectors $\boldsymbol{e}_\rho$, $\boldsymbol{e}_\omega$, $\boldsymbol{e}_\varphi$. Then estimate the forces acting on the surfaces $\mathrm{d}S_\varphi(\varphi)$, $\mathrm{d}S_\varphi(\varphi + \mathrm{d}\varphi)$; $\mathrm{d}S_\omega(\omega)$, $\mathrm{d}S_\omega(\omega + \mathrm{d}\omega)$; and $\mathrm{d}S_\rho(a)$. The sum $\boldsymbol{F}_\varphi$ of forces acting on $\mathrm{d}S_\varphi(\varphi)$ and $\mathrm{d}S_\varphi(\varphi + \mathrm{d}\varphi)$ is given by

$$\begin{aligned} \boldsymbol{F}_\varphi &= -\mathrm{d}\omega\mathrm{d}\varphi \int_0^a \left(T_{\varphi\varphi}\frac{\partial \boldsymbol{e}_\varphi}{\partial \varphi} + T_{\varphi\omega}\frac{\partial \boldsymbol{e}_\omega}{\partial \varphi} + T_{\varphi\rho}\frac{\partial \boldsymbol{e}_\rho}{\partial \varphi}\right)\rho\mathrm{d}\rho \\ &= -\mathrm{d}\omega\mathrm{d}\varphi \int_0^a (T_{\varphi\varphi}{}^0(\boldsymbol{e}_\omega \sin\omega - \boldsymbol{e}_\rho \cos\omega) - T_{\varphi\omega}{}^0\boldsymbol{e}_\varphi \sin\omega)\rho\mathrm{d}\rho. \end{aligned} \tag{7.229}$$

When the forces acting on $\mathrm{d}S_\omega(\omega)$ and $\mathrm{d}S_\omega(\omega + \mathrm{d}\omega)$ are estimated, we must take into account the differences in $\boldsymbol{e}_\omega$, $T_{\omega\alpha}$, $\mathrm{d}S_\omega = (R + \rho\cos\omega)\mathrm{d}\rho\mathrm{d}\varphi$

at ω and $\omega + \mathrm{d}\omega$. The sum $\boldsymbol{F}_\omega$ of forces is

$$\boldsymbol{F}_\omega = -\mathrm{d}\omega\mathrm{d}\varphi \int_0^a \left[T_{\omega\omega}\frac{\partial}{\partial\omega}(\boldsymbol{e}_\omega(R+\rho\cos\omega)) + T_{\omega\varphi}\frac{\partial}{\partial\omega}(\boldsymbol{e}_\varphi(R+\rho\cos\omega)) \right.$$
$$\left. + T_{\omega\rho}\frac{\partial}{\partial\omega}(\boldsymbol{e}_\rho(R+\rho\cos\omega)) + \frac{\partial T_{\omega\omega}}{\partial\omega}R\boldsymbol{e}_\omega + \frac{\partial T_{\omega\varphi}}{\partial\omega}R\boldsymbol{e}_\varphi + \frac{\partial T_{\omega\rho}}{\partial\omega}R\boldsymbol{e}_\rho \right]\mathrm{d}\rho$$
$$= \mathrm{d}\omega\mathrm{d}\varphi\left(R\boldsymbol{e}_\rho \int_0^a T_{\omega\omega}{}^0\mathrm{d}\rho \right)$$
$$+ \mathrm{d}\omega\mathrm{d}\varphi\left[\boldsymbol{e}_\rho\left(\cos\omega \int T_{\omega\omega}{}^0\rho\mathrm{d}\rho + R\int T_{\omega\omega}{}^{(1)}\mathrm{d}\rho \right) \right.$$
$$\left. + \boldsymbol{e}_\omega\left(\sin\omega \int T_{\omega\omega}{}^0\rho\mathrm{d}\rho - R\int \frac{\partial T_{\omega\omega}{}^{(1)}}{\partial\omega}\mathrm{d}\rho \right) \right]$$
$$+ \mathrm{d}\omega\mathrm{d}\varphi\boldsymbol{e}_\varphi\left(\sin\omega \int T_{\omega\varphi}{}^0\rho\mathrm{d}\rho - R\int \frac{\partial T_{\omega\varphi}{}^{(1)}}{\partial\omega}\mathrm{d}\rho \right)$$
$$+ \mathrm{d}\omega\mathrm{d}\varphi\left(-\boldsymbol{e}_\omega R\int T_{\omega\rho}{}^{(1)}\mathrm{d}\rho - \boldsymbol{e}_\rho R\int \frac{\partial T_{\omega\rho}{}^{(1)}}{\partial\omega}\mathrm{d}\rho \right)$$
$$= \mathrm{d}\omega\mathrm{d}\varphi\left(R\boldsymbol{e}_\rho \int_0^a T_{\omega\omega}{}^0\mathrm{d}\rho \right)$$
$$+ \mathrm{d}\omega\mathrm{d}\varphi\boldsymbol{e}_\rho\left(\cos\omega \int T_{\omega\omega}{}^0\rho\mathrm{d}\rho + R\int\left(T_{\omega\omega}{}^{(1)} - \frac{\partial T_{\omega\rho}{}^{(1)}}{\partial\omega} \right)\mathrm{d}\rho \right)$$
$$+ \mathrm{d}\omega\mathrm{d}\varphi\boldsymbol{e}_\omega\left(\sin\omega \int T_{\omega\omega}{}^0\rho\mathrm{d}\rho - R\int\left(T_{\omega\rho}{}^{(1)} + \frac{\partial T_{\omega\omega}{}^{(1)}}{\partial\omega} \right)\mathrm{d}\rho \right)$$
$$+ \mathrm{d}\omega\mathrm{d}\varphi\boldsymbol{e}_\varphi\left(\sin\omega \int T_{\omega\varphi}{}^0\rho\mathrm{d}\rho - R\int \frac{\partial T_{\omega\varphi}{}^{(1)}}{\partial\omega}\mathrm{d}\rho \right). \tag{7.230}$$

As $B_\rho(a) = 0$, the force $\boldsymbol{F}_\rho$ acting on $\mathrm{d}S_\rho(a)$ is

$$\boldsymbol{F}_\rho = -\boldsymbol{e}_\rho T_{\rho\rho}(R + a\cos\omega)a\mathrm{d}\varphi\mathrm{d}\omega$$
$$= \boldsymbol{e}_\rho(-T_{\rho\rho}{}^0 Ra - (T_{\rho\rho}{}^{(1)}Ra + T_{\rho\rho}{}^0 a^2\cos\omega)). \tag{7.231}$$

The equilibrium condition $\boldsymbol{F}_\varphi + \boldsymbol{F}_\omega + \boldsymbol{F}_\rho = 0$ is reduced to

$$\int T_{\omega\omega}{}^0\mathrm{d}\rho = T_{\rho\rho}{}^0(a)a \tag{7.232}$$

$$\frac{\partial}{\partial \omega}\int T_{\omega\varphi}{}^{(1)}\mathrm{d}\rho = \frac{2\sin\omega}{R}\int T_{\omega\varphi}{}^{0}\rho\,\mathrm{d}\rho \tag{7.233}$$

$$\int\left(T_{\omega\rho}{}^{(1)} + \frac{\partial T_{\omega\omega}{}^{(1)}}{\partial \omega}\right)\mathrm{d}\rho = \frac{\sin\omega}{R}\int (T_{\omega\omega}{}^{0} - T_{\varphi\varphi}{}^{0})\rho\,\mathrm{d}\rho \tag{7.234}$$

$$\cos\omega\int (T_{\varphi\varphi}{}^{0} + T_{\omega\omega}{}^{0})\rho\,\mathrm{d}\rho + R\int\left(T_{\omega\omega}{}^{(1)} - \frac{\partial T_{\omega\rho}{}^{(1)}}{\partial\omega}\right)\mathrm{d}\rho$$
$$- T_{\rho\rho}{}^{0}a^2\cos\omega - T_{\rho\rho}{}^{(1)}Ra = 0. \tag{7.235}$$

From $T^{(1)} \propto \sin\omega, \cos\omega$, it follows that $\partial^2 T^{(1)}/\partial\omega^2 = -T^{(1)}$. So eq. (7.234) is

$$\int\left(\frac{\partial T_{\omega\rho}{}^{(1)}}{\partial\omega} - T_{\omega\omega}{}^{(1)}\right)\mathrm{d}\rho = \frac{\cos\omega}{R}\int (T_{\omega\omega}{}^{0} - T_{\varphi\varphi}{}^{0})\rho\,\mathrm{d}\rho. \tag{7.236}$$

Using this relation, we can rewrite eq. (7.235) as

$$T_{\rho\rho}{}^{(1)}(a) = \frac{a}{R}\cos\omega\left(-T_{\rho\rho}{}^{0}(a) + \frac{2}{a^2}\int_0^a T_{\varphi\varphi}{}^{0}\rho\,\mathrm{d}\rho\right). \tag{7.237}$$

$T_{\rho\rho}$ and $T_{\varphi\varphi}$ are given by

$$T_{\rho\rho} = p + \frac{B_\omega{}^2}{2\mu_0} + \frac{B_\varphi{}^2}{2\mu_0} \qquad T_{\varphi\varphi} = p + \frac{B_\omega{}^2}{2\mu_0} - \frac{B_\varphi{}^2}{2\mu_0}. \tag{7.238}$$

From eq. (7.167), B_φ is

$$B_\varphi = \frac{\mu_0 I(\psi)}{2\pi r} = \frac{\mu_0 I(\psi)}{2\pi R}\left(1 - \frac{\rho}{R}\cos\omega + \cdots\right)$$
$$= B_\varphi(\rho)\left(1 - \frac{\rho}{R}\cos\omega + \cdots\right). \tag{7.239}$$

When $B_\omega(a)$ is written as $B_\omega(a) = B_a + B_\omega{}^{(1)}$, eqs. (7.238) and (7.239) yield the expression

$$T_{\rho\rho}{}^{(1)}(a) = \frac{B_a B_\omega{}^{(1)}}{\mu_0} - \frac{B_{\varphi\mathrm{v}}{}^2(a)}{2\mu_0}2\frac{a}{R}\cos\omega. \tag{7.240}$$

On the other hand, eqs. (7.205) and (7.237) give $T_{\rho\rho}{}^{(1)}(a)$ as

$$T_{\rho\rho}{}^{(1)}(a) = \frac{a}{R}\cos\omega\left(-p_a - \frac{B_a{}^2}{2\mu_0} - \frac{B_{\varphi\mathrm{v}}{}^2(a)}{2\mu_0} + \langle p\rangle + \frac{\langle B_\omega{}^2\rangle}{2\mu_0} - \frac{\langle B_\varphi{}^2\rangle}{2\mu_0}\right)$$

$$= \frac{a}{R}\cos\omega\left(\frac{B_a^{\,2}}{2\mu_0}l_{\rm i} + 2(\langle p\rangle - p_a) - \frac{B_a^{\,2}}{\mu_0} - \frac{B_{\varphi {\rm v}}^{\,2}(a)}{\mu_0}\right), \tag{7.241}$$

where

$$l_{\rm i} \equiv \frac{1}{B_a^{\,2}}\frac{2}{a^2}\int_0^a B_\omega^{\,2}\rho\,{\rm d}\rho \tag{7.242}$$

is the normalized internal inductance of the plasma per unit length (the internal inductance $L_{\rm i}$ of the plasma per unit length is given by $\mu_0 l_{\rm i}/4\pi$). Accordingly, $B_\omega^{(1)}$ must be

$$B_\omega^{(1)} = \frac{a}{R}B_a\cos\omega\left(\frac{l_{\rm i}}{2} + \frac{2\mu_0(\langle p\rangle - p_a)}{B_a^{\,2}} - 1\right). \tag{7.243a}$$

B_a is ω component of the magnetic field due to the plasma current $I_{\rm p}$, i.e.,

$$B_a = -\frac{\mu_0 I_{\rm p}}{2\pi a}.$$

When the poloidal beta ratio $\beta_{\rm p}$ (recall that this is the ratio of the plasma pressure p to the magnetic pressure due to B_a) is used, $B_\omega^{(1)}$ is given by

$$B_\omega^{(1)} = \frac{a}{R}B_a\cos\omega\left(\frac{l_{\rm i}}{2} + \beta_{\rm p} - 1\right). \tag{7.243b}$$

7.7c Upper Limit of Beta Ratio

In the previous subsection, the value of B_ω necessary for equilibrium was derived. In this derivation, $(a/R)\Lambda < 1$ was assumed, i.e.,

$$\left(\beta_{\rm p} + \frac{l_{\rm i}}{2}\right) < \frac{R}{a}.$$

The vertical field $B_\perp$ for plasma equilibrium is given by

$$B_\perp = B_a\frac{a}{2R}\left(\ln\frac{8R}{a} + \Lambda - \frac{1}{2}\right).$$

The direction of $B_\perp$ is opposite to that of B_ω due to the plasma current inside the torus, so that the resultant poloidal field becomes zero at some point inside the torus and a separatrix is formed (see fig. 7.10). When the plasma pressure is increased and $\beta_{\rm p}$ becomes large, the necessary amount of $B_\perp$ is increased and the separatrix shifts toward the plasma. For simplicity, let us consider a sharp-boundary model in which the plasma pressure is constant inside the plasma boundary, and in which the boundary encloses a plasma current I_p. Then the pressure-balance equation is

$$\frac{B_\omega^2}{2\mu_0} + \frac{B_{\varphi v}^2}{2\mu_0} \approx p + \frac{B_{\varphi i}^2}{2\mu_0}. \tag{7.244a}$$

The φ components $B_{\varphi v}$, $B_{\varphi i}$ of the field outside and inside the plasma boundary are proportional to $1/r$, according to eq. (7.167). If the values of $B_{\varphi v}$, $B_{\varphi i}$ at $r = R$ are denoted by $B_{\varphi v}{}^0$, $B_{\varphi i}{}^0$ respectively, eq. (7.244a) may be written as

$$B_\omega{}^2 = 2\mu_0 p - ((B_{\varphi v}{}^0)^2 - (B_{\varphi i}{}^0)^2)\left(\frac{R}{r}\right)^2. \tag{7.244b}$$

The upper limit of the plasma pressure is determined by the condition that the resultant poloidal field at $r = r_{\min}$ inside the torus is zero,

$$2\mu_0 p_{\max}\frac{r_{\min}{}^2}{R^2} = (B_{\varphi v}{}^0)^2 - (B_{\varphi i}{}^0)^2.$$

As r is expressed by $r = R + a\cos\omega$, eq. (7.244b) is reduced (with $r_{\min} = R - a$) to

$$B_\omega{}^2 = 2\mu_0 p_{\max}\left(1 - \frac{r_{\min}{}^2}{r^2}\right) = 8\mu_0 p_{\max}\frac{a}{R}\cos^2\frac{\omega}{2}.$$

Here $a/R \ll 1$ is assumed. From the relation $\oint B_\omega a\,d\omega = \mu_0 I_p$, the upper limit $\beta_p{}^c$ of the poloidal beta ratio is

$$\beta_p{}^c = \frac{\pi^2}{16}\frac{R}{a} \approx 0.5\frac{R}{a}. \tag{7.245}$$

Thus the upper limit of $\beta_p{}^c$ is half of the aspect ratio R/a in this simple model; experimentally, the value of β_p is found to be in the range of about 0.5–1 (for ohmically heated plasmas).

When the rotational transform angle ι and the safety factor $q_s = 2\pi/\iota$ defined in sec. 2.4 are introduced, we find from eq. (2.148) that

$$\frac{B_\omega}{B_\varphi} = \frac{a}{R}\left(\frac{\iota}{2\pi}\right) = \frac{a}{Rq_s}, \tag{7.246}$$

so that

$$\beta = \frac{p}{B^2/2\mu_0} \approx \frac{p}{B_\omega{}^2/2\mu_0}\left(\frac{B_\omega}{B_\varphi}\right)^2 = \left(\frac{a}{Rq_s}\right)^2 \beta_p. \tag{7.247}$$

Accordingly, the upper limit of the beta ratio is

$$\beta^c = \frac{0.5}{q_s{}^2}\frac{a}{R}. \tag{7.248}$$

When $\beta_p \approx 0.5$, the beta ratio is

$$\beta \approx 0.5\left(\frac{a}{R}\right)^2 \frac{1}{q_s{}^2}. \tag{7.249}$$

7.8 Stellarator Equilibrium; Upper Limit of the Stellarator Beta Ratio

When the plasma pressure is isotropic, the current $\boldsymbol{j}$ in the plasma is given by eqs. (7.74) and (7.82) as

$$j_\perp = \frac{\boldsymbol{b}}{B} \times \nabla p \tag{7.250}$$

$$\nabla \boldsymbol{j}_\parallel = -\nabla \boldsymbol{j}_\perp = 2\nabla p \cdot \left(\nabla\left(p + \frac{B^2}{2\mu_0}\right) \times \boldsymbol{B}\right)\frac{\mu_0}{B^4}. \tag{7.251}$$

When the beta ratio is small, $j_\parallel$ is

$$\frac{\partial j_\parallel}{\partial s} = 2\nabla p \cdot \frac{(\nabla B \times \boldsymbol{b})}{B^2}, \tag{7.252}$$

where s is length along a line of magnetic force. In the zeroth-order approximation, we can put $B = |B_0|(1 - (r/R)\cos\theta)$, $p = p(r)$, and $\partial/\partial s = (\partial\theta/\partial s)(\partial/\partial\theta) = (\iota/(2\pi R))\partial/\partial\theta$, where ι is the rotational transform angle. When s increases by $2\pi R$, θ increases by ι. Then eq. (7.252) is reduced to

$$\frac{\iota}{2\pi R}\frac{\partial j_\parallel}{\partial\theta} = -\frac{\partial p}{\partial r}\frac{2}{RB_0}\sin\theta, \tag{7.253}$$

i.e.,

$$\frac{\partial j_\parallel}{\partial\theta} = -\frac{4\pi}{\iota B_0}\frac{\partial p}{\partial r}\sin\theta$$

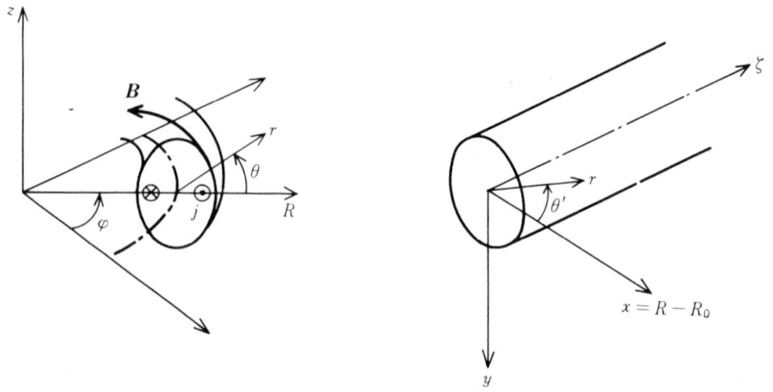

Fig. 7.13
Pfirsch-Schlüter current in a toroidal plasma.

$$j_{\|} = \frac{4\pi}{\iota B_0} \frac{\partial p}{\partial r} \cos\theta. \tag{7.254}$$

This current is called the *Pfirsch-Schlüter current* (fig. 7.13).
These formulas are very important, and will be used to obtain the diffusion coefficient of a toroidal plasma. The Pfirsch-Schlüter factor will be also derived from these formulas. The Pfirsch-Schlüter current is due to the short circuiting, along magnetic-field lines, of toroidal drift polarization charges. The resultant current is inversely proportional to ι.

Let us take the radial variation in plasma pressure p and ι to be

$$p(r) = p_0\left(1 - \left(\frac{r}{a}\right)^m\right) \tag{7.255}$$

$$\iota(r) = \iota_0\left(\frac{r}{a}\right)^{2\iota - 4}, \tag{7.256}$$

respectively; then $j_{\|}$ is

$$j_{\|} = -\frac{4\pi m p_0}{B_0 \iota_0 a}\left(\frac{r}{a}\right)^{m - 2l + 3} \cos\theta. \tag{7.257}$$

Let us estimate the magnetic field $\boldsymbol{B}^\beta$ produced by the current $j_{\|}$. As a/R is small, $\boldsymbol{B}^\beta$ is estimated from the corresponding linear configuration of fig. 7.13. We utilize the coordinates r, θ', ζ and put $\theta = -\theta'$ and $j_{\|} \approx j_\zeta$. Then the vector potential $\boldsymbol{A}^\beta = (0, 0, A_\zeta{}^\beta)$ for $\boldsymbol{B}^\beta$ is given by

$$\frac{1}{r}\frac{\partial}{\partial r}\left(r\frac{\partial A_\zeta{}^\beta}{\partial r}\right) + \frac{1}{r^2}\frac{\partial^2 A_\zeta{}^\beta}{\partial \theta'^2} = -\mu_0 j_\zeta. \tag{7.258}$$

When $A_\zeta{}^\beta(r, \theta') = A^\beta(r)\cos\theta'$, $s = m - 2l + 3$, and $\alpha = 4\pi m p_0 \mu_0/(B_0 \iota_0)$ $= m\beta B_0/(\iota_0/2\pi)$, eq. (7.258) becomes

$$\frac{1}{r}\frac{\partial}{\partial r}\left(r\frac{\partial A^\beta}{\partial r}\right) - \frac{A^\beta}{r^2} = \frac{\alpha}{a}\left(\frac{r}{a}\right)^s.$$

In the plasma region ($r < a$), the vector potential is

$$A_{\text{in}}{}^\beta = \left(\frac{\alpha r^{s+2}}{((s+2)^2 - 1)a^{s+1}} + \beta r\right)\cos\theta' \tag{7.259}$$

and $A_{\text{out}}{}^\beta$ outside the plasma region ($r > a$) is

$$A_{\text{out}}{}^\beta = \frac{\gamma}{r}\cos\theta'. \tag{7.260}$$

Since $\boldsymbol{B}^\beta(r, \theta')$ must be continuous at the boundary $r = a$, the solution for $\boldsymbol{B}^\beta$ inside the plasma is

$$\left.\begin{aligned} B_r{}^\beta &= -\frac{\alpha}{(s+2)^2 - 1}\left(\left(\frac{r}{a}\right)^{s+1} - \frac{s+3}{2}\right)\sin\theta' \\ B_{\theta'}{}^\beta &= -\frac{\alpha}{(s+2)^2 - 1}\left((s+2)\left(\frac{r}{a}\right)^{s+1} - \frac{s+3}{2}\right)\cos\theta' \end{aligned}\right\} \tag{7.261}$$

and the solution outside is

$$\left.\begin{aligned} B_r{}^\beta &= \frac{\alpha}{(s+2)^2 - 1}\,\frac{s+1}{2}\left(\frac{a}{r}\right)^2 \sin\theta' \\ B_\theta{}^\beta &= \frac{-\alpha}{(s+2)^2 - 1}\,\frac{s+1}{2}\left(\frac{a}{r}\right)^2 \cos\theta' \end{aligned}\right\} \tag{7.262}$$

($B_r = r^{-1}\partial A_\zeta/\partial\theta'$, $B_{\theta'} = -\partial A_\zeta/\partial r$). As is clear from eq. (7.261), there is a homogeneous vertical-field component

$$\begin{aligned} B_z &= \frac{-(s+3)\alpha}{2((s+2)^2 - 1)} = \frac{-(m - 2l + 6)m}{2((m - 2l + 5)^2 - 1)}\frac{\beta}{(\iota_0/2\pi)}B_0 \\ &= -f(m, l)\frac{\beta}{(\iota_0/2\pi)}B_0. \end{aligned} \tag{7.263}$$

This field causes the magnetic surface to be displaced by the amount Δ. From eq. (2.160), Δ is given by

$$\frac{\Delta}{R} = \frac{-2\pi B_z}{\iota(\Delta) B_0} = f(m, l)\frac{(2\pi)^2 \beta}{\iota(\Delta)\iota_0} \approx f(m, l)\frac{\beta}{(\iota_0/2\pi)^2}. \tag{7.264}$$

$f(m, l)$ is of the order of 1 and the condition $\Delta < a/2$ gives the upper limit of the beta ratio:

$$\beta_c < \frac{1}{2}\frac{a}{R}\left(\frac{2\pi}{\iota}\right)^2. \tag{7.265}$$

This critical value is the same as that for the tokamak. The reader should refer to ref. 16 for more detailed discussion.

References

1. T. Uchida, K. Miyamoto, J. Fujita, S. Kawasaki, N. Inoue, Y. Suzuki, and K. Adati: Nucl. Fusion **8**, 263(1968)
T. Uchida, K. Miyamoto, J. Fujita, C. Leloup, S. Kawasaki, N. Inoue, Y. Suzuki, and K. Adati: ibid. **9**, 259(1969)

2. F. Waelbroeck, C. Leloup, J. P. Poffe, P. Eward, R. DerAgobian, and D. Veron: Plasma Phys. and Controlled Nucl. Fusion Research (Conf. Proceedings, Salzburg 1961). Also in Nucl. Fusion, suppl., pt. 2, p. 675(1962)

3. Y. Mizuno: Summer school textbook for students of plasma physics, 1970, p. 201. (in Japanese)

4. J. B. Taylor: Phys. of Fluids **6**, 1529(1963)
D. R. Dobrott and J. L. Johnson: Plasma Phys. **11**, 211(1969)

5. M. D. Kruskal and R. M. Kulsrud: Phys. of Fluids **1**, 265(1958)

6. V. D. Shafranov: Rev. of Plasma Phys. **2** (ed. by M. A. Leontovich), Consultants Bureau, New York, 1966

7. V. D. Shafranov and E. I. Yurckenko: Sov. Phys. JETP **26**, 682(1968)

8. S. Hamada: Nucl. Fusion **1**, 23(1962)

9. W. B. Thompson: An Introduction to Plasma Physics, p. 56, Addison-Wesley, Reading, Mass., 1964

10. Y. Suzuki: Nucl. Fusion **14**, 345(1974)

11. G. Laval, H. Luc, E. K. Maschke and C. Mercier: Plasma Phys. and Controlled Nucl. Fusion Research **2**, 507 (Conf. Proceeding, Madison 1971) (IAEA, Vienna, 1971)

12. J. L. Johnson, C. R. Oberman, R. M. Kulsrud, and E. A. Frieman: Phys. of Fluids **1**, 281(1958)

13. S. Chandrasekhar and P. C. Kendall: Astrophys. J. **126**, 457(1957)
P. Barberio-Corsetti: Plasma Phys. **15**, 1131(1973)
D. Correa, D. Lortz: Nucl. Fusion **13**, 127(1973)

14. V. S. Mukhovatov and V. D. Shafranov: ibid. **11**, 605(1971)

14a. See, for example, J. A. Stratton: Electromagnetic Theory, sec. 2.14, McGraw-Hill, New York, 1941.

15. V. D. Shafranov: Plasma Phys., J. of Nucl. Energy pt. C **5**, 251 (1963)

16. L. S. Solovev and V. D. Shafranov: Rev. of Plasma Phys. **5**, 1 (ed. by M. A. Leontovich), Consultants Bureau, New York, 1966

8 Diffusion and Confinement Time

Diffusion and confinement of plasmas are among the most important subjects in fusion research, with theoretical and experimental investigations being carried out concurrently. Although a general discussion of diffusion and confinement requires that the various instabilities (which will be studied in subsequent chapters) be taken into account, it is also important to consider diffusion for the ideal stable cases. A typical example (sec. 8.1) is classical diffusion, in which collisions between electrons and ions are dominant. Sections 8.2 and 8.3 describe the neoclassical diffusion of toroidal plasmas confined in tokamak and stellarator systems, for both the rare-collision and intermediate-collision regions. Sometimes the diffusion of an unstable plasma may be studied in a phenomenological way, without recourse to a detailed knowledge of instabilities. In this manner, diffusion caused by fluctuations in a plasma (Bohm diffusion and pseudo-classical diffusion) are explained in secs. 8.4 and 8.5. In sec. 8.6, diffusion enhanced by impurity ions is discussed; and the bootstrap current in a toroidal plasma, which is considered to be induced by diffusion of the trapped particles, is studied in sec. 8.7.

The *diffusion coefficient* D, density n, and confinement time τ are related by the diffusion equation as follows:

$$\nabla(D\nabla n(\boldsymbol{r}, t)) = \frac{\partial n(\boldsymbol{r}, t)}{\partial t}. \tag{8.1a}$$

When it is possible to assume that $n(\boldsymbol{r}, t) = n(\boldsymbol{r})\exp(-t/\tau)$, eq. (8.1a) yields

$$\nabla(D\nabla n(\boldsymbol{r})) = -\frac{1}{\tau}n(\boldsymbol{r}). \tag{8.1b}$$

When D is constant and the plasma column is a cylinder of radius a, eq. (8.1a) is reduced to

$$\frac{1}{r}\frac{\partial}{\partial r}\left(r\frac{\partial n}{\partial r}\right) + \frac{1}{D\tau}n = 0. \tag{8.1c}$$

The solution satisfying the boundary condition $n(a) = 0$ is

$$n(r) = n_0 J_0\left(\frac{2.4\,r}{a}\right)\exp\left(\frac{-t}{\tau}\right) \tag{8.2}$$

$$\tau = \frac{a^2}{(2.4)^2 D} = \frac{a^2}{5.8\,D}, \tag{8.3}$$

where J_0 is the zeroth-order Bessel function. The relationship (8.3) between τ and D holds generally, with only a slight modification of the numerical factor. This formula is frequently used to obtain the diffusion coefficient from the observed values of the plasma radius and confinement time.

8.1 Classical Diffusion (Collision Dominant)

8.1a Magnetohydrodynamic Treatment

A magnetohydrodynamic treatment is applicable to diffusion when the e-i collision frequency is large and the mean free path is shorter than the *connection length* of the inside regions of good curvature and the outside region of bad curvature of the torus; i.e.,

$$\frac{v_T{}^e}{\nu_{ei}} \lesssim \frac{2\pi R}{\iota}$$

$$\nu_{ei} \gtrsim \frac{1}{R}\frac{\iota}{2\pi}v_T{}^e = \frac{1}{R}\frac{\iota}{2\pi}\left(\frac{T_e}{m_e}\right)^{1/2}, \tag{8.4}$$

where $v_T{}^e$ is the electron thermal velocity and ν_{ei} is the e-i collision frequency. The flow velocity across the magnetic field (eq. (7.45)) is

$$nV_\perp = \frac{1}{B}\left(\left(nE - \frac{T_i}{e}\nabla n\right) \times b\right) - (\rho_B{}^e)^2 \nu_{ei}\left(1 + \frac{T_i}{T_e}\right)\nabla n, \tag{8.5}$$

where $\rho_B{}^e = v_T{}^e/\Omega_e$, $v_T{}^e = (T_e/m_e)^{1/2}$.

If the first term of the right-hand side of eq. (8.5) can be neglected, the diffusion coefficient is given by

$$D = (\rho_B{}^e)^2 \nu_{ei}\left(1 + \frac{T_i}{T_e}\right). \tag{8.6}$$

Here the *classical diffusion coefficient* D_{cl} is defined by

$$D_{cl} \equiv (\rho_B{}^e)^2 \nu_{el} = \frac{nT_e}{\sigma_\perp B^2} = \frac{\beta_e \eta_\parallel}{\mu_0}, \tag{8.7}$$

where $\sigma_\perp = n_e e^2/(m_e \nu_{ei})$, $\eta_\parallel = 1/2\sigma_\perp$.

However, the first term of the right-hand side of eq. (8.5) is not always negligible. In the toroidal configuration, the charge separation due to the

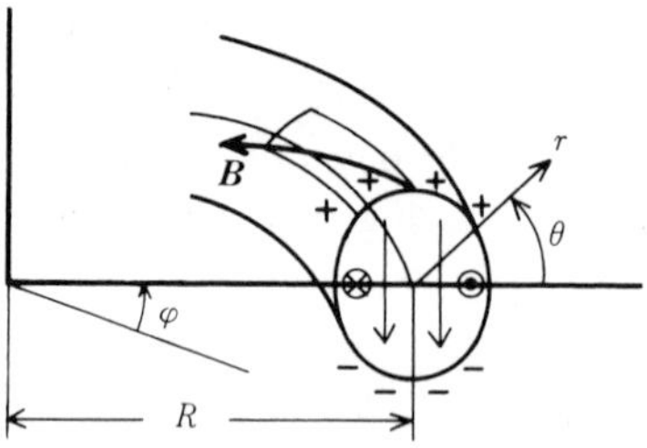

Fig. 8.1
Electric field in a plasma confined in a toroidal field. The symbols ⊗ and ⊙ here show the direction of the Pfirsch-Schlüter current.

toroidal drift is not completely cancelled along the magnetic-field lines and an electric field $\boldsymbol{E}$ arises (fig. 8.1).

Therefore the $\boldsymbol{E} \times \boldsymbol{b}$ term in eq. (8.5) contributes to the diffusion. Let us consider this term. From the equilibrium equation, the diamagnetic current

$$\boldsymbol{j}_\perp = \frac{\boldsymbol{b}}{B} \times \nabla p, \qquad j_\perp = \left|\frac{1}{B}\frac{\partial p}{\partial r}\right|$$

flows in the plasma. From $\nabla \boldsymbol{j}$, we find $\nabla \boldsymbol{j}_\parallel = -\nabla \boldsymbol{j}_\perp$. By means of the equation $B = |B_0|(1 - (r/R)\cos\theta)$, $j_\parallel$ may be written as (see eq. (7.254))

$$j_\parallel = 2\frac{2\pi}{\iota}\frac{1}{B_0}\frac{\partial p}{\partial r}\cos\theta. \tag{8.8}$$

If the plasma conductivity along the lines of magnetic force is $\sigma_\parallel$, the parallel electric field is $E_\parallel = j_\parallel/\sigma_\parallel$. As is clear from fig. 8.1, the relation

$$\frac{E_\theta}{E_\parallel} \approx \frac{B_0}{B_\theta}$$

holds. From $B_\theta/B_0 \approx (r/R)(\iota/2\pi)$, the θ component of the electric field is given by

$$E_\theta = \frac{B_0}{B_\theta}E_\parallel = \frac{R}{r}\frac{2\pi}{\iota}\frac{1}{\sigma_\parallel}j_\parallel = \frac{2}{\sigma_\parallel}\frac{R}{r}\left(\frac{2\pi}{\iota}\right)^2\frac{1}{B_0}\frac{\partial p}{\partial r}\cos\theta. \tag{8.9}$$

Accordingly, eq. (8.5) is reduced to

$$n\boldsymbol{V}_r = -n\frac{E_\theta}{B} - (\rho_B{}^{\mathrm{e}})^2\nu_{\mathrm{ei}}\left(1 + \frac{T_{\mathrm{i}}}{T_{\mathrm{e}}}\right)\frac{\partial n}{\partial r}$$

$$= -\left(\frac{R}{r}\cdot 2\left(\frac{2\pi}{\iota}\right)^2 \frac{nT_e}{\sigma_\parallel B_0{}^2}\cos\theta\left(1+\frac{r}{R}\cos\theta\right)\right.$$

$$\left. + \frac{nT_e}{\sigma_\perp B_0{}^2}\left(1+\frac{r}{R}\cos\theta\right)^2\right)\times\left(1+\frac{T_i}{T_e}\right)\frac{\partial n}{\partial r}.$$

Noting that the area of a surface element is dependent on θ, and taking the average of nV_r over θ, we find that

$$\langle nV_r\rangle = \frac{1}{2\pi}\int_0^{2\pi} nV_r\left(1+\frac{r}{R}\cos\theta\right)\mathrm{d}\theta$$

$$= -\frac{nT_e}{\sigma_\perp B_0{}^2}\left(1+\frac{T_i}{T_e}\right)\left(1+\frac{2\sigma_\perp}{\sigma_\parallel}\left(\frac{2\pi}{\iota}\right)^2\right)\frac{\partial n}{\partial r}. \tag{8.10}$$

Using the relation $\sigma_\perp = \sigma_\parallel/2$ given in sec. 7.3c, we obtain the diffusion coefficient of a toroidal plasma:

$$D_{\mathrm{P.S.}} = \frac{nT_e}{\sigma_\perp B_0{}^2}\left(1+\frac{T_i}{T_e}\right)\left(1+\left(\frac{2\pi}{\iota}\right)^2\right). \tag{8.11}$$

This diffusion coefficient is $(1+(2\pi/\iota)^2)$ times as large as the diffusion coefficient of eq. (8.2). This value is called *Pfirsch-Schlüter factor*[1]. When $\iota/2\pi$ is about 0.3, the Pfirsch-Schlüter factor is about 10.

The foregoing relation can be derived more directly from eq. (7.112) as follows: take

$$\int_\psi \boldsymbol{V}\cdot\boldsymbol{n}\mathrm{d}S = \int_\psi \frac{\eta \boldsymbol{j}^2}{|\nabla p|}\mathrm{d}S, \tag{8.12}$$

where η is the specific resistivity and $\boldsymbol{n}$ is the outward unit vector normal to $\mathrm{d}S$. The right-hand side of this equation is equal to

$$\int \frac{\left(\dfrac{j_\perp{}^2}{\sigma_\perp}+\dfrac{j_\parallel{}^2}{\sigma_\parallel}\right)}{\left|\dfrac{\partial p}{\partial r}\right|}\mathrm{d}S = \int \frac{1}{\sigma_\perp B_0{}^2}\frac{\partial p}{\partial r}\left(1+\frac{\sigma_\perp}{\sigma_\parallel}4\left(\frac{2\pi}{\iota}\right)^2\cos^2\theta\right)\mathrm{d}S$$

$$= \frac{T_e}{\sigma_\perp B_0}\left(1+\frac{T_i}{T_e}\right)\left(1+\left(\frac{2\pi}{\iota}\right)^2\right)\left|\frac{\partial n}{\partial r}\right| S, \tag{8.13}$$

and the left-hand side is equal to $D|\partial n/\partial r|S/n$.

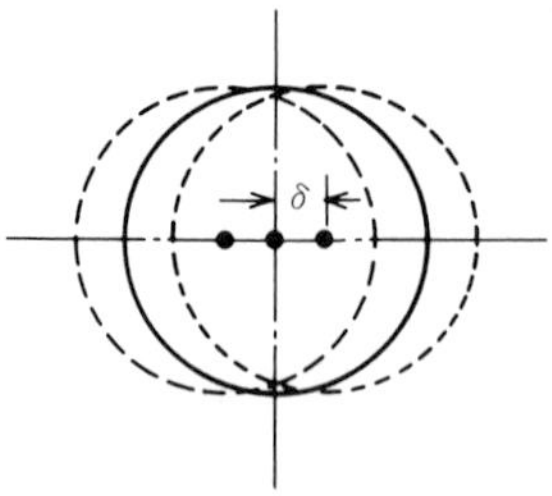

Fig. 8.2
Magnetic surface (solid line) and drift surfaces (dotted lines).

8.1b A Particle Model

The classical diffusion coefficient of electrons

$$D_{\mathrm{cl}} = (\rho_B{}^{\mathrm{e}})^2 \nu_{\mathrm{ei}}$$

is that for electrons which move in a random walk with a step length equal to the Larmor radius. Let us consider a toroidal plasma. For a rotational transform angle ι, the displacement δ of the electron-drift surface from the magnetic surface is (see fig. 8.2)

$$\delta \approx \pm \rho_B{}^{\mathrm{e}} \frac{2\pi}{\iota}. \tag{8.14}$$

The + or − sign is to be taken as the direction of electron motion is parallel or antiparallel to the magnetic field (refer to sec. 3.4c). As an electron can be transferred from one drift surface to the other by collision, the step length across the magnetic field is

$$\Delta = \left(\frac{2\pi}{\iota}\right) \rho_B{}^{\mathrm{e}}. \tag{8.15}$$

Consequently, the diffusion coefficient is given by

$$D_{\mathrm{P.S.}} = \Delta^2 \nu_{\mathrm{ei}} = \left(\frac{2\pi}{\iota}\right)^2 (\rho_B{}^{\mathrm{e}})^2 \nu_{\mathrm{ei}}\,; \tag{8.16}$$

thus the Pfirsch-Schlüter factor has been reduced.

8.2 Neoclassical Diffusion in a Tokamak Field

The magnitude B of the magnetic field of a tokamak is given by

$$B = \frac{RB_0}{R(1 + \varepsilon_t \cos\theta)} = B_0(1 - \varepsilon_t \cos\theta), \tag{8.17}$$

where

$$\varepsilon_t = \frac{r}{R}. \tag{8.18}$$

Consequently, when the perpendicular component $v_\perp$ of a plasma electron's velocity is much larger than the parallel component $v_\parallel$, i.e., when

$$\frac{v_\perp}{v_\parallel} > \frac{1}{\varepsilon_t^{1/2}}, \tag{8.19}$$

the electron is trapped outside of the torus, where the magnetic field is weak. Such an electron drifts in a banana orbit (see fig. 8.3 and sec. 3.4c). In order to complete a circuit of the banana orbit, the effective collision time $\tau_{eff} = 1/\nu_{eff}$ of the trapped electron must be longer than one period τ_b of the banana orbit (see eq. 3.118):

$$\tau_b \approx \frac{R}{v_\perp \varepsilon_t^{1/2}}\left(\frac{2\pi}{\iota}\right). \tag{8.20}$$

The effective collision frequency ν_{eff} of the trapped electron is the frequency in which the condition (8.19) of the trapped electron is violated by collision. As the collision frequency ν_{ei} is the inverse of the diffusion time required to change the direction of velocity by 1 radian, the effective collision frequency is given by

$$\nu_{eff} = \frac{1}{\varepsilon_t}\nu_{ei}. \tag{8.21}$$

Accordingly, if $\nu_{eff} < 1/\tau_b$, i.e.,

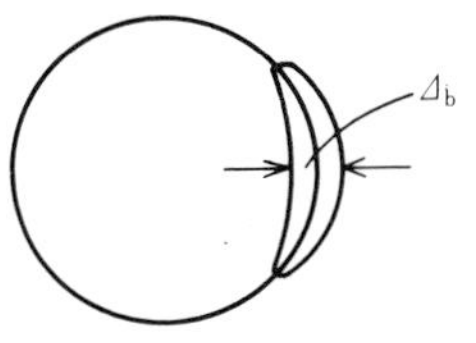

Fig. 8.3
Banana orbit of a trapped particle.

$$\nu_{ei} < \frac{v_\perp \varepsilon_t^{3/2}}{R}\left(\frac{\iota}{2\pi}\right) = \varepsilon_t^{3/2}\frac{1}{R}\left(\frac{\iota}{2\pi}\right)\left(\frac{T_e}{m_e}\right)^{1/2}, \tag{8.22}$$

the trapped electron can travel the entire banana orbit. When the trapped electron collides, it can shift its position by an amount Δ_b, the width of the banana orbit (see eq. 3.100):

$$\Delta_b = \frac{mv_\parallel}{eB_p} \approx \frac{mv_\perp}{eB}\frac{v_\parallel}{v_\perp}\frac{B}{B_p} \approx \rho_B{}^e \varepsilon_t^{1/2}\frac{R}{r}\frac{2\pi}{\iota} = \left(\frac{2\pi}{\iota}\right)\varepsilon_t^{-1/2}\rho_B{}^e. \tag{8.23}$$

As the number of trapped electrons is $\varepsilon^{1/2}$ times the total number of electrons, the trapped-electron contribution to diffusion is

$$D_{G.S.} = \varepsilon_t^{1/2}\Delta_b^2\nu_{eff} = \varepsilon_t^{1/2}\left(\frac{2\pi}{\iota}\right)^2 \varepsilon_t^{-1}(\rho_B{}^e)^2\frac{1}{\varepsilon_t}\nu_{ei}$$

$$= \varepsilon_t^{-3/2}\left(\frac{2\pi}{\iota}\right)^2(\rho_B{}^e)^2\nu_{ei}. \tag{8.24}$$

This diffusion coefficient, introduced by A. A. Galeev and R. Z. Sagdeev,[2] is $\varepsilon_t^{-3/2} = (R/r)^{3/2}$ times as large as the diffusion coefficient for the collision-dominant case.

The collision-frequency region in which the MHD treatment is applicable is found from eq. (8.4) to be

$$\nu_{ei} > \nu_p = \frac{1}{R}\frac{\iota}{2\pi}v_T{}^e = \frac{1}{R}\left(\frac{\iota}{2\pi}\right)\left(\frac{T_e}{m_e}\right)^{1/2}. \tag{8.25}$$

The collision frequency at which electrons can complete a curcuit of the banana orbit is

$$\nu_{ei} < \nu_b = \varepsilon_t^{3/2}\nu_p. \tag{8.26}$$

The diffusion coefficients are expressed by

$$D_{P.S.} = \left(\frac{2\pi}{\iota}\right)^2(\rho_B{}^e)^2\nu_{ei}, \qquad \nu_{ei} > \nu_p \tag{8.27}$$

$$D_{G.S.} = \varepsilon_t^{-3/2}\left(\frac{2\pi}{\iota}\right)^2(\rho_B{}^e)^2\nu_{ei}, \qquad \nu_{ei} < \nu_b = \varepsilon_t^{3/2}\nu_p. \tag{8.28}$$

When $\nu_b < \nu_{ei} < \nu_p$, the electrons cannot complete a banana orbit, so that it is not possible to treat the diffusion phenomena in this region by means of a simple model. MHD treatment is not applicable. In this region we must resort to the drift approximation of Vlasov's equation (see sec. 5.2). For the distribution function $f(r, \theta, z, v_{\parallel}, v_{\perp})$, Vlasov's equation is

$$\frac{\partial f}{\partial t} + \dot{\boldsymbol{r}}\nabla_{\boldsymbol{r}} f + \dot{v}_{\parallel}\frac{\partial f}{\partial v_{\parallel}} + \dot{v}_{\perp}\frac{\partial f}{\partial v_{\perp}} = \left(\frac{\partial f}{\partial t}\right)_{\text{coll}}. \tag{8.29}$$

The zeroth-order terms are

$$f_0(r, v_{\parallel}{}^2 + v_{\perp}{}^2) \qquad \dot{\boldsymbol{r}}_0 = \left(0, v_{\parallel}\frac{\iota}{2\pi}\frac{r}{R}, v_{\parallel}\right), \tag{8.30}$$

and the first-order terms are

$$f_1(r, \theta, z, v_{\parallel}, v_{\perp}) \qquad \dot{\boldsymbol{r}}_1 = (V_{\perp}\sin\theta, V_{\perp}\cos\theta, 0), \tag{8.31}$$

where $V_{\perp}$ is the toroidal-drift velocity

$$V_{\perp} = \frac{-m}{eB_0R}\left(\frac{v_{\perp}{}^2}{2} + v_{\parallel}{}^2\right). \tag{8.32}$$

When the energy and the magnetic moment of the particle are denoted by W and μ_{m}, respectively, the velocity and the acceleration are

$$v_{\parallel} = \left(2\frac{W - \mu_{\mathrm{m}}B(\boldsymbol{r})}{m}\right)^{1/2} \tag{8.33}$$

$$v_{\perp} = \left(2\mu_{\mathrm{m}}\frac{B(\boldsymbol{r})}{m}\right)^{1/2} \tag{8.34}$$

$$\dot{v}_{\parallel} = (\dot{\boldsymbol{r}}\cdot\nabla)v_{\parallel} = -\frac{v_{\perp}{}^2}{2v_{\parallel}}\frac{(\dot{\boldsymbol{r}}\cdot\nabla)B}{B} \tag{8.35}$$

$$\dot{v}_{\perp} = (\dot{\boldsymbol{r}}\cdot\nabla)v_{\perp} = \frac{v_{\perp}}{2}\frac{(\dot{\boldsymbol{r}}\cdot\nabla)B}{B}. \tag{8.36}$$

Since the magnitude of the magnetic field is

$$B \approx |B_0|\left(1 - \frac{r}{R}\cos\theta\right)$$

and the relation $\dot{v}_{\parallel}\partial f_0/\partial v_{\parallel} + \dot{v}_{\perp}\partial f_0/\partial v_{\perp} = 0$ holds for $f_0 = f_0(r, v_{\parallel}{}^2 + v_{\perp}{}^2)$, we find

$$\left(\frac{\iota}{2\pi}\right)\frac{v_{\parallel}}{R}\frac{\partial f_1}{\partial\theta} + V_{\perp}\sin\theta\frac{\partial f_0}{\partial r} = -\nu_{\text{eff}}f_1 \tag{8.37}$$

and the solution of f_1 is

$$f_1 = -V_\perp \frac{\partial f_0}{\partial r} \frac{\nu_{\text{eff}} \sin\theta - \left(\dfrac{\iota}{2\pi}\dfrac{v_\parallel}{R}\right)\cos\theta}{(\nu_{\text{eff}})^2 + \left(\dfrac{\iota}{2\pi}\dfrac{v_\parallel}{R}\right)^2}. \tag{8.38}$$

Accordingly the loss of particles is given by

$$\langle nV_r \rangle = \int V_\perp \sin\theta f_1(r, \theta, v_\parallel, v_\perp)\, \mathrm{d}\theta\, v_\perp\, \mathrm{d}v_\perp\, \mathrm{d}v_\parallel$$

$$= -\pi \int \frac{\nu_{\text{eff}}}{\nu_{\text{eff}}^2 + \left(\dfrac{\iota}{2\pi}\dfrac{v_\parallel}{R}\right)^2} V_\perp{}^2 \frac{\partial f_0}{\partial r} v_\perp\, \mathrm{d}v_\perp\, \mathrm{d}v_\parallel.$$

When the relation

$$\lim_{\varepsilon \to 0} \int_{-\infty}^{\infty} \frac{\varepsilon}{x^2 + \varepsilon^2} \mathrm{d}x = \pi\delta(x)$$

is used, $\langle nV_r \rangle$ is reduced, for the case $\nu_{\text{eff}} < (\iota/2\pi)(v_{\text{T}}{}^{\text{e}}/R)$, to

$$\langle nV_r \rangle \approx -\left(\frac{2\pi}{\iota}\right) R \left(\frac{T_{\text{e}}}{eBR}\right)^2 \left(\frac{m_{\text{e}}}{T_{\text{e}}}\right)^{1/2} \frac{\partial n}{\partial r}$$

$$= -(\rho_B{}^{\text{e}})^2 \left(\frac{2\pi}{\iota}\right)^2 \left(\frac{\iota}{2\pi}\frac{v_{\text{T}}{}^{\text{e}}}{R}\right) \frac{\partial n}{\partial r} = -(\rho_B{}^{\text{e}})^2 \left(\frac{2\pi}{\iota}\right)^2 \nu_{\text{p}} \frac{\partial n}{\partial r}. \tag{8.39}$$

Consequently the diffusion coefficient is[2,3]

$$D_{\text{p}} = \left(\frac{2\pi}{\iota}\right)^2 (\rho_B{}^{\text{e}})^2 \nu_{\text{p}}, \qquad \nu_{\text{p}} > \nu_{\text{ei}} > \nu_{\text{b}} = \varepsilon_{\text{t}}{}^{3/2} \nu_{\text{p}}. \tag{8.40}$$

The dependence of the diffusion coefficient on the collision frequency (eqs. (8.27), (8.28), and (8.40)) is shown in fig. 8.4. The region $\nu_{\text{ei}} > \nu_{\text{p}}$ is called the *MHD region* or *collision region*. The region $\nu_{\text{p}} > \nu_{\text{ei}} > \nu_{\text{b}}$ is the *plateau region* or the *intermediate region*; and the region $\nu_{\text{b}} > \nu_{\text{ei}}$ is the *banana region* or the *collisionless region*. These diffusion processes are collectively called *neoclassical diffusion*. There is an excellent review[3] of neoclassical diffusion. The difference of the diffusions of electrons and ions introduces an electric field into the plasma; the analysis including the effect of this electric field has also been done.[4]

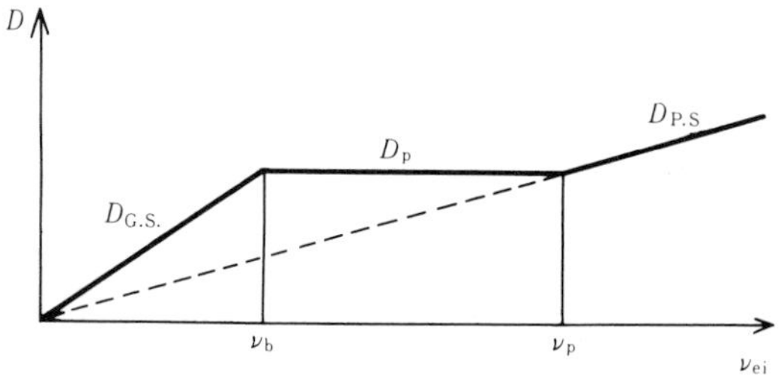

Fig. 8.4
Dependence of the diffusion coefficient on collision frequency in a tokamak field.

*8.3 Neoclassical Diffusion in a Stellarator Field

In a stellarator field, or even in a tokamak toroidal field produced by finite number of coils, there is as asymmetric inhomogeneous term

$$B = |B_0|(1 - \varepsilon_t \cos\theta - \varepsilon_h \cos(l\theta - m\varphi)) \tag{8.41}$$

in addition to the toroidal term $-\varepsilon_t \cos\theta$. The variation of B along lines of magnetic force is shown in fig. 8.5. Particles trapped by the inhomogeneous field $-\varepsilon_h \cos(l\theta - m\varphi)$ drift across the magnetic surfaces and contribute to the particle diffusion. The orbit of the banana center has already been described in sec. 3.5b; the equation of the orbit is expressed, using the coordinates shown in fig. 8.1 (see eqs. (3.126)–(3.131)), as

$$\frac{d\theta}{dt} = \frac{V_\perp}{r}\cos\theta + \omega_h + \omega_E \tag{8.42}$$

$$\frac{dr}{dt} = V_\perp \sin\theta, \tag{8.43}$$

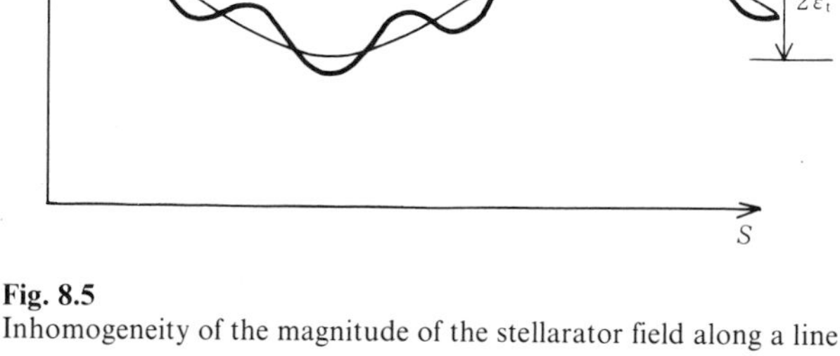

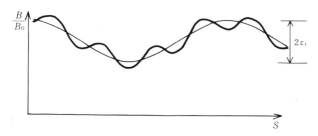

Fig. 8.5
Inhomogeneity of the magnitude of the stellarator field along a line of magnetic force.

where

$$V_\perp = -\frac{m}{eB_0R}\left(\frac{v_\perp{}^2}{2} + v_\parallel{}^2\right) \approx \frac{-T}{eB_0R} \tag{8.44}$$

$$\omega_\mathrm{h} = -2l\left(\frac{E(\kappa)}{K(\kappa)} - \frac{1}{2}\right)\varepsilon_\mathrm{h}\frac{T}{eB_0r^2} = \varepsilon_\mathrm{h}\frac{-T}{eB_0r^2}f(\kappa) \tag{8.45}$$

$$\omega_\mathrm{E} = \frac{E_r}{B_0r} \tag{8.46}$$

$$\kappa^2 = \frac{W - \mu_\mathrm{m}B(1-\varepsilon_\mathrm{h})}{2\varepsilon_\mathrm{h}\mu_\mathrm{m}B} \approx \frac{1}{2\varepsilon_\mathrm{h}}\frac{v_\parallel{}^2}{v_\perp{}^2}. \tag{8.47}$$

E_r is a radial electric field. If we define

$$\omega_0 \equiv \frac{T}{eB_0r^2} \qquad \alpha \equiv -\frac{eE_rr}{T}, \tag{8.48}$$

we find that

$$\frac{V_\perp}{r} : \omega_\mathrm{h} : \omega_\mathrm{E} \approx \varepsilon_\mathrm{t} : \varepsilon_\mathrm{h} : \alpha. \tag{8.49}$$

The dependence of $f(\kappa)/2l = E(\kappa)/K(\kappa) - 1/2$ in eq. (8.45) on κ is shown in fig. 8.6.[5]

***8.3a The Case $\varepsilon_\mathrm{h} \gg \varepsilon_\mathrm{t}, \alpha$**

When the inhomogeneous term of a stellarator field is larger than the toroidal term, eqs. (8.42) and (8.43) are reduced to

$$\frac{\mathrm{d}\theta}{\mathrm{d}t} = \omega_\mathrm{h} \qquad \frac{\mathrm{d}r}{\mathrm{d}t} = V_\perp \sin\theta. \tag{8.50}$$

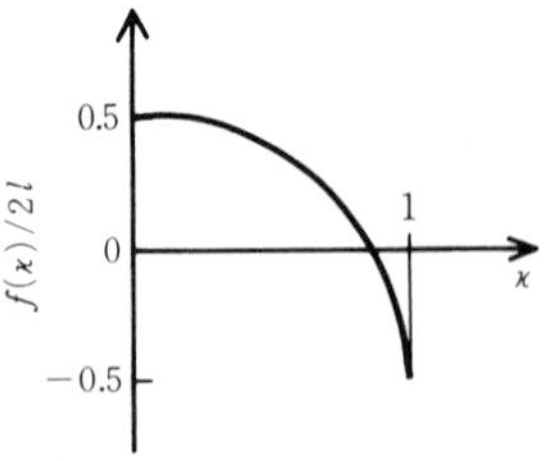

Fig. 8.6
The dependence of $f(\kappa)/2l = E(\kappa)/K(\kappa) - 1/2$ in eq. (8.45) on κ.

The distribution function for the particles trapped in a helical field is obtained by a method similar to that used to derive eq. (8.37), and we have

$$\omega_{\mathrm{h}}\frac{\partial f_1}{\partial\theta} + V_\perp \sin\theta \frac{\partial f_0}{\partial r} = -\nu_{\mathrm{eff}} f_1.$$

The particle flux $\langle nV_r\rangle$ can be estimated from f_1:

$$\langle nV_r\rangle = -\frac{1}{2}\int \frac{\nu_{\mathrm{eff}}}{\nu_{\mathrm{eff}}^2 + \omega_{\mathrm{h}}^2} V_\perp{}^2 \frac{\partial f_0}{\partial r} v_\perp \mathrm{d}v_\perp \mathrm{d}v_\parallel. \tag{8.51}$$

The effective collision frequency for the trapped particles is $\nu_{\mathrm{eff}} = \nu_{\mathrm{ei}}/\varepsilon_{\mathrm{h}}$. If $\nu_{\mathrm{eff}} > \omega_{\mathrm{h}}$, ω_{h} in eq. (8.51) can be neglected and the diffusion coefficient is

$$D_{\mathrm{h}} \approx \frac{1}{\nu_{\mathrm{eff}}}\left(\frac{T}{eBR}\right)^2 \varepsilon_{\mathrm{h}}{}^{1/2} = \varepsilon_{\mathrm{h}}{}^{3/2}\varepsilon_{\mathrm{t}}{}^2 \frac{\omega_0}{\nu_{\mathrm{ei}}}\frac{T}{eB} \tag{8.52}$$

$$(\nu_{\mathrm{ei}} > \varepsilon_{\mathrm{h}}\omega_{\mathrm{h}} \approx \varepsilon_{\mathrm{h}}{}^2\omega_0).$$

This expression may be interpreted as follows. Particles trapped in the helical field move across the magnetic field with the toroidal-drift velocity (sec. 3.5). The (collision) step length of the trapped particles is

$$\Delta_{\mathrm{h}} = \frac{V_\perp}{\nu_{\mathrm{eff}}}. \tag{8.53}$$

As the fraction of particles trapped is $\varepsilon_{\mathrm{h}}{}^{1/2}$, the diffusion coefficient is

$$D_{\mathrm{h}} = \varepsilon_{\mathrm{h}}{}^{1/2}\Delta_{\mathrm{h}}{}^2\nu_{\mathrm{eff}} = \frac{1}{\nu_{\mathrm{eff}}}V_\perp{}^2\varepsilon_{\mathrm{h}}{}^{1/2} = \varepsilon_{\mathrm{h}}{}^{3/2}\varepsilon_{\mathrm{t}}{}^2\frac{\omega_0}{\nu_{\mathrm{ei}}}\frac{T}{eB}. \tag{8.54}$$

This is quite the same as eq. (8.52).[6]

If $\nu_{\mathrm{eff}} < \omega_{\mathrm{h}}$, the term $\nu_{\mathrm{eff}}/(\nu_{\mathrm{eff}}{}^2 + \omega_{\mathrm{h}}{}^2)$ in the integrand of eq. (8.51) can be replaced by $\pi\delta(\omega_{\mathrm{h}})$; then $\langle nV_r\rangle$ is reduced to

$$\langle nV_r\rangle \approx -V_\perp{}^2 \frac{1}{\omega_{\mathrm{h}}}\frac{f(\kappa)\varepsilon_{\mathrm{h}}{}^{1/2}}{f'(\kappa)\kappa}\frac{\partial n}{\partial r} \approx -V_\perp{}^2\frac{\varepsilon_{\mathrm{h}}{}^{1/2}}{\omega_{\mathrm{h}}}\frac{\partial n}{\partial r} \tag{8.55}$$

and the diffusion coefficient is

$$D_{\mathrm{hp}} \approx \frac{\left(\dfrac{T}{eBR}\right)^2}{\varepsilon_{\mathrm{h}}{}^{1/2}\dfrac{T}{eBr^2}} = \frac{\varepsilon_t^2}{\varepsilon_{\mathrm{h}}{}^{1/2}}\frac{T}{eB}. \tag{8.56}$$

If the collision frequency is further reduced, diffusion of particles in superbanana orbits becomes dominant. The particles in a superbanana must satisfy $\mathrm{d}\theta/\mathrm{d}t = 0$ at some point. This condition requires $\omega_\mathrm{h} < V_\perp/r$ as is clear from eq. (8.42). The superbanana region in velocity space becomes $\delta(v_\parallel/v_\perp)^2 < \varepsilon_\mathrm{t}$ (see eqs. 3.123, 3.131, and 3.133), so that the effective collision frequency is $\nu_\mathrm{eff} = \nu_\mathrm{ei}/\varepsilon_\mathrm{t}$. On the other hand, the period T_s of superbanana motion is given by eq. (3.138). Consequently, the existence condition for the superbanana is $\nu_\mathrm{eff} < 1/T_\mathrm{s}$, i.e.,

$$\nu_\mathrm{eff} < \frac{1}{T_\mathrm{s}} \approx (\varepsilon_\mathrm{t}\varepsilon_\mathrm{h})^{1/2}\,\omega_0 \tag{8.57}$$

$$\nu_\mathrm{ei} < \varepsilon_\mathrm{t}{}^{3/2}\varepsilon_\mathrm{h}{}^{1/2}\,\omega_0. \tag{8.58}$$

The width of the superbanana is given from eq. (3.139) as

$$\varDelta_\mathrm{sb} \approx \left(\frac{\varepsilon_\mathrm{t}}{\varepsilon_\mathrm{h}}\right)^{1/2} r. \tag{8.59}$$

This width does not depend on the magnitude of the magnetic field. Then the diffusion coefficient due to the superbanana is

$$D_\mathrm{sb} = \varepsilon_\mathrm{t}{}^{1/2}\varDelta_\mathrm{sb}{}^2\nu_\mathrm{eff} = \varepsilon_\mathrm{t}{}^{1/2}\left(\frac{\varepsilon_\mathrm{t}}{\varepsilon_\mathrm{h}}\right)r^2\frac{\nu_\mathrm{ei}}{\varepsilon_\mathrm{t}} = \frac{\varepsilon_\mathrm{t}{}^{1/2}}{\varepsilon_\mathrm{h}}r^2\nu_\mathrm{ei}. \tag{8.60}$$

This does not become small even if the magnetic field is strong.

The dependence of the diffusion coefficient D on ν_ei is summarized in fig. 8.7.[3,7]

*8.3b The Case $\varepsilon_\mathrm{t} \gg \varepsilon_\mathrm{h}, \alpha$

The condition $\varepsilon_\mathrm{t} > \varepsilon_\mathrm{h}$ is usually satisfied by most toroidal stellarator devices. When $\varepsilon_\mathrm{t} \gg \varepsilon_\mathrm{h}$ and $\varepsilon_\mathrm{t} \gg \alpha$, the first term in the right-hand side of eq. (8.42) is dominant and the particles trapped in the helical field escape from the plasma due to the toroidal drift. Consequently the diffusion coefficient is obtained in the same way used to derive eq. (8.54), so that we have

$$D_\mathrm{h} = \varepsilon_\mathrm{h}{}^{1/2}\left(\frac{V_\perp}{\nu_\mathrm{eff}}\right)^2\nu_\mathrm{eff} = \varepsilon_\mathrm{h}{}^{3/2}\varepsilon_\mathrm{t}{}^2\frac{\omega_0}{\nu_\mathrm{ei}}\frac{T}{eB}.$$

However, the step length $\varDelta_\mathrm{h} = V_\perp/\nu_\mathrm{eff}$ of the trapped particle becomes large when the collision frequency becomes small. When $\nu_\mathrm{ei} < V_\perp\varepsilon_\mathrm{h}/r =$

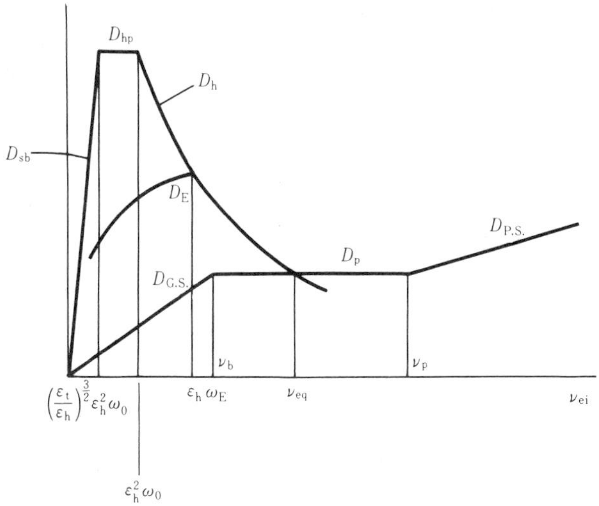

Fig. 8.7
Dependence of the diffusion coefficient on collision frequency ν_{ei} in a stellarator field with $\varepsilon_h > \varepsilon_t$ ($\nu_{eq} = (\varepsilon_h/\varepsilon_t)^{3/2}\nu_b$).

$\varepsilon_t\varepsilon_h\omega_0$, the step length is $\Delta > r$. Once particles are trapped, the trapped particles escape from the plasma region immediately. So the loss of particles is determined by the diffusion of particles to the trapped-particle region in velocity space.

The diffusion equation in velocity space is expressed in spherical coordinates v, γ, δ as follows (see eq. (5.34)):

$$\frac{1}{v^2 \sin\gamma} D_\perp \frac{\partial}{\partial\gamma}\left(\sin\gamma\frac{\partial f}{\partial\gamma}\right) + \frac{1}{v^2}\frac{\partial}{\partial v}\left(v^2\left(D_\parallel\frac{\partial f}{\partial v} - Af\right)\right) = \frac{\partial f}{\partial t}. \tag{8.61}$$

$D_{\perp,\parallel}$ and A are the diffusion tensor and the coefficient of dynamic friction, and they satisfy $mvD_\parallel/T + A = 0$. If the distribution function f is

$$f = \Theta(\gamma)\exp\left(-\frac{mv^2}{2T}\right)\exp\left(-\frac{t}{\tau}\right), \tag{8.62}$$

the diffusion equation becomes

$$\frac{1}{\sin\gamma}\frac{\partial}{\partial\gamma}\left(\sin\gamma\frac{\partial\Theta}{\partial\gamma}\right) = -\frac{v^2}{D_\perp\tau}\Theta. \tag{8.63}$$

The boundary condition in velocity space is $\Theta = 0$ at

$$\sin\gamma = (1 - \varepsilon_h)^{1/2} = 1 - \frac{1}{2}\varepsilon_h. \tag{8.64}$$

Since ε_h is small, we find that $\Theta \propto \cos\gamma$ and

$$\frac{1}{\tau(v)} = \frac{2D_\perp}{v^2} = 2\nu_{ei}\left(1 - \frac{T}{2mv^2}\right) = 2\nu_{ei}\left(1 - \frac{\langle v^2\rangle}{4v^2}\right) \approx 2\nu_{ei}. \tag{8.65}$$

If D is taken to be $D = r^2/5.8\,\tau$, the diffusion coefficient in the region $\nu_{ei} < \varepsilon_t\varepsilon_h\omega_0$ is[8]

$$D_v = \frac{2\nu_{ei}}{5.8}r^2 \approx \frac{r^2}{3}\nu_{ei}. \tag{8.66}$$

When $\nu_{ei} > \varepsilon_t\varepsilon_h\omega_0$, $D \approx r^2\nu_{ei}/3$ gives the upper limit of the diffusion coefficient. At $\nu_{ei} = \nu_v \equiv \varepsilon_t\varepsilon_h\omega_0$ the diffusion coefficient is

$$D_{vp} = \frac{\varepsilon_t\varepsilon_h}{3}\frac{T}{eB}. \tag{8.67}$$

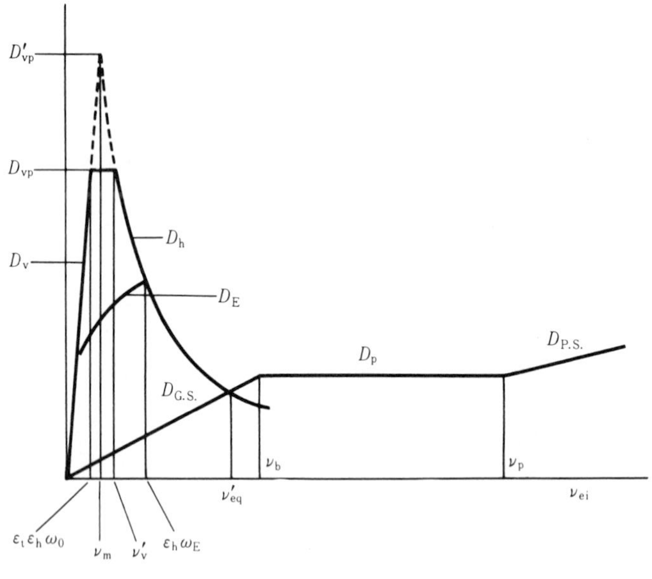

Fig. 8.8
Dependence of the diffusion coefficient D on collision frequency ν_{ei} in a stellarator field with $\varepsilon_t > \varepsilon_h$, where $\nu_{eq}' = (\varepsilon_h/\varepsilon_t)^{3/4}\nu_b$, $\nu_v' = 3\varepsilon_t\varepsilon_h^{1/2}\omega_0$, $\nu_m = 3^{1/2}\varepsilon_t\varepsilon_h^{3/4}\omega_0$, and $D_{vp}' = (\varepsilon_t\varepsilon_h^{3/4}/3^{1/2})(T/eB)$.

It is usually assumed that D does not depend on the collision frequency near $\nu_{ei} \approx \nu_v$. The dependence of the diffusion coefficient on ν_{ei} in a stellarator field with $\varepsilon_t > \varepsilon_h$ is summarized in fig. 8.8.

*8.3c The Case $\alpha \gg \varepsilon_t, \varepsilon_h$

When $\alpha \gg \varepsilon_t, \varepsilon_h$, the term ω_E in the right-hand side of eq. (8.42) is dominant and the bananas of trapped particles can circulate as is shown in fig. 8.9 (This is the untrapped or transit banana discussed in an earlier chapter.) The displacement δr of the orbit of the transit banana from the magnetic surface due to the toroidal drift $V_\perp$ is

$$\delta r = V_\perp/\omega_E. \tag{8.68}$$

In this case the transition between trapped particles and untrapped particles dominates the diffusion processes. The displacement of the orbit of an untrapped particle from the magnetic surface, $(2\pi/\iota)\rho_B{}^e$, is negligible compared with δr. Therefore, the step length of the transition is about δr (see fig. 8.9). In terms of the effective collision frequency ν_{eff} and the fraction of particles undergoing transition δf, the diffusion coefficient is

$$D_E = \delta f \left(\frac{V_\perp}{\omega_E}\right)^2 \nu_{eff},$$

where $\nu_{eff} = \nu_{ei}/(\delta f)^2$. Since ν_{eff} can be considered to be ω_E, we find that $\delta f^2 \approx \nu_{ei}/\omega_E$ and

$$D_E = \varepsilon_t{}^2 \left(\frac{\nu_{ei}}{\omega_E}\right)^{1/2} \frac{1}{\alpha}\frac{T}{eB}. \tag{8.69}$$

This dependence is shown in figs. 8.7 and 8.8.[3]

We have considered the neoclassical diffusion coefficient of a stellarator field for different cases. A more general discussion is given in ref. 9.

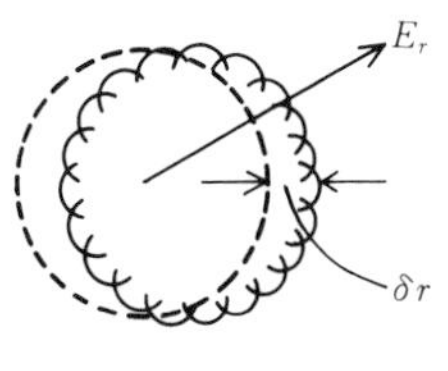

Fig. 8.9
Orbit of the transit banana and of an untrapped particle (dotted line).

8.4 Fluctuation Loss, Bohm Diffusion, and Convective Loss

In the foregoing sections we have discussed diffusion due to binary collision and have derived the confinement times for such diffusion. However, a plasma will be, in many cases, more or less unstable, and fluctuations in the density and electric field will induce collective motions of particles and bring about anomalous losses. We will study such losses here.

Assume the plasma density $n(\boldsymbol{r}, t)$ consists of the zeroth-order term $n_0(\boldsymbol{r}, t)$ and 1st-order perturbation terms $\tilde{n}_k(\boldsymbol{r}, t) = n_k \exp i(\boldsymbol{k}\cdot\boldsymbol{r} - \omega_k t)$:

$$n = n_0 + \sum_k \tilde{n}_k.$$

Since n and n_0 must be real, we have the relations

$$\tilde{n}_{-k} = (\tilde{n}_k)^* \qquad n_{-k} = n_k{}^* \qquad \omega_{-k} = -\omega_k{}^*$$

where * denotes the complex conjugate. ω_k is generally complex ($\omega_k = \omega_{kr} + i\gamma_k$) and

$$\omega_{-kr} = -\omega_{kr} \qquad \gamma_{-k} = \gamma_k.$$

When the plasma velocity due to the perturbation is given in the form

$$\boldsymbol{V}(\boldsymbol{r}, t) = \sum_k \tilde{\boldsymbol{V}}_k = \sum_k \boldsymbol{V}_k \exp i(\boldsymbol{k}\cdot\boldsymbol{r} - \omega_k t),$$

then $\boldsymbol{V}_{-k} = \boldsymbol{V}_k{}^*$ and the equation of continuity

$$\frac{\partial n}{\partial t} + \nabla(n\boldsymbol{V}) = 0$$

may be written as

$$\frac{\partial n_0}{\partial t} + \sum_k \frac{\partial \tilde{n}_k}{\partial t} + \nabla\left(\sum_k n_0 \tilde{\boldsymbol{V}}_k + \sum_{k,k'} \tilde{n}_k \tilde{\boldsymbol{V}}_{k'}\right) = 0.$$

When the 1st- and the 2nd-order terms are separated, then

$$\sum \frac{\partial \tilde{n}_k}{\partial t} + \nabla \sum n_0 \tilde{\boldsymbol{V}}_k = 0 \tag{8.70}$$

$$\frac{\partial n_0}{\partial t} + \nabla\left(\sum_{k,k'} \tilde{n}_k \tilde{\boldsymbol{V}}_{k'}\right) = 0. \tag{8.71}$$

Here we have assumed that the time variation of n_0 is of 2nd order. The time average of the product of eq. (8.70) and $\tilde{n}_{-k}$ becomes

$$\left.\begin{aligned} \gamma_k|n_k|^2 + \nabla n_0 \cdot \mathrm{Re}(n_k \boldsymbol{V}_{-k}) + n_0 \boldsymbol{k} \cdot \mathrm{Im}(n_k \boldsymbol{V}_{-k}) = 0 \\ \omega_{kr}|n_k|^2 + \nabla n_0 \cdot \mathrm{Im}(n_k \boldsymbol{V}_{-k}) - n_0 \boldsymbol{k} \cdot \mathrm{Re}(n_k \boldsymbol{V}_{-k}) = 0. \end{aligned}\right\} \tag{8.72}$$

If the time average of eq. (8.71) is taken, we find that

$$\frac{\partial n_0}{\partial t} + \nabla\left(\sum \mathrm{Re}(n_k \boldsymbol{V}_{-k})\exp(2\gamma_k t)\right) = 0. \tag{8.73}$$

On the other hand, the diffusion equation is

$$\frac{\partial n_0}{\partial t} = \nabla(D \nabla n_0)$$

and the outward particle flux Γ is

$$\Gamma = -D\nabla n_0 = \sum_k \mathrm{Re}(n_k \cdot \boldsymbol{V}_{-k})\exp 2\gamma_k t. \tag{8.74}$$

Equation (8.72) alone is not enough to determine the quantity $\nabla n_0 \cdot \mathrm{Re}(n_k \boldsymbol{V}_{-k})\exp 2\gamma_k t$. Denote $\beta_k = n_0 \boldsymbol{k} \cdot \mathrm{Im}(n_k \boldsymbol{V}_{-k})/\nabla n_0 \cdot \mathrm{Re}(n_k \boldsymbol{V}_{-k})$; then eq. (8.74) is reduced to

$$D|\nabla n_0|^2 = \sum \frac{\gamma_k |n_k|^2 \exp 2\gamma_k t}{1 + \beta_k},$$

and

$$D = \sum_k \gamma_k \frac{|\tilde{n}_k|^2}{|\nabla n_0|^2} \frac{1}{1 + \beta_k}. \tag{8.75}$$

This is the *anomalous diffusion coefficient due to fluctuation loss.*

Let us consider next the case in which the fluctuation $\tilde{\boldsymbol{E}}_k$ of the electric field is electrostatic and can be expressed by a potential $\tilde{\phi}_k$. Then the perturbed electric field is expressed by

$$\tilde{\boldsymbol{E}}_k = -\nabla\tilde{\phi}_k = -i\boldsymbol{k} \cdot \phi_k \exp i(\boldsymbol{k} \cdot \boldsymbol{r} - \omega_k t).$$

The electric field results in an $\boldsymbol{E} \times \boldsymbol{B}$ drift, i.e.,

$$\tilde{\boldsymbol{V}}_k = (\tilde{\boldsymbol{E}}_k \times \boldsymbol{B})/B^2 = -i(\boldsymbol{k} \times \boldsymbol{b})\tilde{\phi}_k/B, \tag{8.76}$$

where $\boldsymbol{b} = \boldsymbol{B}/B$. Substitution of eq. (8.76) into (8.70) yields

$$\tilde{n}_k = \nabla n_0 \cdot (\boldsymbol{b} \times \boldsymbol{k}) \frac{\tilde{\phi}_k}{\omega_k}. \tag{8.77}$$

In general, ∇n_0 and $\boldsymbol{b}$ are orthogonal. Take the z axis in the direction of $\boldsymbol{b}$ and the x axis in the direction of $-\nabla n_0$, i.e., let $\nabla n_0 = -\kappa n_0 \boldsymbol{e}_x$, where κ is the inverse of the scale of the density gradient. Then eq. (8.77) gives

$$\frac{\tilde{n}_k}{n_0} = \frac{\kappa}{B}\frac{k_y}{\omega_k}\tilde{\phi}_k = k_y \frac{\kappa T_\mathrm{e}}{eB\omega_k}\frac{e\tilde{\phi}_k}{T_\mathrm{e}} = \frac{\omega_k{}^*}{\omega_k}\frac{e\tilde{\phi}_k}{T_\mathrm{e}},$$

where $\omega_k{}^* \equiv (k_y \kappa T_\mathrm{e})/eB$ is called the drift frequency. (This quantity will be further discussed in ch. 12.) If the frequency ω_k is real (i.e., if $\gamma_k = 0$), $\tilde{n}_k$ and $\tilde{\phi}_k$ have the same phase, and the fluctuation $\boldsymbol{E}_k$ does not contribute to anomalous diffusion as is clear from eq. (8.75). When $\gamma_k > 0$, so that ω_k is complex, with $\exp(-i\omega_k t) = \exp(-i\omega_{kr} t)\exp \gamma_k t$, there is a phase difference between $\tilde{n}_k$ and $\tilde{\phi}_k$, and the fluctuation in the electric field contributes to anomalous diffusion. (When $\gamma_k < 0$, the amplitude of the fluctuation is damped and does not contribute to diffusion.) Using the real coefficients A_k, α_k, we have

$$\omega_k = \omega_{kr} + i\gamma_k = \omega_k{}^* A_k \exp i\alpha_k \qquad (A_k > 0,\ \alpha_k \text{ is real});$$

and $\boldsymbol{V}_k$ is expressed by

$$\tilde{\boldsymbol{V}}_k = -i(\boldsymbol{k} \times \boldsymbol{b})\frac{\omega_k}{k_y \kappa}\frac{\tilde{n}_k}{n_0} = -i(\boldsymbol{k} \times \boldsymbol{b})\frac{T_\mathrm{e}}{eB}\frac{\tilde{n}_k}{n_0} A_k \exp i\alpha_k .$$

Then the diffusion coefficient may be obtained from eq. (8.74) as follows:[10]

$$\begin{aligned} D &= \sum_k \frac{\omega_k{}^*}{\kappa^2} A_k \sin \alpha_k \left|\frac{n_k}{n_0}\right|^2 \exp 2\gamma_k t \\ &= \left(\sum_k \frac{k_y}{\kappa} A_k \sin \alpha_k \left|\frac{n_k}{n_0}\right|^2 \exp 2\gamma_k t\right)\frac{T_\mathrm{e}}{eB}. \end{aligned} \tag{8.78}$$

The anomalous diffusion coefficient due to fluctuation loss increases with time (from eqs. (8.75) and (8.78)) and eventually the term of maximum growth rate γ_k becomes dominant. However, the amplitude $|\tilde{n}_k|$ will saturates due to nonlinear effects; the saturated amplitude will be of the order of

$$|\tilde{n}_k| \approx |\nabla n_0|\frac{\lambda_{ky}}{2\pi} \approx \frac{\kappa}{k_y} n_0 .$$

Then eq. (8.75) becomes

$$D = \frac{\gamma_k}{\kappa^2}\left|\frac{\tilde{n}_k}{n_0}\right|^2 \approx \frac{\gamma_k}{k_y{}^2}. \tag{8.79}$$

When the nondimensional coefficient inside the parentheses in eq. (8.78) is at its maximum of 1/16, we have the *Bohm diffusion coefficient*

$$D_{\mathrm{B}} = \frac{1}{16}\frac{T_{\mathrm{e}}}{eB}. \tag{8.80}$$

It appears that eqs. (8.79) and (8.80) give the largest possible diffusion coefficient.

When the density and potential fluctuations $\tilde{n}_k$, $\tilde{\phi}_k$ are measured, $\tilde{\boldsymbol{V}}_k$ can be calculated, and the estimated outward particle flux Γ and diffusion coefficient D can be compared to the values obtained by experiment. As the relation of $\tilde{n}_k$ and $\tilde{\phi}_k$ is given by eq. (8.77), the phase difference will indicate whether $\gamma_k = 0$ (oscillatory mode) or $\gamma_k > 0$ (growing mode), so that this equation is very useful in interpreting experimental results.

Next, let us consider convective losses. Even if fluctuations in the density and electric field are not observed at a fixed position, it is possible that the plasma can move across the magnetic field and continuously escape. When a stationary electric field exists and the equipotential surfaces do not coincide with the magnetic surfaces $\psi = \text{const.}$, the $\boldsymbol{E} \times \boldsymbol{B}$ drift is normal to the electric field $\boldsymbol{E}$, which itself is normal to the equipotential surface. Consequently the plasma drifts along the equipotential surfaces (see fig. 8.10) which cross the magnetic surfaces. The resultant loss is called *convective loss.*[11]

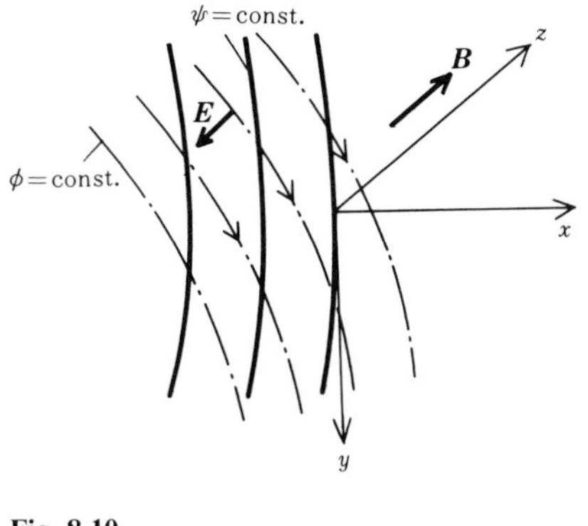

Fig. 8.10
Magnetic surfaces $\psi = \text{const.}$, and electric-field equipotential surfaces $\phi = \text{const.}$ The plasma moves along the equipotential surfaces by virtue of the $\boldsymbol{E} \times \boldsymbol{B}$ drift.

The particle-loss flux Γ is obviously given by

$$\Gamma_x = n_0 \frac{E_y}{B}. \tag{8.81}$$

The experimental results related to this phenomenon will be described in sec. 16.3. The losses due to diffusion by binary collision are proportional to B^{-2}; but fluctuation or convective losses are proportional to B^{-1}. Even if the magnetic field becomes strong, the loss due to fluctuation or convection does not decrease rapidly.

Let us now consider the diffusion process due to the convective motions which are caused by thermally excited convective cells.[12,13] The $\boldsymbol{E} \times \boldsymbol{B}$ drift velocity across the magnetic field is $\boldsymbol{V} = \boldsymbol{E} \times \boldsymbol{B}/B^2$. When $\boldsymbol{V}$ is expanded in Fourier components we find that

$$\boldsymbol{V} = \sum_k \boldsymbol{V}_k \exp(i\boldsymbol{k}\cdot\boldsymbol{r}) \qquad (\boldsymbol{V}_k \cdot \boldsymbol{k}) = 0.$$

The energy which belongs to each mode is

$$W = \mathscr{V} \sum_k \left(\frac{\rho |\boldsymbol{V}_k|^2}{2} + \frac{\varepsilon_0 |\boldsymbol{E}_k|^2}{2} \right) = \rho a^2 l \left(1 + \frac{B^2}{\rho \mu_0 c^2} \right) \sum \frac{|\boldsymbol{V}_k|^2}{2},$$

where a^2 is the plasma cross section and l is the length of the plasma, taken along the magnetic field lines. $\mathscr{V}$ is the volume of the plasma.

When the system is in thermal equilibrium the energy of each mode is equal to $T/2$, and

$$|\boldsymbol{V}_k|^2 = T \left(\rho a^2 l \left(1 + \frac{B^2}{\mu_0 \rho c^2} \right) \right)^{-1}. \tag{8.82}$$

The diffusion coefficient for this process is

$$D = \sum_k |\boldsymbol{V}_k|^2 \tau_{\mathrm{AC},k},$$

where $\tau_{\mathrm{AC},k}$ is the autocorrelation time of the kth mode. Since the autocorrelation time is equal to the duration of the kth mode, it is expressed by

$$\tau_{\mathrm{AC},k} = (k^2 D)^{-1}.$$

From these relations, the diffusion coefficient is found to be given by

$$D = \left(\frac{T \ln(ak_{max}/2\pi)}{\rho l\left(1 + \dfrac{B^2}{\mu_0 \rho c^2}\right)} \right)^{1/2} \tag{8.83}$$

(The mode density is $ak\,dk/2\pi$.) When $B^2/\mu_0\rho > c^2$ the diffusion coefficient is proportional to $T^{1/2}/B$. It can be shown that D is independent of B when $B^2/\mu_0\rho < c^2$.

8.5 Pseudoclassical Diffusion[14,15]

The relation between plasma diffusion and the plasma current has been given in sec. 7.4c:

$$\int V_d \mathrm{d}S = \eta \int \frac{j^2 \mathrm{d}S}{|\nabla p|},$$

where dS is an infinitesimal magnetic-surface element and η is the specific resistivity. Consequently the diffusion coefficient D satisfies

$$\frac{D}{D_{cl}} = \frac{\nu_{eff} j^2}{\nu_{ei} j_{dia}{}^2},$$

where D_{cl} is the classical diffusion and $j_{dia} = |\nabla p|/B$ is

$$j_{dia} = -\frac{T_e}{B}\left(1 + \frac{T_i}{T_e}\right)\frac{n_0}{a}. \tag{8.84}$$

When fluctuations exist, the current flow is turbulent, and satisfies

$$\left.\begin{aligned} &\nabla \cdot \boldsymbol{j} = i\boldsymbol{k} \cdot \boldsymbol{j} = 0 \\ &j_{\parallel} = -\frac{k_{\perp}}{k_{\parallel}} j_{\perp}. \end{aligned}\right\} \tag{8.85}$$

Most low-frequency instability has small $k_{\parallel}$ ($k_{\perp}/k_{\parallel} \gg 1$) so that we find

$$j^2 \approx j_{\parallel}{}^2. \tag{8.86}$$

Ohm's law (6.43b) is used to derive the equation for $j_{\parallel}$, i.e.,

$$0 = -\nabla_{\parallel} n T_e - enE_{\parallel} + \frac{m\nu_{eff}}{e} j_{\parallel} \tag{8.87}$$

$$j_{\parallel} = \frac{e}{m\nu_{eff}}((\nabla_{\parallel} n) T_e + en_0 E_{\parallel} + e\delta n E_{\parallel}). \tag{8.88}$$

When the density and potential fluctuations are expressed by

$$\delta n = n_1 \exp i(-\omega t + k_y y + k_z z) \tag{8.89}$$

$$\delta\phi = \phi_1 \exp i(-\omega t + k_y y + k_z z + \alpha) \tag{8.90}$$

$$\frac{n_1}{n_0} \approx \frac{e\phi_1}{T_e}, \tag{8.91}$$

then $j_\parallel$ is given by

$$j_\parallel = \frac{e}{m\nu_{\mathrm{eff}}}(ik_\parallel \delta n T_e(1 - \exp i\alpha) + e\delta n E_\parallel). \tag{8.92}$$

When the 1st term is larger than the 2nd in the right-hand side of eq. (8.92), $j_\parallel$ is

$$j^2 \approx {j_\parallel}^2 \approx \frac{e^2}{m^2 {\nu_{\mathrm{eff}}}^2} {k_\parallel}^2 {n_1}^2 {T_e}^2 4\sin^2\frac{\alpha}{2}, \tag{8.93}$$

and the diffusion coefficient is

$$D = \frac{T_e}{eB}\frac{1}{\Omega_e}\frac{e^2B^2}{m^2\nu_{\mathrm{eff}}}(k_\parallel a)^2 \frac{1}{\left(1+\dfrac{T_i}{T_e}\right)^2}\left(\frac{n_1}{n_0}\right)^2 4\sin^2\frac{\alpha}{2}. \tag{8.94}$$

This diffusion coefficient (8.94) must be the same as eq. (8.78) ($n_k \exp\gamma_k t = n_1$, $A = 1$, $k = 1/a$ in eq. (8.78)) and it follows that

$$2\tan\frac{\alpha}{2} = \frac{(1 + T_i/T_e)^2}{(k_\parallel a)^2}(ak_y)\frac{\nu_{\mathrm{eff}}}{\Omega_e} \equiv u. \tag{8.95}$$

When u is small and $\nu_{\mathrm{eff}} = \nu_{\mathrm{ei}}$, eq. (8.94) becomes

$$D = \frac{T_e}{eB}\left(\frac{n_1}{n_0}\right)^2 \frac{{k_y}^2}{{k_\parallel}^2}\left(1 + \frac{T_i}{T_e}\right)^2 \frac{\nu_{\mathrm{ei}}}{\Omega_e} = ({\rho_B}^{e})^2 \nu_{\mathrm{ei}}\left(\frac{n_1}{n_0}\right)^2 \frac{{k_y}^2}{{k_\parallel}^2}\left(1 + \frac{T_i}{T_e}\right)^2.$$

It can be assumed that $k_y \approx m/a$, $k_\parallel \approx \iota/2\pi R$ in any toroidal system, so that D is

$$D \approx ({\rho_B}^{e})^2 \left(\frac{R}{a}\frac{2\pi}{\iota}\right)^2 \nu_{\mathrm{ei}}\left(\frac{n_1}{n_0}\right)^2 m^2.$$

The poloidal field B_ω is given by $B_\omega = (\iota/2\pi)(a/R)B_\varphi$. When the Larmor radius corresponding to B_ω is denoted by ${\rho_{B\mathrm{p}}}^{e}$, then the diffusion coefficient D is reduced to

$$D \approx C^2 (\rho_{Bp}{}^{e})^2 \nu_{ei}, \tag{8.96}$$

where $C = (n_1/n_0)m$ and C is of the order of 1. Equation (8.96) is of the same form as that for classical diffusion, the only difference being that $\rho_{Bp}{}^{e}$ replaces $\rho_B{}^{e}$, so that this is called *pseudoclassical diffusion.*

In the derivation of eq. (8.96), the 2nd term in the right-hand side of eq. (8.92) was neglected. This assumption

$$|n_1 T_e(1 - \exp i\alpha)| > |en_1 \phi_1|$$

no longer holds when

$$\alpha \approx u \propto \frac{n_0}{T_e{}^{3/2} B}$$

becomes small (of the order of n_1/n_0). That is to say, we may neglect the 2nd term and assume pseudoclassical diffusion in the region

$$D > \frac{T_e}{eB}(ak_y)\left(\frac{n_1}{n_0}\right)^3 \tag{8.97a}$$

(as may be found by substituting $\alpha \approx n_1/n_0$ in eq. (8.78)). When this condition is not satisfied, the diffusion is not pseudoclassical and the diffusion coefficient is equal to the right-hand side of eq. (8.97a). On the other hand the pseudoclassical region cannot extend into the Bohm region, so that we must have

$$\frac{T_e}{eB}(ak_y)\left(\frac{n_1}{n_0}\right)^2 > D. \tag{8.97b}$$

When this condition is not satisfied, the diffusion coefficient is equal to the left-hand side of (8.97b). The comparison of this analysis with the experimental results will be described in ch. 16.

*8.6 Diffusion by Impurity Ions

The diffusion described in secs. 8.1–8.3 is due to e-i the collisions. The magnitudes of the resulting diffusion coefficients are about $(\rho_B{}^{e})^2 \nu_{ei}$. When impurity ions are contained in a plasma, collisions of the plasma ions with the impurity ions enhance the diffusion. The diffusion coefficient is of the order of $(\rho_B{}^{\mathrm{i}})^2 \nu_{\mathrm{ii}}$, or $(m_{\mathrm{i}}/m_{\mathrm{e}})^{1/2}$ times as large as the classical one. Let us consider diffusion due to the impurity ions in the collision region in which MHD treatment is applicable.[16] Diffusion of this type in the banana region is discussed in ref. 17.

The equation of continuity and the equations of motion with respect some ion α are given by

$$\nabla(n_\alpha \boldsymbol{V}_\alpha) = 0 \tag{8.98}$$

$$\nabla p_\alpha = q_\alpha n_\alpha(\boldsymbol{E} + \boldsymbol{V}_\alpha \times \boldsymbol{B}) + \boldsymbol{R}_\alpha, \tag{8.99}$$

where $\boldsymbol{R}_\alpha$ is the frictional force on the ion α due to Coulomb collision, expressed by (see sec. 4.4)

$$\boldsymbol{R}_\alpha = -\sum_\beta m_{\alpha\beta} n_\alpha \nu_{\alpha\beta}(\boldsymbol{V}_\alpha - \boldsymbol{V}_\beta) \tag{8.100}$$

$$m_{\alpha\beta} \equiv \frac{m_\alpha m_\beta}{m_\alpha + m_\beta}$$

$$\frac{1}{\nu_{\alpha\beta}} = \tau_{\alpha\beta} = \frac{3\pi\varepsilon_0{}^2 (m_{\alpha\beta})^{1/2} T^{3/2}}{(2\pi)^{1/2} q_\alpha{}^2 q_\beta{}^2 n_\beta \ln \Lambda}.$$

Here, for simplicity, it is assumed that the temperature of each component of the plasma is the same. Neglecting the collision term $\boldsymbol{R}_\alpha$ in eq. (8.99), we find

$$(\nabla p_\alpha - q_\alpha n_\alpha \boldsymbol{E}) = q_\alpha n_\alpha(\boldsymbol{V}_\alpha \times \boldsymbol{B}) \tag{8.101}$$

$$(n_\alpha \boldsymbol{V}_\alpha)_\perp = -\frac{1}{q_\alpha B^2}(\nabla p_\alpha - q_\alpha n_\alpha \boldsymbol{E}) \times \boldsymbol{B}, \tag{8.102}$$

i.e.,*

$$(n_\alpha \boldsymbol{V}_\alpha)_\theta = \frac{-1}{q_\alpha B_0}\left(\frac{\partial p_\alpha}{\partial r} - q_\alpha n_\alpha E_r\right). \tag{8.103}$$

By setting $E_r = -\partial\phi/\partial r$ and $p_\alpha{}' = p_\alpha + q_\alpha n_\alpha \phi$, the equation of continuity becomes

$$\nabla_{\|}(n_\alpha \boldsymbol{V}_\alpha)_{\|} = -\nabla_\perp(n_\alpha \boldsymbol{V}_\alpha)_\perp$$

$$\approx \frac{1}{q_\alpha B^2}\nabla_\perp(\nabla p_\alpha{}' \times \boldsymbol{B}) + \frac{1}{q_\alpha}(\nabla p_\alpha{}' \times \boldsymbol{B})\cdot\nabla_\perp\left(\frac{1}{B^2}\right).$$

*The r, $-\theta$, φ coordinates form a right-handed system.

As $\nabla_{\|} \approx (B_\theta/rB_0)\partial/\partial\theta$ in an axially symmetric system, $(n_\alpha V)_{\|}$ is given by

$$(n_\alpha V_\alpha)_{\|} \approx \frac{2r}{q_\alpha B_\theta R}\frac{\partial p_\alpha'}{\partial r}\cos\theta. \tag{8.104}$$

This equation corresponds to eq. (8.8). Using $\nabla_{\|} \approx (B_\theta/rB_0)\partial/\partial\theta$, we obtain the parallel component of eq. (8.99):

$$\frac{1}{r}\frac{\partial p_\alpha}{\partial\theta} \approx q_\alpha n_\alpha E_\theta + R_{\alpha\|}\frac{B_0}{B_\theta},$$

and thus

$$E_\theta = \frac{1}{q_\alpha n_\alpha}\left(\frac{1}{r}\frac{\partial p_\alpha}{\partial\theta} - R_{\alpha\|}\frac{B_0}{B_\theta}\right). \tag{8.105}$$

The θ component of eq. (8.99) is

$$\frac{1}{r}\frac{\partial p_\alpha}{\partial\theta} \approx q_\alpha n_\alpha(E_\theta + V_{\alpha r}B_0) + R_{\alpha\perp}. \tag{8.106}$$

Combining this relation with eq. (8.105), we find the radial flux of particles $n_\alpha V_{\alpha r}$ to be

$$n_\alpha V_{\alpha r} = \frac{-1}{q_\alpha B_0}\left(R_{\alpha\perp} - R_{\alpha\|}\frac{B_0}{B_\theta}\right). \tag{8.107}$$

$R_{\alpha\|}$, $R_{\alpha\perp}$ are derived from eqs. (8.100), (8.103), and (8.104):

$$R_{\alpha\perp} = \sum_\beta m_{\alpha\beta}n_\alpha\nu_{\alpha\beta}\frac{1}{B_0}\left(\frac{1}{q_\alpha n_\alpha}\frac{\partial p_\alpha}{\partial r} - \frac{1}{q_\beta n_\beta}\frac{\partial p_\beta}{\partial r}\right)$$

$$R_{\alpha\|} = -\sum_\beta m_{\alpha\beta}n_\alpha\nu_{\alpha\beta}\frac{2r}{B_\theta R}\left(\frac{1}{q_\alpha n_\alpha}\frac{\partial p_\alpha}{\partial r} - \frac{1}{q_\beta n_\beta}\frac{\partial p_\beta}{\partial r}\right)\cos\theta.$$

Substitution of these equations into eq. (8.107) yields the radial flux Γ_α of the α ions. The average of Γ_α over the plasma surface is

$$\begin{aligned}\langle\Gamma_\alpha\rangle &= \frac{1}{2\pi}\int_0^{2\pi}\left(1 + \frac{r}{R}\cos\theta\right)n_\alpha V_{\alpha r}\,\mathrm{d}\theta \\ &= -\sum_\beta \frac{m_{\alpha\beta}n_\alpha}{q_\alpha B_0^2}\nu_{\alpha\beta}\left(1 + \left(\frac{2\pi}{\iota}\right)^2\right)\left(\frac{1}{q_\alpha n_\alpha}\frac{\partial p_\alpha}{\partial r} - \frac{1}{q_\beta n_\beta}\frac{\partial p_\beta}{\partial r}\right).\end{aligned} \tag{8.108}$$

When the temperature is uniform, $\langle\Gamma_\alpha\rangle$ is reduced to

$$\langle \Gamma_\alpha \rangle = -\sum_\beta \frac{m_{\alpha\beta} n_\alpha}{q_\alpha{}^2 B_0{}^2} T \nu_{\alpha\beta} \left(1 + \left(\frac{2\pi}{\iota} \right)^2 \right) \left(\frac{1}{n_\alpha} \frac{\partial n_\alpha}{\partial r} - \frac{(q_\alpha/q_\beta)}{n_\beta} \frac{\partial n_\beta}{\partial r} \right)$$

$$= -\sum_\beta (\rho_B{}^\alpha)^2 \nu_{\alpha\beta} \left(1 + \left(\frac{2\pi}{\iota} \right)^2 \right) \left(\frac{\partial n_\alpha}{\partial r} - \left(\frac{q_\alpha}{q_\beta} \right) \left(\frac{n_\alpha}{n_\beta} \right) \frac{\partial n_\beta}{\partial r} \right), \tag{8.109}$$

where $\rho_B{}^\alpha$ is the Larmor radius of particles of charge q_α and mass $m_{\alpha\beta}$. This is the equation of the particle flux induced by presence of impurity ions.

When the density of plasma ions is denoted by n_{i} and the charge on a plasma ion by e, and these quantities for impurity ions by n_z and Z_e, respectively, $\langle \Gamma_\alpha \rangle$ becomes zero under the condition

$$\frac{1}{n_{\mathrm{i}}} \frac{\partial n_{\mathrm{i}}}{\partial r} - \frac{1}{Z n_z} \frac{\partial n_z}{\partial r} = 0, \tag{8.110}$$

i.e.,

$$n_z(r) \propto (n_{\mathrm{i}}(r))^z. \tag{8.111}$$

This means that impurity ions are likely to be concentrated at the center of the plasma and enhance the radiative cooling of the plasma core. It is not yet clear that impurity concentration has been observed in experiments.[18,19]

8.7 Bootstrap Current

It is predicted theoretically that radial diffusion induces a current in the toroidal direction and that the current becomes large in the banana region. Although this has not yet been confirmed by experiment, it is an interesting process which may provide the means to sustain the plasma current in tokamaks.

As was described in sec. 8.2, electrons in the collisionless region $\nu_{\mathrm{ei}} < \nu_{\mathrm{b}}$ make complete circuits of the banana orbit. When a density gradient exists, there is a difference in particle numbers on neighboring banana orbits passing through a point A, as shown in fig. 8.11. The difference is $(\mathrm{d}n_{\mathrm{t}}/\mathrm{d}r)\Delta_{\mathrm{b}}$, Δ_{b} being the width of the banana orbit. When the component of the velocity parallel to the magnetic field is denoted by $v_\parallel$, its value for banana particles is $v_\parallel = \varepsilon^{1/2} v_{\mathrm{T}}$, where v_{T} is the thermal velocity. Consequently, the current density due to the trapped particles is

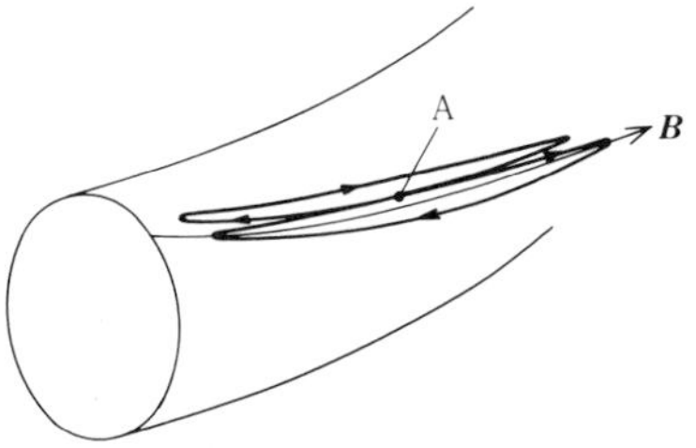

Fig. 8.11
Banana motion of electrons that induces the bootstrap current.

$$j_{\mathrm{bt}} = (ev_{\|})\left(-\frac{\mathrm{d}n_{\mathrm{t}}}{\mathrm{d}r}\Delta_{\mathrm{b}}\right) = -\varepsilon^{3/2}\frac{1}{B_{\mathrm{p}}}\frac{\partial p}{\partial r}.$$

The foregoing equation has been derived in a semiquantitative manner. In a more rigorous analysis, the distribution function f_{t} of the trapped particles is solved by means of Vlasov's equation[2] and the current density due to the trapped particles is obtained from $j_{\mathrm{bt}} = \int ev_{\|} f_{\mathrm{t}} \mathrm{d}\boldsymbol{v}$. The untrapped particles start to drift in the same direction as the trapped particles due to the collisions between them, and the drift becomes steady-state due to the collisions with ions. The drift velocity V_{b} of untrapped particles in the steady state is given by

$$\nu_{\mathrm{ei}} m_{\mathrm{e}} V_{\mathrm{b}} = \left(\frac{\nu_{\mathrm{ee}}}{\varepsilon}\right) m_{\mathrm{e}} \left(\frac{j_{\mathrm{bt}}}{-en_{\mathrm{e}}}\right),$$

where $\nu_{\mathrm{ee}}/\varepsilon$ is the effective collision frequency between trapped and untrapped electrons. The current density due to the drift velocity V_{b} of the untrapped electrons is

$$j_{\mathrm{b}} = -0.33\varepsilon^{1/2}\frac{1}{B_{\mathrm{p}}}\frac{\partial p}{\partial r}. \tag{8.112a}$$

This current is called the *bootstrap current*.[20–22]

This formula for the bootstrap current can be derived more exactly. When the system is axially symmetric and is collision free, the angular momentum of particles is conserved and a particle orbit is described by eq. (3.17) or eq. (3.19). When collisions are taken into account the equation of average particle flow is reduced from eqs. (3.17) and (3.19) to[23]

$$\frac{\mathrm{d}}{\mathrm{d}t}(n_e m_e v_{\parallel}) + n_e e B_p v_r = f_{ei}, \tag{8.113}$$

where B_p is the poloidal field and v_r is the radial-flow velocity. f_{ei} is the force due to the e-i collisions: $f_{ei} = -n_e m_e u_{\parallel} \nu_{ei}$. Electron-electron collisions do not change the total momentum of electrons and this term does not appear in eq. (8.113). In the steady state the 1st term of the right-hand side of eq. (8.113) is zero and the radial particle flow in the banana region is $n_e v_r = -D_{G.S.} \partial n_e/\partial r$. Therefore the flow velocity $u_{\parallel}$ of electrons in the steady state is obtained and the current density due to the electron flow is

$$j_b = -\frac{\varepsilon^{1/2}}{B_p}\frac{\mathrm{d}p}{\mathrm{d}r}. \tag{8.112b}$$

This formula coincides with eq. (8.112a) except for the factor 0.33.

The poloidal field B_p induced by the bootstrap current is

$$B_p \approx \frac{\mu_0(\pi a^2 j_b)}{2\pi a} = \frac{\mu_0 j_b a}{2}.$$

When eq. (8.112b) is substituted into this equation, the Kruskal-Shafranov condition yields the upper limit of the poloidal beta ratio due to the bootstrap current:[21]

$$\beta_p \equiv \frac{p}{B_p{}^2/(2\mu_0)} < 4\left(\frac{R}{a}\right)^{1/2}. \tag{8.114}$$

References

1. D. Pfirsch and A. Schlüter: MPI/PA/7/62, Max-Planck-Institut für Physik und Astrophysik, Munich (1962)

2. A. A. Galeev and R. Z. Sagdeev: Sov. Phys. JETP **26**, 233(1968)

3. B. B. Kadomtsev and O. P. Pogutse: Nucl. Fusion **11**, 67(1971)

4. T. E. Stringer: Phys. of Fluids **3**, 810(1970); Phys. of Fluids **13**, 1586(1970)

5. B. B. Kadomtsev and O. P. Pogutse: Sov. Phys. JETP **24**, 1172(1967)

6. A. Gibson and D. W. Mason: Plasma Physics **11**, 121(1969)

7. A. A. Galeev, R. Z. Sagdeev, H. P. Furth, and M. N. Rosenbluth: Phys. Rev. Lett. **22**, 511(1969)

8. K. Miyamoto: Phys. of Fluids **17**, 1476(1974)

9. H. P. Furth and M. N. Rosenbluth: Plasma Phys. and Controlled Nucl. Fusion Research **1**, 821 (Conf. Proceedings, Novosibirsk 1968) (IAEA, Vienna, 1969)

10. L. Spitzer, Jr.: Phys. of Fluids **3**, 659(1960)
S. Yoshikawa and D. J. Rose: ibid. **5**, 334(1962)

11. S. Yoshikawa, M. Barrault, W. L. Harris, D. Meade, R. Palladino, and S. von Goeler: Plasma Phys. and Controlled Nucl. Fusion Research **1**, 403 (Conf. Proceedings, Novosibirsk 1968) (IAEA, Vienna, 1969)
W. Harris: Phys. of Fluids **13**, 140(1970)

12. J. M. Dawson, H. Okuda, and R. Carlile: Phys. Rev. Lett. **27**, 491(1971)

13. H. Okuda and J. M. Dawson: ibid. **28**, 1625(1972)

14. S. Yoshikawa: Phys. Rev. Lett. **25**, 353(1970)
S. Yoshikawa and N. Christofilos: Plasma Phys. and Controlled Nucl. Fusion Research **2**, 357 (Conf. Proceedings, Madison 1971) (IAEA, Vienna, 1971)

15. S. Yoshikawa: MATT-959, Plasma Phys. Laboratory, Princeton University (1973)

16. T. Tsuda and M. Tanaka: JAERI-M 5376, Aug. 1973 Japan Atomic Energy Research Institute, Tokyo

17. J. W. Connor: Plasma Phys. **15**, 765(1973)

18. V. A. Vershkov and S. V. Mirnov: Nucl. Fusion **14**, 383(1974)

19. N. Bretz, D. Dimock, A. Greenberger, E. Hinnov, E. Meservey, W. Stodiek, and S. von Goeler: Plasma Phys. and Controlled Nucl. Fusion Research **1**, 55(Conf. Proceedings, Tokyo 1974) (IAEA, Vienna, 1975)
T. F. R. Group: ibid., 127

20. R. J. Bickerton, J. W. Connor, and J. B. Taylor: Nature **229**, 110(1971)

21. A. A. Galeev and R. Z. Sagdeev: Plasma Phys. and Controlled Nucl. Fusion Research **1**, 481 (Conf. Proceedings, Madison 1971) (IAEA, Vienna, 1971). English translation in Nucl. Fusion, suppl. p. 45(1972)

22. B. B. Kadomtsev and V. D. Shafranov: ibid. **2**, 479. English translation in Nucl. Fusion, suppl., p. 209(1972)

23. R. Z. Sagdeev and A. A. Galeev: Sov. Phys. Doklady **14**, 1198(1970)

9 Magnetohydrodynamic Instabilities

The stability of plasmas in magnetic fields is one of the primary research subjects in the area of controlled thermonuclear fusion and both theoretical and experimental investigations have been actively pursued. If a plasma is free from all possible instabilities and if the confinement is dominated by neoclassical diffusion in the banana region, then the confinement time τ is given by

$$\tau = \frac{a^2}{5.8 D_{\text{G.S.}}} = \frac{1}{5.8}\left(\frac{\iota}{2\pi}\right)^2 \varepsilon_t^{3/2}\left(\frac{a}{\rho_B{}^{\text{e}}}\right)^2 \frac{1}{\nu_{\text{ei}}},$$

where a is the plasma radius, $\rho_B{}^{\text{e}}$ is the electron Larmor radius, and ν_{ei} is the electron-ion collision frequency. For such an ideal case, a device of a reasonable size satisfies Lawson's condition. For example, with $B = 5\ \text{W/m}^2$, $a = 1$ m, electron temperature $T_{\text{e}} = 20$ keV, $\iota/2\pi = 0.33$, inverse aspect ratio $\varepsilon = a/R = 0.2$, and effective ion charge $Z = 3$, the value of $n\tau$ is approximately $5.8 \times 10^{21}\ \text{cm}^{-3}$ s, which is one or two orders of magnitude larger than that of Lawson's condition.

A plasma consists of many moving charged particles and has many magnetohydrodynamic degrees of freedom as well as degrees of freedom in velocity space. When a certain mode of perturbation grows, it enhances diffusion. Heating a plasma increases the kinetic energy of the charged particles but at the same time can induce fluctuations in the electric and magnetic fields, which in turn augment anomalous diffusion.

Therefore, it is very important to study whether any particular perturbed mode is stable (damping mode) or unstable (growing mode). In the stability analysis, it is assumed that the deviation from the equilibrium state is small so that a linearized approximation can be used. In this chapter we will consider instabilities that can be described by linearized magnetohydrodynamic equations. These instabilities are called the *magnetohydrodynamic (MHD) instabilities* or *macroscopic instabilities.*

A small perturbation $\boldsymbol{F}(r, t)$ of the first order is expanded in terms of its Fourier components

$$\boldsymbol{F}(r, t) = \boldsymbol{F}(r)\exp(-i\omega t) \qquad \omega = \omega_{\text{r}} + i\omega_{\text{i}},$$

and each term can be treated independently in the linearized approximation. The dispersion equation is solved for ω and the stability of the perturbation depends on the sign of the imaginary part ω_{i} (unstable for $\omega_{\text{i}} > 0$ and stable for $\omega_{\text{i}} < 0$). When $\omega_{\text{r}} \neq 0$, the perturbation is oscillatory and when $\omega_{\text{r}} = 0$, it grows or damps monotonically.

In the following sections, typical MHD instabilities are introduced. In sec. 9.1, interchange instability and kink instability are explained in an intuitive manner. Although many mathematical expressions appear in secs. 9.2–9.4, readers should not lose sight of the main thread of the analysis. In sec. 9.2 the magnetohydrodynamic equations are linearized and the boundary conditions are implemented. The stability criterion is deduced from the energy principle, eqs. (9.46)–(9.49). In sec. 9.3, a cylindrical plasma is studied as an important example; and the associated energy integrals, eqs. (9.91)–(9.93), are derived. Furthermore, three important stability conditions are obtained; they are the Kruskal-Shafranov limit, eq. (9.66), the Suydam criterion, eq. (9.96), and the Newcomb theorem. Tokamak and diffuse pinch (Zeta mode) configurations are taken as typical of cylindrical plasmas, and their stabilities are examined. In sec. 9.4, a stability condition of the local modes in a toroidal plasma is derived (eq. (9.133)). This condition is reduced to a more intuitively understandable formula (eq. (9.151)) for the axisymmetric case. Sections 9.5–9.7 explain resistive instability, ballooning instability, and resistive drift instability, respectively.

In this chapter, only the most common and tractable magnetohydrodynamic instabilities are introduced; it should be understood that there are many other instabilities. General reviews of plasma instabilities may be found in refs. 1 and 2.

9.1 Interchange Instability, Sausage Instability, and Kink Instability

Let us study simple examples of instabilities intuitively before discussing the general linear theory of instabilities.

9.1a Interchange Instability

Let $x = 0$ be the boundary between plasma and vacuum and let the z axis be taken in the direction of the magnetic field B. The plasma region is $x < 0$ and the vacuum region is $x > 0$. It is assumed that the acceleration $\boldsymbol{g}$ is applied in the x direction (see fig. 9.1). Ions and electrons drift in opposite directions to each other, due to the acceleration, with drift velocities

$$v_{G,i} = \frac{M}{e}\frac{\boldsymbol{g} \times \boldsymbol{B}}{B^2} \tag{9.1}$$

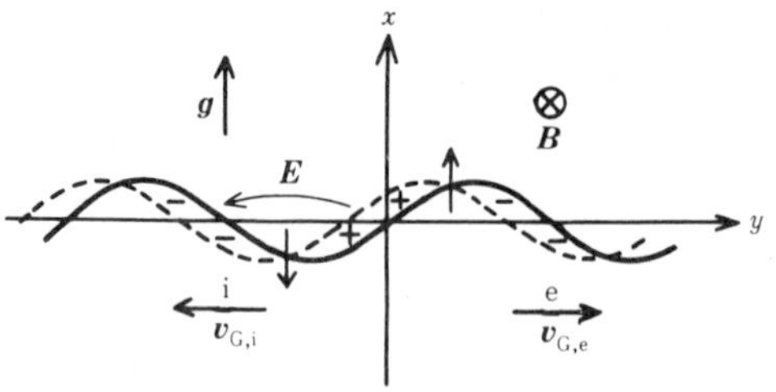

Fig. 9.1
Ion and electron drifts and the resultant electric field for interchange instability.

$$v_{G,e} = -\frac{m}{e}\frac{g \times B}{B^2}. \tag{9.2}$$

Let us assume that, due to a perturbation, the boundary of the plasma is displaced from the surface $x = 0$ in the amount

$$\delta x = a(t)\sin(k_y y). \tag{9.3}$$

The charge separation due to the opposite ion and electron drifts yields an electric field. The resultant $\boldsymbol{E} \times \boldsymbol{B}$ drift enhances the original perturbation if the direction of the acceleration $\boldsymbol{g}$ is outward from the plasma. From fig. 9.3 (or 9.4b) we see that this is the same as saying that the magnetic flux originally inside but near the plasma boundary is displaced so that it is outside the boundary, while the flux outside moves in to fill the depression thus left in the boundary; because of this geometrical picture of the process, this type of instability has come to be called *interchange instability*. As the perturbed plasma boundary is in the form of flutes along the lines of magnetic force (fig. 9.4b), this instability is also called *flute instability*.

The drift due to the acceleration yields a surface charge on the plasma, of charge density

$$\sigma_s = \sigma(t)\cos(k_y y)\delta(x) \tag{9.4}$$

(see fig. 9.1). The electrostatic potential ϕ of the induced electric field $\boldsymbol{E} = -\nabla\phi$ is given by

$$\varepsilon_\perp \frac{\partial^2 \phi}{\partial y^2} + \frac{\partial}{\partial x}\left(\varepsilon_\perp \frac{\partial \phi}{\partial x}\right) = -\sigma_s. \tag{9.5}$$

The boundary condition is

$$\varepsilon_0 \left(\frac{\partial \phi}{\partial x} \right)_{+0} - \left(\varepsilon_\perp \frac{\partial \phi}{\partial x} \right)_{-0} = -\sigma_s$$

$$\phi_{+0} = \phi_{-0}.$$

Under the assumption $k_y > 0$, the solution ϕ is

$$\phi = \frac{\sigma(t)}{k_y(\varepsilon_0 + \varepsilon_\perp)} \cos(k_y y) \exp(-k_y |x|). \tag{9.6}$$

The velocity of the boundary $\mathrm{d}(\delta x)/\mathrm{d}t$ is equal to $\boldsymbol{E} \times \boldsymbol{B}/B^2$ at $x = 0$, with $\boldsymbol{E}$ found from the potential (9.6). The velocity is

$$\frac{\mathrm{d}a(t)}{\mathrm{d}t} \sin(k_y y) = \frac{\sigma(t)}{(\varepsilon_0 + \varepsilon_\perp)B} \sin(k_y y). \tag{9.7}$$

The charge flux in the y direction is

$$ne|\boldsymbol{v}_{\mathrm{G,i}}| = \frac{\rho_{\mathrm{m}} g}{B},$$

where $\rho_{\mathrm{m}} = nM$. Accordingly the time rate of change of charge density is

$$\frac{\mathrm{d}\sigma(t)}{\mathrm{d}t} \cos(k_y y) = \frac{\rho_{\mathrm{m}} g}{B} a(t) \frac{\mathrm{d}}{\mathrm{d}y} \sin(k_y y), \tag{9.8}$$

and

$$\frac{\mathrm{d}^2 a}{\mathrm{d}t^2} = \frac{\rho_{\mathrm{m}} g k_y}{(\varepsilon_0 + \varepsilon_\perp)B^2} a. \tag{9.9}$$

The solution is in the form $a \propto \exp \gamma t$; the growth rate γ is given by

$$\gamma = \left(\frac{\rho_{\mathrm{m}}}{(\varepsilon_0 + \varepsilon_\perp)B^2} \right)^{1/2} (g k_y)^{1/2}. \tag{9.10}$$

In the low-frequency (compared with the ion cyclotron frequency) case the dielectric constant is given by

$$\varepsilon_\perp = \varepsilon_0 \left(1 + \frac{\rho_{\mathrm{m}}}{B^2 \varepsilon_0} \right) \gg \varepsilon_0, \tag{9.11}$$

as will be explained in ch. 10. Accordingly the growth rate γ is[3]

$$\gamma = (g k_y)^{1/2}. \tag{9.12}$$

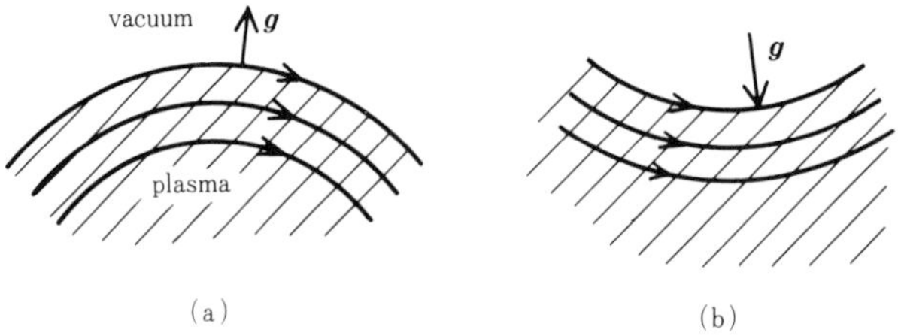

Fig. 9.2
Centrifugal force due to the curvature of magnetic-force lines.

When the acceleration is outward, a perturbation with the propagation vector $\boldsymbol{k}$ normal to the magnetic field $\boldsymbol{B}$ is unstable:

$$(\boldsymbol{k} \cdot \boldsymbol{B}) = 0. \tag{9.13}$$

However, if the acceleration is inward ($g < 0$), γ of eq. (9.12) is imaginary and the perturbation is oscillatory and stable.

The origin of interchange instability is charge separation due to the acceleration. When the lines of magnetic force are curved, as is shown in fig. 9.2, the charged particles are subjected to a centrifugal force. If the lines of magnetic force are convex outward (fig. 9.2a), this centrifugal acceleration induces interchange instability. If the lines are concave outward, the plasma is stable. Accordingly, the plasma is stable when the magnitude B of the magnetic field increases outward. In other words, if B is a minimum at the plasma region the plasma is stable. This is the *minimum-B condition* for stability.

The drift motion of charged particles is expressed by eq. (3.64):

$$\boldsymbol{v}_{\mathrm{g}} = \frac{\boldsymbol{E} \times \boldsymbol{b}}{B} + \frac{\boldsymbol{b}}{\Omega} \times \left(\boldsymbol{g} + \frac{(v_\perp^2/2) + v_\parallel^2}{R} \boldsymbol{n} \right) + v_\parallel \boldsymbol{b},$$

where $\boldsymbol{n}$ is the normal unit vector from the center of curvature to a point on a line of magnetic force. The equivalent acceleration is

$$\boldsymbol{g} = \frac{(v_\perp^2/2) + v_\parallel^2}{R} \boldsymbol{n}. \tag{9.14}$$

For a perturbation with propagation vector $\boldsymbol{k}$ normal to the magnetic field $\boldsymbol{B}$, i.e., $(\boldsymbol{k} \cdot \boldsymbol{B}) = 0$, another mechanism of charge separation may cause the same type of instability. When a plasma rotates with the velocity

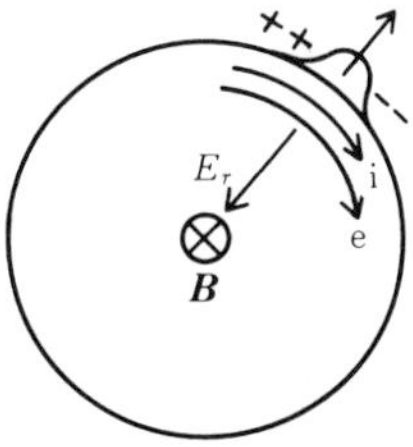

Fig. 9.3
Charge separation due to the difference in velocities of ions and electrons.

$v_\theta = E_r/B$ due to an inward radial electric field, and if the rotation velocity of ions falls below that of electrons, the perturbation is unstable. Several possible mechanisms can retard ion rotation. The collision of ions and neutral particles delays the ion velocity and causes *neutral drag instability*.[4] When the magnitude of the zeroth-order electric field changes radially, the effective electric field which ions feel on the average is different from that felt by electrons due to the different Larmor radii, so that their rotation velocities E_r/B are different. This causes charge separation and may induce the instability.

When the growth rate $\gamma \approx (gk_y)^{1/2}$ is not very large and the ion Larmor radius $\rho_B{}^{\mathrm{i}}$ is large enough to satisfy

$$(k_y \rho_B{}^{\mathrm{i}})^2 > \frac{\gamma}{|\Omega_{\mathrm{i}}|},$$

the perturbation is stabilized.[5] When the ion Larmor radius becomes large, the average perturbation electric field felt by the ions is different that felt by the electrons. The charge separation thus induced has opposite phase from the charge separation due to acceleration and stabilizes the instability. The stabilizing effect of finite Larmor radii will be discussed in detail in ch. 12.

9.1b Stability Condition for Interchange Instability

Let us assume that a line of magnetic force has "good" curvature at one place B and "bad" curvature at another place A (fig. 9.4a). Then the directions of the centrifugal force at A and B are opposite, as is the charge separation. The charge can easily be short circuited along the lines of the magnetic field, so that the problem of stability has a different aspect. Let us here consider perturbations in which the magnetic flux of region 1 is

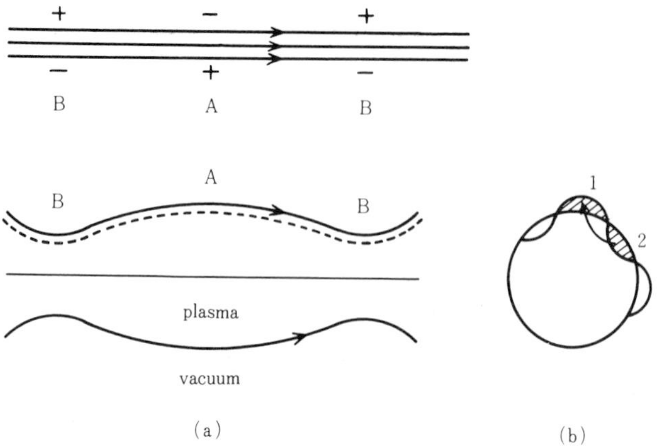

Fig. 9.4
Charge separation in interchange instability. (a) The lower figure shows the unstable part A and the stable part B along a line of magnetic force. The upper figure shows the charge separation due to the acceleration along a flute. (b) Cross section of the perturbed plasma.

interchanged with that of region 2 and the plasma in the region 2 is interchanged wfth the plasma in the region 1 (interchange perturbations, fig. 9.4b). It is assumed that the plasma is low-beta so that the magnetic field is nearly identical to the vacuum field. Any deviation from the vacuum field is accompanied by an increase in the energy of the disturbed field.

It can be shown that the most dangerous perturbations are those which exchange equal magnetic fluxes, as follows.

The energy of the magnetic field inside a magnetic tube is

$$Q_{\mathrm{M}} = \int \mathrm{d}r \frac{B^2}{2\mu_0} = \int \mathrm{d}l S \frac{B^2}{2\mu_0}, \tag{9.15}$$

where l is length taken along a line of magnetic force and S is the cross section of the magnetic tube. As the magnetic flux $\Phi = BS$ is constant, the energy is

$$Q_{\mathrm{M}} = \frac{\Phi^2}{2\mu_0} \int \frac{\mathrm{d}l}{S}. \tag{9.16}$$

The change δQ_{M} in the magnetic energy due to the interchange of the fluxes of regions 1 and 2 is

$$\delta Q_{\mathrm{M}} = \frac{1}{2\mu_0}\left(\left(\Phi_1{}^2\int_2 \frac{\mathrm{d}l}{S} + \Phi_2{}^2\int_1 \frac{\mathrm{d}l}{S}\right) - \left(\Phi_1{}^2\int_1 \frac{\mathrm{d}l}{S} + \Phi_2{}^2\int_2 \frac{\mathrm{d}l}{S}\right)\right) \geq 0. \tag{9.17}$$

If the exchanged fluxes Φ_1 and Φ_2 are the same, the energy change δQ_{M} is zero, so that perturbations resulting in $\Phi_1 = \Phi_2$ are the most dangerous.

The kinetic energy Q_{p} of a plasma of volume $\mathscr{V}$ is

$$Q_{\mathrm{p}} = \frac{nT\mathscr{V}}{\gamma - 1} = \frac{p\mathscr{V}}{\gamma - 1}, \tag{9.18}$$

where γ is the specific-heat ratio. As the perturbation is adiabatic,

$$p\mathscr{V}^{\gamma} = \text{const.}$$

is conserved during the interchange process. The change in the plasma energy is

$$\delta Q_{\mathrm{p}} = \frac{1}{\gamma - 1}\left(p_1\left(\frac{\mathscr{V}_1}{\mathscr{V}_2}\right)^{\gamma}\mathscr{V}_2 - p_1\mathscr{V}_1 + p_2\left(\frac{\mathscr{V}_2}{\mathscr{V}_1}\right)^{\gamma}\mathscr{V}_1 - p_2\mathscr{V}_2\right). \tag{9.19}$$

Setting

$$p_2 = p_1 + \delta p$$

$$\mathscr{V}_2 = \mathscr{V}_1 + \delta\mathscr{V},$$

we can write δQ_{p} as

$$\delta Q_{\mathrm{p}} = \delta p \delta\mathscr{V} + \gamma p\frac{(\delta\mathscr{V})^2}{\mathscr{V}}. \tag{9.20}$$

Since the stability condition is $\delta Q_{\mathrm{p}} + \delta Q_{\mathrm{M}} > 0$, the sufficient condition is

$$\delta p \delta\mathscr{V} > 0.$$

Since the volume is

$$\mathscr{V} = \int \mathrm{d}lS = \Phi\int\frac{\mathrm{d}l}{B},$$

the stability condition for interchange instability is written as

$$\delta p \delta\int\frac{\mathrm{d}l}{B} > 0. \tag{9.21a}$$

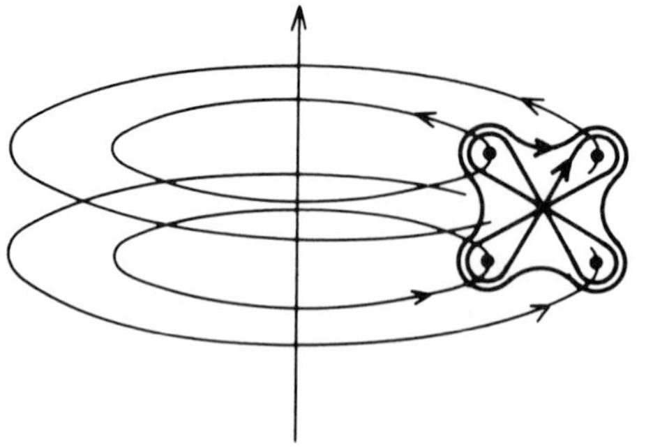

Fig. 9.5
An example of an average minimum-B configuration; toroidal octopole.

Usually the pressure decreases outward ($\delta p < 0$), so that the condition is

$$\delta \int \frac{\mathrm{d}l}{B} < 0 \tag{9.21b}$$

in the outward direction.[6] The integral is to be taken only over the plasma region.

One can define the average magnitude of the magnetic field by

$$\bar{B} \equiv \left(\int \frac{\mathrm{d}l}{B} \bigg/ \int \mathrm{d}l \right)^{-1}.$$

When the total length of the magnetic-field line is denoted by L, the stability condition is $\delta(\bar{B}/L) > 0$. Therefore this condition is called the average minimum-B condition. The magnetic configuration shown in fig. 9.5 satisfies this condition.

9.1c Sausage Instability

Let us consider a cylindrical plasma with a sharp boundary. Only a longitudinal magnetic field B_z exists inside the plasma region and only an azimuthal field $B_\theta = \mu_0 I_z / 2\pi r$ due to the plasma current I_z exists outside the plasma region. We examine an azimuthally symmetric perturbation which constricts the plasma like a sausage (fig. 9.6).

When the plasma radius a is changed by δa, conservation of magnetic flux and the current in the plasma yields

$$\delta B_z = -B_z \frac{2\delta a}{a}$$

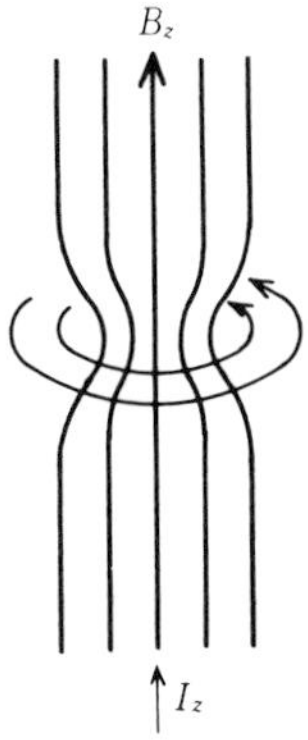

Fig. 9.6
Sausage instability.

$$\delta B_\theta = -B_\theta \frac{\delta a}{a}.$$

The longitudinal magnetic field inside the plasma acts against the perturbation, while the external azimuthal field destabilizes the perturbation. The difference $\delta p_{\rm m}$ in the magnetic pressures is

$$\delta p_{\rm m} = \frac{B_z^{\,2}}{\mu_0}\frac{2\delta a}{a} - \frac{B_\theta^{\,2}}{\mu_0}\frac{\delta a}{a}.$$

The plasma is stable if $\delta p_{\rm m} < 0$ for $\delta a < 0$, so that the stability condition is

$$B_z^{\,2} > \frac{B_\theta^{\,2}}{2}. \tag{9.22}$$

This type of instability is called *sausage instability*.

9.1d Kink Instability

Let us consider a perturbation that kinks the plasma column as shown in fig. 9.7. The configuration of the plasma is the same as that in the previous subsection (sharp boundary, an internal longitudinal field, an external azimuthal field). Denote the characteristic length of the kink by λ and its radius of curvature by R. The longitudinal magnetic field acts as a restoring force on the plasma due to the longitudinal tension; the restoring force on the plasma region of length λ is

$$\frac{B_z^{\,2}}{2\mu_0}\pi a^2 \frac{\lambda}{R}.$$

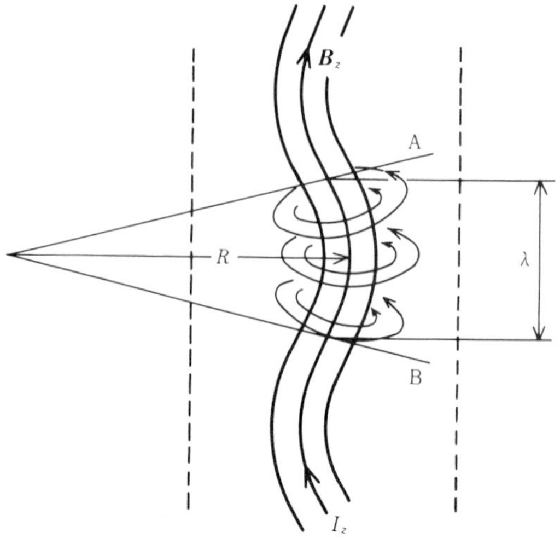

Fig. 9.7
Kink instability.

The azimuthal magnetic field becomes strong at the inner (concave) side of the kink and destabilizes the plasma column. In order to estimate the destabilizing force, we consider a cylindrical lateral surface of radius λ around the plasma and two planes A and B which pass through the center of curvature (see fig. 9.7). Let us compare the contributions of the magnetic pressure on the surfaces enclosing the kink. The contribution of the magnetic pressure on the cylindrical surface is negligible compared with those on the planes A and B. The contribution of the magnetic pressure on the planes A and B is

$$\int_a^\lambda \frac{B_\theta^2}{2\mu_0} 2\pi r dr \times \frac{\lambda}{2R} = \frac{B_\theta^2(a)}{2\mu_0} \pi a^2 \ln \frac{\lambda}{a} \times \frac{\lambda}{R}.$$

Accordingly

$$\frac{B_z^2}{B_\theta^2(a)} > \ln \frac{\lambda}{a} \tag{9.23}$$

is the stability condition.[3] However, the pressure balance

$$p + \frac{B_z^2}{2\mu_0} = \frac{B_\theta^2}{2\mu_0}$$

holds, so that perturbations of large λ are unstable. This type of instability is called *kink instability*.

In this section, a cylindrical sharp-boundary plasma has been analyzed in an intuitive way. The stability of a cylindrical plasma column will be treated in sec. 9.2 in more general and systematic ways.

9.2 General Formulation of Magnetohydrodynamic Instabilities

9.2a Linearization of Magnetohydrodynamic Equations

The stability problems of plasmas can be studied by analysing infinitesimal perturbations of the equilibrium state. If the mass density, pressure, flow velocity, and magnetic field are denoted by ρ_m, p, $\boldsymbol{V}$, and $\boldsymbol{B}$, respectively, the equation of motion, conservation of mass, Ohm's law, and the adiabatic relation are

$$\rho_m \frac{\partial \boldsymbol{V}}{\partial t} = -\nabla p + \boldsymbol{j} \times \boldsymbol{B}$$

$$\frac{\partial \rho_m}{\partial t} + \nabla(\rho_m \boldsymbol{V}) = 0$$

$$\boldsymbol{E} + \boldsymbol{V} \times \boldsymbol{B} = 0$$

$$\left(\frac{\partial}{\partial t} + \boldsymbol{V}\cdot\nabla\right)(p\rho_m{}^{-\gamma}) = 0,$$

respectively. Maxwell's equations are then

$$\nabla \times \boldsymbol{E} = -\frac{\partial \boldsymbol{B}}{\partial t}$$

$$\nabla \times \boldsymbol{B} = \mu_0 \boldsymbol{j}$$

$$\nabla \boldsymbol{B} = 0.$$

These are the magnetohydrodynamic equations of a plasma with zero specific resistivity (see sec. 6.5). The values of ρ_m, p, $\boldsymbol{V}$, and $\boldsymbol{B}$ in the equilibrium state are $\rho_{m0}(\boldsymbol{r})$, $p_0(\boldsymbol{r})$, $\boldsymbol{V}_0 = 0$, and $\boldsymbol{B}_0(\boldsymbol{r})$ respectively. The first-order small quantities are ρ_{m1}, p_1, $\boldsymbol{V}_1 = \boldsymbol{V}$, and $\boldsymbol{B}_1$. The zeroth-order equations are

$$\nabla p_0 = \boldsymbol{j}_0 \times \boldsymbol{B}_0$$

$$\nabla \times \boldsymbol{B}_0 = \mu_0 \boldsymbol{j}_0$$

$$\nabla \cdot \boldsymbol{B}_0 = 0.$$

The first-order linearized equations are

$$\frac{\partial \rho_{m1}}{\partial t} + \nabla(\rho_{m0} \boldsymbol{V}) = 0 \tag{9.24}$$

$$\rho_{m0} \frac{\partial \boldsymbol{V}}{\partial t} + \nabla p_1 = \boldsymbol{j}_0 \times \boldsymbol{B}_1 + \boldsymbol{j}_1 \times \boldsymbol{B}_0 \tag{9.25}$$

$$\frac{\partial p_1}{\partial t} + (\boldsymbol{V} \cdot \nabla) p_0 + \gamma p_0 \nabla \boldsymbol{V} = 0 \tag{9.26}$$

$$\frac{\partial \boldsymbol{B}_1}{\partial t} = \nabla \times (\boldsymbol{V} \times \boldsymbol{B}_0). \tag{9.27}$$

If displacement of the plasma from the equilibrium position $\boldsymbol{r}_0$ is denoted by $\xi(\boldsymbol{r}_0, \mathrm{t})$, it follows that

$$\xi(\boldsymbol{r}_0, t) = \boldsymbol{r} - \boldsymbol{r}_0,$$

$$\boldsymbol{V} = \frac{\mathrm{d}\xi}{\mathrm{d}t} \approx \frac{\partial \xi}{\partial t}.$$

Equation (9.27) is reduced to

$$\frac{\partial \boldsymbol{B}_1}{\partial t} = \nabla \times \left(\frac{\partial \xi}{\partial t} \times \boldsymbol{B}_0\right),$$

and

$$\boldsymbol{B}_1 = \nabla \times (\xi \times \boldsymbol{B}_0). \tag{9.28}$$

From $\mu_0 \boldsymbol{j} = \nabla \times \boldsymbol{B}$, it follows that

$$\mu_0 \boldsymbol{j}_1 = \nabla \times \boldsymbol{B}_1. \tag{9.29}$$

Equations (9.24) and (9.26) yield

$$\rho_{m1} = -\nabla(\rho_{m0} \xi) \tag{9.30}$$

$$p_1 = -\xi \cdot \nabla p_0 - \gamma p_0 \nabla \xi. \tag{9.31}$$

The substitution of these equations into eq. (9.25) gives

$$\rho_{m0}\frac{\partial^2 \boldsymbol{\xi}}{\partial t^2} = \nabla(\boldsymbol{\xi}\cdot\nabla p_0 + \gamma p_0 \nabla\boldsymbol{\xi}) + \frac{1}{\mu_0}(\nabla\times\boldsymbol{B}_0)\times\boldsymbol{B}_1 + \frac{1}{\mu_0}(\nabla\times\boldsymbol{B}_1)\times\boldsymbol{B}_0 \tag{9.32a}$$

$$= -\nabla\left(p_1 + \frac{\boldsymbol{B}_0\cdot\boldsymbol{B}_1}{\mu_0}\right) + \frac{1}{\mu_0}\left((\boldsymbol{B}_0\cdot\nabla)\boldsymbol{B}_1 + (\boldsymbol{B}_1\cdot\nabla)\boldsymbol{B}_0\right). \tag{9.32b}$$

This is the linearized equation of motion in terms of $\boldsymbol{\xi}$.

Next let us consider the boundary conditions. Where the plasma contacts an ideal conductor, the tangential component of the electric field is zero, i.e., $\boldsymbol{n}\times\boldsymbol{E}=0$. This is equivalent to $\boldsymbol{n}\times(\boldsymbol{\xi}\times\boldsymbol{B}_0)=0$, $\boldsymbol{n}$ being taken in the outward direction. The conditions $\boldsymbol{\xi}\cdot\boldsymbol{n}=0$ and $\boldsymbol{B}_1\cdot\boldsymbol{n}=0$ must also be satisfied. At the boundary surface between plasma and vacuum, the total pressure must be continuous and

$$p(\boldsymbol{r}) + \frac{B_{\mathrm{i}}^2(\boldsymbol{r})}{2\mu_0} = \frac{B_{\mathrm{e}}^2(\boldsymbol{r})}{2\mu_0}$$

$$p_0(\boldsymbol{r}_0) + \frac{B_{0\mathrm{i}}^2(\boldsymbol{r}_0)}{2\mu_0} = \frac{B_{0\mathrm{e}}^2(\boldsymbol{r}_0)}{2\mu_0},$$

where $\boldsymbol{B}_{\mathrm{i}}(\boldsymbol{r})$, $\boldsymbol{B}_{0\mathrm{i}}(\boldsymbol{r}_0)$ give the internal magnetic field of the plasma and $\boldsymbol{B}_{\mathrm{e}}(\boldsymbol{r})$, $\boldsymbol{B}_{0\mathrm{e}}(\boldsymbol{r}_0)$ give the external field. The boundary condition is reduced to

$$-\gamma p_0\cdot\nabla\boldsymbol{\xi} + \frac{\boldsymbol{B}_{0\mathrm{i}}\cdot(\boldsymbol{B}_{1\mathrm{i}} + (\boldsymbol{\xi}\cdot\nabla)\boldsymbol{B}_{0\mathrm{i}})}{\mu_0} = \frac{\boldsymbol{B}_{0\mathrm{e}}\cdot(\boldsymbol{B}_{1\mathrm{e}} + (\boldsymbol{\xi}\cdot\nabla)\boldsymbol{B}_{0\mathrm{e}})}{\mu_0}, \tag{9.33}$$

when $\boldsymbol{B}_{\mathrm{i}}(\boldsymbol{r})$, $\boldsymbol{B}_{\mathrm{e}}(\boldsymbol{r})$ and $p(\boldsymbol{r})$ are expanded in $\boldsymbol{\xi}=\boldsymbol{r}-\boldsymbol{r}_0$ ($f(\boldsymbol{r}) = f_0(\boldsymbol{r}_0) + (\boldsymbol{\xi}\cdot\nabla)f_0 + f_1$).

From Maxwell's equations, the boundary conditions are

$$\boldsymbol{n}_0\cdot(\boldsymbol{B}_{0\mathrm{i}} - \boldsymbol{B}_{0\mathrm{e}}) = 0 \tag{9.34}$$

$$\boldsymbol{n}_0\times(\boldsymbol{B}_{0\mathrm{i}} - \boldsymbol{B}_{0\mathrm{e}}) = \mu_0\boldsymbol{K}, \tag{9.35}$$

where $\boldsymbol{K}$ is the surface current. Ohm's law and eq. (9.34) yield

$$\boldsymbol{n}_0\times(\boldsymbol{E}_{\mathrm{i}} - \boldsymbol{E}_{\mathrm{e}}) = (\boldsymbol{n}_0\cdot\boldsymbol{V})(\boldsymbol{B}_{0\mathrm{i}} - \boldsymbol{B}_{0\mathrm{e}}). \tag{9.36a}$$

The boundary condition can be written as

$$\boldsymbol{E}_{\mathrm{t}} + (\boldsymbol{V}\times\boldsymbol{B}_{0\mathrm{e}})_{\mathrm{t}} = 0, \tag{9.36b}$$

where the subscript t indicates the tangential component. As the electric field $\boldsymbol{E}^*$ in coordinates moving with the plasma is $\boldsymbol{E}^* = \boldsymbol{E} + \boldsymbol{V}\times\boldsymbol{B}_{0\mathrm{i}}$ and the specific resistivity of the plasma is zero, the tangential component of the electric field is naturally zero. Since the normal component of $\boldsymbol{B}$ is given by the tangential component of $\boldsymbol{E}$ by the relation $\nabla \times \boldsymbol{E} = -\partial \boldsymbol{B}/\partial t$, eq. (9.36b) is reduced to

$$(\boldsymbol{n}_0 \cdot \boldsymbol{B}_{\text{ie}}) = \boldsymbol{n}_0 \cdot \nabla \times (\xi \times \boldsymbol{B}_{0\text{e}}). \tag{9.36c}$$

The electric field $\boldsymbol{E}_{\text{e}}$ and the magnetic field $\boldsymbol{B}_{\text{e}}$ in the external (vacuum) region can be expressed in terms of a vector potential:

$$\boldsymbol{E}_{\text{e}} = -\frac{\partial \boldsymbol{A}}{\partial t} \tag{9.37}$$

$$\boldsymbol{B}_{\text{ie}} = \nabla \times \boldsymbol{A} \tag{9.38}$$

$$\nabla \boldsymbol{A} = 0. \tag{9.39}$$

If no current flows in the vacuum region, $\boldsymbol{A}$ satisfies

$$\nabla \times \nabla \times \boldsymbol{A} = 0. \tag{9.40}$$

Using the vector potential, we may express eq. (9.36b) as

$$\boldsymbol{n}_0 \times \left(-\frac{\partial \boldsymbol{A}}{\partial t} + \boldsymbol{V} \times \boldsymbol{B}_{0\text{e}} \right) = 0.$$

For $\boldsymbol{n}_0 \cdot \boldsymbol{B}_{0\text{i}} = \boldsymbol{n}_0 \cdot \boldsymbol{B}_{0\text{e}} = 0$, the boundary condition is

$$\boldsymbol{n}_0 \times \boldsymbol{A} = -\xi_n \boldsymbol{B}_{0\text{e}}. \tag{9.36d}$$

At the wall of an ideal conductor, eq. (9.37) becomes

$$\boldsymbol{n} \times \boldsymbol{A} = 0. \tag{9.41}$$

The stability problem now becomes one of solving eqs. (9.32) and (9.40) under the boundary conditions (9.33), (9.36d), and (9.41). When a normal mode $\xi(\boldsymbol{r}, t) = \xi(\boldsymbol{r}) \exp(-i\omega t)$ is considered, the problem is reduced to the eigenvalue problem $\rho_{\text{m}0}\omega^2 \xi = \boldsymbol{F}(\xi)$. If any eigenvalue is negative, the plasma is unstable; if all the eigenvalues are positive, the plasma is stable.

9.2b Energy Principle[7]

The eigenvalue problem is complicated and difficult to solve in general. When we introduce a potential energy associated with the displacement

$\boldsymbol{\xi}$, the stability problem can be simplified. The equation of motion has the form

$$\rho_{m0}\frac{\partial^2 \boldsymbol{\xi}}{\partial t^2} = \boldsymbol{F}(\boldsymbol{\xi}) = -\hat{\boldsymbol{K}}\cdot\boldsymbol{\xi}, \tag{9.42}$$

where $\hat{\boldsymbol{K}}$ is a Hermitian, i.e., a linear self-adjoint, operator as will be shown in the following paragraph. This equation can be integrated:

$$\frac{1}{2}\int \rho_{m0}\left(\frac{\partial \boldsymbol{\xi}}{\partial t}\right)^2 \mathrm{d}\boldsymbol{r} + \frac{1}{2}\int \boldsymbol{\xi}\cdot\hat{\boldsymbol{K}}\boldsymbol{\xi}\mathrm{d}\boldsymbol{r} = \text{const.}$$

The kinetic energy T and the potential energy W are

$$T \equiv \frac{1}{2}\int \rho_{m0}\left(\frac{\partial \boldsymbol{\xi}}{\partial t}\right)^2 \mathrm{d}\boldsymbol{r}$$

$$W \equiv \frac{1}{2}\int \boldsymbol{\xi}\cdot\hat{\boldsymbol{K}}\boldsymbol{\xi}\mathrm{d}\boldsymbol{r} = \frac{-1}{2}\int \boldsymbol{\xi}\cdot\boldsymbol{F}(\boldsymbol{\xi})\mathrm{d}\boldsymbol{r},$$

respectively. Accordingly if

$$W > 0$$

for all possible displacements, the system is stable. This is the stability condition for the *energy principle*. W is called the *energy integral*.

Let us prove that $\hat{\boldsymbol{K}}$ is Hermitian. For this purpose, a displacement $\boldsymbol{\eta}$ and a vector potential $\boldsymbol{Q}$ are introduced which satisfy the same boundary conditions as $\boldsymbol{\xi}$ and $\boldsymbol{A}$, i.e.,

$$\boldsymbol{n}_0 \times \boldsymbol{Q} = -\eta_n \boldsymbol{B}_{0\mathrm{e}}$$

at the plasma-vacuum boundary and

$$\boldsymbol{n}_0 \times \boldsymbol{Q} = 0$$

at the conducting wall. If

$$\int \boldsymbol{\eta}\cdot\hat{\boldsymbol{K}}\boldsymbol{\xi}\mathrm{d}\boldsymbol{r} = \int \boldsymbol{\xi}\cdot\hat{\boldsymbol{K}}\boldsymbol{\eta}\mathrm{d}\boldsymbol{r}, \tag{9.43}$$

is satisfied, $\hat{\boldsymbol{K}}$ is a Hermitian operator. By substitution of eq. (9.32a), the integral in the plasma region V_i is seen to be

$$\int_{V_\mathrm{i}} \boldsymbol{\eta}\cdot\hat{\boldsymbol{K}}\boldsymbol{\xi}\mathrm{d}\boldsymbol{r} = \int_{V_\mathrm{i}} \Big(\gamma p_0(\nabla\boldsymbol{\eta})(\nabla\boldsymbol{\xi}) + (\nabla\eta)(\boldsymbol{\xi}\cdot\nabla p_0)$$

$$+\frac{1}{\mu_0}(\nabla\times(\boldsymbol{\eta}\times\boldsymbol{B}_0))\cdot(\nabla\times(\boldsymbol{\xi}\times\boldsymbol{B}_0))$$

$$-\frac{1}{\mu_0}(\boldsymbol{\eta}\times(\nabla\times\boldsymbol{B}_0))\cdot\nabla\times(\boldsymbol{\xi}\times\boldsymbol{B}_0)\Bigg)\mathrm{d}\boldsymbol{r}$$

$$+\int_S \mathrm{d}S\boldsymbol{n}_0\cdot\boldsymbol{\eta}\left(\frac{\boldsymbol{B}_{0\mathrm{i}}\cdot\nabla\times(\boldsymbol{\xi}\times\boldsymbol{B}_{0\mathrm{i}})}{\mu_0}-\gamma p_0(\nabla\cdot\xi)-(\xi\cdot\nabla p_0)\right). \tag{9.44}$$

First let us consider the volume integral in this equation. The contribution of $\boldsymbol{\eta}_{\parallel}=\alpha\boldsymbol{B}$ to the 2nd and 4th terms of the integral is

$$\begin{aligned}
&(\nabla\cdot\boldsymbol{\eta}_{\parallel})(\xi\cdot\nabla p_0)-\frac{1}{\mu_0}(\boldsymbol{\eta}_{\parallel}\times(\nabla\times\boldsymbol{B}_0))\cdot\nabla\times(\xi\times\boldsymbol{B}_0)\\
&\quad=(\xi\cdot\nabla p_0)(\boldsymbol{B}_0\cdot\nabla\alpha)+\alpha\nabla p_0\cdot\nabla\times(\xi\times\boldsymbol{B}_0)\\
&\quad=(\xi\cdot\nabla p_0)(\boldsymbol{B}_0\cdot\nabla\alpha)+\alpha\nabla\cdot((\xi\times\boldsymbol{B}_0)\times\nabla p_0)\\
&\quad=(\xi\cdot\nabla p_0)(\boldsymbol{B}_0\cdot\nabla\alpha)+\alpha\nabla\cdot((\xi\cdot\nabla p_0)\cdot\boldsymbol{B}_0)\\
&\quad=\nabla\cdot((\xi\cdot\nabla p_0)\alpha\boldsymbol{B}_0).
\end{aligned}$$

The volume integral becomes a surface integral, which is zero by virtue of the boundary condition $\boldsymbol{n}_0\cdot\boldsymbol{\eta}_{\parallel}=0$. Let us express $\boldsymbol{\xi}$ and $\boldsymbol{\eta}$ by the three vectors $\boldsymbol{B}_0$, $\nabla\times\boldsymbol{B}_0$, and $\boldsymbol{e}=\nabla p_0/|\nabla p_0|$, i.e.,

$$\begin{aligned}
\boldsymbol{\xi}&=\xi_1\nabla\times\boldsymbol{B}_0+\xi_2\boldsymbol{e}+\xi_3\boldsymbol{B}_0=\xi'+\xi_3\boldsymbol{B}_0\\
\boldsymbol{\eta}&=\eta_1\nabla\times\boldsymbol{B}_0+\eta_2\boldsymbol{e}+\eta_3\boldsymbol{B}_0=\boldsymbol{\eta}'+\eta_3\boldsymbol{B}_0.
\end{aligned}$$

As the contribution of $\boldsymbol{\eta}_{\parallel}$ to the integral is zero, only the contribution of $\boldsymbol{\eta}'$ must be taken into account:

$$\begin{aligned}
&-\frac{1}{\mu_0}(\boldsymbol{\eta}'\times(\nabla\times\boldsymbol{B}_0))\cdot\nabla\times(\boldsymbol{\xi}'\times\boldsymbol{B}_0)\\
&\quad=-\frac{\eta_2}{\mu_0}(\boldsymbol{e}\times(\nabla\times\boldsymbol{B}_0))\cdot\nabla\times(\mu_0\xi_1\nabla p_0+\xi_2(\boldsymbol{e}\times\boldsymbol{B}_0))\\
&\quad=-\eta_2(\boldsymbol{e}\times(\nabla\times\boldsymbol{B}_0))\cdot(\nabla\xi_1\times\nabla p_0)\\
&\qquad-\frac{\eta_2}{\mu_0}(\boldsymbol{e}\times(\nabla\times\boldsymbol{B}_0))\cdot[\xi_2(\boldsymbol{B}_0\cdot\nabla)\boldsymbol{e}-\xi_2(\boldsymbol{e}\cdot\nabla)\boldsymbol{B}_0-\xi_2\boldsymbol{B}_0(\nabla\cdot\boldsymbol{e})\\
&\qquad\qquad+\nabla\xi_2\times(\boldsymbol{e}\times\boldsymbol{B}_0)]
\end{aligned}$$

$$= (\eta_2 e\cdot\nabla p_0)(\nabla\times \boldsymbol{B}_0\cdot\nabla\xi_1)$$

$$-\frac{\eta_2}{\mu_0}(e\times(\nabla\times \boldsymbol{B}_0))\cdot[-(\nabla(\xi_2 e))\boldsymbol{B}_0 - \xi_2((\boldsymbol{B}_0\cdot\nabla)e - (e\cdot\nabla)\boldsymbol{B}_0)]$$

$$= (\boldsymbol{\eta}\cdot\nabla p_0)(\nabla\times \boldsymbol{B}_0\cdot\nabla\xi_1) + (\boldsymbol{\eta}\cdot\nabla p_0)\nabla(\xi_2 e)$$

$$+\frac{\xi_2\eta_2}{\mu_0}(e\times(\nabla\times \boldsymbol{B}_0))\cdot((e\cdot\nabla)\boldsymbol{B}_0 - (\boldsymbol{B}_0\cdot\nabla)e).$$

Accordingly, it follows that

$$(\nabla\boldsymbol{\eta}')(\boldsymbol{\xi}\cdot\nabla p_0) - \frac{1}{\mu_0}(\boldsymbol{\eta}'\times(\nabla\times \boldsymbol{B}_0))\cdot\nabla\times(\boldsymbol{\xi}\times \boldsymbol{B}_0)$$

$$= (\boldsymbol{\xi}\cdot\nabla p_0)(\nabla\boldsymbol{\eta}') + (\boldsymbol{\eta}\cdot\nabla p_0)(\nabla\boldsymbol{\xi}')$$

$$+\frac{(\boldsymbol{\eta}\cdot e)(\boldsymbol{\xi}\cdot e)}{\mu_0}(e\times(\nabla\times \boldsymbol{B}_0))\cdot((e\cdot\nabla)\boldsymbol{B} - (\boldsymbol{B}\cdot\nabla)e).$$

This equation is symmetric with respect to $\boldsymbol{\xi}$ and $\boldsymbol{\eta}$. Thus it has been shown that the volume integral of eq. (9.44) is symmetric with respect to $\boldsymbol{\xi}$ and $\boldsymbol{\eta}$.

Next let us consider the surface integral in eq. (9.44). Due to the boundary condition $\boldsymbol{n}_0\times \boldsymbol{Q} = -\eta_n \boldsymbol{B}_{0\mathrm{e}}$, we find that

$$\int \eta_n \boldsymbol{B}_{0\mathrm{e}}\cdot(\nabla\times A)\mathrm{d}S = -\int_S (\boldsymbol{n}_0\times \boldsymbol{Q})\cdot\nabla\times A\mathrm{d}S$$

$$= -\int_S \boldsymbol{n}_0\cdot(\boldsymbol{Q}\times(\nabla\times A))\mathrm{d}S$$

$$= \int_{V_\mathrm{e}} \nabla(\boldsymbol{Q}\times(\nabla\times A))\mathrm{d}\boldsymbol{r}$$

$$= \int_{V_\mathrm{e}} ((\nabla\times \boldsymbol{Q})\cdot(\nabla\times A) - \boldsymbol{Q}\cdot\nabla\times(\nabla\times A))\mathrm{d}\boldsymbol{r}$$

$$= \int_{V_\mathrm{e}} (\nabla\times \boldsymbol{Q})\cdot(\nabla\times A)\mathrm{d}\boldsymbol{r}.$$

By means of this and the boundary condition (9.33), the surface integral of eq. (9.44) may be reduced to

$$\int_S \left(\eta_n \frac{\boldsymbol{B}_{0\mathrm{e}} \cdot \boldsymbol{B}_{1\mathrm{e}}}{\mu_0} + \eta_n (\boldsymbol{\xi} \cdot \nabla) \left(\frac{B_{0\mathrm{e}}{}^2}{2\mu_0} - \frac{B_{0\mathrm{i}}{}^2}{2\mu_0} - p_0 \right) \right) \mathrm{d}S$$

$$= \int_{V_\mathrm{e}} (\nabla \times \boldsymbol{Q}) \cdot (\nabla \times \boldsymbol{A}) \mathrm{d}\boldsymbol{r} - \int_S \eta_n \xi_n \frac{\partial}{\partial n} \left(p_0 + \frac{B_{0\mathrm{i}}{}^2}{2\mu_0} - \frac{B_{0\mathrm{e}}{}^2}{2\mu_0} \right) \mathrm{d}S$$

(notice that $\boldsymbol{B}_{1\mathrm{e}} = \nabla \times \boldsymbol{A}$). The integral region V_e is the region outside the plasma. Finally, the energy integral is reduced to

$$\int_{V_\mathrm{i}} \boldsymbol{\eta} \cdot \hat{\boldsymbol{K}} \boldsymbol{\xi} \mathrm{d}\boldsymbol{r} = \int_{V_\mathrm{i}} \left(\gamma p_0 (\nabla \boldsymbol{\eta})(\nabla \boldsymbol{\xi}) + \frac{1}{\mu_0} (\nabla \times (\boldsymbol{\eta} \times \boldsymbol{B}_0)) \cdot (\nabla \times (\boldsymbol{\xi} \times \boldsymbol{B}_0)) \right.$$

$$\left. + (\nabla \boldsymbol{\eta})(\boldsymbol{\xi} \cdot \nabla p_0) - \frac{1}{\mu_0} (\boldsymbol{\eta} \times (\nabla \times \boldsymbol{B}_0)) \cdot \nabla \times (\boldsymbol{\xi} \times \boldsymbol{B}_0) \right) \mathrm{d}\boldsymbol{r}$$

$$+ \frac{1}{\mu_0} \int_{V_\mathrm{e}} (\nabla \times \boldsymbol{Q}) \cdot (\nabla \times \boldsymbol{A}) \mathrm{d}\boldsymbol{r}$$

$$+ \int_S \eta_n \xi_n \frac{\partial}{\partial n} \left(\frac{B_{0\mathrm{e}}{}^2}{2\mu_0} - \frac{B_{0\mathrm{i}}{}^2}{2\mu_0} - p_0 \right) \mathrm{d}S. \tag{9.45}$$

Thus it is proved that $\hat{\boldsymbol{K}}$ is a Hermitian operator.[8]

The energy integral W is divided into three parts W_P, W_S, and W_V, the contributions of the plasma internal region V_i, the boundary region S, and the external vacuum region V_e, i.e.,

$$W = \frac{1}{2} \int_{V_\mathrm{i}} \boldsymbol{\xi} \cdot \hat{\boldsymbol{K}} \boldsymbol{\xi} \mathrm{d}\boldsymbol{r} \tag{9.46}$$

$$= W_\mathrm{P} + W_\mathrm{S} + W_\mathrm{V}$$

$$W_\mathrm{P} = \frac{1}{2} \int_{V_\mathrm{i}} \left(\gamma p_0 (\nabla \boldsymbol{\xi})^2 + \frac{1}{\mu_0} (\nabla \times (\boldsymbol{\xi} \times \boldsymbol{B}_0))^2 + (\nabla \boldsymbol{\xi})(\boldsymbol{\xi} \cdot \nabla p_0) \right.$$

$$\left. - \frac{1}{\mu_0} (\boldsymbol{\xi} \times (\nabla \times \boldsymbol{B}_0)) \cdot \nabla \times (\xi \times \boldsymbol{B}_0) \right) \mathrm{d}\boldsymbol{r}$$

$$= \frac{1}{2} \int_{V_\mathrm{i}} \left(\frac{\boldsymbol{B}_1{}^2}{\mu_0} - p_1 (\nabla \boldsymbol{\xi}) - \boldsymbol{\xi} \cdot (\boldsymbol{j}_0 \times \boldsymbol{B}_1) \right) \mathrm{d}\boldsymbol{r} \tag{9.47}$$

$$W_\mathrm{S} = \frac{1}{2} \int_S \xi_n{}^2 \frac{\partial}{\partial n} \left(\frac{B_{0\mathrm{e}}{}^2}{2\mu_0} - \frac{B_{0\mathrm{i}}{}^2}{2\mu_0} - p_0 \right) \mathrm{d}S \tag{9.48}$$

$$W_{\mathrm{V}} = \frac{1}{2\mu_0}\int_{V_{\mathrm{e}}} (\nabla \times \boldsymbol{A})^2 \mathrm{d}\boldsymbol{r} = \int_{V_{\mathrm{e}}} \frac{\boldsymbol{B}_1{}^2}{2\mu_0} \mathrm{d}\boldsymbol{r}. \tag{9.49}$$

The stability condition is $W > 0$ for all possible $\boldsymbol{\xi}$. The frequency or growth rate of a perturbation can be obtained from the energy integral. When the perturbation varies as $\exp(-i\omega t)$, the equation of motion is

$$\omega^2 \rho_{\mathrm{m}0} \boldsymbol{\xi} = \hat{\boldsymbol{K}} \boldsymbol{\xi}. \tag{9.50}$$

The solution of the eigenvalue problem is the same as the solution based on the calculus of variations,

$$\omega^2 = \frac{\displaystyle\int \boldsymbol{\xi} \cdot \hat{\boldsymbol{K}} \boldsymbol{\xi} \mathrm{d}\boldsymbol{r}}{\displaystyle\int \rho_{\mathrm{m}0} \boldsymbol{\xi}^2 \mathrm{d}\boldsymbol{r}}. \tag{9.51}$$

As $\hat{\boldsymbol{K}}$ is a Hermitian operator, ω^2 is real. In the MHD analysis of an ideal plasma with zero resistivity, the perturbation either increases or decreases monotonically, or else the perturbed plasma oscillates with constant amplitude.

9.3 Instabilities of a Cylindrical Plasma

9.3a Instabilities of Sharp-Boundary Configurations; the Kruskal-Shafranov Condition

Let us consider a sharp-boundary plasma of radius a, with a longitudinal magnetic field $B_{\mathrm{i}z}$ inside the boundary and a longitudinal magnetic field $B_{\mathrm{e}z}$ and an azimuthal magnetic field $B_\theta = \mu_0 I/(2\pi r)$ outside. $B_{\mathrm{i}z}$ and $B_{\mathrm{e}z}$ are assumed to be constant.

We can consider the displacement

$$\boldsymbol{\xi}(\boldsymbol{r}) \exp(im\theta + ikz), \tag{9.52}$$

since any displacement may be expressed by a superposition of such modes. Since the term in $\nabla \boldsymbol{\xi}$ in the energy integral is positive, incompressible perturbation is the most dangerous. We examine only the worst mode,

$$\nabla \boldsymbol{\xi} = 0. \tag{9.53}$$

The perturbation of the magnetic field $\boldsymbol{B}_1 = \nabla \times (\xi \times \boldsymbol{B}_{\rm i})$ is

$$\boldsymbol{B}_1 = ikB_{iz}\xi. \tag{9.54}$$

The equation of motion (9.32b) becomes

$$\left(-\omega^2\rho_{\rm m0} + \frac{k^2 B_{\rm iz}{}^2}{\mu_0}\right)\xi = -\nabla\left(p_1 + \frac{\boldsymbol{B}_{\rm i}\cdot\boldsymbol{B}_1}{\mu_0}\right) \equiv -\nabla p^*. \tag{9.55}$$

As $\nabla\cdot\xi = 0$, it follows that $\Delta p^* = 0$, i.e.,

$$\left(\frac{{\rm d}^2}{{\rm d}r^2} + \frac{1}{r}\frac{\rm d}{{\rm d}r} - \left(k^2 + \frac{m^2}{r^2}\right)\right)p^*(r) = 0. \tag{9.56}$$

The solution without singularity at $r = 0$ is given by the modified Bessel function $I_m(kr)$, so that $p^*(r)$ is

$$p^*(r) = p^*(a)\frac{I_m(kr)}{I_m(ka)}.$$

Accordingly, we find

$$\xi_r(a) = \frac{\dfrac{kp^*(a)}{I_m(ka)}}{\omega^2\rho_{\rm m0} - \dfrac{k^2 B_{\rm i}^2}{\mu_0}} I_m'(ka). \tag{9.57}$$

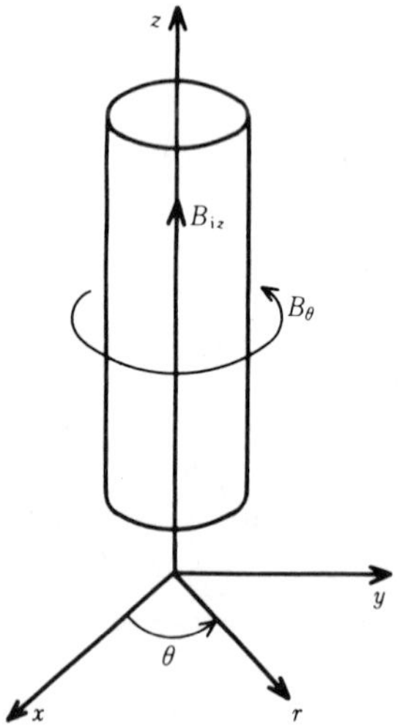

Fig. 9.8
Sharp-boundary plasma.

As the perturbation of the vacuum magnetic field B_{ie} satisfies $\nabla \times \boldsymbol{B} = 0$ and $\nabla \boldsymbol{B} = 0$, $\boldsymbol{B}_{1\mathrm{e}}$ is expressed by $\boldsymbol{B}_{1\mathrm{e}} = \nabla\psi$. The scalar magnetic potential ψ satisfies $\nabla\psi = 0$ and $\psi \to 0$ as $r \to \infty$. Then

$$\psi = C\frac{K_m(kr)}{K_m(ka)}\exp(im\theta + ikz). \tag{9.58a}$$

The boundary condition (9.33) is

$$p^* \equiv p_1 + \frac{1}{\mu_0}\boldsymbol{B}_\mathrm{i}\cdot\boldsymbol{B}_1 = \frac{1}{\mu_0}\boldsymbol{B}_\mathrm{e}\cdot\boldsymbol{B}_{1\mathrm{e}} + (\boldsymbol{\xi}\cdot\nabla)\left(\frac{B_\mathrm{e}^2}{2\mu_0} - \frac{B_\mathrm{i}^2}{2\mu_0} - p_0\right)$$

$$= \frac{1}{\mu_0}\boldsymbol{B}_\mathrm{e}\cdot\boldsymbol{B}_{1\mathrm{e}} + (\boldsymbol{\xi}\cdot\nabla)\left(\frac{B_\theta^2}{2\mu_0}\right).$$

As $B_\theta \propto 1/r$, $p^*(a)$ is given by

$$p^*(a) = \frac{i}{\mu_0}\left(kB_{\mathrm{e}z} + \frac{m}{a}B_\theta\right)C - \frac{B_\theta^2}{\mu_0 a}\xi_r(a). \tag{9.59}$$

The boundary condition (9.36c) is reduced to

$$Ck\frac{K_m'(ka)}{K_m(ka)} = i\left(kB_{\mathrm{e}z} + \frac{m}{a}B_\theta\right)\xi_r(a). \tag{9.60}$$

From eqs. (9.57), (9.59), and (9.60), the dispersion equation is

$$\frac{\omega^2}{k^2} = \frac{B_{\mathrm{i}z}^2}{\mu_0\rho_{\mathrm{m}0}} - \frac{(kB_{\mathrm{e}z} + (m/a)B_\theta)^2}{\mu_0\rho_{\mathrm{m}0}k^2}\frac{I_m'(ka)}{I_m(ka)}\frac{K_m(ka)}{K_m'(ka)}$$

$$- \frac{B_\theta^2}{\mu_0\rho_{\mathrm{m}0}}\frac{1}{(ka)}\frac{I_m'(ka)}{I_m(ka)}. \tag{9.61a}$$

The 1st and 2nd terms represent the stabilizing effect of $B_{\mathrm{i}z}$ and $B_{\mathrm{e}z}$ ($K_m/K_m' < 0$). If the propagation vector $\boldsymbol{k}$ is normal to the magnetic field, i.e., if

$$(\boldsymbol{k}\cdot\boldsymbol{B}_\mathrm{e}) = kB_{\mathrm{e}z} + \frac{m}{a}B_\theta = 0,$$

the 2nd term of eq. (9.61a) becomes zero, so that a flutelike perturbation is indicated. The 3rd term is the destabilizing term.

(i) The $m = 0$ Mode with $B_{\mathrm{e}z} = 0$ Let us consider the $m = 0$ mode with $B_{\mathrm{e}z} = 0$. This configuration corresponds to that of the sausage instability described in sec. 9.1c. Equation (9.61) reduces to

$$\omega^2 = \frac{B_{iz}^2 k^2}{\mu_0 \rho_{m0}}\left(1 - \frac{B_\theta^2}{B_{iz}^2}\frac{I_0'(ka)}{(ka)I_0(ka)}\right). \tag{9.62}$$

Since $I_0'(x)/x\, I_0(x) < 1/2$, the stability condition is

$B_{iz}^2 > B_\theta^2/2.$

(ii) The $m = 1$ Mode with $B_{ez} = 0$ For the $m = 1$ mode with $B_{ez} = 0$, eq. (9.61) is

$$\omega^2 = \frac{B_{iz}^2 k^2}{\mu_0 \rho_{m0}}\left(1 + \frac{B_\theta^2}{B_{iz}^2}\frac{1}{(ka)}\frac{I_1'(ka)}{I_1(ka)}\frac{K_1(ka)}{K_1'(ka)}\right). \tag{9.63a}$$

For perturbations with long characteristic length, eq. (9.63a) becomes

$$\omega^2 = \frac{B_{iz}^2 k^2}{\mu_0 \rho_{m0}}\left(1 - \left(\frac{B_\theta}{B_{iz}}\right)^2 \ln\frac{1}{ka}\right). \tag{9.63b}$$

This dispersion equation corresponds to kink instability, which is unstable for the perturbation with long wavelength.

(iii) Instability in the Case of $|B_{ez}| \gg |B_e|$ When $|B_{ez}| \gg |B_\theta|$, the case $|ka| \ll 1$ predominates. Expanding the modified Bessel function ($m > 0$ is assumed), we find

$$\mu_0 \rho_{m0} \omega^2 = k^2 B_{iz}^2 + \left(k B_{ez} + \frac{m}{a} B_\theta\right)^2 - \frac{m}{a^2} B_\theta^2. \tag{9.64a}$$

ω^2 becomes minimum at $\partial\omega/\partial k = 0$, i.e., $k(B_{iz}^2 + B_{ez}^2) + (m/a)B_\theta B_{ez} = 0$. In this case ω^2 is

$$\omega_{min}^2 = \frac{B_\theta^2}{\mu_0 \rho_{m0} a^2}\left(\frac{m^2 B_{iz}^2}{B_{ez}^2 + B_{iz}^2} - m\right) = \frac{B_\theta^2}{\mu_0 \rho_{m0} a^2} m\left(m\frac{1-\beta}{2-\beta} - 1\right), \tag{9.65}$$

where β is the beta ratio. Accordingly, the plasma is unstable when $0 < m < (2-\beta)/(1-\beta)$. For a low-beta plasma only the modes $m = 1$ and $m = 2$ become unstable. However, if

$$\left(\frac{B_\theta}{B_z}\right)^2 < (ka)^2 \tag{9.66a}$$

is satisfied the plasma is stable even for $m = 1$. Usually the length of the plasma is finite, so that k cannot be smaller than $2\pi/L$. Accordingly, when

$$\left|\frac{B_\theta}{B_z}\right| < \frac{2\pi a}{L} \tag{9.66b}$$

the plasma is stable. This stability condition is called the *Kruskal-Shafranov condition.*[9, 10]

When a cylindrical conducting wall of radius b surrounds the plasma, the scalar magnetic potential of the external magnetic field is

$$\psi = \left(c_1 \frac{K_m(kr)}{K_m(ka)} + c_2 \frac{I_m(kr)}{I_m(ka)}\right)\exp(im\theta + ikz) \tag{9.58b}$$

instead of eq. (9.58a). The boundary condition $B_{1\mathrm{e}r} = 0$ at $r = b$ yields

$$\frac{c_1}{c_2} = -\frac{I_m{}'(kb)\,K_m(ka)}{K_m{}'(kb)\,I_m(ka)}.$$

The dispersion equation becomes

$$\frac{\omega^2}{k^2} = \frac{B_{\mathrm{i}z}{}^2}{\mu_0\rho_{\mathrm{m}0}} - \frac{(kB_{\mathrm{e}z} + (m/a)B_\theta)^2}{\mu_0\rho_{\mathrm{m}0}k^2}\frac{I_m{}'(ka)}{I_m(ka)}\left(\frac{K_m(ka)I_m{}'(kb) - I_m(ka)K_m{}'(kb)}{K_m{}'(ka)I_m{}'(kb) - I_m{}'(ka)K_m{}'(kb)}\right)$$
$$- \frac{B_\theta^2}{\mu_0\rho_{\mathrm{m}0}}\frac{1}{(ka)}\frac{I_m{}'(ka)}{I_m(ka)}. \tag{9.61b}$$

Expanding the modified Bessel functions under the conditions $ka \ll 1$, $kb \ll 1$, we find

$$\mu_0\rho_{\mathrm{m}0}\omega^2 = k^2 B_{\mathrm{i}z}{}^2 + \frac{1 + (a/b)^{2m}}{1 - (a/b)^{2m}}\left(kB_{\mathrm{e}z} + \frac{m}{a}B_\theta\right)^2 - \frac{m}{a^2}B_\theta^2. \tag{9.64b}$$

The closer the wall to the plasma boundary, the more effective is the wall stabilization.

In toroidal systems, the propagation constant is $k = n/R$ where n is an integer and R is the major radius of the torus. If the *safety factor* q_a at the plasma boundary $r = a$

$$q_a = \frac{a}{R}\frac{B_{\mathrm{e}z}}{B_\theta} \tag{9.67}$$

is introduced, $(\boldsymbol{k}\cdot\boldsymbol{B})$ may be written as

$$(\boldsymbol{k}\cdot\boldsymbol{B}) = \left(kB_{\mathrm{e}z} + \frac{m}{a}B_\theta\right) = \frac{B_\theta}{a}(m + q_a).$$

The Kruskal-Shafranov condition (9.66b) can then be expressed in terms of the safety factor as

$$q_a > 1. \tag{9.66c}$$

This is why q_a is called the safety factor.

9.3b Instabilities of Diffuse-Boundary Configurations

The sharp-boundary configuration treated in sec. 9.3a is a special case; in most cases the plasma current decreases gradually at the boundary. Let us consider the case of a diffuse-boundary plasma whose parameters in the equilibrium state are

$$p_0(r), \qquad \boldsymbol{B}_0(r) = (0,\, B_\theta(r),\, B_z(r)).$$

The perturbation $\boldsymbol{\xi}$ is assumed to be

$$\boldsymbol{\xi} = \boldsymbol{\xi}(r)\exp(im\theta + ikz).$$

The perturbation of the magnetic field $\boldsymbol{B}_1 = \nabla \times (\boldsymbol{\xi} \times \boldsymbol{B}_0)$ is

$$B_{1r} = i(\boldsymbol{k}\cdot\boldsymbol{B}_0)\xi_r \tag{9.68}$$

$$B_{1\theta} = ikA - \frac{\mathrm{d}}{\mathrm{d}r}(\xi_r B_\theta) \tag{9.69}$$

$$B_{1z} = -\left(\frac{imA}{r} + \frac{1}{r}\frac{\mathrm{d}}{\mathrm{d}r}(r\xi_r B_z)\right), \tag{9.70}$$

where

$$(\boldsymbol{k}\cdot\boldsymbol{B}_0) = kB_z + \frac{m}{r}B_\theta \tag{9.71}$$

$$A \equiv \xi_\theta B_z - \xi_z B_\theta = (\boldsymbol{\xi} \times \boldsymbol{B}_0)_r. \tag{9.72}$$

Since the pressure terms $\gamma p_0(\nabla\boldsymbol{\xi})^2 + (\nabla\boldsymbol{\xi})(\boldsymbol{\xi}\cdot\nabla p_0) = (\gamma - 1)p_0(\nabla\boldsymbol{\xi})^2 + (\nabla\boldsymbol{\xi})(\nabla p_0\,\boldsymbol{\xi})$ in the energy integral are nonnegative, we examine the incompressible displacement $\nabla\boldsymbol{\xi} = 0$ again, i.e.,

$$\frac{1}{r}\frac{\mathrm{d}}{\mathrm{d}r}(r\xi_r) + \frac{im}{r}\xi_\theta + ik\xi_z = 0. \tag{9.73}$$

From this and eq. (9.72) for A, ξ_θ and ξ_z are expressed in terms of ξ_r and A as

$$i(\boldsymbol{k}\cdot\boldsymbol{B})\xi_\theta = ikA - \frac{B_\theta}{r}\frac{\mathrm{d}}{\mathrm{d}r}(r\xi_r) \tag{9.74}$$

$$-i(\boldsymbol{k}\cdot\boldsymbol{B})\xi_z = \frac{imA}{r} + \frac{B_z}{r}\frac{\mathrm{d}}{\mathrm{d}r}(r\xi_r). \tag{9.75}$$

From $\mu_0\boldsymbol{j}_0 = \nabla\times\boldsymbol{B}_0$, it follows that

$$\mu_0 j_{0\theta} = -\frac{\mathrm{d}B_z}{\mathrm{d}r} \tag{9.76}$$

$$\mu_0 j_{0z} = \frac{\mathrm{d}B_\theta}{\mathrm{d}r} + \frac{B_\theta}{r} = \frac{1}{r}\frac{\mathrm{d}}{\mathrm{d}r}(rB_\theta). \tag{9.77}$$

The terms of the energy integral are given by

$$W_\mathrm{P} = \frac{1}{4}\int_{V_\mathrm{i}}\left(\gamma p_0|\nabla\boldsymbol{\xi}|^2 + (\nabla\boldsymbol{\xi}^*)(\boldsymbol{\xi}\cdot\nabla p_0) + \frac{1}{\mu_0}|\boldsymbol{B}_1|^2 - \boldsymbol{\xi}^*(\boldsymbol{j}_0\times\boldsymbol{B}_1)\right)\mathrm{d}\boldsymbol{r} \tag{9.78}$$

$$W_\mathrm{S} = \frac{1}{4}\int_\mathrm{S}|\boldsymbol{\xi}_n|^2\frac{\partial}{\partial n}\left(\frac{B_{0\mathrm{e}}{}^2}{2\mu_0} - \frac{B_{0\mathrm{i}}{}^2}{2\mu_0} - p_0\right)\mathrm{d}S \tag{9.79}$$

$$W_\mathrm{V} = \frac{1}{4\mu_0}\int_{V_\mathrm{e}}|\boldsymbol{B}_1|^2\mathrm{d}\boldsymbol{r}. \tag{9.80}$$

ξ_θ and ξ_z can be eliminated by means of eqs. (9.74) and (9.75) and $\mathrm{d}B_z/\mathrm{d}r$ and $\mathrm{d}B_\theta/\mathrm{d}r$ can be eliminated by means of eqs. (9.76) and (9.77) in eq. (9.78). Then W_P becomes

$$\begin{aligned} W_\mathrm{P} = \frac{1}{4}\int_{V_\mathrm{i}}\Bigg[&\frac{(\boldsymbol{k}\cdot\boldsymbol{B}_0)^2}{\mu_0}|\xi_r|^2 + \left(k^2 + \frac{m^2}{r^2}\right)\frac{|A|^2}{\mu_0} \\ &+ \frac{1}{\mu_0}\left|B_\theta\frac{\mathrm{d}\xi_r}{\mathrm{d}r} + \xi_r\left(\mu_0 j_{0z} - \frac{B_\theta}{r}\right)\right|^2 + \frac{1}{\mu_0}\left|\frac{\xi_r B_z}{r} + B_z\frac{\mathrm{d}\xi_r}{\mathrm{d}r}\right|^2 \\ &+ \frac{2}{\mu_0}\mathrm{Re}\left(ikA^*\left(B_\theta\frac{\mathrm{d}\xi_r}{\mathrm{d}r} + \left(\mu_0 j_{0z} - \frac{B_\theta}{r}\right)\xi_r\right)\right. \\ &\left. - \frac{imA^*}{r^2}\left(\xi_r B_z + rB_z\frac{\mathrm{d}\xi_r}{\mathrm{d}r}\right)\right) \\ &+ 2\mathrm{Re}\left(\xi_r^* j_{0z}\left(-B_\theta\frac{\mathrm{d}\xi_r}{\mathrm{d}r} - \frac{\xi_r\mu_0 j_{0z}}{2} + ikA\right)\right)\Bigg]\mathrm{d}\boldsymbol{r}. \end{aligned}$$

The integrand of W_P is reduced to

$$\frac{1}{\mu_0}\left(k^2+\frac{m^2}{r^2}\right)\left|A+\frac{ikB_\theta[(\mathrm{d}\xi_r/\mathrm{d}r)-\xi_r/r]-im(B_z/r)[(\mathrm{d}\xi_r/\mathrm{d}r)+(\xi_r/r)]}{k^2+(m^2/r^2)}\right|^2$$
$$+\left(\frac{(\boldsymbol{k}\cdot\boldsymbol{B})^2}{\mu_0}-\frac{2j_zB_\theta}{r}\right)|\xi_r|^2+\frac{B_z^2}{\mu_0}\left|\frac{\mathrm{d}\xi_r}{\mathrm{d}r}+\frac{\xi_r}{r}\right|^2+\frac{B_\theta^2}{\mu_0}\left|\frac{\mathrm{d}\xi_r}{\mathrm{d}r}-\frac{\xi_r}{r}\right|^2$$
$$-\frac{|ikB_\theta[(\mathrm{d}\xi_r/\mathrm{d}r)-(\xi_r/r)]-im(B_z/r)[(\mathrm{d}\xi_r/\mathrm{d}r)+(\xi_r/r)]|^2}{\mu_0(k^2+(m^2/r^2))}.$$

Accordingly, the integrand is a minimum when

$$A=\xi_\theta B_z-\xi_z B_\theta$$
$$=-\frac{i}{k^2+(m^2/r^2)}\left(\left(kB_\theta-\frac{m}{r}B_z\right)\frac{\mathrm{d}\xi_r}{\mathrm{d}r}-\left(kB_\theta+\frac{m}{r}B_z\right)\frac{\xi_r}{r}\right). \tag{9.81}$$

Then W_P is reduced to

$$W_P=\frac{\pi}{2\mu_0}\int_0^a\left(\frac{|(\boldsymbol{k}\cdot\boldsymbol{B}_0)(\mathrm{d}\xi_r/\mathrm{d}r)+h(\xi_r/r)|^2}{k^2+(m/r)^2}\right.$$
$$\left.+\left((\boldsymbol{k}\cdot\boldsymbol{B}_0)^2-\frac{2\mu_0 j_z B_\theta}{r}\right)|\xi_r|^2\right)r\mathrm{d}r, \tag{9.82}$$

where

$$h\equiv kB_z-\frac{m}{r}B_\theta.$$

Let us next determine W_S. From eq. (7.7b), it follows that $(\mathrm{d}/\mathrm{d}r)(p_0+(B_z^2+B_\theta^2)/2\mu_0)=-B_\theta^2/(r\mu_0)$. B_θ^2 is continuous across the boundary $r=a$, so that

$$\frac{\mathrm{d}}{\mathrm{d}r}\left(p_0+\frac{B_z^2+B_\theta^2}{2\mu_0}\right)=\frac{\mathrm{d}}{\mathrm{d}r}\left(\frac{B_{\mathrm{e}z}^2+B_{\mathrm{e}\theta}^2}{2\mu_0}\right).$$

Accordingly we find

$$W_S=0, \tag{9.83}$$

as is clear from eq. (9.79).

The expression for W_V can be obtained when the quantities in eq. (9.82) for W_P are replaced as follows: $\boldsymbol{j}\to 0$, $B_z\to B_{\mathrm{e}z}=B_\mathrm{s}$ (=const.), $B_\theta\to B_{\mathrm{e}\theta}=B_a a/r$, $B_{1r}=i(\boldsymbol{k}\cdot\boldsymbol{B}_0)\xi_r\to B_{\mathrm{e}1r}=i(\boldsymbol{k}\cdot\boldsymbol{B}_0)\eta_r$. This replacement yields

$$W_{\mathrm{V}} = \frac{\pi}{2\mu_0}\int_a^b \left(\frac{|[kB_{\mathrm{s}} + (m/r)(B_a a/r)](\mathrm{d}\eta_r/\mathrm{d}r) + [kB_{\mathrm{s}} - (m/r)(B_a a/r)]\eta_r/r|^2}{k^2 + (m/r)^2} + \left(kB_{\mathrm{s}} + \frac{m}{r}\frac{B_a a}{r}\right)^2 |\eta_r|^2\right) r\mathrm{d}r. \tag{9.84}$$

By partial integration, W_{P} is seen to be

$$W_{\mathrm{P}} = \frac{\pi}{2\mu_0}\int_0^a \left(\frac{r(\boldsymbol{k}\cdot\boldsymbol{B}_0)^2}{k^2 + (m/r)^2}\left|\frac{\mathrm{d}\xi_r}{\mathrm{d}r}\right|^2 + g|\xi_r|^2\right)\mathrm{d}r + \frac{\pi}{2\mu_0}\frac{k^2 B_{\mathrm{s}}^2 - (m/a)^2 B_a^2}{k^2 + (m/a)^2}|\xi_r(a)|^2 \tag{9.85}$$

$$g = \frac{1}{r}\frac{(kB_z - (m/r)B_\theta)^2}{k^2 + (m/r)^2} + r(\boldsymbol{k}\cdot\boldsymbol{B}_0)^2 - \frac{2B_\theta}{r}\frac{\mathrm{d}(rB_\theta)}{\mathrm{d}r} - \frac{\mathrm{d}}{\mathrm{d}r}\left(\frac{k^2 B_z^2 - (m/r)^2 B_\theta^2}{k^2 + (m/r)^2}\right). \tag{9.86}$$

Using the notation $\zeta \equiv rB_{\mathrm{e1}r} = ir(\boldsymbol{k}\cdot\boldsymbol{B}_{\mathrm{e0}})\eta_r$, we find that

$$W_{\mathrm{V}} = \frac{\pi}{2\mu_0}\int_a^b \left(\frac{1}{r(k^2 + (m/r)^2)}\left|\frac{\mathrm{d}\zeta}{\mathrm{d}r}\right|^2 + \frac{1}{r}|\zeta|^2\right)\mathrm{d}r. \tag{9.87a}$$

The functions ξ_r or ζ that will minimize W_{P} or W_{V} are the solutions of Euler's equation:

$$\frac{\mathrm{d}}{\mathrm{d}r}\left(\frac{r(\boldsymbol{k}\cdot\boldsymbol{B}_0)^2}{k^2 + (m/r)^2}\frac{\mathrm{d}\xi_r}{\mathrm{d}r}\right) - g\xi_r = 0 \qquad r \le a \tag{9.88}$$

$$\frac{\mathrm{d}}{\mathrm{d}r}\left(\frac{1}{r(k^2 + (m/r)^2)}\frac{\mathrm{d}\zeta}{\mathrm{d}r}\right) - \frac{1}{r}\zeta = 0 \qquad r > a. \tag{9.89}$$

There are two independent solutions, which tend to $\xi_r \propto r^{m-1}, r^{-m-1}$ as $r \to 0$. As ξ_r is finite at $r = 0$, the solution must satisfy the conditions

$$r \to 0 \qquad \xi_r \propto r^{m-1}$$

$$r = a \qquad \zeta(a) = ia\left(kB_{\mathrm{s}} + \frac{m}{a}B_a\right)\xi_r(a)$$

$$r = b \qquad \zeta(b) = 0.$$

Using the solution of eq. (9.89), we obtain

$$W_{\mathrm{V}} = \frac{\pi}{2\mu_0}\left|\frac{1}{r(k^2 + (m/r)^2)}\frac{\mathrm{d}\zeta}{\mathrm{d}r}\zeta^*\right|_a^b. \tag{9.87b}$$

The solution of eq. (9.89) is

$$\zeta = i\frac{I_m'(kr)K_m'(kb) - K_m'(kr)I_m'(kb)}{I_m'(ka)K_m'(kb) - K_m'(ka)I_m'(kb)}r\left(kB_s + \frac{m}{a}B_a\right)\xi_r(a). \tag{9.90}$$

The stability problem is now reduced to one of examining the sign of $W_P + W_V$. For this we use

$$\left.\begin{aligned} W_P &= \frac{\pi}{2\mu_0}\int_0^a\left(f\left|\frac{d\xi_r}{dr}\right|^2 + g|\xi_r|^2\right)dr + W_a \\ W_a &= \frac{\pi}{2\mu_0}\frac{k^2B_s^2 - (m/a)^2B_a^2}{k^2 + (m/a)^2}|\xi_r(a)|^2 \\ W_V &= \frac{\pi}{2\mu_0}\frac{-1}{a(k^2 + (m/a)^2)}\left|\frac{d\zeta}{dr}\zeta^*\right|_{r=a}, \end{aligned}\right\} \tag{9.91}$$

where

$$f = \frac{r(kB_z + (m/r)B_\theta)^2}{k^2 + (m/r)^2} \tag{9.92}$$

$$g = \frac{1}{r}\frac{(kB_z - (m/r)B_\theta)^2}{k^2 + (m/r)^2} + r\left(kB_z + \frac{m}{r}B_\theta\right)^2 - \frac{2B_\theta}{r}\frac{d(rB_\theta)}{dr} - \frac{d}{dr}\left(\frac{k^2B_z^2 - (m/r)^2B_\theta^2}{k^2 + (m/r)^2}\right) \tag{9.93a}$$

$$= \frac{2k^2}{k^2 + (m/r)^2}\mu_0\frac{dp_0}{dr} + r\left(kB_z + \frac{m}{r}B_\theta\right)^2\frac{k^2 + (m/r)^2 - (1/r)^2}{k^2 + (m/r)^2} + \frac{(2k^2/r)(k^2B_z^2 - (m/r)^2B_\theta^2)}{(k^2 + (m/r)^2)^2}. \tag{9.93b}$$

It can be easily proved from eq. (7.7b) that eq. (9.93b) is equal to eq. (9.93a).

9.3c Suydam's Criterion and Newcomb's Theorem

(i) Suydam's Criterion[11] The function f in the integrand of W_P in the previous section is always $f \geq 0$, so that the term in f is a stabilizing term. The 1st and 2nd terms in eq. (9.86) for g are stabilizing terms, but the 3rd and 4th terms may contribute to the instabilities. When a singular point

$$f \propto (\boldsymbol{k}\cdot\boldsymbol{B}_0)^2 = 0$$

of Euler's equation (9.88) is located at some point $r = r_0$ within the plasma region, the contribution of the stabilizing term becomes small near $r = r_0$, so that a local mode near the singular point is dangerous. In terms of the notation

$$r - r_0 = x$$

$$f = \alpha x^2$$

$$g = \beta$$

$$\alpha = \frac{r_0}{k^2 r_0^2 + m^2}\left(kr\frac{\mathrm{d}B_z}{\mathrm{d}r} + kB_z + m\frac{\mathrm{d}B_\theta}{\mathrm{d}r}\right)^2_{r=r_0} = \frac{r_0 B_\theta^2 B_z^2}{B^2}\left(\frac{\tilde{\mu}'}{\tilde{\mu}}\right)^2_{r=r_0}$$

$$\tilde{\mu} \equiv \frac{B_\theta}{rB_z}$$

$$\beta = \frac{2B_\theta^2}{B^2}\mu_0 \frac{\mathrm{d}p_0}{\mathrm{d}r}\bigg|_{r=r_0},$$

Euler's equation is reduced to

$$\alpha\frac{\mathrm{d}}{\mathrm{d}x}\left(x^2\frac{\mathrm{d}\xi_r}{\mathrm{d}x}\right) - \beta\xi_r = 0.$$

The solution is

$$\xi_r = c_1 x^{-n_1} + c_2 x^{-n_2}, \tag{9.94}$$

where n_1 and n_2 are given by

$$\left.\begin{aligned} n_i^2 - n_i - \frac{\beta}{\alpha} &= 0 \\ n_i &= \frac{1 \pm (1 + 4\beta/\alpha)^{1/2}}{2}. \end{aligned}\right\} \tag{9.95}$$

When $\alpha + 4\beta \geq 0$, n_1 and n_2 are real. The relation $n_1 + n_2 = 1$ holds always. For $n_1 < n_2$, we have the solution x^{-n_1}, called a *small solution.* When n is complex ($n = \gamma \pm i\delta$), ξ_r is in the form $\exp((-\gamma \mp i\delta)\ln x)$ and ξ_r is oscillatory.

Let us consider a local mode ξ_r, which is nonzero only in the neighborhood ε around $r = r_0$ and set

$$r - r_0 = \varepsilon t \qquad \xi_r(r) = \xi(t) \qquad \xi(1) = \xi(-1) = 0.$$

Then W_P becomes

$$W_P = \frac{\pi}{2\mu_0}\varepsilon\int_{-1}^{1}\left(\alpha t^2\left|\frac{d\xi}{dt}\right|^2 + \beta|\xi|^2\right)dt + O(\varepsilon^2).$$

Since Schwartz's inequality yields

$$\int_{-1}^{1} t^2|\xi'|^2 dt \int_{-1}^{1}|\xi|^2 dt \geq \left|\int_{-1}^{1} t\xi'\xi^* dt\right|^2 = \left(\frac{1}{2}\int_{-1}^{1}|\xi|^2 dt\right)^2,$$

W_P is

$$W_P > \frac{\pi}{2\mu_0}\frac{1}{4}(\alpha + 4\beta)\int_{-1}^{1}|\xi|^2 dt.$$

The stability condition is $\alpha + 4\beta > 0$, i.e.,

$$\frac{r}{4}\left(\frac{\tilde{\mu}'}{\tilde{\mu}}\right)^2 + \frac{2\mu_0}{B_z^2}\frac{dp_0}{dr} > 0. \tag{9.96}$$

Usually the 2nd term is negative, since, most often, $dp_0/dr < 0$. The 1st term $(\tilde{\mu}'/\tilde{\mu})^2$ represents the stabilizing effect of shear. This condition is called *Suydam's criterion.* This is a necessary condition for stability; but it is not always a sufficient condition, as Suydam's criterion is derived from consideration of local-mode behavior only.

(ii) Newcomb's Theorems[12] Newcomb derived the necessary and sufficient conditions for the stability of a cylindrical plasma. His twelve theorems are described in ref. 12; only some of them are introduced here.

Assume a situation of increasing m, while keeping the ratio of k and m/a constant in the $m \neq 0$ modes. For such a case only the 2nd term in g changes; this term is positive, and its increase is proportional to m^2. Accordingly if the plasma is stable for the mode $m = 1$, $-\infty < k < \infty$, it is also stable for modes with $m \geq 2$. When $m = 0$, W_P and W_V are

$$W_P = \frac{\pi}{2\mu_0}\int_0^a\left[\left(rB_z^2\left|\frac{d\xi_r}{dr}\right|^2 + \left(\frac{B_z^2}{r} + 2\mu_0\frac{dp_0}{dr}\right)|\xi_r|^2\right) + k^2 rB_z^2|\xi_r|^2\right]dr$$

$$+ \frac{\pi}{2\mu_0}B_z^2|\xi_r(a)|^2$$

$$W_V = \frac{\pi}{2\mu_0}\int_a^b\left(\left(rB_s^2\left|\frac{d\eta_r}{dr}\right|^2 + \frac{B_s^2}{r}|\eta_r|^2\right) + k^2 rB_s^2|\eta_r|^2\right)dr$$

$$-\frac{\pi}{2\mu_0}B_s^2|\eta_r(a)|^2.$$

The terms in k^2 are stabilizing, so that if the mode $m = 0$, $k \to 0$ is stable, the mode $m = 0$, $k^2 > 0$ is also stable.

[Theorem 1] All modes are stable if and only if the modes $m = 0$, $k \to 0$ and $m = 1$, $-\infty < k < \infty$ are stable.

The function η_r was introduced in the expression $B_{1er} \equiv i(\boldsymbol{k} \cdot \boldsymbol{B}_{e0})\eta_r$ for the perturbation $\boldsymbol{B}_{1e}$ of the external magnetic field, and eq. (9.84) for W_V was reduced by using η_r. Therefore if we set $\eta_r = \xi_r$ even in $a \le r \le b$, the energy integral is formally reduced to

$$W = W_P + W_V = \frac{\pi}{2\mu_0}\int_0^b \left(f\left|\frac{d\xi_r}{dr}\right|^2 + g|\xi_r|^2 \right) dr. \tag{9.97}$$

Euler's equation for W is

$$\frac{d}{dr}\left(f\frac{d\xi_r}{dr} \right) - g\xi_r = 0. \tag{9.98}$$

Let the singular points $f = 0$ be $r_1, r_2, \ldots, r_n$, and divide the interval $[0, b]$ into the subintervals $[0, r_1]; [r_1, r_2]; \ldots; [r_n, b]$. Then we have:

[Theorem 3] The plasma is stable if and only if it is stable in each of the subintervals.

The solution of Euler's equation near a singular point r_s is given by eq. (9.94). The small solution is defined by the following:

if r_i is a singular point $\qquad \xi_r(r_i) \propto x^{-n_1} \qquad (n_1 < n_2)$

at $r = 0$ $\qquad \xi_r(r) \propto r^{m-1} \qquad (m \ge 0)$.

Here it is assumed that Suydam's criterion is satisifed, so that n_1 and n_2 are real (if Suydam's criterion is not satisfied, the plasma is unstable). Then we have:

[Theorem 7] Let $[r_1, r_2]$ be any subinterval and let ξ_r be the small solution at $r = r_1$. The plasma is stable in the subinterval if and only if the small solution does not become zero anywhere in this subinterval.

If ξ_r is the solution of Euler's equation (it is assumed here that ξ_r is real), then eq. (9.97) for W becomes

$$
\begin{aligned}
W(c, d; \xi_r) &= \frac{\pi}{2\mu_0} \int_c^d \mathrm{d}r \left(f \left(\frac{\mathrm{d}\xi_r}{\mathrm{d}r} \right)^2 + g\xi_r^2 \right) \mathrm{d}r \\
&= \frac{\pi}{2\mu_0} \int_c^d \mathrm{d}r \left(f \left(\frac{\mathrm{d}\xi_r}{\mathrm{d}r} \right)^2 + \xi_r \frac{\mathrm{d}}{\mathrm{d}r} \left(f \frac{\mathrm{d}\xi_r}{\mathrm{d}r} \right) \right) \\
&= \frac{\pi}{2\mu_0} \int_c^d \mathrm{d}r \frac{\mathrm{d}}{\mathrm{d}r} \left(f \xi_r \frac{\mathrm{d}\xi_r}{\mathrm{d}r} \right) \\
&= \frac{1}{2\mu_0} f \xi_r \frac{\mathrm{d}\xi_r}{\mathrm{d}r} \bigg|_c^d .
\end{aligned}
$$

The contribution of $\xi_r\, \mathrm{d}\xi_r/\mathrm{d}r$ at a singular point is zero, if ξ_r is a small solution. Thus we may state:

[Theorem 12] Let r_0 be a point inside $[r_1, r_2]$ and let $\xi_1(r)$ be the small solution at r_1 and $\xi_2(r)$ be the small solution at r_2. The plasma is stable and if and only if $\xi_1(r)$ has no zero in $[r_1, r_0]$ and $\xi_2(r)$ has no zero in $[r_0, r_2]$ and these solutions satisfy

$$
\frac{\xi_1'(r_0)}{\xi_1(r_0)} > \frac{\xi_2'(r_0)}{\xi_2(r_0)}.
$$

The necessary and sufficient conditions enunciated in the various statements of Newcomb's theorem provide the analytical means to study the stability of the equilibrium solution $p_0(r)$, $B_0(r)$. If small solutions can be obtained, analytically or numerically, one can predict stability by means of the theorems 7 or 12.

9.3d Tokamak Configuration and Diffuse-Pinch (Zeta) Configuration

(i) Tokamak Configuration In this case the longitudinal magnetic field B_s is much larger than the poloidal magnetic field B_a. The plasma region is $r \le a$ and the vacuum region is $a \le r \le b$ and an ideal conducting wall is at $r = b$.

The function ζ in eq. (9.87b) for W_V is

$$
\zeta = i \frac{(mB_a + kaB_s)}{1 - (a/b)^{2m}} \xi_r(a) \frac{a^m}{b^m} \left(\frac{b^m}{r^m} - \frac{r^m}{b^m} \right)
$$

(from eq. (9.90)), and W_V becomes

$$
W_V = \frac{\pi}{2\mu_0} \frac{(mB_a + kaB_s)^2}{m} \xi_r^2(a) \lambda
$$

$$\lambda \equiv \frac{1 + (a/b)^{2m}}{1 - (a/b)^{2m}}.$$

From the periodic condition for a torus, it follows that

$$\frac{2\pi n}{k} = -2\pi R$$

(n an integer), so that $(\boldsymbol{k} \cdot \boldsymbol{B})$ is given by

$$a(\boldsymbol{k} \cdot \boldsymbol{B}) = mB_a + kaB_s = mB_a\left(1 - \frac{nq_a}{m}\right)$$

in terms of the safety factor. The W_a term in eq. (9.91) is reduced to

$$k^2 B_s^2 - \left(\frac{m}{a}\right)^2 B_a^2 = \left(kB_s + \frac{m}{a} B_a\right)^2 - 2\frac{m}{a} B_a\left(kB_s + \frac{m}{a} B_a\right)$$

$$= \left(\frac{mB_a}{a}\right)^2\left(\left(1 - \frac{nq_a}{m}\right)^2 - 2\left(1 - \frac{nq_a}{m}\right)\right).$$

Accordingly, the energy integral becomes

$$W_P + W_V = \frac{\pi}{2\mu_0} B_a^2 \xi_r^2(a)\left(\left(1 - \frac{nq_a}{m}\right)^2 (1 + m\lambda) - 2\left(1 - \frac{nq_a}{m}\right)\right)$$

$$+ \frac{\pi}{2\mu_0} \int\left(f\left(\frac{d\xi_r}{dr}\right)^2 + g\xi_r^2\right)dr. \tag{9.99}$$

The 1st term of eq. (9.99) is negative when

$$1 - \frac{2}{1 + m\lambda} < \frac{nq_a}{m} < 1. \tag{9.100}$$

The assumption $nq_a/m \approx 1$ corresponds to $ka \approx mB_a/B_s$. As $B_a/B_s \ll 1$, this is consistent with the assumption $ka \ll 1$. When $m = 1$, $(m^2 - 1)/m^2$ in the 2nd term of eq. (9.93) for g is zero. The magnitude of g is of the order of $k^2 r^2$, which is very small since $kr \ll 1$. The term in $f(d\xi_r/dr)^2$ can be very small if ξ_r is nearly constant. Accordingly the contribution of the integral term in eq. (9.99) is negligible. When $m = 1$ and $a^2/b^2 < nq_a < 1$, the energy integral becomes negative ($W < 0$). The mode $m = 1$ is unstable in the region specified by eq. (9.100) irrespective of the current distribution. The Kruskal-Shafranov condition for the mode $m = 1$ derived from the sharp-boundary configuration is also applicable to the diffuse-boundary plasma. The growth rate $\gamma^2 = -\omega^2$ is

$$\gamma^2 = \frac{-W}{\int (\rho_{m0}|\xi|^2/2)\mathrm{d}\boldsymbol{r}} \approx \frac{1}{\langle \rho_m \rangle} \frac{B_a^2}{\mu_0 a^2} \left(2(1 - nq_a) - \frac{2(1 - nq_a)^2}{1 - a^2/b^2} \right) \tag{9.101}$$

$$\langle \rho_m \rangle = \frac{\int \rho_{m0} |\xi|^2 2\pi r \mathrm{d}r}{\pi a^2 \xi_r^2(a)}.$$

The maximum growth rate is $\gamma_{max}^2 \approx (1 - a^2/b^2) B_a^2 / (\mu \langle \rho_m \rangle a^2)$. When $m \neq 1$, $(m^2 - 1)/m^2$ in the 2nd term of eq. (9.93b) for g is large, and $g \approx 1$. Accordingly, the contribution of the integral term to W_P must be checked. The region $g < 0$ is given by $\chi_1 < \chi < \chi_2$, when $\chi \equiv -krB_z/B_\theta = nq_s(r)$ and

$$\chi_{1,2} = m - \frac{2}{m(m^2 - 1)} k^2 r^2 \pm \frac{2k^2 r^2}{m(m^2 - 1)} \left(1 - \frac{m^2(m^2 - 1)}{2k^2 r^2} \frac{\mu_0 r p_0'}{B_\theta^2} \right)^{1/2}. \tag{9.102}$$

Since $kr \ll 1$, the region $g < 0$ is narrow and close to the singular point $nq(r) = m$ and the contribution of the integral term to W_P can be neglected. Therefore if nq_a/m is in the range given by eq. (9.100), the plasma is unstable due to the displacement $\xi_r(a)$ of the plasma boundary. When the current distribution is $j(r) = j_0 \exp(-k^2 r^2/a^2)$ and the conducting wall is at infinity ($b = \infty$), γ^2 can be calculated from eq. (9.101), using the solution of Euler's equation; and the dependence of γ^2 on q_a can be estimated. The result is shown in fig. 9.9.[13]

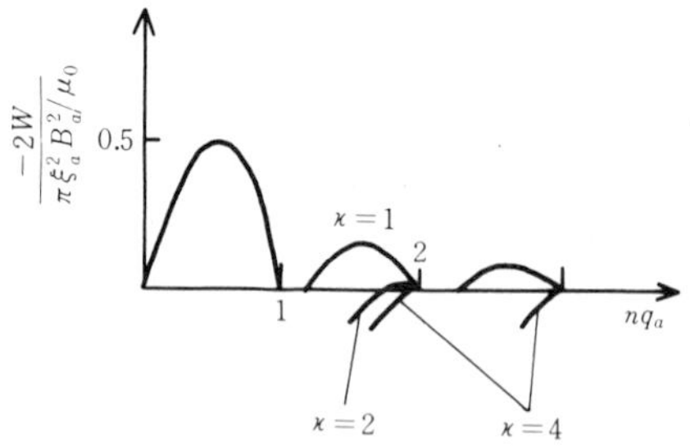

Fig. 9.9
The relation of the growth rate and nq_a for kink instability $(-2W/(\pi\xi_a^2 B_a^2/\mu_0) = \gamma^2 a^2(\langle \rho_m \rangle \mu_0 / B_a^2))$. After V. D. Shafranov: Sov. Phys. Tech. Phys. **15**, 175 (1970).

When the value of nq_a/m is outside the region given by eq. (9.100), the effect of the displacement of the plasma boundary is not great and the contribution of the integral term in W_P is dominant. However, the growth rate γ^2 is k^2r^2 times as small as that given by eq. (9.101), as is clear from consideration of eq. (9.102).

(ii) Diffuse-Pinch (Zeta) Configuration[14] The characteristics of the Zeta configuration is that B_s and B_a are of the same order of magnitude, so that the approximation based upon $ka \ll 1$ or $B_a \ll B_s$ can no longer be used. As is clear from the expression (9.82) for W_p, the plasma is stable if

$$2\mu_0 j_z \frac{B_\theta}{r} = 2\frac{B_\theta}{r^2}\frac{\mathrm{d}}{\mathrm{d}r}(rB_\theta) = \frac{1}{r^3}\frac{\mathrm{d}}{\mathrm{d}r}(rB_\theta)^2 < 0 \tag{9.103}$$

is satisfied everywhere. This is a sufficient condition; however, it can never be satisfied in real cases. When the expression (9.93b) for g is rewritten in terms of $P \equiv rB_z/B_\theta$ ($2\pi P$ is the pitch of the magnetic-field lines), we find

$$\begin{aligned} g = {} & \frac{2(kr)^2\mu_0}{m^2+(kr)^2}\frac{\mathrm{d}p_0}{\mathrm{d}r} \\ & + \frac{B_\theta{}^2/r}{(m^2+(kr)^2)^2}(kP+m) \\ & \quad \times \left[kP((m^2+k^2r^2)^2-(m^2-k^2r^2)) + m((m^2+k^2r^2)^2 \right. \\ & \qquad \left. -(m^2+3k^2r^2))\right]. \end{aligned} \tag{9.104a}$$

When $m = 1$, g becomes

$$g = \frac{2(kr)^2\mu_0}{1+(kr)^2}\frac{\mathrm{d}p_0}{\mathrm{d}r} + \frac{(kr)^2B_\theta{}^2/r}{(1+(kr)^2)^2}(kP+1)(kP(3+k^2r^2)+(k^2r^2-1)). \tag{9.104b}$$

The 2nd term in eq. (9.104b) is quadratic with respect to P and has the minimum value

$$g(r) > 2\mu_0\frac{\mathrm{d}p_0}{\mathrm{d}r}\frac{k^2r^2}{1+k^2r^2} - \frac{4B_\theta{}^2}{r}\frac{k^2r^2}{(1+k^2r^2)^2(3+k^2r^2)}.$$

The condition $g(r) > 0$ is reduced to

$$\frac{r\mu_0}{B_\theta{}^2}\frac{\mathrm{d}p_0}{\mathrm{d}r} > \frac{2}{(1+k^2r^2)(3+k^2r^2)} \tag{9.105}$$

(dp_0/dr must be positive). Accordingly if the equilibrium solution is found to satisfy the condition (9.105) near the plasma center and also to satisfy the condition (9.103) at the plasma boundary, the positive contribution of the integral term may dominate the negative contribution from the plasma boundary and this equilibrium configuration may be stable.

Let us consider the 2nd term of eq. (9.104b):

$$g_j = \frac{k^2 r B_\theta{}^2}{(1 + k^2r^2)^2}(kP + 1)(kP(3 + k^2r^2) + (k^2r^2 - 1)). \tag{9.106}$$

This term is positive when

$$kP < -1 \quad \text{or} \quad kP > (1 - k^2r^2)/(3 + k^2r^2). \tag{9.107}$$

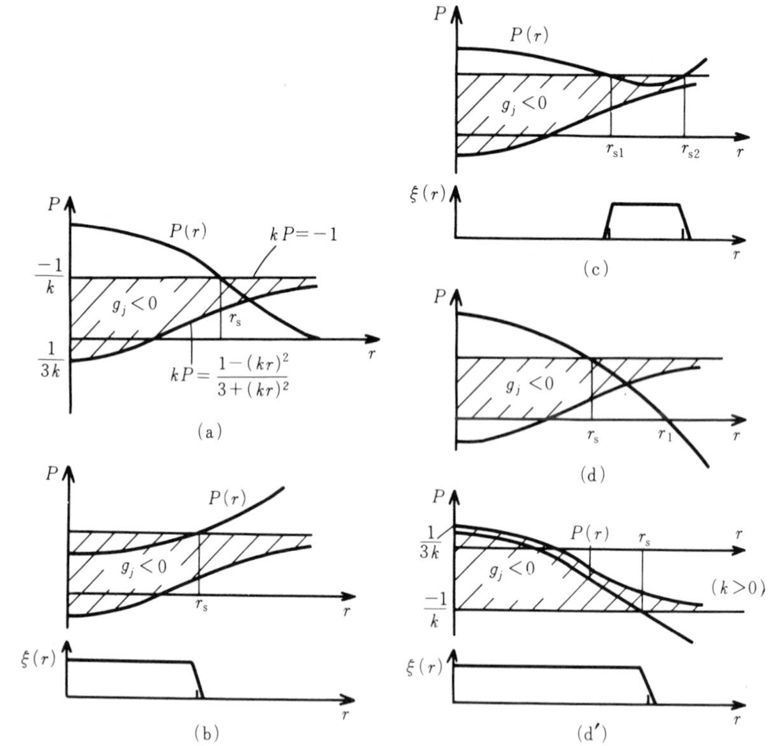

Fig. 9.10
Dependence of pitch $P(r)$ on r, and the region $g_j(P, r) < 0$. The displacements $\xi_r(r)$ of the unstable modes are also shown. Parts a–d are for $k < 0$.

The point $kP = -1$ is a singular point. The region $g_j < 0$ is shown in the P, r diagrams of fig. 9.10a–d for a given $k(<0)$. r_s is a singular point. Several typical examples of $P(r)$ are shown in the figure. It is clear that $\xi_r(r)$ shown in (b) and (c) makes W negative. The example (d) is the case where the longitudinal magnetic field B_z is reversed at $r = r_1$. If $k(>0)$ is chosen so that $kP(0) < 1/3$ and if the singular point r_s satisfying $kP(r_s) = -1$ is smaller than $r = b$ (i.e., r_s does not lie on the conducting wall), the plasma is unstable for the displacement $\xi_r(r)$ shown in fig. 9.10d′.
The necessary condition for the stability of the reverse-field configuration is

$$-P(b) < 3P(0). \tag{9.108}$$

This means that B_θ cannot be very small compared with B_z and that the value of the reversed B_z cannot be too large.

When $m = 1$, eq. (9.82) for W_P yields the sufficient condition of stability:

$$2\mu_0 j_z \frac{B_\theta}{r} < \frac{B_\theta{}^2}{r^2}(1 + kP)^2. \tag{9.109}$$

The most dangerous mode is $k = -1/P(a)$. With the assumption $B_\theta > 0$, the stability condition becomes

$$\mu_0 j_z < \frac{1}{2}\frac{B_\theta}{r}\left(-\frac{P(r)}{P(a)} + 1\right)^2. \tag{9.110}$$

Accordingly if the condition

$$\left|\frac{P(r)}{P(a)}\right| > 1 \tag{9.111}$$

is satisfied at small r ($\mu_0 j_z \approx 2B_\theta/r$) and j_z is negative at r near the boundary (the reverse-current configuration), this configuration may be stable. This is the configuration used in Zeta and HBTX, which machines will be discussed in sec. 16.5.

Let us consider limitations on the beta ratio from the standpoint of stability. For this purpose the dangerous mode $kP(a) = -1$ is examined, using eq. (9.82) for W_P. The substitution

$$\xi_r(r) = \xi \qquad 0 \le r < a - \varepsilon$$
$$\xi_r(r) = 0 \qquad r > a + \varepsilon$$

into eq. (9.82) yields

$$W_{\mathrm{P}} = \frac{\pi}{2\mu_0}\xi^2 \int_0^a \frac{\mathrm{d}r}{r}\left(-2B_\theta\frac{\mathrm{d}}{\mathrm{d}r}(rB_\theta) + (krB_z + mB_\theta)^2 + \frac{(krB_z - mB_\theta)^2}{m^2 + k^2r^2}\right).$$

When $m = 1$, then W_{P} is

$$W_{\mathrm{P}} = \frac{\pi}{2\mu_0}\xi^2 \int_0^a \frac{\mathrm{d}r}{r}\left(-2B_\theta\frac{\mathrm{d}}{\mathrm{d}r}(rB_\theta) + 2k^2r^2B_z^2 + 2B_\theta^2 - \frac{k^2r^2(krB_z - B_\theta)^2}{1 + k^2r^2}\right).$$

Since the last term in the integrand is always negative, the integration of the other three terms must be positive, i.e.,

$$\frac{\pi}{2\mu_0}\xi^2\left(-B_\theta^2(a) + 2k^2\int_0^a rB_z^2\mathrm{d}r\right) > 0.$$

Using the equilibrium equation of sec. 7.7b,

$$\frac{2}{a^2}\int_0^a\left(\mu_0 p_0 + \frac{B_z^2}{2}\right)r\mathrm{d}r = \left(\mu_0 p_0 + \frac{B_z^2 + B_\theta^2}{2}\right)_{r=a},$$

we can convert the necessary condition for stability to

$$a^2B_\theta^2(a) > 4\mu_0\int_0^a rp_0\mathrm{d}r,$$

i.e.,

$$\bar{\beta}_\theta \equiv \frac{2\pi}{\pi a^2}\int p_0 r\mathrm{d}r\left(\frac{2\mu_0}{B_\theta^2}\right) < 1. \tag{9.112}$$

Next let us study the stability of the mode $m = 0$ in the reverse-field configuration. It is assumed that B_z reverses at $r = r_1$. The substitution

$$\xi_r(r) = \lambda r \qquad 0 \le r < r_1 - \varepsilon$$
$$\xi_r(r) = 0 \qquad r > r_1 + \varepsilon$$

into eq. (9.82) yields

$$W_{\mathrm{P}} = \frac{\pi}{2\mu_0}\lambda^2\int_0^{r_1} r\mathrm{d}r\left(4B_z^2 - 2B_\theta\frac{\mathrm{d}}{\mathrm{d}r}(rB_\theta) + k^2r^2B_z^2\right).$$

Using the equilibrium equation, we obtain the necessary condition for stability:

$$\mu_0^{-1} r_1^2 B_\theta^2(r_1) > 8 \int_0^{r_1} r p_0 \mathrm{d}r - 4 r_1^2 p_0(r_1).$$

If $p_0(r_1) \approx 0$, we have the condition

$$\bar{\beta}_\theta < \frac{1}{2}. \tag{9.113}$$

*9.4 Local-Mode Stability in Toroidal Systems

*9.4a Energy Integral for Toroidal Systems

The energy integral is complicated even in the foregoing simple case of axially symmetric cylindrical plasmas. The energy integral for toroidal systems is much more difficult to analyze. The integration will be simplified if it is carried out in the natural coordinates described in sec. 7.5. If the natural coordinates are denoted by V, θ', φ', then V satisfies

$$\boldsymbol{B} \cdot \nabla V = 0,$$

where $V =$ const. is the magnetic surface and V is the volume inside the magnetic surface. We set the periods of θ' and φ' equal to 1. The natural coordinates are a special case of general curvilinear coordinates (see app. 2), with

$$u^1 = V \qquad u^2 = \theta' \qquad u^3 = \varphi'.$$

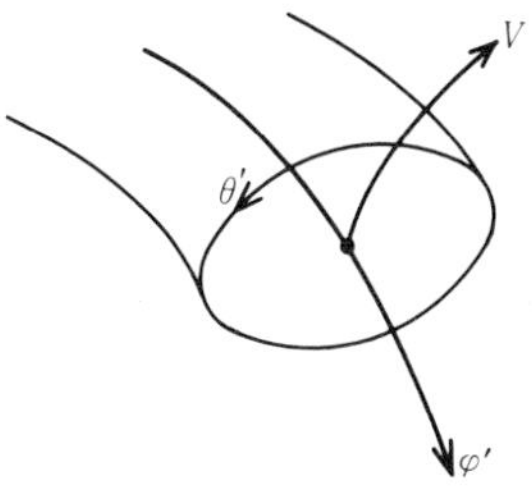

Fig. 9.11
Natural coordinates.

By definition of the coordinate vectors

$$\boldsymbol{a}^j \equiv \nabla u^j \qquad \boldsymbol{a}_j \equiv \frac{\partial \boldsymbol{r}}{\partial u^j},$$

any vector is expressed by

$$A = \sum A^i a_i = \sum A_j a^j,$$

where

$$A^i = A \cdot a^i \qquad A_j = A \cdot a_j.$$

The A^i are the contravariant components and the A_j are the covariant components. As was explained in sec. 7.5, the magnetic field $\boldsymbol{B}$ and the current density $\boldsymbol{j}$ are expressed by the contravariant components

$$\boldsymbol{B} = (0, \dot{X}(V), \dot{\Phi}(V)) \tag{9.114}$$

$$\boldsymbol{j} = (0, \dot{I}(V), \dot{J}(V)), \tag{9.115}$$

where $\Phi(V)$ and $J(V)$ are the magnetic flux and the current, respectively, in the $\nabla\varphi'$ direction within the magnetic surface V, while $X(V)$ and $I(V)$ are the magnetic flux and the current in the $\nabla\theta'$ direction. The dot here represents differentiation with respect to the volume V.

Let us use eq. (9.47) for the energy integral,

$$W_{\rm P} = \frac{1}{2\mu_0}\int\Big((\mu_0\gamma p(\nabla\boldsymbol{\xi})^2 + \mu_0(\boldsymbol{\xi}\cdot\nabla p)(\nabla\boldsymbol{\xi}) + (\nabla\times(\boldsymbol{\xi}\times\boldsymbol{B}))^2$$
$$- (\boldsymbol{\xi}\times(\nabla\times\boldsymbol{B}))\cdot\nabla\times(\boldsymbol{\xi}\times\boldsymbol{B}))\mathrm{d}\boldsymbol{r}. \tag{9.47}$$

When $D^j = (0, \partial B^2/\partial V, \partial B^3/\partial V)$, $e^j = (1, 0, 0)$ $(\boldsymbol{e} = \boldsymbol{a}_1)$ and

$$\boldsymbol{f} \equiv \nabla\times(\boldsymbol{\xi}\times\boldsymbol{B}) + (\nabla\boldsymbol{\xi})\boldsymbol{B} + \xi^1\boldsymbol{D}$$
$$\boldsymbol{F} \equiv \boldsymbol{j}\times\boldsymbol{e}$$

are introduced, we find

$$\int((\boldsymbol{\xi}\cdot\boldsymbol{F})f^1 + (\boldsymbol{f}\cdot\boldsymbol{F})\xi^1)\mathrm{d}\boldsymbol{r} = \int\nabla((\boldsymbol{\xi}\cdot\boldsymbol{F})\xi^1\boldsymbol{B})\mathrm{d}\boldsymbol{r} = 0. \tag{9.116}$$

Accordingly, the energy integral (9.47) is reduced to

$$W_{\rm P} = \frac{1}{2\mu_0}\int\Big(((\nabla\times(\boldsymbol{\xi}\times\boldsymbol{B}) + (\mu_0\boldsymbol{j}\times\boldsymbol{e})\xi^1)^2 + \mu_0\gamma p(\nabla\boldsymbol{\xi})^2$$
$$+ (\mu_0\boldsymbol{j}\times\boldsymbol{e})(\boldsymbol{D} - (\mu_0\boldsymbol{j}\times\boldsymbol{e}))(\xi^1)^2)\mathrm{d}\boldsymbol{r}. \tag{9.117}$$

The reduction of eqs. (9.116) and (9.117) will be described at the end of this section. The equilibrium condition is

$$\dot{p} = \dot{I}\dot{\Phi} - \dot{J}\dot{X}. \tag{9.118}$$

For convenience, the following quantities are defined:

$$S \equiv \dot{X}\ddot{\Phi} - \dot{\Phi}\ddot{X} = -\dot{\Phi}^2 \frac{\mathrm{d}}{\mathrm{d}V}\left(\frac{\dot{X}}{\dot{\Phi}}\right) \tag{9.119}$$

$$\Omega \equiv \dot{I}\ddot{\Phi} - \dot{J}\ddot{X}. \tag{9.120}$$

S represents shear as is clear from the definition. When S is zero, Ω is

$$Q = \dot{p}\frac{\ddot{\Phi}}{\dot{\Phi}} = -\dot{p}\frac{\mathrm{d}^2 V}{\mathrm{d}\Phi^2}\dot{\Phi}^2,$$

so that Ω is a measure of the average minimum B. In terms of Ω, the energy integral becomes

$$W_{\mathrm{P}} = \frac{1}{2\mu_0}\int((\nabla \times (\boldsymbol{\xi} \times \boldsymbol{B}) + (\mu_0\boldsymbol{j} \times \boldsymbol{e})\xi^1)^2 + \mu_0\gamma p(\nabla\boldsymbol{\xi})^2$$
$$+ (\mu_0\Omega + (\mu_0\boldsymbol{j} \times \boldsymbol{e})^2)(\xi^1)^2)\mathrm{d}\boldsymbol{r}.$$

The sufficient condition for stability is

$$\mu_0\Omega + (\mu_0\boldsymbol{j} \times \boldsymbol{e})^2 \le 0. \tag{9.121}$$

When it is assumed that the magnetic field is quasi-uniform and $B = \partial B/\partial V$ is a first-order small quantity, $\boldsymbol{j}$, $\dot{p}$, S are of the order of ε and Ω is of the order of ε^2. In this case, $\boldsymbol{e} = \boldsymbol{a}_1 = \partial\boldsymbol{r}/\partial V$ is nearly normal to the magnetic surface and $\boldsymbol{e} \approx \nabla V/|\nabla V|^2$. Then eq. (9.121) is

$$\mu_0\Omega + \mu_0{}^2 j^2/|\nabla V|^2 \le 0. \tag{9.122}$$

*9.4b Local-Mode Stability Condition[15]

Let us study the stability of a local mode of the displacement $\boldsymbol{\xi}$ located near a rational magnetic surface. The contravariant components of $\boldsymbol{\xi}$ are denoted by ξ^1, ξ^2, ξ^3 and the following quantities are introduced;

$$\zeta = \dot{\Phi}\xi^2 - \dot{X}\xi^3 = (\xi \times \boldsymbol{B})\cdot\boldsymbol{e} \tag{9.123}$$

$$\eta = \dot{J}\xi^2 - \dot{I}\xi^3 = (\xi \times \boldsymbol{j})\cdot\boldsymbol{e}. \tag{9.124}$$

The vectors $\boldsymbol{B}$, $\boldsymbol{j}$, and $\boldsymbol{\xi}$ are expressed in terms of the contravariant components as follows:

$$\boldsymbol{B} = \dot{X}\boldsymbol{a}_2 + \dot{\Phi}\boldsymbol{a}_3 \tag{9.125}$$

$$\boldsymbol{j} = \dot{I}\boldsymbol{a}_2 + \dot{J}\boldsymbol{a}_3 \tag{9.126}$$

$$\boldsymbol{\xi} = \xi^1\boldsymbol{a}_1 + \frac{\dot{I}\zeta - \dot{X}\eta}{\dot{\Phi}\dot{I} - \dot{X}\dot{J}}\boldsymbol{a}_2 + \frac{\dot{J}\zeta - \dot{\Phi}\eta}{\dot{\Phi}\dot{I} - \dot{X}\dot{J}}\boldsymbol{a}_3 \tag{9.127}$$

$$\nabla \times (\boldsymbol{\xi} \times \boldsymbol{B}) = \left(\dot{X}\frac{\partial \xi^1}{\partial \theta'} + \dot{\Phi}\frac{\partial \xi^1}{\partial \varphi'}\right)\boldsymbol{a}_1 + \left(\frac{\partial \zeta}{\partial \varphi'} - \frac{\partial}{\partial V}\xi^1\dot{X}\right)\boldsymbol{a}_2 + \left(\frac{\partial}{\partial V}(-\xi^1\dot{\Phi}) - \frac{\partial \zeta}{\partial \theta'}\right)\boldsymbol{a}_3. \tag{9.128}$$

$\xi^V \equiv \xi^1$ and ζ can be expanded in terms of a small parameter of the field inhomogeneity:

$$\xi^1 = \xi_0{}^V + \varepsilon\xi_1{}^V + \cdots \qquad \zeta = \zeta_0 + \varepsilon\zeta_1 + \cdots.$$

Set $V = V_0 + \varepsilon v$ and expand $\dot{X}(V)$ and $\dot{\Phi}(V)$ near V_0; then

$$\begin{aligned}\nabla \times (\boldsymbol{\xi} \times \boldsymbol{B}) = {} & (\boldsymbol{B}_0\cdot\nabla\xi_0{}^V + \varepsilon(\boldsymbol{B}_0\cdot\nabla\xi_1{}^V + v\boldsymbol{D}_0\cdot\nabla\xi_0{}^V))\boldsymbol{a}_1 \\ & + \left(\frac{\partial\zeta_0}{\partial\varphi'} - \dot{X}_0\dot{\xi}_0{}^V + \varepsilon\left(\frac{\partial\zeta_1}{\partial\varphi'} - \dot{X}_0\dot{\xi}_1{}^V\right) - \ddot{X}_0\frac{\partial}{\partial v}(v\xi_0{}^V)\right)\boldsymbol{a}_2 \\ & - \left(\frac{\partial\zeta_0}{\partial\theta'} + \dot{\Phi}_0\dot{\xi}_0{}^V + \varepsilon\left(\frac{\partial\zeta_1}{\partial\theta'} + \dot{\Phi}_0\dot{\xi}_1{}^V\right) + \ddot{\Phi}_0\frac{\partial}{\partial v}(v\xi_0{}^V)\right)\boldsymbol{a}_3.\end{aligned}$$

If $\nabla \times (\boldsymbol{\xi} \times \boldsymbol{B}) \neq 0$ is of zeroth order, the energy integral W_P is positive and the equilibrium is always stable. Accordingly, it is necessary to consider only the case of $\nabla \times (\boldsymbol{\xi} \times \boldsymbol{B}) = 0$ at $V = V_0$ in the zeroth order, i.e., the equilibrium is given by

$$\dot{X}_0\dot{\xi}_0{}^V = \partial\zeta_0/\partial\varphi'$$

$$\dot{\Phi}_0\dot{\xi}_0{}^V = -\partial\zeta_0/\partial\theta'$$

$$\boldsymbol{B}_0\cdot\nabla\xi_0{}^V = 0.$$

This condition is satisfied if $n\dot{\Phi}_0 = m\dot{X}_0$ and $\xi_0{}^V = \xi_0{}^V(u, a)$, u being $u = m\theta' - n\varphi'$. This condition says that the line of magnetic force must be closed and the magnetic surface must be rational.

If the vectors $\boldsymbol{B}$, $\boldsymbol{J}$, and $\nabla \times (\xi \times \boldsymbol{B})$ are represented by

$$\boldsymbol{B} = \alpha_1(\boldsymbol{e} \times \boldsymbol{B}) + \beta_1(\boldsymbol{e} \times \boldsymbol{j}) + \gamma_1\boldsymbol{e} \tag{9.129}$$

$$j = \alpha_2(e \times B) + \beta_2(e \times j) + \gamma_2 e \tag{9.130}$$

$$\nabla \times (\xi \times B) = aB + bj + ce, \tag{9.131}$$

the coefficients are

$$\alpha_1 = -\frac{(e \times B)\cdot(e \times j)}{\dot{p}|e|^2} \qquad \beta_1 = \frac{(e \times B)^2}{\dot{p}|e|^2} \qquad \gamma_1 = \frac{(e \cdot B)}{|e|^2}$$

$$\alpha_2 = -\frac{(e \times j)^2}{\dot{p}|e|^2} \qquad \beta_2 = \frac{(e \times B)\cdot(e \times j)}{\dot{p}|e|^2} \qquad \gamma_2 = \frac{(e \cdot j)}{|e|^2}.$$

When the vector product of e with eq. (9.129) is taken, the scalar product of this vector product with $(e \times B)$ yields β_1. From eqs. (9.129)–(9.131) it follows that

$$(e \times j) = -\frac{\alpha_1}{\beta_1}(e \times B) + \frac{1}{\beta_1|e|^2}(e \times (B \times e))$$

$$\begin{aligned}\nabla \times (\xi \times B) &+ \xi^V(\mu_0 j \times e)\\ &= (a\gamma_1 + b\gamma_2 + c)e + (a\alpha_1 + b\alpha_2)(e \times B)\\ &\quad + (a\beta_1 + b\beta_2 - \mu_0\xi^V)(e \times j)\\ &= (a\gamma_1 + b\gamma_2 + c)e + \left(b\alpha_2 - \frac{(b\beta_2 - \mu_0\xi^V)\alpha_1}{\beta_1}\right)(e \times B)\\ &\quad + \left(a + \frac{b\beta_2 - \mu_0\xi^V}{\beta_1}\right)\frac{(e \times (B \times e))}{|e|^2}.\end{aligned}$$

Since the three terms of the foregoing formula are mutually orthogonal, we find

$$(\nabla \times (\xi \times B) + \xi^V(\mu_0 j \times e))^2 \geq \left(\frac{b(\alpha_2\beta_1 - \beta_2\alpha_1) + \mu_0\xi^V\alpha_1}{\beta_1}\right)^2 (e \times B)^2.$$

From eqs. (9.125), (9.126), the vectors a_j are given by

$$a_1 = e \qquad a_2 = \frac{\dot{\Phi}j - \dot{J}B}{\dot{p}} \qquad a_3 = \frac{-\dot{X}j + \dot{I}B}{\dot{p}}.$$

The substitution of these relations into eq. (9.128) yields

$$b = \frac{1}{\dot{p}}\left(\varepsilon(B_0\cdot\nabla)\zeta_1 + S_0\frac{\partial}{\partial v}(v\xi_0{}^V)\right).$$

We accordingly find

$$(\nabla \times (\boldsymbol{\xi} \times \boldsymbol{B}) + \xi_0{}^V(\mu_0 \boldsymbol{j} \times \boldsymbol{e}))^2$$

$$\geq \frac{1}{(\boldsymbol{e} \times \boldsymbol{B}_0)^2}\left(\varepsilon(\boldsymbol{B}_0 \cdot \nabla)\zeta_1 + S_0 \frac{\partial}{\partial v}(v\xi_0{}^V) + \xi_0{}^V(\boldsymbol{e} \times \boldsymbol{B}_0)\cdot(\boldsymbol{e} \times \mu_0 \boldsymbol{j}_0)\right)^2,$$

and the inequality regarding the energy integral W_{P} becomes

$$W_{\mathrm{P}} \geq \frac{1}{2\mu_0}\int\left[\frac{1}{(\boldsymbol{e} \times \boldsymbol{B}_0)^2}\left(\varepsilon(\boldsymbol{B}_0 \cdot \nabla)\zeta_1 + S_0 \frac{\partial}{\partial v}(v\xi_0{}^V) + \xi_0{}^V(\boldsymbol{e} \times \boldsymbol{B}_0)\cdot(\boldsymbol{e} \times \mu_0 \boldsymbol{j}_0)\right)^2 - (\mu_0 \Omega + (\mu_0 \boldsymbol{j}_0 \times \boldsymbol{e})^2)(\xi_0{}^V)^2\right]\mathrm{d}\boldsymbol{r}.$$

(Here the positive contribution of $\gamma p_0(\nabla \xi)^2$ has been dropped.) The average of f along the closed line of magnetic force is defined by

$$\langle f \rangle = \oint f \frac{\mathrm{d}l}{B} \Big/ \oint \frac{\mathrm{d}l}{B}.$$

Schwarz's inequality $\langle x^2 \rangle \langle y^2 \rangle \geq \langle xy \rangle^2$ applied to

$$x = \left(\varepsilon(\boldsymbol{B}_0 \cdot \nabla)\zeta_1 + S_0 \frac{\partial}{\partial v}(v\xi_0{}^V) + \xi_0{}^V(\boldsymbol{e} \times \mu_0 \boldsymbol{j}_0)\cdot(\boldsymbol{e} \times \boldsymbol{B}_0)\right)\frac{1}{|\boldsymbol{e} \times \boldsymbol{B}_0|}$$

$$y = |\boldsymbol{e} \times \boldsymbol{B}_0|$$

yields the stability condition

$$\frac{1}{2\mu_0}\int\left(\langle(\boldsymbol{e} \times \boldsymbol{B}_0)^2\rangle^{-1}\left\langle S_0 \frac{\partial}{\partial v}(v\xi_0{}^V) + \xi_0{}^V(\boldsymbol{e} \times \boldsymbol{B}_0)(\boldsymbol{e} \times \mu_0 \boldsymbol{j}_0)\right\rangle^2 - \langle \mu_0 \Omega + (\boldsymbol{e} \times \mu_0 \boldsymbol{j}_0)^2\rangle(\xi_0{}^V)^2\right)\mathrm{d}v \geq 0.$$

When $\xi_0{}^V = 0$ at the boundary, it follows that

$$\int \frac{\partial}{\partial v}(v\xi_0{}^V)\xi_0{}^V \,\mathrm{d}v = \int(\xi_0{}^V)^2 \,\mathrm{d}v + \frac{1}{2}\int v \frac{\partial}{\partial v}(\xi_0{}^V)^2 \,\mathrm{d}v$$

$$= \frac{1}{2}\int(\xi_0{}^V)^2 \,\mathrm{d}v.$$

From Schwarz's inequality

$$\int (\xi_0{}^V)^2 \mathrm{d}v \int \left(\frac{\partial}{\partial v}(v\xi_0{}^V)\right)^2 \mathrm{d}v \geq \left(\int \xi_0{}^V \frac{\partial}{\partial v}(v\xi_0{}^V)\mathrm{d}v\right)^2 = \frac{1}{4}\left(\int (\xi_0{}^V)^2 \mathrm{d}v\right)^2,$$

the stability condition is expressed by

$$\left\langle \frac{S}{2} + (\boldsymbol{B} \times \boldsymbol{e})\cdot(\mu_0 \boldsymbol{j} \times \boldsymbol{e}) \right\rangle^2 - \langle (\boldsymbol{B} \times \boldsymbol{e})^2 \rangle \langle \mu_0 \Omega + (\mu_0 \boldsymbol{j} \times \boldsymbol{e})^2 \rangle \geq 0. \quad (9.132)$$

When the field inhomogeneity is small, $\boldsymbol{e}$ is nearly equal to $\boldsymbol{e} = \boldsymbol{a}_1 = \partial \boldsymbol{r}/\partial V \approx \nabla V/|\nabla V|^2$, and the stability condition for a local mode near a rational magnetic surface is*

$$\left\langle \frac{S}{2} + \frac{\mu_0 \boldsymbol{j}\cdot\boldsymbol{B}}{|\nabla V|^2} \right\rangle^2 - \left\langle \frac{\boldsymbol{B}^2}{|\nabla V|^2} \right\rangle \left\langle \mu_0 \Omega + \frac{\mu_0{}^2 \boldsymbol{j}^2}{|\nabla V|^2} \right\rangle \geq 0. \quad (9.133)$$

The quantities appearing in this formula are all physical quantities, so that this expression holds irrespective of the coordinate system. It has been shown that this sufficient condition is also necessary.[15] Equation (9.133) can be rewritten as

$$\frac{S^2}{4} + \mu_0 \left(S \left\langle \frac{\boldsymbol{j}\cdot\boldsymbol{B}}{|\nabla V|^2} \right\rangle - \Omega \left\langle \frac{\boldsymbol{B}^2}{|\nabla V|^2} \right\rangle \right) - \mu_0{}^2 \left(\left\langle \frac{\boldsymbol{j}^2}{|\nabla V|^2} \right\rangle \left\langle \frac{\boldsymbol{B}^2}{|\nabla V|^2} \right\rangle - \left\langle \frac{\boldsymbol{j}\cdot\boldsymbol{B}}{|\nabla V|^2} \right\rangle^2 \right) \geq 0. \quad (9.134)$$

The 1st term represents shear stabilization and the Ω term within the 2nd term contributes to the stabilization if the field configuration is average minimum-B ($\Omega < 0$). The 3rd term is always negative due to Schwartz's inequality (μ_0 is the permeability of vacuum).

*9.4c Local-Mode Stability for Axially Symmetric Systems

Let us apply the local-mode stability condition to an axially symmetric toroidal system. In pseudotoroidal coordinates ρ, ω, φ the infinitesimal element of length ds is given by

$$\mathrm{d}s^2 = \mathrm{d}\rho^2 + \rho^2 \mathrm{d}\omega^2 + R^2(1 + \kappa\rho\cos\omega)^2 \mathrm{d}\varphi^2 \qquad \kappa \equiv \frac{1}{R}.$$

R is the radius of curvature of the magnetic axis. Let the deviation of the center of the magnetic surface $a =$ const. from the magnetic axis be

* Some caution is necessary for the use of the approximation $\boldsymbol{e} \approx \nabla V/|\nabla V|^2$.

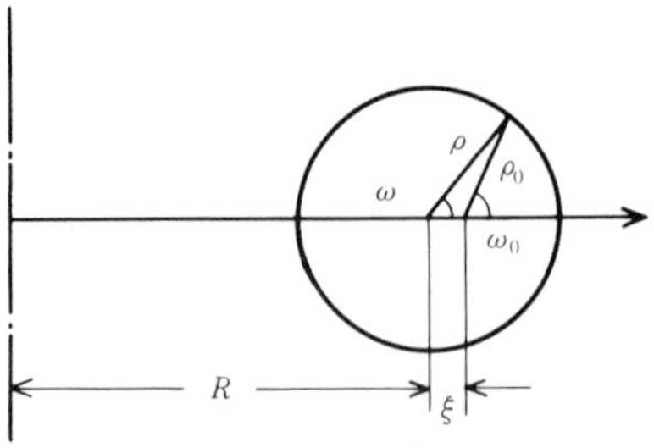

Fig. 9.12
Pseudotoroidal coordinates.

denoted by $\xi(a)$, and let us introduce coordinates ρ_0, ω_0 that satisfy

$$\left.\begin{aligned} \rho\cos\omega &= \rho_0\cos\omega_0 + \xi(a) \\ \rho\sin\omega &= \rho_0\sin\omega_0 \end{aligned}\right\} \tag{9.135}$$

(see fig. 9.12). The magnetic surface $\rho_0 = \rho_0(a, \omega_0)$ can be expanded in terms of a and θ to give

$$\left.\begin{aligned} \rho_0 &= a + \alpha(a)\cos 2\theta \\ \omega_0 &= \theta + \lambda(a)\sin\theta + \mu(a)\sin 2\theta + \cdots. \end{aligned}\right\} \tag{9.136}$$

We assume that the plasma cross section is nearly circular and that ka, λ are 1st-order small quantities and α/a, μ are 2nd-order small quantities. Then ds is given by

$$\mathrm{d}s^2 = g_{11}\mathrm{d}a^2 + 2g_{12}\mathrm{d}a\mathrm{d}\theta + g_{22}\mathrm{d}\theta^2 + g_{33}\mathrm{d}\varphi^2,$$

where

$$\begin{aligned} g_{11} &= 1 + 2\xi'\cos\theta + \xi'^2 \\ &\quad + (a^2\lambda'^2/2 - \xi'\lambda - \xi'\lambda'a)(1 - \cos 2\theta) + 2\alpha'\cos 2\theta + \cdots \end{aligned}$$

$$\begin{aligned} g_{12} &= (a^2\lambda' - a\xi')\sin\theta \\ &\quad + (a^2\mu' - 2\alpha - a\xi'\lambda + a^2\lambda\lambda'/2)\sin 2\theta + \cdots \end{aligned}$$

$$\begin{aligned} g_{22} &= a^2(1 + 2\lambda)\cos\theta + \lambda^2/2 \\ &\quad + (2\alpha/a + 4\mu + \lambda^2/2)\cos 2\theta + \cdots \end{aligned}$$

$$g_{33} = R^2(1 + \kappa a\cos\theta + \kappa\xi - \kappa a\lambda\sin^2\theta + \cdots)^2$$

$$g_{13} = g_{23} = 0$$

$$g^{1/2} = aR\left(1 + (\lambda + \xi' + \kappa a)\cos\theta + \left(\kappa\xi + \frac{\kappa a\xi'}{2}\right)\right.$$
$$\left. + (\alpha' + \lambda\xi' + 2\mu + \alpha/a + \kappa a\lambda + \kappa a\xi'/2)\cos 2\theta \cdots\right). \tag{9.137}$$

Equations for $g_{ik}/g^{1/2}$ are

$$\frac{g_{11}}{g^{1/2}} = \frac{1}{aR}(1 + (\xi' - \lambda - \kappa a)\cos\theta + \cdots)$$

$$\frac{g_{12}}{g^{1/2}} = \frac{1}{R}((a\lambda' - \xi')\sin\theta + \cdots)$$

$$\frac{g_{22}}{g^{1/2}} = \frac{a}{R}(1 + (\lambda - \xi' - \kappa a)\cos\theta + \cdots)$$

$$\frac{g_{33}}{g^{1/2}} = \frac{R}{a}\left(1 - (-\kappa a + \lambda + \xi')\cos\theta\right.$$
$$\left. - \left(\kappa a\lambda + \frac{\kappa a\xi'}{2} - \frac{(\lambda + \xi')^2}{2} - \kappa\xi\right) + O(2)\cos 2\theta \cdots\right). \tag{9.138}$$

Natural coordinates a, θ, φ were introduced in sec. 7.5. The contravariant components of the magnetic field are given by

$$B^1 = 0 \qquad B^2 = \frac{X'(a)}{2\pi g^{1/2}} \qquad B^3 = \frac{\Phi'(a)}{2\pi g^{1/2}} \tag{9.139}$$

in these coordinates. The prime here means differentiation with respect to a. The contravariant components of the current density are given by (see eq. (7.148))

$$j^1 = 0 \qquad j^2 = \frac{I'(a) - \dfrac{\partial v}{\partial \varphi}}{2\pi g^{1/2}} \qquad j^3 = \frac{J'(a) + \dfrac{\partial v}{\partial \theta}}{2\pi g^{1/2}}. \tag{9.140}$$

The covariant components of $\boldsymbol{B}$ are $B_2 = g_{22}B^2 + g_{23}B^3$, $B_3 = g_{23}B^2 + g_{33}B^3$; these are related to $J(a)$ and $I(a)$ by (see eq. (7.150))

$$\mu_0 J(a) = \int_0^{2\pi} B_2 \mathrm{d}\theta \qquad \mu_0 I(a) = -\int_0^{2\pi} B_3 \mathrm{d}\varphi. \tag{9.141}$$

The equilibrium equation is

$$\left.\begin{aligned} & p'V' = I'\Phi' - J'X' \\ & X'\frac{\partial v}{\partial \theta} + \Phi'\frac{\partial v}{\partial \varphi} = p'(V' - 4\pi^2 g^{1/2}). \end{aligned}\right\} \tag{9.142}$$

The equilibrium equation and the local-mode stability condition are given in terms of natural coordinates. Therefore it is necessary to know the explicit form of the natural coordinates. We start with pseudotoroidal coordinates ρ, ω, φ and introduce a, θ, φ with eqs. (9.135) and (9.136). Here we examine the condition for the coordinates a, θ, φ to be equivalent to the natural coordinates. Denote the average of f over θ by $(f)_0$:

$$(f)_0 \equiv \frac{1}{2\pi}\int_0^{2\pi} f\,\mathrm{d}\theta.$$

From eq. (9.141), it follows that

$$\left.\begin{aligned} & \mu_0 J(a) = X'(a)(g_{22}/g^{1/2})_0 \\ & \mu_0 I(a) = -\Phi'(a)(g_{33}/g^{1/2})_0. \end{aligned}\right\} \tag{9.143}$$

The parameter λ is chosen so that the $\cos\theta$ term of $g_{33}/g^{1/2}$ is zero, i.e., we choose

$$\lambda = -\xi' + \kappa a.$$

As we are here discussing axially symmetric systems, we set $\partial/\partial\varphi = 0$. From the relation $\mu_0 \boldsymbol{j} = \nabla \times \boldsymbol{B}$, we find that

$$\mu_0 j^3 = \frac{1}{g^{1/2}}\left(\frac{\partial B_2}{\partial a} - \frac{\partial B_1}{\partial \theta}\right) = \frac{1}{2\pi g^{1/2}}\left(\left(\frac{g_{22}}{g^{1/2}}X'\right)' - X'\frac{\partial}{\partial \theta}\left(\frac{g_{12}}{g^{1/2}}\right)\right).$$

By means of the relations (9.140), (9.142), and (9.143), the foregoing equation is reduced to

$$\left(J\frac{g_{22}/g^{1/2}}{(g_{22}/g^{1/2})_0}\right)' - J' - J\frac{\partial(g_{12}/g^{1/2})/\partial\theta}{(g_{22}/g^{1/2})_0} = \frac{4\pi^2 p'}{\mu_0 J}\left(\frac{g_{22}}{g^{1/2}}\right)_0((g^{1/2})_0 - g^{1/2}), \tag{9.144}$$

In the following we will adopt the notation

$$V'(a) = \iint_0^{2\pi} g^{1/2}\mathrm{d}\theta\mathrm{d}\varphi = 4\pi^2(g^{1/2})_0.$$

When the definitions

$$B_\theta \equiv \frac{\mu_0 J}{2\pi a} \qquad B_t \equiv \frac{\mu_0 I}{2\pi R}$$

are used, eq. (9.144) can be rewritten as

$$\xi'' + \xi'\left(\frac{1}{a} + \frac{2B_\theta'}{B_\theta}\right) = -\kappa\left(1 - \frac{ap'}{B_\theta^2/2\mu_0}\right). \tag{9.145}$$

The specific volume $U \equiv \mathrm{d}V/\mathrm{d}\Phi = V'/\Phi'$ is

$$U = \frac{V'}{\Phi'} = \frac{-4\pi^2 (g^{1/2})_0}{\mu_0 I}\left(\frac{g_{33}}{g^{1/2}}\right)_0 = \frac{-4\pi^2 R^2}{\mu_0 I}(1-u) \tag{9.146}$$

$$u \equiv 1 - \left(\frac{g_{33}}{g^{1/2}}\right)_0 \frac{(g^{1/2})_0}{R^2} = -2\kappa\xi - \kappa a \xi' + \frac{(\kappa a)^2}{2}. \tag{9.147}$$

By using eq. (9.145) for ξ'', we find that

$$u' = -\frac{2\mu_0 p'}{B_\theta^2}\kappa^2 a^2 + 2\kappa^2 a - 2\kappa a \xi'\left(\frac{q_s'}{q_s} - \frac{B_t'}{B_t}\right), \tag{9.148}$$

where q_s is the safety factor. Since S and Ω are expressed in terms of $\tilde{\mu} \equiv X'/\Phi'$ by

$$S = -\left(\frac{\mathrm{d}\Phi}{\mathrm{d}V}\right)^2 \frac{\mathrm{d}}{\mathrm{d}V}\left(\frac{\mathrm{d}X}{\mathrm{d}\Phi}\right) = -\frac{(\Phi')^2}{(V')^3}\left(\frac{X'}{\Phi'}\right)' = -\frac{X'\Phi'}{(V')^3}\left(\frac{\tilde{\mu}'}{\tilde{\mu}}\right)$$

$$\Omega = \dot{I}\ddot{\Phi} - \dot{J}\ddot{X} = \frac{1}{(V')^3}(I'\Phi'' - J'X'' - p'V''),$$

the local-mode stability condition is

$$\frac{1}{4}\left(\frac{\tilde{\mu}'}{\tilde{\mu}}\right)^2 + \frac{(V')^3\mu_0}{(X')^2(\Phi')^2}\left(\left\langle\frac{\boldsymbol{B}^2}{|\nabla V|^2}\right\rangle(J'X'' - I'\Phi'' + p'V'') - \left\langle\frac{\boldsymbol{j}\cdot\boldsymbol{B}}{|\nabla V|^2}\right\rangle\tilde{\mu}'\Phi'^2\right)$$
$$+ \frac{(V')^6\mu_0^2}{(X')^2(\Phi')^2}\left(\left\langle\frac{\boldsymbol{j}\cdot\boldsymbol{B}}{|\nabla V|^2}\right\rangle^2 - \left\langle\frac{\boldsymbol{B}^2}{|\nabla V|^2}\right\rangle\left\langle\frac{\boldsymbol{j}^2}{|\nabla V|^2}\right\rangle\right) \geq 0. \tag{9.149}$$

This condition is reduced to the following equation after a long calculation, which will be given at the end of this section:

$$\left(\frac{\tilde{\mu}'}{2\tilde{\mu}} + \frac{p'}{B_\theta^2/2\mu_0}\kappa a \xi'\right)^2 + \frac{p'/a}{B_t^2/2\mu_0} - \frac{p'u'/2}{B_\theta^2/2\mu_0} - \frac{1}{2}\left(\frac{p'}{B_\theta^2/2\mu_0}\right)^2\kappa^2 a^2 > 0. \tag{9.150}$$

This condition is the same as that derived by Kadomtsev and Pogutse[16] except for the term in ξ'. When $\tilde{\mu}' = 0$ (when shear is zero), the term in u' may stabilize the local mode due to the average minimum-B effect. When eq. (9.148) for u' is substituted into eq. (9.150) and the term in $(\xi')^2$ within the 1st term of eq. (9.150) is neglected, the local-mode stability condition is simplified to

$$\frac{1}{4}\left(\frac{\tilde{\mu}'}{\tilde{\mu}}\right)^2 + \frac{p'/a}{B_t^2/2\mu_0}(1 - q_s^2) > 0, \qquad (9.151)$$

where q_s is the safety factor $q_s = (a/R)(B_\theta/B_t)$. This formula is very important in the study of toroidal systems (see sec. 16.5) and may be regarded as a generalization of Suydam's criterion for a toroidal system.[17] When $q_s^2 > 1$, the local mode is stable. When $q_s^2 < 1$, the local mode is unstable at $\tilde{\mu}' = 0$. In other words if the pitch of the magnetic-field line is maximum or minimum, the local mode is unstable when $q_s^2 < 1$.

The reduction of eqs. (9.116) and (9.117) will now be described. The vectors $\boldsymbol{f}$ and $\boldsymbol{F}$ are defined by

$$\boldsymbol{f} \equiv \nabla \times (\boldsymbol{\xi} \times \boldsymbol{B}) + (\nabla\boldsymbol{\xi})\boldsymbol{B} + \xi^1 \boldsymbol{D}$$
$$\boldsymbol{F} \equiv \boldsymbol{j} \times \boldsymbol{e} \qquad (F_1, F_2, F_3) = (0, \dot{J}, -\dot{I}).$$

From eq. (9.128) it follows that

$$\begin{aligned}\boldsymbol{F}\cdot\boldsymbol{f} &= \sum F_i f^i \\ &= \dot{J}\left(\frac{\partial \zeta}{\partial \varphi'} - \frac{\partial}{\partial V}(\xi^1 \dot{X}) + (\nabla\boldsymbol{\xi})\dot{X} + \xi^1 \ddot{X}\right) \\ &\quad - \dot{I}\left(-\frac{\partial \zeta}{\partial \theta'} + \frac{\partial}{\partial V}(-\xi^1 \dot{\Phi}) + (\nabla\boldsymbol{\xi})\dot{\Phi} + \xi^1 \ddot{\Phi}\right) \\ &= \left(\dot{I}\frac{\partial}{\partial \theta'} + \dot{J}\frac{\partial}{\partial \varphi'}\right)\zeta + \dot{\xi}^1 \dot{p} - \dot{p}(\nabla\boldsymbol{\xi}).\end{aligned}$$

Since eqs. (9.123)–(9.127) yield

$$\nabla\boldsymbol{\xi} = \nabla\left(\frac{\zeta \boldsymbol{j} - \eta \boldsymbol{B}}{\dot{p}} + \xi^1 \boldsymbol{e}\right) = \frac{1}{\dot{p}}(\boldsymbol{j}\cdot\nabla\zeta - \boldsymbol{B}\cdot\nabla\eta) + \dot{\xi}^1,$$

$\boldsymbol{F}\cdot\boldsymbol{f}$ becomes

$$\boldsymbol{F}\cdot\boldsymbol{f} = (\boldsymbol{B}\cdot\nabla)\eta = (\boldsymbol{B}\cdot\nabla)(\dot{J}\xi^2 - \dot{I}\xi^3).$$

Using eq. (9.128), we find

$$\begin{aligned}\nabla((\boldsymbol{\xi}\cdot\boldsymbol{F})\xi^1\boldsymbol{B}) &= (\boldsymbol{\xi}\cdot\boldsymbol{F})\nabla(\xi^1\boldsymbol{B}) + \xi^1(\boldsymbol{B}\cdot\nabla)(\boldsymbol{\xi}\cdot\boldsymbol{F})\\ &= (\boldsymbol{\xi}\cdot\boldsymbol{F})f^1 + \xi^1(\boldsymbol{B}\cdot\nabla)(\dot{J}\xi^2 - \dot{I}\xi^3)\\ &= (\boldsymbol{\xi}\cdot\boldsymbol{F})f^1 + \xi^1(\boldsymbol{f}\cdot\boldsymbol{F}),\end{aligned}$$

and the relation (9.116) is proved.

To prove that eqs. (9.47) and (9.117) are equivalent, we must show that

$$\int [2\nabla\times(\boldsymbol{\xi}\times\boldsymbol{B})\cdot(\boldsymbol{j}\times\boldsymbol{e})\xi^1 + (\boldsymbol{j}\times\boldsymbol{e})\cdot\boldsymbol{D}(\xi^1)^2$$
$$- (\boldsymbol{\xi}\cdot\nabla p)(\nabla\boldsymbol{\xi}) - (\boldsymbol{j}\times\boldsymbol{\xi})\cdot\nabla\times(\boldsymbol{\xi}\times\boldsymbol{B})]\,\mathrm{d}\boldsymbol{r} = 0.$$

From eq. (9.116), it follows that

$$\int [\boldsymbol{\xi}\cdot(\boldsymbol{j}\times\boldsymbol{e})f^1 + \xi^1(\boldsymbol{j}\times\boldsymbol{e})\cdot\nabla\times(\boldsymbol{\xi}\times\boldsymbol{B})$$
$$+ \xi^1(\boldsymbol{j}\times\boldsymbol{e})\cdot\boldsymbol{B}(\nabla\boldsymbol{\xi}) + (\xi^1)^2(\boldsymbol{j}\times\boldsymbol{e})\cdot\boldsymbol{D}]\,\mathrm{d}\boldsymbol{r} = 0.$$

When we compare the foregoing two equations, the problem becomes one of proving that

$$\nabla\times(\boldsymbol{\xi}\times\boldsymbol{B})\cdot(\boldsymbol{j}\times(\xi^1\boldsymbol{e} - \boldsymbol{\xi})) - \boldsymbol{\xi}\cdot(\boldsymbol{j}\times\boldsymbol{e})f^1 = 0.$$

This is clear from

$$\begin{aligned}\nabla\times(\boldsymbol{\xi}\times\boldsymbol{B})\cdot(\dot{J}\xi^2 - \dot{I}\xi^3)\boldsymbol{a}^1 &= f^1(\dot{J}\xi^2 - \dot{I}\xi^3)\\ &= f^1\boldsymbol{e}\cdot(\boldsymbol{\xi}\times\boldsymbol{j}).\end{aligned}$$

Let us examine the derivation of eq. (9.150) from (9.149). When the relations

$$\langle f\rangle \equiv \frac{\oint f\dfrac{\mathrm{d}l}{B}}{\oint\dfrac{\mathrm{d}l}{B}} = \frac{2\pi}{V'}\int_0^{2\pi} f\cdot g^{1/2}\,\mathrm{d}\theta = \frac{(2\pi)^2}{V'}(f\cdot g^{1/2})_0$$

$$|\nabla V|^2 = g^{11}(V')^2 = \frac{(V')^2(g_{22}g_{33} - g_{23}{}^2)}{g} = \frac{(V')^2 g_{22}g_{33}}{g}$$

$$\left\langle\frac{\boldsymbol{B}^2}{|\nabla V|^2}\right\rangle = \left\langle\frac{(X'/2\pi g^{1/2})^2 g_{22} + (\Phi'/2\pi g^{1/2})^2 g_{33}}{V'^2 g_{22}g_{33}/g}\right\rangle$$

$$= \frac{X'^2\left\langle\frac{1}{g_{33}}\right\rangle + \Phi'^2\left\langle\frac{1}{g_{22}}\right\rangle}{4\pi^2 V'^2}$$

are used, eq. (9.149) may be rewritten as

$$\frac{1}{4}\left(\frac{\tilde{\mu}'}{\tilde{\mu}}\right)^2 + \frac{\mu_0 p' V'}{4\pi^2}\left(\left\langle\frac{1}{g_{33}}\right\rangle\frac{X'}{\Phi'^2}\left(\frac{V'}{X'}\right)' + \left\langle\frac{1}{g_{22}}\right\rangle\frac{\Phi'}{X'^2}\left(\frac{V'}{\Phi'}\right)'\right)$$

$$- \frac{\mu_0 \Phi' V' \tilde{\mu}'}{4\pi^2 X'^2}\left\langle\frac{1}{g_{22}}\frac{\partial v}{\partial\theta}\right\rangle + \frac{\mu_0{}^2 (V')^2}{16\pi^4 (X')^2 (\Phi')^2}\left(-\left\langle\frac{1}{g_{22}}\right\rangle\left\langle\frac{1}{g_{33}}\right\rangle p'^2 V'^2\right.$$

$$+ 2\left\langle\frac{1}{g_{22}}\frac{\partial v}{\partial\theta}\right\rangle\left\langle\frac{1}{g_{33}}\right\rangle X' p' V' + \left\langle\frac{1}{g_{22}}\frac{\partial v}{\partial\theta}\right\rangle^2 \Phi'^2 - \left\langle\frac{1}{g_{22}}\right\rangle\left\langle\frac{1}{g_{22}}\left(\frac{\partial v}{\partial\theta}\right)^2\right\rangle \Phi'^2$$

$$\left. - \left\langle\frac{1}{g_{33}}\right\rangle\left\langle\frac{1}{g_{22}}\left(\frac{\partial v}{\partial\theta}\right)^2\right\rangle X'^2\right) > 0.$$

Combine the 1st, 4th, and 7th terms, express $p'^2 V'$ of the 5th term by $p'^2 V' = p'(I'\Phi' - J'X')$, and combine the 2nd and 3rd terms. Use eqs. (9.143) and $\mu_0 J' = (X'(g_{22}/g^{1/2})_0)' = X''(g_{22}/g)_0{}^{1/2} + X'((g_{22}/g)_0{}^{1/2})'$. Then the foregoing relation is reduced to

$$\left(\frac{\tilde{\mu}'}{2\tilde{\mu}} - \left\langle\frac{1}{g_{22}}\frac{\partial v}{\partial\theta}\right\rangle\frac{V'\mu_0}{4\pi^2 X'}\right)^2$$

$$+ \frac{\mu_0 p' V'}{\Phi'^2}\left(\frac{g^{1/2}}{g_{33}}\right)_0\left[\frac{V''}{V'} + \left(\frac{g^{1/2}}{g_{22}}\right)_0\left(\frac{g_{22}}{g^{1/2}}\right)_0' + \frac{X''}{X'}\left(\left(\frac{g^{1/2}}{g_{22}}\right)_0\left(\frac{g_{22}}{g^{1/2}}\right)_0 - 1\right)\right]$$

$$+ \frac{\mu_0 p' V'}{X'^2}\left(\frac{g^{1/2}}{g_{22}}\right)_0\left(\frac{U'}{U} + \frac{I'}{I}\right)$$

$$- \frac{\mu_0{}^2}{4\pi^2}\left\langle\frac{1}{g_{22}}\left(\frac{\partial v}{\partial\theta}\right)^2\right\rangle\left(\frac{V'}{X'^2}\left(\frac{g^{1/2}}{g_{22}}\right)_0 + \frac{V'}{\Phi'^2}\left(\frac{g^{1/2}}{g_{33}}\right)_0\right)$$

$$+ \frac{4\pi^2\mu_0{}^2}{V'}\left(\frac{g^{1/2}}{g_{33}}\right)_0\left\langle\frac{1}{g_{22}}\frac{\partial v}{\partial\theta}\right\rangle\frac{V'^2}{4\pi^2\Phi'^2}\frac{2p'V'}{4\pi^2 X'} > 0. \tag{9.152}$$

Here the relation

$$\frac{U'}{U} + \frac{I'}{I} = (\log UI)' = \left(\log\frac{V'}{\Phi'}\left(-\frac{\Phi' g_{33}}{\mu_0 g^{1/2}}\right)\right)' = \frac{V''}{V'} + \left(\frac{g_{33}}{g^{1/2}}\right)' \Big/ \left(\frac{g_{33}}{g^{1/2}}\right)$$

was used. Equation (9.142) yields

$$\frac{\partial v}{\partial \theta} = -\frac{p'}{X'} 8\pi^2 a^2 \cos\theta = -4\pi \frac{a^2}{R}\frac{p'}{B_\theta}\cos\theta.$$

From eq. (9.137) it follows that

$$\left\langle \frac{1}{g_{22}} \frac{\partial v}{\partial \theta} \right\rangle = -4\pi^2 \xi' \frac{2p'}{X'} = \frac{-2\pi\xi'}{R}\frac{2p'}{B_\theta}$$

$$\left\langle \frac{1}{g_{22}} \left(\frac{\partial v}{\partial \theta}\right)^2 \right\rangle = \frac{2\pi^2 a^2}{R^2} \left(\frac{2p'}{B_\theta}\right)^2.$$

The last term of eq. (9.152) is of the same order as the 2nd one in the term with the factor $(\partial v/\partial \theta)^2$ and the 2nd one is $(B_\theta/B_t)^2$ times as small as the 1st one in the same term. If these terms are neglected, we obtain eq. (9.150).

*9.5 Resistive Instability[18]

In the preceding sections we have discussed instabilities of plasmas with zero resistivity. In such a case the conducting plasma is frozen to the line of magnetic force. However, the resistivity of a plasma is not generally zero and the plasma may hence deviate from the line of magnetic force. Modes which are stable in the ideal case may in some instances become unstable if a finite resistivity is introduced.

Ohm's law is

$$\eta \boldsymbol{j} = \boldsymbol{E} + \boldsymbol{V} \times \boldsymbol{B}. \tag{9.153}$$

For simplicity we here assume that $\boldsymbol{E}$ is zero. The current density is $\boldsymbol{j} = \boldsymbol{V} \times \boldsymbol{B}/\eta$ and the $\boldsymbol{j} \times \boldsymbol{B}$ force $\boldsymbol{F}_s$ is

$$\boldsymbol{F}_s = \boldsymbol{j} \times \boldsymbol{B} = \frac{\boldsymbol{B}(\boldsymbol{V}\cdot\boldsymbol{B}) - \boldsymbol{V}B^2}{\eta}. \tag{9.154}$$

When η tends to zero, this force becomes infinite and prevents the deviation of the plasma from the line of magnetic force. When the magnitude B of magnetic field is small, this force does not become large, even if η is small, and the plasma can deviate from the line of magnetic force. When we consider a perturbation with the propagation vector $\boldsymbol{k}$, only the parallel (to $\boldsymbol{k}$) component of the zeroth-order magnetic field $\boldsymbol{B}$ affects the perturbation, as will be shown later. Even if shear exists, we can choose a propagation vector $\boldsymbol{k}$ perpendicular to the magnetic field $\boldsymbol{B}$:

$$(\boldsymbol{k}\cdot\boldsymbol{B}) = 0. \tag{9.155}$$

Accordingly, if there is any force $\boldsymbol{F}_{\mathrm{dr}}$ driving the perturbation, this driving force may easily exceed the force $\boldsymbol{F}_{\mathrm{s}}$, which is very small for a perturbation where $(\boldsymbol{k} \cdot \boldsymbol{B}) = 0$, and the plasma becomes unstable. This type of instability is called *resistive instability*.

*9.5a Basic Relations

Let us consider a plane model in which the zeroth-order magnetic field $\boldsymbol{B}_0$ depends on only x:

$$\boldsymbol{B}_0 = B_{y0}(x)\boldsymbol{e}_y + B_{z0}(x)\boldsymbol{e}_z. \tag{9.156}$$

Ohm's law (9.153) becomes

$$\frac{\partial \boldsymbol{B}}{\partial t} = -\nabla \times \boldsymbol{E} = \nabla \times (\boldsymbol{V} \times \boldsymbol{B}) - \nabla \times \left(\frac{\eta}{\mu_0}\nabla \times \boldsymbol{B}\right). \tag{9.157}$$

It is assumed that the plasma is incompressible. Since the growth rate of the resistive instability is small compared with the MHD characteristic time and the movement is slower than the velocity of sound, the assumption of incompressibility is justified and it follows that

$$\nabla \boldsymbol{V} = 0. \tag{9.158}$$

The equation of motion is

$$\nabla \times \left(\rho_{\mathrm{m}}\frac{\mathrm{d}\boldsymbol{V}}{\mathrm{d}t}\right) = \nabla \times \left(\frac{1}{\mu_0}(\nabla \times \boldsymbol{B}) \times \boldsymbol{B} + \rho_{\mathrm{m}}\boldsymbol{g}\right). \tag{9.159}$$

ρ_{m} is the mass density and $\boldsymbol{g}$ is the acceleration. It is assumed that the specific resistivity changes only because of the displacement of the plasma fluid, i.e.,

$$\frac{\partial \eta}{\partial t} + \boldsymbol{V}\nabla\eta = 0. \tag{9.160}$$

This assumption does not hold if the temperature is high and the heat conductivity is good. It is also assumed that

$$\frac{\partial}{\partial t}(\rho_{\mathrm{m}}\boldsymbol{g}) + \boldsymbol{V}\nabla(\rho_{\mathrm{m}}\boldsymbol{g}) = 0. \tag{9.161}$$

Since the zeroth-order $\boldsymbol{V}$ is zero in an equilibrium state, we find from eq. (9.157) that

$$\nabla \times (\eta_0 \nabla \times \boldsymbol{B}_0) = 0. \tag{9.162}$$

In the following, denote 1st-order quantities by means of a subscript 1. Let us consider a perturbation of the form

$$f_1(r,t) = f_1(x)\exp(i(k_y y + k_z z) + st).$$

From eq. (9.157), it follows that

$$s\boldsymbol{B}_1 = \nabla \times (\boldsymbol{V}_1 \times \boldsymbol{B}_0) - \frac{1}{\mu_0}\nabla \times (\eta_0 \nabla \times \boldsymbol{B}_1 + \eta_1 \nabla \times \boldsymbol{B}_0). \tag{9.163}$$

From eqs. (9.159)–(9.161) we find

$$s\nabla \times (\rho_{\mathrm{m}}\boldsymbol{V}_1) = \nabla \times \left(\frac{1}{\mu_0}((\boldsymbol{B}_0 \cdot \nabla)\boldsymbol{B}_1 + (\boldsymbol{B}_1 \cdot \nabla)\boldsymbol{B}_0) + (\rho_{\mathrm{m}}\boldsymbol{g})_1\right) \tag{9.164}$$

$$\nabla \boldsymbol{V}_1 = \nabla \boldsymbol{B}_1 = 0$$

$$s\eta_1 + (\boldsymbol{V}_1 \cdot \nabla)\eta_0 = 0$$

$$s(\rho_{\mathrm{m}}\boldsymbol{g})_1 + (\boldsymbol{V}_1 \cdot \nabla)(\rho_{\mathrm{m}}\boldsymbol{g})_0 = 0.$$

The zeroth-order quantities of $\boldsymbol{B}$, ρ_{m}, $\boldsymbol{g}$, and η are given by ($B_{x0} = 0$, $B_{y0}(x)$, $B_{z0}(x)$), $\rho_0(x)$, $\boldsymbol{g}_0(x)$, and $\eta_0(x)$, respectively. With the relations

$$\frac{\partial}{\partial x}B_{1x} + ik_y B_{1y} + ik_z B_{1z} = 0$$

$$\frac{\partial}{\partial x}V_x + ik_y V_y + ik_z V_z = 0,$$

the x component of eq. (9.163) reduces to

$$\frac{\partial^2 B_{1x}}{\partial x^2} = \left(k^2 + \frac{\mu_0 s}{\eta}\right)B_{1x} - i(\boldsymbol{k}\cdot\boldsymbol{B}_0)V_x\frac{\mu_0}{\eta_0}\left(1 + \frac{\eta_0{}'}{\mu_0 s}\frac{(\boldsymbol{k}\cdot\boldsymbol{B}_0)'}{(\boldsymbol{k}\cdot\boldsymbol{B}_0)}\right). \tag{9.165}$$

Multiply k_y and the z component of eq. (9.164) and multiply k_z and the y component and take their difference. Eliminate the term in $\partial^2 B_{1x}/\partial x^2$ by means of eq. (9.165); then we find

$$\mu_0 si\left[\left(k^2\rho_0 + \frac{k^2}{s^2}(\rho_0 \boldsymbol{g})' + \frac{(\boldsymbol{k}\cdot\boldsymbol{B}_0)^2}{\eta_0 s}\left(1 + \frac{\eta_0{}'}{\mu_0 s}\frac{(\boldsymbol{k}\cdot\boldsymbol{B}_0)'}{(\boldsymbol{k}\cdot\boldsymbol{B}_0)}\right)\right)V_x - (\rho_0 V_x{}')'\right]$$

$$= (\boldsymbol{k}\cdot\boldsymbol{B}_0)\left(\frac{\mu_0 s}{\eta_0} - \frac{(\boldsymbol{k}\cdot\boldsymbol{B}_0)''}{(\boldsymbol{k}\cdot\boldsymbol{B}_0)}\right)B_{1x}, \tag{9.166}$$

where the prime means differentiation in x. We introduce the dimensionless quantities

$$\psi \equiv \frac{B_{1x}}{B_0} \qquad W \equiv -iV_x k\tau_R \qquad F \equiv \frac{(\boldsymbol{k}\cdot\boldsymbol{B}_0)}{kB_0} \qquad \alpha = kL_s \quad \xi = \frac{x}{L_s},$$

where $B_0 = |\boldsymbol{B}_0| \approx$ const. and $k = |\boldsymbol{k}| = (k_y^2 + k_z^2)^{1/2}$. τ_R is the diffusion time, given by

$$\tau_R = \frac{\mu_0 L_s^2}{\langle\eta_0\rangle},$$

and τ_H is the Alfvén-wave transit time

$$\tau_H = \frac{L_s}{B/(\mu_0\langle\rho_0\rangle)^{1/2}}.$$

L_s is a characteristic length in the x direction with respect to $F = (\boldsymbol{k}\cdot\boldsymbol{B})/kB$, so that $L_s \approx (\partial F/\partial x)^{-1}$. L_s is called the shear length. Furthermore we define

$$S \equiv \frac{\tau_R}{\tau_H} \qquad p \equiv s\tau_R \qquad G \equiv \tau_H^2(-g/\langle\rho_0\rangle)\frac{\partial\rho_0}{\partial x}$$

$$\tilde{\rho} = \frac{\rho_0}{\langle\rho_0\rangle} \qquad \tilde{\eta} = \frac{\eta_0}{\langle\eta_0\rangle},$$

where $\langle\rho_0\rangle$ and $\langle\eta_0\rangle$ are average values of ρ_0 and η_0. Equations (9.165) and (9.166) are then

$$\frac{\psi''}{\alpha^2} = \left(1 + \frac{p}{\tilde{\eta}\alpha^2}\right)\psi + \left(\frac{F}{\tilde{\eta}} + \frac{F'}{p}\frac{\tilde{\eta}'}{\tilde{\eta}}\right)\frac{1}{\alpha^2}W \tag{9.167}$$

$$\frac{(\tilde{\rho}W')'}{\alpha^2} = \left(\tilde{\rho} - \frac{S^2 G}{p^2} + \frac{FS^2}{p}\left(\frac{F}{\tilde{\eta}} + \frac{\tilde{\eta}'F'}{\tilde{\eta}p}\right)\right)W + S^2\left(\frac{F}{\tilde{\eta}} - \frac{F''}{p}\right)\psi. \tag{9.168}$$

From now on the prime will mean differentiation in $\xi = x/L_s$. The parameter S is the magnetic Reynolds number, given by

$$S = 2.6 \times 10^3 \frac{L_s B(T_e/e)^{3/2}}{ZA^{1/2}(n/10^{20})^{1/2}} \tag{9.169}$$

in mks units (T_e/e is in eV). S is usually very large. The growth rate $p = s\tau_R$ is measured in units of the diffusion time. It must be noticed that *the zeroth-order magnetic field* $\boldsymbol{B}_0$ *appears only in the form of* $(\boldsymbol{k}\cdot\boldsymbol{B}_0)$ *in eqs.* (9.165) *and* (9.166) *or eqs.* (9.167) *and* (9.168). The zeroth-order relation (9.162) reduces to

$$\frac{\partial}{\partial x}\left(\eta_0 \frac{\partial}{\partial x}(\boldsymbol{k}\cdot\boldsymbol{B}_0)\right) = 0, \tag{9.170}$$

and

$$\left(\frac{\tilde{\eta}'}{\tilde{\eta}}\right)F' = -F''; \tag{9.171}$$

i.e.,

$$\tilde{\eta}F' = \text{const.} \tag{9.172}$$

Since F' is of the order of 1, $\langle \eta_0 \rangle$ can be adjusted to be $\tilde{\eta}F' = 1$. Accordingly, eqs. (9.167) and (9.168) are expressed by

$$p\psi'' - p\psi\left(a^2 + \frac{F''}{F}\right) = (p\psi + WF)\left(pF' - \frac{F''}{F}\right) \tag{9.173}$$

$$\frac{p^2}{\alpha^2 S^2 F}\left((\tilde{\rho}W')' + \alpha^2 W\left(\frac{S^2 G}{p^2} - \tilde{\rho}\right)\right) = (p\psi + WF)\left(pF' - \frac{F''}{F}\right). \tag{9.174}$$

When $S \to \infty$ $(p \to \infty)$, we find

$$p\psi = -FW\left(1 - \frac{G}{F^2 F' p}\right) \approx -FW \tag{9.175}$$

except near $F = 0$. The substitution of this relation into eq. (9.173) yields

$$\psi'' - \psi\left(\alpha^2 + \frac{F''}{F} - \frac{G}{F^2}\right) = 0 \tag{9.176}$$

except near $F = 0$.

Let us examine the properties of the functions ψ and W. As the singular point $(\boldsymbol{k}\cdot\boldsymbol{B}_0) = 0$ is important, we choose $\xi = x/L_s = 0$ at the singular point. When the magnetic field $\boldsymbol{B}_0$ has shear, $F = (\boldsymbol{k}\cdot\boldsymbol{B}_0)/\boldsymbol{k}\boldsymbol{B}_0$ changes from zero to ± 1 according to $\xi = 0 \to \pm 1$ as is shown in fig. 9.13. ψ and W must be zero at $\xi = \pm\infty$ or at ξ_1 and ξ_2 (the conducting walls). Accordingly the problem of instability is reduced to the problem of solving the differential equations (9.173) and (9.174) under the given boundary conditions. The growth rate p can be estimated from the solution of the eigenvalue problem.

For the eigenvalue problem, we use the approximation of eqs. (9.175) and (9.176) to solve for ψ and W in the region excluding a neighborhood

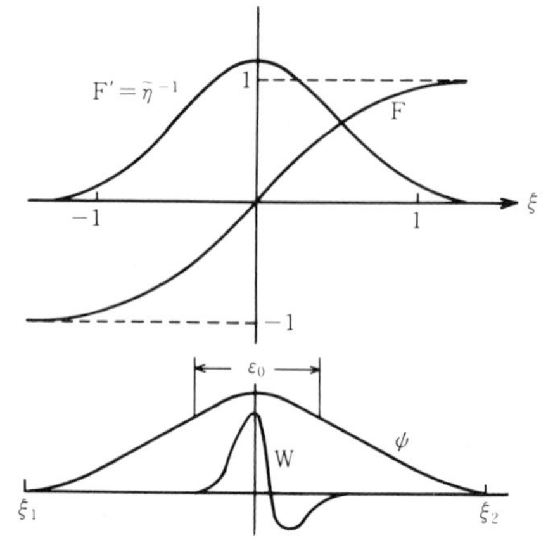

Fig. 9.13
The dependence of F on ζ and the dependence of ψ and W on ζ.

around $F = 0$. In the region near $F = 0$, we find solutions ψ and W by using the approximation $F = F'\xi$. The eigenvalue problem is solved by combining both solutions (this method is the same as the WKB approximation to the Schrödinger equation in quantum mechanics).

Let ψ_1 be the solution of eq. (9.176) which satisfies the boundary condition $\psi = 0$ at $\xi = \xi_1$ and ψ_2 be the solution under the condition $\psi = 0$ at $\xi = \xi_2$. In order to connect the solutions, the gradient of ψ must change rapidly near $F = 0$, as is shown in fig. 9.13. In other words, ψ'' must be large near $F = 0$. ψ'' can be large only near $\xi = 0$ or $F = 0$ as is clear from eq. (9.173). When ψ_1 and ψ_2 are normalized so that $\psi_1 = \psi_2$ at $\xi = 0$, the difference of the gradients of ψ_1 and ψ_2 at $\xi = 0$ is given from eq. (9.176) as follows (assuming $\xi_1 \to -\infty$, $\xi_2 \to \infty$):

$$\Delta' \equiv \frac{\psi_2' - \psi_1'}{\psi_1} \approx -2\alpha - \frac{1}{\psi_1}\int_{-\infty}^{\infty} \mathrm{d}\xi e^{-\alpha|\xi|}\psi\frac{F''}{F} + O\left(\frac{G}{(F')^2\varepsilon_0}\right). \qquad (9.177a)$$

The quantity ε_0 is the range of ξ in which ψ' changes rapidly. The solutions of $d^2\psi/d\xi^2 - \alpha^2\psi = f$ are

$$\psi_1 = e^{\alpha\xi}\left(\int_{\infty}^{\xi} e^{-2\alpha x}\mathrm{d}x\int_{\infty}^{x} fe^{\alpha y}\mathrm{d}y + A\right)$$

$$\psi_2 = e^{-\alpha\xi}\left(\int_{-\infty}^{\xi} e^{2\alpha x}\mathrm{d}x\int_{\infty}^{x} fe^{-\alpha y}\mathrm{d}y + B\right),$$

and thus eq. (9.177a) has been obtained. In order to check the dependence of Δ' on $\alpha = kL_s$, we choose a trial function

$$F = \tanh \xi.$$

Then eq. (7.176) is solved in terms of Legendre functions and Δ' is given by

$$\Delta' = 2\left(\frac{1}{\alpha} - \alpha\right). \tag{9.177b}$$

When we choose another trial function

$$F = \xi \qquad |\xi| < 1$$

$$F = \xi \qquad \xi > 1$$

$$F = -1 \qquad \xi < -1,$$

Δ' is

$$\Delta' = 2\alpha\left(\frac{(1-\alpha) - \alpha \tanh \alpha}{\alpha - (1-\alpha)\tanh \alpha}\right). \tag{9.177c}$$

From these examples we find that Δ' changes from ∞ to $-\infty$ monotonically as α changes from 0 to ∞.

Now let us consider eqs. (9.173) and (9.174) near $\xi = 0$. It may be assumed that $F = F'\xi$ and F', F'', $\tilde{\eta}$, $\tilde{\eta}'$, G, and $\tilde{\rho}$ are constant near $\xi = 0$.

It is also assumed that $\tilde{\rho}' W'$ is negligible compared with $\tilde{\rho} W''$. When

$$\theta = (1/\varepsilon)\left(\xi + \frac{\tilde{\eta}'}{2p}\right),$$

eqs. (9.173) and (9.174) are reduced to

$$\frac{d^2\psi}{d\theta^2} - \varepsilon^2\alpha^2\psi = \varepsilon\Omega(4\psi + U(\theta + \delta_1)) \tag{9.178}$$

$$\frac{d^2 U}{d\theta^2} + \left(\Lambda - \frac{1}{4}\theta^2\right)U = (\theta - \delta)\psi, \tag{9.179}$$

where

$$\varepsilon = \left(\frac{p\tilde{\eta}\tilde{\rho}}{4\alpha^2 S^2 (F')^2}\right)^{1/4} \tag{9.180}$$

$$U = W\left(\frac{4\varepsilon F'}{p}\right) \tag{9.181}$$

$$\Omega = \frac{p\varepsilon}{4\tilde{\eta}} \tag{9.182}$$

$$\delta = \left(\frac{1}{4\Omega}\right)\left(\frac{F''}{F'} + \frac{\tilde{\eta}'}{2\tilde{\eta}}\right) \tag{9.183}$$

$$\delta_1 = \left(\frac{1}{8\Omega}\right)\left(\frac{\tilde{\eta}'}{\tilde{\eta}}\right) \tag{9.184}$$

$$\Lambda = \frac{(\tilde{\eta}')^2}{16\varepsilon^2 p^2} + \frac{S^2\alpha^2\varepsilon^2 G}{p^2\tilde{\rho}} - \alpha^2\varepsilon^2. \tag{9.185}$$

From eq. (9.171), it follows that $\delta \approx -\delta_1$.
When U is expanded in Hermite functions u_n, we find

$$U = \sum_{n=0}^{\infty} a_n u_n$$

$$\frac{d^2 u_n}{d\theta^2} + \left(n + \frac{1}{2} - \frac{1}{4}\theta^2\right)u_n = 0 \tag{9.186}$$

$$u^n = \frac{(-1)^n}{(2\pi)^{1/4}(n!)^{1/2}} e^{\theta^2/4} \frac{d^n}{d\theta^n} e^{-\theta 2/2}. \tag{9.187}$$

Accordingly, the coefficients a_n are given by

$$a_n = \frac{1}{\Lambda - \left(n + \frac{1}{2}\right)} \int_{-\infty}^{\infty} d\theta_1 u_n \psi(\theta_1 - \delta).$$

Since $W \approx -(p/F)\psi$ and $F \approx F'\xi$, W and U change more rapidly than ψ near $\xi = 0$. Accordingly, the condition $\varepsilon < \varepsilon_0$ must be satisfied.

Let us calculate the difference in the gradients of ψ at both sides of $\xi = 0$. By means of eq. (9.178) we find

$$\Delta' \equiv \frac{\psi_2' - \psi_1'}{\psi_1} = \frac{1}{\psi_1}\int_{-\infty}^{\infty} \psi'' d\xi$$

$$\approx \frac{\Omega}{\psi_1}\int_{-\infty}^{\infty} (4\psi + (\theta + \delta_1)U) d\theta$$

$$\approx \Omega \int_{-\infty}^{\infty} \left(4 + (\theta + \delta_1)\frac{U}{\psi_1}\right) d\theta$$

$$= \Omega \int_{-\infty}^{\infty} \sum_n \left(\left(\int_{-\infty}^{\infty} 4u_n d\theta\right) u_n + (\theta + \delta_1) \times \frac{a_n}{\psi_1} u_n\right) d\theta$$

$$\approx \Omega \sum_n \left(4\left(\int_{-\infty}^{\infty} u_n d\theta\right)^2 + \int_{-\infty}^{\infty} (\theta + \delta_1) u_n d\theta \frac{\int_{-\infty}^{\infty} (\theta_1 - \delta) u_n d\theta_1}{\Lambda - \left(n + \frac{1}{2}\right)}\right).$$

Here we have assumed that ψ is nearly constant in the region $|\xi| < \varepsilon$, i.e., $\varepsilon|\psi'/\psi| \sim \varepsilon|\Delta'| \ll 1$. From eq. (9.177) we may rewrite this condition as

$2\varepsilon < \alpha < 1/2\varepsilon$.

When the relations

$$\int_{-\infty}^{\infty} u_n d\theta \equiv 2^{3/4} \left(\frac{\Gamma\left(\frac{1}{2}n + \frac{1}{2}\right)}{\Gamma\left(\frac{1}{2}n + 1\right)}\right)^{1/2} \qquad (n \text{ even})$$

$$= 0 \qquad (n \text{ odd})$$

$$\int_{-\infty}^{\infty} u^n \theta d\theta = 0 \qquad (n \text{ even})$$

$$= 2^{9/4} \left(\frac{\Gamma\left(\frac{1}{2}n + 1\right)}{\Gamma\left(\frac{1}{2}n + \frac{1}{2}\right)}\right)^{1/2} \qquad (n \text{ odd})$$

are used, we find

$$\Delta' = 2^{7/2}\Omega \sum_{m=0}^{\infty} \frac{\Gamma\left(m + \frac{1}{2}\right)}{\Gamma(m+1)} \left(\frac{\Lambda - \frac{1}{2}}{\Lambda - \left(2m + \frac{3}{2}\right)} - \frac{\frac{\delta\delta_1}{4}}{\Lambda - \left(2m + \frac{1}{2}\right)}\right). \qquad (9.188)$$

This value of Δ' must be equal to Δ' of eq. (9.177). From this relation, the eigenvalue Λ is obtained. The growth rate $p = s\tau_R$ can be estimated from

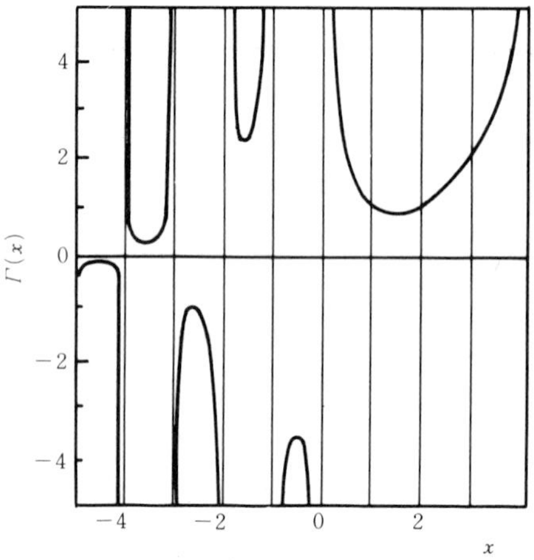

Fig. 9.14
The dependence of $\Gamma(x)$ on x.

eqs. (9.180) and (9.185). By means of the hypergeometric series of argument 1, the sum in eq. (9.188) can be reduced to

$$\Delta' = 2^{7/2}\pi\Omega\left(\frac{\Gamma\left(\frac{3}{4}-\frac{1}{2}\Lambda\right)}{\Gamma\left(\frac{1}{4}-\frac{1}{2}\Lambda\right)} + \frac{\delta\delta_1}{8}\frac{\Gamma\left(\frac{1}{4}-\frac{1}{2}\Lambda\right)}{\Gamma\left(\frac{3}{4}-\frac{1}{2}\Lambda\right)}\right). \tag{9.189}$$

The Γ function depends on the argument as is shown in fig. 9.14. Usually $\delta\delta_1$ is negative ($\delta\delta_1 < 0$). We can observe that:

1. If $\delta\delta_1 < 0$, then Δ'/Ω changes from ∞ to $-\infty$ as Λ changes from 1/2 to 3/2. Hence for any given Δ' there is an infinite sequence of eigenvalues $\Lambda \approx 1, 2, 3, \ldots$
2. If $\delta\delta_1 = 0$, then Δ'/Ω changes from ∞ to $-\infty$ when Λ changes from $2m + 3/2$ to $2m + 7/2$. The sequence of eigenvalues is $\Lambda \approx 2, 4, 6, \ldots$.
3. If $0 < \delta\delta_1 \ll \Lambda$, then Δ'/Ω covers almost the entire range from ∞ to $-\infty$ as Λ goes from $2m + 3/2$ to $2m + 7/2$. The excluded interval is

$$\left|\frac{\Delta'}{\Omega}\right| < 4\pi(\delta\delta_1)^{1/2}.$$

For the case outside the excluded interval, the sequence of eigenvalues is similar to the case 2.

4. When $|\Lambda| \ll 1/2$, eq. (9.189) reduces to

$$\Delta' = \Omega(12 + 13\delta\delta_1). \tag{9.190}$$

In this case, Ω is to be determined by the value of Δ' given by eq. (9.177); and the condition $\Lambda \ll 1$ is to be verified by means of eqs. (9.180) and (9.185).

Since eq. (9.179) is very similar to the differential equation (9.186), it could be expected that the eigenvalue is $\Lambda \approx 1, 2, 3\ldots$ in the cases 1–3. Thus the eigenvalue problem has been solved. We can now identify a number of basic modes.

*9.5b Rippling, Gravitational Interchange, and Tearing Modes

(i) Rippling Mode The rippling mode is characterized by the finiteness of Λ and the predominance of the $(\eta')^2$ term in

$$\Lambda = (\tilde{\eta}')^2/16\varepsilon^2 p^2 + S^2\alpha^2\varepsilon^2 G/p^2\tilde{\rho} - \alpha^2\varepsilon^2.$$

The gradient of specific resistivity or of the electron temperature plays a dominant role in the rippling mode. From eqs. (9.180) and (9.185) the growth rate $p = s\tau_R$ and the interval ε of $\xi = x/L_s$ in which $U \propto W \propto v_y$ is large, are given by

$$p = \left(\frac{(\tilde{\eta}')^2 \alpha S|F'|}{8\Lambda\tilde{\eta}^{1/2}\tilde{\rho}^{1/2}}\right)^{2/5} \approx \left(\frac{1}{8\Lambda}\right)^{2/5}\left(\frac{L_s}{\delta x}\right)^{4/5}\left(kL_s\frac{\tau_R}{\tau_H}\right)^{2/5} \tag{9.191}$$

$$\varepsilon = \frac{|\tilde{\eta}'|}{4p\Lambda^{1/2}} \approx \frac{(8\Lambda)^{2/5}}{4\Lambda^{1/2}}\left(\frac{L_s}{\delta x}\right)^{1/5}\left(kL_s\frac{\tau_R}{\tau_H}\right)^{-2/5}. \tag{9.192}$$

The length δx is defined by $\delta x \equiv \eta/(\partial\eta/\partial x)$. τ_R and τ_H are the diffusion time and Alfvén-wave transit times. When $\alpha \gg 1$, eqs. (9.177a–c) yield a large negative Δ', and $\Lambda \approx 1/2, 3/2\ldots$. When $\alpha \ll 1$, Δ' is large and positive, and $\Lambda \approx 3/2, 5/2\ldots$. The mode with the maximum growth rate corresponds to $\Lambda \approx 1/2$, which is a mode with large α, i.e., short wavelength.

(ii) Gravitational Interchange Mode This mode is characterized by the finiteness of Λ and the predominance of the G term on the right-hand side of eq. (9.185). Instability occurs when $G > 0$. Equations (9.180) and (9.185) yield

$$p = \left(\frac{S\alpha G\tilde{\eta}^{1/2}}{2\Lambda|F'|\tilde{\rho}^{1/2}}\right)^{2/3} \approx \left(\frac{1}{2\Lambda}\right)^{2/3}\left(GkL_s\frac{\tau_R}{\tau_H}\right)^{2/3} \tag{9.193}$$

$$\varepsilon = \left(\frac{(\tilde{\rho}G)^{1/2}\tilde{\eta}}{2^2\Lambda^{1/2}|F'|^2\alpha S}\right)^{1/3} \approx \left(\frac{1}{2^2\Lambda^{1/2}}\right)^{2/3} G^{1/6}\left(kL_s\frac{\tau_R}{\tau_H}\right)^{-1/3}, \tag{9.194}$$

where $G = -(g/\langle\rho_0\rangle)(\partial\rho_0/\partial x)\tau_H{}^2$. Since this mode is dependent upon the density gradient, it is related to the drift wave which will be described in ch. 12. The growth rate $\gamma = p\tau_R{}^{-1}$ can be put into the form

$$\gamma \approx \left(\frac{L_s}{a}\right)^{2/3}\left(\frac{kg}{\Omega_i}\right)^{2/3}\left(\frac{m_e}{m_i}\nu_{ei}\right)^{1/3}, \tag{9.195}$$

by using the relations $\eta = m_e\nu_{ei}/ne^2$ and $-(\partial\rho_0/\partial x)/\rho_0 \equiv a^{-1}$. The growth rate is proportional to $(\nu_{ei})^{1/3}$. However, if the temperature becomes high and the growth rate becomes smaller than the drift frequency ω^*, the MHD treatment is no longer valid and the growth rate becomes $\gamma \propto \eta \propto \nu_{ei}$ (see sec. 12.1).

(iii) Tearing Mode The tearing mode is characterized by $|\Lambda| \ll 1/2$. Equations (9.180) and (9.182) yield

$$p = 4(\alpha S\Omega^2\tilde{\eta}^{3/2}\tilde{\rho}^{-1/2}|F'|)^{2/5} \approx 4\left(\Omega^2 kL_s\frac{\tau_R}{\tau_H}\right)^{2/5} \tag{9.196}$$

$$\varepsilon = \left(\frac{\Omega^{1/2}\tilde{\eta}\tilde{\rho}^{1/2}}{\alpha S|F'|}\right)^{2/5} \approx \left(\frac{\Omega^{1/2}}{kL_s\tau_R/\tau_H}\right)^{2/5}, \tag{9.197}$$

and the condition on Λ can be expressed by

$$\Lambda = \frac{1}{\Omega^2}\left(\left(\frac{\tilde{\eta}'}{16\tilde{\eta}}\right)^2 + \frac{pG}{64\tilde{\eta}(F')^2}\right) \ll 1.$$

If the current layer is symmetric, $\tilde{\eta}'$ and G are zero and this condition is satisfied. From eq. (9.190), we have the instability condition

$$\Omega \approx \Delta'/12 > 0. \tag{9.198}$$

It is clear from eq. (9.182) that the condition $\Delta' > 0$ is necessary for instability. The tearing mode is more unstable for longer wavelength ($\Delta' \approx 2/\alpha$ for small α), although there is a lower limit of $\alpha > 2\varepsilon$ due to the constant-ψ approximation. As the terms Λ and δ are small, the solution for U is antisymmetric as is clear from eq. (9.179).

The characteristic time τ_R for resistive instability is given by $\tau_R = \mu_0 L_s^2/\eta$. Since the classical diffusion coefficient is given by $D_{cl} = \beta_e \eta/\mu_0$, τ_R may be rewritten as

$$\tau_R = \frac{L_s^2}{D_{cl}} \beta_e \approx \beta_e \tau_{cl},$$

where τ_{cl} is the classical diffusion time of a plasma of size L_s.

First-order perturbations of the magnetic field and fluid velocity are shown in figs. 9.15a–c for rippling, gravitational interchange, and tearing modes, respectively.

To understand the basic character of resistive instabilities, simple models are used. When a resistivity gradient exists, Ohm's law in its linearized form is

$$\eta_0 \boldsymbol{j}_1 = -\eta_1 \boldsymbol{j}_0 + \boldsymbol{V} \times \boldsymbol{B}$$
$$(\eta_1 = -\boldsymbol{V} \cdot \nabla \eta_0 / s),$$

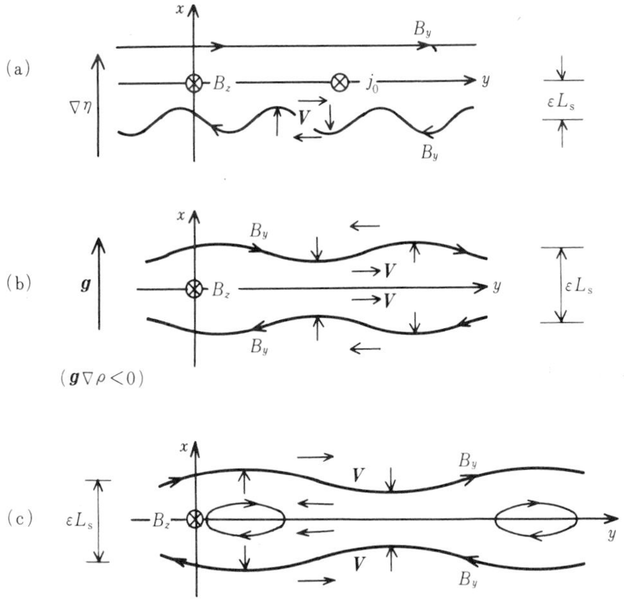

Fig. 9.15
Perturbed magnetic-force lines and velocities: (a) rippling mode, (b) gravitational interchange mode, and (c) tearing mode.

where $\boldsymbol{E}$ is assumed to be zero. Due to the term $\boldsymbol{j}_1 = (-\eta_1/\eta_0)\boldsymbol{j}_0$, the force

$$\boldsymbol{F}_{\mathrm{dr}} = \boldsymbol{j}_1 \times \boldsymbol{B} = ((\boldsymbol{V}\cdot\nabla\eta_0)/s\eta_0)\boldsymbol{j}_0 \times \boldsymbol{B} \qquad (9.199)$$

acts on the plasma. As the sign of B_y changes as we proceed across $F = 0$, this component generates a stabilizing force on the side of higher resistivity (the direction of the force is opposite to that of $\boldsymbol{V}$). However the force is destabilizing on the other side (fig. 9.15a) and causes the rippling mode.

In the gravitational interchange instability, the force

$$\boldsymbol{F}_{\mathrm{dr}} = \rho_1 \boldsymbol{g} = -(V_x \rho_0'/s)\boldsymbol{g} \qquad (9.200)$$

acts on the plasma. When $\rho_0' g < 0$, $\boldsymbol{F}_{\mathrm{dr}}$ enhances the perturbation of V_x (fig. 9.15b).

The tearing mode (fig. 9.15c) differs from the other two modes in that it is typically of long wavelength relative to the shear length. An electric field induced by the perturbation in the magnetic field must be taken into account in Ohm's law. $\nabla \times \boldsymbol{E} = -\partial \boldsymbol{B}/\partial t$ yields

$$\left.\begin{aligned} E_s &\approx \frac{-sB_x}{ik_y}. \\ j_1 &\approx \frac{E_z}{\eta_0} \\ (F_{\mathrm{dr}})_x &\approx (\boldsymbol{j}_1 \times \boldsymbol{B})_x \approx \frac{sB_y B_x}{ik_y \eta_0}. \end{aligned}\right\} \qquad (9.201)$$

Since this mode is likely to break up the current sheet into a set of parallel pinches, this mode is called the tearing mode. This tearing mode is closely related to disruptive instability in tokamak plasmas.

9.6 Ballooning Modes

In interchange instability, the parallel component $k_{\parallel} = (\boldsymbol{k}\cdot\boldsymbol{B})/B$ of the propagation vector of the magnetic field is zero and an average minimum-B condition may stabilize such an instability. Suydam's condition and the toroidal-system local-mode stability condition are involved in perturbations with $k_{\parallel} = 0$. Resistive instabilities may occur near $k_{\parallel} = 0$ even if shear exists. In this section we will study perturbations where $k_{\parallel} \neq 0$. Although the interchange instability is stabilized by an average minimum-

B configuration, it is possible that the perturbation can grow locally in the bad region of the average minimum-B field. This type of instability is called the *ballooning mode*. When the resistivity of the plasma is zero there is an upper limit β_c of the beta ratio under which the ballooning mode will not be excited. When the plasma resistivity is finite, the plasma can move across lines of magnetic force and a *resistive ballooning mode* may be excited even if $\beta < \beta_c$. The growth rate is $\gamma \approx (\beta/\beta_c)\eta k^2/\mu_0 \approx 1/(\tau_{cl}\beta_c)$. τ_{cl} is the classical diffusion time for a plasma of size equal to the wavelength. As the value of β_c increases with the magnetic well depth, the average minimum-B property tends to suppress the growth rate of resistive ballooning. Shear can also stabilize this mode. Since the average minimum-B property and shear also stabilize drift instability, as will be described in ch. 12, these parameters are important figures of merit for the magnetic field configurations used in plasma confinement.[19–22]

9.6a Collisionless Ballooning Mode

The energy integral δW is given by

$$\delta W = \frac{1}{2\mu_0}\int((\nabla \times (\boldsymbol{\xi} \times \boldsymbol{B}_0))^2 - (\boldsymbol{\xi} \times (\nabla \times \boldsymbol{B}_0))\cdot\nabla \times (\boldsymbol{\xi} \times \boldsymbol{B}_0)$$
$$+ \gamma\mu_0 p_0(\nabla\cdot\xi)^2 + \mu_0(\nabla\cdot\xi)(\xi\cdot\nabla p_0))\mathrm{d}\boldsymbol{r}.$$

Assuming a low-beta plasma, $\boldsymbol{\xi}$ may be expressed by

$$\boldsymbol{\xi} = \frac{\boldsymbol{B}_0 \times \nabla\phi}{{B_0}^2}, \tag{9.202}$$

where ϕ is considered to be the time integral of the scalar electrostatic potential of the perturbed electric field. Because

$$\boldsymbol{\xi} \times \boldsymbol{B}_0 = \nabla_\perp\phi,$$

the energy integral is reduced to

$$\delta W = \frac{1}{2\mu_0}\int\left((\nabla \times \nabla_\perp\phi)^2 - \left(\frac{(\boldsymbol{B}_0 \times \nabla_\perp\phi) \times \mu_0\boldsymbol{j}_0}{{B_0}^2}\right)\nabla \times \nabla_\perp\phi\right.$$
$$\left. + \gamma\mu_0 p_0(\nabla\cdot\xi)^2 + \mu_0(\nabla\cdot\xi)(\xi\cdot\nabla p_0)\right)\mathrm{d}\boldsymbol{r}.$$

$\nabla\cdot\xi$ is given by

$$\nabla\xi = \nabla\left(\frac{\boldsymbol{B}_0 \times \nabla\phi}{B_0{}^2}\right) = \nabla\phi\cdot\nabla\times\left(\frac{\boldsymbol{B}_0}{B_0{}^2}\right)$$
$$= \nabla\phi\cdot\left(\left(\nabla\frac{1}{B_0{}^2}\right)\times\boldsymbol{B}_0 + \frac{1}{B_0{}^2}\nabla\times\boldsymbol{B}_0\right).$$

The 2nd term in ( ) is negligible compared with the 1st term in the low-beta case. By means of $\nabla p_0 = \boldsymbol{j}_0 \times \boldsymbol{B}_0$, δW is expressed by

$$\delta W = \frac{1}{2\mu_0}\iint\left[(\nabla\times\nabla_\perp\phi)^2 + \frac{\mu_0\nabla p_0\cdot(\nabla_\perp\phi\times\boldsymbol{B}_0)}{B_0{}^2}\frac{\boldsymbol{B}_0\cdot\nabla\times\nabla_\perp\phi}{B_0{}^2}\right.$$
$$-\frac{\mu_0(\boldsymbol{j}_0\cdot\boldsymbol{B}_0)}{B_0{}^2}\nabla_\perp\phi\cdot\nabla\times\nabla_\perp\phi + \gamma\mu_0 p_0\left(\nabla\left(\frac{1}{B_0{}^2}\right)\cdot(\boldsymbol{B}_0\times\nabla_\perp\phi)\right)^2$$
$$\left.+\frac{\mu_0\nabla p_0\cdot(\boldsymbol{B}_0\times\nabla_\perp\phi)}{B_0{}^2}\left(\nabla\left(\frac{1}{B_0{}^2}\right)\cdot(\boldsymbol{B}_0\times\nabla_\perp\phi)\right)\right]\mathrm{d}\boldsymbol{r}.$$

Let us use cylindrical coordinates. The r, θ, z components of ∇p_0, $\boldsymbol{B}$, and $\nabla\phi$ are given by

$$\nabla p_0 = (p_0', 0, 0) \qquad \boldsymbol{B} = (0, B_\theta(r), B_0(1 - rR_c^{-1}(z)))$$
$$\nabla\phi = (\partial\phi/\partial r, \partial\phi/r\partial\theta, \partial\phi/\partial z) \qquad \phi(r, \theta, z) = \phi(r, z)\mathrm{Re}(\exp im\theta).$$

$R_c(z)$ is the radius of curvature of the line of magnetic force:

$$\frac{1}{R_c(z)} = \frac{1}{R_0}\left(-w + \cos 2\pi\frac{z}{L}\right). \tag{9.203}$$

When $R_c(z) < 0$, the curvature is said to be good. If the configuration is average minimum-B, w and R_0 must be $1 > w > 0$ and $R_0 > 0$ in eq. (9.203). Since B_θ/B_0, r/R, r/L are all small quantities, we find

$$\nabla_\perp\phi = \nabla\phi - \nabla_\parallel\phi \simeq \mathrm{Re}\left(\frac{\partial\phi}{\partial r}, \frac{im}{r}\phi, 0\right)$$

$$\nabla\times(\nabla_\perp\phi) \approx \mathrm{Re}\left(\frac{-im}{r}\frac{\partial\phi}{\partial z}, \frac{\partial^2\phi}{\partial z\partial r}, 0\right)$$

$$\boldsymbol{B}_0\times\nabla_\perp\phi \approx \mathrm{Re}\left(-\frac{im}{r}B_0\phi, B_0\frac{\partial\phi}{\partial r}, 0\right),$$

and δW is reduced to

$$\delta W = \frac{1}{2\mu_0}\int\frac{m^2}{r^2}\left(\left(\frac{\partial\phi(r, z)}{\partial z}\right)^2 - \frac{\beta}{r_p R_c(z)}(\phi(r, z))^2\right)2\pi r\mathrm{d}r\mathrm{d}z, \tag{9.204}$$

where $-p_0/p_0' = r_p$ and $\beta = p_0/(B_0^2/2\mu_0)$. The 2nd term of eq. (9.204) contributes to stability in the region $R_c(z) < 0$ and contributes to instability in the region of $R_c(z) > 0$. Euler's equation is given by

$$\frac{d^2\phi}{dz^2} + \frac{\beta}{r_p R_c(z)}\phi = 0. \tag{9.205}$$

R_c is nearly equal to $B/|\nabla B|$. Equation (9.205) is a Mathieu differential equation, whose eigenvalue is

$$w = F\left(\frac{\beta L^2}{2\pi^2 r_p R_0}\right).$$

Since

$$F(x) = x/4 \qquad x \ll 1$$
$$F(x) = 1 - x^{-1/2} \qquad x \gg 1,$$

we find the approximate relation

$$\beta_c \approx \frac{4w}{(1+3w)(1-w)^2}\frac{2\pi^2 r_p R_0}{L^2}. \tag{9.206a}$$

Since w is of the order of $r_p/2R_0$ and the connection length is

$$L \approx 2\pi R_0(2\pi/\iota)$$

(ι being the rotational transform angle), the critical beta ratio β_c is

$$\beta_c \approx \left(\frac{\iota}{2\pi}\right)^2\left(\frac{r_p}{R_0}\right)^2. \tag{9.206b}$$

If β is smaller than the critical beta ratio β_c, then $\delta W > 0$, and the plasma is stable. The stability condition for the ballooning mode in the shearless case is given by[20,22]

$$\beta < \beta_c. \tag{9.207}$$

9.6b Resistive Ballooning Mode[22]

In the resistive ballooning mode, the plasma diffuses across the magnetic field and nonequilibrium of pressure occurs along magnetic-force lines, thus exciting acoustic waves. Let us consider the plane model of a plasma. The x, y, z components of the equilibrium magnetic field are

$$\boldsymbol{B}_0 = \left(0, \; -2\pi\frac{x}{L}B_0, \; B_0\right).$$

The equation of motion which takes the curvature of field lines into account is

$$\rho_m\frac{\partial \boldsymbol{V}}{\partial t} = -\nabla p + \boldsymbol{j}\times\boldsymbol{B}_0 + \frac{2p}{R_0}\left(-w + \cos 2\pi\frac{z}{L}\right)\boldsymbol{e}_x \tag{9.208}$$

($\boldsymbol{e}_x$ is the unit vector in the x direction). Ohm's law, the equation of state, and the divergence equation for the current are

$$-\nabla\phi + \boldsymbol{V}\times\boldsymbol{B}_0 = \eta\boldsymbol{j}_\parallel \tag{9.209}$$

$$\frac{\partial p}{\partial t} + \boldsymbol{V}\cdot\nabla p + \gamma_H p\nabla\boldsymbol{V} = 0 \tag{9.210}$$

$$\nabla\boldsymbol{j} = 0, \tag{9.211}$$

where γ_H is the ratio of specific heats. For simplicity only the term $\eta\boldsymbol{j}_\parallel$ is taken into account in Ohm's law. We consider a quasi-mode ϕ:

$$\phi = \exp(st)\exp ik\left(y - \frac{2\pi x}{L_s}z\right)\phi(z). \tag{9.212}$$

Let p_1 be the deviation of the pressure from the equilibrium value. Equations (9.209) and (9.208) yield $V_\perp$ and $V_\parallel$, respectively. Equations (9.208) and (9.209) yield $j_\perp$ and $j_\parallel$, respectively. From eqs. (9.210) and (9.211), we find

$$-s\rho_0 k^2\left(1 + \left(\frac{2\pi z}{L_s}\right)^2\right)\frac{\phi}{B_0} + \frac{2ki}{R_0}p_1\left(-w + \cos\left(2\pi\frac{z}{L}\right)\right)$$

$$+\frac{B_0{}^2}{\eta}\frac{\partial^2}{\partial z^2}\left(\frac{\phi}{B_0}\right) = 0 \tag{9.213}$$

$$sp_1 + \frac{ikp_0}{r_p}\frac{\phi}{B_0} - \frac{c_s{}^2}{s}\frac{\partial^2 p_1}{\partial z^2} = 0, \tag{9.214}$$

where c_s is the velocity of sound in the plasma ($c_s{}^2 = \gamma_H p_0/\rho_{m0}$) and $-p_0/(\partial p_0/\partial x) \equiv r_p$. Let us take Fourier transform of $p_1(z)$ and $\phi(z)$ and eliminate p_1:

$$p_1(\alpha) = \frac{1}{L}\int_0^L p_1(z)\exp\left(-\frac{2\pi i\alpha}{L}z\right)dz$$

$$\phi(\alpha) = \frac{1}{L}\int_0^L \phi(z)\exp\left(-\frac{2\pi i\alpha}{L}z\right)dz.$$

Since $\partial^2\phi(\alpha)/\partial\alpha^2 = -(2\pi z/L)^2\phi(\alpha)$, we find

$$\frac{L^2}{L_s^2}\frac{d^2\phi(\alpha)}{d\alpha^2} - \phi(\alpha) - \frac{\alpha^2\phi(\alpha)}{s'\varepsilon}$$

$$-G\left\{\frac{w\phi(\alpha)}{s'^2+\alpha^2} - \frac{1}{2}\left(\frac{\phi(\alpha+1)}{s'^2+(\alpha+1)^2} + \frac{\phi(\alpha-1)}{s'^2+(\alpha-1)^2}\right)\right\} = 0, \tag{9.215}$$

where

$$s' = \frac{L}{2\pi c_s}s$$

$$G = \frac{L^2}{2\pi^2 r_p R_0 \gamma_H}$$

$$\varepsilon = \frac{\beta\gamma_H L k^2}{4\pi\mu_0 c_s}\eta.$$

Let us consider the shearless case ($L_s \to \infty$). Equation (9.215) is then linear in $\phi(\alpha)$, $\phi(\alpha+1)$, and $\phi(\alpha-1)$. The value of s is obtained from the determinant of the coefficients. When η is small, the diagonal element is large except in the case $\alpha = 0$. Accordingly it is sufficient to take the 3×3 determinant in coefficients of $\phi(0)$, $\phi(\pm 1)$, i.e.,

$$(s'^2 + Gw)(s'^2 + 1 + \varepsilon Gws') = \frac{1}{2}\varepsilon s' G^2. \tag{9.216}$$

With the assumption of small ε, two solutions are given by

$$s_1' = \pm i(Gw)^{1/2} + \frac{1}{4}\varepsilon\frac{G^2}{1-Gw} \tag{9.217}$$

$$s_2' = \pm i + \frac{1}{4}\frac{\varepsilon G^2}{Gw-1} - \frac{\varepsilon Gw}{2}. \tag{9.218}$$

The average minimum-B configuration has $w > 0$ and usually $Gw > 1$. The mode corresponding to s_1' is stable while the mode corresponding to s_2' is unstable. This latter mode corresponds to an acoustic wave of wavelength L. The growth rate γ is proportional to η and

$$\gamma \approx \frac{2\pi c_s}{L}\frac{\varepsilon G}{w} \approx \frac{1}{\beta_c \tau_{c1}} \tag{9.219}$$

$$\tau_{cl} = \frac{\mu_0}{\beta \eta k^2}.$$

(The classical diffusion coefficient is $D_{cl} = \rho_e{}^2 \nu_{ei} = \beta_e \eta / \mu_0$.)

When shear exists, the mode $\phi = \exp(st) \times \exp ik(y - (2\pi x/L_s)z)\phi(z)$ propagates a distance c_s/γ during the time interval $1/\gamma$. If $c_s/(L_s\gamma) > 1$, phase mixing occurs, so that

$$\frac{\beta_c \tau_{c1} c_s}{L_s} > 1 \tag{9.220}$$

is the shear-stabilizing condition for the resistive ballooning mode.

9.7 Resistive Drift Instability[23-25]

A finite density gradient always exists at a plasma boundary. Configurations including a gradient may be unstable under certain conditions. Let us consider a plane model. The direction of the uniform magnetic field is taken in the z direction and $\boldsymbol{B}_0 = (0, 0, B_0)$. The x axis is taken in the direction of the density gradient with the positive direction outward from the plasma, so that the pressure is $p_0 = p_0(x)$. The zeroth-order plasma current $\boldsymbol{j}_0$ is $(0, p_0'/B_0, 0)$, and we assume that the flow velocity and the electric field are zero ($\boldsymbol{V}_0 = 0$, $\boldsymbol{E}_0 = 0$) in the zeroth order. The flow velocity due to classical diffusion is neglected here. Electron inertia and ion motion along magnetic-force lines are also neglected.

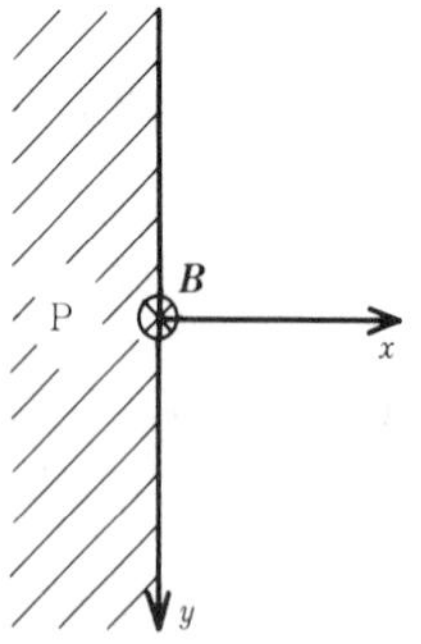

Fig. 9.16
Plane model of resistive drift wave.

The usual relations in this configuration are

$$Mn\frac{\partial \boldsymbol{V}}{\partial t} = \boldsymbol{j} \times \boldsymbol{B} - \nabla p \tag{9.221}$$

$$\boldsymbol{E} + \boldsymbol{V} \times \boldsymbol{B} = \eta \boldsymbol{j} + \frac{1}{en}(\boldsymbol{j} \times \boldsymbol{B} - \nabla p_e) \tag{9.222}$$

$$\frac{\partial n}{\partial t} + \nabla(n\boldsymbol{V}) = 0 \tag{9.223}$$

$$\nabla \boldsymbol{j} = 0, \tag{9.224}$$

where M is the ion mass. In this configuration, electrostatic perturbations can be considered. The 1st-order electric field $\boldsymbol{E}_1$ is expressed by the electrostatic potential ($\boldsymbol{E}_1 = -\nabla\phi_1$) and the 1st-order magnetic field is zero ($\partial \boldsymbol{B}_1/\partial t = -\nabla \times \boldsymbol{E} = 0$). The characteristics of electrostatic perturbation will be explained in ch. 10 in detail. For simplicity the ion temperature is assumed to be zero ($T_i = 0$). Let us consider the mode

$$n_1 = n_1(x)\exp i(ky + k_{\parallel} z - \omega t)$$

$$\phi_1 = \phi_1(x)\exp i(ky + k_{\parallel} z - \omega t).$$

n_1 is the 1st-order density perturbation. From eqs. (9.221) and (9.222), it follows that

$$-i\omega M n_0 \boldsymbol{V}_1 = \boldsymbol{j}_1 \times \boldsymbol{B}_0 - T_e \nabla n_1 \tag{9.225}$$

$$\boldsymbol{j}_1 \times \boldsymbol{B}_0 - T_e \nabla n_1 = e n_0(-\nabla\phi_1 + \boldsymbol{V}_1 \times \boldsymbol{B}_0 - \eta \boldsymbol{j}_1) \tag{9.226a}$$

$$i\omega\left(\frac{M}{e}\right)\boldsymbol{V}_1 = \nabla\phi_1 - \boldsymbol{V}_1 \times \boldsymbol{B}_0 + \eta \boldsymbol{j}_1. \tag{9.226b}$$

When η is small ($\nu_{ei} \ll \Omega_e$), the contribution of $\eta \boldsymbol{j}$ can be neglected in eq. (9.226b), i.e., we may write

$$V_x = -ik\frac{\phi_1}{B_0}$$

$$V_y = \left(\frac{\omega}{\Omega_i}\right)\frac{k\phi_1}{B_0}$$

$$V_z = \left(-\frac{\Omega_i}{\omega}\right)\frac{k_{\parallel}\phi_1}{B_0}.$$

Ω_i is, as usual, the ion Larmor frequency ($\Omega_i = -eB/M$). The wave frequency ω is assumed to be low $(\omega/\Omega_i)^2 \ll 1$. The x, y components of eq. (9.225) and the z component of eq. (9.226a) are

$$j_x = -ik\frac{T_e n_1}{B_0}$$

$$j_y = \frac{\partial n_1}{\partial x}\frac{T_e}{B_0} + kn_0\left(\frac{\omega}{\Omega_i}\right)\frac{e\phi_1}{B_0}$$

$$j_z = \frac{ik_\parallel}{e\eta}\left(T_e\frac{n_1}{n_0} - e\phi_1\right).$$

Since eq. (9.224) is $j_x' + ikjy + ik_\parallel j_z = 0$ and eq. (9.223) is $-i\omega n_1 + n_0 ikV_y + n_0 ik_\parallel V_z = 0$, we find

$$\frac{k_\parallel{}^2 T_e}{e\eta}\frac{n_1}{n_0} + \left(-\frac{k_\parallel{}^2 B}{\eta} - ik^2 en_0\frac{\omega}{\Omega_i}\right)\frac{\phi_1}{B_0} = 0 \tag{9.227}$$

$$\frac{n_1}{n_0} + \left(\frac{-k^2}{\Omega_i} + \frac{k_\parallel{}^2\Omega_i}{\omega^2} + \frac{n_0'}{n_0}\frac{k}{\omega}\right)\frac{\phi_1}{B_0} = 0. \tag{9.228}$$

The dispersion equation is given by the determinant of the coefficients of eqs. (9.227) and (9.228):

$$\left(\frac{\omega}{\Omega_i}\right)^2 - i\left(\frac{\omega}{\Omega_i}\right)\left(\frac{k_\parallel}{k}\right)^2\frac{B_0}{n_0 e\eta}\left(1 - \frac{T_e k^2}{eB_0\Omega_i} - \frac{T_e}{M}\frac{k_\parallel{}^2}{\omega^2}\right)$$
$$- i\left(\frac{k_\parallel}{k}\right)^2\frac{B_0}{n_0 e\eta}\frac{T_e}{eB_0}\frac{k}{\Omega_i}\frac{n_0'}{n_0} = 0, \tag{9.229}$$

where $\eta = m_e\nu_{ei}/ne^2$, $B_0/(n_0 e\eta) = \Omega_e/\nu_{ei}$. *The drift velocities* v_{di}, v_{de} of ions and electrons due to the density gradient are given by

$$v_{di} = \frac{-(T_i\nabla n_0/n_0)\times b}{eB_0} = \frac{T_i}{eB_0}\frac{n_0'}{n_0}e_y$$

$$v_{de} = \frac{(T_e\nabla n_0/n_0)\times b}{eB_0} = \frac{T_e}{eB_0}\left(\frac{-n_0'}{n_0}\right)e_y.$$

The *drift frequencies* of ions and electrons are defined by $\omega_e{}^* \equiv kv_{de}$ and $\omega_i{}^* = kv_{de}$. As $-n_0'/n_0 > 0$, then $\omega_e{}^* > 0$ and $\omega_i{}^* = -(T_i/T_e)\omega_e{}^* < 0$. Since $\omega_e{}^*$ is also expressed by $\omega_e{}^* = k(-n_0'/n_0)(T_e/m\Omega_e)$, the dispersion equation is reduced to

$$\left(\frac{\omega}{\omega_e{}^*}\right)^2 - i\left(1 + (k\rho_B)^2 - \frac{T_e}{M}\frac{k_{\parallel}{}^2}{\omega^2}\right)\frac{\Omega_e\Omega_i}{\nu_{ei}\omega_e{}^*}\left(\frac{k_{\parallel}}{k}\right)^2\left(\frac{\omega}{\omega_e{}^*}\right) + i\frac{\Omega_e\Omega_i}{\nu_{ei}\omega_e{}^*}\left(\frac{k_{\parallel}}{k}\right)^2 = 0. \tag{9.230}$$

ρ_B is the ion Larmor radius when the ions are assumed to have the electron temperature T_e. Denote $\omega/\omega_e{}^* = x + iz$ and $-(\Omega_e\Omega_i/\nu_{ei}\omega_e{}^*)(k_{\parallel}/k)^2 = y^2$ and assume $|(k\rho_B)^2 - (T_e/M)(k_{\parallel}{}^2/\omega^2)| \ll 1$. The dispersion equation is then

$$(x + iz)^2 + iy^2(x + iz) - iy^2 = 0. \tag{9.231}$$

The dependence of the two solutions $x_1(y)$, $z_1(y)$ and $x_2(y)$, $z_2(y)$ on $y \propto (k_{\parallel}/k)$ is shown in fig. 9.17. As $z_2(y) < 0$, the mode corresponding to x_2, z_2 is stable. This wave propagates in the direction of the ion drift. The solution x_1, $z_1 > 0$ propagates in the direction of the electron drift and it is unstable. If the value of $(k_{\parallel}/k)$ is adjusted to be $y \approx 1.3$, the value of z_1 becomes maximum $z_1 \approx 0.25$ and the growth rate is $\mathrm{Im}\,\omega \approx 0.25\omega_e{}^*$. If η is small, the wavelength of the most unstable wave becomes long and the necessary number of collision to interrupt the electron motion along the line of magnetic field is maintained. If the lower limit of $k_{\parallel}$ is fixed by an appropriate method, the growth rate is

$$\mathrm{Im}(\omega/\omega_e{}^*) \approx y^{-2} = \frac{\nu_{ei}\omega_e{}^*}{\Omega_e|\Omega_i|}\left(\frac{k}{k_{\parallel}}\right)^2$$

and the growth rate is proportional to $\eta \propto \nu_{ei}$. This instability is called

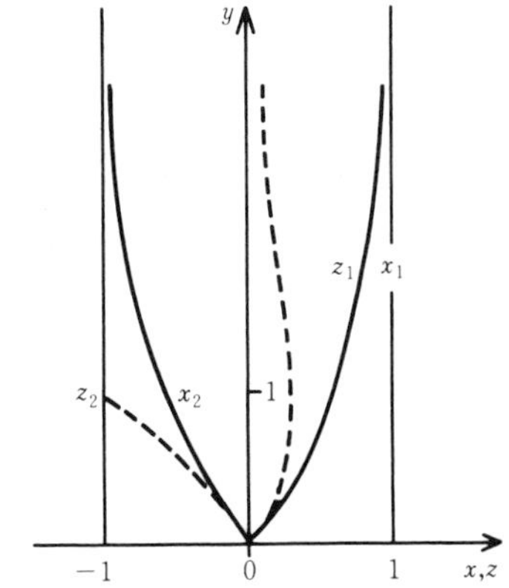

Fig. 9.17
Dependence of $\omega/\omega_e{}^* = x + iz$ on $y \propto k_{\parallel}/k$ for the resistive drift wave.

resistive drift instability or *dissipative drift instability*. This instability originates in the charge separation between electrons and ions due to ion inertia. The charge separation thus induced is neutralized by electron motion along the line of magnetic force. However, if the parallel motion of electrons is interrupted by collision, i.e., resistivity, the charge separation grows and the wave becomes unstable.[24] This instability is therefore also called *collisional drift instability*.

The mechanism of charge separation between ions and electrons can be more easily understood if the equations of motion of the two components, ion and electron, are used. The equations of motion are

$$\left.\begin{aligned} 0 &= -ikn_1 T_e + ik\phi_1 en_0 - en_0(\boldsymbol{V}_e \times \boldsymbol{B}) - n_0 m_e \nu_{ei}(\boldsymbol{V}_e - \boldsymbol{V}_i)_{\|} \\ M\frac{n_0}{Z}(-i\omega)\boldsymbol{V}_i &= -ikn_1\frac{T_i}{Z} - ik\phi_1 en_0 + en_0(\boldsymbol{V}_i \times \boldsymbol{B}) \\ &\qquad + n_0 m_e \nu_{ei}(\boldsymbol{V}_e - \boldsymbol{V}_i)_{\|}. \end{aligned}\right\} \tag{9.232}$$

From these equations we find

$$\boldsymbol{V}_{e\perp} = \frac{-iT_e}{eB}(\boldsymbol{b} \times \boldsymbol{k})\left(\frac{n_1}{n_0} - \frac{e\phi_1}{T_e}\right) \tag{9.233}$$

$$\boldsymbol{V}_{e\|} = \frac{-ik_{\|} T_e}{m_e \nu_{ei}}\left(\frac{n_1}{n_0} - \frac{e\phi_1}{T_e}\right) \tag{9.234}$$

$$\boldsymbol{V}_{i\perp} = \frac{iT_i}{ZeB}(\boldsymbol{b} \times \boldsymbol{k})\left(\frac{n_1}{n_0} + \frac{Ze\phi_1}{T_i}\right) + \left(\frac{-M}{BZe}\right)(i\omega)(\boldsymbol{b} \times \boldsymbol{V}_{i\perp}) \tag{9.235}$$

$$\boldsymbol{V}_{i\|} = \frac{k_{\|} Z c_s^2}{\omega}\frac{n_1}{n_0}, \tag{9.236}$$

where it has been assumed that $|\boldsymbol{V}_{e\|}| \gg |\boldsymbol{V}_{i\|}|$ and where $c_s \equiv T_e/M$ has been used. The continuity equations $\partial n/\partial t + \nabla(n\boldsymbol{V}) = 0$ for ions and electrons yield

$$-i\omega n_1 + \nabla\left(\frac{\boldsymbol{b} \times i\boldsymbol{k}}{B} n_0 \phi_1\right) + (\boldsymbol{b}\cdot\nabla)(n_0 \boldsymbol{V}_{e\|}) = 0$$

$$-i\omega n_1 + \nabla\left(\frac{\boldsymbol{b} \times i\boldsymbol{k}}{B} n_0 \phi_1\right) + \frac{M(-i\omega)k^2 T_i}{ZeB \quad ZeB}\left(\frac{n_1}{n_0} + \frac{Ze\phi_1}{T_i}\right)n_0$$

$$+ \frac{ik_{\|}^2 c_s^2}{\omega} Z n_1 = 0.$$

From the equations for electrons, it follows that

$$\frac{n_1}{n_0} = \frac{\omega_e^* + ik_{\parallel}^2 T_e/(m_e \nu_{ei})}{\omega + ik_{\parallel}^2 T_e/(m_e \nu_{ei})} \left(\frac{e\phi_1}{T_e} \right); \tag{9.237}$$

and from the equation for ions, it follows that

$$\frac{n_1}{n_0} = \frac{\omega_i^* + b\omega}{\omega(1+b) - k_{\parallel}^2 c_s^2 Z/\omega} \left(\frac{-Ze\phi_1}{T_i} \right), \tag{9.238}$$

where $b = k^2(\rho_B)^2$. Charge neutrality gives the dispersion equation, which is equivalent to eq. (9.230):

$$\frac{\omega_e^* + ik_{\parallel}^2 T_e/(m_e \nu_{ei})}{\omega + ik_{\parallel}^2 T_e/(m_e \nu_{ei})} = \frac{\omega_e^* - (ZT_e/T_i)b\omega}{\omega(1+b) - k_{\parallel}^2 c_s^2 Z/\omega}. \tag{9.239}$$

The electron motion perpendicular to the magnetic field is in the x direction; but the ion motion has a y component (the 2nd term of eq. (9.235)), due to the ion inertia, in addition to its x component.

When the plasma is collisionless, eq. (9.230) is reduced to[26]

$$(1 + (k\rho_B)^2)\omega^2 - \omega_e^* \omega - c_s^2 k_{\parallel}^2 = 0. \tag{9.240}$$

The dispersion relation ($k_{\parallel}$ vs. ω) is shown in fig. 9.18. The instability does not appear in the collisionless case in the MHD approximation. However, the instability occurs even in a collisionless plasma analyzed from the standpoint of kinetic theory. The collisionless drift wave will be further discussed in ch. 12.

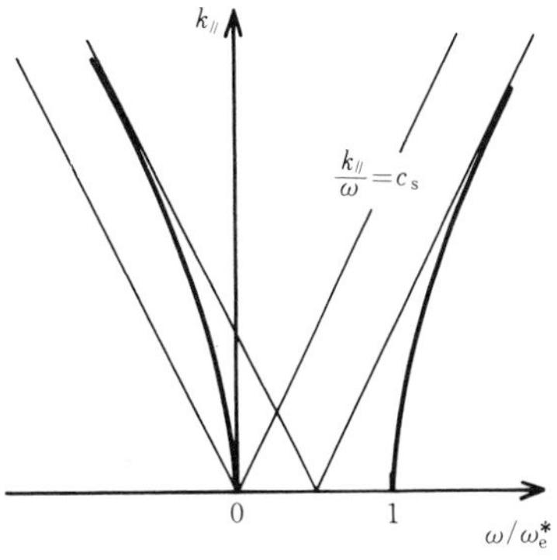

Fig. 9.18
Dispersion relation for the drift wave in the collisionless case (MHD treatment).

References

1. A. A. Vedenov, E. P. Velikhov, and P. Z. Sagdeev: Sov. Phys. Uspekhi **4**, 332(1961)
J. D. Jukes: Rep. Prog. Phys. **30**, 333(1967)
M. Sato: IPPJ-7 (March 1963), Inst. of Plasma Physics Nagoya University
A. B. Mikhailovskii: Theory of Plasma Instabilities **1**, **2**, Consultants Bureau, New York, 1974

2. B. Lehnert: Plasma Phys. **9**, 301(1967)

3. M. Kruskal and M. Schwarzschild: Proc. Roy. Soc. A**223**, 348(1954)

4. F. C. Hoh: Phys. of Fluids **6**, 1184(1963)

5. M. N. Rosenbluth, N. A. Krall, and N. Rostoker: Nucl. Fusion, suppl., pt. 1, p. 143 (1962)

6. M. N. Rosenbluth and C. L. Longmire: Annals of Phys. **1**, 120(1957)

7. I. B. Bernstein, E. A. Frieman, M. D. Kruskal, and R. M. Kulsrud: Proc. Roy. Soc. A**244**, 17(1958)

8. B. B. Kadomtsev: Rev. of Plasma Phys. **2**, 153 (ed. by M. A. Leontovich), Consultant Bureau. New York, 1966

9. M. D. Kruskal, J. L. Johnson, M. B. Gottlieb, and L. M. Goldman: Phys. of Fluids **1**, 421(1958)

10. V. D. Shafranov: Sov. Phys. JETP **6**, 545(1958); Zh. Exp. Teor. Fyz. **33**, 710(1957)

11. B. R. Suydam: Proc. 2nd UN International Conf. on Peaceful Uses of Atomic Energy **31**, 157 (Geneva, 1958)

12. W. A. Newcomb: Annals of Phys. **10**, 232(1960)

13. V. D. Shafranov: Sov. Phys. Tech. Phys. **15**, 175(1970)
T. Amano and M. Wakatani: J. Phys. Soc. Japan **33**, 782(1972)

14. D. C. Robinson: Plasma Phys. **13**, 439(1971)

15. L. S. Solovev: Sov. Phys. JETP **26**, 1167(1968)
J. M. Green and J. L. Johnson: Phys. of Fluids **5**, 510(1962)
C. Mercier: Plasma Phys. and Controlled Nucl. Fusion Research (Conf. Proceedings, Salzburg 1961); Nucl. Fusion, suppl., pt. 2, p. 801(1962)

16. B. B. Kadomtsev and O. P. Pogutse: Sov. Phys. Dokl. **11**, 858(1967)

17. V. D. Shafranov and E. I. Yurchenko: Sov. Phys. JETP **26**, 682(1968)

18. H. P. Furth, J. Killeen, and M. N. Rosenblutch: Phys. of Fluids **6**, 459(1963)

19. H. P. Furth: Advances in Plasma Physics vol. 1, p. 69 (ed. by A. Simon and W. B. Thompson), Interscience, New York, 1968

20. R. M. Kulsrud: Plasma Phys. and Controlled Nucl. Fusion Research **1**, 127 (Conf. Proceedings, Culham 1965) (IAEA, Vienna, 1966)

21. B. Coppi and M. N. Rosenbluth: ibid., 617

22. H. P. Furth, J. Killeen, M. N. Rosenbluth, and B. Coppi: ibid., 103

23. S. S. Moiseev and R. Z. Sagdeev: Sov. Phys. JETP **17**, 515(1963), Sov. Phys. Tech. Phys. **9**, 196(1964)

24. F. F. Chen: Phys. of Fluids **8**, 912 and 1323(1965)

25. H. W. Hendel and T. K. Chu: Methods of Experimental Physics vol. 9, part A, p. 345 (ed. by H. R. Griem and R. H. Loveberg), Academic Press, New York, 1970

26. B. B. Kadomtsev: Plasma Turbulence, Academic Press, New York, 1965

III KINETIC DESCRIPTIONS OF WAVES AND INSTABILITIES

10 Waves in a Cold Plasma

A plasma is an ensemble of an enormous number of moving ions and electrons interacting with each other. In order to describe the behavior of such an ensemble, the distribution function was introduced in ch. 5; and Boltzmann's and Vlasov's equations were obtained with respect to the distribution function. A plasma viewed as an ensemble of a large number of particles has a large number of degrees of freedom; thus the mathematical description of plasma behavior is feasible only for simplified analytical models.

In ch. 6, statistical averages in velocity space, such as mass density, flow velocity, pressure, etc., were introduced and the magnetohydrodynamic equations for these averages were derived. We have thus obtained a mathematical description of the magnetohydrodynamic fluid model; and we have studied the equilibrium conditions, stability problems, etc., for this model in chs. 7–9. Since the fluid model considers only average quantities in velocity space, it is not capable of describing instabilities or damping phenomena, in which the profile of the distribution function plays a significant role. The phenomena which can be handled by means of the fluid model are of low frequency (less than the ion or electron cyclotron frequency); high-frequency phenomena are not describable in terms of it.

In this chapter, we will focus on a model which allows us to study wave phenomena while retaining the essential features of plasma dynamics, at the same time maintaining relative simplicity in its mathematical form. Such a model is given by a homogeneous plasma of ions and electrons at 0°K in a uniform magnetic field. In the unperturbed state, both the ions and electrons of this plasma are motionless. Any small deviation from the unperturbed state induces an electric field and a time-dependent component of the magnetic field, and consequently movements of ions and electrons are excited. The movements of the charged particles induce electric and magnetic fields which are themselves consistent with the previously induced small perturbations. This is called the kinetic model of a cold plasma. We will use it in this chapter to derive the dispersion relation which characteristizes wave phenomena in the cold plasma.

In sec. 10.7, the interactions of particles and waves are discussed. These interactions can only occur in a hot plasma, i.e., one with a nonzero temperature distribution in velocity space. One of the most important examples of such interaction is Landau damping.

Waves and the instabilities for plasmas of nonzero temperature will be discussed in chs. 11–13 using generalized mathematical analysis applied to the kinetic model of a hot plasma. References 1–4 give detailed descriptions of wave phenomena in plasmas.

10.1 Dispersion Equation of Waves in a Cold Plasma

In an unperturbed cold plasma, the particle density n and the magnetic field $\boldsymbol{B}_0$ are both homogeneous in space and constant in time. The ions and electrons are motionless.

Now assume that the first-order perturbation term $\exp i(\boldsymbol{k}\cdot\boldsymbol{r} - \omega t)$ is applied. The ions and electrons are forced to move by the perturbed electric field $\boldsymbol{E}$ and the induced magnetic field $\boldsymbol{B}_1$. Let us denote velocity by $\boldsymbol{v}_k$, where k indicates the species of particle (electrons, or ions of various kinds). The current $\boldsymbol{j}$ due to the particle motion is given by

$$\boldsymbol{j} = \sum_k n_k q_k \boldsymbol{v}_k. \tag{10.1}$$

n_k and q_k are the density and charge of the kth species, respectively. The electric displacement $\boldsymbol{D}$ is

$$\boldsymbol{D} = \varepsilon_0 \boldsymbol{E} + \boldsymbol{P} \tag{10.2}$$

$$\boldsymbol{j} = \frac{\partial \boldsymbol{P}}{\partial t} = -i\omega \boldsymbol{P}, \tag{10.3}$$

where $\boldsymbol{E}$ is the electric intensity, $\boldsymbol{P}$ is the electric polarization, and ε_0 is the dielectric constant of vacuum. Consequently $\boldsymbol{D}$ is expressed by

$$\boldsymbol{D} = \varepsilon_0 \boldsymbol{E} + \frac{i}{\omega}\boldsymbol{j} \equiv \varepsilon_0 \boldsymbol{K}\boldsymbol{E}. \tag{10.4}$$

$\boldsymbol{K}$ is called the *dielectric tensor*. The equation of motion of a single particle of the kth kind is

$$m_k \frac{\mathrm{d}\boldsymbol{v}_k}{\mathrm{d}t} = q_k(\boldsymbol{E} + \boldsymbol{v}_k \times \boldsymbol{B}). \tag{10.5}$$

Here $\boldsymbol{B}$ consists of $\boldsymbol{B}_0 + \boldsymbol{B}_1$, where $\boldsymbol{v}_k$, $\boldsymbol{E}$, $\boldsymbol{B}_1$ are first-order quantities. The linearized equation in these quantities is

$$-i\omega m_k \boldsymbol{v}_k = q_k(\boldsymbol{E} + \boldsymbol{v}_k \times \boldsymbol{B}_0). \tag{10.6}$$

When the z axis is taken along the direction of $\boldsymbol{B}_0$, the solution is given by

$$\left.\begin{aligned} v_{k,x} &= \frac{-iE_x}{B_0}\frac{\Omega_k\omega}{(\omega^2-\Omega_k^2)} - \frac{E_y}{B_0}\frac{\Omega_k^2}{(\omega^2-\Omega_k^2)} \\ v_{k,y} &= \frac{E_x}{B_0}\frac{\Omega_k^2}{\omega^2-\Omega_k^2} - \frac{iE_y}{B_0}\frac{\Omega_k\omega}{(\omega^2-\Omega_k^2)} \\ v_{k,z} &= \frac{-iE_z}{B_0}\Omega_k\frac{1}{\omega}, \end{aligned}\right\} \tag{10.7}$$

where Ω_k is the cyclotron frequency of the charged particle of the kth kind:

$$\Omega_k = \frac{-q_k B_0}{m_k} \tag{10.8}$$

($\Omega_e > 0$ for electrons and $\Omega_i < 0$ for ions). The components of $\boldsymbol{v}_k$ are the linear functions of $\boldsymbol{E}$ given by eq. (10.7); and the dielectric tensor $\boldsymbol{K}$ is obtained from eqs. (10.4) and (10.1):

$$\boldsymbol{KE} = \begin{pmatrix} K_\perp & -iK_\times & 0 \\ iK_\times & K_\perp & 0 \\ 0 & 0 & K_\parallel \end{pmatrix}\begin{pmatrix} E_x \\ E_y \\ E_z \end{pmatrix}, \tag{10.9}$$

where

$$K_\perp \equiv 1 - \sum_k \frac{\Pi_k^2}{\omega^2-\Omega_k^2} \tag{10.10}$$

$$K_\times \equiv -\sum_k \frac{\Pi_k^2}{(\omega^2-\Omega_k^2)}\left(\frac{\Omega_k}{\omega}\right) \tag{10.11}$$

$$K_\parallel \equiv 1 - \sum_k \frac{\Pi_k^2}{\omega^2}, \tag{10.12}$$

with

$$\Pi_k^2 \equiv \frac{n_k q_k^2}{\varepsilon_0 m_k}. \tag{10.13}$$

For convenience the following quantities are introduced:[1]

$$\left.\begin{aligned} R &\equiv 1 - \sum_k \frac{\Pi_k^2}{\omega^2}\left(\frac{\omega}{\omega-\Omega_k}\right) = K_\perp + K_\times \\ L &\equiv 1 - \sum_k \frac{\Pi_k^2}{\omega^2}\left(\frac{\omega}{\omega+\Omega_k}\right) = K_\perp - K_\times \end{aligned}\right\} \tag{10.14}$$

From Maxwell's equations,

$$\left.\begin{aligned} \nabla \times \boldsymbol{E} &= -\frac{\partial \boldsymbol{B}}{\partial t} \\ \nabla \times \boldsymbol{H} &= \boldsymbol{j} + \varepsilon_0 \frac{\partial \boldsymbol{E}}{\partial t} = \frac{\partial \boldsymbol{D}}{\partial t}, \end{aligned}\right\} \tag{10.15}$$

it follows that

$$\boldsymbol{k} \times \boldsymbol{E} = \omega \boldsymbol{B}_1$$
$$\boldsymbol{k} \times \boldsymbol{H}_1 = -\omega\varepsilon_0 \boldsymbol{K}\boldsymbol{E}$$

and

$$\boldsymbol{k} \times (\boldsymbol{k} \times \boldsymbol{E}) + \frac{\omega^2}{c^2} \boldsymbol{K}\boldsymbol{E} = 0. \tag{10.16}$$

Let us define a dimensionless vector

$$\boldsymbol{N} \equiv \frac{\boldsymbol{k}c}{\omega} \tag{10.17}$$

(c is the velocity of light). The absolute value $N = |\boldsymbol{N}|$ is the ratio of the velocity of light to the phase velocity of the wave, i.e., N is the refractive index. Using $\boldsymbol{N}$, we may write eq. (1.16) as

$$\boldsymbol{N} \times (\boldsymbol{N} \times \boldsymbol{E}) + \boldsymbol{K}\boldsymbol{E} = 0. \tag{10.18}$$

If the angle between $\boldsymbol{N}$ and $\boldsymbol{B}_0$ is denoted by θ (see fig. 10.1) and the y axis is taken so that $\boldsymbol{N}$ lies in the y, z plane, then eq. (10.18) may be expressed by

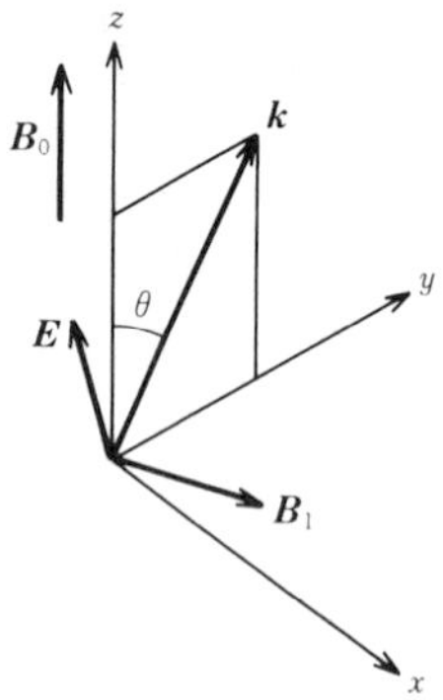

Fig. 10.1
Propagation vector $\boldsymbol{k}$ and x, y, z coordinates.

$$\begin{pmatrix} K_\perp - N^2 & -iK_x & 0 \\ iK_x & K_\perp - N^2\cos^2\theta & N^2\sin\theta\cos\theta \\ 0 & N^2\sin\theta\cos\theta & K_\parallel - N^2\sin^2\theta \end{pmatrix}\begin{pmatrix} E_x \\ E_y \\ E_z \end{pmatrix} = 0. \tag{10.19}$$

For a nontrivial solution to exist, the determinant of the matrix must be zero, or

$$AN^4 - BN^2 + C = 0, \tag{10.20}$$

where

$$A = K_\perp \sin^2\theta + K_\parallel \cos^2\theta \tag{10.21}$$
$$B = (K_\perp{}^2 - K_x{}^2)\sin^2\theta + K_\parallel K_\perp(1 + \cos^2\theta) \tag{10.22}$$
$$C = K_\parallel(K_\perp{}^2 - K_x{}^2) = K_\parallel RL. \tag{10.23}$$

Equation (10.20) determines the relationship between the propagation vector $\boldsymbol{k}$ and the frequency ω, and it is thus the dispersion equation. The solution of (10.20) is

$$N^2 = \frac{B \pm (B^2 - 4AC)^{1/2}}{2A}$$
$$= \frac{(K_\perp{}^2 - K_x{}^2)\sin^2\theta + K_\parallel K_\perp(1+\cos^2\theta) \pm [(K_\perp{}^2 - K_x{}^2 - K_\parallel K_\perp)^2\sin^4\theta + 4K_\parallel{}^2K_x{}^2\cos^2\theta]^{1/2}}{2(K_\perp\sin^2\theta + K_\parallel\cos^2\theta)}. \tag{10.24}$$

When the wave propagates along the line of magnetic force, i.e., when $\theta = 0$, the dispersion equation (10.20) is

$$K_\parallel(N^4 - 2K_\perp N^2 + (K_\perp{}^2 - K_x{}^2)) = 0 \tag{10.25}$$

and the solutions are

$$K_\parallel = 0 \qquad N^2 = K_\perp + K_x = R \qquad N^2 = K_\perp - K_x = L. \tag{10.26}$$

For the wave propagating in the direction perpendicular to the magnetic field ($\theta = \pi/2$), the dispersion equation and the solutions are given by

$$K_\perp N^4 - (K_\perp{}^2 - K_x{}^2 + K_\parallel K_\perp)N^2 + K_\parallel(K_\perp{}^2 - K_x{}^2) = 0 \tag{10.27}$$

$$N^2 = \frac{K_\perp{}^2 - K_x{}^2}{K_\perp} = \frac{RL}{K_\perp} \qquad N^2 = K_\parallel. \tag{10.28}$$

10.2 Properties of the Waves

10.2a Polarization and Particle Movement

The dispersion relation for waves in a cold plasma was derived in the previous section. We consider here the electric field of the waves and the resultant particle motion. The x component of eq. (10.19) is

$$(K_\perp - N^2)E_x - iK_x E_y = 0,$$

or

$$\frac{iE_x}{E_y} = \frac{K_x}{N^2 - K_\perp}. \tag{10.29}$$

The relation between the components of the particle velocity is

$$\frac{iv_{k,x}}{v_{k,y}} = \frac{i\left(\dfrac{-iE_x}{E_y}\dfrac{\omega}{\omega^2 - \Omega_k^2} - \dfrac{\Omega_k}{\omega^2 - \Omega_k^2}\right)}{\left(\dfrac{E_x}{E_y}\dfrac{\Omega_k}{\omega^2 - \Omega_k^2} - i\dfrac{\omega}{\omega^2 - \Omega_k^2}\right)}$$

$$= \frac{(\omega + \Omega_k)(N^2 - L) - (\omega - \Omega_k)(N^2 - R)}{(\omega + \Omega_k)(N^2 - L) + (\omega - \Omega_k)(N^2 - R)}. \tag{10.30}$$

The wave satisfying $N^2 = R$ at $\theta = 0$ has $iE_x/E_y = 1$ and the electric field is right-circularly polarized. In other word, the electric field rotates in the direction of the electron Larmor motion. The motion of ions and electrons is also right-circular motion. In the wave satisfying $N^2 = L$ at $\theta = 0$, the relation $iE_x/E_y = -1$ holds and the electric field is left-circularly polarized. The motion of ions and electrons is also left-circular motion. The waves with $N^2 \to R$ and $N^2 \to L$ as $\theta \to 0$, are called the R *wave* and the L *wave*, respectively. The solution of the dispersion equation (10.24) at $\theta = 0$ is

$$N^2 = \frac{1}{2}\left(R + L \pm \frac{|K_\parallel|}{K_\parallel}|R - L|\right), \tag{10.31}$$

so that R and L waves are exchanged when $K_\parallel$ changes sign. When $R - L = K_\times$ changes sign, i.e., when $R = \infty$ or $L = \infty$, R and L waves are also exchanged.

When $\theta = \pi/2$, the electric field of the wave satisfying $N^2 = K_\parallel$ is $E_x = E_y = 0$, $E_z \neq 0$. For the wave satisfying $N^2 = RL/K_\perp$, the electric field satisfies the relations $iE_x/E_y = -(R+L)/(R-L) = -K_\perp/K_\times$, $E_x = 0$. The waves with $N^2 \to K_\parallel$ and $N^2 \to RL/K_\perp$ as $\theta \to \pi/2$ are called the *ordinary wave* (O) and the *extraordinary wave* (X), respectively. It should be pointed out that the electric field of the extraordinary wave at $\theta = \pi/2$ is perpendicular to the magnetic field and the electric field of the ordinary wave at $\theta = \pi/2$ is parallel to the magnetic field. The dispersion relation (10.24) at $\theta = \pi/2$ is

$$\begin{aligned} N^2 &= \frac{1}{2K_\perp}(K_\perp{}^2 - K_\times{}^2 + K_\parallel K_\perp + |K_\perp{}^2 - K_\times{}^2 - K_\parallel K_\perp|) \\ &= \frac{1}{2K_\perp}(RL + K_\parallel K_\perp \pm |RL - K_\parallel K_\perp|), \end{aligned} \tag{10.32}$$

so that the ordinary wave and the extraordinary wave are exchanged at $RL - K_\parallel K_\perp = 0$.

Besides the classification into R and L and O and X waves, there is another classification, namely, that of *fast wave* and *slow wave*, following the difference in phase velocity at $\theta = 0$ and $\theta = \pi/2$. Since the term inside the square root of the equation $N^2 = (B \pm (B^2 - 4AC)^{1/2})/2A$ is always positive, as is clear from eq. (10.24), the fast wave and slow wave do not exchange between $\theta = 0$ and $\theta = \pi/2$.

10.2b Cutoff and Resonance

The refractive index (10.24), the solution of the dispersion equation, may become infinity or zero. When $N^2 = 0$, the wave is said to be at *cutoff*; at cutoff the phase velocity

$$v_{\mathrm{ph}} = \frac{\omega}{k} = \frac{c}{N} \tag{10.33}$$

becomes infinite. As is clear from eqs. (10.20) and (10.23), cutoff occurs when

$$K_\parallel = 0 \qquad R = 0 \qquad L = 0. \tag{10.34}$$

When $N^2 = \infty$, the wave is said to be at *resonance*; here the phase velocity becomes zero. The wave will be absorbed by the plasma at resonance. The resonance condition is

$$\tan^2\theta = -\frac{K_\parallel}{K_\perp}. \tag{10.35}$$

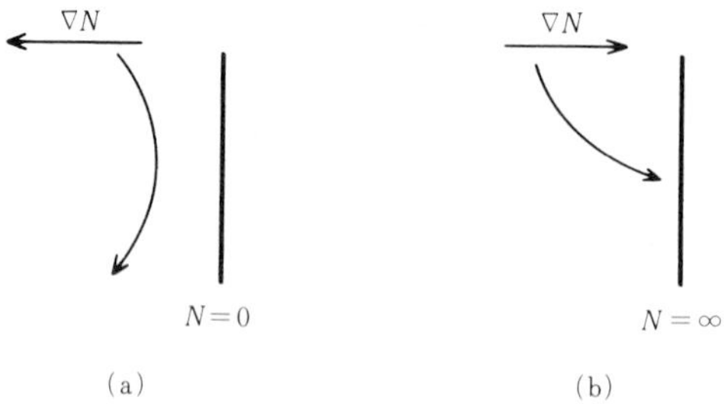

Fig. 10.2
Wave propagation (a) near a cutoff region and (b) near a resonance region.

When $\theta = 0$, the resonance condition is $K_\perp = (R + L)/2 \to \pm\infty$. The condition $R \to \pm\infty$ is satisfied at $\omega = \Omega_e$, Ω_e being the electron cyclotron frequency. This is called *electron cyclotron resonance.* The condition $L \to \pm\infty$ holds when $\omega = |\Omega_i|$, and this is called *ion cyclotron resonance.*

When $\theta = \pi/2$, $K_\perp = 0$ is the resonance condition. This is called *hybrid resonance.* When waves approach a cutoff region, the wave path is curved as is shown in fig. 10.2a and the waves are reflected. When waves approach a resonance region, the waves propagate perpendicularly toward the resonance region. The phase velocities tend to zero and the wave energy will be absorbed.

10.3 Waves in a Two-Component Plasma

Let us consider a plasma which consists of electrons and of one kind of ion. Charge neutrality is

$$n_i Z_i = n_e. \tag{10.36}$$

A dimensionless parameter is introduced for convenience:

$$\delta = \frac{\mu_0 (n_i m_i + n_e m_e) c^2}{B_0{}^2}. \tag{10.37}$$

The quantity defined by eq. (10.13), with respect to the plasma electrons,

$$\Pi_e{}^2 = n_e e^2/(\varepsilon_0 m_e) \tag{10.38}$$

is called the *electron plasma frequency.* We have the relations

$$\Pi_e{}^2/\Pi_i{}^2 = m_i/m_e Z \gg 1$$

$$\frac{\Pi_i^2 + \Pi_e^2}{|\Omega_i|\Omega_e} = \delta \approx \frac{\Pi_i^2}{\Omega_i^2}. \tag{10.39}$$

$K_\perp$, $K_\times$, $K_\parallel$, R, and L are given by

$$K_\perp = 1 - \frac{\Pi_i^2}{\omega^2 - \Omega_i^2} - \frac{\Pi_e^2}{\omega^2 - \Omega_e^2}$$

$$K_\times = -\frac{\Pi_i^2}{(\omega^2 - \Omega_i^2)}\frac{\Omega_i}{\omega} - \frac{\Pi_e^2}{(\omega^2 - \Omega_e^2)}\frac{\Omega_e}{\omega} \tag{10.40}$$

$$K_\parallel = 1 - \frac{(\Pi_e^2 + \Pi_i^2)}{\omega^2} \approx 1 - \frac{\Pi_e^2}{\omega^2}$$

$$R = 1 - \frac{\Pi_e^2 + \Pi_i^2}{(\omega - \Omega_i)(\omega - \Omega_e)} \approx \frac{\omega^2 - (\Omega_i + \Omega_e)\omega + \Omega_i\Omega_e - \Pi_e^2}{(\omega - \Omega_i)(\omega - \Omega_e)} \tag{10.41}$$

$$L = 1 - \frac{\Pi_e^2 + \Pi_i^2}{(\omega + \Omega_i)(\omega + \Omega_e)} \approx \frac{\omega^2 + (\Omega_i + \Omega_e)\omega + \Omega_i\Omega_e - \Pi_e^2}{(\omega + \Omega_i)(\omega + \Omega_e)}. \tag{10.42}$$

The dispersion relations for the waves propagating parallel to $\boldsymbol{B}_0$ are found by setting $K_\parallel = 0$, $N^2 = R$, and $N^2 = L$. Then

$$\omega^2 = \Pi_e^2 \tag{10.43}$$

$$\frac{\omega^2}{c^2 k_\parallel^2} = \frac{1}{R} = \frac{(\omega - \Omega_i)(\omega - \Omega_e)}{\omega^2 - \omega\Omega_e + \Omega_e\Omega_i - \Pi_e^2} = \frac{(\omega + |\Omega_i|)(\omega - \Omega_e)}{(\omega - \omega_R)(\omega + \omega_L)} \tag{10.44}$$

$$\frac{\omega^2}{c^2 k_\parallel^2} = \frac{1}{L} = \frac{(\omega + \Omega_i)(\omega + \Omega_e)}{\omega^2 + \omega\Omega_e + \Omega_e\Omega_i - \Pi_e^2} = \frac{(\omega - |\Omega_i|)(\omega + \Omega_e)}{(\omega - \omega_L)(\omega + \omega_R)}, \tag{10.47}$$

where

$$\omega_R = \frac{\Omega_e}{2} + \left(\left(\frac{\Omega_e}{2}\right)^2 + \Pi_e^2 + |\Omega_e\Omega_i|\right)^{1/2} > 0 \tag{10.45}$$

$$\omega_L = -\frac{\Omega_e}{2} + \left(\left(\frac{\Omega_e}{2}\right)^2 + \Pi_e^2 + |\Omega_e\Omega_i|\right)^{1/2} > 0. \tag{10.46}$$

Note that $\Omega_e > 0$, $\Omega_i < 0$. We have $\omega_R > \Omega_e > \omega_L > |\Omega_i|$ when $2\Omega_e^2 > \Pi_e^2 > \Omega_i^2$. Plots of the dispersion relations ω vs. $ck_\parallel$ are shown in fig. 10.3.

The dispersion relations for waves propagating perpendicular to $\boldsymbol{B}_0$ are found by setting $N^2 = K_\parallel$ (ordinary wave) and $N^2 = (K_\perp^2 - K_\times^2)/K_\perp$ (extraordinary wave). Then

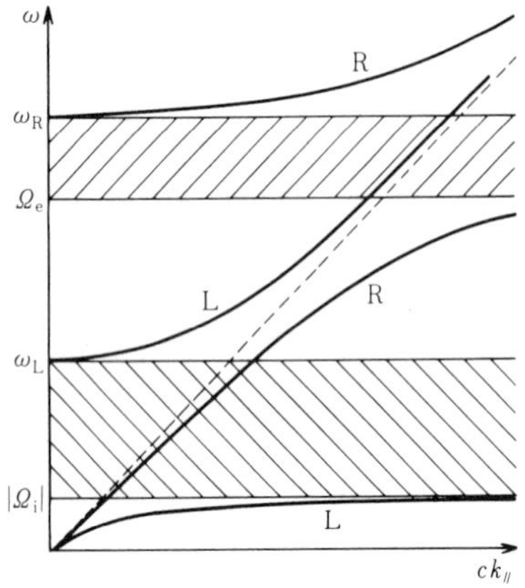

Fig. 10.3
Dispersion relations of R and L waves propagating parallel to the magnetic field (ω vs. $ck_{\parallel}$, $\theta = 0$).

$$\text{O wave:}\quad \frac{\omega^2}{c^2 k_{\perp}^2} = \frac{1}{K_{\parallel}} = \left(1 - \frac{\Pi_e^2}{\omega^2}\right)^{-1} = \left(1 + \frac{\Pi_e^2}{c^2 k_{\perp}^2}\right) \tag{10.48}$$

$$\text{X wave:}\quad \frac{\omega^2}{c^2 k_{\perp}^2} = \frac{K_{\perp}}{K_{\perp}^2 - K_{\times}^2} = \frac{K_{\perp}}{RL}$$

$$= \frac{2(\omega^2 - \Omega_i^2)(\omega^2 - \Omega_e^2) - \Pi_e^2((\omega + \Omega_i)(\omega + \Omega_e) + (\omega - \Omega_i)(\omega - \Omega_e))}{2(\omega^2 - \omega_L^2)(\omega^2 - \omega_R^2)}$$

$$= \frac{\omega^4 - (\Omega_i^2 + \Omega_e^2 + \Pi_e^2)\omega^2 + \Omega_i^2\Omega_e^2 - \Pi_e^2\Omega_i\Omega_e}{(\omega^2 - \omega_L^2)(\omega^2 - \omega_R^2)}. \tag{10.49}$$

Equation (10.48) is the dispersion equation for the *electron plasma wave* (*Langmuir wave*). Let us define ω_{UH} and ω_{LM} by

$$\omega_{UH}^2 \equiv \Omega_e^2 + \Pi_e^2 \tag{10.50}$$

$$\frac{1}{\omega_{LH}^2} \equiv \frac{1}{\Omega_i^2 + \Pi_i^2} + \frac{1}{|\Omega_i|\Omega_e}. \tag{10.51}$$

ω_{UH} is called the *upper hybrid resonant frequency* and ω_{LH} is called the *lower hybrid resonant frequency*. Using these, we may write eq. (10.49) as

$$\frac{\omega^2}{c^2 k_{\perp}^2} = \frac{(\omega^2 - \omega_{LH}^2)(\omega^2 - \omega_{UH}^2)}{(\omega^2 - \omega_L^2)(\omega^2 - \omega_R^2)}. \tag{10.52}$$

We have $\omega_R > \omega_{UH} > \Omega_e$, $\omega_{LH} > |\Omega_i|$, and $\omega_{UH} > \Pi_e$. Plots of the dispersion relation ω vs. $ck_\perp$ are shown in fig. 10.4. The gradient ω/ck in a ω, ck diagram is the ratio of the phase velocity v_{ph} to the velocity of light c. The steeper the gradient, the greater the phase velocity. The regions (in terms of ω) of R and L waves at $\theta = 0$, O and X waves at $\theta = \pi/2$, and F (fast) and S (slow) waves are shown in fig. 10.5, for $\omega_L < \Pi_e < \Omega_e$.

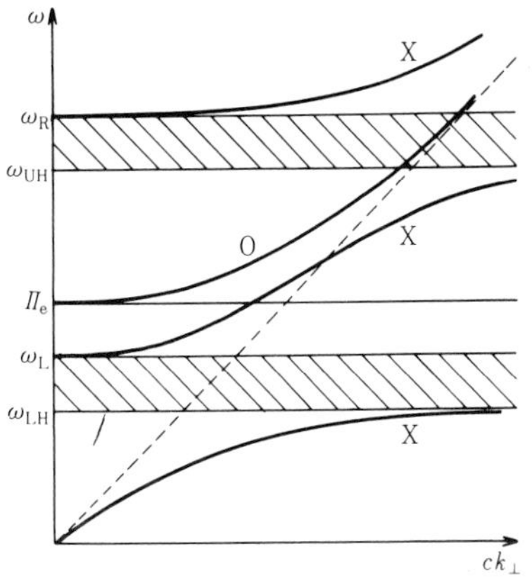

Fig. 10.4
Dispersion relations of O and X waves propagating perpendicular to the magnetic field (ω vs. $ck_\perp$, $\theta = \pi/2$).

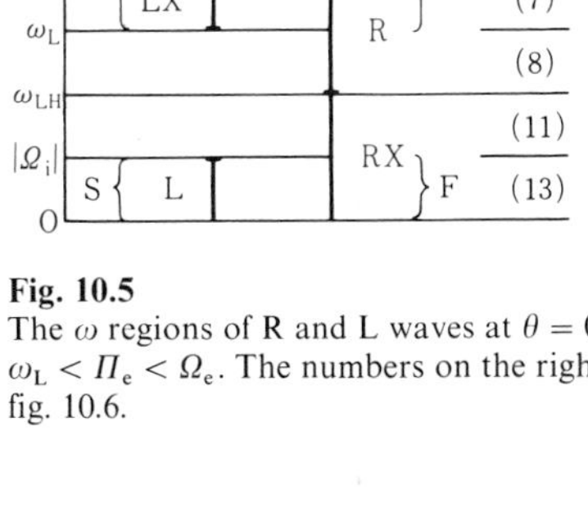

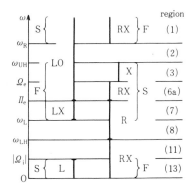

Fig. 10.5
The ω regions of R and L waves at $\theta = 0$; O, X waves at $\theta = \pi/2$; and F and S waves, for $\omega_L < \Pi_e < \Omega_e$. The numbers on the right identify regions shown in the CMA diagram, fig. 10.6.

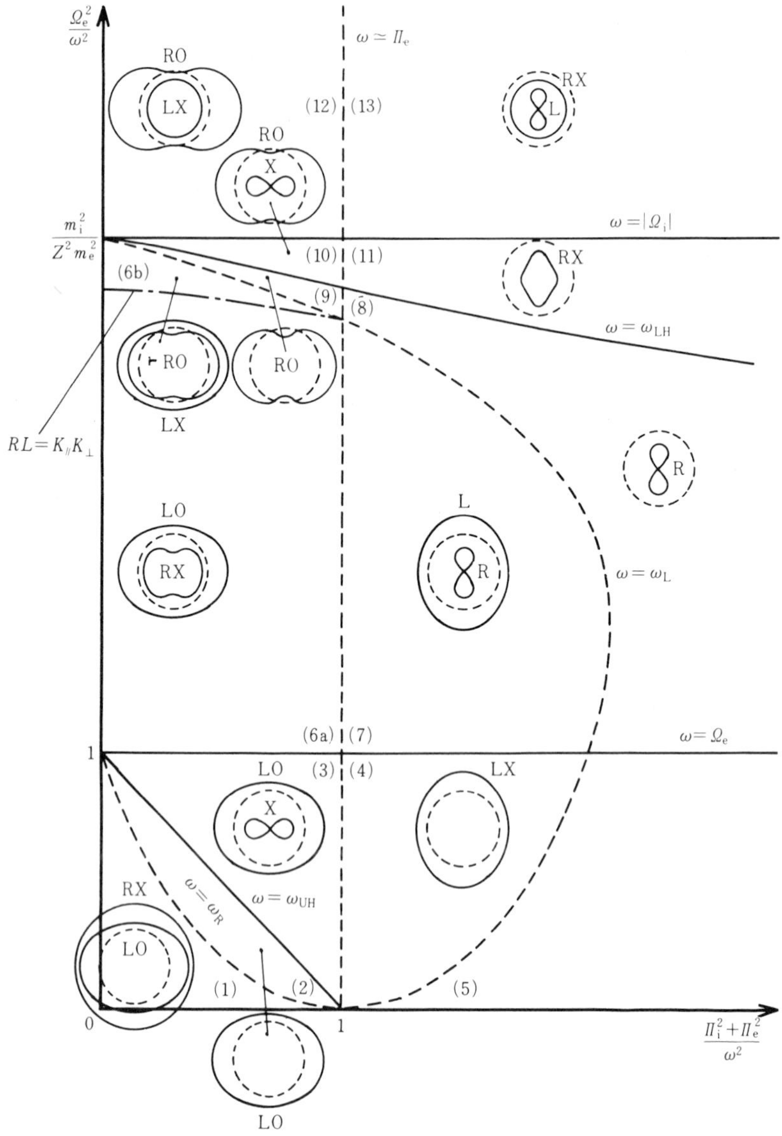

Fig. 10.6
CMA diagram of a two-component plasma. The surfaces of constant phase are drawn in each region. The dotted circles give the wave front in vacuum. The magnetic field is directed toward the top of the diagram.

We explain here the *CMA diagram* (fig. 10.6), which was introduced by P. C. Clemmon and R. F. Mullaly and later modified by W. P. Allis. The quantities Ω_e^2/ω^2 and $(\Pi_i^2 + \Pi_e^2)/\omega^2$ are plotted along the vertical and horizontal ordinates, respectively. The cutoff conditions $R = 0$ ($\omega = \omega_R$), $L = 0$ ($\omega = \omega_L$), and $K_\parallel = 0$ ($\omega \approx \Pi_e$) are shown by the dotted lines, and the resonance conditions $R = \infty$ ($\omega = \Omega_e$), $L = \infty$ ($\omega = |\Omega_i|$), and $K_\perp = 0$ ($\omega = \omega_{LH}$, $\omega = \omega_{UH}$) are shown by the solid lines. The cutoff and resonance contours form the boundaries of the different regions. The boundary $RL = K_\parallel K_\perp$, at which O wave and X wave are exchanged, is also shown in fig. 10.6. The surfaces of constant phase for R, L and O, X waves are shown for the different regions. As the vertical and horizontal ordinates correspond to the magnitude of $\boldsymbol{B}_0$ and the density n_e, one can easily assign waves to the corresponding regions simply by giving their frequencies ω.

10.4 Various Waves

10.4a Alfvén Wave

When the frequency ω is smaller than the ion cyclotron frequency ($\omega \ll |\Omega_i|$), the dielectric tensor K is expressed by

$$\left.\begin{aligned} K_\perp &= 1 + \delta \\ K_\times &= 0 \\ K_\parallel &= 1 - \frac{\Pi_e^2}{\omega^2}, \end{aligned}\right\} \tag{10.53}$$

where $\delta = \mu_0 n_i m_i c^2/B_0^2$. As $\Pi_e^2/\omega^2 = (m_i/m_e)(\Omega_i^2/\omega^2)\delta$, we find $\Pi_e^2/\omega^2 \gg \delta$. Assuming that $\Pi_e^2/\omega^2 \gg 1$, we have $|K_\parallel| > |K_\perp|$; then A, B, and C of eq. (10.20) are given by

$$\left.\begin{aligned} A &\approx -\frac{\Pi_e^2}{\omega^2}\cos^2\theta \\ B &\approx -\frac{\Pi_e^2}{\omega^2}(1 + \delta)(1 + \cos^2\theta) \\ C &\approx -\frac{\Pi_e^2}{\omega^2}(1 + \delta)^2, \end{aligned}\right\} \tag{10.54}$$

and the dispersion relations are

$$\frac{c^2}{N^2} = \frac{\omega^2}{k^2} = \frac{c^2}{1+\delta} = \frac{c^2}{1 + \dfrac{\mu_0 \rho_m c^2}{B_0{}^2}} \approx \frac{B_0{}^2}{\mu_0 \rho_m} \tag{10.55}$$

$$\frac{c^2}{N^2} = \frac{\omega^2}{k^2} = \frac{c^2}{1+\delta} \cos^2\theta \tag{10.56}$$

(ρ_m is the mass density). The wave satisfying this dispersion relation is called the *Alfvén wave*. We define the *Alfvén velocity* by

$$v_A{}^2 = \frac{c^2}{1+\delta} = \frac{c^2}{1 + \dfrac{\mu_0 \rho_m c^2}{B_0{}^2}} \approx \frac{B_0{}^2}{\mu_0 \rho_m}. \tag{10.57}$$

Equations (10.55) and (10.56) correspond to modes appearing in region (13) of the CMA diagram. Substitution of eqs. (10.55) and (10.56) into (10.19) shows that E_z for either mode is $E_z = 0$; E_y of the mode (10.55) (R wave, F wave, X wave) is $E_y = 0$; and E_x of the mode (10.56) (L wave, S wave) is $E_x = 0$. From eq. (10.6), we find for $\omega \ll |\Omega_i|$ that

$$\boldsymbol{E} + \boldsymbol{v}_i \times \boldsymbol{B}_0 = 0 \tag{10.58}$$

and $\boldsymbol{v}_i = (\boldsymbol{E} \times \boldsymbol{B}_0)/B_0{}^2$, so that $\boldsymbol{v}_i$ of the mode (10.55) is

$$\boldsymbol{v}_i \approx \hat{\boldsymbol{y}} \cos(k_y y + k_z z + \omega t); \tag{10.59}$$

and $\boldsymbol{v}_i$ of the mode (10.56) is

$$\boldsymbol{v}_i \approx \hat{\boldsymbol{x}} \cos(k_y y + k_z z - \omega t), \tag{10.60}$$

where $\hat{\boldsymbol{x}}$ and $\hat{\boldsymbol{y}}$ are unit vectors of along the x and y axes, respectively. From these last equations, the fast mode (10.55) is called the compressional mode and the slow mode (10.56) is called the torsional or shear mode. The R wave (10.55) still exists, though it is deformed in the transition from region (13) to regions (11) and (8), but the L wave (10.56) disappears in these transitions.

As is clear from eq. (10.58), the plasma is frozen to the magnetic field. There is tension $B_0{}^2/2\mu_0$ along the magnetic-field lines and a pressure $B_0{}^2/2\mu_0$ exerted perpendicularly to the magnetic field. As the plasma, of mass density ρ_m, sticks to the field lines, the wave propagation speed in the direction of the field is $B_0{}^2/(\mu_0 \rho_m)$.

10.4b Ion Cyclotron Wave

Let us consider the case where the frequency ω is near the ion cyclotron frequency and $\Pi_e^2/\omega^2 \gg 1$. The corresponding waves are located in regions (13) and (11) of the CMA diagram. When $|\omega| \ll \Omega_e$, $\delta \gg 1$, and $\Pi_e^2/\omega^2 \gg 1$, the values of $K_\perp$, $K_\times$, and $K_\parallel$ are

$$K_\perp = \frac{-\delta\Omega_i^2}{\omega^2 - \Omega_i^2} \qquad K_\times = \frac{-\delta\omega\Omega_i}{\omega^2 - \Omega_i^2} \qquad K_\parallel = -\frac{\Pi_e^2}{\omega^2}. \tag{10.61}$$

The coefficiencies A, B, and C are

$$\left.\begin{aligned} A &= -\frac{\Pi_e^2}{\omega^2}\cos^2\theta \\ B &= \frac{\Pi_e^2}{\omega^2}\frac{\delta\Omega_i^2}{\omega^2 - \Omega_i^2}(1 + \cos^2\theta) \\ C &= \frac{\Pi_e^2}{\omega^2}\frac{\delta^2\Omega_i^2}{\omega^2 - \Omega_i^2}, \end{aligned}\right\} \tag{10.62}$$

since $\Pi_e^2/\omega^2 = (m_i/m_e)(\Omega_i^2/\omega^2)\delta \gg \delta$. The dispersion equation becomes

$$N^4\cos^2\theta - N^2\frac{\delta\Omega_i^2}{\Omega_i^2 - \omega^2}(1 + \cos^2\theta) + \frac{\delta^2\Omega_i^2}{\Omega_i^2 - \omega^2} = 0. \tag{10.63}$$

Setting $N^2\cos^2\theta = c^2k_\parallel^2/\omega^2$ and $N^2\sin^2\theta = c^2k_\perp^2/\omega^2$, we may write eq. (10.63) as

$$k_\perp^2c^2 = \frac{\omega^4\delta^2\Omega_i^2 - \omega^2(2\delta\Omega_i^2k_\parallel^2c^2 + k_\parallel^4c^4) + \Omega_i^2k_\parallel^4c^4}{\omega^2(\delta\Omega_i^2 + k_\parallel^2c^2) - \Omega_i^2k_\parallel^2c^2}. \tag{10.64}$$

Therefore resonance occurs at

$$\omega^2 = \Omega_i^2\frac{k_\parallel^2c^2}{k_\parallel^2c^2 + \delta\Omega_i^2} = \Omega_i^2\frac{k_\parallel^2c^2}{k_\parallel^2c^2 + \Pi_i^2}. \tag{10.65}$$

The dispersion equation (10.63) approaches

$$N^2 \approx \frac{\delta}{1 + \cos^2\theta} \tag{10.66}$$

$$N^2\cos^2\theta \approx \delta(1 + \cos^2\theta)\frac{\Omega_i^2}{\Omega_i^2 - \omega^2} \tag{10.67}$$

as $|\omega|$ approaches $|\Omega_i|$. The mode (10.66) extends to the Alfvén compressional mode and is not affected by the ion cyclotron resonance. The dispersion relation (10.67) is that of the *ion cyclotron wave*, and can be expressed by

$$\omega^2 = \Omega_i^2 \left(1 + \frac{\Pi_i^2}{k_\parallel^2 c^2} + \frac{\Pi_i^2}{k_\parallel^2 c^2 + k_\perp^2 c^2}\right)^{-1}. \tag{10.68}$$

Note that here ω^2 is always less than Ω_i^2. The ions move in a right circular motion (i.e., in the direction of the ion Larmor motion) at $\omega \approx |\Omega_i|$.

The mode (10.66) satisfies $iE_x/E_y = 1$, i.e., it is circularly polarized, with the electric field rotating opposite to the ion Larmor motion.

The ion cyclotron wave satisfies

$$\frac{iE_x}{E_y} \approx -\frac{\omega}{|\Omega_i|}\frac{1}{\left(1 + \dfrac{k_\perp^2}{k_\parallel^2}\right)}, \tag{10.69}$$

i.e., the electric field is elliptically polarized, rotating in the same direction as the ion Larmor motion.

10.4c Lower Hybrid Resonance

The frequency at lower hybrid resonance at $\theta = \pi/2$ is given by

$$\omega^2 = \omega_{LH}^2$$
$$\frac{1}{\omega_{LH}^2} = \frac{1}{\Omega_i^2 + \Pi_i^2} + \frac{1}{|\Omega_i|\Omega_e} \qquad \frac{\omega_{LH}^2}{|\Omega_i|\Omega_e} = \frac{\Pi_i^2 + \Omega_i^2}{\Pi_i^2 + |\Omega_i|\Omega_e + \Omega_i^2}. \tag{10.70}$$

When the density is high and $\Pi_i^2 \gg |\Omega_i|\Omega_e$, it follows that $\omega_{LH}^2 \to |\Omega_i|\Omega_e$. When $\Pi_i^2 \ll |\Omega_i|\Omega_e$, then $\omega_{LH}^2 \to \Pi_i^2 + \Omega_i^2$ (see fig. 10.7). At lower hybrid resonance, we have $E_x = E_z = 0$ and $E_y \neq 0$.

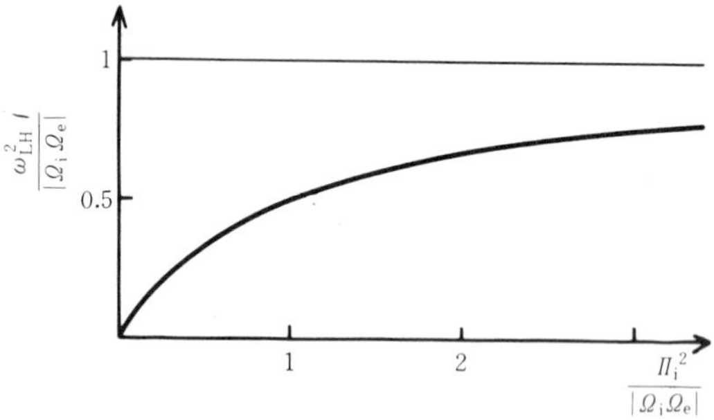

Fig. 10.7
Dependence of lower hybrid resonant frequency on the density.

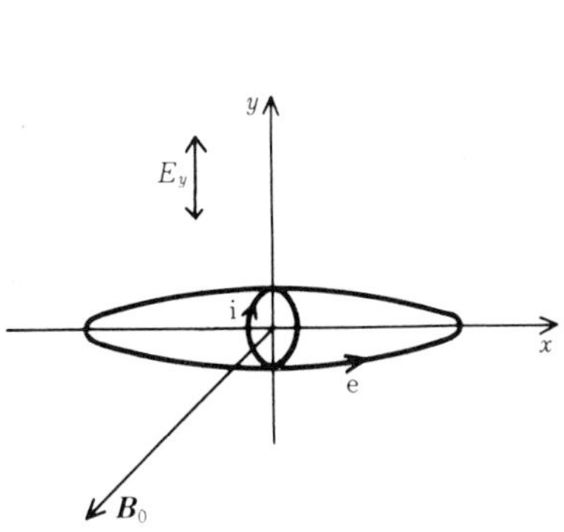

Fig. 10.8
Orbits of ions and electrons at lower hybrid resonance.

When the density is high (say $\Pi_i^2 \gtrsim |\Omega_i|\Omega_e$), then $|\Omega_i| \ll \omega \ll \Omega_e$ and the analysis of the motions of ions and electrons becomes simple. From eq. (10.7), the velocity is given by

$$v_{k,y} = \frac{i\varepsilon_k E_y}{B_0}\left(\frac{\omega|\Omega_k|}{\omega^2 - \Omega_k^2}\right) \tag{10.71}$$

and $v_{k,y} = \mathrm{d}y_k/\mathrm{d}t = -i\omega y_k$ yields

$$y_k = \frac{-\varepsilon_k E_y}{B_0}\left(\frac{|\Omega_k|}{\omega^2 - \Omega_k^2}\right). \tag{10.72}$$

At $\omega^2 = |\Omega_i|\Omega_e$, we find that $y_i \approx -E_y/B_0\Omega_e$ and $y_e \approx -E_y/B_0\Omega_e$, or $y_i \approx y_e$ (see fig. 10.8). Consequently, charge separation does not occur and the lower hybrid wave can exist.

We have been discussing lower hybrid resonance at $\theta = \pi/2$. Let us consider the case in which θ is slightly different from $\pi/2$. The resonant condition is obtained from eq. (10.24) as follows. Set

$$K_\perp \sin^2\theta + K_\parallel \cos^2\theta = 0. \tag{10.73}$$

Using eqs. (10.46), (10.50), and (10.51), reduce eq. (10.73) to

$$\frac{(\omega^2 - \omega_{LH}^2)(\omega^2 - \omega_{UH}^2)}{(\omega^2 - \Omega_i^2)(\omega^2 - \Omega_e^2)}\sin^2\theta + \left(1 - \frac{\Pi_e^2}{\omega^2}\right)\cos^2\theta = 0. \tag{10.74}$$

When θ is near $\pi/2$ and ω is not much different from ω_{LH}, we find that

$$\omega^2 - \omega_{LH}^2 = \frac{(\omega_{LH}^2 - \Omega_e^2)(\omega_{LH}^2 - \Omega_i^2)}{(\omega_{LH}^2 - \omega_{UH}^2)} \times \frac{\Pi_e^2 - \omega_{LH}^2}{\omega_{LH}^2}\cos^2\theta$$

$$\approx \frac{\Omega_e^2 \Pi_e^2}{\omega_{UH}^2}\left(1 - \left(\frac{\Omega_i}{\omega_{LH}}\right)^2\right)\left(1 - \left(\frac{\omega_{LH}}{\Pi_e}\right)^2\right)\cos^2\theta.$$

As $\omega_{UH}^2 \omega_{LH}^2 = \Omega_i^2 \Omega_e^2 + \Pi_e^2 |\Omega_i| \Omega_e$, ω^2 is expressed by

$$\omega^2 = \omega_{LH}^2 \left[1 + \frac{m_i}{Z m_e}\cos^2\theta \frac{\left(1 - \left(\frac{\Omega_i}{\omega_{LH}}\right)^2\right)\left(1 - \left(\frac{\omega_{LH}}{\Pi_e}\right)^2\right)}{\left(1 + \frac{|\Omega_i|\Omega_e}{\Pi_e^2}\right)}\right]. \tag{10.75}$$

When $\Pi_e^2/|\Omega_i|\Omega_e \approx \delta = c^2/v_A^2 \gg 1$, eq. (10.75) becomes

$$\omega^2 = \omega_{LH}^2 \left(1 + \frac{m_i}{Z m_e}\cos^2\theta\right). \tag{10.76}$$

Even if θ is different from $\pi/2$ only by the slight amount $(Z m_e/m_i)^{1/2}$, ω^2 becomes $\omega^2 \approx 2\omega_{LH}^2$, which contradicts the assumption that ω is not much different from ω_{LH}. Consequently, the relation (10.76) holds only in the region very near $\theta = \pi/2$.

10.4d Upper Hybrid Resonance

The upper hybrid resonant frequency ω_{UH} is given by

$$\omega_{UH}^2 = \Pi_e^2 + \Omega_e^2. \tag{10.77}$$

Since this frequency is much larger than $|\Omega_i|$, ion motion can be neglected. The electrons at $r = r_0$ in a cylindrical plasma expand by Δr, and a radial electric field of magnitude

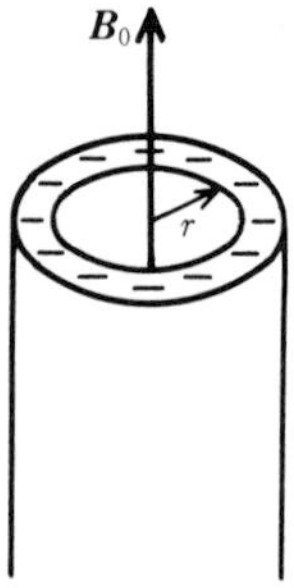

Fig. 10.9
Motion of electrons at upper hybrid resonance.

$$E_r = \frac{n_e e \Delta r}{\varepsilon_0}$$

is induced (see fig. 10.9). The motion of electrons is given by

$$m_e \frac{d^2}{dt^2} \Delta \boldsymbol{r} = -e\left(\frac{n_e e}{\varepsilon_0}(\hat{\boldsymbol{r}} \cdot \Delta \boldsymbol{r})\hat{\boldsymbol{r}} + \frac{d}{dt}\Delta \boldsymbol{r} \times \boldsymbol{B}_0\right), \tag{10.78}$$

where $\hat{\boldsymbol{r}}$ is the unit vector in the radial direction. Under the assumption $\Delta r \propto \exp(-i\omega t)$, the frequency ω becomes the upper hybrid resonant frequency. When $B_0 = 0$, ω becomes $\omega = \Pi_e$, i.e., that of the electron plasma wave.

10.4e Electron Cyclotron Wave (Whistler Wave)

Let us consider high-frequency waves, so that ion motion can be neglected. When $\omega \gg |\Omega_i|$, we find

$$\left.\begin{aligned} K_\perp &\approx 1 - \frac{\Pi_e^2}{\omega^2 - \Omega_e^2} \\ K_\times &\approx -\frac{\Pi_e^2}{(\omega^2 - \Omega_e^2)}\left(\frac{\Omega_e}{\omega}\right) \\ K_\parallel &= 1 - \frac{\Pi_e^2}{\omega^2}. \end{aligned}\right\} \tag{10.79}$$

The solution

$$N^2 = \frac{B \pm (B^2 - 4AC)^{1/2}}{2A}$$

of the dispersion equation may be modified to

$$N^2 - 1 = \frac{-2(A - B + C)}{2A - B \pm (B^2 - 4AC)^{1/2}}$$

$$= \frac{-2\Pi_e^2(1 - \Pi_e^2/\omega^2)}{2\omega^2(1 - \Pi_e^2/\omega^2) - \Omega_e^2 \sin^2\theta \pm \Omega_e \Delta}, \tag{10.80}$$

$$\Delta = \left(\Omega_e^2 \sin^4\theta + 4\omega^2\left(1 - \frac{\Pi_e^2}{\omega^2}\right)^2 \cos^2\theta\right)^{1/2} \tag{10.81}$$

The ordinary wave and the extraordinary wave will be obtained by taking the plus and minus sign, respectively, in eq. (10.80).

When the 1st term in eq. (10.81) is much larger than the 2nd term,

$$\Omega_e^{\,2}\sin^4\theta \gg 4\omega^2\left(1-\frac{\Pi_e^{\,2}}{\omega^2}\right)^2\cos^2\theta, \tag{10.82}$$

we find

$$N^2 = \frac{1-\Pi_e^{\,2}/\omega^2}{1-(\Pi_e/\omega)^2\cos^2\theta} \tag{10.83}$$

$$N^2 = \frac{(1-\Pi_e^{\,2}/\omega^2)^2\omega^2-\Omega_e^{\,2}\sin^2\theta}{(1-\Pi_e^{\,2}/\omega^2)\omega^2-\Omega_e^{\,2}\sin^2\theta}. \tag{10.84}$$

Equation (10.83) becomes $N^2 = K_{\parallel} = 1-\Pi_e^{\,2}/\omega^2$ at $\theta \approx \pi/2$ and does not depend on the magnitude of the magnetic field. This wave is used for density measurements by microwave interferometry (see ch. 15).

When the 1st term in eq. (10.81) is much smaller than the 2nd term,

$$\Omega_e^{\,2}\sin^4\theta \ll 4\omega^2\left(1-\frac{\Pi_e^{\,2}}{\omega^2}\right)\cos^2\theta, \tag{10.85}$$

and the additional condition

$$\Omega_e^{\,2}\sin^2\theta \ll \left|2\omega^2\left(1-\frac{\Pi_e^{\,2}}{\omega^2}\right)\right| \tag{10.86}$$

is assumed, the dispersion relations become

$$N^2 = 1-\frac{\Pi_e^{\,2}}{(\omega+\Omega_e\cos\theta)\omega} \tag{10.87}$$

$$N^2 = 1-\frac{\Pi_e^{\,2}}{(\omega-\Omega_e\cos\theta)\omega}. \tag{10.88}$$

Equation (10.87) corresponds to the L wave, and eq. (10.88) to the R wave. R-wave resonance occurs near the electron cyclotron frequency. This wave can propagate in regions (7) and (8) of the CMA diagram, where the frequency is less than the plasma frequency. This wave is called the *electron cyclotron wave*. It must be noticed that the assumptions (10.85) and (10.86) are not satisfied near $K_{\parallel} = 1-\Pi_e^{\,2}/\omega^2 \approx 0$.

The electron cyclotron wave is also called the *whistler wave*. Electromagnetic disturbances initiated by lighting flashes propagate through the ionosphere along magnetic-field lines. The frequency of a lightning-induced whistler wave falls in the audio region, and its group velocity increases with frequency; so that this wave is perceived as a whistle of descending tone. This is why it is called the whistler wave.

10.5 Energy Flow and Accessibility

Energy transport and the propagation of waves in the plasma medium are very important in the wave heating of plasmas. The equation of energy flow is derived from the Poynting equation, i.e.,

$$\nabla \cdot (\boldsymbol{E} \times \boldsymbol{H}) + \boldsymbol{E} \cdot \frac{\partial \boldsymbol{D}}{\partial t} + \boldsymbol{H} \cdot \frac{\partial \boldsymbol{B}}{\partial t} = \boldsymbol{j} \cdot \boldsymbol{E}. \tag{10.89}$$

We also have the relations $\boldsymbol{D} = \varepsilon_0 \boldsymbol{K E}$, $\boldsymbol{B} = \mu_0 \boldsymbol{H}$. Dielectric tensor $\boldsymbol{K}$ is associated with the kinetic energy of coherent motion of plasma particles.

The quasi-periodic functions A, B may be expressed by $A = A_0 \exp(-i\phi)$, $B = B_0 \exp(-i\phi)$, where A_0, B_0 are constant in time, and where $\phi = \phi_r + i\phi_i$ such that $\exp(-i\phi_r)$ is a periodic function of time, compared to which ϕ_i changes quite slowly. When the average of the multiplication of the real part of A with the real part of B is denoted by $\overline{AB}$, then $\overline{AB}$ is given by

$$\begin{aligned}\overline{AB} &= \frac{1}{4}\langle (A_0 \exp(-i\phi) + A_0{}^* \exp(i\phi^*))(B_0 \exp(-i\phi) + B_0{}^* \exp(i\phi^*))\rangle \\ &= \frac{1}{4}(A_0 B_0{}^* + A_0{}^* B_0)\exp 2\phi_i = \frac{1}{2}\mathrm{Re}((A_0 \exp \phi_i)(B_0 \exp \phi_i)^*). \end{aligned} \tag{10.90}$$

Re( ) is the real part of ( ). When ϕ is expressed by

$$\phi = \int_{-\infty}^{t} \omega \mathrm{d}t' = \int_{-\infty}^{t} (\omega_r + i\omega_i)\mathrm{d}t',$$

the average of the Poynting equation becomes

$$\nabla \cdot \boldsymbol{P} + \frac{\partial W}{\partial t} = \boldsymbol{j} \cdot \boldsymbol{E} \tag{10.91}$$

$$\boldsymbol{P} = \frac{1}{2\mu_0} \mathrm{Re}(E_0 \times B_0{}^*) \exp 2 \int_{-\infty}^{t} \omega_i \mathrm{d}t'$$

$$\frac{\partial W}{\partial t} = \frac{1}{2}\left(\omega_i \frac{\boldsymbol{B}_0{}^* \cdot \boldsymbol{B}_0}{\mu_0} + \omega_i \varepsilon_0 \boldsymbol{E}_0{}^* \cdot \frac{\boldsymbol{K} + \boldsymbol{K}^\dagger}{2} \boldsymbol{E}_0 - i\omega_r \varepsilon_0 \boldsymbol{E}_0{}^* \cdot \frac{\boldsymbol{K} - \boldsymbol{K}^\dagger}{2} \boldsymbol{E}_0\right) \times \exp 2 \int_{-\infty}^{t} \omega_i \mathrm{d}t'. \tag{10.92}$$

$\boldsymbol{K}^\dagger$ is the complex conjugate of the transpose of $\boldsymbol{K}$, i.e., $\boldsymbol{K}^\dagger \equiv (\tilde{\boldsymbol{K}})^*$. $(\boldsymbol{E}_0 \cdot (\boldsymbol{K E}_0)^* = \boldsymbol{E}_0 \cdot (\boldsymbol{K}^* \boldsymbol{E}_0{}^*) = (\boldsymbol{K}^\dagger \boldsymbol{E}_0) \cdot \boldsymbol{E}_0{}^* = \boldsymbol{E}_0{}^* \cdot (\boldsymbol{K}^\dagger \boldsymbol{E}_0).)$

A Hermitian matrix

$$K_{\mathrm{h}}(\omega_r) \equiv \frac{1}{2}(K(\omega_r) + K^{\dagger}(\omega_r)) \qquad (10.93)$$

is defined. For $|\omega_i| \ll |\omega_r|$ in $\omega = \omega_r + i\omega_i$, we can write

$$-iK_{\mathrm{h}}(\omega) + iK_{\mathrm{h}}{}^{\dagger}(\omega) = 2\omega_i\left(\frac{\partial K_{\mathrm{h}}}{\partial \omega}\right)_{\omega=\omega r}. \qquad (10.94)$$

Substitution of eq. (10.94) into eq. (10.92) gives

$$W = W_0 + [\text{loss term}]$$

$$W_0 = \frac{1}{2}\mathrm{Re}\left(\frac{B_0{}^* \cdot B_0}{2\mu_0} + \frac{\varepsilon_0}{2}E_0{}^* \cdot \left(\frac{\partial}{\partial\omega}(\omega K_{\mathrm{h}})\right)E_0\right)\exp 2\int_{-\infty}^{t}\omega_i \mathrm{d}t'. \qquad (10.95)$$

The 1st term is the energy density of the magnetic field and the 2nd term is the energy density of the electric field which takes account of the kinetic energy of coherent motion associated with the wave. Equation (10.95) gives the energy density of the wave in a dispersive medium.

Let us consider the velocity of the wave packet

$$F(r, t) = \int_{-\infty}^{\infty} f(k)\exp i(k\cdot r - \omega(k)t)\mathrm{d}k \qquad (10.96)$$

when the dispersion equation $\omega = \omega(k)$ is given. If $f(k)$ varies slowly, the position of the maximum of $F(r, t)$ is the position of the stationary phase of $(k\cdot r - \omega t)$, i.e., the position where

$$\frac{\partial}{\partial k}(k\cdot r - \omega(k)t) = 0.$$

Consequently the velocity of the maximum is

$$v_{\mathrm{gr}} = \frac{\partial\omega(k)}{\partial k} \equiv \left(\frac{\partial\omega}{\partial k_x}, \frac{\partial\omega}{\partial k_y}, \frac{\partial\omega}{\partial k_z}\right). \qquad (10.97)$$

This velocity is called the group velocity. This is the velocity of the energy flow.

Finally we will discuss the *accessibility* of the inner region of a plasma to a wave. When a wave is transmitted from the outside into an inhomogeneous plasma, whose density is high at the center and zero at the boundary, it is important to know whether the wave will propagate inward, or whether it will be reflected before reaching the inner part of the plasma.

Let us consider the accessibility at ion cyclotron resonance. The dispersion relation (10.64) is

$$k_y^2 c^2 = \frac{\omega^2 \left(\dfrac{\delta_0 |\Omega_i|}{|\Omega_i| - \omega} - \delta \right) \left(\dfrac{\delta_0 |\Omega_i|}{|\Omega_i| + \omega} - \delta \right)}{\delta - \delta_0}, \tag{10.98}$$

where

$$\delta_0 \equiv \frac{k_z^2 c^2 (\Omega_i^2 - \omega^2)}{\Omega_i^2 \omega^2} \tag{10.99}$$

is the value of δ at resonance. The frequency of the ion cyclotron wave is $|\omega| < |\Omega_i|$ and $k_y^2 c^2$ is positive when

$$\frac{\delta_0 |\Omega_i|}{|\Omega_i| + |\omega|} < \delta < \delta_0. \tag{10.100}$$

Since δ is proportional to the electron density, the left-hand side of eq. (10.100) indicates that $k_y^2 c^2$ is positive only when the density is larger than the critical value; otherwise k_y^2 is negative and we have an *evanescent wave*, damping according to $\exp(-|k_y|y)$. Equation (10.98) can be rewritten as

$$k_y^2 c^2 = -k_z^2 c^2 + \omega^2 \delta + \frac{\omega^2 \delta k_z^2 c^2}{\Omega_i^2 (\delta_0 - \delta)}. \tag{10.101}$$

The 2nd and 3rd terms of eq. (10.101) are both positive when $0 < \delta < \delta_0$, so that the damping rate is less than that of the wave in vacuum $k_y^2 c^2 = -k_z^2 c^2$. If the range between the plasma boundary ($\delta = 0$) and the region where $\delta = \delta_0 |\Omega_i| / (|\Omega_i + |\omega|)$ is not large compared with $1/|k_y|$, the damping rate is not large and the accessibility for the wave is not disturbed very much.

The accessibility of the resonant positions of the lower hybrid wave and two-ion hybrid waves (for plasmas containing two kinds of ions) are discussed in ch. 3 of ref. 1.

10.6 Conditions for Electrostatic Waves

When the electric field $\boldsymbol{E}$ can be expressed by an electrostatic potential

$$\boldsymbol{E} = -\nabla\phi = -i\boldsymbol{k}\phi, \tag{10.102}$$

the resultant wave is called an *electrostatic wave*. The electric field $\boldsymbol{E}$ is always parallel to the propagation vector $\boldsymbol{k}$, so that the electrostatic wave is longitudinal. The magnetic field $\boldsymbol{B}_1$ of the electrostatic wave is always zero:

$$\boldsymbol{B}_1 = \boldsymbol{k} \times \boldsymbol{E}/\omega = 0. \tag{10.103}$$

Alfvén waves are not electrostatic waves. We will here discuss the conditions for electrostatic waves. Since the dispersion relation is

$$\boldsymbol{N} \times (\boldsymbol{N} \times \boldsymbol{E}) + \boldsymbol{K}\cdot\boldsymbol{E} = 0,$$

the scalar product with $\boldsymbol{N}$ becomes

$$\boldsymbol{N}\cdot\boldsymbol{K}(\boldsymbol{E}_{\parallel} + \boldsymbol{E}_{\perp}) = 0,$$

where $\boldsymbol{E}_{\parallel}$ and $\boldsymbol{E}_{\perp}$ are the components of the electric field parallel and perpendicular to $\boldsymbol{k}$. If $|\boldsymbol{E}_{\parallel}| \gg |\boldsymbol{E}_{\perp}|$, the wave is electrostatic and the dispersion relation becomes

$$\boldsymbol{N}\cdot\boldsymbol{K}\boldsymbol{N} = 0. \tag{10.104}$$

Rewriting the dispersion relation as

$$(N^2 - \boldsymbol{K})\boldsymbol{E}_{\perp} = \boldsymbol{K}\boldsymbol{E}_{\parallel}$$

shows that $|\boldsymbol{E}_{\parallel}| \gg |\boldsymbol{E}_{\perp}|$ holds when

$$|N^2| \gg |K_{ij}| \tag{10.105}$$

is satisfied for all K_{ij}. The dispersion relation (10.104) for the electrostatic wave is then

$$k_y{}^2 K_{yy} + 2k_y k_z K_{yz} + k_z{}^2 K_{zz} = 0. \tag{10.106}$$

The condition (10.105) for the electrostatic wave indicates that the phase velocity $\omega/k = c/N$ of this wave is low. The K_{ij} have already been given by eqs. (10.9)–(10.12) for cold plasmas, and the general formula for hot plasmas will be discussed in chs. 11–13. We have stated (eq. (10.103)) that the magnetic field $\boldsymbol{B}_1$ of the electrostatic wave is zero. Disturbances of the magnetic field propagate with the Alfvén velocity $v_A^2 \approx B_0^2/\mu_0 n_i m_i$. If the phase velocity of the wave is much lower than v_A, the disturbance of the magnetic field will be damped within a few cycles of the wave and the propagated magnetic-field disturbance becomes $\boldsymbol{B}_1 = 0$, as stated. When the electron thermal velocity v_T^e is taken as a typical phase velocity for electrostatic waves, then

$$\frac{B_0^2}{\mu_0 n_i m_i (v_T^e)^2} = \frac{2m_e}{m_i \beta_e} > 1$$

$$\beta_e < \frac{2m_e}{m_i}. \tag{10.107}$$

This is a condition that a wave is electrostatic.

At resonance the refractive index N becomes infinite. As the K_{ij} are finite for lower hybrid and upper hybrid resonance, the condition (10.105) is satisfied so that these hybrid waves are electrostatic. Since some of the K_{ij} become infinite for the ion or electron cyclotron waves, these cyclotron waves are not always electrostatic.

10.7 Interactions of Waves and Particles in (Hot) Plasma with Velocity Distribution

The existence of a damping mechanism by which plasma particles absorb wave energy even in a collisionless plasma was found by L. D. Landau,[5] under the condition that the plasma is not cold and the velocity distribution is of finite extent. Energy-exchange processes between particles and wave are possible even in a collisionless plasma and play important roles in plasma heating by waves and in the mechanism of instabilities. These processes will be described systematically by analysis of wave phenomena in hot plasmas with finite velocity distributions in chs. 11–13. In the present section, however, we attempt to explain these important processes in terms of simplified physical models. When the plasma has a finite velocity distribution, pressure and particle-wave interaction terms appear in the dielectric tensor that are absent in the tensor for a cold plasma.

10.7a Landau Damping

Let us assume that many particles drift with different velocities in the direction of the lines of magnetic force. When an electrostatic wave (a

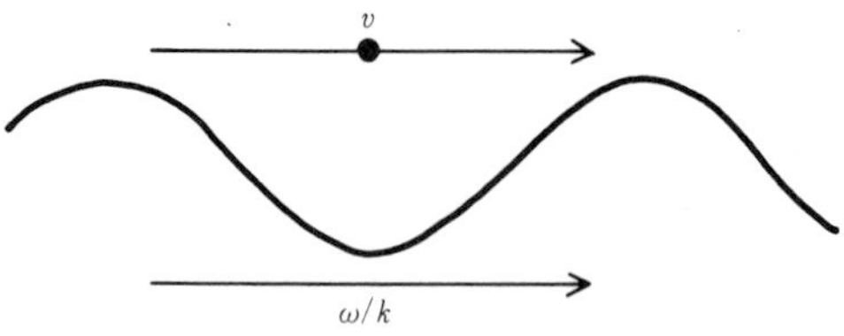

Fig. 10.10
Propagation of wave and motion of particles in the process of Landau damping.

longitudinal wave with $\boldsymbol{k}/\!/\boldsymbol{E}$) propagates along the lines of magnetic force, there appears an interaction between the wave and a group of particles (see fig. 10.10). Take the z axis in the direction of the magnetic field and denote the unit vector in this direction by $\hat{z}$. Then the electric field and the velocity $\boldsymbol{v} = v\hat{z}$ satisfy

$$E = \hat{z}E\cos(kz - \omega t) \tag{10.108}$$

$$m\frac{dv}{dt} = qE\cos(kz - \omega t). \tag{10.109}$$

The electric field E is a quantity of the 1st order. The zeroth-order solution of eq. (10.109) is

$$z = v_0 t + z_0$$

and the 1st-order equation is

$$m\frac{dv_1}{dt} = qE\cos(kz_0 + kv_0 t - \omega t). \tag{10.110}$$

The solution of eq. (10.110) for the initial condition $v_1 = 0$ at $t = 0$ is

$$v_1 = \frac{qE}{m}\frac{\sin(kz_0 + kv_0 t - \omega t) - \sin kz_0}{kv_0 - \omega}. \tag{10.111}$$

The kinetic energy of the particle becomes

$$\frac{d}{dt}\frac{mv^2}{2} = v\frac{d}{dt}mv = v_1\frac{d}{dt}mv_1 + v_0\frac{d}{dt}mv_2 + \cdots. \tag{10.112}$$

We have the relation

$$m\frac{d(v_1 + v_2)}{dt} = qE\cos(k(z_0 + v_0 t + z_1) - \omega t)$$

$$= qE\cos(kz_0 + \alpha t) - qE\sin(kz_0 + \alpha t)kz_1$$

$$z_1 = \int_0^t v_1 \mathrm{d}t = \frac{qE}{m}\left(\frac{-\cos(kz_0 + \alpha t) + \cos kz_0}{\alpha} - \frac{t\sin kz_0}{\alpha}\right),$$

where

$$\alpha \equiv kv_0 - \omega.$$

Using these, we may put eq. (10.112) into the form

$$\frac{\mathrm{d}}{\mathrm{d}t}\frac{mv^2}{2} = \frac{q^2E^2}{m}\left(\frac{\sin(kz_0 + \alpha t) - \sin kz_0}{\alpha}\right)(\cos(kz_0 + \alpha t))$$

$$+ \frac{kv_0 q^2 E^2}{m}\left(\frac{-\cos(kz_0 + \alpha t) + \cos kz_0}{\alpha^2} - \frac{t\sin kz_0}{\alpha}\right)(-\sin(kz_0 + \alpha t)).$$

The average of the foregoing quantity with respect to the initial position z_0 is

$$\left\langle \frac{\mathrm{d}}{\mathrm{d}t}\frac{mv^2}{2} \right\rangle_{z_0} = \frac{q^2E^2}{2m}\left(\frac{-\omega\sin\alpha t}{\alpha^2} + t\cos\alpha t + \frac{\omega t\cos\alpha t}{\alpha}\right). \tag{10.113}$$

When we take the velocity average of eq. (10.113) over v_0 with the weighting factor or distribution function

$$f(v_0) = f\left(\frac{\alpha + \omega}{k}\right) = g(\alpha),$$

the rate of increase of the kinetic energy of the particles is obtained. The distribution function is normalized:

$$\int_{-\infty}^{\infty} f(v_0)\mathrm{d}v_0 = \frac{1}{k}\int g(\alpha)\mathrm{d}\alpha = 1.$$

The integral of the 2nd term of eq. (10.113),

$$\frac{1}{k}\int g(\alpha)t\cos\alpha t\,\mathrm{d}\alpha = \frac{1}{k}\int g\left(\frac{x}{t}\right)\cos x\,\mathrm{d}x, \tag{10.114}$$

approaches zero as $t \to \infty$. The integral of the 3rd term of eq. (10.113) becomes

$$\frac{\omega}{k}\int \frac{g(\alpha)t\cos\alpha t}{\alpha}\mathrm{d}\alpha = \frac{\omega}{k}\int \frac{t}{x}g\left(\frac{x}{t}\right)\cos x\,\mathrm{d}x. \tag{10.115}$$

The function $g(\alpha)$ can be considered to be the sum of an even and an odd function. The even function does not contribute to the integral. The contribution of the odd function approaches zero when $t \to \infty$ (see eq. (10.114)) if $g(\alpha)$ is continuous at $\alpha = 0$. Therefore, only the contribution of the 1st term in eq. (10.113) remains and we find

$$\left\langle \frac{\mathrm{d}}{\mathrm{d}t} \frac{mv^2}{2} \right\rangle_{z_0, v_0} = -\frac{\omega q^2 E^2}{2mk} P \int \frac{g(\alpha) \sin \alpha t}{\alpha^2} \mathrm{d}\alpha. \tag{10.116}$$

The main contribution to the integral comes from near $\alpha = 0$, so that $g(\alpha)$ may be expanded around $\alpha = 0$:

$$g(\alpha) = g(0) + \alpha g'(0) + \frac{\alpha^2}{2} g''(0) + \cdots.$$

As $(\sin \alpha t)/\alpha^2$ is an odd function, only the 2nd term of the foregoing equation contributes to the integral and we find for large t that

$$\begin{aligned} \left\langle \frac{\mathrm{d}}{\mathrm{d}t} \frac{mv^2}{2} \right\rangle_{z_0, v_0} &= -\frac{\omega q^2 E^2}{2mk} \int \frac{g'(0) \sin \alpha t}{\alpha} \mathrm{d}x \\ &= \frac{-\pi q^2 E^2}{2m|k|} \left(\frac{\omega}{k}\right) \left(\frac{\partial f(v_0)}{\partial v_0}\right)_{v_0 = \omega/k}. \end{aligned} \tag{10.117}$$

If the number of particles slightly slower than the phase velocity of the wave is larger than the number slightly faster, i.e., if $v_0 \; \partial f_0/\partial v_0 < 0$, the group of particles as a whole gains energy from the wave and the wave is damped. On the contrary, when $v_0 \; \partial f_0/\partial v_0 > 0$ at $v_0 = \omega/k$ the particles gives their energy to the wave and the amplitude of the wave increases (fig. 10.11). This mechanism is called Landau damping or amplification. Experimental verification of Landau damping of waves in

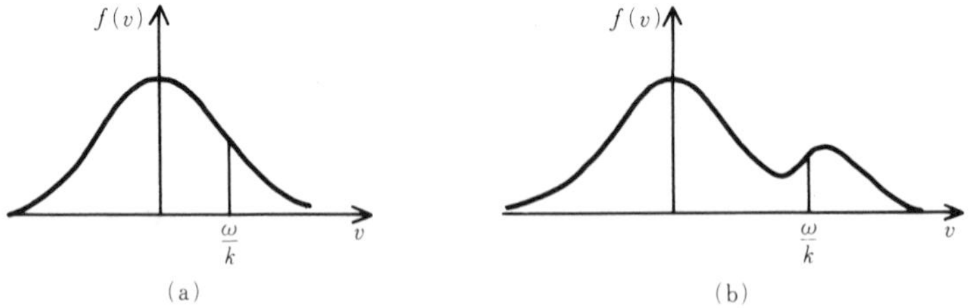

Fig. 10.11
Landau damping (a) and Landau amplification (b).

a collisionless plasma was demonstrated by J. M. Malemberg and C. B. Wharton[6] in 1965, twenty years after Landau's prediction.[5]

There is a restriction on the applicability of linear Landau damping. When this phenomenon runs its course before the particle orbit deviates from the linear-approximation solution, the reductions leading to linear Landau damping are justified. The period of oscillation in the potential well of the electric field of the wave gives the time for the particle orbit to deviate from the linear approximation ($m\omega^2 x = eE$ and $\omega^2 \approx eEk/m$). The period of oscillation is

$$\tau_{\mathrm{osc}} = \frac{1}{\omega_{\mathrm{osc}}} \approx \left(\frac{m}{ekE}\right)^{1/2}. \tag{10.118}$$

Consequently the condition for the applicability of linear Landau damping is that the Landau damping time $1/\omega_{\mathrm{i}}$ is shorter than τ_{osc} or the collision time $1/\nu_{\mathrm{coll}}$ is shorter than τ_{osc}:

$$|\omega_{\mathrm{i}}\tau_{\mathrm{osc}}| \gtrsim 1 \tag{10.119}$$

$$|\nu_{\mathrm{coll}}\tau_{\mathrm{osc}}| \gtrsim 1. \tag{10.120}$$

On the other hand, it was assumed that particles are collisionless. The condition that the collision time $1/\nu_{\mathrm{coll}}$ is longer than λ/v_{rms} is necessary for the asymptotic approximation of the integral (10.116) as $t \to \infty$, where λ is the wavelength of the wave and v_{rms} is the spread in the velocity distribution:

$$\frac{1}{\nu_{\mathrm{coll}}} > \frac{2\pi}{kv_{\mathrm{rms}}}. \tag{10.121}$$

10.7b Transit-Time Damping

We have already described, the properties of Alfvén waves in cold plasmas. There are compressional and torsional modes. The compressional mode becomes magnetosonic in hot plasmas, as will be described in ch. 13. In the low-frequency region, the magnetic moment μ_{m} is conserved and the equation of motion along the field lines is

$$m\frac{\mathrm{d}v_z}{\mathrm{d}t} = -\mu_{\mathrm{m}}\frac{\partial B_{1z}}{\partial z}. \tag{10.122}$$

This equation is the same as that for Landau damping if $-\mu_{\mathrm{m}}$ and $\partial B_{1z}/\partial z$

are replaced by the electric charge and the electric field, respectively. The rate of change of the kinetic energy is derived similarly, and is

$$\left\langle \frac{\mathrm{d}}{\mathrm{d}t}\frac{mv^2}{2} \right\rangle_{z_0, v_0} = -\frac{\pi {\mu_\mathrm{m}}^2 |k|}{2m} |B_{1z}|^2 \left(\frac{\omega}{k}\right)\left(\frac{\partial f(v_0)}{\partial v_0}\right)_{v_0 = \omega/k}. \tag{10.123}$$

This phenomena is called *transit-time damping.*

10.7c Cyclotron Damping

The mechanism of cyclotron damping is different from that of Landau damping. Here the electric field of the wave is perpendicular to the direction of the magnetic field and the particle drift and accelerates the particle perpendicularly to the drift direction. Let us consider a simple case in which the thermal energy of particles perpendicular to the magnetic field is zero and the velocity of particles parallel to the magnetic field is V. The equation of motion is

$$m\frac{\partial v}{\partial t} + mV\frac{\partial v}{\partial z} = q(E_1 + v \times \hat{z}B_0 + V\hat{z} \times B_1). \tag{10.124}$$

As our interest is in the perpendicular acceleration we assume $\boldsymbol{E}_1 \cdot \hat{z} = 0$. $\boldsymbol{B}_1$ is given by $\boldsymbol{B}_1 = (\boldsymbol{k} \times \boldsymbol{E})/\omega$. With the definitions $v^\pm = v_x \pm iv_y$, $E^\pm = E_x \pm iE_y$, the solution for the initial condition $\boldsymbol{v} = 0$ at $t = 0$ is

$$v^\pm = \frac{iqE^\pm(\omega - kV)\exp(ikz - i\omega t)}{m\omega}\frac{1 - \exp(i\omega t - ikVt \pm i\Omega t)}{\omega - kV \pm \Omega} \tag{10.125}$$

$$\Omega = -\frac{iB}{m}.$$

The macroscopic value of $\boldsymbol{v}_\perp$ is obtained by taking the average of $\boldsymbol{v}_\perp$ weighed by the distribution function $f_0(V)$ as follows:

$$\langle \boldsymbol{v}_\perp \rangle = \frac{iq\exp(ikz - i\omega t)}{2m}[(c^+ + c^-)\boldsymbol{E}_\perp + i(c^+ - c^-)\boldsymbol{E}_\perp \times \hat{z}] \tag{10.126}$$

$$c^\pm = \alpha^\pm - i\beta^\pm \tag{10.127}$$

$$\alpha^\pm = \int_{-\infty}^{\infty} \mathrm{d}V \frac{f_0(V)(1 - kV/\omega)(1 - \cos(\omega - kV \pm \Omega)t)}{\omega - kV \pm \Omega} \tag{10.128}$$

$$\beta^\pm = \int_{-\infty}^{\infty} \mathrm{d}V \frac{f_0(V)(1 - kV/\omega)\sin(\omega - kV \pm \Omega)t}{\omega - kV \pm \Omega}. \tag{10.129}$$

As the time t becomes large we find that

$$\alpha^{\pm} \to P \int_{-\infty}^{\infty} \mathrm{d}V \frac{f_0(V)(1 - kV/\omega)}{\omega - kV \pm \Omega} \tag{10.130}$$

$$\beta^{\pm} \to \frac{\mp \pi\Omega}{\omega|k|} f_0\left(\frac{\omega \pm \Omega}{k}\right). \tag{10.131}$$

When

$$t \gg \frac{2\pi}{kV_{\mathrm{rms}}}, \tag{10.132}$$

where V_{rms} is the spread of the velocity distribution $V_{\mathrm{rms}} = \langle V^2 \rangle^{1/2}$, the approximations (10.130) and (10.131) are justified. The absorption of the wave energy by the plasma particles is given by

$$\begin{aligned}&\langle \mathrm{Re}(q\boldsymbol{E}\exp(ikz - i\omega t)) \mathrm{Re}(\langle \boldsymbol{v}_{\perp} \rangle \exp(ikz - i\omega t))\rangle_z \\ &\quad = \frac{q^2}{4m}(\beta^+ |E_x + iE_y|^2 + \beta^- |E_x - iE_y|^2).\end{aligned} \tag{10.133}$$

Consequently, absorption may occur when β^+ and β^- are positive. The *cyclotron velocity* V_{c} is defined so that the Doppler shift $|\omega - kV|$ is equal to the cyclotron frequency $|\Omega|$, i.e.,

$$\omega - kV_{\mathrm{c}} \pm \Omega = 0 \tag{10.134}$$

$$V_{\mathrm{c}} = \frac{\omega \pm \Omega}{k} = \frac{\omega}{k}\left(1 \pm \frac{\Omega}{\omega}\right). \tag{10.135}$$

As is clear from the sign of $\beta^{\pm}$ corresponding to $V_{\mathrm{c}} = (\omega/k)(1 \pm \Omega/\omega)$, we find $\beta^{\pm} < 0$ when $|V_{\mathrm{c}}| > |\omega/k|$ and $\beta^{\pm} > 0$ when $|V_{\mathrm{c}}| < |\omega/k|$. Therefore when the absolute value of the cyclotron velocity is smaller than the phase velocity of the wave, the particles absorb wave energy. This phenomena is called *cyclotron damping*.

We have discussed the case in which the perpendicular thermal energy is zero. When the perpendicular thermal energy is larger than the parallel thermal energy, so-called cyclotron instability may occur. The mutual interaction between particles and wave will be discussed again in sec. 14.3c in connection with heating phenomena.

Let us consider the change in the kinetic energy of the particles in the case of cyclotron damping. Then the equation of motion is

$$m\frac{\mathrm{d}\boldsymbol{v}}{\mathrm{d}t} - q(\boldsymbol{v} \times \boldsymbol{B}_0) = q\boldsymbol{E}_\perp + q(\boldsymbol{v} \times \boldsymbol{B}_1).$$

Since $\boldsymbol{B}_1 = (\boldsymbol{k} \times \boldsymbol{E})/\omega$ and $E_z = 0$, we have

$$m\frac{\mathrm{d}v_z}{\mathrm{d}t} = \frac{qk_z}{\omega}(\boldsymbol{v}_\perp \cdot \boldsymbol{E}_\perp),$$

$$m\frac{\mathrm{d}\boldsymbol{v}_\perp}{\mathrm{d}t} - q(\boldsymbol{v}_\perp \times \boldsymbol{B}_0) = q\boldsymbol{E}_\perp\left(1 - \frac{k_z v_z}{\omega}\right)$$

so that

$$m\boldsymbol{v}_\perp \cdot \frac{\mathrm{d}\boldsymbol{v}_\perp}{\mathrm{d}t} = q(\boldsymbol{v}_\perp \cdot \boldsymbol{E}_\perp)\left(1 - \frac{k_z v_z}{\omega}\right).$$

Then

$$\frac{\mathrm{d}}{\mathrm{d}t}\left(\frac{mv_z^2}{2}\right) = \frac{k_z v_z}{\omega - k_z v_z}\frac{\mathrm{d}}{\mathrm{d}t}\left(\frac{mv_\perp^2}{2}\right),$$

$$\frac{1}{2}\left[v_\perp^2 + \left(v_z - \frac{\omega}{k_z}\right)^2\right] = \text{const.}$$

In the analysis of cyclotron damping, we assumed that $v_z = V$ is constant; the condition for the validity of the linearized theory is[1]

$$\frac{k_z^2 q^2 E_\perp^2 |\omega - k_z v_z| t^3}{24\omega^2 m^2} < 1.$$

References

1. T. H. Stix: The Theory of Plasma Waves, McGraw-Hill, New York, 1962

2. W. P. Allis, S. J. Buchsbaum, and A. Bers: Waves in Anisotropic Plasmas, MIT Press, Cambridge, Mass. 1963

3. G. Bekefi: Radiation Processes in Plasma, Willey, New York, 1966

4. V. L. Ginzburg: Propagation of Electromagnetic Waves in Plasma, Gordon and Breach, New York, 1961

5. L. D. Landau: J. Phys. (USSR) **10**, 45(1946)

6. J. H. Malmberg, C. B. Wharton, and W. E. Drummond: Plasma Phys. and Controlled Nucl. Fusion Research **1**, 485 (Conf. Proceedings, Culham 1965) (IAEA, Vienna, 1966)

11 Electrostatic Waves in a Hot Plasma

In the preceding chapter we discussed the wave phenomena in cold plasmas, where ions and electrons were assumed to be immobile in the unperturbed state. In hot plasmas, the charged particles move along spiral trajectories even in the unperturbed state. The motion of charged particles in a uniform magnetic field $\boldsymbol{B}_0 = B_0\hat{z}$ and a uniform gravitational field $\boldsymbol{G} = G\hat{x}$ may be described by

$$\frac{d\boldsymbol{r}'}{dt'} = \boldsymbol{v}' \qquad \frac{d\boldsymbol{v}'}{dt'} = \frac{q}{m}\boldsymbol{v}' \times \boldsymbol{B}_0 + G\hat{\boldsymbol{x}}. \tag{11.1}$$

Assuming that $\boldsymbol{r}' = \boldsymbol{r}$, $\boldsymbol{v}' = \boldsymbol{v}$ at $t' = t$, the solution of the eqs. (11.1) is obtained as follows:

$$\left.\begin{aligned} v'_x(t') &= v_\perp \cos(\theta + \Omega(t' - t)) \\ v'_y(t') &= v_\perp \sin(\theta + \Omega(t' - t)) + \frac{G}{\Omega} \\ v'_z(t') &= v_z \end{aligned}\right\} \tag{11.2}$$

$$\left.\begin{aligned} x'(t') - x &= \frac{v_\perp}{\Omega}(\sin(\theta + \Omega(t' - t)) - \sin\theta) \\ y'(t') - y &= -\frac{v_\perp}{\Omega}(\cos(\theta + \Omega(t' - t)) - \cos\theta) + \frac{G}{\Omega}(t' - t) \\ z'(t') - z &= v_z(t' - t), \end{aligned}\right\} \tag{11.3}$$

where $\boldsymbol{v} = (v_\perp \cos\theta,\ v_\perp \sin\theta,\ v_z)$ and $\Omega = -qB_0/m$. The analysis of the behavior due to a perturbation of this system must be based on Boltzmann's equation. Previously the magnetohydrodynamic equations were derived using the fluid model of a plasma; the equations were written in terms of statistical averages of macroscopic quantities such as mass density, flow velocity, etc. On the other hand, the kinetic model of a plasma considers the plasma as an ensemble of particles. The zeroth-order approximation of the particle motion has already been obtained, and Boltzmann's equation can be solved for the distribution function in velocity space. Based on this solution, various physical phenomena may be investigated. In this manner, it becomes possible to analyze those wave-particle interactions that are strongly dependent on the shape of the distribution function, although the mathematical procedure is somewhat complicated. The kinetic model is also general enough to enable us to analyse high-frequency wave phenomena.

The first part of this chapter (sec. 11.1) introduces the dispersion equation of waves in hot plasmas by solving the linearized collisionless Boltzmann equation. In sec. 11.2, a simplified dispersion relation is obtained for the electrostatic wave, whose perturbed electric field can be expressed by means of a scalar potential. Subsequently (sec. 11.3), the dispersion relation is applied to a homogeneous plasma to study the characteristics of ion acoustic, electron acoustic, and Bernstein waves. Two-stream instabilities are discussed in sec. 11.4. The Nyquist criterion applied to the instabilities and the difference between absolute and convective instabilities are explained in secs. 11.5 and 11.6.

11.1 Dispersion Equation of Waves in a Hot Plasma

Boltzmann's or Vlasov's equation for a collisionless plasma with the velocity distribution function $f_k(\boldsymbol{r}, \boldsymbol{v}, t)$ for the kth kind of particles is given by

$$\frac{\partial f_k(\boldsymbol{r}, \boldsymbol{v}, t)}{\partial t} + \boldsymbol{v} \cdot \nabla_r f_k(\boldsymbol{r}, \boldsymbol{v}, t) + \left(\frac{q_k}{m_k}\boldsymbol{E} + \boldsymbol{G} + \frac{q_k \boldsymbol{v}}{m_k} \times \boldsymbol{B}\right) \cdot \nabla_v f_k(\boldsymbol{r}, \boldsymbol{v}, t) = 0. \tag{11.4}$$

Maxwell's equations are

$$\nabla \cdot \boldsymbol{E} = \frac{1}{\varepsilon_0} \sum_k q_k \int f_k \mathrm{d}\boldsymbol{v} \tag{11.5}$$

$$\frac{1}{\mu_0} \nabla \times \boldsymbol{B} = \varepsilon_0 \frac{\partial \boldsymbol{E}}{\partial t} + \sum_k q_k \int \boldsymbol{v} f_k \mathrm{d}\boldsymbol{v} \tag{11.6}$$

$$\nabla \times \boldsymbol{E} = -\frac{\partial \boldsymbol{B}}{\partial t} \tag{11.7}$$

$$\nabla \cdot \boldsymbol{B} = 0. \tag{11.8}$$

As usual, we indicate zeroth-order quantities (the unperturbed state) by a subscript 0 and the 1st-order perturbation terms by a subscript 1. The 1st-order terms are expressed in the form of exp $i(\boldsymbol{k} \cdot \boldsymbol{r} - \omega t)$. Using

$$f_k = f_{k0}(\boldsymbol{r}, \boldsymbol{v}) + f_{k1} \tag{11.9}$$

$$\boldsymbol{B} = \boldsymbol{B}_0(\boldsymbol{r}) + \boldsymbol{B}_1 \tag{11.10}$$

$$\boldsymbol{E} = 0 + \boldsymbol{E}_1, \tag{11.11}$$

we may linearize eqs. (11.4)–(11.8) as follows:

$$\boldsymbol{v}\cdot\nabla_r f_{k0} + \left(\boldsymbol{G} + \frac{q_k \boldsymbol{v}}{m_k} \times \boldsymbol{B}_0\right)\nabla_v f_{k0} = 0 \tag{11.12}$$

$$\sum_k q_k \int f_{k0}\,\mathrm{d}\boldsymbol{v} = 0 \tag{11.13}$$

$$\frac{1}{\mu_0}\nabla \times \boldsymbol{B}_0 = \sum_k q_k \int \boldsymbol{v} f_{k0}\,\mathrm{d}\boldsymbol{v} \tag{11.14}$$

$$\frac{\partial f_{k1}}{\partial t} + \boldsymbol{v}\cdot\nabla_r f_{k1} + \left(\boldsymbol{G} + \frac{q_k \boldsymbol{v}}{m_k} \times \boldsymbol{B}_0\right)\nabla_v f_{k1} = -\frac{q_k}{m_k}(\boldsymbol{E}_1 + \boldsymbol{v} \times \boldsymbol{B}_1)\nabla_v f_{k0} \tag{11.15}$$

$$i\boldsymbol{k}\cdot\boldsymbol{E}_1 = \frac{1}{\varepsilon_0}\sum_k q_k \int f_{k1}\,\mathrm{d}\boldsymbol{v} \tag{11.16}$$

$$\frac{1}{\mu_0}(\boldsymbol{k} \times \boldsymbol{B}_1) = -\omega\left(\varepsilon_0 \boldsymbol{E}_1 + \frac{i}{\omega}\sum_k q_k \int \boldsymbol{v} f_{k1}\,\mathrm{d}\boldsymbol{v}\right) \tag{11.17}$$

$$\boldsymbol{B}_1 = \frac{1}{\omega}(\boldsymbol{k} \times \boldsymbol{E}_1). \tag{11.18}$$

The right-hand side of eq. (11.15) is a linear equation in $\boldsymbol{E}_1$ as is clear from eq. (11.18), so that f_{k1} is given as a linear equation in $\boldsymbol{E}_1$. The *dielectric tensor* of the hot plasma is defined by

$$\boldsymbol{E}_1 + \frac{i}{\varepsilon_0 \omega}\sum_k q_k \int \boldsymbol{v} f_{k1}\,\mathrm{d}\boldsymbol{v} \equiv \boldsymbol{K}\boldsymbol{E}_1. \tag{11.19}$$

The dispersion equation is derived from eqs. (11.17) and (11.18):

$$\boldsymbol{k} \times (\boldsymbol{k} \times \boldsymbol{E}_1) + \frac{\omega^2}{c^2}\boldsymbol{K}\boldsymbol{E}_1 = 0. \tag{11.20}$$

Consequently if f_{k1} can be solved from eq. (11.15), then $\boldsymbol{K}$ can be obtained. As for cold plasmas, the properties of waves in hot plasmas can be studied by solving the dispersion equation (11.20). When the right-hand side of eq. (11.15) is integrated along the particle orbit (eqs. (11.2), (11.3)) in the unperturbed state, we find that

$$f_{k1}(\boldsymbol{r}, \boldsymbol{v}, t) = -\frac{q_k}{m_k}\int_{-\infty}^{t}\left(\boldsymbol{E}_1(\boldsymbol{r}'(t'), t') + \frac{1}{\omega}\boldsymbol{v}'(t') \times (\boldsymbol{k} \times \boldsymbol{E}_1(\boldsymbol{r}'(t'), t'))\right)$$

$$\cdot\nabla_{v'}f_{k0}(\boldsymbol{r}'(t'), \boldsymbol{v}'(t'))\mathrm{d}t'. \tag{11.21}$$

This is the solution of eq. (11.15), because the following relation holds:[1-3]

$$\begin{aligned}\frac{\mathrm{d}}{\mathrm{d}t}f_{k1}(\boldsymbol{r}, \boldsymbol{v}, t) &= \frac{\partial}{\partial t}f_{k1} + \frac{\mathrm{d}\boldsymbol{r}}{\mathrm{d}t}\cdot\nabla_r f_{k1} + \frac{\mathrm{d}\boldsymbol{v}}{\mathrm{d}t}\cdot\nabla_v f_{k1}\\ &= \frac{\partial f_{k1}}{\partial t} + \boldsymbol{v}\cdot\nabla_r f_{k1} + \left(\frac{q_k}{m_k}\boldsymbol{v} \times \boldsymbol{B}_0 + \boldsymbol{G}\right)\cdot\nabla_v f_{k1}\\ &= -\frac{q_k}{m_k}(\boldsymbol{E}_1 + \boldsymbol{v} \times \boldsymbol{B}_1)\cdot\nabla_v f_{k0}.\end{aligned} \tag{11.22}$$

Since the 1st-order terms vary as $\exp(-i\omega t)$, the integral (11.21) converges when the imaginary part of ω is positive. When the imaginary part of ω is negative, the solution can be given by analytic continuation from the region of the positive imaginary part.

When waves are electrostatic i.e., when the electric field can be expressed by the electrostatic potential ϕ_1, we find

$$\boldsymbol{E}_1 = -\nabla\phi_1 \qquad \frac{\partial \boldsymbol{B}_1}{\partial t} = -\nabla \times \boldsymbol{E}_1 = 0 \tag{11.23}$$

and the dispersion relation becomes simple:

$$-\nabla^2\phi_1 = \frac{1}{\varepsilon_0}\sum_k q_k \int f_{k1}\mathrm{d}\boldsymbol{v} \tag{11.24}$$

$$f_{k1} = \frac{q_k}{m_k}\int_{-\infty}^{t} \nabla_{r'}\phi_1(\boldsymbol{r}', t')\cdot\nabla_{v'}f_{k0}(\boldsymbol{r}', \boldsymbol{v}')\mathrm{d}t'. \tag{11.25}$$

We will discuss the simple electrostatic-wave case in this chapter. A condition for the electrostatic approximation was described in sec. 10.6. Waves of low phase velocity in low-beta plasmas are electrostatic in many cases, so that the analysis of electrostatic waves is important.

11.2 Dispersion Equation of Electrostatic Waves

11.2a Dispersion Equation

Let us take the z axis in the direction of the magnetic field and x axis in the direction of the density gradient. It is assumed that the acceleration lies in the direction of the x axis. The zeroth-order distribution function $f_0(\boldsymbol{r}, \boldsymbol{v})$ must satisfy eq. (11.12), or

$$v_x \frac{\partial f_0}{\partial x} + G \frac{\partial f_0}{\partial v_x} - \Omega\left(v_y \frac{\partial}{\partial v_x} - v_x \frac{\partial}{\partial v_y}\right) f_0 = 0. \tag{11.26}$$

The solutions of the equations for particle motion are

$$v_x^2 + v_y^2 - 2Gx = \alpha \tag{11.27}$$

$$(v_z - V)^2 = \beta \tag{11.28}$$

$$x - \frac{v_y}{\Omega} = \gamma. \tag{11.29}$$

Consequently the general solution of eq. (11.26) is

$$f_0(v_x^2 + v_y^2 - 2Gx, (v_z - V)^2, x - v_y/\Omega) = f_0(\alpha, \beta, \gamma). \tag{11.30}$$

f_1 can be obtained in terms of the integral (11.25):

$$f_1 = \frac{q}{m} \int_{-\infty}^{t} \nabla_{r'} \phi_1(\boldsymbol{r}', t') \cdot \left(2(\boldsymbol{v}' - v_z \hat{\boldsymbol{z}}) \frac{\partial f_0}{\partial \alpha'} + 2(v_z - V) \frac{\partial f_0}{\partial \beta'} \hat{\boldsymbol{z}} - \frac{1}{\Omega} \frac{\partial f_0}{\partial \gamma'} \hat{\boldsymbol{y}}\right) dt', \tag{11.31}$$

where $\hat{\boldsymbol{x}}$, $\hat{\boldsymbol{y}}$, and $\hat{\boldsymbol{z}}$ are unit vectors for the x, y, and z axes, respectively. Let us consider

$$\phi_1(\boldsymbol{r}, t) = \phi_1(x) \exp(ik_y y + ik_z z - i\omega t). \tag{11.32}$$

Then the integrand becomes

$$\nabla_{r'} \phi_1 \cdot \nabla_{v'} f_0 = (\boldsymbol{v}' \cdot \nabla_{r'} \phi_1) 2 \frac{\partial f_0}{\partial \alpha'} + \left(2ik_z(v_z' - V) \frac{\partial f_0}{\partial \beta'} - 2ik_z v_z' \frac{\partial f_0}{\partial \alpha'} - \frac{ik_y}{\Omega} \frac{\partial f_0}{\partial \gamma'}\right) \phi_1.$$

Using

$$\frac{d\phi_1}{dt'} = \frac{\partial \phi_1}{\partial t'} + \boldsymbol{v}' \cdot \nabla_{r'} \phi_1 = -i\omega\phi_1 + \boldsymbol{v}' \cdot \nabla_{r'} \phi_1,$$

$$\int^t \boldsymbol{v}' \cdot \nabla_{r'} \phi_1 dt' = \phi_1 + i\omega \int^t \phi_1 dt', \qquad \alpha' = \alpha, \qquad \beta' = \beta \qquad \gamma' = \gamma,$$

we find that

$$f_1 = \frac{q}{m}\left(2\frac{\partial f_0}{\partial \alpha}\phi_1 + \left(2i\omega\frac{\partial f_0}{\partial \alpha} + 2ik_z(v_z - V)\frac{\partial f_0}{\partial \beta} - 2ik_z v_z\frac{\partial f_0}{\partial \alpha} - \frac{ik_y}{\Omega}\frac{\partial f_0}{\partial \gamma}\right)\right.$$
$$\left.\times \int^t \phi_1(x')\exp(i(k_y y' + k_z z') - i\omega t')\mathrm{d}t'\right). \quad (11.33)$$

The substitution of eq. (11.3) for y' and z' in the integrand of eq. (11.33) yields

$$\int_{-\infty}^{t} \phi_1(x')\exp(i(k_y y' + k_z z') - i\omega t')\mathrm{d}t'$$
$$= \phi_1(x)\exp(i(k_y y + k_z z) - i\omega t)\exp\left(i\frac{k_y v_\perp}{\Omega}\cos\theta\right)$$
$$\times \int_{-\infty}^{0} \exp\left(-i\frac{k_y v_\perp}{\Omega}\cos(\theta + \Omega\tau) + i\left(k_z v_z + \frac{k_y G}{\Omega} - \omega\right)\tau\right)\mathrm{d}\tau.$$

Here it is assumed that $\phi_1(x')$ changes slowly. Using the expansions

$$\exp(-ia\cos\theta) = \sum_{n=-\infty}^{\infty} (-i)^n J_n(a)\exp(-in\theta)$$

$$\exp(ia\cos\theta) = \sum_{m=-\infty}^{\infty} (i)^m J_m(a)\exp im\theta$$

$$(J_{-m}(a) = (-1)^m J_m(a)),$$

we write the integral as

$$\int_{-\infty}^{t} \phi_1(\boldsymbol{r}', t')\mathrm{d}t'$$
$$= \phi_1(\boldsymbol{r}, t)\sum_{n=-\infty}^{\infty}\frac{i}{(\omega - k_z v_z - k_y G/\Omega) + n\Omega}$$
$$\times (J_n{}^2(a) + i J_n(a)J_{n+1}(a)\exp i\theta$$
$$- i J_n(a)J_{n-1}(a)\exp(-i\theta) + \cdots), \quad (11.34)$$

where $a = k_y v_\perp/\Omega$. The substitution of the foregoing equation into eq. (11.33) gives

$$f_1(\boldsymbol{r}, \boldsymbol{v}, t)$$
$$= \frac{q}{m}\phi_1(\boldsymbol{r}, t)\left[2\frac{\partial f_0}{\partial \alpha} - \left(2(\omega - k_z v_z)\frac{\partial f_0}{\partial \alpha} + 2k_z(v_z - V)\frac{\partial f_0}{\partial \beta} - \frac{k_y}{\Omega}\frac{\partial f_0}{\partial \gamma}\right)\right.$$
$$\left. \times \sum_{n=-\infty}^{\infty} \frac{J_n^{\ 2} + iJ_nJ_{n+1}\exp i\theta - iJ_nJ_{n-1}\exp(-i\theta) + \cdots}{(\omega - k_z v_z - k_y G/\Omega) + n\Omega}\right]. \tag{11.35}$$

When this expression for f_1 is substituted into the dispersion equation (11.24), we find

$$\left(k_y^{\ 2} + k_z^{\ 2} - \frac{\partial^2}{\partial x^2}\right)\phi_1 = \phi_1 \sum_j \frac{q_j^{\ 2}}{\varepsilon_0 m_j}$$
$$\times \iiint \left[2\frac{\partial f_0}{\partial \alpha} - \left(2(\omega - k_z v_z)\frac{\partial f_0}{\partial \alpha} + 2k_z(v_z - V)\frac{\partial f_0}{\partial \beta} - \frac{k_y}{\Omega}\frac{\partial f_0}{\partial \gamma}\right)\right.$$
$$\left. \times \sum_{n=-\infty}^{\infty} \frac{J_n^{\ 2} + iJ_nJ_{n+1}\exp i\theta - iJ_nJ_{n-1}\exp(-i\theta) + \cdots}{(\omega - k_z v_z - k_y G/\Omega) + n\Omega}\right]_j$$
$$\times \mathrm{d}\theta v_\perp \mathrm{d}v_\perp \mathrm{d}v_z. \tag{11.36}$$

For $|(k_y^{\ 2} + k_z^{\ 2})\phi_1| \gg |\partial^2 \phi_1/\partial x^2|$, eq. (11.36) is reduced to

$$k_y^{\ 2} + k_z^{\ 2} - \sum_j \Pi_j^{\ 2}\frac{1}{n_j}\iiint [\quad]_j \mathrm{d}\theta v_\perp \mathrm{d}v_\perp \mathrm{d}v_z = 0, \tag{11.37}$$

where $[\quad]_j$ is the same as the integrand of eq. (11.36).

In order to take the variation of $\phi_1(x')$ into account in $\phi_1(\boldsymbol{r}', t) = \phi_1(x')\exp(ik_y y' + ik_z z' - i\omega t)$, we expand $\phi_1(x')$ around $x' = x$ as $\phi_1(x') = \phi_1(x) + (x' - x)\mathrm{d}\phi_1/\mathrm{d}x + \frac{1}{2}(x' - x)^2 \mathrm{d}^2\phi_1/\mathrm{d}x^2 + \cdots$. The more accurate expression of eq. (11.34) is then

$$\int_0^t \phi_1(\boldsymbol{r}', t')\exp(i(k_y y' + k_z z') - i\omega t')\mathrm{d}t'$$
$$= \phi_1(\boldsymbol{r}, t)\sum_{n=-\infty}^{\infty} \frac{i(J_n^{\ 2}(a) + \cdots)}{(\omega - k_z v_z - k_y G/\Omega) + n\Omega} + \frac{1}{2}\frac{\mathrm{d}^2\phi_1(\boldsymbol{r}, t)}{\mathrm{d}x^2} \times$$
$$\sum_{n=-\infty}^{\infty} \frac{i(v_\perp^{\ 2}/\Omega^2)(J_n^{\ 2} + (J_{n+2} + J_{n-2})J_n/2 - (J_{n+1} + J_{n-1})^2/2) + \cdots}{(\omega - k_z v_z - k_y G/\Omega) + n\Omega}$$
$$+ \cdots.$$

Let us consider the zeroth-order distribution function

$$f_0\left(v_\perp{}^2 - 2Gx, (v_z - V)^2, x - \frac{v_y}{\Omega}\right)$$
$$= n_0\left(1 + \left(-\varepsilon + \delta_\perp \frac{v_\perp{}^2 - 2Gx}{2v_{\mathrm{T}\perp}{}^2} + \delta_z \frac{(v_z - V)^2}{2v_{\mathrm{T}z}{}^2}\right)\left(x - \frac{v_y}{\Omega}\right)\right)$$
$$\times \left(\frac{1}{2\pi v_{\mathrm{T}\perp}{}^2}\right)\left(\frac{1}{2\pi v_{\mathrm{T}z}{}^2}\right)^{1/2} \exp\left(-\frac{1}{2v_{\mathrm{T}\perp}{}^2}(v_\perp{}^2 - 2Gx) - \frac{1}{2v_{\mathrm{T}z}{}^2}(v_z - V)^2\right) \tag{11.38}$$

where $v_{\mathrm{T}\perp}{}^2 = T_\perp/m$ and $v_{\mathrm{T}z}{}^2 = T_z/m$. $T_\perp$ and T_z are temperatures perpendicular and parallel to the magnetic field respectively. This distribution function represents a plasma with finite density and temperature gradients, and the shifted velocity V. The density and temperature gradients are

$$\frac{1}{n_0}\frac{\mathrm{d}n_0}{\mathrm{d}x} = -\varepsilon + \delta_\perp + \frac{\delta_z}{2} + \frac{G}{v_{\mathrm{T}\perp}{}^2} \tag{11.39}$$

$$\frac{1}{T_\perp}\frac{\mathrm{d}T_\perp}{\mathrm{d}x} = \delta_\perp \tag{11.40}$$

$$\frac{1}{T_z}\frac{\mathrm{d}T_z}{\mathrm{d}x} = \delta_z, \tag{11.41}$$

respectively.

From the expression (11.38) for f_0, we have

$$2\frac{\partial f_0}{\partial \alpha} = -\frac{n_0}{v_{\mathrm{T}\perp}{}^2}\left(1 + ((-\varepsilon') - \delta_\perp)\left(x - \frac{v_y}{\Omega}\right)\right)F_\perp F_z$$

$$2\frac{\partial f_0}{\partial \beta} = -\frac{n_0}{v_{\mathrm{T}z}{}^2}\left(1 + ((-\varepsilon') - \delta_z)\left(x - \frac{v_y}{\Omega}\right)\right)F_\perp F_z$$

$$\frac{\partial f_0}{\partial \gamma} = n_0(-\varepsilon')F_\perp F_z$$

$$(-\varepsilon') \equiv \left(-\varepsilon + \delta_\perp \frac{v_\perp{}^2 - 2Gx}{2v_{\mathrm{T}\perp}{}^2} + \delta_z \frac{(v_z - V)^2}{2v_{\mathrm{T}z}{}^2}\right)$$

$$F_\perp = \left(\frac{1}{2\pi v_{\mathrm{T}\perp}{}^2}\right)\exp\left(-\frac{1}{2v_{\mathrm{T}\perp}{}^2}(v_\perp{}^2 - 2Gx)\right)$$

$$F_z = \left(\frac{1}{2\pi v_{\mathrm{T}z}{}^2}\right)^{1/2} \exp\left(-\frac{1}{2v_{\mathrm{T}z}{}^2}(v_z - V)^2\right).$$

In calculating the dispersion equation (11.37), the following notations and relations are used:

$$v_y = v_\perp \sin\theta = v_\perp \frac{\mathrm{e}^{i\theta} - \mathrm{e}^{-i\theta}}{2i}$$

$$\frac{J_{n+1}(x) + J_{n-1}(x)}{2} = \frac{n}{x} J_n(x)$$

$$\int_0^\infty J_n{}^2(b^{1/2}x)\exp\left(-\frac{x^2}{2\alpha}\right)\cdot x\mathrm{d}x = \alpha I_n(\alpha b)\mathrm{e}^{-\alpha b}$$

$$\int_0^\infty J_n{}^2(b^{1/2}x)\exp\left(-\frac{x^2}{2}\right)\cdot\frac{x^2}{2} x\mathrm{d}x = f_n(b) I_n(b)\mathrm{e}^{-b}$$

$$f_n(b) \equiv (1 - b) + bI_n'(b)/I_n(b)$$

$$\sum_{n=-\infty}^{\infty} J_n{}^2(x) = 1$$

$$\sum_{n=-\infty}^{\infty} I_n(b)\mathrm{e}^{-b} = 1$$

$$Z(\zeta) = \frac{1}{\pi^{1/2}} \int_{-\infty}^{\infty} \frac{\exp(-\beta^2)}{\beta - \zeta}\mathrm{d}\beta$$

$$b \equiv \left(\frac{k_y v_{\mathrm{T}\perp}}{\Omega}\right)^2$$

$$\omega_n \equiv \omega - k_z V - \frac{k_y G}{\Omega} + n\Omega$$

$$\zeta_n \equiv \frac{\omega_n}{2^{1/2} k_z v_{\mathrm{T}z}}.$$

We obtain

$$\int_{-\infty}^{\infty} \frac{F_z}{\omega_n - k_z(v_z - V)}\mathrm{d}v_z = \frac{1}{\pi^{1/2}} \int_{-\infty}^{\infty} \frac{\exp(-\beta^2)\mathrm{d}\beta}{\omega_n - 2^{1/2}k_z v_{\mathrm{T}z}\beta}$$

$$= \frac{1}{\omega_n}(-\zeta_n) Z(\zeta_n)$$

$$\int_{-\infty}^{\infty} \frac{k_z(v_z - V)}{\omega_n - k_z(v_z - V)} F_z \mathrm{d}v_z = -(1 + \zeta_n Z(\zeta_n))$$

$$\int_{-\infty}^{\infty} \frac{(k_z(v_z - V))^2}{\omega_n - k_z(v_z - V)} F_z \mathrm{d}v_z = -\omega_n(1 + \zeta_n Z(\zeta_n))$$

$$\int_{-\infty}^{\infty} \frac{(k_z(v_z - V))^3}{\omega_n - k_z(v_z - V)} F_z \mathrm{d}v_z = -\frac{k_z^2 T_z}{m} - \omega_n^2(1 + \zeta_n Z(\zeta_n))$$

$$\int_0^{\infty} J_n^2\left(\frac{k_y v_\perp}{\Omega}\right) F_\perp 2\pi v_\perp \mathrm{d}v_\perp = \exp(-b) I_n(b).$$

Finally, the dispersion equation is reduced to

$$\begin{aligned} k_y^2 + k_z^2 + \sum_j \Pi_j^2 \Bigg[& \frac{1}{v_{\mathrm{T}z}^2} \\ & + \sum_{n=-\infty}^{\infty} I_n(b) \mathrm{e}^{-b} \Bigg\{ \frac{1}{v_{\mathrm{T}z}^2} \zeta_n Z(\zeta_n) + \frac{1}{v_{\mathrm{T}\perp}^2} \frac{k_y G/\Omega - n\Omega}{\omega_n} \zeta_n Z(\zeta_n) \\ & + \frac{1}{v_{\mathrm{T}\perp}^2} \frac{n}{k_y} \Bigg[(\varepsilon + \delta_\perp - f_n(b)\delta_\perp) \left(1 - \frac{k_y G/\Omega - n\Omega}{\omega_n} \zeta_n Z(\zeta_n) \right) \\ & \qquad - \frac{\delta_z}{2} \left(1 - \frac{(k_y G/\Omega - n\Omega)\omega_n}{k_z^2 v_{\mathrm{T}z}^2} (1 + \zeta_n Z(\zeta_n)) \right) \Bigg] \\ & - \frac{1}{v_{\mathrm{T}z}^2} \frac{n}{k_y} \Bigg[(\varepsilon + \delta_z - f_n(b)\delta_\perp)(1 + \zeta_n Z(\zeta_n)) \\ & \qquad - \frac{\delta_z}{2} \left(1 + \frac{\omega_n^2}{k_z^2 v_{\mathrm{T}z}^2} (1 + \zeta_n Z(\zeta_n)) \right) \Bigg] \\ & - \frac{k_y}{\Omega} \Bigg[(\varepsilon - f_n(b)\delta_\perp) \frac{1}{\omega_n} \zeta_n Z(\zeta_n) - \frac{\delta_z}{2} \frac{\omega_n}{k_z^2 v_{\mathrm{T}z}^2} (1 + \zeta_n Z(\zeta_n)) \Bigg] \Bigg\} \Bigg]_j = 0. \end{aligned} \tag{11.42}$$

The function $Z(\zeta)$ is called the plasma dispersion function and its properties will be described in sec. 11.2b.

When the plasma is homogeneous, i.e., when $\varepsilon = \delta_\perp = \delta_z = 0$, the dispersion equation becomes

$$k_y^2 + k_z^2 + \sum_j \Pi_j^2 m_j \left(\frac{1}{T_z} + \sum_{n=-\infty}^{\infty} I_n(b) \mathrm{e}^{-b} \left(\frac{1}{T_z} + \frac{1}{T_\perp} \frac{(k_y G/\Omega - n\Omega)}{\omega_n} \right) \times \zeta_n Z(\zeta_n) \right)_j = 0. \tag{11.43}$$

When the frequency ω is higher than the cyclotron frequency Ω, a particle can be considered to move along a straight-line segment during the characteristic time $1/\omega$. When $G = 0$ and $v_{\mathrm{T}\perp}^2 = v_{\mathrm{T}z}^2$, the dispersion equation is

$$k_y^2 + k_z^2 + \sum_j \frac{1}{\lambda_{\mathrm{d}j}^2}(1 + \zeta Z(\zeta))_j = 0 \tag{11.44}$$

$$\frac{1}{\lambda_{\mathrm{d}}^2} = \frac{n_0 q^2}{\varepsilon_0 T} \qquad \zeta = \frac{\omega - \boldsymbol{k}\cdot\boldsymbol{V}}{2^{1/2}k(T/m)^{1/2}}.$$

This is also the exact dispersion equation in the case of zero magnetic field.

For the low-frequency case $\omega \ll |\Omega|$, we have

$$\zeta_n = (\omega - k_z V - k_y G/\Omega + n\Omega)/(2^{1/2}k_z v_{\mathrm{T}z}) \gg 1$$

for $n \neq 0$. When the relation

$$\zeta_n Z(\zeta_n) \approx -(1 + 1/(2\zeta_n^2) + \cdots)$$

is used (this approximation will be derived later), the low frequency dispersion equation becomes[4]

$$k_y^2 + k_z^2 + \sum_j \Pi_j^2 \left\{ \frac{m}{T_\perp} + I_0 e^{-b} \left[\frac{m}{T_\parallel}(1 + \zeta_0 Z(\zeta_0)) - \frac{m}{T_\perp}\left(1 - \frac{k_y G/\Omega}{\omega_0}\zeta_0 Z(\zeta_0)\right) - \frac{k_y}{\Omega\omega_0}(\varepsilon - f_0(b)\delta_\perp)\zeta_0 Z(\zeta_0) + \frac{\delta_z}{2}\frac{\omega_0}{k_z^2(T_z/m)}\frac{k_y}{\Omega}(1 + \zeta_0 Z(\zeta_0)) \right] \right\}_j = 0. \tag{11.45}$$

This dispersion equation explains the properties of the drift wave, as will be seen in the next chapter.

11.2b Plasma Dispersion Function

The *plasma dispersion function* $Z_{\mathrm{p}}(\zeta)$ is defined by

$$Z_{\mathrm{p}}(\zeta) = \frac{1}{\pi^{1/2}}\int_{-\infty}^{\infty} \frac{\exp(-\beta^2)}{\beta - \zeta}\,\mathrm{d}\beta \tag{11.46}$$

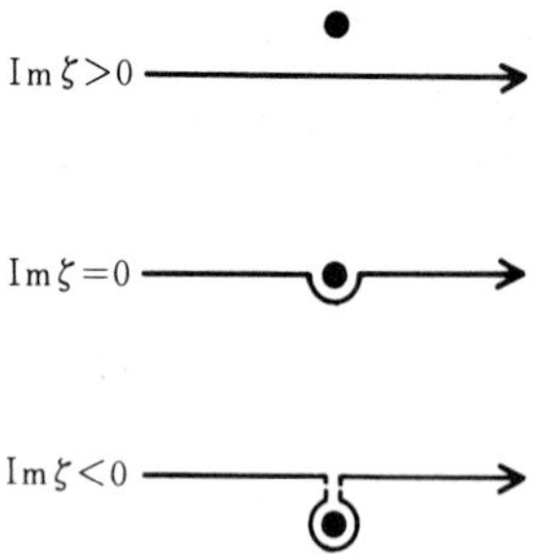

Fig. 11.1
Integration paths for the dispersion function.

when Im $\zeta > 0$. When Im $\zeta < 0$, analytic continuation must be used. The integral paths for Im $\zeta \gtreqless 0$ are shown in fig. 11.1. When ζ is real and is equal to x, $Z_p(x)$ is given by

$$Z_p(x) = \frac{1}{\pi^{1/2}} \mathrm{P} \int_{-\infty}^{\infty} \frac{\exp(-\beta^2)}{\beta - x} \mathrm{d}\beta + i\pi^{1/2} \exp(-x^2) \tag{11.47}$$

where $\mathrm{P} \int_{-\infty}^{\infty}$ is Cauchy's principal value $\lim_{\varepsilon \to 0} \left(\int_{-\infty}^{x-\varepsilon} + \int_{x+\varepsilon}^{\infty} \right)$.

Setting $\beta - \zeta = \gamma$ in eq. (11.46) and using the relation

$$\int_{-\infty}^{\infty} (\exp(-\gamma^2 - 2\zeta\gamma)/\gamma)\mathrm{d}\gamma = -2\pi^{1/2} \int_{+i\infty}^{\zeta} \exp(t^2)\mathrm{d}t$$

(if both sides are differentiated, it is clear that they are equal), we find

$$Z_p(\zeta) = 2i \exp(-\zeta^2) \int_{-\infty}^{i\zeta} \exp(-t^2)\mathrm{d}t$$

$$= i\pi^{1/2} \exp(-\zeta^2)(1 + \mathrm{erf}(i\zeta)) \tag{11.48}$$

$$\mathrm{erf}(\eta) \equiv \left(\frac{2}{\pi^{1/2}} \int_{-\infty}^{\eta} \exp(-t^2)\mathrm{d}t - 1 \right) = 2\Phi(2^{1/2}\eta) - 1 \tag{11.49}$$

$$\Phi(\alpha) \equiv \frac{1}{(2\pi)^{1/2}} \int_{-\infty}^{\alpha} \exp\left(-\frac{t^2}{2}\right)\mathrm{d}t.$$

This expression is valid for all cases, Im $\zeta \gtreqless 0$. The series expansion of $Z_p(\zeta)$ is

$$Z_p(\zeta) = i\pi^{1/2}\exp(-\zeta^2) - \zeta \sum_{n=0}^{\infty} \frac{(-\zeta^2)^n \pi^{1/2}}{\Gamma(n+3/2)}$$

$$= i\pi^{1/2}\exp(-\zeta^2) - 2\zeta\left(1 - \frac{2\zeta^2}{3} + \frac{4\zeta^4}{15} - \cdots\right), \qquad (11.50a)$$

and the asymptotic expansion of $Z_p(\zeta)$ is given by[5,6]

$$Z_p(\zeta) \approx i\pi^{1/2}\sigma\exp(-\zeta^2) - \frac{1}{\zeta}\left(1 + \frac{1}{2\zeta^2} + \frac{3}{4\zeta^4} + \cdots\right)$$

$$\approx i\pi^{1/2}\sigma\exp(-\zeta^2) - \sum_{n=0}^{\infty} \zeta^{-(2n+1)} \frac{\Gamma(n+1/2)}{\pi^{1/2}} \qquad (11.51a)$$

$$\sigma = \begin{cases} 0 & \mathrm{Im}\zeta > 0, \quad |\mathrm{Im}\,\zeta| > |\mathrm{Re}\,\zeta| \\ 1 & |\mathrm{Im}\,\zeta| < |\mathrm{Re}\,\zeta| \\ 2 & \mathrm{Im}\,\zeta < 0, \quad |\mathrm{Im}\,\zeta| > |\mathrm{Re}\,\zeta|. \end{cases}$$

The curves for real and imaginary parts of $Z_p(x)$, when ζ is real and equal to x, are shown in fig. 11.2.

In the derivation of the dispersion equation (11.42), we have used

$$Z\left(\frac{\omega_n}{k_z 2^{1/2} v_{Tz}}\right) = \frac{1}{\pi^{1/2}} \int_{-\infty}^{\infty} \frac{\exp(-\beta^2)\mathrm{d}\beta}{\beta - \omega_n/(k_z 2^{1/2} v_{Tz})}.$$

This function Z in the dispersion equation is defined for Im $\omega > 0$, and it can be analytically continued for Im $\omega < 0$. Accordingly, the foregoing function is the same as the plasma dispersion function $Z_p(\zeta)$ itself for

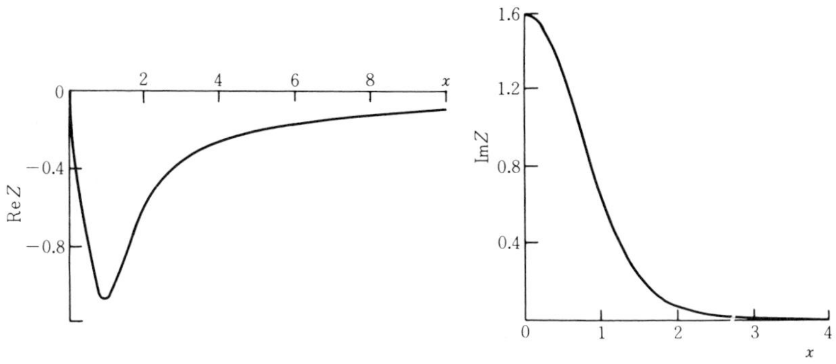

Fig. 11.2
Real and imaginary parts of $Z_p(x)$.

$k_z > 0$:

$$Z(\zeta) = Z_{\mathrm{p}}(\zeta) \qquad (k_z > 0).$$

Let us consider the case $k_z < 0$. As the dispersion equation was deduced for $\mathrm{Im}\,\omega > 0$, the function Z appearing therein was defined for $\mathrm{Im}(\omega_n/2^{1/2}k_z v_{\mathrm{T}z}) < 0$ when $k_z < 0$. Analytic continuation must be used for $\mathrm{Im}(\omega_n/2^{1/2}k_z v_{\mathrm{T}z}) > 0$ in this case. The series expansion of $Z(\zeta)$ is

$$Z(\zeta) = i\frac{k_z}{|k_z|}\pi^{1/2}\exp(-\zeta^2) - 2\zeta(1 - 2\zeta^2/3 + 4\zeta^4/15 - \cdots), \tag{11.50b}$$

and the asymptotic expansion is

$$Z(\zeta) \approx i\pi^{1/2}\sigma\frac{k_z}{|k_z|}\exp(-\zeta^2) - \frac{1}{\zeta}\left(1 + \frac{1}{2\zeta^2} + \frac{3}{4\zeta^4} + \cdots\right). \tag{11.51b}$$

11.3 Ion Acoustic, Electron Acoustic, and Bernstein Waves

Let us consider low-frequency waves in a homogeneous plasma without gravity ($\varepsilon = \delta_\perp = \delta_z = 0$, $G = 0$, $\omega \ll |\Omega|$). The dispersion equation (eq. (11.45)) is

$$k_y^2 + k_z^2 + \sum_j \Pi_j^2 m_j\left(\frac{1}{T_z}I_0(b)\mathrm{e}^{-b}(1 + \zeta_0 Z(\zeta_0))\right)_j = 0 \tag{11.52}$$

$$b = \left(k_y\frac{v_{\mathrm{T}\perp}}{\Omega}\right)^2 \tag{11.53}$$

$$\zeta_0 = \frac{\omega - k_z V}{2^{1/2}k_z v_{\mathrm{T}z}} \tag{11.54}$$

$$v_{\mathrm{T}\perp}^2 = \frac{T_\perp}{m} \qquad v_{\mathrm{T}z}^2 = \frac{T_z}{m}. \tag{11.55}$$

When the drift velocity V_{e} of an electron is smaller than the thermal velocity v_{Tze} and the phase velocity of the wave is also much smaller than the thermal velocity of the electron, then ($|\zeta_{0\mathrm{e}}| \ll 1$) and

$$k_y^2 + k_z^2 + \Pi_{\mathrm{e}}^2\frac{m_{\mathrm{e}}}{T_{z\mathrm{e}}}\left(1 + i\sigma\frac{k_z}{|k_z|}\pi^{1/2}\zeta_{0\mathrm{e}}\exp(-\zeta_{0\mathrm{e}}^2)\right)$$

$$+ \Pi_{\mathrm{i}}^2\frac{m_{\mathrm{i}}}{T_{z\mathrm{i}}}I_0(b_{\mathrm{i}})\mathrm{e}^{-b_{\mathrm{i}}}(1 + \zeta_{0\mathrm{i}}Z(\zeta_{0\mathrm{i}})) = 0. \tag{11.56}$$

The electron Larmor radius is assumed to be zero ($b_e = 0$). Let us consider the case where the wave propagates along magnetic lines of force; then $k_y = 0$ and $b_i = 0$. If the function $W(\zeta) \equiv -(1 + \zeta Z(\zeta))$ is introduced, eq. (11.56) becomes

$$\frac{k_z^2}{\Pi_i^2}\frac{T_{zi}}{m_i} + \frac{T_{zi}}{ZT_{ze}} = W(\zeta_{0i}) - \frac{T_{zi}}{ZT_{ze}} i\sigma \frac{k_z}{|k_z|}\pi^{1/2}\zeta_{0e}\exp(-\zeta_{0e}^2). \tag{11.57}$$

Consider the stationary wave, for which the imaginary part of the right-hand side of eq. (11.57) is zero. It is given by

$$i\frac{T_{zi}}{Z}\pi^{1/2}\left(-\frac{Z}{T_{zi}}\frac{(\omega/k_z)m_i^{1/2}}{2^{1/2}T_{zi}^{1/2}}\exp\left(-\frac{m_i(\omega/k_z)^2}{2T_{zi}}\right)\right.$$
$$\left.-\frac{1}{T_{ze}}\frac{(\omega/k_z - V_e)m_e^{1/2}}{2^{1/2}T_{ze}^{1/2}}\exp\left(-\frac{m_e(\omega/k_z - V_e)^2}{2T_{ze}}\right)\right). \tag{11.58}$$

(The ion drift velocity V_i is assumed to be zero.) The first term is the ion Landau damping term and the second is the electron Landau damping or amplifying term. In order to make the imaginary part (11.58) zero, V_e must be larger than $|\omega/k_z|$. The real part of $W(\zeta_{0i})$ is shown in fig. 11.3. $W(\zeta_{0i})$ has a maximum value of 0.285 (at $|\zeta_{0i}| = 1.502$), so that the left-hand side of eq. (11.57) must be smaller than 0.285. Accordingly, we have a necessary condition for the existence of the wave:

$$T_{ze} > \frac{1}{Z \times 0.285} T_{zi}. \tag{11.59}$$

When the solution ζ_{0i} of eq. (11.57) is obtained, ω/k_z is found from eq. (11.54) to be

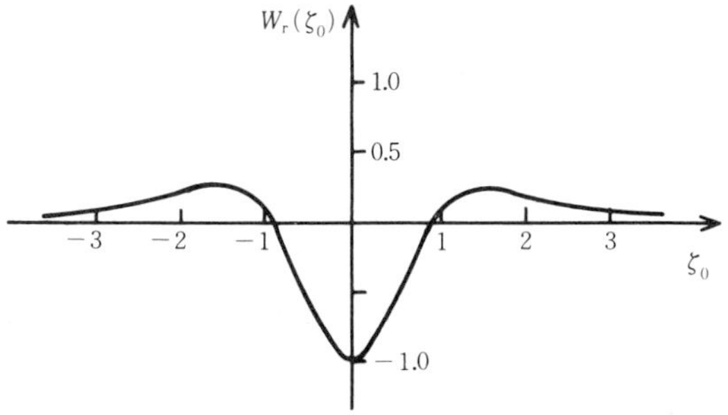

Fig. 11.3
Real part of $W(\zeta_0)$.

$$\frac{\omega}{k_z} = 2^{1/2}\zeta_{0\mathrm{i}}\left(\frac{T_{z\mathrm{i}}}{m_\mathrm{i}}\right)^{1/2}. \tag{11.60}$$

This wave is called the *ion acoustic wave.*[6] The minimum wavelength is of the order of the Debye length as is clear from eq. (11.57).

When the phase velocity of the wave is much larger than the ion thermal velocity, the dispersion equation becomes

$$k_y^2 + k_z^2 + \Pi_\mathrm{e}^2\frac{m_\mathrm{e}}{T_{z\mathrm{e}}}(1 + \zeta_{0\mathrm{e}}Z(\zeta_{0\mathrm{e}})) - k_z^2 I_0(b_\mathrm{i})\mathrm{e}^{-b_\mathrm{i}}\frac{\Pi_\mathrm{i}^2}{\omega^2}$$

$$+ i\pi^{1/2}\frac{k_z}{|k_z|}\frac{\Pi_\mathrm{i}^2 m_\mathrm{i}}{T_{z\mathrm{i}}} I_0(b_\mathrm{i})\mathrm{e}^{-b_\mathrm{i}}\exp(-\zeta_{0\mathrm{i}}^2) = 0, \tag{11.61}$$

since $|\zeta_{0\mathrm{i}}| \gg 1$. Here we assume that the imaginary part of the dispersion equation is zero and $\zeta_{0\mathrm{e}}$ and $\zeta_{0\mathrm{i}}$ are real. The dispersion equation of the wave propagating along a field line is

$$\left(1 - \frac{\Pi_\mathrm{i}^2}{\omega^2}\right)\frac{k_z^2}{\Pi_\mathrm{e}^2}\frac{T_{z\mathrm{e}}}{m_\mathrm{e}} = \mathrm{Re}(W(\zeta_{0\mathrm{e}})). \tag{11.62}$$

The right-hand side has a maximum value of 0.285 (again at $|\zeta_{0\mathrm{e}}| = 1.502$). The wavelength at this point is of the order of the Debye length. The phase velocity is

$$\frac{\omega}{k_z} - V_\mathrm{e} = 2^{1/2}\zeta_{0\mathrm{e}}\left(\frac{T_{z\mathrm{e}}}{m_\mathrm{e}}\right)^{1/2}, \tag{11.63}$$

which is of the order of the electron thermal velocity. This wave is called the *electron acoustic wave.*[6]

Let us next consider the high-frequency waves in a homogeneous plasma. When we assume $G = 0$ and $|\zeta_n| \gg 1$ for $n \neq 0$, eq. (11.43) becomes

$$k_y^2 + k_z^2 + \sum_j \Pi_j^2\frac{m_j}{T_{\perp j}}\left(\sum_{n=1}^{\infty} I_n(b)\mathrm{e}^{-b}\frac{2n^2\Omega^2}{n^2\Omega^2 - \omega^2}\right)_j$$

$$+ \sum_j \Pi_j^2\frac{m_j}{T_{zj}}\left(I_0(b)\mathrm{e}^{-b}(1 + \zeta_0 Z(\zeta_0))\right)_j = 0. \tag{11.64}$$

The imaginary part of the last term expresses the Landau damping. For the wave propagating perpendicularly to the magnetic field, the Landau damping term disappears and $k_z \to 0$ ($\zeta_0 \to \infty$), i.e., $(1 + \zeta_0 Z(\zeta_0)) \to 0$. The dispersion equation is reduced to[7]

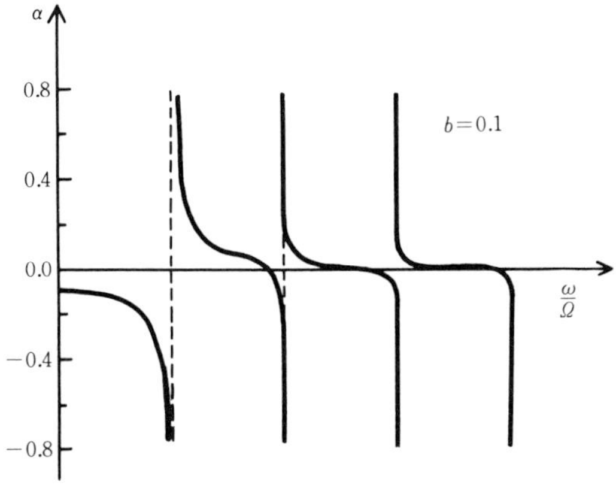

Fig. 11.4
Dependence of $\alpha(\omega/\Omega, b)$ on ω/Ω. After I. B. Bernstein: Phys. Rev. **109**, 10 (1958).

$$k_y^2 = \sum_j \Pi_j^2 \frac{m_j}{T_{\perp j}} \alpha\left(\frac{\omega}{\Omega_j}, b_j\right) \tag{11.65}$$

$$\alpha\left(\frac{\omega}{\Omega}, b\right) = -2 \sum_{n=1}^{\infty} I_n(b) e^{-b} \frac{n^2}{n^2 - (\omega/\Omega)^2}.$$

A plot of $\alpha(\omega/\Omega, b)$ is shown in fig. 11.4.

When the frequency of the wave is larger than the electron cyclotron frequency, the contribution of ion terms can be neglected. In this case the positive part of $\alpha(\omega/\Omega, b)$ (see fig. 11.4) expresses the right-hand side of the dispersion equation. This wave is called the *Bernstein wave.*

11.4 Two-Stream Instability

The interaction between beam and plasma is important. Let us consider an excited wave which propagates along lines of magnetic force ($k_y = 0$) in a homogeneous plasma without gravity. The dispersion equation (eq. (11.37)) is

$$1 = -\frac{1}{k} \sum_j \frac{\Pi_j^2}{n_j} \int_{-\infty}^{\infty} \frac{\dfrac{\partial f_j(v_z)}{\partial v_z}}{\omega - k v_z} dv_z. \tag{11.66}$$

Partial integration of eq. (11.66) gives

$$1 = \sum_j \frac{\Pi_j^{\,2}}{n_j} \int_{-\infty}^{\infty} \frac{f_j(v_z)}{(\omega - kv_z)^2} \mathrm{d}v_z. \tag{11.67}$$

When the jth particles drift with the velocity V_j, and the spread of the velocity is zero, i.e., when the temperature is zero, the distribution function is

$$f_j(v_z) = n_j \delta(v_z - V_j),$$

and eq. (11.67) becomes

$$1 = \sum_j \frac{\Pi_j^{\,2}}{(\omega - kV_j)^2}. \tag{11.68}$$

This simple dispersion equation can be directly derived by the linearization of the equation of motion and the equation of continuity:

$$-i\omega v_j + i(kV_j)v_j = \frac{q_j}{m_j}E \tag{11.69}$$

$$-\omega n_{j\perp} + n_{jo}(kv_j) + (kV_j)n_{j1} = 0. \tag{11.70}$$

Combining these equations with the linearized Poisson equation

$$ikE = \frac{1}{\varepsilon_0}\sum_j q_j n_{j1}, \tag{11.71}$$

we find eq. (11.68).

In the special case $\Pi_1^{\,2} = \Pi_2^{\,2} (n_1^{\,2} q_1^{\,2}/m_1 = n_2^{\,2} q_2^{\,2}/m_2)$ the dispersion equation is quadratic:

$$(\omega - k\bar{V})^2 = \Pi_t^{\,2}\left(\frac{1 + 2x^2 \pm (1 + 8x^2)^{1/2}}{2}\right), \tag{11.72}$$

where

$$\Pi_t^{\,2} = \Pi_1^{\,2} + \Pi_2^{\,2} \qquad x = \frac{k(V_1 - V_2)}{2\Pi_t} \qquad \bar{V} = \frac{(V_1 + V_2)}{2}.$$

If the positive sign is taken in eq. (11.72), the dispersion equation is

$$(\omega - k\bar{V})^2 = \Pi_t^{\,2}\left(1 + \frac{3(k(V_1 - V_2))^2}{4\Pi_t^{\,2}} + \cdots\right). \tag{11.73}$$

For the negative sign, the dispersion equation is

$$(\omega - k\bar{V})^2 = \Pi_t^{\,2}(-x^2 + x^3 + \cdots), \tag{11.74}$$

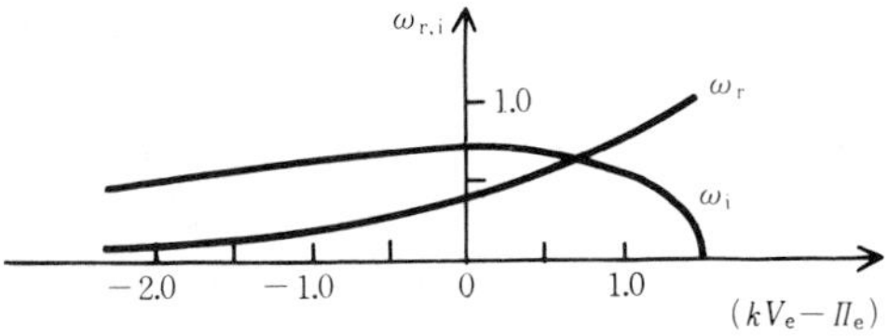

Fig. 11.5
Dispersion relation for Buneman instability. The ordinate and abscissa are plotted in units of $(\pi_i^2\pi_e)^{1/3}$. After 0. Buneman: Phys. Rev. **115**, 503(1959).

and the wave is unstable when $x < 1$, or

$$(k(V_1 - V_2))^2 < 4\Pi_t^2. \tag{11.75}$$

When $x^2 = 3/8$, the growth rate is maximum and $\operatorname{Im}\omega = \Pi_t^2/2^{3/2}$.

The energy to excite this instability comes from the zeroth-order kinetic energy of beam motion. When some disturbance occurs in the beam motion, charged particles may be bunched and an electric field induced. If this electric field acts to amplify the bunching, the disturbance grows. This instability is called *two-stream instability*.

Let us consider a two-component plasma in which the ions are motionless and the electrons move with velocity V_e. The dispersion equation is

$$\frac{\Pi_i^2}{\omega^2} + \frac{\Pi_e^2}{(\omega - kV_e)^2} = 1. \tag{11.76}$$

The dependence of the solution $\omega = \omega_r \pm i\omega_i$ on kV_e is shown in fig. 11.5, which is taken from a paper by Buneman.[8]
Equation (11.76) is written as

$$kV_e = \omega + \Pi_e\left(1 - \frac{\Pi_i^2}{\omega^2}\right)^{-1/2} \approx \Pi_e + \omega + \frac{\Pi_e\Pi_i^2}{2\omega^2}. \tag{11.77}$$

When

$$\omega = (\Pi_e\Pi_i^2\cos\theta)^{1/3}e^{i\theta}, \tag{11.78}$$

then kV_e is real. The growth rate is maximum at $\theta = \pi/3$ and $\omega_i = \Pi_e(m/2M)^{1/3}3^{1/2}/2$, $kV_e = \Pi_e$. For atomic hydrogen, the imaginary part of ω is $\omega_i = 0.0562\,\Pi_e$. From the relation

$$\frac{d(kV_e)}{d\omega} = 1 - \Pi_e\Pi_i^2/\omega^3 \tag{11.79}$$

it follows that $d(kV_e)/d\omega = 3$ at $\theta = \pi/3$, so that the group velocity is $V_e/3$. The phase velocity is $\omega_r/k = V_e(m/2M)^{1/3}/2$, which is smaller than V_e. We may consider the following model for the excitation of this instability. An electron beam (plasma electrons) runs with the velocity V_e through the ion wave propagating with the velocity Π_i/k, so that an electron plasma wave of (low) velocity $V_e - \Pi_e/k$ interacts with the ion wave. This instability is called *Buneman instability*.

Next, let us consider the interation of a weak beam of velocity V_0 with a plasma which consists of cold ions and hot electrons. The dispersion equation (eq. (11.67)) is

$$\frac{\Pi_b{}^2}{(\omega - kV_0)^2} = 1 - \frac{\Pi_i{}^2}{\omega^2} - \Pi_e{}^2 \int_{-\infty}^{\infty} \frac{f_e(v_z)dv_z}{(\omega - kv_z)^2} \equiv K_{\parallel}(\omega, k). \tag{11.80}$$

$K_{\parallel}(\omega, k)$ is nearly equal to zero for the weak beam ($\Pi_b \to 0$) if $\omega \neq kV_0$. This is the dispersion equation of the plasma without the beam. The dispersion equation including the effect of the beam must be in the form

$$\omega - kV_0 = \delta_\omega(k) \qquad (\delta_\omega(k) \ll kV_0). \tag{11.81}$$

Using δ_ω, we reduce eq. (11.80) to

$$\frac{\Pi_b{}^2}{\delta_\omega{}^2} = K_{\parallel}(\omega = kV_0, k) + \left(\frac{\partial K_{\parallel}}{\partial \omega}\right)_{\omega = kV_e} \delta_\omega. \tag{11.82}$$

If $K_{\parallel} \neq 0$, the second term of the right-hand side of eq. (11.82) can be neglected. The expression for $K_{\parallel}$ is

$$K_{\parallel}(\omega = kV_0, k) = K_r + iK_i \tag{11.83}$$

$$K_r = 1 - \frac{\Pi_i{}^2}{\omega^2} + \frac{\Pi_e{}^2}{k} \mathrm{P} \int_{-\infty}^{\infty} \frac{\dfrac{\partial f_e(v_z)}{\partial v_z}}{\omega - kv_z} dv_z \tag{11.84}$$

$$K_i = i\sigma\pi \frac{\Pi_e{}^2}{|k|k} \left(\frac{\partial f_e}{\partial v_z}\right)_{v_z = \omega/k} \tag{11.85}$$

The K_i term is the Landau damping. When V_0 is in a region where Landau damping is ineffective, then $|K_i| \ll |K_r|$. Therefore if the condition

$$K_r < 0 \tag{11.86}$$

is satisfied, δ_ω is complex and the wave is unstable. When the *dielectric constant is negative*, electric charges are likely to be bunched and we can safely predict the occurence of this instability.

Let us consider the case in which $K_r > 0$ and the Landau term K_i is not small. The energy density of a wave propagating through a dispersive medium is given by eq. (10.95):

$$\frac{\varepsilon_0}{4} \boldsymbol{E}^* \frac{\partial}{\partial \omega}(\omega K_h) \boldsymbol{E}. \tag{11.87}$$

When the wave energy density is negative, the right-hand side of eq. (11.82) is negative and the wave is unstable. When energy is lost from the wave, the amplitude of the wave increases because the *wave energy density is negative.*[9]

Here we have only briefly discussed interactions between beam and plasma. Readers may refer to ref. 9 for more detailed analysis.

11.5 The Nyquist Criterion[10]

The dispersion equation for a homogeneous plasma without gravity is

$$H\left(\frac{\omega}{k}, k\right) = 1 - \frac{1}{k^2} \sum_j \frac{\Pi_j^2}{n_j} \int_{-\infty}^{\infty} \frac{f_j'(v)}{v - \omega/k} \mathrm{d}v = 0 \tag{11.88}$$

for $\omega_i = \mathrm{Im}(\omega) > 0$. When $f_j'(v)$ is absolute integrable, i.e., when

$$\int_{-\infty}^{\infty} |f_j'(v)| \, \mathrm{d}v = M,$$

then $H(\omega/k, k)$ is analytic. The number N_0 of points where $H(\omega/k, k) = 0$ in the upper half of the ω_r, ω_i plane is given by the residue theorem:

$$N_0 = \frac{1}{2\pi i} \int_C \frac{H'(\omega)}{H(\omega)} \mathrm{d}\omega. \tag{11.89}$$

The integration contour C is shown in fig. 11.6a. The contour C in the ω_r, ω_i plane maps to the contour D in the H_r, H_i plane. This contour D is called the *Nyquist diagram* (fig. 11.6b).

The arc in the ω_r, ω_i plane of fig. 11.6a maps to the point $H_r = 1$, while the chord maps to the rest of the contour D in fig. 11.6b. The integral (11.89) can be rewritten along the contour D in the H_r, H_i plane as

$$N_0 = \frac{1}{2\pi i} \int_D \frac{\mathrm{d}H}{H}. \tag{11.90}$$

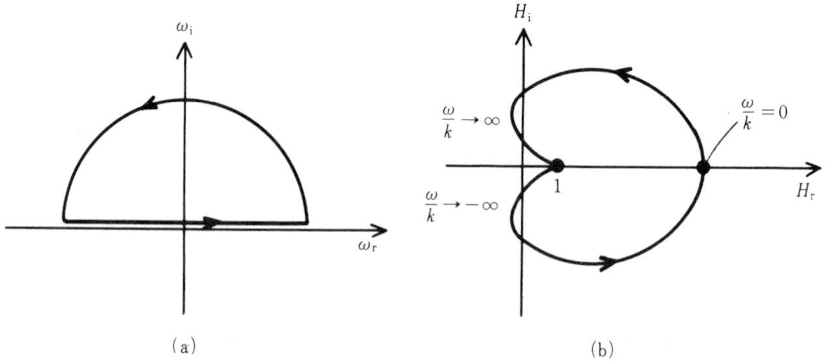

Fig. 11.6
Nyquist diagram. The contour in the ω_r, ω_i plane (a) maps to the contour in the H_r, H_i plane (b).

The number of unstable modes which satisfy $H = 0$ is the same as the number of turns encircling the origin $H = 0$ of the H_r, H_i plane. In other word, if the contour does not encircle the origin, the system is stable. This is called the *Nyquist criterion.*

*11.6 Absolute Instability and Convective Instability[11–13]

It has been assumed that small perturbations in plasmas can be expressed in the form $\exp i(\boldsymbol{k}\cdot\boldsymbol{r} - \omega t)$; for such perturbations we have derived dispersion relations $\boldsymbol{k}$ in the form

$$D(\omega, \boldsymbol{k}) = 0. \tag{11.91}$$

For a given real propagation vector $\boldsymbol{k}$, ω can be solved as a function of $\boldsymbol{k}$. When the imaginary part ω_i of $\omega = \omega_r + \omega_i i$ is positive, the wave is unstable. In general there are two types of instability, absolute and convective. The growth of a disturbance in time and space is shown in fig. 11.7. In fig. 11.7a, the disturbance grows at all points; instability of this type is called *absolute instability*. In fig. 11.7b, the disturbance propagates while the amplitude increases. At any fixed position, the disturbance grows at first, but it damps out after the disturbance has passed. Instability of this type is called *convective instability*.

Let us consider both of these instabilities analytically. It is assumed that the perturbation is described by a linear differential equation with constant coefficients:

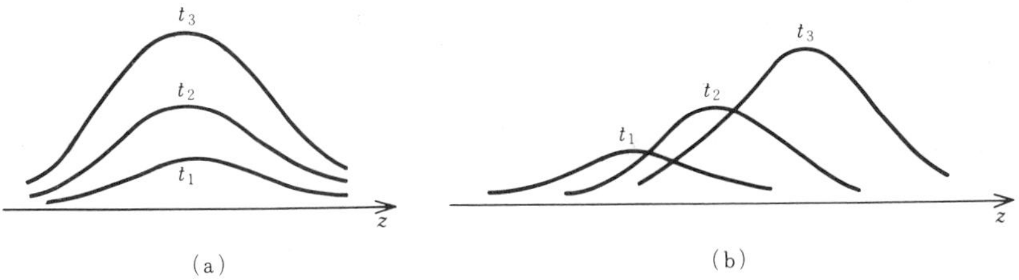

Fig. 11.7
Absolute instability (a) and convective instability (b).

$$D\left(i\frac{\partial}{\partial t}, -i\nabla\right)u \equiv \left(\left(i\frac{\partial}{\partial t}\right)^n + p_1(-i\nabla)\left(i\frac{\partial}{\partial t}\right)^{n-1} + \cdots + p_n(-i\nabla)^n\right)u = 0. \tag{11.92}$$

When u is in the form $\exp i(\boldsymbol{k}\cdot\boldsymbol{r} - \omega t)$, the dispersion equation is, as we have stated,

$$D(\omega, \boldsymbol{k}) = 0. \tag{11.93}$$

Let us solve the differential equation with the initial conditions

$$i^n\frac{\partial^l u(0, \boldsymbol{r})}{\partial t^l} = \begin{cases} 0 & (l = 0, 1, 2, \cdots, n-2) \\ \psi(\boldsymbol{r}) & (l = n-1). \end{cases} \tag{11.94}$$

When Fourier and Laplace transforms of eq. (11.92) are taken with respect to the space coordinates and the time, respectively, we find

$$D(\omega, \boldsymbol{k})U(\omega, \boldsymbol{k}) = \varphi(\boldsymbol{k}), \tag{11.95}$$

where

$$U(\omega, \boldsymbol{k}) = \int_0^\infty \left(\exp(i\omega t)\int_{-\infty}^\infty u(t, \boldsymbol{r})\exp(-i\boldsymbol{k}\cdot\boldsymbol{r})\mathrm{d}\boldsymbol{r}\right)\mathrm{d}t. \tag{11.96}$$

$\varphi(\boldsymbol{k})$ is the Fourier transform of $\psi(\boldsymbol{r})$. The inverse transform of eq. (11.96) is

$$\begin{aligned} u(t, \boldsymbol{r}) &= \frac{1}{2\pi}\int_{-\infty+i\sigma}^{+\infty+i\sigma}\left(\exp(-i\omega t)\int_{-\infty}^\infty U(\omega, \boldsymbol{k})\exp(i\boldsymbol{k}\cdot\boldsymbol{r})\frac{\mathrm{d}\boldsymbol{k}}{(2\pi)^3}\right)\mathrm{d}\omega \\ &= \frac{1}{2\pi}\int_{-\infty+i\sigma}^{+\infty+i\sigma}\left(\exp(-i\omega t)\int_{-\infty}^\infty \frac{\varphi(\boldsymbol{k})}{D(\omega, \boldsymbol{k})}\exp(i\boldsymbol{k}\cdot\boldsymbol{r})\frac{\mathrm{d}\boldsymbol{k}}{(2\pi)^3}\right)\mathrm{d}\omega. \end{aligned} \tag{11.97}$$

For simplicity, we analyse only the one-dimensional case. Equations (10.96) and (10.97) become

$$u(t, z) = \int_{-\infty + i\sigma}^{\infty + i\sigma} F(\omega, z) \exp(-i\omega t) \frac{\mathrm{d}\omega}{2\pi} \tag{11.98}$$

$$F(\omega, z) = \int_{-\infty}^{\infty} \frac{\varphi(k)}{D(\omega, k)} \exp(ikz) \frac{\mathrm{d}k}{2\pi}. \tag{11.99}$$

The integral path is shown in fig. 11.8. When $z > 0$, $\exp i(k_{\mathrm{r}} + ik_{\mathrm{i}})z = \exp ik_{\mathrm{r}}z \cdot \exp(-k_{\mathrm{i}}z)$ approaches zero as $k_{\mathrm{i}} \to \infty$ and the integral path of eq. (11.99) in the k_{r}, k_{i} plane can be a closed loop ABCA. Accordingly, eq. (11.99) can be expressed by the sum of residues of the integrand of eq. (11.99) inside the contour of the integration. The singular points of the integrand are given by eq. (11.93), and the solutions of k are denoted by $k_{n+}(\omega)$. For $z > 0$, eq. (11.99) is reduced to

$$F(\omega, z) = \sum_n \frac{i\varphi(k_{n+}(\omega))}{\left(\dfrac{\partial}{\partial k} D(\omega, k)\right)_{k = k_{n+}(\omega)}} \exp ik_{n+}(\omega) z. \tag{11.100}$$

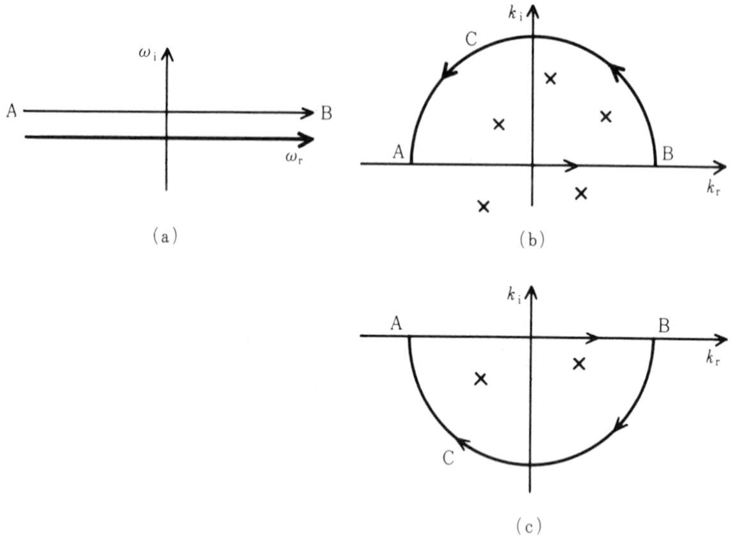

Fig. 11.8
(a) Contour of the integral (11.98). (b) Contour of the integral (11.99) for $z > 0$. (c) Contour of the integral (11.99) for $z < 0$.

Accordingly $u(t, z)$ is given by

$$u(t, z) = \int_{-\infty + i\sigma}^{\infty + i\sigma} \sum_n \frac{i\varphi(k_{n+}(\omega))}{\left(\dfrac{\partial}{\partial k} D(\omega, k)\right)_{k = k_{n+}(\omega)}} \exp(-i\omega t + ik_{n+}(\omega) z) \mathrm{d}\omega. \tag{11.101}$$

When σ of $\omega = \omega_r + i\sigma$ approaches zero, then $k_{n+}(\omega)$ also changes. The orbit of $k_{n+}(\omega)$ in the k_r, k_i plane is shown in fig. 11.9. If $k_{n+}(\omega)$ does not cross the k_r axis, then $k_i > 0$ and the wave is evanescent, damping exponentially for $z > 0$. If $k_{n+}(\omega)$ crosses the k_r axis, then $k_i < 0$ and the amplitude of the wave grows exponentially for $z > 0$. This amplifying wave is convectively unstable.

We also have the case in which two solutions $k_1(\omega)$ and $k_2(\omega)$ coincide for some σ. In this case the contour for eq. (11.101) must be chosen as is indicated in fig. 11.10.

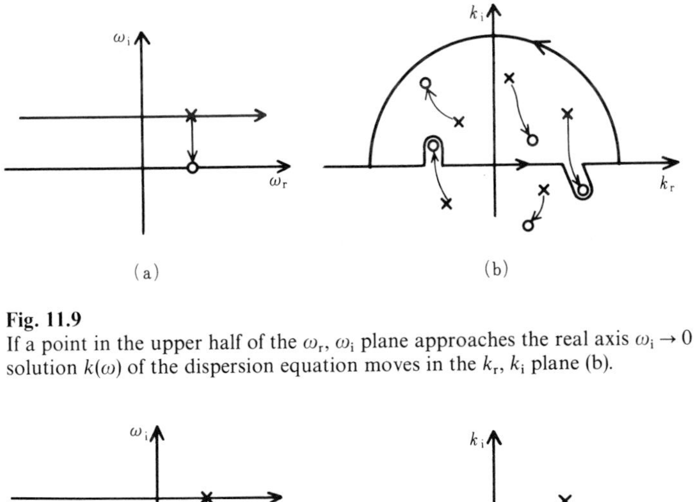

Fig. 11.9
If a point in the upper half of the ω_r, ω_i plane approaches the real axis $\omega_i \to 0$ (a), the solution $k(\omega)$ of the dispersion equation moves in the k_r, k_i plane (b).

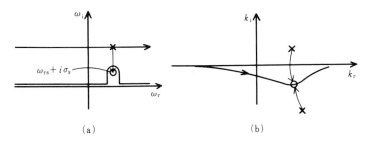

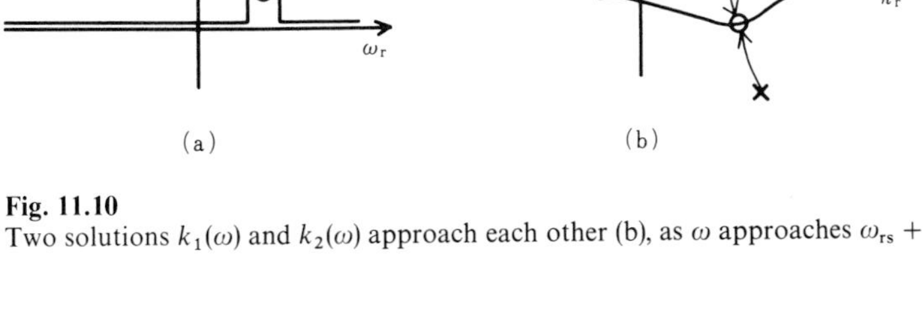

Fig. 11.10
Two solutions $k_1(\omega)$ and $k_2(\omega)$ approach each other (b), as ω approaches $\omega_{rs} + i\sigma_s$.

When the values of ω, k at the encounter are denoted by ω_s and k_s, the dispersion equation can be expanded near ω_s and k_s:

$$D(\omega, k) \approx \left(\frac{\partial D}{\partial \omega}\right)_s (\omega - \omega_s) + \frac{1}{2}\left(\frac{\partial^2 D}{\partial k^2}\right)_s (k - k_s)^2. \tag{11.102}$$

When $\omega = \omega_s$, then $k = k_s$ is a double root. $F(\omega, z)$ near $\omega = \omega_s$ is given by

$$F(\omega, z) = \frac{\pm \varphi(k_s)\exp(ik_s z)}{\left(2\left(\frac{\partial D}{\partial \omega}\right)\left(\frac{\partial^2 D}{\partial k^2}\right)\right)^{1/2}_{w_s,\, k_s}} \frac{1}{(\omega - \omega_s)^{1/2}}, \tag{11.103}$$

and $u(t, z)$ is reduced to

$$u(t, z) = \int F(\omega, z)\exp(-i\omega t)\mathrm{d}\omega$$

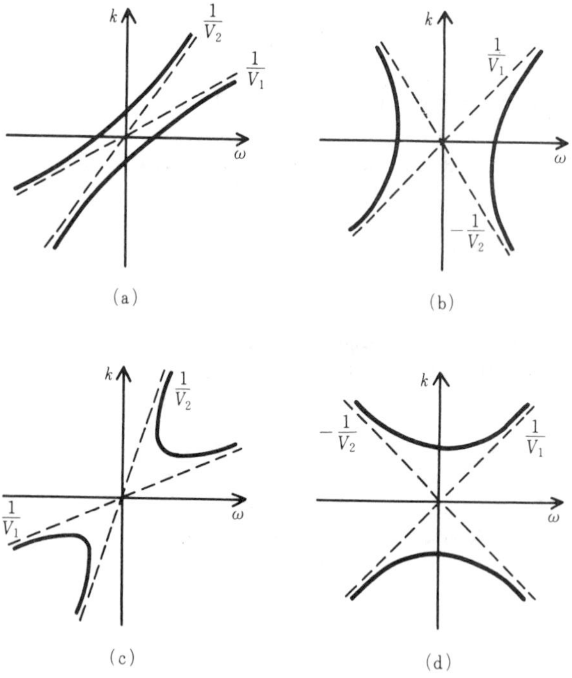

Fig. 11.11
The curves (a)–(d) show k vs. ω given by eqs. (11.105), 1)–4) respectively.

$$= \pm \frac{\varphi(k_s)\exp(-i(\omega_s t - k_s z))}{\left(2\left(\frac{\partial D}{\partial \omega}\right)\left(\frac{\partial^2 D}{\partial k^2}\right)\right)^{1/2}_{w_s,\, k_s}} \int \frac{\exp(-i(\omega - \omega_s)t)}{(\omega - \omega_s)^{1/2}} \frac{d\omega}{2\pi}$$

$$= \pm \frac{\varphi(k_s)\exp(-i(\omega_s t - k_s z))}{\left(2\pi i\left(\frac{\partial D}{\partial \omega}\right)\left(\frac{\partial^2 D}{\partial k^2}\right)\right)^{1/2}_{w_s,\, k_s} t^{1/2}}. \tag{11.104}$$

Since $\mathrm{Im}(\omega_s) > 0$, $u(t, z)$ increases with time. Suppose that there are two solutions $k = k_r + k_i i$ of $D(\omega, k) = 0$ with $\mathrm{Im}(k) > 0$ or $\mathrm{Im}(k) < 0$ for a specified ω with a large positive imaginary part ω_i. Suppose that the two solutions meet before $\mathrm{Im}(\omega)$ falls to zero from this large positive value. Then the wave amplitude grows with time at every point. This wave is absolutely unstable.

Let us consider the following simple examples of dispersion equations for two combined waves:

$$\left.\begin{aligned}
&1)\quad \left(k - \frac{\omega}{V_1}\right)\left(k - \frac{\omega}{V_2}\right) = k_0^2\\
&2)\quad \left(k - \frac{\omega}{V_1}\right)\left(k + \frac{\omega}{V_2}\right) = -k_0^2\\
&3)\quad \left(k - \frac{\omega}{V_1}\right)\left(k - \frac{\omega}{V_2}\right) = -k_0^2\\
&4)\quad \left(k - \frac{\omega}{V_1}\right)\left(k + \frac{\omega}{V_2}\right) = k_0^2,
\end{aligned}\right\} \tag{11.105}$$

where V_1 and V_2 are positive and k_0 is positive and real. These dispersion relations are shown for real k, ω in fig. 11.11a–d.

In fig. 11.11a there are real solutions of k for any real ω and there are real solutions of ω for any real k, so that there is no instability.

In fig. 11.11b there are real solutions of ω for any real k but there are complex solutions of k for some real ω. It is easy to show that the corresponding waves are evanescent.

In fig. 11.11c, there are two solutions $k \to \omega/V_1$, $k \to \omega/V_2$ for $|\omega| \to \infty$. The imaginary part of k is positive if $\omega_i \to \infty$. If a solution for k has $k_i < 0$ for real ω, the wave is amplifying and is convectively unstable. In this case the imaginary parts of both solutions are positive for large ω_i and there is no possibility that two solutions encounter as in fig. 11.10b. Accordingly, absolute instability cannot occur.

In fig. 11.11d there are two solutions $k \to \omega/V_1, k \to -\omega/V_2$ for $|\omega| \to \infty$, and one solution has $k_i > 0$ and the other solution has $k_i < 0$. When ω is

$$\omega = \omega_s = 2ik_0\left(\frac{1}{V_1} + \frac{1}{V_2}\right)^{-1},$$

there is a double root $k = k_s$:

$$k = k_s = ik_0\left(\frac{V_2 - V_1}{V_1 + V_2}\right).$$

This is a case of absolute instability.

References

1. J. E. Drummond: Phys. Rev. **110**, 293(1958)
2. R. Z. Sagdeev and V. D. Shafranov: Proc. 2nd Intern. Conf. Geneva **31**, 118(1958)
3. M. N. Rosenbluth and N. Rostoker: ibid., 144
4. N. A. Krall and M. N. Rosenbluth: Phys. of Fluids **8**, 1488 (1965)
5. B. D. Fried and S. D. Conte: The Plasma Dispersion Function, Academic Press, New York, 1961
6. T. H. Stix: The Theory of Plasma Waves, McGraw-Hill, New York, 1962
7. I. B. Bernstein: Phys. Rev. **109**, 10(1958)
8. O. Buneman: ibid. **115**, 503(1959)
9. R. J. Briggs: Electron-Stream Interaction with Plasmas, MIT Press, Cambridge, Mass., 1964
10. H. Nyquist: Bell System Tech. J. **11**, 126(1932)
11. P. A. Sturrock: Phys. Rev. **112**, 1488(1958)
12. H. Derfler: Phys. Lett. A**24**, 763(1967)
13. K. B. Dysthe: Nucl. Fusion **6**, 215(1966)

12 Velocity-Space Instabilities (Electrostatic)

Besides the magnetohydrodynamic instabilities discussed in ch. 9, there is another type of instability, caused by deviations of the velocity distribution function from the stable Maxwellian form. Instabilities which are dependent on the shape of the velocity distribution function are called velocity-space instabilities or *microscopic instabilities*. However, the distinction between microscopic and macroscopic or MHD instability is not always clear, and it is sometimes difficult to classify a given instability.

In this chapter, the characteristics of the perturbation (wave) will be analyzed based on the dispersion equations derived in the preceding chapter. Discussion will be limited here to electrostatic perturbation, which is simple while providing a good description of perturbation in low-β plasmas. In sec. 12.1, it will be seen that drift instability is induced by a pressure gradient within a plasma. Trapped particle instabilities in toroidal plasmas are discussed in sec. 12.2. The frequencies of these perturbations are much lower than the ion cyclotron frequency. Harris instability and loss-cone instability induced by anisotropic velocity-space distribution functions are the subjects of sec. 12.3. The frequencies of these two types of instabilities are much higher than those of the former ones. In sec. 12.4, the cross-field instabilities are discussed.

12.1 Instabilities due to Pressure Gradient

12.1a Dispersion Equation

Let us take the z axis to be in the direction of the static magnetic field and assume that the density gradient is perpendicular to the magnetic field (fig. 12.1). The positive x axis is taken in the direction of the density gradient. The direction of the acceleration $\boldsymbol{G}$ on the plasma is assumed to be in the x direction. When the frequency is lower than the ion cyclotron

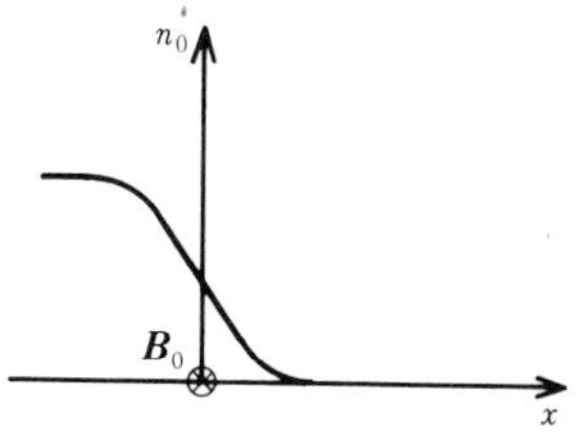

Fig. 12.1
Density gradient and coordinates ($dn_0/dx < 0$).

frequency $|\Omega_i|$, the dispersion equation for low-frequency waves is given by eqs. (11.39)–(11.41), (11.45) as follows:[1]

$$
\begin{aligned}
&k_y^2 + k_z^2 \\
&\quad + \sum_j \Pi_j^2 \Bigg[\frac{m}{T_\perp} + I_0 \mathrm{e}^{-b} \left(\frac{m}{T_z} - \frac{m}{T_\perp} \right) \\
&\quad + \frac{\delta_z}{2} \frac{\omega_0}{k_z^2 (T_z/m)} \frac{k_y}{\Omega} I_0 \mathrm{e}^{-b} (1 + \zeta_0 Z(\zeta_0)) \\
&\quad + I_0 \mathrm{e}^{-b} \zeta_0 Z(\zeta_0) \left(\frac{m}{T_z} + \frac{k_y}{\omega_0 \Omega} \left(\frac{1}{n_0} \frac{\mathrm{d}n_0}{\mathrm{d}x} + (f_0 - 1)\delta_\perp - \frac{\delta_z}{2} \right) \right) \Bigg] = 0,
\end{aligned}
\tag{12.1}
$$

where

$$\omega_0 = \omega - k_z V - \frac{k_y G}{\Omega}$$

$$\zeta_0 = \frac{\omega_0}{2^{1/2} k_z (T_z/m)^{1/2}}$$

$$\delta_\perp = \frac{1}{T_\perp} \frac{\mathrm{d}T_\perp}{\mathrm{d}x}$$

$$\delta_z = \frac{1}{T_z} \frac{\mathrm{d}T_z}{\mathrm{d}x}$$

$$b = k_y^2 \left(\frac{T}{m\Omega^2} \right) = (k_y \rho_B)^2.$$

Where a line of magnetic force is of radius of curvature R, the plasma accelerates with an acceleration

$$G = \frac{(T_z + T_\perp)/m}{R}, \tag{12.2}$$

due to the centrifugal force. When the magnetic-force lines are convex toward the outside of the plasma column, then $R > 0$; when they are concave toward the outside, $R < 0$.

For convenience, we define the inverse of the density gradient:

$$\kappa \equiv -\frac{1}{n_0}\frac{dn_0}{dx} > 0. \tag{12.3}$$

The drift velocity is $\kappa(T/m)/\Omega$, and the drift frequency is defined by

$$\omega_0^* \equiv k_y\kappa\frac{T_{\perp e}/m_e}{\Omega_e} = k_y\kappa\frac{T_{\perp e}}{eB_0} > 0 \tag{12.4}$$

$$\omega_i^* \equiv k_y\kappa\frac{T_{\perp i}/m_i}{\Omega_i} = -k_y\kappa\frac{T_{\perp i}}{ZeB_0} < 0. \tag{12.5}$$

12.1b Interchange Instability (Finite Larmor Radius Effect)

When k_z is very small, interchange instability must be considered. In this case,

$$|\zeta_0| = \left|\frac{\omega_0}{2^{1/2}k_z(T_z/m)_e^{1/2}}\right| > 1 \qquad \left(\left|\frac{\omega_0}{k_z}\right| > \left(\frac{T_z}{m}\right)_e^{1/2}\right).$$

The electron Larmor radius is assumed to be negligible. It is assumed that the fluid velocity V along the lines of magnetic force is zero and that the plasma is isotropic ($T_\perp = T_z$) and homogeneous ($\delta_\perp = \delta_z = 0$). Using the asymptotic expansion of the dispersion function

$$Z(\zeta_0) \approx -\zeta_0^{-1}(1 + (2\zeta_0^2)^{-1} + \cdots) \qquad (\zeta_0 \gg 1),$$

we can write the dispersion equation as

$$1 - I_0(b)e^{-b}\left(1 - \frac{\omega_i^*}{\omega - k_yG_i/\Omega_i}\right) = \frac{T_i}{Z_iT_e}\frac{(-\omega_e^*)}{\omega - k_yG_e/\Omega_e}, \tag{12.6}$$

where the Debye length has been assumed to be much smaller than the wavelength. Since $k_yG/\Omega = \omega^*(\kappa R)$ and $e^{-b}I_0(b) = 1 - b$, $b \equiv b_i \ll 1$, the dispersion equation becomes

$$\omega^2 - \omega\omega_i^*\left(1 + \frac{1}{\kappa R}\left(1 - \frac{Z_iT_e}{T_i}\right)\right) - \frac{(\omega_i^*)^2}{\kappa R}\left(\frac{Z_iT_e}{T_i}\left(1 + \frac{1}{\kappa R}\right) - \frac{1}{b}\left(1 + \frac{Z_iT_e}{T_i}\right)\right) = 0,$$

i.e.,

$$\frac{\omega}{\omega_i^*} = \frac{1}{2}\left(1 + \frac{1}{\kappa R}\left(1 - \frac{Z_i T_e}{T_i}\right)\right)$$
$$\pm \frac{1}{2}\left[\left(1 + \frac{1}{\kappa R}\left(1 - \frac{Z_i T_e}{T_i}\right)\right)^2\right.$$
$$\left. + \frac{4}{\kappa R}\left(\frac{Z_i T_e}{T_i}\left(1 + \frac{1}{\kappa R}\right) - \frac{1}{b}\left(1 + \frac{Z_i T_e}{T_i}\right)\right)\right]^{1/2}. \quad (12.7a)$$

When $T_e \approx T_i$, and $\kappa R \gg 1$, it follows that

$$\frac{\omega}{\omega_i^*} = \frac{1}{2} \pm \frac{1}{2}\left(1 - \frac{4}{b\kappa R}\left(1 + \frac{Z_i T_e}{T_i}\right)\right)^{1/2}. \quad (12.7b)$$

When b is very small, the case $R > 0$ is unstable. But if the Larmor radius becomes large and

$$k_y^2 (\rho_B^i)^2 > \frac{4}{\kappa R}\left(1 + \frac{Z_i T_e}{T_i}\right)\right)^{1/2} \quad (12.8)$$

the plasma becomes stable.[2] Thus the finite Larmor radius has a stabilizing effect upon interchange instability. When the ratio of Larmor radius to the wavelength becomes larger than the critical value (12.8), the average electric field acting on ions is somewhat different from the local electric field acting on electrons, so that charge separation occurs. This charge separation tends to cancel the charge separation due to the acceleration. From the real part ω_r of the frequency ω given by eq. (12.7b), the propagation velocity ω_r/k_y is the order of $\omega_{*i}/k_y \approx \kappa T_i/(Z_i eB)$ and the direction is that of the ion drift velocity due to the pressure gradient. (If $R > 0$ and $Z_i T_e/T_i > (\kappa R + 1)$, the direction will be that of the electron drift velocity as is clear from eq. (12.7a).)

12.1c Drift Instability

In the previous subsection we have discussed the disturbances of small k_z satisfying $|\omega_0/k_z| > (T_z/m)_e^{1/2}$, i.e., the interchange instabilities propagating perpendicularly to the magnetic field. In this subsection we discuss disturbances of slightly larger k_z. Assume an isotropic ($T_\perp = T_z$) plasma without temperature gradient ($\delta_z = \delta_\perp = 0$) and zero electron Larmor radius. The dispersion equation is

$$0 = (k_y{}^2 + k_z{}^2)\lambda_d{}^2 + \frac{Z_i T_e}{T_i}\left(1 + e^{-b} I_0(b)\zeta_i Z(\zeta_i)\left(1 - \frac{\omega_i{}^*}{\omega_{0i}}\right)\right)$$

$$+ \left(1 + \zeta_e Z(\zeta_e)\left(1 - \frac{\omega_e{}^*}{\omega_{0e}}\right)\right). \quad (12.9)$$

For $\omega = \omega_r + i\gamma$ and $|\omega_r| \gg |\gamma|$, the real and imaginary parts of the dispersion function are given by

$$\zeta Z(\zeta) \to \zeta Z_r(\zeta) + i\frac{k_z}{|k_z|}\pi^{1/2}\zeta \exp(-\zeta^2).$$

The solution for $G = 0$ is[1]

$$\frac{\omega_e{}^*}{\omega_r} = \frac{(k_y{}^2 + k_z{}^2)\lambda_d{}^2 + (Z_i T_e/T_i)(1 + e^{-b} I_0 \zeta_i Z_r(\zeta_i)) + 1 + \zeta_e Z_r(\zeta_e)}{\zeta_e Z_r(\zeta_e) - e^{-b} I_0 \zeta_i Z_r(\zeta_i)} \quad (12.10)$$

$$\frac{\gamma}{\omega_r} = \pi^{1/2}\frac{k_z}{|k_z|} \frac{\zeta_e\left(1 - \dfrac{\omega_r}{\omega_e{}^*}\right)\exp(-\zeta_e{}^2) - e^{-b} I_0 \zeta_i\left(1 + \dfrac{\omega_r}{\omega_e^*}\dfrac{Z_i T_e}{T_i}\right)\exp(-\zeta_i{}^2)}{\zeta_e Z_r(\zeta_e) - e^{-b} I_0 \zeta_i Z_r(\zeta_i)}. \quad (12.11)$$

Here $|\omega_r| \gg |\gamma|$ is assumed. The 2nd term in the numerator of γ/ω_r is the contribution of ion Landau damping. The 1st term of the numerator contributes to instability when $\omega_r/\omega_e{}^* < 1$.

When k_z is in the region given by

$$\left(\frac{T_i}{m_i}\right)^{1/2} < \left|\frac{\omega_0}{k_z}\right| < \left(\frac{T_e}{m_e}\right)^{1/2},$$

it follows that $|\zeta_e| < 1$, $|\zeta_i| > 1$ and the contribution of ion Landau damping decreases, and the 1st term in the numerator of eq. (12.11) becomes dominant. For $b \ll 1$, eqs. (12.10) and (12.11) become

$$\frac{\omega_e{}^*}{\omega_r} = \frac{1 + (Z_i T_e/T_i)b}{1 - b} \approx 1 + \left(1 + \frac{Z_i T_e}{T_i}\right)b \quad (12.12)$$

$$\frac{\gamma}{\omega_e^*} = \frac{k_z}{|k_z|}\pi^{1/2} b(1 + Z_i T_e/T_i)\zeta_e \exp(-\zeta_e^2) \approx \frac{b\omega_e^*}{|k_z|(T_e/m_e)^{1/2}} \tag{12.13}$$

($\zeta_i Z(\zeta_i) \approx -1$, $\zeta_e Z(\zeta_e) \approx -2\zeta_e^2$). The growth rate of drift instability of a collisionless plasma becomes large when the ion Larmor radius is large. When $|\omega_0/k_z| < (T_i/m_i)^{1/2}$ and $|\omega_0/k_z| < (T_e/m_e)^{1/2}$, then $|\zeta_e|, |\zeta_i| < 1$ and the ion Landau damping term stabilizes the disturbance.

As this instability is caused by an electron drift due to the density gradient, it is called *drift instability*. Since a density gradient must exist in any plasma of finite extent, this is also called *universal instability*.[3] There are several reviews of drift instability.[4–7]

From the relations (11.35), (11.36), and (11.45), the relation between $n_{e1} = \int f_{e1}\, dv$ and the electrostatic potential ϕ_1 can be obtained. From eqs. (12.9), (12.12), and (12.13), it follows that

$$\frac{n_{e1}}{n_0} \approx \frac{e\phi_1}{T_e}\left(1 + \zeta_e Z(\zeta_e)\left(1 - \frac{\omega_e^*}{\omega_{0e}}\right)\right)$$

$$\approx \frac{e\phi_1}{T_e}\left(1 + i\frac{\gamma}{\omega_e^*}\zeta_e Z(\zeta_e)\right).$$

The diffusion coefficient due to the fluctuation loss in drift instability (see the discussion of sec. 8.4) is given by

$$D = \frac{\gamma}{\kappa^2}\left|\frac{n_{e1}}{n_0}\right|^2 (-\zeta_e Z(\zeta_e)) \approx \frac{\gamma}{\kappa^2}\left|\frac{n_{e1}}{n_0}\right|^2.$$

Resistive drift instability was treated in sec. 9.7. The maximum growth rate reaches $\gamma/\omega_e^* \approx 0.3$ for an appropriate value of (k_z/k_y). When k_z is fixed, the growth rate γ_R of resistive drift instability is given by

$$\frac{\gamma_R}{\omega_e^*} \approx \frac{\nu_{ei}\omega_e^*}{\Omega_e(-\Omega_i)}\left(\frac{k_y}{k_z}\right)^2.$$

The ratio of γ_R to γ (this latter for collisionless drift instability) is

$$\frac{\gamma_R}{\gamma} \approx \frac{Z_i T_e}{T_i}\frac{\omega_e^*}{|k_z|(T_e/m_e)^{1/2}}\frac{\nu_{ei}}{\omega_e^*}.$$

Here $\zeta_e = \omega_e^*/(|k_z|(2T_e/m_e)^{1/2}) < 1$ is assumed. In many cases, resistive drift instability is the more dangerous of the two. The mechanism of

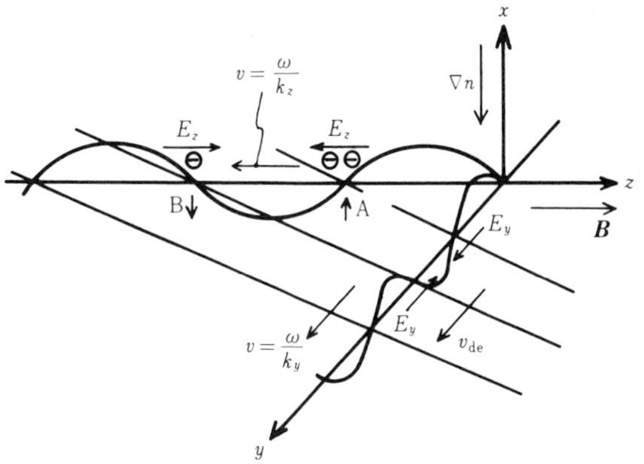

Fig. 12.2
Mechanism of collisionless drift instability.

resistive drift instability was explained in sec. 7.9. It is easy to explain the mechanism of collisionless drift instability if one uses fig. 12.2.[8] The $\boldsymbol{E} \times \boldsymbol{B}$ motion due to the potential fluctuation is directed upward at A and downward at B. Because of the density gradient, the density increases at A and decreases at B. Particles of velocity $v = \omega/k_z$ interact resonantly with the wave. Electrons at A move against the electric field and give up energy to the wave, while electrons at B extract energy from the wave. As a result the wave is amplified. The 1st term of the numerator of eq. (12.11) represents these phenomena.

We have been discussing the cases where only a density gradient exists. When a temperature gradient ($\delta_\perp \neq 0$, $\delta_z \neq 0$) and a current ($V \neq 0$) also exist, the case can be treated by means of the dispersion equation (12.1).

12.1d Stability Conditions for Drift Instability

(i) Stabilization by Good Curvature of Magnetic-Force Lines Good curvature of magnetic-force lines is one of the stability conditions for the drift wave. The origin of the instability, as we have seen, is electron drift due to the density gradient. When the drift caused by the centrifugal force due to the curvature of the lines of force cancels the drift due to the density gradient, the curvature is said to be "good;" the curvature is said to be "bad" if the resultant drift reinforces the gradient-induced drift.

Because of the sign convention in the definition (12.2), convex-outward curvature ($R > 0$) is bad, and the concave-outward curvature ($R < 0$) is good. When $\delta_\perp = \delta_z = 0$ and $G \neq 0$, the dispersion equation is

$$0 = (k_y{}^2 + k_z{}^2)\lambda_d{}^2 + \frac{Z_i T_e}{T_i}\left(1 + e^{-b} I_0(b)\zeta_i Z(\zeta_i)\left(1 - \frac{\omega_i{}^*}{\omega_i{}'}\right)\right)$$
$$+ \left(1 + \zeta_e Z(\zeta_e)\left(1 - \frac{\omega_e{}^*}{\omega_e{}'}\right)\right), \qquad (12.14)$$

where $\omega_i{}' \equiv \omega - k_y G_i/\Omega_i$, $\omega_e{}' \equiv \omega - k_y G_e/\Omega_e$, and $G_i = (2T/m)_i/R$, $G_e = (2T/m)_e/R$, i.e., $\omega_i{}' \equiv \omega + 2k_y T_i/(ZeB_0R)$ and $\omega_e{}' \equiv \omega - 2k_y T_e(eB_0R)$. ζ_i and ζ_e are given by $\zeta_i = \omega_i{}'/(k_z(2T_i/m_i)^{1/2})$, $\zeta_e = \omega_e{}'/(k_z(2T_e/m_e)^{1/2})$. When $|\zeta_i| > 1$, $|\zeta_e| < 1$, $b \ll 1$, the dispersion equation is

$$\frac{\omega_{*e}}{\omega_i{}'} \approx 1 + \left(\frac{Z_i T_e}{T_i} + 1\right)b \qquad (12.15)$$

$$\frac{\gamma}{\omega_i{}'} \approx \pi^{1/2}\frac{k_z}{|k_z|}\frac{\zeta_e\left(\dfrac{\omega_i{}'}{\omega_e{}'} - \dfrac{\omega_i{}'}{\omega_e{}^*}\right)\exp(-\zeta_e{}^2) - e^{-b} I_0 \zeta_i\left(1 + \dfrac{\omega_i{}'}{\omega_e{}^*}\dfrac{Z_i T_e}{T_i}\right)\exp(-\zeta_i{}^2)}{\zeta_e Z(\zeta_e)(\omega_i{}'/\omega_e{}') - e^{-b} I_0 \zeta_i Z(\zeta_i)}. \qquad (12.16)$$

The 2nd term of the numerator of eq. (12.16) represents ion Landau damping and is the stabilizing term. When $R < 0$, $\zeta_i = \omega_i{}'/(k_z(2T_i/m_i)^{1/2})$ becomes small and Landau damping is then more effective. If the first term of the numerator is

$$\frac{\omega_i{}'}{\omega_e{}'}\left(1 - \frac{\omega_e{}'}{\omega_e{}^*}\right) < 0 \qquad (12.17)$$

the drift wave is stable, i.e.,

$$\omega_e{}' = \omega - \frac{2k_y T_e}{(eB_0R)} > \omega_e{}^*$$

$$\omega_i{}' - \frac{2(1 + T_i/(Z_i T_e))}{\kappa R}\omega_e{}^* > \omega_e{}^*.$$

When $|\omega_i{}'/\omega_e{}^*| \ll 1$, it follows that[1]

$$2\left(1 + \frac{T_i}{Z_i T_e}\right)\frac{1}{\kappa} > -R > 0. \qquad (12.18)$$

Unless T_i/T_e is large, the radius of curvature must be comparable to the scale of the density gradient. This condition is very difficult to satisfy in practice.

(ii) Stabilization due to Short Connection Length The unstable condition is given by the imaginary part of ω, eq. (12.11):

$$\left(1 - \frac{\omega_r}{\omega_e^*}\right)\exp(-\zeta_e^2) - \left(\frac{m_i T_e}{T_i m_e}\right)^{1/2} e^{-b} I_0 \left(1 + \frac{\omega_r T_e}{\omega_e^* T_i}\right)\exp(-\zeta_i^2) > 0.$$

Instability is satisfied when $\omega_r < \omega_e^*$. However if the relation (12.10) of the real part of ω does not hold, an unstable wave cannot exist. ω_e^* can be expressed by ($k_y > 0$)

$$\frac{\omega_e^*}{\omega} = \left(\frac{b}{2}\right)^{1/2} \frac{Z_i T_e}{T_i} \kappa \frac{1}{k_z \zeta_i}. \tag{12.19}$$

The substitution of eq. (12.19) into eq. (12.10) gives

$$\frac{2\pi}{|k_z|}\kappa = 2\pi \left(\frac{2}{b}\right)^{1/2} \frac{k_z}{|k_z|} \zeta_i \frac{(T_i/Z_i T_e)(1 + (k_y^2 + k_z^2)\lambda_d^2) + 1 + e^{-b} I_0 \zeta_i Z(\zeta_i)}{(-\zeta_i Z(\zeta_i) e^{-b} I_0)}, \tag{12.20}$$

where $\zeta_e Z(\zeta_e) \approx -2\zeta_e^2$ is neglected ($|\zeta_e| \ll 1$). The right-hand side of eq. (12.20) has a minimum at $\zeta_i \approx 1, b \approx 1$. From the relations $|-\zeta_i Z(\zeta_i)| < 1.3$, $I_0 e^{-b} < 1$, $I_0 e^{-b} \approx (2\pi b)^{-1/2}$ for $b \gg 1$, we have

$$(\text{right-hand side of eq. (12.20)}) \gtrsim 20\left(1 + \frac{T_i}{Z_i T_e}\left(1 + \frac{\lambda_d^2}{(\rho_B^{i})^2}\right)\right).$$

Accordingly when the length of a plasma device along a line of magnetic force L satisfies

$$L < 20\left(1 + \frac{T_i}{Z_i T_e}(1 + \lambda_d^2/(\rho_B^{i})^2)\right)\left|\frac{n_0}{n_0'}\right|, \tag{12.21}$$

the left-hand side of eq. (12.20), $2\pi\kappa/k_z \approx |n_0'/n_0| L$, cannot be equal to the right-hand side. The mode which satisfies the dispersion equation (12.20) cannot exist and the unstable drift wave does not occur.[1] L may be regarded as the connection length of adjacent regions of good curvature. Accordingly if the connection length is shorter than, say, 20 times the scale of the density gradient, the drift wave is stabilized.

(iii) Stabilization by Shear To stabilize the drift wave, it is necessary to reduce the 1st term of the numerator of eqs. (12.11) or (12.16) and to increase the 2nd (Landau damping) term. Stabilization by good curvature is due to the reduction of $|\omega_i'|$ and $\zeta_i{}^2 = \omega_i'/(k_z(T_z/m)_i)^{1/2}$ so that Landau damping becomes more effective. Stabilization by shear is due to the enhancement of the ion Landau damping by virtue of increasing $|k_z|$ and decreasing $\zeta_i{}^2$. When L_s is the shear length, the magnetic field is given by

$$\boldsymbol{B} = B_0\left(\hat{z} + \frac{x}{L_s}\hat{y}\right). \tag{12.22}$$

The parallel component $k_\parallel$ of the propagation vector is

$$k_\parallel = \frac{(\boldsymbol{k}\cdot\boldsymbol{B})}{B_0} = k_z + \frac{x}{L_s}k_y. \tag{12.23}$$

The drift wave $\exp(i(k_y y + k_z z) - i\omega t)$ is expressed by

$$\begin{aligned}
&\exp i(k_y y + k_z z) \\
&\quad = \exp i\left(\left(k_y - \frac{x}{L_s}k_z\right)\left(y - \frac{x}{L_s}z\right) + \left(k_z + \frac{x}{L_s}k_y\right)\left(z + \frac{x}{L_s}y\right)\right) \\
&\quad = \exp i(k_\perp y' + k_\parallel z'),
\end{aligned} \tag{12.24}$$

where $y' = y - (x/L_s)z$, $z' = z + (x/L_s)y$. Usually $|k_y| \gg |k_z|$ and $k_\perp \approx k_y$. If k_z is replaced by $k_\parallel$ and the term in $\partial^2/\partial x^2$ is added to eq. (12.1) (eq. (12.1) comes from Poisson's equation $\nabla\phi_1 = (-k_y{}^2 - k_z{}^2 + \partial^2/\partial x^2)\phi_1$), the dispersion equation, from eq. (11.36), is

$$\left(\frac{d^2}{dx^2} - Q(x)\right)\phi_1(x) = 0 \tag{12.25}$$

$$\begin{aligned}
Q(x) &= k_y{}^2 + k_z{}^2 \\
&\quad + \frac{1}{\lambda_d{}^2}\left[\frac{Z_i T_e}{T_i}\left(1 + e^{-b}I_0\zeta_i Z(\zeta_i)\left(1 - \frac{\omega_i^*}{\omega_i'}\right)\right) + 1 + \zeta_0 Z(\zeta_e)\left(1 - \frac{\omega_e^*}{\omega_e'}\right)\right],
\end{aligned} \tag{12.26}$$

where

$$\zeta_i = \frac{\omega_i'}{(k_z + (x/L_s)k_y)(2T/m)_i{}^{1/2}}$$

$$\zeta_e = \frac{\omega_e'}{(k_z + (x/L_s)k_y)(2T/m)_e^{1/2}}.$$

As the unstable region for the drift wave is $|\zeta_e| < 1, |\zeta_i| > 1$, the dispersion function can be expanded and eq. (12.25) is solved approximately by the WKB[9] method (the potential is expressed by $\phi_1 \propto A_\pm \exp(\pm Q^{1/2}x)$). Assume that $k_y^2 + k_z^2 \ll \lambda_d^{-2}$ and $G = 0$; then

$$\begin{aligned} Q(x) &= \frac{1}{\lambda_d^2}\frac{Z_i T_e}{T_i}\left[1 - e^{-b}I_0\left(1 - \frac{\omega_i^*}{\omega}\right)\left(1 + \frac{T_i/m_i}{\omega^2}\left(\frac{x}{L_s}k_y\right)^2\right)\right. \\ &\qquad \left. + \frac{T_i}{Z_i T_e} - i\frac{T_i}{Z_i T_e}\left(1 - \frac{\omega_e^*}{\omega}\right)\frac{\pi^{1/2}\omega}{(2T/m)_e^{1/2}|k_\parallel|}\right] \\ &= \frac{1}{\lambda_d^2}\frac{Z_i T_e}{T_i}\left[1 + \frac{T_i}{Z_i T_e} - e^{-b}I_0\left(1 - \frac{\omega_i^*}{\omega}\right)\right. \\ &\qquad - e^{-b}I_0 x^2\left(\frac{k_y^2 T_i/m_i}{\omega^2 L_s^2}\left(1 - \frac{\omega_i^*}{\omega}\right) - \frac{\omega_i^*}{\omega}\frac{(n'/n)''}{n'/n}\right) \\ &\qquad \left. - i\frac{T_i}{Z_i T_e}\left(1 - \frac{\omega_e^*}{\omega}\right)\pi^{1/2}\zeta_e\frac{k_\parallel}{|k_\parallel|}\right]. \end{aligned} \tag{12.27}$$

If the coefficient of x^2 is positive, $Q(x)$ is negative near $x \approx 0$ and $Q(x)$ becomes positive at $x_1(< 0)$ and at $x_2(> 0)$, as is shown in fig. 12.3a. ϕ_1 is oscillatory in $x_1 < x < x_2$ and decays exponentially for $x < x_1, x > x_2$. The solution is essentially the same as the solution without shear ($L_s \to \infty$), and the drift wave is unstable.

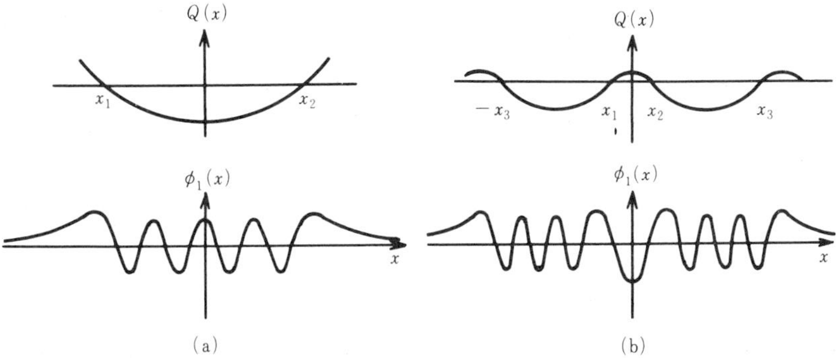

Fig. 12.3
Behavior of perturbation potential $\phi_1(x)$ satisfying eq. (12.25): (a) the case $Q(0) < 0$; (b) the case $Q(0) > 0$.

When the shear length becomes short (shear becomes large) and the coefficient of x^2 is negative, i.e., for

$$\frac{k_y{}^2 T_i/m_i}{\omega^2 L_s{}^2}\left(1-\frac{\omega_i^*}{\omega}\right) > \frac{\omega_i^*}{\omega}\frac{(n'/n)''}{n'/n}, \tag{12.28}$$

then $\phi_1(x)$ behaves differently. $Q(x)$ is positive near $x \approx 0$ and $Q(x) = 0$ at $x = x_1, x = x_2$. At first $Q(x)$ is negative outside the region. $x_1 < x < x_2$, but becomes positive again in the region $|x| > x_3$ as is seen from eq. (12.26). This behavior of $Q(x)$ and $\phi_1(x)$ is shown in fig. 12.3b.

Let us prove that the solution ϕ_1 is stable in this case. Multiplication of eq. (12.25) by ϕ_1^* and integration over x gives

$$\int_{-\infty}^{\infty}\left|\frac{d\phi_1}{dx}\right|^2 dx = -\int_{-\infty}^{\infty} Q(x)|\phi_1|^2 dx.$$

Accordingly we find

$$\int_{-\infty}^{\infty} \operatorname{Im} Q |\phi_1|^2 dx = 0.$$

The imaginary part is expressed by

$$\operatorname{Im} Q \approx \frac{1}{\lambda_d{}^2}\left[\frac{Z_i T_e}{T_i} e^{-b} I_0 \zeta_i Z(\zeta_i) \frac{\omega_i^*}{\omega_r}\frac{\gamma}{\omega_r} + \left(1-\frac{\omega_e^*}{\omega_r}\right)\pi^{1/2}\frac{\omega_r}{|k_\parallel|(2T_e/m_e)^{1/2}} \right.$$
$$+ \frac{Z_i T_e}{T_i} e^{-b} I_0 \left(1-\frac{\omega_i^*}{\omega_r}\right)\pi^{1/2}$$
$$\left.\times \frac{\omega_r}{|k_y(x/L_s)|(2T_i/m_i)^{1/2}} \exp\left(\frac{-\omega_r{}^2}{k_y{}^2(x/L_s)^2\, 2T_i/m_i}\right)\right]. \tag{12.29}$$

The ion Landau damping term, the last term of $\operatorname{Im} Q$, becomes dominant near $x \approx \pm x_3$. As the function $\phi_1(x)$ is oscillatory in the region $|x| < x_3$ and $\phi_1(x)$ decreases exponentially in the region $|x| > x_3$, the following relation is necessary in the region $|x| < x_3$ to satisfy $\int \operatorname{Im} Q |\phi_1{}^2| dx = 0$:

$$\frac{\gamma}{\omega_r} = -\frac{\omega_r}{|k_y(x/L_s)|(2T/m_i)^{1/2}} O(1),$$

where 0(1) is a positive number of the order of 1. The imaginary part γ of ω is negative and the drift wave is stabilized. Using $\operatorname{Re} Q(x) \approx 0$ ($\operatorname{Re} Q(x) = 0$ at $x = x_1, x_2$), we have

$$\frac{\omega_i^*}{\omega} \approx -\frac{(1 + (T_i/Z_i T_e) - e^{-b} I_0)}{e^{-b} I_0};$$

the stability condition (12.28) is then

$$\frac{1}{L_s^2} > \frac{(n'/n)'}{(-n'/n)} \frac{(\omega_i^*)^2}{k_y^2 T_i/m_i} \frac{(e^{-b} I_0)^2}{(1 + (T_i/Z_i T_e) - e^{-b} I_0)(1 + (T_i/Z_i T_e))}. \qquad (12.30a)$$

As $(\omega_i^*)^2/(k_y^2 T_i/m_i) = \kappa^2 (\rho_B^{\ i})^2$, $-n_0'/n_0 = \kappa$, $(n_0'/n_0)'' \approx \kappa^3$, $e^{-b} I_0 \approx 1$, the stability condition with respect to shear length is[1]

$$\frac{1}{L_s} > \kappa(\kappa \rho_B^{\ i}) \frac{1}{\left(\frac{T_i}{Z_i T_e}\left(1 + \frac{T_i}{Z_i T_e}\right)\right)^{1/2}}. \qquad (12.30b)$$

If the shear length is shorter than $r/\rho_B^{\ i}$ times the scale of the density gradient $r = \kappa^{-1}$, the drift wave is stabilized. This stability condition is more likely to be satisfied for large values of $T_i/Z_i T_e$.

(iv) Stabilization by Finite Beta Ratio When $\beta > m_e/m_i$, the Alfvén velocity v_A is smaller than the electron thermal velocity $v_{Te} = (T_e/m_e)^{1/2}$, since $(v_A/v_{Te})^2 = 2m_e/(m_i \beta_e)$.

In this case electrostatic treatment is no longer applicable (see sec. 10.6) and the dispersion equation (12.1) cannot be used. One must use the dispersion equation for the electromagnetic plasma wave, which will be introduced in ch. 13. In the present subsection the effect of finite β is discussed qualitatively. The reader may refer to refs. 6 and 10 for detailed analyses.

When β becomes large, the wave with Alfvén velocity $\omega/k_z \approx v_A$ is likely to be unstable. The ion Landau damping term is expressed by $\zeta_i \exp(-\zeta_i^2)$, where $\zeta_i^2 = \omega^2/k_z^2(2T_i/m_i)$. In the case of $\omega/k_z \approx v_A$, we have $\zeta_i^2 \approx \beta^{-1}$; the ion Landau damping term $\beta^{-1/2} \exp(-1/\beta)$ becomes effective if β becomes large.

The equilibrium condition in terms of the plasma pressure and the magnetic pressure is

$$\varepsilon_B \equiv \frac{1}{B_0}\frac{dB_0}{dx} = \frac{1}{2}\beta\kappa \qquad \kappa = -\frac{1}{n_0}\frac{dn_0}{dx}.$$

The magnetic-pressure gradient is associated with the density gradient and one must take account of the drift induced by the B gradient. The velocity of the B-gradient drift is

$$v_B = \frac{-\varepsilon_B T}{m_i \Omega_i} = \frac{-\beta}{2} \frac{\omega_i^*}{k_y};$$

this forces $\omega_i' = \omega - k_y v_B$ to become smaller, enhancing the ion Landau damping. Accordingly, drift instability due to the density gradient does not occur when β is larger than some value.[10]

*12.1e Drift Cyclotron Instability

The drift frequency due to the density gradient is expressed by

$$\omega_e^* = k_y \kappa \frac{T_e}{eB} = \frac{ZT_e}{T_i}(k_y \kappa)(\rho_B{}^i)^2(-\Omega_i)$$

$$\omega_i^* = (k_y \kappa)(\rho_B{}^i)^2 \Omega_i.$$

If k_y is large or ρ_i is large, the drift frequency may become $\omega_e^* \approx |\omega_i^*| \approx |l\Omega_i|$, where l is an integer. Then the drift wave couples with the ion cyclotron wave and becomes unstable. For simplicity, it is assumed that the temperature gradient, the acceleration, and the fluid velocity are zero ($\delta_\perp = \delta_z = 0$, $G = 0$, $V = 0$). The dispersion equation (eq. (11.42)) is

$$k_y{}^2 + k_z{}^2 + \sum_j \Pi_j{}^2 \left\{ \frac{m}{T_z} + \sum_{n=-\infty}^{\infty} I_n e^{-b} \left[\left(\frac{m}{T_z} - \frac{m}{T_\perp} \frac{n\Omega}{\omega + n\Omega} - \frac{k_y \kappa}{\Omega(\omega + n\Omega)} \right) \right. \right.$$
$$\left. \left. \times \zeta_n Z(\zeta_n) + \frac{m}{T_\perp} \frac{n}{k_y} \kappa \left(1 + \frac{n\Omega}{\omega + n\Omega} \zeta_n Z(\zeta_n) - \frac{T_\perp}{T_z}(1 + \zeta_n Z(\zeta_n)) \right) \right] \right\}_j$$
$$= 0. \tag{12.31}$$

When the temperature is isotropic, the equation is

$$k_y{}^2 + k_z{}^2 + \sum_{ei} \frac{1}{\lambda_d{}^2} \left[1 + \sum_{n=-\infty}^{\infty} I_n e^{-b} \frac{\zeta_n Z(\zeta_n)}{\omega + n\Omega} \left(\omega - \omega^* \left(1 + \frac{n\omega}{b\Omega} \right) \right) \right] = 0. \tag{12.32}$$

Consider the case $|\Omega_i| \gg |\omega + n\Omega_i| \gg |k_z|(T_i/m_i)^{1/2}$, $|k_z|(T_e/m_e)^{1/2} \gg \Omega_e \gg |\omega|$, $b_e = 0$: the zeroth-order electron term and the nth-order ion term are dominant, i.e.,

$$(k_y{}^2 + k_z{}^2)(\lambda_d{}^e)^2 + 1 + \frac{\omega - \omega_e^*}{\omega} i\pi^{1/2} \frac{\omega}{|k_z|(2T/m)_e{}^{1/2}}$$
$$+ \frac{Z_i T_e}{T_i}\left(1 - I_n e^{-b} \frac{\omega - \omega_i^*}{\omega + n\Omega_i} \right) = 0 \tag{12.33}$$

and

$$\omega + n\Omega_{\rm i} = \frac{(\omega - \omega_{\rm i}^*)(2\pi b)^{-1/2}}{1 + \dfrac{T_{\rm i}}{Z_{\rm i}T_{\rm e}}(1 + k^2\lambda_{\rm d}^2) - \dfrac{T_{\rm i}}{Z_{\rm i}T_{\rm e}} \dfrac{i\pi^{1/2}(n\Omega_{\rm i} + \omega_{\rm e}^*)}{|k_z|(2T/m)_{\rm e}^{1/2}}}. \quad (12.34)$$

Here the approximation $I_n(b) \approx e^b(2\pi b)^{-1/2}$ is used for $b \gg 1$. The sign of the imaginary part γ of ω is the same as the sign of

$$(\omega_{\rm r} - \omega_{\rm i}^*)(\omega_{\rm e}^* + n\Omega_{\rm i}) = (\omega_{\rm r} - \omega_{\rm i}^*)(\omega_{\rm e}^* - \omega_{\rm r}).$$

When $\omega_{\rm r}/\omega_{\rm i}^* > 0$ (when the wave propagates in the same direction as the ion drift), the condition for instability is $\omega_{\rm r}/\omega_{\rm i}^* < 1$. Since

$$\omega_{\rm i}^* = k_y \rho_B{}^{\rm i} \kappa (T_{\rm i}/m_{\rm i})^{1/2},$$

the instability condition is reduced to

$$\kappa \rho_B{}^{\rm i} > \frac{1}{\rho_B{}^{\rm i}} \left| \frac{1}{k_y} \right|. \quad (12.35)$$

When $\omega/\omega_{\rm e}^* > 0$ (when the wave propagates in the direction of the electron drift), the instability condition $\omega_{\rm r}/\omega_{\rm e}^* < 1$ is

$$\kappa \rho_B{}^{\rm i} > \frac{T_{\rm i}}{Z_{\rm i} T_{\rm e} \rho_B{}^{\rm i}} \left| \frac{n}{k_y} \right|. \quad (12.36)$$

The instabilities (12.35) or (12.36) are called drift cyclotron instabilities.[6]

12.1f Effect of Collisions on Drift Instability

The effect of collisions has not been considered in the foregoing subsections. Collision processes introduce damping effects in many cases, but sometimes a collision process will interrupt some other stabilizing process and actually induce instability.

Boltzmann's equation in the collisional case is

$$\frac{\partial f_j}{\partial t} + \boldsymbol{v} \cdot \nabla_r f_j + \left(\frac{q_j}{m_j} \boldsymbol{E} + \boldsymbol{G} + \frac{q_j}{m_j} \boldsymbol{v} \times \boldsymbol{B} \right) \nabla_v f_j = -\nu_j (f_j - f_{jL}). \quad (12.37)$$

The right-hand side represents the change of the distribution function f_j due to collisions. f_{jL} is determined so that particle number and momentum are conserved. This term is called Krook's collision term or the BGK collision term.[11] Boltzmann's equation (12.37) is linearized by putting $f = f_{j0} + f_{j1}$, $\boldsymbol{B} = \boldsymbol{B}_0 + \boldsymbol{B}_1$. The quantity ν_j is the 1st-order small quantity. The zeroth-order distribution function f_0 is

$$f_0 = n_0\left(1 - \varepsilon\left(x - \frac{v_y}{\Omega}\right)\right)\left(\frac{m}{2\pi T}\right)^{3/2} \exp\left(-\frac{m}{2T}((v_\perp{}^2 - 2Gx) + v_\parallel{}^2)\right).$$

In a manner similar to that of sec. 10.1, the linearized Boltzmann equation is reduced to

$$\frac{\partial f_{j1}}{\partial t} + v \cdot \nabla_r f_{j1} + \left(\boldsymbol{G}_j + \frac{q_j}{m_j}\boldsymbol{v} \times \boldsymbol{B}_0\right)\nabla_v f_{j1}$$

$$= -\frac{q_j}{m_j}(\boldsymbol{E}_1 + \boldsymbol{v} \times \boldsymbol{B}_1)\nabla_v f_{j0} - \nu_j(f_{j1} - f_{jL}'). \tag{12.38}$$

The collision frequency ν_j consists of ν_{jj} between particles of the same kind and ν_{jl} between particles of different kind:

$$\nu_j = \nu_{jj} + \sum_{l \neq j} \nu_{jl}.$$

f_{jL}' is given by

$$f_{jL}' = \left(\frac{n_{j1}}{n_{j0}} + \frac{m_j}{T_j}\frac{1}{\nu_j}\boldsymbol{v}\cdot(\nu_{jj}\boldsymbol{u}_j + \sum_{l \neq j} \nu_{jl}\boldsymbol{u}_l)\right) f_{j0}$$

$$n_{j1} = \int f_{j1}\, \mathrm{d}\boldsymbol{v}$$

$$\boldsymbol{u}_j = \frac{\int \boldsymbol{v} f_{j1}\, \mathrm{d}\boldsymbol{v}}{n_{j0}}.$$

The change $-\nu_j(\mathrm{f}_{j1} - \mathrm{f}_{jL}')$ of the distribution function due to collisions satisfies the conservation of particle number and momentum, i.e., we can write

$$-\nu_j \int (f_{j1} - f_{jL}')\, \mathrm{d}\boldsymbol{v} = \nu_j(n_{j1} - n_{j1}) = 0$$

$$-\sum_j m_j \nu_j \int \boldsymbol{v}(f_{j1} - f_{jL}')\mathrm{d}\boldsymbol{v} = -\sum_j m_j\left(\nu_j \boldsymbol{u}_j n_{j0} - \nu_{jj}\boldsymbol{u}_j n_{j0} - \sum_{l \neq j} \nu_{jl}\boldsymbol{u}_l n_{j0}\right)$$

$$= -\sum_j \sum_l m_j n_{j0} \nu_{jl}(\boldsymbol{u}_j - \boldsymbol{u}_l)$$

$$= 0.$$

The solution of the linear differential equation (12.38) is

$$f_{j1} = e^{-\nu_j t} \int_{-\infty}^{t} dt' e^{\nu_j t'} \left(\left(\frac{-q_j}{m_j} \right) (\boldsymbol{E} + \boldsymbol{v} \times \boldsymbol{B}_1) \nabla_v f_{j0} + \nu_j f_{jL}' \right). \tag{12.39a}$$

When the wave is electrostatic, then $\boldsymbol{E} = -\nabla\phi$, $\boldsymbol{B}_1 = 0$, and f_{j1} is

$$f_{j1} = e^{-\nu_j t} \int_{-\infty}^{t} dt' e^{\nu_j t'} \left(\frac{q_j}{m_j} \nabla\phi \cdot \nabla_v f_{j0} + \nu_j f_{jL}' \right). \tag{12.39b}$$

Let us consider the simple case of a low-frequency wave $|\omega| \ll |\Omega_i|$. Only the 1st term in Krook's collision term $f_L' = \frac{n_1}{n_0} f_0$ is taken. When we apply a process similar to that used in the derivation of eqs. (11.36) and (11.45), we find

$$\int f_1 dv = \frac{-n_0 q^\phi}{T} \left(1 + e^{-b} I_0(b) \left(1 - \frac{\omega^*}{\omega + i\nu} \right) \zeta Z(\zeta) \right)$$
$$\times \left(1 + \frac{i\nu}{\omega + i\nu} e^{-b} I_0(b) \zeta Z(\zeta) \right)^{-1}.$$

The factor $(\ )^{-1}$ comes from $\nu f_L'$ of eq. (12.39b). Here ζ is

$$\zeta = \frac{\omega + i\nu}{2^{1/2} k_z (T/m)^{1/2}}.$$

The substitution of the equation $\int f_1 dv$ into Poisson's equation (11.24) gives the dispersion equation

$$k_y^2 + k_z^2 + \sum_j \frac{\Pi_j^2 \frac{m_j}{T_j} \left(1 + \left(1 - \frac{\omega_j^*}{\omega + i\nu_j} \right) e^{-b_j} I_0(b_j) \zeta_j Z(\zeta_j) \right)}{1 + \frac{i\nu_j}{\omega + i\nu_j} e^{-b_j} I_0(b_j) \zeta_j Z(\zeta_j)} = 0. \tag{12.40}$$

This equation becomes the same as eq. (12.9) for the collisionless case in the limit $\nu_j \to 0$. When $|\zeta_e| \gg 1$, $|\zeta_i| \gg 1$, $b_e \to 0$, $b_i \equiv b \ll 1$, $\nu_i = 0$, $|\nu_e| \ll |\omega|$, then the dispersion equation is

$$k^2 + \frac{1}{(\lambda_d^{\,i})^2} \frac{1}{\omega} (\omega_i^* + (\omega - \omega_i^*) b) + \frac{1}{(\lambda_d^{\,e})^2} \frac{\omega_e^* - \dfrac{(\omega - \omega_e^* + i\nu_e) k_z^2 T_e}{m_e (\omega + i\nu_e)^2}}{\omega - \dfrac{i\nu_e k_z^2 T_e}{m_e (\omega + i\nu_e)^2}} = 0. \tag{12.41}$$

For the case $(k_z^2 T_e/m_e)/(\omega_e^*)^2 \gg b$, it follows that

$$\frac{\omega}{\omega_e^*} = 1 + i\frac{\nu_e}{\omega_e^*}\frac{(\omega_e^*)^2}{k_z^2 T_e/m_e}\left(b\left(1 + \frac{T_e Z}{T_i}\right) + k^2(\lambda_d^e)^2\right). \tag{12.42}$$

Accordingly, the growth rate is proportional to the collision frequency. This growth rate is the same as the rate for resistive drift instability that was derived by means of the MHD equation in sec. 9.7, as is expected. The resistive drift instability is an example which can be treated by means of a kinetic model as well as a MHD model.

12.2 Trapped Particle Instabilities

Magnetic-field configurations for plasma confinement were discussed in ch. 2. In the mirror configuration, velocity-space anisotropy appears as a consequence of the existence of the loss cone; and this anisotropy induces instabilities, as will be explained in sec. 12.3. In the toroidal system, the distribution function is nearly Maxwellian and the velocity-space anisotropy appears to be unimportant. However, this is not true. The magnitudes of the magnetic fields in toroidal systems are approximately equal to

$$B \approx |B_0|\frac{R}{R + r\cos\theta} \approx |B_0|\left(1 - \frac{r}{R}\cos\theta\right)$$

in the coordinates shown in fig. 12.4. The magnetic field is weak in the region outside the torus. Accordingly, there exist local mirrors and the particles with the parallel velocity $v_\parallel$ such that $(v_\parallel/v)^2 < r/R$ are trapped in the local mirrors. The number of trapped particles is $(r/R)^{1/2}$ of the total. When $r/R \approx 0.2$, 40% of the particles are trapped. The trapped

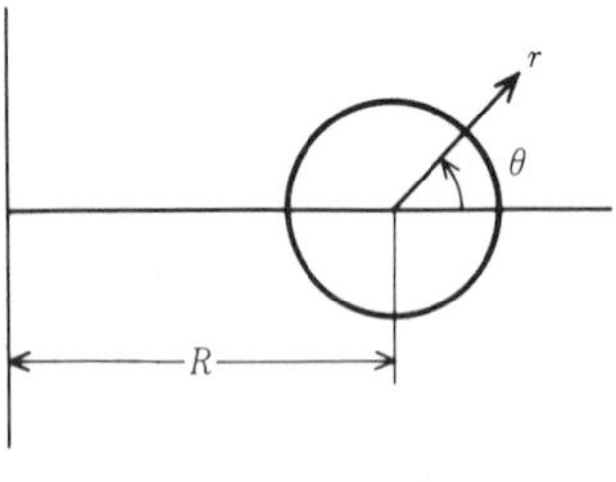

Fig. 12.4
Coordinate system for torus.

particles remain in the regions of bad curvature and may induce interchange instability. There is a difference between this and the linear-mirror case, however. In toroidal systems, transit particles are not trapped and play the role of a medium of large dielectric constant (see eq. (13.36)):

$$K_{zz} = 1 + \frac{1}{k_z{}^2 \lambda_{\mathrm{d}}{}^2}.$$

Hence the electric-field perturbation due to the trapped particles is nearly cancelled, so that the growth rate of the instability is reduced. The growth rate of interchange instability in a linear mirror is

$$\gamma \approx (gk_y)^{1/2} \approx \left(\frac{T/m}{Ra}\right)^{1/2} \approx \frac{v_{\mathrm{T}}}{(Ra)^{1/2}} \approx \varepsilon^{1/2}\frac{v_{\mathrm{T}}}{a}.$$

On the contrary, the growth rate of the trapped particle instability is

$$\gamma \approx \frac{\rho_B^{\mathrm{i}}}{a}\varepsilon^{1/4}\varepsilon^{1/2}\frac{v_{\mathrm{T}}}{a},$$

as will be shown later. g is the acceleration, a is the plasma radius, and ε is the inverse aspect ratio $\varepsilon = a/R$.

The existence of trapped particles is not only the origin of the neoclassical diffusion that has been described in sec. 8.5, but is also the source of various instabilities. In the present section collisionless trapped particle instabilities and dissipative trapped particle instabilities will be discussed. The latter are more dangerous in actual toroidal plasmas. Reviews of trapped particle instabilities will be found in refs. 12–15.

12.2a Collisionless Trapped Particle Instability

The dispersion equation which has been used assumes the homogeneity of the zeroth-order magnetic field. For the analysis of the behavior of trapped particles, one must take account of the spatial variation of the zeroth-order magnetic field. Therefore one must employ processes similar to those used in ch. 11 to derive the dispersion equation.

Let us call the coordinate along the line of magnetic force ζ. The coordinate directed radially outward in the plasma column is denoted by ξ. The coordinate η is normal to both the foregoing, so that ξ, η, ζ form a right-handed system (fig. 12.5). We assume that the magnetic field is nearly constant within the Larmor radius and that the charged particles are accelerated only in the direction ξ by the amount G.

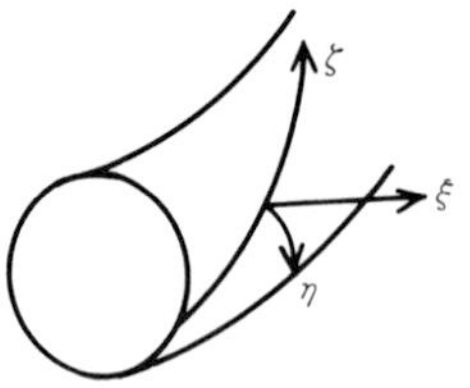

Fig. 12.5
Coordinates system ξ, η, ζ.

The orbit of a charged particle is given by

$$\left.\begin{aligned} \xi'(t') - \xi &= \frac{v_\perp}{\Omega}(\sin(\theta + \Omega(t'-t)) - \sin\theta) \\ \eta'(t') - \eta &= \frac{v_\perp}{\Omega}(\cos(\theta + \Omega(t'-t)) - \cos\theta) + \int_t^{t'} \left(\frac{G}{\Omega}\right) \mathrm{d}t'' \\ \zeta'(t') - \zeta &= \int_t^{t'} v_\parallel \, \mathrm{d}t''. \end{aligned}\right\} \quad (12.43)$$

For simplicity, the drift velocity due to the acceleration is denoted by

$$v_G = \frac{G}{\Omega}.$$

If the zeroth-order distribution function is denoted by

$$\begin{aligned} f_0 &= n_0\left(1 - \varepsilon\left(\xi - \frac{v_\eta}{\Omega}\right)\right)\left(\frac{m}{2\pi T}\right)^{3/2} \exp\left(-\frac{m}{2T}((v_\perp{}^2 - 2Gx) + v_\parallel{}^2)\right) \\ &\equiv n_0\left(1 - \varepsilon\left(\xi - \frac{v_\eta}{\Omega}\right)\right) F_0, \end{aligned}$$

the 1st-order perturbation of the distribution function is given by eq. (11.33) as

$$\begin{aligned} f_1 &= \frac{q n_0 F_0}{m}\left(-\frac{m}{T}\phi_1 - \frac{m}{T} i \left(\omega - \frac{k_\eta \varepsilon T}{\Omega m}\right) \int_{-\infty}^{t} \phi_1 \mathrm{d}t'\right) \\ &= -\frac{q n_0 F_0}{T}\left(\phi_1 + i(\omega - \omega^* - k_\eta v_G) \int_{-\infty}^{t} \phi_1(\xi', \eta', \zeta', t')\mathrm{d}t'\right). \end{aligned} \quad (12.44)$$

The integration is carried out along the orbit equations (12.43) of the particle; ω^* is the drift frequency and

$$\frac{1}{n_0}\frac{\mathrm{d}n_0}{\mathrm{d}x} = -\varepsilon + \frac{Gm}{T} \equiv -\kappa$$

$$\omega^* = \frac{k_\eta \kappa T}{\Omega m}.$$

Let us consider the perturbation

$$\phi_1 = \phi_1(\xi', \xi')\exp(i(k_\eta \eta' - \omega t')) \approx \phi_1(\zeta')\exp(i(k_\eta \eta' - \omega t')).$$

Substitute ϕ_1 into eq. (12.44) and expand $\exp(ik_\eta(v_\perp/\Omega)\cos\theta)$ in terms of Bessel functions. As we are interested in perturbations of a frequency low compared with the cyclotron frequency, only the term in the zeroth-order Bessel function is taken into account. Accordingly eq. (12.44) is reduced to

$$f_1 = -\frac{qn_0}{T}F_0\Bigg[\phi_1 + i(\omega - \omega^* - k_\eta v_{\mathrm{G}})\,J_0(a)$$
$$\times \int_{-\infty}^{t}\left(J_0(a')\phi_1(\zeta'(t'))\exp\left(ik_\eta\Big(\eta + \int_t^{t'} v_{\mathrm{G}}\mathrm{d}t''\Big) - i\omega t'\right)\right)\mathrm{d}t'\Bigg], \tag{12.45}$$

where the integrations are carried out along the orbit of the guiding center and where $a = k_\eta v_\perp/\Omega$.

The substitution of $n_1 = \int f_1 \,\mathrm{d}\boldsymbol{v}$ into Poisson's equation

$$-\nabla^2\phi_1 = \frac{e}{\varepsilon_0}(Z_\mathrm{i} n_{\mathrm{i}1} - n_{\mathrm{e}1})$$

gives

$$Z_\mathrm{i} n_{\mathrm{i}1} - n_{\mathrm{e}1} = \frac{\varepsilon_0 k^2}{e}\phi_1 \approx 0.$$

Usually the product of k^2 and the square of the Debye length $(\varepsilon_0 T/n_\mathrm{e} e^2)$ is small and charge neutrality holds. This quasi-neutrality condition yields the dispersion equation.

Trapped particles are confined in the local mirrors, and move back and forth within the local mirror fields. The frequency of this motion is called the bounce frequency and is denoted by ω_b. If the length of a mirror is L, then the average value of ω_b is

$$\omega_b \approx \frac{2\pi}{L} v_{\parallel}$$

$$\approx \frac{2\pi}{L} \varepsilon^{1/2} v_T,$$

where v_T is the thermal velocity and $\varepsilon = r/R$. The bounce frequencies of ions and electrons are different, and usually $\omega_{bi} \ll \omega_{be}$. When the frequency ω of the perturbation is such that

$$\omega \gg \omega_{be} \gg \omega_{bi},$$

the effect of trapped particles does not appear. When the frequency ω is such that

$$\omega_{be} \gg \omega \gg \omega_{bi},$$

the bounce motion of trapped electrons is the source of wave excitation; this type of perturbation is called the *electron mode*. When the frequency ω is such that

$$\omega_{be} \gg \omega_{bi} \gg \omega,$$

the bounce motions of both trapped ions and electrons play roles in the perturbation; this is called the *ion mode*.

When the perturbed field $\phi_1(\zeta)\exp(ik_\eta \eta - i\omega t)$ does not change very much along the line of magnetic force, the perturbation is called *flutelike*. When part of the variation $\tilde{\phi}_1(\zeta)$ is dominant compared with the average value $\bar{\phi}_1(\zeta)$, the perturbation is said to be *ballooning*.

When the wave couples resonantly with particles whose velocities match the wave phase velocity, the mode is called *resonant*. When the wave interacts with entire collection of plasma particles, we have the *fluid mode*.

(i) Ion Mode ($\omega \ll \omega_{bi} \ll \omega_{be}$) When we integrate the 1st-order velocity distribution function f_1, we must treat the distributions for untrapped and trapped particles separately. Let us consider the contribution of the untrapped particles. The orbit is given by

$$\zeta' - \zeta \approx v_{\parallel}(t' - t).$$

If the perturbed field is taken to be in the form $\phi_1(\xi, \eta, \zeta) = \phi_1 \exp i(k_\eta \eta + k_{\parallel}\zeta - \omega t)$, then the integral in eq. (12.44) is

$$\int_{-\infty}^{t} J_0(a')\phi_1(\xi)\exp i\left(k_\eta\left(\eta + \int_t^{t'} v_G \mathrm{d}t''\right) + k_\parallel \zeta' - \omega t'\right)\mathrm{d}t'$$

$$\approx J_0(a)\phi_1(\xi)\exp i(k_\eta \eta + k_\parallel \zeta - \omega t)$$

$$\times \int_{-\infty}^{0} \exp i\left(k_\eta \int_0^\tau v_G(t+\tau')\mathrm{d}\tau' + (k_\parallel v_\parallel - \omega)\tau\right)\mathrm{d}\tau$$

$$\approx J_0(a)\phi_1 \frac{1}{i(k_\eta v_G + k_\parallel v_\parallel - \omega)}. \qquad (12.46)$$

Next we consider the contribution of the trapped particles. The integral becomes

$$\int_{-\infty}^{t} J_0(a')\phi_1(\zeta'(t'))\exp i\left(k_\eta\left(\eta + \int_t^{t'} v_G \mathrm{d}t''\right) - \omega t'\right)\mathrm{d}t'$$

$$\approx J_0(a)\exp i(k_\eta \eta - \omega t)$$

$$\times \int_{-\infty}^{0} \phi_1(\zeta'(\tau + t))\exp i\left(k_\eta \int_0^\tau \tilde{v}_G(\tau' + t)\mathrm{d}\tau' + (k_\eta \bar{v}_G - \omega)\tau\right)\mathrm{d}\tau. \qquad (12.47)$$

Here v_G is taken to be composed of an average part $\bar{v}_G$ and variable part $\tilde{v}_G$. The term $\phi_1(\zeta'(\tau + t))\exp(ik_\eta \int_0^\tau \tilde{v}_G \mathrm{d}\tau')$ is a periodic function of period $\tau_b = 2\pi/\omega_b$ for trapped particles. When $h(\tau)$ is a periodic function of period T, it follows that

$$\int_{-\infty}^{0} h(\tau)\exp(-i\omega\tau)\mathrm{d}\tau = \frac{1}{1 - \exp(i\omega T)} \int_{-T}^{0} h(\tau)\exp(-i\omega\tau)\mathrm{d}\tau$$

$$\approx \frac{i}{\omega}\frac{1}{T}\int_{-T}^{0} h(\tau)\exp(-i\omega\tau)\mathrm{d}\tau \qquad (\omega T \ll 1).$$

Accordingly, eq. (12.47) is reduced to

$$J_0(a)\exp i(k_\eta \eta - \omega t)\frac{i}{(\omega - k_\eta \bar{v}_G)}\frac{1}{\tau_b}\int_{-\tau_b}^{0}\left(\phi_1(\zeta'(\tau + t))\exp ik_\eta \int_0^\tau \tilde{v}_G \mathrm{d}\tau'\right)\mathrm{d}\tau. \qquad (12.48)$$

Let us compare the contributions of untrapped and trapped particles to

$n_1 = \int f_1 \mathrm{d}v$. Where the radius of curvature of magnetic-force lines is R, the drift velocity v_G due to the centrifugal force is

$$v_G = \frac{-T}{qBR} = \frac{-v_T \rho_B}{R},$$

where $v_T = T/m$ and ρ_B is the Larmor radius. When

$$k_\parallel \gg \frac{k_\eta \rho_B}{r},$$

it follows that

$$\left|\frac{1}{k_\parallel v_\parallel}\right| \ll \left|\frac{\varepsilon}{k_\eta v_G}\right|.$$

Accordingly, the contribution of untrapped particles is negligible compared to that of trapped particles. When the rotational transform angle is ι, the mirror length L is given by $2\pi R/\iota$. Accordingly if

$$\frac{\iota}{R} \gg |k_\parallel|,$$

then $\phi_1(\zeta')$ can be considered to be constant along the line of magnetic force. As a result, when the condition

$$\frac{\iota}{R} \gg |k_\parallel| \gg \frac{\rho_B}{r}|k_\eta|$$

is satisfied, then eq. (12.48) becomes

$$\frac{iJ_0(a)\phi_1}{\omega - k_\eta \bar{v}_G}.$$

Using the notation $\omega_G \equiv k_\eta \bar{v}_G$, we can express the dispersion equation by[15]

$$\frac{1}{T_e} + \frac{Z_i}{T_i} = \varepsilon^{1/2}\left(\frac{1}{T_e}\frac{\omega - \omega_e^* - \omega_{Ge}}{\omega - \omega_{Ge}} + \frac{Z_i}{T_i}\frac{\omega - \omega_i^* - \omega_{Gi}}{\omega - \omega_{Gi}}\right). \tag{12.49}$$

This is the dispersion equation for *collisionless trapped particle instability*. As $\omega^* = -k_\eta \kappa T/qB$, it follows that $\omega^* \approx R\kappa\,\omega_G$. When $T_e = T_i/Z_i$, then $\omega_i^* = -\omega_e^*$ and $\omega_{Gi} = -\omega_{Ge}$ and the dispersion equation may be simplified to

$$\omega^2 = (\omega_{Ge})^2 - \varepsilon^{1/2}\,\omega_e{}^*\omega_{Ge}. \tag{12.50}$$

The condition $\omega_e{}^*\,\omega_{Ge} < 0$ is a sufficient condition for stability. We have $\omega_e{}^* = -k_\eta(T/eB)(\partial n_0/\partial r)/n_0 = -k_\eta(\partial p/\partial r)(eBn_0)^{-1}$. In terms of the longitudinal adiabatic invariant J, v_{Ge} is given by (see sec. 3.5)

$$v_{Ge} = \frac{-1}{qB_0}\frac{\partial J}{\partial r}\bigg/\frac{\partial J}{\partial W},$$

where W is the energy of the particle. (The sign is opposite to that of eq. (3.119) as the direction of η has been reversed.) The stability condition is

$$\nabla p \nabla J + \frac{n_0}{\varepsilon^{1/2}}\frac{(\nabla J)^2}{\left(\dfrac{\partial J}{\partial W}\right)} > 0. \tag{12.51a}$$

When $R\kappa \gg 1$, the first term of eq. (12.51a) is large and the stability condition is

$$\nabla p \nabla J > 0. \tag{12.51b}$$

By analogy to the stability condition for interchange instability $\nabla p \cdot \nabla U < 0$ ($U = \oint \mathrm{d}l/B$), this condition is called the *maximum-J condition.*

When the electron and ion temperatures are different and their ratio is in the region

$$\frac{1}{4\varepsilon^{1/2}}\frac{\omega_{Ge}}{\omega_e} < \frac{T_i}{T_e} < 4\varepsilon^{1/2}\frac{\omega_e{}^*}{\omega_{Ge}}, \tag{12.52}$$

the situation is unstable, as is clear from the dispersion equation (12.49). This is a trapped ion mode as well as a fluid and flutelike mode.

(ii) Electron Mode ($\omega_{be} \gg \omega \gg \omega_{bi}$) Let us consider the electron mode of trapped particle instability. The 1st-order perturbation of the ion distribution function is derived from eqs. (12.45) and (12.46) as follows:

$$f_{i1} = -\frac{Z_i e n_0}{T_i}F_{0i}\Bigg(\phi_1 - \frac{\omega - \omega_i{}^*}{\omega - \omega_{Gi}}J_0\int_{-\infty}^{0}\frac{\partial}{\partial\tau}\exp(-i(\omega-\omega_{Gi})\tau)A(\zeta'(\tau+t))\mathrm{d}\tau \times \exp i(k_\eta\eta - \omega t)\Bigg), \tag{12.53}$$

where

$$A(\zeta'(\tau + t)) = J_0\phi_1(\zeta'(\tau + t))\exp ik_\eta \int_0^\tau \tilde{v}_G(\tau' + t)d\tau'.$$

The integral term in eq. (12.53) is modified by partial integration:

$$J_0\phi_1(\zeta) - \frac{iv_\parallel}{\omega - \omega_{Gi}}\frac{\partial}{\partial\zeta}(J_0\phi_1(\zeta)) - \frac{1}{(\omega - \omega_{Gi})^2}\frac{\partial^2}{\partial t^2}(J_0\phi_1(\zeta)) + \cdots.$$

The integral of f_{i1} over v is the ion density perturbation n_{i1}, and

$$\frac{n_{i1}}{n_0} = -\frac{Z_i e}{T_i}\left(\phi_1(\zeta) - \left\langle\frac{\omega - \omega_i^*}{\omega - \omega_{Gi}} J_0\left(J_0\phi_1(\zeta) - \frac{1}{(\omega - \omega_{Gi})^2}\frac{\partial^2}{\partial t^2}(J_0\phi_1(\zeta))\right)\right\rangle\right). \quad (12.54)$$

Since $\int v_\parallel F_0 dv_\parallel = 0$, the term in $\partial(J_0\phi_1(\zeta))/\partial\zeta$ disappears. The notation $\langle h\rangle$ is the average of h in velocity space with the weighting factor F_0. The term $\exp i(k_\eta\eta - \omega t)$ is omitted.

Next we must obtain the electron density perturbation n_{e1}. Using the notation $v_G = \bar{v}_G + \tilde{v}_G$, let us consider the term

$$\phi_1(\zeta'(t'))\exp\left(ik_\eta \int_t^{t'} \tilde{v}_G dt''\right) \quad (12.55)$$

in eq. (12.45). This is a periodic function of t' of fundamental frequency ω_{be}, which can be Fourier expanded as

$$\sum_n \phi_e^{(n)} \exp(-in\omega_{be}t').$$

The substitution of Fourier series into eq. (12.45) yields

$$f_{e1} = \frac{en_0}{T_e}F_{0e}\left(\phi_1 - (\omega - \omega_e^* - k_\eta v_{Ge}) \times \sum_n \frac{\phi_e^{(n)}\exp(ik_\eta\eta + i(k_\eta v_{Ge} - \omega - n\omega_{be})t)}{\omega + n\omega_{be} - k_\eta\bar{v}_{Ge}}\right), \quad (12.56)$$

and the perturbed electron density n_{e1} is then given by

$$\frac{n_{e1}}{n_0} = \frac{e}{T_e}\left(\phi_1 - (\omega - \omega_e^*)\left\langle\sum_n \frac{\phi_e^{(n)}\exp(-in\omega_{be}t)}{\omega + n\omega_{be} - k_\eta \bar{v}_{Ge}}\right\rangle\right). \tag{12.57}$$

It is assumed that $|k_\eta v_{Ge}| \ll \omega_e^*$, and also that $J_0 = 1$ because of the small electron Larmor radius. The term $\exp i(k_\eta \eta - \omega t)$ is omitted. The dispersion equation $n_{e1} = n_{i1}$ is[13]

$$\begin{aligned}\phi_1(\zeta)\left(1 - \left\langle\frac{\omega - \omega_i^*}{\omega - \omega_{Gi}} J_0{}^2\right\rangle\right)\frac{Z_i T_e}{T_i} \\ + \left(1 - \frac{\omega_i^*}{\omega}\right)\frac{1}{\omega^2}\left\langle J_0 \frac{\partial^2}{\partial t^2}(J_0\phi_1(\zeta))\right\rangle\frac{Z_i T_e}{T_i} \\ + \phi_1(\zeta) - (\omega - \omega_e^*)\left\langle\sum_n \frac{\phi_e^{(n)}\exp(-in\omega_{be}t)}{\omega + n\omega_{be} - \omega_{Ge}}\right\rangle = 0.\end{aligned} \tag{12.58}$$

Multiplication by $\phi_1^*(\zeta)$ and volume integration within a tube of magnetic flux give

$$\iiint \mathrm{d}\xi\, \mathrm{d}\eta\, \mathrm{d}\zeta\, \phi_1^*(\zeta) \propto \int \frac{\mathrm{d}\zeta}{B_0}\phi_1^*(\zeta) = \int \frac{v_\parallel}{B_0}\mathrm{d}t\, \phi_1^*(\zeta(t))$$

and

$$\begin{aligned}\int \frac{\mathrm{d}\zeta}{B_0}|\phi_1|^2\left(1 + \frac{Z_i T_e}{T_i}\left(1 - \left\langle\frac{\omega - \omega_i^*}{\omega - \omega_{Gi}} J_0{}^2\right\rangle\right)\right) \\ - \frac{Z_i T_e}{T_i}\left(1 - \frac{\omega_i^*}{\omega}\right)\frac{1}{\omega^2}\left\langle\frac{v_\parallel}{B_0}\int\left|\frac{\partial}{\partial t}J_0\phi_1\right|^2 \mathrm{d}t\right\rangle \\ - (\omega - \omega_e^*)\left\langle\frac{v_\parallel}{B_0}\frac{2\pi}{\omega_{be}}\sum_n \frac{|\phi_e^{(n)}|^2}{\omega + n\omega_{be} - \omega_{Ge}}\right\rangle = 0.\end{aligned} \tag{12.59}$$

Let us consider $\phi(\zeta)$ to be an odd function with respect to the center of a local mirror. Then $\phi_e^{(0)} = 0$, and eq. (12.59) becomes

$$\begin{aligned}\oint \frac{\mathrm{d}\zeta}{B_0}|\phi_1|^2\left(1 + \frac{Z_i T_e}{T_i}\left(1 - \left\langle\frac{\omega - \omega_i^*}{\omega - \omega_{Gi}} J_0{}^2\right\rangle\right)\right) \\ + \left(1 - \frac{\omega_e^*}{\omega}\right)\frac{\omega^2}{\langle\omega_{be}\rangle^2}\left\langle\left(\frac{\langle\omega_{be}\rangle}{\omega_{be}}\right)^2 \frac{v_\parallel}{B_0}\left(\frac{2\pi}{\omega_{be}}\right)\sum_{n\neq 0}\frac{|\phi_e^{(n)}|^2}{n^2}\right\rangle = 0.\end{aligned} \tag{12.60}$$

When $|\omega_{\mathrm{Ge}}| \ll |\omega| \ll \omega_{\mathrm{e}}{}^*$, eq. (12.60) is reduced to

$$\left(1+\frac{Z_{\mathrm{i}}T_{\mathrm{e}}}{T_{\mathrm{i}}}\right)\left(\oint\frac{d\zeta}{B_0}|\phi_1|^2\right)-\frac{\omega_{\mathrm{e}}^*}{\omega}\left(J_0{}^2\oint\frac{d\zeta}{B_0}|\phi_1|^2\right)-\frac{\omega_{\mathrm{e}}{}^*\omega}{\langle\omega_{\mathrm{be}}\rangle_2}\oint\frac{d\zeta}{B_0}|\phi_1|^2\frac{1}{\Xi}=0$$

where

$$\Xi=\frac{\displaystyle\oint\frac{d\zeta}{B_0}|\phi_1|^2}{\left\langle\left(\frac{\langle\omega_{\mathrm{be}}\rangle}{\omega_{\mathrm{be}}}\right)^2\frac{v_\parallel}{B_0}\left(\frac{2\pi}{\omega_{\mathrm{be}}}\right)\sum_{n\neq0}\frac{|\phi_{\mathrm{e}}{}^{(n)}|^2}{n^2}\right\rangle}.$$

The solution is

$$\frac{\omega}{\langle\omega_{\mathrm{be}}\rangle}=\frac{1}{2}\left(1+\frac{Z_{\mathrm{i}}T_{\mathrm{e}}}{T_{\mathrm{i}}}\right)\frac{\langle\omega_{\mathrm{be}}\rangle}{\omega_{\mathrm{e}}^*}\Xi\pm\left(\frac{1}{4}\left(1+\frac{Z_{\mathrm{i}}T_{\mathrm{e}}}{T_{\mathrm{i}}}\right)^2\left(\frac{\langle\omega_{\mathrm{be}}\rangle}{\omega_{\mathrm{e}}^*}\right)^2\Xi^2-J_0{}^2\Xi\right)^{1/2}, \tag{12.61}$$

and the instability condition is

$$\frac{\omega_{\mathrm{e}}^*}{\langle\omega_{\mathrm{be}}\rangle}>\frac{1}{2}\left(1+\frac{Z_{\mathrm{i}}T_{\mathrm{e}}}{T_{\mathrm{i}}}\right)\frac{\Xi^{1/2}}{J_0}.$$

It must be remarked that we have assumed $|\omega| \ll \omega_{\mathrm{be}}$, $|\omega| \ll \omega_{\mathrm{e}}{}^*$.

Trapped particle instability is reviewed in refs. 12 and 13. The reader should also refer to refs. 16 and 17 on this topic. Instabilities arising from deformations of the velocity-space distribution function by trapped particles is analyzed in ref. 18.

12.2b Dissipative Trapped Particle Instability

In the previous subsection the effects of collisions were neglected. This is justified if the collision frequency is less than the characteristic frequency of the plasma. For trapped particle instability, the characteristic frequency is the drift frequency $\omega^* \approx T/(eBr^2) = v_{\mathrm{T}}\rho_B/r^2$. The e-i collision frequency must be smaller than ω^*. This condition $\omega^*/\nu_{\mathrm{ei}} = \lambda_{\mathrm{e}}\rho_B{}^{\mathrm{e}}/r^2 \gg 1$ is expressed by

$$S\equiv\frac{\lambda_{\mathrm{e}}\rho_B}{r^2}\gg\left(\frac{m_{\mathrm{i}}}{m_{\mathrm{e}}}\right)^{1/2},$$

where $\lambda_{\mathrm{e}} = v_{\mathrm{Te}}/\nu_{\mathrm{ei}}$ is the electron mean free path. The value of S is on the order of 1 in many experiments as well as for most predicted reactors, so that collisions cannot be neglected. Therefore dissipative trapped particle instability, induced by the collisions, is of great practical importance. Dissipative trapped ion and electron instabilities are described in this subsection.

(i) Dissipative Trapped Ion Instability Collision have been taken into account in the discussion of drift instability in sec. 12.1f. The same analytical method can be applied to trapped ion instability. The frequency ω in the denominator of the right-hand side of eq. (12.49) must be replaced by $\omega + i\nu_{\mathrm{eff}}$. The effective collision frequency is $\nu_{\mathrm{eff}} = \nu/\varepsilon$ for the trapped particles. Furthermore the $\boldsymbol{B}$ curvature drift term is neglected ($\omega_{\mathrm{G}} \approx 0$). Then the dispersion equation is expressed by[12,15]

$$\frac{1}{T_{\mathrm{e}}} + \frac{Z_{\mathrm{i}}}{T_{\mathrm{i}}} = \varepsilon^{1/2}\left(\frac{1}{T_{\mathrm{e}}}\frac{\omega - \omega_{\mathrm{e}}^{*}}{\omega + i\nu_{\mathrm{e}}/\varepsilon} + \frac{Z_{\mathrm{i}}}{T_{\mathrm{i}}}\frac{\omega - \omega_{\mathrm{i}}^{*}}{\omega + i\nu_{\mathrm{i}}/\varepsilon}\right). \tag{12.62}$$

The same formula can be deduced from the two-dimensional fluid model averaged along the lines of magnetic force. The equation of continuity for the trapped particles is

$$\frac{\partial n_{\mathrm{t}}}{\partial t} + v_{\xi}\frac{\partial}{\partial \xi}n_{\mathrm{t}} = -\nu_{\mathrm{eff}}\left(n_{\mathrm{t}} - \varepsilon^{1/2}n_0\left(1 - \frac{q\phi}{T}\right)\right).$$

The average of the density n_{t} of trapped particle is $n_{\mathrm{t}0} = \varepsilon^{1/2}n_0\exp(-q\phi/T) \approx \varepsilon^{1/2}n_0(1 - q\phi/T)$. The collision term must be proportional to the deviation of n_{t} from $n_{\mathrm{t}0}$. The ξ component of $\boldsymbol{v}$ is

$$v_{\xi} = \left(\frac{\boldsymbol{E}\times\boldsymbol{b}}{B_0}\right)_{\varepsilon} = \frac{-ik_{\eta}\phi}{B_0}.$$

Introducing $n_{\mathrm{t}j}' \equiv n_{\mathrm{t}j} - \varepsilon^{1/2}n_0$, we can write the equations of continuity for trapped ions or electrons as

$$-i\omega n_{\mathrm{t}j}' - \frac{ik_{\eta}\phi}{B_0}\frac{\partial\varepsilon^{1/2}n_0}{\partial\xi} = -\frac{\nu_j}{\varepsilon}\left(n_{\mathrm{t}j}' + \varepsilon^{1/2}n_0\frac{q_j\phi}{T_j}\right), \tag{12.63}$$

i.e., as

$$\frac{n_{\mathrm{t}j}'}{n_0} = \frac{-\varepsilon^{1/2}(\omega_j^{*} + i\nu_j/\varepsilon)}{\omega + i\nu_j/\varepsilon}\frac{q_j\phi}{T_j}. \tag{12.64}$$

A quasi-neutrality condition gives the dispersion equation

$$\frac{n_{ti}'}{n_0} - \frac{Z_i e\phi}{T_i}(1 - \varepsilon^{1/2}) = \frac{n_{te}'}{n_0} + \frac{e\phi}{T_e}(1 - \varepsilon^{1/2}).$$

The substitution of eq. (12.64) into this equation yields eq. (12.62).

The dispersion equation is here quadratic. When

$$\nu_e/\varepsilon > \varepsilon^{1/2}\omega_e^*$$

the equation in the square root of the solution can be expanded, and the solution is given by

$$\omega = \frac{\varepsilon^{1/2}\omega_e^*}{(1 + Z_i T_e/T_i)} - \frac{\nu_i}{\varepsilon} i + \frac{\varepsilon^2}{(1 + Z_i T_e/T_i)} \frac{(\omega_e^*)^2}{\nu_e} i. \tag{12.65}$$

e-e and e-i collisions are destabilizing and i-i collisions are stabilizing. The growth rate γ is a maximum ($\gamma \approx \omega \approx \varepsilon^{1/2}\omega_e^*$) at $\nu_e/\varepsilon \approx \varepsilon^{1/2}\omega_e^*$. As is clear from eq. (12.65), the growth rate is larger for larger values of $k_\eta^2 \propto (\omega_e^*)^2$. When the collision frequency becomes large, the destabilizing electron collision term becomes small and the stabilizing ion collision term increases. Finally the instability is suppressed.

(ii) Dissipative Trapped Electron Instability When the collision frequency increases, a modified dissipative drift wave is induced by the trapped electrons. The mechanism of dissipative (resistive) drift waves was explained in sec. 9.7. Particles drift across the magnetic field under the influence of the electric perturbation field; however, the ion drift velocity is different from the electron drift velocity due to the inertia of the ions. The charge separation thus caused is cancelled by the motion of electrons along the lines of magnetic force. If the motion of the electrons is disturbed by collisions, charge separation occurs and the plasma becomes unstable. In contrast to this, dissipative trapped particle instability is simply a result of trapped electrons becoming untrapped electrons by virtue of collision.

As the ion drift velocity across the magnetic field is $v_\xi = -ik_\eta\phi/B_0$, the 1st-order perturbation of the ion density is given by the equation of continuity:

$$-i\omega n_{1i} + \frac{\partial n_0}{\partial \xi}\left(-ik_\eta \frac{\phi}{B_0}\right) = 0$$

$$\frac{n_{1i}}{n_0} = \frac{-\omega_i^* Z_i e\phi}{\omega \quad T_i} = \frac{\omega_e^*}{\omega}\frac{e\phi}{T_e}.$$

The 1st-order perturbation of the electron density is, from eq. (12.64),

$$\frac{n_{1e}}{n_0} = \frac{n_{te}'}{n_0} + \frac{e\phi}{T_e}(1 - \varepsilon^{1/2})$$

$$\frac{n_{te}'}{n_0} = \left\langle \frac{\omega_e{}^+ + i\nu_{eff}}{\omega + i\nu_{eff}} \right\rangle \frac{\varepsilon^{1/2} e\phi}{T_e},$$

where the effective collision frequency is $\nu_{eff} = (\nu_{ei}/\varepsilon)(v_{Te}/v)^3$, v_{Te} being the electron thermal velocity. $\langle\ \rangle$ means the average in velocity space with the Maxwell distribution as weighting factor. $\omega_e{}^+$ is defined by*

$$\omega_e{}^+ \equiv \omega_e^*\left(1 + \eta\left(\frac{m_e v^2}{2T_e} - \frac{3}{2}\right)\right) \qquad \left(\eta \equiv \frac{d(\ln T_e)}{d(\ln n_0)}\right).$$

The quasi-neutrality condition yields

$$\frac{\omega_e^*}{\omega} = 1 + \varepsilon^{1/2}\left\langle \frac{\omega_e{}^+ - \omega}{\omega + i\nu_{eff}} \right\rangle \tag{12.66}$$

and

$$\omega \approx \omega_e^* + i\omega_e^* \varepsilon^{1/2} \left\langle \frac{(\omega_e{}^+ - \omega_e^*)}{(\omega_e^*)^2 + (\nu_{eff})^2} \nu_{eff} \right\rangle.$$

When $\nu_{eff} = \nu_{ei}/\varepsilon > \omega_e^*$, the growth rate is

$$\gamma \approx \varepsilon^{3/2} \frac{\omega_e^* \omega_{eT}^*}{\nu_e}, \tag{12.67}$$

where $\omega_{eT}^* = -k_\eta(dT_e/dr)/eB$. When the temperature gradient is zero, the growth rate becomes zero according to eq. (12.67). However the effect of the curvature drift was neglected in the derivation of eq. (12.67). The drift velocity due to the curvature of the lines of magnetic force is $v_\eta = T_e/eBR$. If the effect is taken into account, eq. (12.66) is

$$\frac{\omega_e^*}{\omega} = 1 + \varepsilon^{1/2}\left\langle \frac{\omega_e{}^+ - \omega + \omega_G}{\omega - \omega_G + i\nu_{eff}} \right\rangle$$

*The distribution function including density and temperature gradients is given by

$$f(v) = n_0\left(1 + \left(\frac{1}{n_0}\frac{dn_0}{dx} + \frac{1}{T}\frac{dT}{dx}\left(\frac{mv^2}{2T} - \frac{3}{2}\right)\right)\left(x - \frac{v_\eta}{\Omega}\right)\right)\frac{1}{(2\pi v_T)^{3/2}}\exp\left(-\frac{v^2}{2v_T{}^2}\right),$$

from eqs. (11.38) and (11.41).

and

$$\gamma \approx \varepsilon^{3/2}\frac{\omega_e^*\omega_G}{\nu_e} \approx \varepsilon^{5/2}\frac{(\omega_e^*)^2}{\nu_e}.$$

12.2c Diffusion due to Trapped Particle Instability

Let us consider the anomalous diffusion caused by dissipative trapped particle instability.

For dissipative trapped electron instability the growth rate is a maximum, $\gamma \approx \nu_e/\varepsilon^{1/2}$, at $\nu_e/\varepsilon \approx \omega_e^*$ (from eq. (12.67)). The energy loss due to the instability is given (see sec. 8.4) by

$$\mathrm{Re}(n_1 T_e v_{Er}^*) = -\chi_e \frac{\partial T_e}{\partial r} = \kappa T_e \chi_e,$$

where χ_e is the electron thermal conductivity, and $\kappa \equiv T_e/(\partial T_e/\partial r)$. v_{Er} is the radial component of the velocity due to the fluctuating electric field,

$$v_{Er} = -i\frac{k\phi}{B} = -\frac{i\omega_e^*}{\kappa}\frac{e\phi}{T_e}.$$

The anomalous diffusion coefficient $D_{Te} \equiv \chi_e/n_0$ caused by the instability is

$$\begin{aligned} D_{Te} &= -\frac{\omega_e^*}{\kappa^2}\mathrm{Im}\left(\frac{n_{e1}}{n_0}\frac{e\phi^*}{T_e}\right) = +\frac{\omega_e^*}{\kappa^2}\left|\frac{n_{i1}}{n_0}\right|^2 \mathrm{Im}\left(\frac{T_i\omega}{-\omega_i^* Z_i T_e}\right) \\ &= \frac{1}{\kappa^2}\left|\frac{n_1}{n_0}\right|^2 \gamma \approx \frac{\gamma}{k_\perp^2} = \varepsilon^{3/2}\frac{\omega_e^*\omega_{eT}^*}{\nu_e k_\perp^2} \approx \varepsilon^{3/2}\frac{r^2\omega_*^2}{\nu_e}. \end{aligned} \tag{12.68}$$

Here the quantity

$$\omega_* \equiv \frac{T}{eBra}$$

($k_\perp \approx r^{-1}$, $\kappa \approx a^{-1}$) has been introduced.

When the collision frequency is smaller than $\nu_e < \omega_*$, the electron is trapped by the electric field of the wave; this electron trapping by the electric field plays a stabilizing role, and the wave amplitude soon saturates. When the potential ϕ_1 of the wave reaches $e\phi_1 \approx \varepsilon T_e$, the number of electrons trapped by the electric field (E banana) is comparable to the number trapped by the magnetic field. The random-walk step length δr

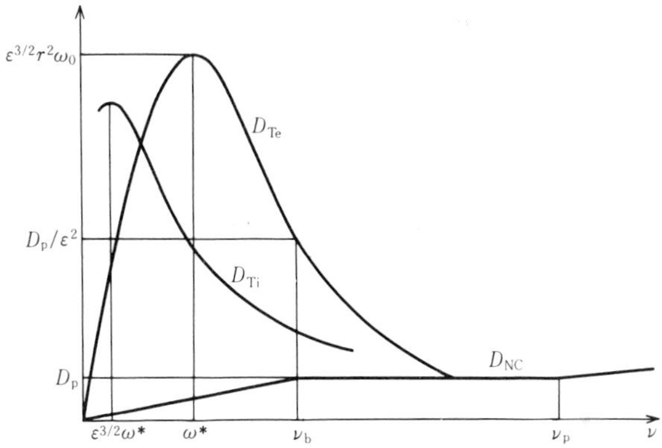

Fig. 12.6
The dependence of dissipative trapped particle instability diffusion coefficients on the collision frequency. D_{NC}, D_{Te}, and D_{Ti} are the neoclassical, dissipative trapped electron, and dissipative trapped ion coefficients, respectively.

is of the order of εr ($(\partial\phi_0/\partial r)\delta r \approx (T_e/er)\delta r \approx \phi_1 \approx \varepsilon T_e/e$). Then the diffusion coefficient is

$$D_{Te} \approx \varepsilon^{1/2}(\delta r)^2 \nu_{eff} \approx \varepsilon^{3/2} r^2 \nu_e \qquad (\nu_e < \omega_*). \tag{12.69}$$

The formula which bridges eq. (12.68) and eq. (12.69) is[12]

$$D_{Te} \approx \varepsilon^{3/2} r^2 \frac{\nu_e}{1 + (\nu_e/\omega_*)^2}. \tag{12.70}$$

The magnitude of the diffusion coefficient at the collision frequency $\nu_b = \varepsilon^{3/2}(\iota/2\pi R)v_T{}^e$ is ε^{-2} times as large as that predicted by neoclassical theory, ν_b being the frequency at the boundary between the plateau and banana regions. The dependence of D_{Te} on ν is shown in fig. 12.6.

Let us now consider the case of dissipative trapped ion instability. The anomalous diffusion coefficient D_{Ti} of particles is

$$\begin{aligned} D_{Ti} &= -\frac{\omega_e^*}{\kappa^2}\,\mathrm{Im}\left(\frac{n_{e1}}{n_0}\frac{e\phi^*}{T_e}\right) = -\frac{\omega_e^*}{\kappa^2}\,\mathrm{Im}\left(\frac{n_{i1}}{n_0}\frac{e\phi^*}{T_e}\right) \\ &= \frac{\omega_e^*}{\kappa^2}\left|\frac{Ze\phi}{T_i}\right|^2 \frac{T_i}{T_e Z}\,\mathrm{Im}\left(1 - \varepsilon^{1/2}\frac{\omega - \omega_i^*}{\omega}\right) \\ &= \frac{\varepsilon^{1/2}(\omega_e^*)^2}{\kappa^2}\left|\frac{e\phi}{T_e}\right|^2 \mathrm{Im}\left(\frac{-1}{\omega}\right). \end{aligned}$$

The ion-collision term is neglected. The relation between the density and potential (eq. (12.64)) is

$$\frac{n_{\rm ti}}{n_0} \approx \varepsilon^{1/2}\frac{(-\omega_{\rm i}^*)}{\omega}\frac{e\phi}{T_{\rm i}} = \varepsilon^{1/2}\frac{\omega_{\rm e}^*}{\omega}\frac{e\phi}{T_{\rm e}}.$$

Substitution of this relation into the equation for $D_{\rm Ti}$ gives[12]

$$D_{\rm Ti} = \varepsilon^{1/2}\frac{\gamma}{\kappa^2}\varepsilon^{-1}\left|\frac{n_{\rm ti}}{n_0}\right|^2 \approx \varepsilon^{1/2}\frac{\gamma}{\kappa^2}\left|\frac{n_1}{n_0}\right|^2 \approx \varepsilon^{1/2}\frac{\gamma}{k_\perp^2}$$

$$\approx \left(\frac{T_{\rm i}/Z_{\rm i}}{T_{\rm e}+T_{\rm i}/Z_{\rm i}}\right)^2 \varepsilon^{5/2}\frac{r^2\omega_*^2}{\nu_{\rm e}}. \tag{12.71}$$

The value of γ is maximum at $\nu_{\rm e} \approx \varepsilon^{3/2}\omega_{\rm e}^*$. When the collision frequency is less than this, $D_{\rm Ti}$ decreases and collisionless trapped particle instability becomes dominant. The dependence of $D_{\rm Ti}$ on the collision frequency is shown in fig. 12.6.

12.3 Instabilities due to Anisotropy of the Velocity-Space Distribution Function

When a plasma is confined in a mirror field, the particles in the loss cone ($(v_\perp/v)^2 < 1/R_{\rm m}$, $R_{\rm m}$ being the mirror ratio) escape, so that anisotropy appears in the velocity-space distribution function. The temperature perpendicular to the magnetic field may be different from that parallel to the field in certain cases. In these cases the anisotropic distribution function may result in instability. We consider here the Harris instability, which appears near the ion cyclotron harmonic frequencies, and loss-cone instability.

12.3a Harris Instability

Let us consider the case where the distribution function is bi-Maxwellian. However, the density and temperature are uniform ($\varepsilon = \delta_\perp = \delta_z = 0$) and the acceleration and fluid flow are zero ($G = 0$, $V = 0$). In this case the dispersion equation (11.43) is

$$k_y^2 + k_z^2 + \sum_j \Pi_j^2 m_j\left(\frac{1}{T_z} + \sum_{l=-\infty}^{\infty} I_l(b)e^{-b}\left(\frac{1}{T_z} + \frac{1}{T_\perp}\frac{(-l\Omega)}{\omega + l\Omega}\right)\zeta_l Z(\zeta_l)\right)_j$$
$$= 0 \tag{12.72}$$

$$\zeta_l = \frac{\omega + l\Omega}{(2T_z/m)^{1/2}k_z}.$$

We denote the real and imaginary parts of the right-hand side of eq. (12.72) for real $\omega = \omega_r$ by $K(\omega_r) = K_r(\omega_r) + iK_i(\omega_r)$. When the solution of eq. (12.72) is $\omega = \omega_r + i\gamma$, i.e., for $K(\omega_r + i\gamma) = 0$, ω_r and γ are given by

$$K_r(\omega_r) = 0 \qquad \gamma = K_i(\omega_r)(\partial K_r(\omega_r)/\partial\omega_r)^{-1}$$

when $|\gamma| \ll |\omega_r|$. Accordingly, we find

$$k_y^2 + k_z^2 + \sum_j \Pi_j^2 m_j \left(\frac{1}{T_z} + \frac{1}{(2T_z/m)^{1/2}k_z} \sum_{l=-\infty}^{\infty} I_l e^{-b} \left(\frac{\omega + l\Omega}{T_z} - \frac{l\Omega}{T_\perp} \right) Z_r(\zeta_l) \right)_j = 0 \tag{12.73}$$

$$\gamma = \frac{-\pi^{1/2} \sum_j \Pi_j^2 m_j \left(\sum_{l=-\infty}^{\infty} I_l e^{-b} \frac{1}{(2T_z/m)^{1/2}|k_z|} \left(\frac{\omega + l\Omega}{T_z} - \frac{l\Omega}{T_\perp} \right) \exp(-\zeta_l^2) \right)_j}{\sum_j \Pi_j^2 m_j \sum_{l=-\infty}^{\infty} \left[I_l e^{-b} \left(\frac{Z_r(\zeta_l)/T_z}{(2T_z/m)^{1/2}k_z} + \frac{1}{(2T_z/m)k_z^2} \left(\frac{\omega + l\Omega}{T_z} - \frac{l\Omega}{T_\perp} \right) Z_r'(\zeta_l) \right) \right]_j}.$$

$Z_r(\zeta_1)$ is the real part of $Z(\zeta_1)$ and the prime $'$ indicates differentiation with respect to ζ_1.

Let us assume that the electron is cold ($b_e \approx 0$, $|\zeta_{0e}| \gg 1$) and the ion is hot. Then the contribution of the electron term is dominant in eq. (12.73) and the ion term can be neglected. Equation (12.73) becomes

$$k^2 - \Pi_e^2 \frac{k_z^2}{\omega_r^2} = 0, \tag{12.75}$$

i.e.,

$$\omega_r = \pm \Pi_e \frac{k_z}{k}.$$

Substitution of ω_r into eq. (12.74) yields

$$\gamma = \frac{\pi^{1/2}}{2k^2} \sum_j \Pi_j^2 \left(\frac{m}{T_z} \right)_j \frac{1}{(2T_z/m)_j^{1/2}|k_z|}$$
$$\times \left(\sum_{l=-\infty}^{\infty} I_l e^{-b} \omega_r \left(-(\omega_r + l\Omega) + \frac{T_z}{T_\perp} l\Omega \right) \exp(-\zeta_l^2) \right)_j. \tag{12.76}$$

The dependence of γ on $\omega_r/|\Omega_i|$ is shown in fig. 12.7. The dip near $\omega_r \approx 0$ is due to the contribution of the electron term. The dotted curves near

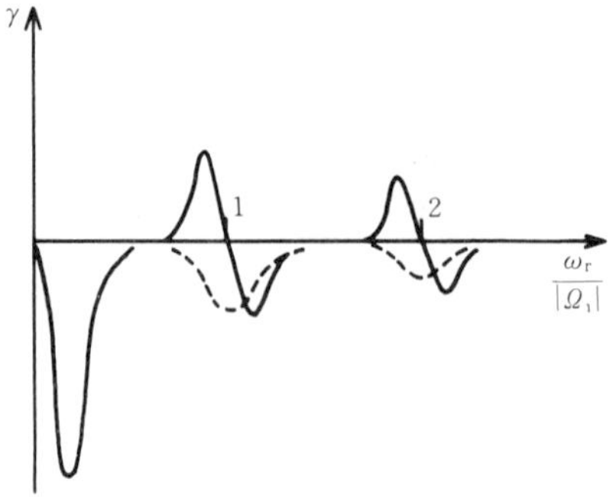

Fig. 12.7
Qualitative plots of the contributions of the 1st term (solid curves) and the 2nd term (dotted curves) of eq. (12.76) for the growth rate γ.

$\omega_r/(-\Omega_i) \approx n(\geq 1)$ are stabilizing contributions of the 2nd term in the ion term (the ion cyclotron damping term). The solid curves are the destabilizing contribution of the 1st term (it must be noticed that the sign of Ω_i is negative). Accordingly, the necessary condition for instability is

$$\Pi_e > l(-\Omega_i) \tag{12.77}$$

(see eq. 12.75). When the plasma density increases to the point that Π_e approaches $|\Omega_i|$, then plasma oscillations couple to the ion Larmor motion, causing the instability. When the plasma density increases further, an oblique Langmuir wave couples with an ion cyclotron harmonic wave $l|\Omega_i|$ and an instability with $\omega_r \approx \Pi_e |k_z/k| \approx l|\Omega_i|$ is induced.

In the region near $\omega_r \approx -l\Omega_i$, the stabilizing 2nd term in eq. (12.76) has the coefficient $l(T_z/T_\perp)_i$, so that the instability is less probable for large l. From eq. (12.76), the instability condition is

$$\left(\frac{T_z}{T_\perp}\right) < \frac{1}{2l}. \tag{12.78}$$

We have discussed the cold-electron case. For nonzero electron temperature, the region of the electron stabilizing term near $\omega_r \approx 0$ is broad and the wave of lower harmonic frequency $-l\Omega_i$ is stabilized. If the electron temperature is increased further, the stabilized region due to the electron term is broadened so widely that the stabilizing term itself becomes small and does not contribute to the suppression of the instability. When

$$\left(\frac{T_z}{m}\right)_{\mathrm{i}} \ll \left(\frac{\omega_{\mathrm{r}}}{k_z}\right)^2 \ll \left(\frac{T_z}{m}\right)_{\mathrm{e}},$$

i.e., for $|\zeta_{0\mathrm{i}}| \gg 1$, $|\zeta_{0\mathrm{e}}| \ll 1$, the real part of the dispersion equation (12.73) is

$$k^2 + \frac{1}{(\lambda_{\mathrm{d}}^{\mathrm{e}})^2} - \Pi_{\mathrm{i}}^2 \frac{k_z^2}{\omega_{\mathrm{r}}^2} = 0$$

$$\omega_{\mathrm{r}} = \frac{\Pi_{\mathrm{i}} k_z}{\left(k^2 + \dfrac{1}{(\lambda_{\mathrm{d}}^{\mathrm{e}})^2}\right)^{1/2}} \approx \left(\frac{T_{\mathrm{e}}}{m_{\mathrm{i}}}\right)^{1/2} k_z. \qquad (12.79)$$

This is the frequency of the ion acoustic wave. The maximum value is Π_{i}, so that the condition

$$\Pi_{\mathrm{i}} > l(-\Omega_{\mathrm{i}}) \qquad (12.80)$$

must be satisfied to excite the wave with $\omega_{\mathrm{r}} \approx -l\Omega_{\mathrm{i}}$. This condition means that the plasma density must be $m_{\mathrm{i}}/m_{\mathrm{e}}$ times as large as that of eq. (12.77).

In summary, the instabilities with ion cyclotron harmonic frequencies appear one after another in a cold-electron plasma under the anisotropic condition (12.78) when the electron density satisfies

$$n_{\mathrm{e}} > l^2 \frac{Z^2 m_{\mathrm{e}}}{m_{\mathrm{i}}} \left(\frac{B^2}{\mu_0 m_{\mathrm{i}} c^2}\right) \qquad (l = 1, 2, 3, \cdots). \qquad (12.81)$$

This instability is called the *Harris instability*.[19,20] When the electrons are hot, the instability occurs when

$$n_{\mathrm{e}} > l^2 Z \left(\frac{B^2}{\mu_0 m_{\mathrm{i}} c^2}\right) \qquad (l = 1, 2, 3, \cdots). \qquad (12.82)$$

12.3b Loss-Cone Instability[21,22]

The velocity-space distribution function of a plasma confined in a mirror field is zero for $(v_\perp/v)^2 < 1/R_{\mathrm{m}}$ due to the loss cone. The instability associated with this is called *loss cone instability*.

Assume that the plasma is uniform and the magnetic field constant. We consider a wave propagating *nearly* perpendicularly to $\boldsymbol{B}$, with frequency in the region

$$\left.\begin{aligned} &|\Omega_{\mathrm{i}}| \ll |\omega| \ll \Omega_{\mathrm{e}} \\ &|k_z| \ll |k_y|. \end{aligned}\right\} \tag{12.83}$$

Under these conditions the ion Larmor motion is so slow that the ion motion can be considered linear over the time scale of the perturbation. Accordingly, the magnetic field can be neglected for ions. The contribution of the ion term to the dispersion equation is given by eq. (11.33) as follows:

$$2\Pi_{\mathrm{i}}^2 \int \mathrm{d}v_x \mathrm{d}v_y \mathrm{d}v_z \left[\frac{\partial f_{\mathrm{i}}}{\partial v_\perp{}^2} - \left((\omega - k_z v_z) \frac{\partial f_{\mathrm{i}}}{\partial v_\perp{}^2} + k_z v_z \frac{\partial f_{\mathrm{i}}}{\partial v_z{}^2} \right) \frac{1}{\omega - k_z v_z - k_y v_y} \right]$$

$$\approx 2\Pi_{\mathrm{i}}^2 \int \frac{-k_y v_y}{\omega - k_y v_y} \frac{\partial f_{\mathrm{i}}}{\partial v_\perp{}^2} \mathrm{d}v_x \mathrm{d}v_y \mathrm{d}v_z .$$

The contribution of the electron term, obtained from eq. (11.36) under the assumption of small electron Larmor radius ($|k_y v_\perp / \Omega_{\mathrm{e}}| \ll 1$), is

$$2\Pi_{\mathrm{e}}^2 \int \mathrm{d}v_x \mathrm{d}v_y \mathrm{d}v_z \frac{-k_z v_z}{\omega - k_z v_z} \frac{\partial f_{\mathrm{e}}}{\partial v_z{}^2} .$$

Only the term in $n = 0$ is retained here, because of the low frequency ($|\omega| \ll \Omega_{\mathrm{e}}$). Under the assumptions eq. (12.83), the dispersion equation becomes

$$k^2 + 2\Pi_{\mathrm{i}}^2 \int_{-\infty}^{\infty} \mathrm{d}v_y \frac{k_y v_y}{\omega - k_y v_y} \iint_{-\infty}^{\infty} \mathrm{d}v_x \mathrm{d}v_z \frac{\partial f_{\mathrm{i}}}{\partial v_\perp{}^2}$$

$$+ 2\Pi_{\mathrm{e}}^2 \int_{-\infty}^{\infty} \mathrm{d}v_z \frac{k_z v_z}{\omega - k_z v_z} \iint_{-\infty}^{\infty} \mathrm{d}v_x \mathrm{d}v_y \frac{\partial f_{\mathrm{e}}}{\partial v_z{}^2} = 0. \tag{12.84}$$

Here the distribution-function normalizations are $\int f_{\mathrm{i,e}} \mathrm{d}\boldsymbol{v} = 1$. If $|\omega| > |k_z| (T_z/m)_{\mathrm{e}}^{1/2}$, electron Landau damping is not effective. When the denominator of the electron term is expanded, the term in $k_z{}^2 \Pi_{\mathrm{e}}{}^2/\omega^2$ appears. The average distribution function over v_z is introduced for convenience:

$$g_0(v_\perp{}^2) \equiv \pi \int \mathrm{d}v_z f_{\mathrm{i}}(v_\perp{}^2, v_z{}^2).$$

In the notation $v_x = v_\perp \sin\phi$, $v_y = v_\perp \cos\phi$, the ion term becomes

$$\iint_{-\infty}^{\infty} \mathrm{d}v_x \mathrm{d}v_y \frac{k_y v_y}{\omega - k_y v_y} \int_{-\infty}^{\infty} \mathrm{d}v_z \frac{\partial f_{\mathrm{i}}}{\partial v_\perp{}^2}$$

$$= \int_0^\infty \frac{dv_\perp^2}{2\pi}\frac{\partial g_0}{\partial v_\perp^2}\int_0^{2\pi}\frac{k_y v_\perp \cos\phi}{\omega - k_y v_\perp \cos\phi}d\phi$$

$$= \int_0^\infty dv_\perp^2 \frac{\partial g_0}{\partial v_\perp^2 (1-(k_y v_\perp/\omega)^2)^{1/2}}.$$

The function $g_0(v_\perp^2)$ is zero at $v_\perp^2 = 0$ because of the loss cone. If the real part ω_r of $\omega = \omega_r + i\gamma$ is $|\omega_r| > |k_y v_\perp|$, we can then write

$$\frac{1}{2\pi}\int_0^{2\pi}\frac{k_y v_\perp \cos\phi}{\omega_r - k_y v_\perp\cos\phi}d\phi = -1 + \frac{1}{\left(1-\left(\frac{k_y v_\perp}{\omega_r}\right)^2\right)^{1/2}}. \quad (12.85)$$

On the other hand, if $|\omega_r| < |k_y v_\perp|$ the integral is

$$\lim_{r\to+0}\frac{1}{2\pi}\int_0^{2\pi}\frac{k_y v_\perp\cos\phi}{\omega - k_y v_\perp \cos\phi}d\phi = -1 + \frac{\frac{\omega_r}{k_y v_\perp}i}{\left(1-\left(\frac{\omega_r}{k_y v_\perp}\right)^2\right)^{1/2}}. \quad (12.86)$$

Using $\bar{v}_\perp^2 \equiv \int v_\perp^2 g_0 dv_\perp^2$, $v_\perp^2 = \bar{v}_\perp^2\alpha$, and $g_0 = \psi/(\bar{v}_\perp^2)(\int \psi d\alpha = 1)$, we can reduce the dispersion equation to

$$1 + \frac{\Pi_e^2}{\Omega_e^2} - \frac{\Pi_e^2}{\omega^2}\frac{k_z^2}{k^2} - \frac{\Pi_i^2}{k^2(\bar{v}_\perp)^2}F\left(\frac{\omega}{k_y\bar{v}_\perp}\right) \equiv H(\omega) = 0 \quad (12.87)$$

$$F(\zeta) = -2\int\frac{(d\psi/d\alpha)d\alpha}{(1-\alpha/\zeta^2)^{1/2}}. \quad (12.88)$$

The 2nd term in eq. (12.87) results from the finite electron Larmor radius. Let us examine the behavior of the function $F(\zeta)$. $F(\zeta)$ can be written as

$$F(\zeta) = -2\int_0^\infty \frac{d\psi}{(1-\alpha/\zeta^2)^{1/2}}.$$

When $\alpha > \zeta_r^2$, we consider the integrand to be

$$\frac{1}{(1-\alpha/\zeta_r^2)^{1/2}} = \frac{\zeta_r i}{(\alpha - \zeta_r^2)^{1/2}}.$$

The function $\psi(\alpha)$ is shown in fig. 12.8. The real part F_r and the imaginary part F_i of $F(\zeta)$ are also shown in fig. 12.8. ζ_1 and ζ_2 in fig. 12.8 are about 1.

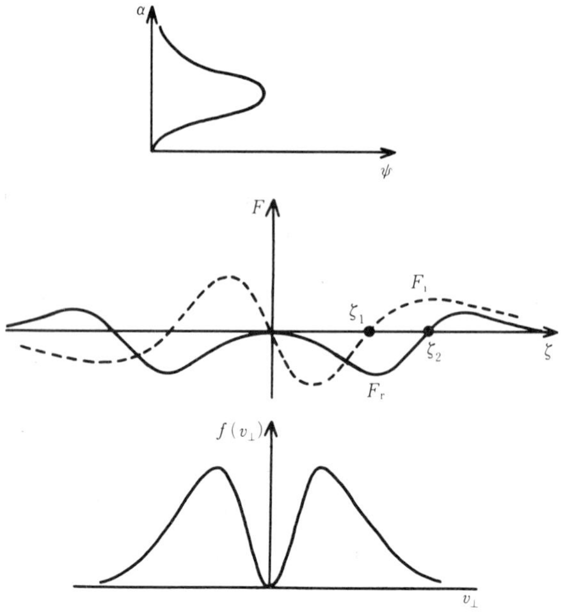

Fig. 12.8
$F(\zeta) = F_r(\zeta) + iF_i(\zeta)$ in the dispersion equation for loss-cone instability.

If the solution $\omega = \omega_r + i\gamma$ with $|\gamma| \ll \omega_r$ exists, γ is given by

$$\frac{\gamma}{\omega_r} \approx \frac{\omega_r^2}{2\Pi_e^2}\frac{k^2}{k_z^2}\frac{\Pi_i^2}{k_y^2(\bar{v}_\perp)^2}F_i\left(\frac{\omega_r}{k_y\bar{v}_\perp}\right) = \frac{Zm_e}{2m_i}\left(\frac{k}{k_z}\right)^2\left(\frac{\omega_r}{k_y\bar{v}_\perp}\right)^2 F_i\left(\frac{\omega_r}{k_y\bar{v}_\perp}\right). \tag{12.89}$$

In the general case we must resort to the Nyquist diagram. For the Nyquist diagram to encircle the origin $H(\omega) = 0$ under the condition $|\zeta| < \zeta_1$, the condition

$$1 + \frac{\Pi_e^2}{\Omega_e^2} - \frac{\Pi_e^2}{(k_y\bar{v}_\perp\zeta_1)^2}\frac{k_z^2}{k^2} + \frac{\Pi_i^2}{k^2(\bar{v}_\perp)^2}|F(\zeta_1)| > 0 \tag{12.90}$$

must be satisfied. This is the instability condition for loss-cone instability. From eq. (12.87) the wave with the maximum growth rate is given roughly by

$$\omega_r \approx \gamma \approx k\bar{v}_\perp \approx \Pi_i \approx k_z\bar{v}_\perp\left(\frac{m_i}{m_e}\right)^{1/2} \tag{12.91}$$

($\zeta_r = \omega_r/k_y\bar{v}_\perp \approx 1$). These relations can be satisfield only under the condition

$$\omega_r \approx \Pi_i > |\Omega_i|. \tag{12.92}$$

(we have assumed that $\Omega_e \gg |\omega_r| \gg |\Omega_i|$).

As has been stated, the magnetic field has no effect upon the ion motion over times characteristic of the perturbation. The ion distribution function has double peaks with respect to $v_\perp$, as is shown in fig. 12.8; it is of similar structure to that of two-stream instability. The 4th term in the left-hand side of eq. (12.87) is very similar to the right-hand side of eq. (11.66). Therefore, we can say that the mechanism of loss-cone instability is essentially the same as that of two-stream instability. The instability condition $\Pi_i > |\Omega_i|$ for loss-cone instability can be satisfied for much higher densities than the condition for Harris instability.

We have been discussing this instability in the case of an infinite plasma, but it must be pointed out that loss-cone instability is convective. As the wave approaches the boundary, then $\Pi_i^2 \to 0$ and $k_z^2 \to \infty$ (see eq. (12.93)) and the wave is absorbed at the boundary. The propagation constant k_z is given in terms of ω by the dispersion equation (12.87):

$$k_z^2 = k^2\left(\omega^2\left(\frac{1}{\Pi_e^2} + \frac{1}{\Omega_e^2}\right) - \frac{Zm_e}{m_i}\zeta^2 F(\zeta)\right). \tag{12.93}$$

Expanding the square root under the condition $k^2\bar{v}_\perp^2 > \Pi_i^2$, we find

$$k_z = k\omega\frac{(\Pi_e^2 + \Omega_e^2)^{1/2}}{\Pi_e\Omega_e} - \frac{Z}{2}\frac{m_e}{m_i}\frac{\Pi_e\Omega_e}{(\Pi_e^2 + \Omega_e^2)^{1/2}}\frac{1}{v_\perp}\zeta F(\zeta).$$

Accordingly, the imaginary part $\mathrm{Im}(k_z)$ of k_z becomes

$$\mathrm{Im}(k_z) \approx -\frac{1}{2}\left(\frac{m_e}{m_i}\right)^{1/2}\left(\frac{\Pi_i}{\bar{v}_\perp}\right)\left(1 + \left(\frac{\Pi_e}{\Omega_e^2}\right)^2\right)^{-1/2}\mathrm{Im}(\zeta F(\zeta));$$

thus the amplitude of the wave increases as the wave propagates. However, if the plasma length along the lines of magnetic force is less than 10 times $(\mathrm{Im}(k_z))^{-1}$, the perturbation can be considered not to be dangerous. That is to say, if the plasma length L_p is

$$L_p \le 20\left(\frac{m_i}{m_e}\right)^{1/2}\left(\frac{\bar{v}_\perp}{\Pi_i}\right)\left(1 + \frac{\Pi_e^2}{\Omega_e^2}\right)^{1/2}|\mathrm{Im}\,\zeta F(\zeta)|^{-1}, \tag{12.94}$$

then the plasma is stable. When $\Pi_e \gg \Omega_e$, this condition becomes simply

$$L_p \leq 10^4 \left(\frac{m_e}{m_i}\right)^{1/2} \rho_B{}^i, \tag{12.95}$$

where $\rho_B{}^i$ is the ion Larmor radius.

12.4 Cross-Field Instability

When the electron flow velocity across the magnetic field is different from the ion flow velocity, instabilities called *cross-field instabilities*[23,24] may occur. These phenomena are most likely to occur where the frequency of the wave is in the region

$$|\Omega_i| \ll |\omega| \ll \Omega_e.$$

In this case the ions move without being affected by the magnetic field but the electrons are completely magnetized and they drift across the magnetic field under the influence of electric fields, the gradient of B, etc.

For simplicity, the plasma is assumed to be isotropic and homogeneous. The ion term in the dispersion equation is that without magnetic field, because $|\omega| \gg |\Omega_i|$, and it is given by eq. (11.44). The electron term is that for low frequency, because $|\omega| \ll \Omega_e$, and it is given by eq. (11.45). Accordingly the dispersion equation is

$$k_y{}^2 + k_z{}^2 + \Pi_i{}^2 \frac{m_i}{T_i}(1 + \zeta_i Z(\zeta_i))$$

$$+ \Pi_e{}^2 \left[\frac{m}{T} + \zeta Z(\zeta) I_0(b) e^{-b} \left(\frac{m}{T} + \frac{k_y}{\omega' \Omega}\left(\frac{m}{T} G - \varepsilon\right)\right)\right]_e = 0, \tag{12.96a}$$

where

$$\frac{G_e}{\Omega_e} = -\frac{E_x}{B_0} - \frac{T_e}{m_e}\frac{\partial B_0}{\partial x}\frac{1}{B_0} \equiv v_d$$

is the y component of the electron drift velocity due to the electric field E_x and the gradient of B. $\omega_e{}'$ is

$$\omega_e{}' = \omega - k_z V - k_y v_d.$$

The density gradient is given from eq. (11.39) as

$$-\kappa \equiv \frac{1}{n_0}\frac{\partial n_0}{\partial x} = -\varepsilon + \frac{m}{T} G.$$

Using the electron drift frequency $\omega_e{}^* = k_y \kappa T_e / eB$, we write the dispersion equation as

$$k_y{}^2 + k_z{}^2 + \Pi_i{}^2 \frac{m_i}{T_i}(1 + \zeta_i Z(\zeta_i))$$

$$+ \Pi_e{}^2 \frac{m_e}{T_e}\left(1 + I_0 e^{-b}\left(1 - \frac{\omega_e{}^*}{\omega_e{}'}\right)\zeta_e Z(\zeta_e)\right) = 0 \qquad (12.96b)$$

$$\zeta_i = \frac{\omega - kV_i}{(2T/m)_i{}^{1/2} k} \qquad (\omega_i{}' = \omega - kV_i)$$

$$\zeta_e = \frac{\omega - k_z V_e - k_y v_d}{(2T/m)_e{}^{1/2} k_z}.$$

V_i is the ion drift velocity in the direction of $\boldsymbol{k}$ and V_e is the z component, i.e., that aligned with $\boldsymbol{B}$, of the electron drift velocity. When the phase velocity of the wave is faster than the thermal velocities of ion and electron, i.e., when $|\zeta_i| \gg 1$, $|\zeta_e| \gg 1$, then eq. (12.96b) is reduced to ($\zeta Z(\zeta) = -(1 + (2\zeta^2)^{-1} + \cdots)$)

$$\frac{\Pi_i{}^2}{(\omega - kV_i)^2} + \frac{\Pi_e{}^2}{(\omega - k_y v_d - k_z V_e)^2}\frac{k_z{}^2}{k^2}$$

$$- \frac{\Pi_e{}^2}{\Omega_e(\omega - k_y v_d - k_z V_e)}\frac{k_y \kappa}{k^2} = 1 + \frac{\Pi_e{}^2}{\Omega_e{}^2}. \qquad (12.97)$$

This dispersion equation is similar to that for two-stream instability. We call the case when the 3rd term is smaller than the 2nd term in eq. (12.97) *modified two-stream instability*. When $|k_z|$ is small and the 2nd term is smaller than the 3rd term, we refer to *flutelike drift instability*. When $V_i = 0$, $v_d = 0$, and the 2nd term is smaller than the 3rd term in eq. (12.97), we have *drift two-stream instability*. Its dispersion equation is

$$\frac{\Pi_i{}^2}{\omega^2} - \frac{\Pi_e{}^2}{\Omega_e(\omega - k_z V_e)}\frac{k_y \kappa}{k^2} = 1 + \frac{\Pi_e{}^2}{\Omega_e{}^2}. \qquad (12.98)$$

12.4a Modified Two-Stream Instability

In this case the dispersion equation is

$$\frac{\Pi_e{}^2}{(\omega - k_y v_d)^2}\frac{k_z{}^2}{k^2} + \frac{\Pi_i{}^2}{\omega^2} = 1 + \frac{\Pi_e{}^2}{\Omega_e{}^2}, \qquad (12.99)$$

where it is assumed that $V_i = 0$, $V_e = 0$. This equation is of the same type as eq. (11.76) for Buneman instability. The maximum growth rate of the instability occurs at

$$\omega_r = \frac{1}{2 \cdot 2^{1/3}} (\Pi_e \Pi_i)^{1/2} \left(\frac{m_i}{m_e}\right)^{1/6} \left(\frac{k_z}{k}\right)^{1/3},$$

and γ is

$$\gamma = 3^{1/2} \omega_r.$$

In the region $k_z/k \approx (m_e/m_i)^{1/2}$, ω_r and γ are $\omega_r \approx \gamma \approx (\Pi_e \Pi_i)^{1/2}$.

12.4b Flutelike Drift Instability

From eq. (12.97), the dispersion equation is

$$\frac{\omega_{LH}^2}{\omega^2} - \frac{\omega_{LH}^2}{|\Omega_i|(\omega - k_y v_d)} \frac{\kappa}{k_y} = 1, \tag{12.100}$$

where we assume $V_i = 0$, $V_e = 0$, $k_z = 0$. ω_{LH} is the lower hybrid resonant frequency,

$$\omega_{LH}^2 = \frac{\Pi_i^2 + \Omega_i^2}{1 + \Pi_e^2/\Omega_e^2}$$

$$\approx \frac{\Pi_i^2}{1 + \Pi_e^2/\Omega_e^2}.$$

Since eq. (12.100) reduces to

$$\left(1 - \frac{\omega_{LH}^2}{\omega^2}\right)(\omega - k_y v_d) = \frac{-\omega_{LH}^2}{|\Omega_i|} \left(\frac{\kappa}{k_y}\right),$$

it can be considered that the interaction of the lower hybrid wave and the wave with $\omega = k_y v_d$ induces the instability. A parameter

$$\alpha = \left|\frac{\kappa v_d}{\Omega_i}\right|$$

is introduced for convenience. When $\alpha \ll 1$ and $k_y v_d \approx \omega_{LH}$, then

$$\omega \approx \omega_{LH} + \left(\frac{\alpha}{2}\right)^{1/2} \omega_{LH} i.$$

When $\alpha \gg 1$ and $k_y v_d \approx \alpha^{1/2} \omega_{LH}$, then

$$\omega \approx \alpha^{1/6}\left(\frac{1 + 3^{1/2}i}{2}\right)\omega_{\mathrm{LH}}.$$

12.4c Drift Two-Stream Instability

In this case the dispersion equation is given eq. (12.98) as

$$\left(1 - \frac{\omega_{\mathrm{LH}}^{2}}{\omega^{2}}\right)(\omega - k_z V_{\mathrm{e}}) = -\frac{\omega_{\mathrm{LH}}^{2}}{|\Omega_{\mathrm{i}}|}\left(\frac{k_y \kappa}{k^2}\right). \qquad (12.101)$$

The plasma becomes unstable as in the case of flutelike drift instability.

These cross-field instabilities appear in the region $|\Omega_{\mathrm{i}}| \ll \omega \ll \Omega_{\mathrm{e}}$; their properties do not depend strongly upon the shape of the velocity-space distribution function or upon the ratio of $T_{\mathrm{e}}/T_{\mathrm{i}}$. The growth rate γ is of the same order of ω_{r} and is very large.

In this chapter we have discussed various electrostatic instabilities. Each instability has its characteristic frequency. The characteristic real part of the frequency of interchange instability is the ion drift frequency and the propagation direction is that of the ion drift due to the density gradient as is clear from the equation $\omega = \omega_{\mathrm{i}}{}^* = -k_y \kappa T_{\mathrm{i}}/(ZeB)$.

Instabilities characterized by the electron drift frequency $\omega_{\mathrm{e}}{}^*$ are drift instability, resistive drift instability (sec. 9.7), and dissipative trapped electron instability. Dissipative trapped ion instability has the characteristic frequency $\varepsilon^{1/2}\omega_{\mathrm{e}}{}^*$. These waves propagate in the direction of the electron drift due to the density gradient.

The characteristic frequencies of Harris instability are the ion cyclotron frequency and its harmonics. Loss-cone instability has a characteristic frequency lying between ion and electron cyclotron frequencies. When the ion plasma frequency Π_{i} is in the region $|\Omega_{\mathrm{i}}| < \Pi_{\mathrm{i}} < \Omega_{\mathrm{e}}$, then the real part of ω_{r} is $\omega_{\mathrm{r}} \approx \Pi_{\mathrm{i}}$. The characteristic frequency for cross-field instability lies in the range $|\Omega_{\mathrm{i}}| < \omega < \Omega_{\mathrm{e}}$.

When the fluctuations of the density and plasma potential are measured in space and in time, the frequency ω and the propagation vector $\boldsymbol{k}$ can be obtained. The phase relation between the density fluctuation n_1 and the potential fluctuation ϕ_1 can also be measured. These results are then compared with the dispersion relation, the predicted phase difference between n_1 and ϕ_1, and so on. Thus the fluctuations observed in experiments can be related to the appropriate instabilities.

References

1. N. A. Krall and M. N. Rosenbluth: Phys. of Fluids **8**, 1488(1965)

2. M. N. Rosenbluth, N. A. Krall, and N. Rostocker: Nucl. Fusion, suppl., pt. 1, p. 143 (1962)

3. A. B. Mikhailovskii and L. I. Rudakov: Sov. Phys. JETP **17**, 621(1963)

4. S. Yoshikawa: Methods of Experimental Physics vol. 9, pt. A, Plasma Physics, p. 305 (ed. by H. R. Griem, R. Loveberg), Academic Press, New York, 1970

5. N. A. Krall: Advances in Plasma Physics vol. 1, p. 153 (ed. by A. Simon and W. B. Thompson), Interscience, New York, 1968

6. A. B. Mikhailovskii: Rev. of Plasma Phys. **3**, 159 (ed. by M. A. Leontovich), Consultants Bureau, New York, 1967

7. M. Sato and Y. Terashima: JAERI-memo 3758, Oct. 1968, Japan Atomic Energy Research Institute (in Japanese)

8. Y. Terashima: Prog. Theor. Phys. **37**, 775(1967)
M. Meade: Phys. of Fluids **12**, 947(1969)

9. L. I. Schiff: Quantum Mechanics, p. 184, McGraw-Hill, New York, 1955

10. L. V. Mikhailovskaya and A. B. Mikhailovskii: Sov. Phys. Tech. Phys. **8**, 896(1964)

11. P. L. Bhatnagar, E. P. Gross, and M. Krooks: Phys. Rev. **94**, 511(1954)
E. P. Gross and M. Krooks: Phys. Rev. **102**, 593 (1956)

12. B. B. Kadomtsev, and O. P. Pogutse: Nucl. Fusion **11**, 67(1971)

13. B. Coppi: Nuovo Cimento **1**, 357(1969)

14. M. Sato and Y. Terashima JAERI-memo 3759, Oct. 1969, Japan Atomic Energy Research Institute (in Japanese)

15. B. B. Kadomtsev and O. P. Pogutse: Sov. Phys. JETP **24**, 1172(1967)

16. P. H. Rutherford and E. A. Frieman: Phys. of Fluids **11**, 569(1968)

17. P. Rutherford, M. Rosenbluth, W. Horton, E. Frieman, and B. Coppi: Plasma Physics and Controlled Nucl. Fusion Research **1**, 367 (Conf. Proceedings, Novosibirsk 1968) (IAEA Vienna, 1969)

18. H. L. Berk and A. A. Galeev: Phys. of Fluids **10**, 441(1967)

19. E. G. Harris: Phys. Rev. Lett. **2**, 34(1959)

20. E. G. Harris: Phys. of Hot Plasmas, p. 145 (ed. by B. J. Rye and J. C. Taylor) Oliver & Boyd, Edinburgh, 1970

21. M. N. Rosenbluth and R. F. Post: Phys. of Fluids **8**, 547(1965)

22. R. F. Post and M. N. Rosenbluth: ibid. **9**, 730(1966)

23. N. A. Krall and P. C. Liewer: Phys. Rev. A**4**, 2094(1971)

24. V. V. Demchenko: Nucl. Fusion **11**, 245(1971)

13 Electromagnetic Waves in a Hot Plasma

Electrostatic waves and electrostatic instabilities have been discussed in chs. 11 and 12. However, the electrostatic approximation is no longer valid if the beta ratio of the plasma is large and if the perturbation of the magnetic field is no longer negligible. The analysis of electromagnetic plasma waves in general, described briefly in sec. 11.1, is much more complicated than the analysis of the electrostatic approximation. To simplify the analysis somewhat, only homogeneous plasmas will be considered here. Take the zeroth-order distribution function to be

$$f_0(\boldsymbol{r}, \boldsymbol{v}) = f(v_\perp, v_z) \qquad v_\perp{}^2 = v_x{}^2 + v_y{}^2.$$

In sec. 13.1, the dispersion equation for electromagnetic plasma waves is derived for a homogeneous plasma. The dispersion equation becomes simple for propagation parallel to the magnetic field. The fire-hose and whistler instabilities are discussed in sec. 13.2 for this case. When the frequencies of the waves are much lower than the cyclotron frequency, the dispersion relation also becomes simple. Such is the case for the fire-hose and mirror instabilities, as is described in sec. 13.3. In sec. 13.4 magnetosonic waves are discussed, starting from a linear approximation to the magnetohydrodynamic equations.

13.1 Dispersion Equation of Electromagnetic Plasma Waves

In order to derive the dispersion equation of electromagnetic waves, the dielectric tensor $\boldsymbol{K}$ must first be derived. For the purpose, the 1st-order term f_1 of the distribution function (11.21) is necessary. Using the vector formula

$$\boldsymbol{v} \times (\boldsymbol{k} \times \boldsymbol{E}_1) = (\boldsymbol{v}\cdot\boldsymbol{E}_1)\boldsymbol{k} - (\boldsymbol{v}\cdot\boldsymbol{k})\boldsymbol{E}_1, \tag{13.1}$$

we obtain

$$f_1(\boldsymbol{r}, \boldsymbol{v}, t) = \frac{-q}{m}\int_{\infty}^{t}\left(\left(1 - \frac{\boldsymbol{k}\cdot\boldsymbol{v}'(t')}{\omega}\right)\boldsymbol{E}_1(\boldsymbol{r}', t') + (\boldsymbol{v}'(t')\cdot\boldsymbol{E}_1(\boldsymbol{r}', t'))\frac{\boldsymbol{k}}{\omega}\right)\nabla_{v'} f_0(\boldsymbol{r}', \boldsymbol{v}')\,\mathrm{d}t'. \tag{13.2}$$

The integral is carried out along the zeroth-order orbit (eqs. (11.2), (11.3)). It is assumed that the electric field is expressed by

$$\boldsymbol{E}_1(\boldsymbol{r}', t') = \boldsymbol{E}\exp i(\boldsymbol{k}\cdot\boldsymbol{r}' - \omega t'), \tag{13.3}$$

and the z axis is taken in the direction of the magnetic field $\boldsymbol{B}_0$. The y axis is taken to be in the plane fixed by $\boldsymbol{B}_0$ and $\boldsymbol{k}$, so that $k_x = 0$. Then eq. (13.2) is

$$f_1(\boldsymbol{r}, \boldsymbol{v}, t) = \frac{-q}{m} \exp i(k_y y + k_z z - \omega t) \times \int_{-\infty}^{0} \left(\left(1 - \frac{\boldsymbol{k} \cdot \boldsymbol{v}'(t')}{\omega} \right) \boldsymbol{E} + (\boldsymbol{v}'(t') \cdot \boldsymbol{E}) \frac{\boldsymbol{k}}{\omega} \right) \cdot \nabla_{v'} f_0 \times \exp\left(i \frac{k_y v_\perp}{\Omega} \cos\theta - i \frac{k_y v_\perp}{\Omega} \cos(\theta + \Omega\tau) + i(k_z v_z - \omega)\tau \right) \mathrm{d}\tau. \tag{13.4}$$

The term exp( ) in the integrand is expanded in Bessel functions:

$$\exp(\ ) = \sum_{m=-\infty}^{\infty} \sum_{n=-\infty}^{\infty} (i)^m J_m \exp im\theta \cdot (i)^{-n} J_n \exp(-in(\theta + \Omega\tau)) \exp i(k_z v_z - \omega)\tau.$$

The other part of the integrand may be written as

$$\left(\left(1 - \frac{\boldsymbol{k} \cdot \boldsymbol{v}'}{\omega} \right) \boldsymbol{E} + (\boldsymbol{v}' \cdot \boldsymbol{E}) \frac{\boldsymbol{k}}{\omega} \right) \cdot \nabla_{v'} f_0$$

$$= \frac{\partial f_0}{\partial v_z} \left(\left(1 - \frac{k_y v_y'}{\omega} \right) E_z + (v_x' E_x + v_y' E_y) \frac{k_z}{\omega} \right) + \frac{\partial f_0}{\partial v_\perp} \left(\left(1 - \frac{k_z v_z}{\omega} \right) \left(E_x \frac{v_x'}{v_\perp} + E_y \frac{v_y'}{v_\perp} \right) + v_z E_z \frac{k_y v_y'}{\omega v_\perp} \right)$$

$$= \left(\frac{\partial f_0}{\partial v_\perp} \left(1 - \frac{k_z v_z}{\omega} \right) + \frac{\partial f_0}{\partial v_z} \frac{k_z v_\perp}{\omega} \right) \times \left(\frac{E_x}{2} (\mathrm{e}^{i(\theta + \Omega\tau)} + \mathrm{e}^{-i(\theta + \Omega\tau)}) + \frac{E_y}{2i} (\mathrm{e}^{i(\theta + \Omega\tau)} - \mathrm{e}^{-i(\theta + \Omega\tau)}) \right)$$

$$+ \left(\frac{\partial f_0}{\partial v_\perp} \frac{k_y v_z}{\omega} - \frac{\partial f_0}{\partial v_z} \frac{k_y v_\perp}{\omega} \right) \frac{E_z}{2i} (\mathrm{e}^{i(\theta + \Omega\tau)} - \mathrm{e}^{-i(\theta + \Omega\tau)}) + \frac{\partial f_0}{\partial v_z} E_z;$$

and eq. (13.4) is reduced to

$$f_1(\boldsymbol{r}, \boldsymbol{v}, t) = \frac{-q}{m} \exp i(k_y y + k_z z - \omega t)$$
$$\times \sum_{m,n} \left[U\left(J_n'(a) i E_x - \frac{n J_n(a)}{a} E_y \right) + \left(-W \frac{n J_n(a)}{a} + \frac{\partial f_0}{\partial v_z} J_n(a) \right) E_z \right]$$
$$\times \frac{i^{m-n} J_m(a) \exp i(m-n)\theta}{i(k_z v_z - \omega - n\Omega)}, \tag{13.5}$$

where U and W are

$$U = \left(1 - \frac{k_z v_z}{\omega}\right) \frac{\partial f_0}{\partial v_\perp} + \frac{k_2 v_\perp}{\omega} \frac{\partial f_0}{\partial v_z} \tag{13.6}$$

$$W = \frac{k_y v_z}{\omega} \frac{\partial f_0}{\partial v_\perp} - \frac{k_y v_\perp}{\omega} \frac{\partial f_0}{\partial v_z} \tag{13.7}$$

$$a = \frac{k_y v_\perp}{\Omega} \qquad \Omega = \frac{-q\boldsymbol{B}}{m}. \tag{13.8}$$

The relations

$$\frac{J_{n-1}(a) + J_{n+1}(a)}{2} = \frac{n}{a} J_n(a) \qquad \frac{J_{n-1}(a) - J_{n+1}(a)}{2} = \frac{\mathrm{d}}{\mathrm{d}a} J_n(a)$$

have been used to obtain eq. (13.5) from eq. (13.4).

Since $f_1(\boldsymbol{r}, \boldsymbol{v}, t)$ is given by eq. (13.5), the dielectric tensor $\boldsymbol{K}$ can be calculated by means of

$$(\boldsymbol{K} - \boldsymbol{I})\boldsymbol{E} = \frac{i}{\varepsilon_0 \omega} \sum_j q_j \int \boldsymbol{v} f_{j1} \, \mathrm{d}v. \tag{13.9}$$

As $v_x = v_\perp \cos\theta$, $v_y = v_\perp \sin\theta$, and $v_z = v_z$, the contribution to the x and y components of eq. (13.9) comes from the terms in $m - n = \pm 1$ in f_1 given by eq. (13.5) and the contribution to the z component of eq. (13.9) comes from the terms in $m - n = 0$, i.e.,[1-3]

$$\boldsymbol{K} = \boldsymbol{I} - \sum_j \frac{\Pi_j^2}{\omega n_{j0}} \sum_{n=-\infty}^{+\infty} \int \mathrm{d}v \frac{\boldsymbol{S}_{j,n}}{(k_z v_z - \omega - n\Omega_j)} \tag{13.10a}$$

$$\boldsymbol{S}_{j,n} = \begin{bmatrix} v_\perp (J_n')^2 U & i v_\perp J_n' \left(n\frac{J_n}{a} \right) U & -i v_\perp J_n' J_n \left(\frac{\partial f_0}{\partial v_z} - \frac{n}{a} W \right) \\ -i v_\perp J_n' \left(n \frac{J_n}{a} \right) U & v_\perp \left(n^2 \frac{J_n^2}{a^2} \right) U & -v_\perp \left(n \frac{J_n^2}{a} \right) \left(\frac{\partial f_0}{\partial v_z} - \frac{n}{a} W \right) \\ i v_z J_n' J_n U & -v_z \left(n \frac{J_n^2}{a} \right) U & v_z J_n^2 \left(\frac{\partial f_0}{\partial v_z} - \frac{n}{a} W \right) \end{bmatrix}_j, \tag{13.11}$$

where Π_j is the plasma frequency for particles of the jth kind. Using the relations

$$\frac{v_z U - v_\perp \left(\dfrac{\partial f_0}{\partial v_z} - \dfrac{n\Omega}{k_y v_\perp} W \right)}{k_z v_z - \omega - n\Omega} = \frac{-v_z}{\omega} \frac{\partial f_0}{\partial v_\perp} + \frac{v_\perp}{\omega} \frac{\partial f_0}{\partial v_z}$$

$$\sum_{n=-\infty}^{\infty} J_n^2 = 1 \qquad \sum_{n=-\infty}^{\infty} J_n J_n' = 0 \qquad \sum_{n=-\infty}^{\infty} n J_n^2 = 0 \qquad (J_{-n} = (-1)^n J_n),$$

K_{xy}, K_{yz}, and K_{zz} can be rewritten. Then the dielectric tensor for the hot plasma is given by[4]

$$\boldsymbol{K} = \boldsymbol{I} - \sum_j \frac{\Pi_j^2}{\omega} \sum_{n=-\infty}^{\infty} \int \boldsymbol{T}_{jn} \frac{U_j (v_\perp n_{j0})^{-1}}{k_z v_z - \omega - n\Omega_j} \mathrm{d}\boldsymbol{v}$$
$$- \boldsymbol{L} \sum_j \frac{\Pi_j^2}{\omega^2} \left(1 + \frac{1}{n_{j0}} \int \frac{v_z^2}{v_\perp} \frac{\partial f_{j0}}{\partial v_\perp} \mathrm{d}\boldsymbol{v} \right) \tag{13.10b}$$

$$\boldsymbol{T}_{jn} = \begin{bmatrix} v_\perp^2 (J_n')^2 & i v_\perp^2 n \frac{J_n J_n'}{a} & -i v_\perp v_z J_n J_n' \\ -i v_\perp^2 n \frac{J_n J_n'}{a} & v_\perp^2 n^2 \frac{J_n^2}{a^2} & -v_\perp v_z \left(n \frac{J_n^2}{a} \right) \\ i v_\perp v_z J_n J_n' & -v_\perp v_z n \frac{J_n^2}{a} & v_z^2 J_n^2 \end{bmatrix}, \tag{13.12}$$

where the only nonzero component of the matrix is $L_{zz} = 1$. From the relations

$$\frac{U}{k_z v_z - \omega - n\Omega_j} = -\frac{1}{\omega} \frac{\partial f_0}{\partial v_\perp} + \frac{1}{\omega (k_z v_z - \omega - n\Omega_j)} \left(-n\Omega_j \frac{\partial f_0}{\partial v_\perp} + k_z v_\perp \frac{\partial f_0}{\partial v_z} \right)$$

$$\sum_{n=-\infty}^{\infty}(J_n')^2 = \sum_{n=-\infty}^{\infty}\frac{n^2 J_n{}^2(z)}{z^2} = \frac{1}{2},$$

another expression of the dielectric tensor is obtained:

$$\boldsymbol{K} = \left(1 - \frac{\Sigma \Pi_j{}^2}{\omega^2}\right)\boldsymbol{I}$$
$$- \sum_{j,n}\frac{\Pi_j{}^2}{\omega^2}\int\frac{\boldsymbol{T}_{jn}}{k_z v_z - \omega - n\Omega_j}\left(-\frac{n\Omega_j}{v_\perp}\frac{\partial f_{j0}}{\partial v_\perp} + k_z\frac{\partial f_{j0}}{\partial v_z}\right)\frac{1}{n_{j0}}\mathrm{d}\boldsymbol{v}. \qquad (13.13)$$

The dispersion relation $\boldsymbol{k} \times (\boldsymbol{k} \times \boldsymbol{E}_1) + (\omega^2/c^2)\boldsymbol{K}\cdot\boldsymbol{E}_1 = 0$ is given by

$$\begin{bmatrix} K_{xx} - N^2 & K_{xy} & K_{xz} \\ K_{yx} & K_{yy} - N^2\cos^2\theta & K_{yz} + N^2\sin\theta\cos\theta \\ K_{zx} & K_{zy} + N^2\sin\theta\cos\theta & K_{zz} - N^2\sin^2\theta \end{bmatrix}\begin{bmatrix} E_x \\ E_y \\ E_z \end{bmatrix} = 0.$$

(13.14)

Since the formula for the dielectric tensor is complicated, it is not easy to analyze further without making simplifying assumptions. One of the simplifications is the electrostatic approximation, which was discussed in sec. 10.6 and chs. 11 and 12. Substitution of eq. (13.13) into eq. (10.106) gives the dispersion equation for the electrostatic wave immediately.

When the Larmor radius is much smaller than the wavelength perpendicular to the magnetic field ($(k_y\rho_B)^2 \ll 1$), the Bessel functions in the matrix $\boldsymbol{T}_{jn}$ of eq. (13.13) can be expanded in Taylor series, i.e.,

$$\boldsymbol{T}_{j0} = \begin{bmatrix} v_\perp{}^2\left(\dfrac{a^2}{4} + \cdots\right) & 0 & iv_\perp v_z\left(\dfrac{a}{2} + \cdots\right) \\ 0 & 0 & 0 \\ -iv_\perp v_z\left(\dfrac{a}{2} + \cdots\right) & 0 & v_z{}^2\left(1 - \dfrac{a^2}{2} + \cdots\right) \end{bmatrix}_j$$

$$\boldsymbol{T}_{j,\pm l} = \begin{bmatrix} v_\perp{}^2(1 + \cdots) & \pm iv_\perp{}^2(1 + \cdots) & -iv_\perp v_z\left(\dfrac{a}{l} + \cdots\right) \\ \mp iv_\perp{}^2(1 + \cdots) & v_\perp{}^2(1 + \cdots) & \mp v_\perp v_z\left(\dfrac{a}{l} + \cdots\right) \\ iv_\perp v_z\left(\dfrac{a}{l} + \cdots\right) & v_\perp v_z\left(\dfrac{a}{l} + \cdots\right) & v_z{}^2\left(\dfrac{a^2}{l^2} + \cdots\right) \end{bmatrix}_j$$

$$\times\left(\frac{(a/2)^{l-1}}{2(l-1)!}\right)_j^2, \tag{13.15}$$

where $l \geq 1$ and $a = k_y v_\perp/\Omega$.

When the zeroth-order distribution function is bi-Maxwellian,

$$f_0(v_\perp, v_z) = n_0 F_\perp(v_\perp) F_z(v_z) \tag{13.16}$$

$$F_\perp(v_\perp) = \frac{m}{2\pi T_\perp}\exp\left(-\frac{m v_\perp{}^2}{2T_\perp}\right)$$

$$F_z(v_z) = \left(\frac{m}{2\pi T_z}\right)^{1/2}\exp\left(-\frac{m(v_z - V)^2}{2T_z}\right),$$

we find

$$\left(-\frac{n\Omega_j}{v_\perp}\frac{\partial f_0}{\partial v_\perp} + k_z\frac{\partial f_0}{\partial v_z}\right)\frac{1}{n_0} = m\left(\frac{n\Omega_j}{T_\perp} - \frac{k_z(v_z - V)}{T_z}\right)F_\perp(v_\perp)F_z(v_z).$$

Integration over v_z can be done using

$$\int_{-\infty}^{\infty}\frac{F_z}{k_z(v_z - V) - \omega_n}\mathrm{d}v_z = \frac{1}{\omega_n}\zeta_n Z(\zeta_n)$$

$$\int_{-\infty}^{\infty}\frac{k_z(v_z - V)F_z}{k_z(v_z - V) - \omega_n}\mathrm{d}v_z = 1 + \zeta_n Z(\zeta_n)$$

$$\int_{-\infty}^{\infty}\frac{(k_z(v_z - V))^2 F_z}{k_z(v_z - V) - \omega_n}\mathrm{d}v_z = \omega_n(1 + \zeta_n Z(\zeta_n))$$

$$\int_{-\infty}^{\infty}\frac{(k_z(v_z - V))^3 F_z}{k_z(v_z - V) - \omega_n}\mathrm{d}v_z = \frac{k_z{}^2 T_z}{m} + \omega_n{}^2(1 + \zeta_n Z(\zeta_n))$$

$$\omega_n = \omega - k_z V + n\Omega$$

$$\zeta_n = \frac{\omega - k_z V + n\Omega}{k_z(2T_z/m)^{1/2}}.$$

Integration of T_j over $v_\perp$ can be carried out using

$$\int_0^{\infty} v_\perp{}^{2n} F_\perp(v_\perp) 2\pi v_\perp \mathrm{d}v_\perp = \left(\frac{2T_\perp}{m}\right)^n n!.$$

When

$$\int_0^\infty J_n^2(b^{1/2}x)\exp\left(-\frac{x^2}{2\alpha}\right)\cdot x\mathrm{d}x = \alpha I_n(\alpha b)\mathrm{e}^{-b\alpha}$$

$$\sum_{n=-\infty}^{\infty} I_n(b) = \mathrm{e}^b \qquad \sum_{n=-\infty}^{\infty} nI_n(b) = 0 \qquad \sum_{n=-\infty}^{\infty} n^2 I_n(b) = b\mathrm{e}^b$$

are used, the formula for the dielectric tensor of a bi-Maxwellian plasma is obtained from eq. (13.10b):

$$\boldsymbol{K} = \boldsymbol{I} + \sum_j \frac{\Pi^2}{\omega^2}\sum_n\left[\zeta_0 Z(\zeta_n) - \left(1 - \frac{1}{\lambda_{\mathrm{T}}}\right)(1 + \zeta_n Z(\zeta_n))\right]\mathrm{e}^{-b}\,\boldsymbol{X}_n$$

$$+ \sum_j \frac{\Pi^2}{\omega^2} 2\eta_0^2 \lambda_{\mathrm{T}} \boldsymbol{L} \tag{13.17a}$$

$$\boldsymbol{X}_n = \begin{bmatrix} \left(\frac{n^2}{b} + 2b\right)I_n - 2bI_n' & in(I_n' - I_n) & -i(2\lambda_{\mathrm{T}})^{1/2}\eta_n\alpha(I_n' - I_n) \\ -in(I_n' - I_n) & \frac{n^2 I_n}{b} & -(2\lambda_{\mathrm{T}})^{1/2}\eta_n\frac{n}{\alpha}I_n \\ i(2\lambda_{\mathrm{T}})^{1/2}\eta_n\alpha(I_n' - I_n) & -(2\lambda_{\mathrm{T}})^{1/2}\eta_n\frac{n}{\alpha}I_n & 2\lambda_{\mathrm{T}}\eta_n^2 I_n \end{bmatrix}, \tag{13.17b}$$

where

$$\eta_n \equiv \frac{\omega + n\Omega}{2^{1/2}k_z v_{\mathrm{T}z}} \qquad \lambda_{\mathrm{T}} \equiv \frac{T_z}{T_\perp} \qquad \alpha \equiv \frac{k_y v_{\mathrm{T}\perp}}{\Omega} \qquad b \equiv \alpha^2.$$

When $T_\perp$ and T_z approach zero and ζ_n, η_n approach infinity, the dielectric tensor of a hot plasma reduces to that of a cold plasma. The propagation vector $\boldsymbol{k}$ is expressed by $\boldsymbol{k} = k_y\hat{\boldsymbol{y}} + k_z\hat{\boldsymbol{z}}$ in the coordinates used here, and $\Omega = -qB/m$. If coordinates where $\boldsymbol{k} = k_{x'}\hat{\boldsymbol{x}}' + k_{z'}\hat{\boldsymbol{z}}'$ is used, the transformations

$$K_{x'x'} = K_{yy} \qquad K_{y'y'} = K_{xx} \qquad K_{z'z'} = K_{zz}$$

$$K_{x'y'} = -K_{yx} \qquad K_{x'z'} = K_{yz} \qquad K_{y'z'} = -K_{xz}$$

are necessary.

13.2 Waves Propagating along Lines of Magnetic Force (Whistler Instability)

When waves propagate along magnetic-field lines, it follows that $k_y = 0$ and $a = 0$, i.e., that $J_n(0) = \delta_{n0}$. Then the dielectric tensor is reduced to

$$K_{xx} = K_{yy} = 1 - \sum_j \frac{\Pi_j^2}{\omega n_{j0}} \int \mathrm{d}\boldsymbol{v} \frac{v_\perp}{4}\left(\left(1 - \frac{kv_z}{\omega}\right)\frac{\partial f_0}{\partial v_\perp} + \frac{kv_\perp}{\omega}\frac{\partial f_0}{\partial v_z}\right)_j \times \left(\frac{1}{kv_z - \omega - \Omega} + \frac{1}{kv_z - \omega + \Omega}\right)_j \tag{13.18}$$

$$K_{zz} = 1 - \sum_j \frac{\Pi_j^2}{\omega n_{j0}} \int \mathrm{d}\boldsymbol{v} \left(\frac{v_z \partial f_0/\partial v_z}{kv_z - \omega}\right)_j \tag{13.19}$$

$$K_{xy} = -K_{yx} = -i\sum_j \frac{\Pi_j^2}{\omega n_{j0}} \int \mathrm{d}\boldsymbol{v} \frac{v_\perp}{4}\left(\left(1 - \frac{kv_z}{\omega}\right)\frac{\partial f_0}{\partial v_\perp} + \frac{kv_\perp}{\omega}\frac{\partial f_0}{\partial v_z}\right)_j \times \left(\frac{1}{kv_z - \omega - \Omega} - \frac{1}{kv_z - \omega + \Omega}\right)_j; \tag{13.20}$$

the other four components are zero. The dispersion equation for the wave with $k_\perp = 0$ is

$$\left.\begin{aligned} \left(K_{xx} - \frac{k^2c^2}{\omega^2}\right)E_x + K_{xy}E_y &= 0 \\ -K_{xy}E_x + \left(K_{xx} - \frac{k^2c^2}{\omega^2}\right)E_y &= 0 \\ K_{zz}E_z &= 0. \end{aligned}\right\} \tag{13.21}$$

One solution,

$$K_{zz} = 0 \qquad E_z \neq 0 \qquad E_x = E_y = 0, \tag{13.22}$$

is the longitudinal electrostatic wave. This wave has already been treated in ch. 11. The other two solutions,

$$E_z = 0 \qquad E_y = \pm iE_x \qquad K_{xx} - \frac{k^2c^2}{\omega^2} = \mp iK_{xy} \tag{13.23}$$

are transverse, circularly polarized waves. Equations (13.23) are satisfied by

$$
\begin{aligned}
K &\equiv K_{xx} \pm iK_{xy} - \frac{k^2c^2}{\omega^2} \\
&= 1 - \frac{k^2c^2}{\omega^2} - \sum_j \frac{\Pi_j^2}{\omega n_j} \int \mathrm{d}\boldsymbol{v} \frac{v_\perp}{2} \left(\left(1 - \frac{kv_z}{\omega}\right) \frac{\partial f_0}{\partial v_\perp} + \frac{kv_\perp}{\omega} \frac{\partial f_0}{\partial v_z} \right)_j \frac{1}{kv_z - \omega \pm \Omega_j} \\
&= 1 - \frac{k^2c^2}{\omega^2} + \sum_j \frac{\Pi_j^2}{\omega n_j} \int \mathrm{d}\boldsymbol{v} \left(\left(1 - \frac{kv_z}{\omega}\right) f_0 - \frac{kv_\perp^2}{2\omega} \frac{\partial f_0}{\partial v_z} \right)_j \frac{1}{kv_z - \omega \pm \Omega_j} \\
&= 0.
\end{aligned} \tag{13.24}
$$

Here partial integration has been used.

When the plasma is cold, the distribution functions of electrons and ions are $f_{j0} = n_{j0}\delta(\boldsymbol{v})$ and the dispersion equation is

$$
1 - \frac{k^2c^2}{\omega^2} - \frac{\Pi_\mathrm{i}^2}{\omega(\omega \mp \Omega_\mathrm{i})} - \frac{\Pi_\mathrm{e}^2}{\omega(\omega \mp \Omega_\mathrm{e})} = 0. \tag{13.25}
$$

When $|\omega| \ll |\Omega_\mathrm{i}|$, the dispersion equation is

$$
\begin{aligned}
1 - \frac{k^2c^2}{\omega^2} \pm \frac{1}{\omega}\left(\frac{\Pi_\mathrm{i}^2}{\Omega_\mathrm{i}} + \frac{\Pi_\mathrm{e}^2}{\Omega_\mathrm{e}}\right) + \left(\frac{\Pi_\mathrm{i}^2}{\Omega_\mathrm{i}^2} + \frac{\Pi_\mathrm{e}^2}{\Omega_\mathrm{e}^2}\right) &= 1 - \frac{k^2c^2}{\omega^2} + \frac{c^2}{(B_0^2/\mu_0\rho_\mathrm{m})} \\
&= 0.
\end{aligned} \tag{13.26}
$$

This is the dispersion equation for the Alfvén wave (eqs. (10.55) and (10.56)).

When the minus sign of eq. (13.25) is taken, the dispersion equation for the electron cyclotron wave (whistler wave) is obtained (eq. 10.88).

Let us take into account the effect of finite temperature on the Alfvén wave. When $|\omega|$, $|kv_z| \ll |\Omega_\mathrm{i}|$, one can expand the denominator of the integrand of eq. (13.24). The dispersion equation for the Alfvén wave becomes

$$
1 + \frac{c^2}{(B_0^2/\mu_0\rho_\mathrm{m})} - \frac{k^2c^2}{\omega^2}\left(1 - \frac{p_\parallel - p_\perp}{B_0^2/\mu_0}\right) = 0, \tag{13.27}
$$

where $p_\parallel$, $p_\perp$ are the pressures parallel and perpendicular to the magnetic field:

$$
\left.
\begin{aligned}
p_\parallel &= \sum_j \int \mathrm{d}\boldsymbol{v} m_j v_z^2 f_{j0} \\
p_\perp &= \sum_j \int \mathrm{d}\boldsymbol{v} m_j \frac{v_\perp^2}{2} f_{j0}.
\end{aligned}
\right\} \tag{13.28}
$$

Finally, the dispersion equation is reduced to

$$\omega^2 = k^2 v_A{}^2\left(1 - \frac{p_\parallel - p_\perp}{B_0{}^2/\mu_0}\right). \tag{13.29}$$

When

$$\left.\begin{aligned} p_\parallel - p_\perp &> \frac{B_0{}^2}{\mu_0} \\ \beta_\parallel &> \beta_\perp + 2, \end{aligned}\right\} \tag{13.30}$$

the Alfvén wave becomes unstable. This is called *fire-hose instability*.[3] This name comes from the observation that a fire-hose moves in a zigzag pattern when water flows too quickly through it.[5] This instability will be treated again in the following section.

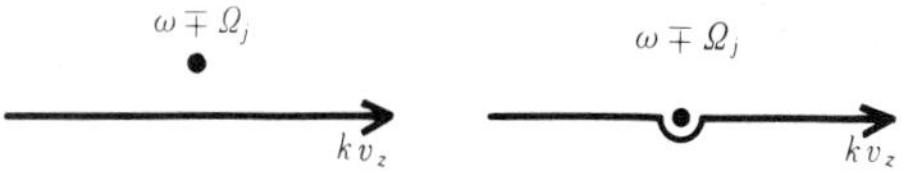

Fig. 13.1
Integration path of kv_z in eq. (13.24) for $k > 0$.

Let us analyze eq. (13.24). The contribution from the poles to the integral of eq. (13.24) is

$$\begin{aligned} K(k, \omega) &\equiv K_r(k, \omega) + iK_i(k, \omega) \\ &= K_r(k, \omega) + i\pi \sum_j \frac{\Pi_j{}^2}{\omega|k|}\left[\left(\frac{\pm\Omega_j}{\omega}\right)F_j\left(\frac{\omega \mp \Omega_j}{k}\right)\right. \\ &\qquad\qquad \left. - \frac{k\langle v_\perp{}^2\rangle}{2\omega}F_j'\left(\frac{\omega \mp \Omega_j}{k}\right)\right] = 0, \end{aligned} \tag{13.31}$$

where $K_r(k, \omega)$ is the principal value of the integral taken over v_z. $F_j(v_z)$ and $\langle v_\perp{}^2\rangle$ are defined by

$$n_{j0}F_j(v_z) \equiv \int f_{j0}(v_\perp, v_z)2\pi v_\perp \mathrm{d}v_\perp$$

$$n_{j0}\langle v_\perp{}^2\rangle F_j(v_z) \equiv \int v_\perp{}^2 f_{j0}(v_\perp, v_z)2\pi v_\perp \mathrm{d}v_\perp.$$

It is assumed that the absolute value of the imaginary part K_i of $K(k, \omega)$

is much smaller than the absolute value of the real part K_r, and that the solution $\omega_r(k)$ of $K_r(k, \omega) = 0$ is real. The solution $\omega = \omega_r + i\gamma$ of $K_r(\omega) + iK_i(\omega) = 0$ is given by

$$K_r(\omega_r) + i\gamma \frac{\partial}{\partial \omega_r} K_r(\omega_r) + iK_i(\omega_r) = 0,$$

so that

$$\gamma = -\left(\frac{K_i(k, \omega)}{\partial K_r / \partial \omega}\right)_{\omega = \omega_r(k)}. \tag{13.32}$$

When $\gamma > 0$, the unstable wave exists.

Let us consider a bi-Maxwellian distribution function

$$f_{j0} = \frac{n_{j0}}{\pi^{3/2}\left(\dfrac{2T_\perp}{m}\right)\left(\dfrac{2T_z}{m}\right)^{1/2}} \exp\left(-\frac{mv_\perp^2}{2T_\perp} - \frac{mv_z^2}{2T_z}\right). \tag{13.33}$$

The value of $K_i(k, \omega)$ of eq. (13.31) is

$$K_i(k, \omega) = \pi \sum_j \frac{\Pi_j^2}{\omega|k|} \frac{1}{\pi^{1/2}\left(\dfrac{2T_z}{m}\right)_j^{1/2}} \exp\left(-\frac{m(\omega \mp \Omega)^2}{2T_z k^2}\right)_j \left(\frac{T_\perp}{T_z} \mp \frac{\Omega}{\omega}\left(\frac{T_\perp}{T_z} - 1\right)\right)_j. \tag{13.34}$$

The whistler wave corresponds to the minus sign of eq. (13.34). When $|\Omega_i| \ll \omega \ll \Omega_e$, the dispersion equation is

$$K_r(k, \omega) \approx 1 - \frac{k^2 c^2}{\omega^2} + \frac{\Pi_e^2}{\omega \Omega_e}.$$

The last two terms are much larger than 1, and thus

$$\omega_r \approx \frac{c^2 k^2 \Omega_e}{\Pi_e^2}.$$

The sign of $\partial K_r / \partial \omega$ at $\omega = \omega_r$ is

$$\left(\frac{\partial K_r}{\partial \omega}\right)_{\omega = \omega_r} \approx \frac{1}{\omega_r^2} \frac{\Pi_e^2}{\Omega_e} > 0.$$

From eq. (13.32) and (13.34), the instability condition for the whistler wave is given by[3]

$$\frac{T_{\perp e}}{T_{ze}} - \frac{\Omega_e}{\omega}\left(\frac{T_{\perp e}}{T_{ze}} - 1\right) < 0. \tag{13.35}$$

When $T_{\perp e}/T_{ze} > 1 + \omega/\Omega_e$, i.e., when the perpendicular temperature is larger than the parallel one by a factor of $1 + \omega/\Omega_e$, the whistler wave is unstable.

13.3 Fire-Hose and Mirror Instabilities

When the Larmor radius of a plasma particle is smaller than the perpendicular wavelength, the dielectric tensor can be simplified by means of eq. (13.17). When the thermal velocity is less than the phase velocity, $|\zeta_n|$ is much larger than 1. When $|\omega| \ll |\Omega|$, then $|k_z \rho_B| \ll 1$ and $|\zeta_n|$ is large ($|\zeta_n| \gg 1$) for $n \neq 0$. In this case the dispersion function is

$$1 + \zeta_n Z(\zeta_n) \approx -\frac{1}{2\zeta_n^2} \qquad (|\zeta_n| > 2).$$

When the thermal velocity is larger than the phase velocity, then $|\zeta_0| \ll 1$ and

$$1 + \zeta_0 Z(\zeta_0) \approx 1 - 2\zeta_0^2 \qquad (|\zeta_0| < 0.5).$$

It is assumed that $b = (k\rho_B)^2$ and ζ_0^2 (in the hot case) or ζ_0^{-2} (in the cold case) are small quantities of the 1st order. The dielectric tensor of a plasma composed of cold ions and hot electrons and $V = 0$ is given to the 1st order by

$$K_{xx} = -\frac{\Pi_i^2}{\omega^2}\frac{2k_y^2(T_\perp/m)_i}{\Omega_i^2} - \frac{\Pi_e^2}{\omega^2}\frac{2k_y^2(T_\perp/m)_e}{\Omega_e^2}\left(1 - \frac{T_{\perp e}}{T_{ze}}\right) + K_{yy}$$

$$K_{yy} = 1 + \frac{\Pi_i^2}{\Omega_i^2} + \frac{\Pi_e^2}{\Omega_e^2} + \frac{\Pi_e^2}{\omega^2}\left(1 - \frac{T_{\perp e}}{T_{ze}}\right)\frac{k_z^2(T_z/m)_e}{\Omega_e^2}$$

$$K_{zz} = 1 - \frac{\Pi_i^2}{\omega^2}\left(1 - \frac{k_y^2(T_\perp/m)_i}{\Omega_i^2}\right) + \frac{\Pi_e^2}{k_z^2(T_z/m)_e}$$

$$K_{xy} = -\left(\frac{\Pi_i^2}{\omega\Omega_i} + \frac{\Pi_e^2}{\omega\Omega_e}\right)i \approx 0$$

$$K_{xz} = \left(-\frac{\Pi_i^2}{\omega\Omega_i}\frac{k_y k_z(T_\perp/m)_i}{\omega^2} + \frac{\Pi_e^2}{\omega\Omega_e}\frac{T_{\perp e}}{T_{ze}}\frac{k_y}{k_z}\right)i$$

$$K_{yz} = -\frac{\Pi_i^2}{\omega^2}\left(1 - \frac{T_{\perp i}}{T_{zi}}\right)\frac{k_y k_z (T_z/m)_i}{\Omega_i^2} - \frac{\Pi_e^2}{\omega^2}\left(1 - \frac{T_{\perp e}}{T_{ze}}\right)\frac{k_y k_z (T_z/m)_e}{\Omega_e^2}. \quad (13.36)$$

In some cases both ion and electron are cold or hot; the other possible case involves cold electrons and hot ions. The appropriate expression, drawn from the hot or cold case, must be used for each term.

When the frequency of the wave is low ($|\omega| \ll |\Omega|$) and the Larmor radius small ($k^2(T/m)/\Omega^2 \ll 1$), the dispersion equation is obtained by using the dielectric tensor (13.36). As the component K_{xy} is small, its contribution can be neglected and the dispersion equation written as[6]

$$K_{zz}(K_{yy} - N_z^2)(K_{xx} - N^2)$$
$$= (K_{xx} - N^2)(K_{yy}N_y^2 + K_{yz}^2 + 2N_y N_z K_{yz}) + (K_{yy} - N_z^2)K_{xz}K_{zx},$$

i.e.,

$$(K_{yy} - N_z^2)\left(K_{xx} - N^2 - \frac{K_{xz}K_{zx}}{K_{zz}}\right)$$
$$= \frac{(K_{xx} - N^2)}{K_{zz}}(K_{yy}N_y^2 + K_{yz}^2 + 2N_y N_z K_{yz}), \quad (13.37)$$

where $\boldsymbol{N} = \boldsymbol{k}c/\omega$. Since $|K_{zz}| \gg |K_{yy}|, |K_{yz}|$, the right-hand side of eq. (13.37) is small compared with the left-hand side and the following dispersion equations are derived:

$$K_{yy} - N_z^2 \approx 0 \quad (13.38)$$

$$K_{xx} - N^2 - \frac{K_{xz}K_{zx}}{K_{zz}} \approx 0. \quad (13.39)$$

From eq. (13.36), K_{yy} is given by

$$K_{yy} = 1 + \delta + \frac{\beta_z - \beta_\perp}{2}N_z^2 \qquad \delta = \frac{(n_i m_i + n_e m_e)c^2}{B^2/\mu_0}$$

and the dispersion equation (13.38) is

$$\frac{k_z^2 c^2}{\omega^2}\left(1 + \frac{\beta_\perp}{2} - \frac{\beta_z}{2}\right) = 1 + \delta. \quad (13.40)$$

This is the same as eq. (13.29). When $\beta_z > \beta_\perp + 2$, *fire-hose instability* occurs[5] (fig. 13.2).

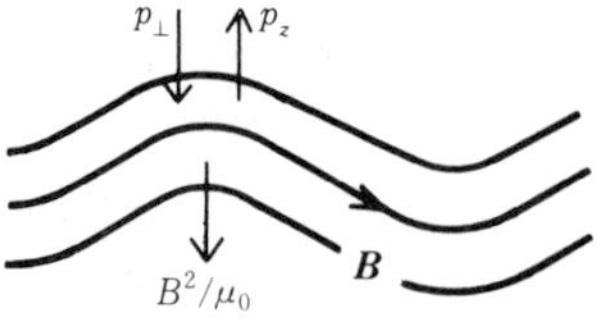

Fig. 13.2
Fire-hose instability.

The other dispersion equation (13.39) may be reduced to

$$N^2\left(1+\left(\beta_\perp-\frac{(\beta_{\perp e})^2}{2\beta_{ze}}\right)\sin^2\theta+(\beta_\perp-\beta_z)\frac{\cos^2\theta}{2}\right)=1+\delta. \tag{13.41}$$

This is the case of cold ions and hot electrons. When

$$\frac{{\beta_{\perp e}}^2}{\beta_{ze}}>2(1+\beta_\perp), \tag{13.42}$$

the plasma becomes unstable at $\sin^2\theta\approx 1$. This is called *mirror instability*.[5] Fire-hose instability is associated with perturbations propagating along the lines of magnetic force, but mirror instability is related to perturbations propagating perpendicularly to the field lines. Where $\beta_\perp$ becomes large, as at the center of a mirror field, the magnetic field within the plasma becomes small due to the diamagnetism of the plasma. The plasma further constricts around the center, $\beta_\perp$ increases, and thus the plasma becomes unstable.

In eqs. (13.40) and (13.41), the Alfvén velocity ${v_A}^2=c^2/(1+\delta)$ and the sound velocity $\beta {v_A}^2\approx(T_e+T_i)/m_i$ appear as characteristic parameters. These are unstable modes of the magnetosonic wave due to temperature anisotropy, as will be explained in the next section.

13.4 Magnetosonic Wave (MHD Treatment)

For frequencies smaller than the ion cyclotron frequency and when the resonant particles (those with velocities near the phase velocity) do not play an important role, MHD treatment provides a good approximation. The dispersion equation derived from the MHD equations must coincide with the result derived in sec. 13.3. The linearization of a small perturbation of MHD equations has already been carried out in sec. 9.2a. From eqs. (9.32) and (9.28), the small displacement $\xi(\boldsymbol{r},t)$ satisfies

$$\rho_{m0}\frac{\partial^2 \boldsymbol{\xi}}{\partial t^2} = \nabla(\boldsymbol{\xi}\cdot\nabla p_0 + \gamma p_0 \nabla \boldsymbol{\xi}) + \frac{1}{\mu_0}(\nabla \times \boldsymbol{B}_0) \times \boldsymbol{B}_1 + \frac{1}{\mu_0}(\nabla \times \boldsymbol{B}_1) \times \boldsymbol{B}_0. \tag{13.43}$$

γ is the specific heat ratio. The small perturbation of the magnetic field $\boldsymbol{B}_1$ is given by

$$\boldsymbol{B}_1 = \nabla \times (\xi \times \boldsymbol{B}_0). \tag{13.44}$$

Assuming that $\boldsymbol{\xi}(\boldsymbol{r}, t) = \boldsymbol{\xi}_1 \exp i(\boldsymbol{k}\cdot\boldsymbol{r} - \omega t)$, $\nabla \times \boldsymbol{B}_0 = 0$, $\nabla p_0 = 0$, we can write eq. (13.43) as

$$-\rho_m \omega^2 \xi_1 = -\boldsymbol{k}(\gamma p_0(\boldsymbol{k}\cdot\xi_1)) - \mu_0{}^{-1}[\boldsymbol{k} \times (\boldsymbol{k} \times (\xi_1 \times \boldsymbol{B}_0))] \times \boldsymbol{B}_0. \tag{13.45}$$

Using the vector formula $\boldsymbol{A} \times (\boldsymbol{B} \times \boldsymbol{C}) = \boldsymbol{B}(\boldsymbol{A}\cdot\boldsymbol{C}) - \boldsymbol{C}(\boldsymbol{A}\cdot\boldsymbol{B})$, we can write eq. (13.45) as

$$((\boldsymbol{k}\cdot\boldsymbol{B}_0)^2 - \mu_0\omega^2\rho_m)\xi_1 + [(\boldsymbol{B}_0{}^2 + \mu_0\gamma p_0)\boldsymbol{k} - (\boldsymbol{k}\cdot\boldsymbol{B}_0)\boldsymbol{B}_0](\boldsymbol{k}\cdot\xi_1)$$
$$- (\boldsymbol{k}\cdot\boldsymbol{B}_0)(\boldsymbol{B}_0\cdot\xi_1)\boldsymbol{k} = 0. \tag{13.46}$$

If the unit vectors of $\boldsymbol{k}$, $\boldsymbol{B}_0$ are denoted by $\hat{\boldsymbol{k}}$, $\boldsymbol{b}$, respectively, eq. (13.46) is reduced to

$$\left(\cos^2\theta - \frac{V^2}{v_A{}^2}\right)\xi_1 + \left(\left(1 + \frac{\gamma\beta}{2}\right)\hat{\boldsymbol{k}} - \cos\theta\boldsymbol{b}\right)(\hat{\boldsymbol{k}}\cdot\xi_1) - \cos\theta\hat{\boldsymbol{k}}(\boldsymbol{b}\cdot\xi_1) = 0, \tag{13.47}$$

where $\omega/k = V$, $v_A{}^2 \equiv B_0{}^2/\mu_0\rho_m$, $\beta \equiv p_0/(B_0{}^2/2\mu_0)$, and $(\hat{\boldsymbol{k}}\cdot\boldsymbol{b}) = \cos\theta$. The scalar products of eq. (13.47) with $\hat{\boldsymbol{k}}$ and $\boldsymbol{b}$, and the scalar product of $\boldsymbol{b}$ and the vector product of $\hat{\boldsymbol{k}}$ and eq. (13.47), are

$$\left.\begin{aligned}
&\left(1 + \frac{\gamma\beta}{2} - \frac{V^2}{v_A{}^2}\right)(\hat{\boldsymbol{k}}\cdot\xi_1) - \cos\theta(\boldsymbol{b}\cdot\xi_1) = 0\\
&\frac{\gamma\beta}{2}\cos\theta(\hat{\boldsymbol{k}}\cdot\xi_1) - \frac{V^2}{v_A{}^2}(\boldsymbol{b}\cdot\xi_1) = 0\\
&\left(\cos^2\theta - \frac{V^2}{v_A{}^2}\right)\boldsymbol{b}\cdot(\hat{\boldsymbol{k}} \times \xi_1) = 0,
\end{aligned}\right\} \tag{13.48}$$

respectively. The solution of eqs. (13.48) is

$$V^2 = v_A{}^2 \cos^2\theta \qquad (\boldsymbol{\xi}_1 \cdot \boldsymbol{k}) = 0 \qquad (\boldsymbol{\xi}_1 \cdot \boldsymbol{B}_0) = 0 \tag{13.49}$$

or

$$\left(\frac{V}{v_A}\right)^4 - \left(1 + \frac{\gamma\beta}{2}\right)\left(\frac{V}{v_A}\right)^2 + \frac{\gamma\beta}{2}\cos^2\theta = 0 \tag{13.50}$$

$$\boldsymbol{B}_0 \cdot (\boldsymbol{k} \times \boldsymbol{\xi}_1) = 0.$$

Equation (13.49) is a torsional Alfvén wave where the displacement $\boldsymbol{\xi}_1$ is normal to $\boldsymbol{k}$ and $\boldsymbol{B}_0$. Equation (13.50) corresponds to a compressional Alfvén wave where the displacement $\boldsymbol{\xi}_1$, $\boldsymbol{k}$, and $\boldsymbol{B}_0$ are coplaner. If the velocity of sound is denoted by $c_s{}^2 = \gamma p_0/\rho_m$, eq. (13.50) becomes

$$V^4 - (v_A{}^2 + c_s{}^2)V^2 + v_A{}^2 c_s{}^2 \cos^2\theta = 0 \tag{13.51}$$

and

$$V_F{}^2 = \frac{1}{2}[v_A{}^2 + c_s{}^2 + ((v_A{}^2 + c_s{}^2)^2 - 4v_A{}^2 c_s{}^2 \cos^2\theta)^{1/2}] \tag{13.52}$$

$$V_S{}^2 = \frac{1}{2}[v_A{}^2 + c_s{}^2 - ((v_A{}^2 + c_s{}^2)^2 - 4v_A{}^2 c_s{}^2 \cos^2\theta)^{1/2}]. \tag{13.53}$$

The two solutions V_F and V_S correspond to a fast wave and a slow wave. These waves are called *magnetosonic waves*. The dependence of the phase velocity on the direction of propagation is shown in fig. 13.3.

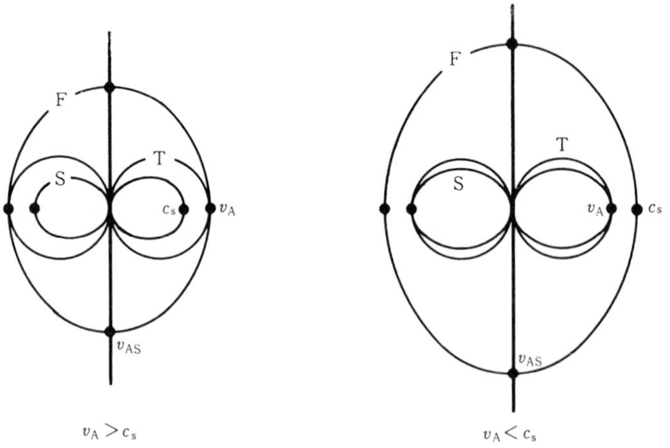

Fig. 13.3
Magnetosonic waves (fast wave F and slow wave S and the torsional Alfvén wave T).

References

1. V. D. Shafranov: Rev. of Plasma Phys. **3**, 1 (ed. by M. A. Leontovich), Consultants Bureau, New York, 1967

2. G. Bekefi: Radiation Processes in Plasma, Wiley, 1966

3. E. D. Harris: Physics of Hot Plasmas, ch. 4 (ed. by B. J. Bye and J. C. Taylor), Oliver & Boyd, Edinburgh, 1970

4. A. I. Akhiezer, I. A. Akhiezer, R. V. Polovin, and A. G. Sitenko: Collective Oscillations in a Plasma, translated by H. S. H. Massey and R. J. Taylor, Pergamon Press, Oxford, 1967

5. A. A. Vedenov, E. P. Velikov, and R. Z. Sagdeev: Sov. Phys. Uspekhi **4**, 332 (1961)
W. B. Thompson: An Introduction to Plasma Physics, Addison-Wesley, Reading, Mass., 1962

6. T. H. Stix: The Theory of Plasma Waves, McGraw-Hill, New York, 1962

IV HEATING, DIAGNOSTICS, AND CONFINEMENT

14 Plasma Heating

In order to raise the temperature of a plasma, it is necessary to inject energy while holding the energy dissipation to a minimum. Define the heating rate H as the energy absorbed by a unit volume of plasma in unit time, and let τ_E be the energy confinement time. Then the change of plasma temperature over time is given by

$$\frac{3}{2}\frac{d(nT)}{dt} = H - \frac{(3/2)nT}{\tau_E}.$$

The heating rate and the energy confinement time are in general dependent on the temperature T, density n, and other parameters of the confinement device. When the temperature is raised, there is a tendency for the heating rate to decrease and the energy loss rate nT/τ_E to increase. An equilibrium state is obtained at $d(nT)/dt = 0$, at which point

$$\tfrac{3}{2}nT = H\tau_E.$$

Accordingly, the heating rate and the energy confinement time are the essential factors to be considered in raising the plasma temperature.

Joule heating and heating by the injection of high-energy neutral beams are considered in sec. 14.1. Heating by compression is described in sec. 14.2. In sec. 14.3, heating due to the interactions between excited waves and plasma particles is discussed and its mechanism is analyzed based upon the theory developed in chs. 11–13.

The heating of plasmas is related to various aspects of plasma physics, and it is found to be closely connected to the nonlinear and stochastic characteristics of plasmas—i.e., to the essential features of plasma kinematics. Since a detailed discussion of this subject is beyond the scope of this book, only a brief introduction is given in sec. 14.4.

14.1 Joule Heating and Neutral Beam Injection

14.1a Joule Heating

When an electric field E is applied to a plasma, the plasma electrons are accelerated, i.e., their kinetic energies are increased. These kinetic energies are thermalized by means of collisions with the plasma ions. This is the mechanism of Joule or ohmic heating. The heating rate for Joule heating is given by

$$j \cdot E = \eta j^2 = \frac{E^2}{\eta}, \tag{14.1}$$

where j is the current density. As was explained in sec. 4.6, the specific resistivity η of the plasma is

$$\eta = \frac{m_e \nu_{ei}}{n_e e^2} \approx 2 \times 10^{-9} Z \ln \Lambda (T_e^*)^{-3/2} \quad \text{(ohm-m)},$$

where T_e^* is expressed in keV. In toroidal systems, the magnitude of the current density j is limited by the Kruskal-Shafranov limit. On the other hand the resistivity is inversely proportional to the 3/2 power of the electron temperature. The resistivity decreases as the electron temperature increases and the heating rate decreases. Runaway electrons are likely to appear in a hot plasma. Accordingly, the electron temperature in a toroidal system is limited to several keV at the most, due to the balance between bremsstrahlung loss and Joule heating, even if the confinement is perfect. (This point will be discussed in sec. 16.2.) When the confinement time is longer than the relaxation time ($\approx (m_i/m_e)\tau_{ei}$), the ions are heated via collisions with the electrons and the ion temperature is $T_i \approx T_e$. From this it is clear that Lawson's condition cannot be realized by Joule heating alone in the tokamak and stellarator configurations. Further heating is necessary in order to realize a practical reactor.

However, the cost of Joule heating is much less than that of other methods and the heating rate can be measured relatively easily; so, even though it is insufficient of itself, Joule heating is frequently used in toroidal devices.

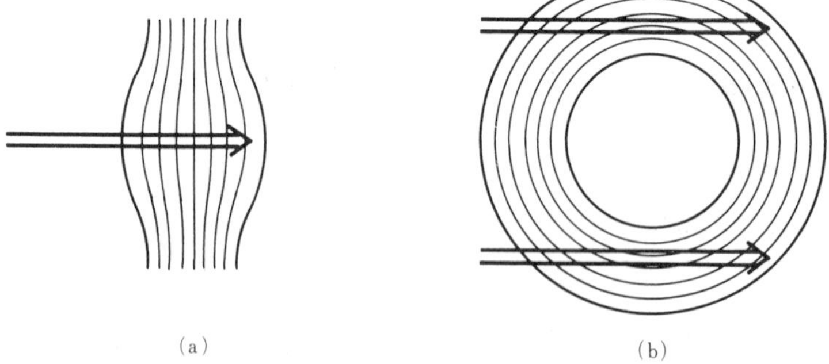

Fig. 14.1
Fast neutral beam injection: (a) normal injection, mirror field; (b) tangential injection, toroidal field.

14.1b Heating by Neutral Beam Injection

High-energy neutral particle beams can be injected into plasmas across strong magnetic fields (fig. 14.1). The neutral particles are converted to high-energy ions by means of charge exchange with plasma ions or ionization. The high-energy ions running through the plasma are slowed down by Coulomb collisions with the plasma ions and electrons and the beam energy is thus transferred to the plasma. This method is called heating by fast neutral beam injection. As was noted in sec. 4.2, the rate of change of the energy ε_b of a fast ion is given by

$$\frac{1}{v_b}\frac{d\varepsilon_b}{dt} = -\frac{Z^2 e^4 \ln \Lambda}{8\pi\varepsilon_0^{\,2}}\left(\frac{A}{\varepsilon_b}\sum\frac{n_i Z_i^{\,2}}{A_i} + \frac{4n_e}{3\pi^{1/2}}\left(\frac{A_e}{A}\right)^{1/2}\frac{\varepsilon_b^{\,1/2}}{T_e^{\,3/2}}\right) = -\frac{\alpha}{\varepsilon_b} - \beta\varepsilon_b^{\,1/2} \tag{14.2}$$

when beam ion's velocity v_b is much less (say 1/3) than the plasma electron thermal velocity $(T_e/m_e)^{1/2}$ and is much larger (say 2 times) than the plasma ion thermal velocity $(T_i/m_i)^{1/2}$. The symbols A, A_i, and A_e give the atomic weights of the injected ion, plasma ion, and electron, respectively. Ze and $Z_i e$ are the charges of the injected and the plasma ion. The first term on the right $(-\alpha/\varepsilon_b)$ is due to beam–ion collisions, and the second term $(-\beta\varepsilon_b^{\,1/2})$ is due to beam–electron collisions. A critical energy ε_{cr} of the beam ion, at which the plasma ions and electrons are heated at equal rates, was given by eq. (4.42):

$$\varepsilon_{cr} = 15T_e\left(A^{3/2}\sum\frac{n_i Z_i^{\,2}}{n_e}\frac{1}{A_i}\right)^{2/3}. \tag{14.3}$$

When the energy of a fast ion is thermalized completely, the ratio G of the energy transferred to plasma ions to the initial energy ε of the beam particle is[1]

$$G = \frac{1}{\varepsilon}\int_v^{\infty}\frac{\alpha}{\varepsilon_b}\,dx = \frac{1}{\varepsilon}\int_0^{\varepsilon}\frac{\alpha}{\alpha + \beta\varepsilon_b^{\,3/2}}\,d\varepsilon_b = \frac{\varepsilon_{cr}}{\varepsilon}\int_0^{\varepsilon/\varepsilon_{cr}}\frac{dy}{1 + y^{3/2}}. \tag{14.4}$$

The dependence of G on $(\varepsilon/\varepsilon_{cr})$ is shown in fig. 14.2.

When $\varepsilon = \varepsilon_{cr} \approx 15T_e$, 74.7% of ε is transferred to the plasma ions. If $\varepsilon = 3\varepsilon_{cr}$, only 44.1% of the energy of the fast ions is transferred to the plasma ions and the rest of the energy heats the plasma electrons. To improve the efficiency of ion heating, the electron temperature must be raised.

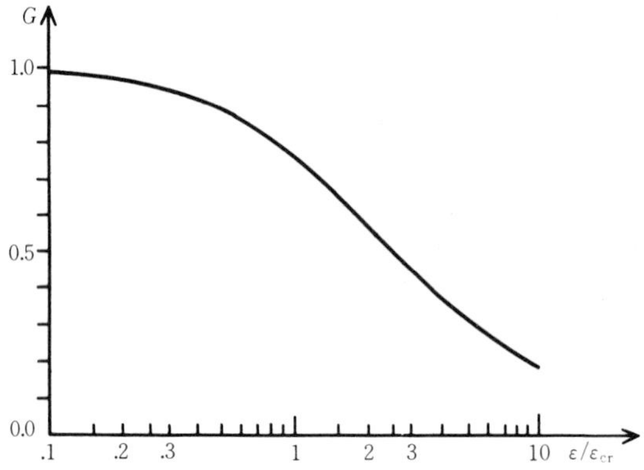

Fig. 14.2
The ratio G of the energy transferred to plasma ions to the initial energy ε of a beam ion versus $\varepsilon/\varepsilon_{cr}$. After T. H. Stix: Plasma Physics **14**, 367 (1972).

Fast neutral beam injection is one of the most promising method of ion heating, but the following conditions must be observed:

1. Most of the fast neutral particles must be ionized during their passage through the plasma.
2. The beam-ion confinement time must be greater than the slowing down time.
3. The ion beam must not induce other losses.

The trajectories of the fast ions in the plasma and the contribution of the fast ions to the current or electric-field distribution have been examined.[1-3] Experiments in fast neutral beam injection have been carried out in Tokamak devices such ORMAK and ATC,[4] and the predicted results have been obtained.

A more ambitious scheme has been proposed.[5] If a neutral 200 keV deuterium beam is injected into a D-T plasma, the high-energy deuterium should ionize and fusion reactions should occur before the ions slow down. This process may relax Lawson's condition.

14.2 Adiabatic Compression and Shock Heating

14.2a Adiabatic Compression

The relation between the plasma pressure p and the volume $\mathscr{V}$ in adiabatic compression is

$$p\mathscr{V}^{\gamma} = \text{const.},$$

where the specific heat ratio γ is given by $\gamma = (2 + \delta)/\delta$, δ being the number of degrees of freedom, so that $\gamma = 5/3, 2, 3$ for $\delta = 3, 2, 1$, respectively. As $p = nT$ and n is inversely proportional to $\mathscr{V}$, it follow that

$$T\mathscr{V}^{\gamma - 1} = \text{const.} \qquad Tn^{-(\gamma - 1)} = \text{const.} \tag{14.5}$$

The compression time $t_{\mathrm{ac}} = B/(\mathrm{d}B/\mathrm{d}t)$ for adiabatic compression must be longer than the Larmor period $(\Omega t_{\mathrm{ac}} > 1)$ or the transit time L/v_{T} $(\Omega t_{\mathrm{ac}} > L/\rho_B)$ (L and v_{T} are the characteristic length and the thermal velocity, respectively). On the other hand t_{ac} must be shorter than the energy confinement time τ_{E}: $\tau_{\mathrm{E}} > t_{\mathrm{ac}}$, $\Omega t_{\mathrm{ac}} > L/\rho_B, 1$. If the collision time is shorter than t_{ac}, then γ is 5/3.

A good example of adiabatic compression in a toroidal plasma is the ATC (Adiabatic Toroidal Compressor) experiment.[6,7]
Let the plasma minor and toroidal radii and the toroidal and poloidal magnetic fields be denoted by a, R and B_{t}, B_{p}, respectively. Under the assumption that the plasma is conductive during the compression time, the conservation of magnetic flux is given by

$$a^2 B_{\mathrm{t}} = \text{const.} \qquad aRB_{\mathrm{p}} = \text{const.}$$

Using $Tn^{-2/3} = \text{const.}$, we find the following relations:

$$\left.\begin{aligned} n &\propto a^{-2}R^{-1} \\ T &\propto a^{-4/3}R^{-2/3} \\ B_{\mathrm{t}} &\propto a^{-2} \\ B_{\mathrm{p}} &\propto a^{-1}R^{-1} \\ \beta &\propto a^{2/3}R^{-5/3}. \end{aligned}\right\} \tag{14.6}$$

β is the beta ratio. There are several possible compression schemes. For example: (i) compression of the plasma radius a keeping $R = \text{const.}$; (ii) compression of a and R keeping $a/R = \text{const.}$; or (iii) compression of a and R keeping $a^2/R = \text{const.}$ The third case is convenient because then the toroidal field $B_{\mathrm{t}} \propto 1/R$ itself need not change in time (fig. 14.3). In the ATC experiment compression of this type is carried out by varying the vertical field, which determines the equilibrium position of the plasma while keeping $a^2/R = \text{const.}$

Adiabatic compression of plasmas between mirrors is also effective and

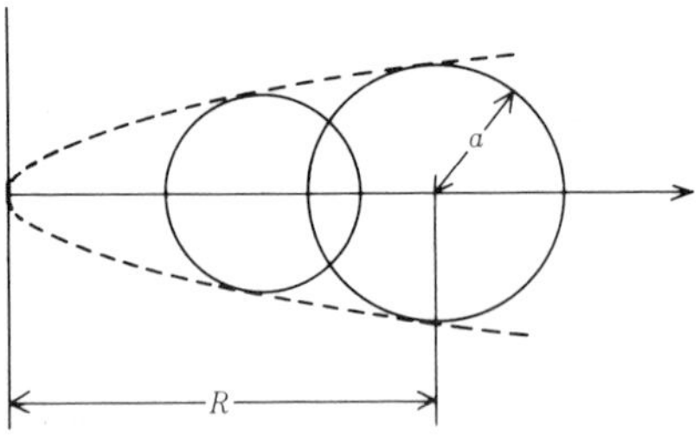

Fig. 14.3
Adiabatic compression of a toroidal plasma.

ion heating in such a configuration is demonstrated in the 2X-II experiment.[8] Electron heating due to adiabatic compression is observed in the hot electron plasma of the TPM experiment.[9]

14.2b Magnetic Pumping

Let us assume that the confining magnetic field B is changed periodically,

$$B = B_0(1 + \varepsilon \cos \omega t), \tag{14.7}$$

where ω is much smaller than the cyclotron frequency. As the magnetic moment is conserved, the kinetic energy $U_\perp$ perdendicular to the magnetic field is proportional to the magnitude of the field. However there is a relaxation process between $U_\perp$ and $U_\parallel$ (the kinetic energy parallel to the magnetic field) due to collision. Accordingly we have the relations

$$\left.\begin{aligned} \frac{\mathrm{d}U_\perp}{\mathrm{d}t} &= \frac{U_\perp}{B}\frac{\mathrm{d}B}{\mathrm{d}t} - \nu\left(\frac{U_\perp}{2} - U_\parallel\right) \\ \frac{\mathrm{d}U_\parallel}{\mathrm{d}t} &= \nu\left(\frac{U_\perp}{2} - U_\parallel\right), \end{aligned}\right\} \tag{14.8}$$

where ν is the collision frequency. The substitution of eq. (14.7) into eqs. (14.8) gives the increment ΔU of the total kinetic energy during a period $2\pi/\omega$ (providing $\varepsilon \ll 1$):

$$\Delta U = \left(\frac{2\pi}{\omega}\right)\varepsilon^2 U \frac{\omega^2 \nu}{6\left(\frac{9}{4}\nu^2 + \omega^2\right)}.$$

The heating rate is

$$\frac{\Delta U}{\Delta t} = \frac{\varepsilon^2}{6} \frac{\omega^2 \nu}{(9\nu^2/4 + \omega^2)} U = \alpha U. \tag{14.9}$$

When $\omega \ll \nu$, then $\alpha \approx (2/27)a^2\varepsilon^2/\nu$. When $\omega \gg \nu$, then $\alpha \approx \varepsilon^2\nu/6$. Since $\nu \propto U^{-3/2}$, the heating rate $\Delta U/\Delta t \propto \varepsilon^2 n U^{-1/2}$ decreases as the temperature increases. This heating method is called (*collisional*) *magnetic pumping*[10] or *gyro-relaxation heating.*[11]

14.2c Theta Pinch, Shock Heating

(i) Heating by Theta Pinch In theta pinch experiments, a large current is fed rapidly to a cylindrical coil by means of capacitor discharge, so that the magnetic field B_z within the cylinder is raised very rapidly (fig. 14.4). If the plasma is weakly ionized, an electric field in the θ direction $E_\theta = \dot{B}_z r/2$ is induced. The plasma, which assumes a shell-like structure near the tube wall, is compressed radially by the magnetic pressure. This is called the *theta pinch.*[12] The entire process is analyzed by putting the MHD equations into a computer; the numerical results thus obtained are compared in detail with the experimental ones.[13] In this section, the so-called *snowplow model* is used to describe the theta pinch. It is assumed that the magnetic field does not penetrate the plasma shell of zero thickness, and the field inside the plasma shell is zero. The equation of motion is reduced to

$$\frac{\mathrm{d}}{\mathrm{d}t}\left(m\frac{\mathrm{d}r}{\mathrm{d}t}\right) = -\left(\frac{B^2}{2\mu_0} - p\right)2\pi r. \tag{14.10}$$

Here m is the mass of the plasma shell per unit length, and p is the plasma pressure. The neutral gas within the region from r to a is ionized in the plasma shell, which then gathers in the ionized particles, so that m is[14]

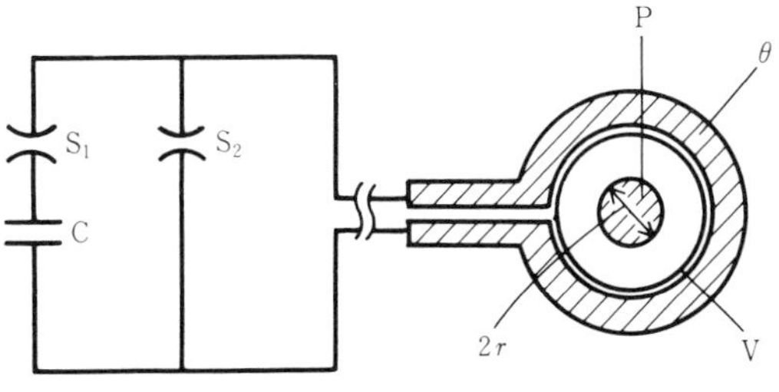

Fig. 14.4
Theta pinch coil and the circuit for capacitor discharge. C, capacitor; S_1, start switch; S_2, crowbar switch; θ, theta pinch coil; V, vacuum wall; P, plasma.

$$m = \pi(a^2 - r^2)\rho_0,$$

where ρ_0 is the mass density of neutral gas and a is the radius of the vacuum tube and the initial radius of the plasma shell. We assume that the plasma pressure changes adiabatically, i.e., that

$$p(\pi r^2)^{5/3} = p_0(\pi a^2)^{5/3}$$

and that the magnitude B of the magnetic field increases linearly in time, $B = \dot{B}t$. Then the equation of motion is

$$\frac{\mathrm{d}}{\mathrm{d}t}\left((a^2 - r^2)\frac{\mathrm{d}r}{\mathrm{d}t}\right) = -\frac{\dot{B}^2}{\mu_0\rho_0}rt^2 + \frac{2p_0}{\rho_0}\left(\frac{a}{r}\right)^{10/3} r.$$

Using the dimensionless parameters $x = r/a$, $\tau = t/t_1$, $\alpha = 2p_0 t_1{}^2/(\rho_0 a^2)$, and a characteristic time $t_1{}^4 = \mu_0\rho_0 a^2/\dot{B}^2$, we may reduce the foregoing equation to

$$\frac{\mathrm{d}}{\mathrm{d}\tau}\left((1 - x^2)\frac{\mathrm{d}x}{\mathrm{d}\tau}\right) = -x\tau^2 + \alpha x^{-7/3}. \tag{14.11}$$

The initial conditions are $x = 1$, $\mathrm{d}x/\mathrm{d}\tau = 0$ at $\tau = \alpha^{1/2}$ (The right-hand side of eq. (14.11) is positive only at $\tau > \alpha^{1/2}$).
The solution of eq. (14.11) is shown qualitatively in fig. 14.5. In the simple case $\alpha = 0$, the time τ_1 at which x becomes zero is a characteristic time for the pinch. From eq. (14.11) it follows that

$$-\left.\frac{\mathrm{d}x}{\mathrm{d}\tau}\right|_{\tau_1} = \int_0^{\tau_1} x\tau^2 \mathrm{d}\tau = \langle x \rangle \frac{\tau_1{}^3}{3}.$$

As $-(\mathrm{d}x/\mathrm{d}\tau)_{\tau 1} \approx 3/\tau_1$ and $\langle x \rangle \approx 0.5$, τ_1 is of the order of $\tau_1 \approx 2$. The characteristic time of the pinch is about $2t_1$. Assuming that there exists processes by which the kinetic energy of the particles may be thermalized,

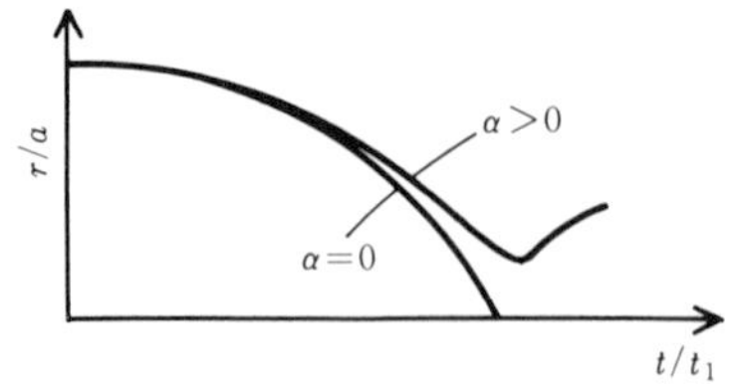

Fig. 14.5
Time variation of radius for a plasma undergoing theta pinch.

the temperature is

$$T \approx m\left(\frac{a}{t_1}\right)^2 = \frac{m\dot{B}a}{(\mu_0 \rho_0)^{1/2}}. \tag{14.12}$$

Taking a typical case, $\dot{B} = 10^6$ W/m^{2}s, $a \approx 0.05$ m, and $p_0 \approx 25_{\mathrm{mTorr}}$ in deuterium gas, then one has $\rho_0 \approx 5 \times 10^{-6}$ Kg/m^3, $2t_1 \approx 0.6$ μs, and $T \approx$ 400 eV. Since the temperature is proportional to the mass, the theta pinch is not an effective method of heating electrons.

When the magnetic field still increases after the pinched plasma reaches equilibrium, the plasma is still compressed adiabatically and is heated. However, the ends of the plasma contact the wall of the discharge tube, so that the electron temperature is further limited by conduction loss.[15] Radiation losses due to impurities also limit the attainable electron temperature.

(ii) Shock Heating Let us call the mass density ρ_{m}, the pressure p, and the flow velocity V of a fluid before a shock front, and ρ_{m}', p', and V' behind the front. Then the Rankine-Hugoniot relations are written as

$$\frac{\rho_{\mathrm{m}}'}{\rho_{\mathrm{m}}} = \frac{V}{V'} = \frac{4}{1 + 3/M^2} \tag{14.13}$$

$$\left(\frac{T'}{T}\right)^2 = 1 + \frac{5M^2}{16}\left(1 - \frac{1}{M^2}\right)\left(1 + \frac{3}{5M^2}\right), \tag{14.14}$$

as has been described in sec. 7.2. The Mach number M is given by $M = V/c_{\mathrm{s}}$, where the velocity of sound is given by $c_{\mathrm{s}}{}^2 = \gamma p/\rho_{\mathrm{m}} = \gamma T/m_{\mathrm{i}}$. When the Mach number is very large, then $\rho_{\mathrm{m}}'/\rho_{\mathrm{m}} \approx 4$, $V'/V \approx 1/4$, and $T'/T \approx 0.56\,M$. It is clear that the temperature is raised. Collisionless shocks (i.e., shocks in rarefied plasmas) cannot be treated by means of a simple fluid model. The readers may refer to refs. 16–18 for further discussion of collisionless shocks. In theta pinch experiments using high voltage discharges, collisionless shocks of large M propagate through the plasma, and the plasma ions are heated to several keV.

14.3 Heating by High-Frequency Waves

For a plasma to be heated by high-frequency waves, three processes must proceed adequately: (1) excitation of the waves, (2) propagation of the waves through the plasma, and (3) absorption and thermalization of the wave energy.[19]

Commonly used methods involve heating by compressional Alfvén waves, ion or electron cyclotron waves, and lower or upper hybrid waves. The electrostatic approximation is applicable to hybrid resonant heating but the others are generally electromagnetic waves. The characteristic frequencies are the ion cyclotron frequency $(2\pi)^{-1}|\Omega_i| \approx 75\,\mathrm{MHz}$ (for H^+, $B = 5\,\mathrm{Wb/m^2}$), the lower hybrid resonant frequency $(2\pi)^{-1}\omega_{LH} \approx 1\,\mathrm{GHz}$, and the electron cyclotron frequency $(2\pi)^{-1}\Omega_e \approx 140\,\mathrm{GHz}\,(B = 5\,\mathrm{Wb/m^2})$. Megawatt power sources for generation of the high-frequency waves have been developed up to the lower hybrid resonant frequency region. Power sources at the electron cyclotron frequency develop from 10 to 100 kW at the present time. However with the recent developement of the gyrotron, the disadvantage of the unavailability of high microwave power may disappear.

14.3a Excitation of Waves

When the electric field $\boldsymbol{E}$ of the excited wave is parallel to the confining magnetic field of the plasma, the electrons, which can move along the magnetic field, may cancel the electric field. However, if the frequency of the wave is larger than the plasma frequency the electrons cannot follow the change in the electric field, and the wave then propagates through the plasma.

When the electric field of the excited wave is perpendicular to the magnetic field, the electrons move in the direction of $\boldsymbol{E} \times \boldsymbol{B}$ (under the condition $\omega < \Omega_e$) and thus they cannot cancel the electric field. Accordingly, in this case the wave can propagate through the plasma even if the wave frequency is smaller than the plasma frequency.

Excitation consists of pumping the high-frequency electromagnetic wave into the plasma through a coupling system. If the structure of the coupling system has the same periodicity as the eigenmode wave, the wave can be excited resonantly. The efficiency of wave excitation is not high except for such *resonant excitation.* The coupling systems for ion cyclotron and Alfvén waves are explained in detail in ch. 5 of ref. 20. The eigenmodes of a cylindrical plasma can be matched by the configuration shown in fig. 14.6a. The exciting coil must be designed to satisfy the boundary condition of the eigenmode. Experiments in ion cyclotron resonant heating have already been carried out.[21,22] A diagram of the exciting coil

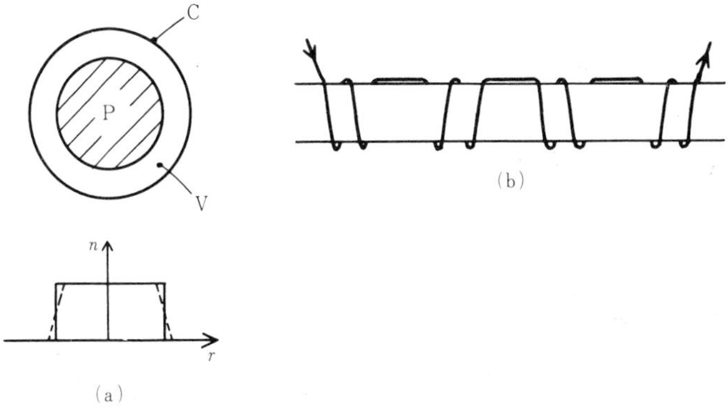

Fig. 14.6
(a) Ion cyclotron resonant heating. C, coil; P, plasma; V, vacuum region. The plot shows the plasma density n vs. r. (b) Stix coil covering 2 wavelengths of an eigenmode. The directions of adjacent coil windings are opposite to each other.

for ion cyclotron resonant heating is shown in fig. 14.6b. It is called a *Stix coil.* Heating experiments using lower hybrid, electron cyclotron, and upper hybrid waves have also been carried out[23–25] and the appropriate coupling systems extensively analyzed.

14.3b Accessibility, Propagation, and Mode Conversion

The dispersion relations of waves discussed in ch. 10 have been deduced for homogeneous and infinitely extended plasmas. However, real plasmas are inhomogeneous and finite. The density and temperature become low in the boundary region. Accordingly, it is very important to examine whether or not an externally excited wave can propagate into an inhomogeneous plasma. This is the problem of the *accessibility* of the resonant region to the wave.

The accessibility of the ion cyclotron resonant region was discussed in sec. 10.5. When the quantity $\delta \equiv \mu_0 n_{\rm i} m_{\rm i} c^2/B_0{}^2$ is less than δ_0,

$$\delta < \delta_0 \equiv \frac{k_z{}^2 c^2(\Omega_{\rm i}{}^2 - \omega^2)}{\Omega_{\rm i}{}^2 \omega^2}, \tag{14.15}$$

the wave is evanescent. But if the distance from the boundary to the region where $\delta = \delta_0$ is not very large compared to the wavelength, the damping is not serious. After the wave propagates into the high-density region, $\delta > \delta_0$, the damping ceases (fig. 14.7).

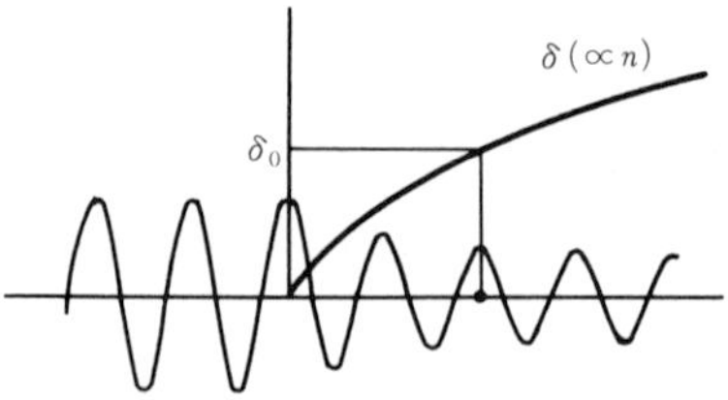

Fig. 14.7
Accessibility of the ion cyclotron resonant region to an externally excited electromagnetic wave which propagates as an ion cyclotron wave in a plasma.

The accessibility to lower hybrid waves is analyzed in refs. 26–28. The accessibility condition is

$$N_z^2 \equiv \left(\frac{ck_z}{\omega}\right)^2 > 2\left(1 + \frac{\Pi_e^2}{\Omega_e^2}\right) \tag{14.16}$$

according to Stix,[20] where k_z is the component of the propagation vector parallel to the magnetic field. The accessibility condition according to Golant[26] is more generous:

$$N_z^2 > \left(1 + \frac{\Pi_e^2}{\Omega_e^2}\right). \tag{14.17}$$

The numerical result obtained by computer methods is close to eq. (14.17).[28]

The electron cyclotron wave has accessibility along the magnetic field even if the frequency ω is lower than the plasma frequency Π_e (provided $\omega < \Omega_e \cos\theta$).

Next let us consider wave propagation and mode conversion. Generally, the density, temperature, and magnetic field are different at different points within the plasma. If these parameters change gradually and can be considered to be constant over a wavelength, the WKB approximation can be applied.

The ion cyclotron wave can propagate under the condition

$$\omega < |\Omega_i|, \tag{14.18}$$

as is clear from eqs. (10.67) and (10.68). When the magnitude of the magnetic field decreases in the direction of propagation, the ion Larmor frequency $|\Omega_i|$ approaches ω and $N = kc/\omega$ approaches ∞. The phase velocity of the wave decreases and the wave energy is thermalized by

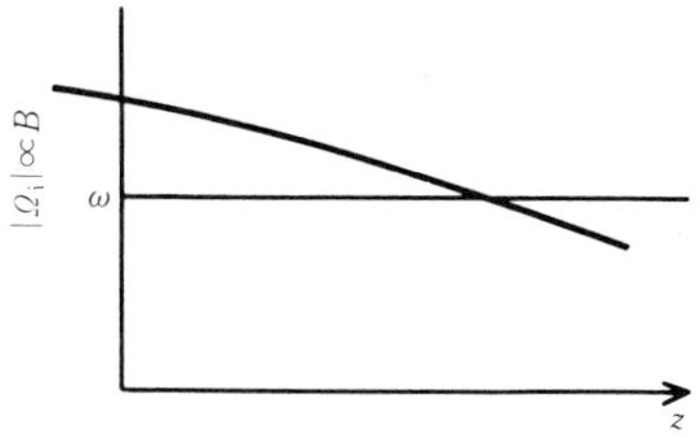

Fig. 14.8
Magnetic beach for an ion cyclotron wave.

cyclotron damping. The region where $\omega = |\Omega_i|$ is called the magnetic beach by analogy to waves breaking on a beach (fig. 14.8).

The electron cyclotron wave can propagate in the region $\omega < \Omega_e$, as is shown by eq. (10.88). When the magnitude of the magnetic field decrease in the direction of propagation, the kinetic energy of electrons perpendicular to the magnetic field increases at the region where $\omega = \Omega_e$. The magnetic moment μ_m increases and the electrons are heated. The force $-\mu_m \nabla B$ acts on the electrons and accelerates them along the lines of magnetic force.[29]

When a lower hybrid wave propagates into a plasma, mode conversion can be expected. The dispersion equation for a cold plasma (eq. (10.20)) is

$$AN^4 - BN^2 + C = 0,$$

where A, B, and C are expressed by eqs. (10.21)–(10.23). The absolute value N of $\boldsymbol{N} \equiv \boldsymbol{k}c/\omega$ is the refractive index. When the dispersion equation is solved with respect to N_y^2 (use the relation $N^2 = N_y^2 + N_z^2$), one finds

$$N_y^2 = \frac{K_\perp \tilde{K}_\perp - K_\times^2 + K_\parallel \tilde{K}_\perp}{2K_\perp} \pm \left(\left(\frac{K_\perp \tilde{K}_\perp - K_\times^2 + K_\parallel \tilde{K}_\perp}{2K_\perp} \right)^2 + \frac{K_\parallel}{K_\perp}(K_\times^2 - \tilde{K}_\perp^2) \right)^{1/2} \quad (14.19)$$

where $\tilde{K}_\perp \equiv K_\perp - N_z^2$. $K_\perp$ and $K_\times$ are defined by eqs. (10.7) and (10.8). The larger N_y^2 (taking the + sign in eq. 14.19) corresponds to the slow wave and the smaller N_y^2 corresponds to the fast wave. There are two solution $N_y = \pm(N_y^2)^{1/2}$ in each case. As was described in ch. 10, the lower hybrid wave is the extraordinary (slow) wave, so that this wave corresponds to the solution with larger N_y^2. When the refractive index becomes large

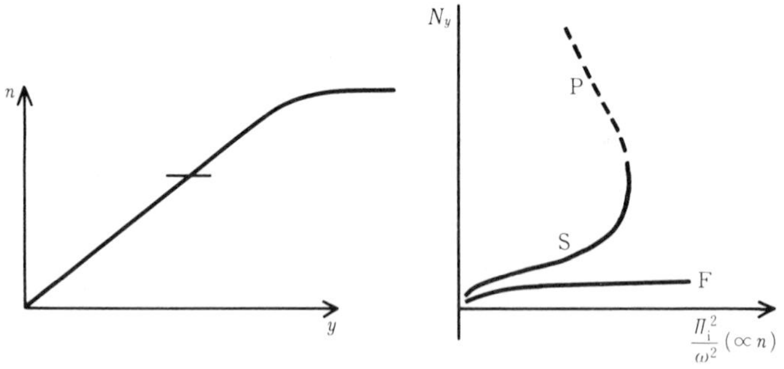

Fig. 14.9
Mode conversion of slow (S) to plasma (P) wave near the region of lower hybrid resonance in the presonce of a density gradient.

and the phase velocity becomes low, one must take account of the effect of the nonzero plasma temperature. The electrostatic approximation can be applied to the lower hybrid wave near resonance and the dispersion equation derived in ch. 11 can be used. According to ref. 27, there are two modes in a plasma of nonzero temperature corresponding to the slow wave of a cold plasma. The first mode is essentially the same as the slow cold-plasma wave and the second is a plasma wave of much lower phase velocity. These two modes are usually independent, but it can be shown that if the magnitude of the magnetic field is stronger than the critical value at which $|\Omega_i \Omega_e|^{1/2} > \omega$, then the first mode will be converted to the second (fig. 14.9). After conversion, the wave with the low phase velocity is absorbed by the plasma by means of the mechanism to be explained in the next subsection.

This mode conversion is a linear phenomenon. There is also a nonlinear case in which an excited wave is converted into two waves. A high-frequency wave may be converted to lower-frequency waves and the low-frequency wave of low phase velocity can be effectively absorbed by the plasma. The mutual interactions between waves will be briefly discussed in sec. 14.4.

14.3c Thermalization

There are two kinds of processes for the absorption of wave energy by plasmas. One is a collision process and the other includes all collisionless processes. The collision process is essentially that of Joule heating. This process becomes less effective as the temperature is increased, and the collisionless processes become dominant. The important collisionless absorption mechanisms are Landau damping and cyclotron damping. These are due to the interaction of resonant particles of velocity v_z and the wave:

$$\omega - k_z v_z - n\Omega = 0 \qquad n = 0, 1, 2, \cdots. \tag{14.20}$$

When one observes the wave in the coordinate system moving with the resonant particles, the electric field of the wave becomes static or changes only with the cyclotron frequency or its harmonics. The case $n = 0$ corresponds to Landau damping and the cases $n = 1, 2, \cdots$ correspond to cyclotron damping and cyclotron-harmonic damping.

The general analysis of thermalization due to the collisionless processes can be carried out in terms of the dielectric tensor of the plasma. The energy absorbed during unit time by a unit volume of the plasma is given by $\boldsymbol{j}\cdot\boldsymbol{E}$. When $\boldsymbol{j}$ and $\boldsymbol{E}$ are complex, the average absorbed power P during a period is $(\boldsymbol{j}\cdot\boldsymbol{E} + \boldsymbol{j}\cdot\boldsymbol{E}^*)/4$. The dielectric tensor $\boldsymbol{K}$ is as defined by eq. (10.4) or (13.9), and the current density $\boldsymbol{j}$ is given by

$$\boldsymbol{j} = \sum_j n_j q_j \langle v \rangle_j = -i\varepsilon_0\omega(\boldsymbol{K} - 1)\boldsymbol{E}. \tag{14.21}$$

Accordingly the average absorbed power P is

$$P = \frac{-i\varepsilon_0\omega}{4}\boldsymbol{E}^{*}\cdot(\boldsymbol{K} - 1)\boldsymbol{E} + \mathrm{cc}, \tag{14.22}$$

where cc represents the complex conjugate of the 1st term.

When $\boldsymbol{k}$ and ω are real, the contribution to the absorbed power comes from the imaginary part of $\boldsymbol{K}$. When the plasma is Maxwellian, the imaginary part of $\boldsymbol{K}$ comes from the imaginary part of the plasma dispersion function $\zeta_n Z(\zeta_n)$. The imaginary part of $\zeta_n Z(\zeta_n)$ in the case of real $\boldsymbol{k}$, ω is

$$\pi^{1/2}\left(\frac{\omega - k_z V + n\Omega}{|k_z|2^{1/2}v_{\mathrm{T}z}}\right)\exp\left(-\frac{(\omega - k_z V + n\Omega)^2}{k_z{}^2\, 2v_{\mathrm{T}z}{}^2}\right). \tag{14.23}$$

The imaginary part of $\zeta_n Z(\zeta_n)$ becomes maximum at $\zeta_n = 2^{-1/2}$, i.e., when

$$\omega - k_z V + n\Omega - k_z v_{\mathrm{T}z} = 0, \tag{14.24}$$

where $v_{\mathrm{T}z}{}^2 = T_z/m$.

The power absorbed from an electrostatic wave with $E_z \neq 0$, $E_x = E_y$

$= 0$, and $k_x = k_y = 0$ by a Maxwellian plasma can be estimated by eq. (14.22); we obtain

$$P = \frac{-i\varepsilon_0\omega}{4} E_z^*(K_{zz} - 1)E_z + \text{cc}$$

$$= \sum_j \left(\frac{\pi}{2}\right)^{1/2} \frac{\Pi^2}{|k_z|} \left(\frac{m}{T_z}\right)^{3/2} \left(\frac{\omega(\omega - k_z V)}{k_z^2}\right) \frac{\varepsilon_0 |E_z|^2}{2} \exp\left(-\frac{m}{2T_z}\left(\frac{\omega - k_z V}{k_z}\right)^2\right). \tag{14.25}$$

This expression is the same as eq. (10.117), the Landau damping term.

Let us express the dielectric tensor given by eqs. (13.12) and (13.13) as follows:

$$\boldsymbol{K} = \boldsymbol{K}_0 + \boldsymbol{K}_1 + \boldsymbol{K}_{-1} + \cdots \tag{14.26}$$

$$\boldsymbol{K}_0 = \left(1 - \frac{\Pi^2}{\omega^2}\right)\boldsymbol{I} - \sum_j \frac{\Pi_j^2}{\omega^2} \int \frac{\boldsymbol{T}_{j0}}{k_z v_z - \omega} k_z \frac{\partial f_{j0}}{\partial v_z} \frac{1}{n_{j0}} \mathrm{d}\boldsymbol{v}$$

$$\boldsymbol{K}_n = -\sum_j \frac{\Pi_j^2}{\omega^2} \int \frac{\boldsymbol{T}_{jn}}{(k_z v_z - \omega - n\Omega_j)} \left(-\frac{n\Omega_j}{v_\perp}\frac{\partial f_{j0}}{\partial v_\perp} + k_z \frac{\partial f_{j0}}{\partial v_z}\right) \frac{1}{n_{j0}} \mathrm{d}\boldsymbol{v} \qquad (|n| \geq 1).$$

When the wave is of low frequency, $\boldsymbol{K}_0$ is dominant. The magnetic and electric fields of the compressional Alfvén wave are

$$B_{z1}(z, t) = B_{z1} \exp i(k_y y + k_z z - \omega t)$$
$$E_x = -\omega(B_{z1}/k_y) \exp i(k_y y + k_z z - \omega t).$$

For the Alfvén wave, the $\boldsymbol{K}_0$ term is dominant and the absorbed power is

$$P = \frac{-i\varepsilon_0\omega}{4} E_x^*(K_{xx} - 1)E_x + \text{cc} \tag{14.27}$$

$$= \sum \left(\frac{\pi}{2}\right)^{1/2} \beta_\perp \left(\frac{m}{T_z}\right)^{1/2} \frac{T_\perp}{T_z} \frac{\omega(\omega - k_z V)}{|k_z|} \frac{|B_{z1}|^2}{2\mu_0} \exp\left(-\frac{m}{2T_z}\left(\frac{\omega - k_z V}{k_z}\right)^2\right).$$

This equation is the same as eq. (10.123), the transit-time damping term, where $\beta_\perp = nT_\perp/(B_0^2/2\mu_0)$. Accordingly this damping is more effective for high-beta plasmas. This heating process is called *transit-time magnetic pumping*.

Lastly, let us estimate the power absorbed from an ion cyclotron wave. In this case the $\boldsymbol{K}_1$ term is dominant and we have the relations $K_{xx} \approx K_{yy} \approx K_{\perp,\pm 1}$, $K_{xy} \approx -K_{yx} \approx \pm iK_{\perp,\pm 1}$. The absorbed power is

$$P = \sum_j \left(\frac{-i\varepsilon_0 \omega}{4} (K_{\perp,1}|E_x + iE_y|^2 + K_{\perp,-1}|E_x - iE_y|^2) + \mathrm{cc} \right)$$

$$= \sum_j \frac{1}{2}\left(\frac{\pi}{2}\right)^{1/2} \frac{\Pi^2}{|k_z|}\left(\frac{m}{T_z}\right)^{1/2}$$

$$\times \left[\left(\frac{(\omega - k_z V_z + \Omega) T_\perp - \Omega T_z}{\omega T_z} \right) \times \exp\left(-\frac{m}{2T_z}\left(\frac{\omega - k_z V + \Omega}{k_z} \right)^2 \right) \frac{\varepsilon_0 |E_x + iE_y|^2}{2} \right.$$

$$\left. + \left(\frac{(\omega - k_z V_z - \Omega) T_\perp + \Omega T_z}{\omega T_z} \right) \times \exp\left(-\frac{m}{2T_z}\left(\frac{\omega - k_z V - \Omega}{k_z} \right)^2 \right) \frac{\varepsilon_0 |E_x - iE_y|^2}{2} \right]. \tag{14.28}$$

If $V_z = 0$ and $T_\perp = 0$, this equation is the same as eq. (10.133) for cyclotron damping.[20]

Let us compare Landau damping P_L, transit-time damping P_T, and cyclotron damping P_C. If the electric field E_z of the electrostatic wave is of the same magnitude as the electric field E_x of the compressional Alfvén field, the ratio P_T/P_L is

$$\frac{P_T}{P_L} \approx \frac{\beta_\perp \omega^2}{\Pi^2} \frac{|B_{zi}|^2}{2\mu_0} \left(\frac{\varepsilon_0 |E_z|^2}{2} \right)^{-1} \approx (k_y \rho_B)^2 = b, \tag{14.29}$$

where ρ_B is the Larmor radius, $\rho_B{}^2 = T/m\Omega^2$. Usually $b = (k_y \rho_B)^2$ is less than 1.[20] The effect of collisions on heating via transit-time magnetic pumping is discussed in ref. 30. If the magnitude of the electric fields for Landau and cyclotron damping are the same, the values of exp( ) are the same, and $V = 0$, then the ratio P_C/P_L is

$$\frac{P_C}{P_L} \approx \frac{(\omega^2/k^2)}{(T_z/m)} \approx 1. \tag{14.30}$$

The cyclotron harmonic damping term is the term $\boldsymbol{K}_n$ of eq. (14.26). As is clear from the expression for $\boldsymbol{K}_n$ given in ch. 13, $\boldsymbol{K}_n$ is proportional to b^n and the contribution of cyclotron harmonic damping becomes small as n increases for $b < 1$.

The absorbed power $P_{\pm n}$ due to the nth harmonic cyclotron damping in the case of a bi-Maxwellian plasma with $b \ll 1$ is derived in a way similar to the fundamental cyclotron damping, by expansion in b of eq. (13.17b), as follows:

$$P_{\pm n} = \omega \left(\frac{\Pi}{\omega}\right)^2 \frac{n^2}{2 \cdot n!} \left(\frac{b}{2}\right)^{n-1} G_{\pm n} \frac{\varepsilon_o}{2} |E_x \pm iE_y|^2 \tag{14.31}$$

$$G_{\pm n} \equiv \mathrm{Im}\left[\zeta_0 - \left(1 - \frac{1}{\lambda_{\mathrm{T}}}\right)\zeta_{\pm n}\right] Z(\zeta_{\pm n})$$

$$= \pi^{1/2}\left[\zeta_0 - \left(1 - \frac{T_\perp}{T_z}\right)\zeta_{\pm n}\right] \exp(-\zeta_{\pm n}^2). \tag{14.32}$$

The physical concepts underlying heating processes in collisionless plasmas have been explained in section 10.7. In spite of the simple model used there, we derived the same results as those discussed in the present section.

When the plasma is Maxwellian ($T_z = T_\perp$, $V = 0$), eqs. (14.25), (14.27), (14.28), and (14.31) are all positive and the waves are damped. The plasmas are heated. However, the waves can be amplified in some cases.

When a group of particles is traveling in the same direction and faster than the wave phase velocity ($T_\perp = T_z$), that is, when

$$\frac{\omega - k_z V}{\omega} < 0,$$

the absorbed power P of eq. (14.28) becomes negative, so that cyclotron amplification of the wave may occur in this case.

Let us consider the case where $T_\perp \neq T_z$ and $V = 0$. When the anisotropy of the temperature is

$$\frac{T_{\perp \mathrm{i}}}{T_{z\mathrm{i}}} > \frac{1}{1 - \omega/|\Omega_i|} \quad \text{and} \quad \omega < |\Omega_i|$$

for an ion cyclotron wave, the absorbed power P of eq. (14.28) becomes negative and instability occurs. When

$$\frac{T_{\perp \mathrm{e}}}{T_{z\mathrm{e}}} > \frac{1}{1 - \omega/\Omega_{\mathrm{e}}} \quad \text{and} \quad \omega < \Omega_{\mathrm{e}}$$

for an electron cyclotron wave, instability occurs.

Cyclotron amplification can occur for electrons drifting through an ion cyclotron wave with $\omega = |\Omega_i|$ and $E_x = iE_y$. When we consider a group of electrons with drift velocity V such that

$$\omega - k_z V + \Omega_e \approx 0,$$

then eq. (14.28) becomes[1]

$$P_{\rm e} \simeq -\frac{1}{2}\left(\frac{\pi}{2}\right)^{1/2}\left(\frac{m_{\rm e}}{T_{z\rm e}}\right)^{1/2}\frac{\Pi_{\rm e}}{|k_z|}\frac{\Omega_{\rm e}}{|\Omega_{\rm i}|}\frac{\varepsilon_0}{2}|E_x + iE_y|^2$$

$$= -\frac{1}{2}\left(\frac{\pi}{2}\right)^{1/2}\left(\frac{m_{\rm i}}{Zm_{\rm e}}\right)\frac{\Pi_{\rm e}}{|k_z|}\left(\frac{m_{\rm e}}{T_{z\rm e}}\right)^{1/2}\frac{\varepsilon_0}{2}|E_x + iE_y|^2,$$

and $P_{\rm e}$ is negative.

*14.4 Nonlinear Phenomena and Stochastic Processes

We have discussed wave phenomena and instabilities under the assumption that perturbations are small enough so that zeroth-order quantities are not affected by the perturbations. However, the instabilities grow and the perturbations become large in real cases. The zeroth-order quantities themselves are modified and the growth rate of the instabilities decreases. The plasma shifts to a new steady state in which the magnitude of the perturbed electric fields is saturated. It is important to know the saturation levels of the electric fields in order to estimate the heating efficiency of waves or the confinement time. Another typical example of a nonlinear phenomenon is the interaction between waves. One wave may be converted to two waves by parametric nonlinear phenomena. These nonlinear phenomena are being intensively studied, theoretically as well as experimentally. The injections of ion or electron beams into a plasma may excite waves of large amplitude, which then heat the plasma. Large currents applied to a plasma may induce turbulence in which various waves are excited.[31–33] Plasma heating as a consequence of cross-field instability (sec. 12.4) has been investigated.[34] Relativistic electron beams[35,36] or high-power laser light[37,38] concentrated upon a small target, resulting in heating mechanisms involving extremely high energy densities, are under investigation. In this section such nonlinear phenomena will be briefly described.

Stochastic processes are also an essential feature of real plasmas. The density and the temperature as well as the magnetic field confining the plasma are not uniform in general. (Recall that most of the dispersion equations discussed in the previous chapters have been derived only for homogeneous plasmas.) Accordingly, one must be careful in applying these dispersion equation to real cases. The region of cyclotron resonance becomes local for mirror or toroidal fields. When particles traverse local resonance regions many times, the particles can gain or lose energy, and stochastic processes become important. When a plasma is in a turbulent state, the accelerations of particles are random and stochastic analysis is necessary.

*14.4a Quasi-Linear Theory of Change in the Distribution Function

When an electron beam of nonzero energy distribution is injected into a plasma, a hump appears in the plasma electron distribution function $f(v)$. Denote the velocity of the electron beam by $V_b > 0$. As the gradient of $f(v)$ near $v = V_b$ is $\partial f/\partial v > 0$, Landau amplification occurs and the plasma is unstable. Let us consider a one-dimensional case for simplicity. Vlasov's equation is

$$\frac{\partial f}{\partial t} + v\frac{\partial f}{\partial x} - \frac{e}{m}E\frac{\partial f}{\partial v} = 0. \tag{14.33}$$

Let f be divided into two parts, $f = f^0 + f^1$, where f^0 is a slowly changing function and f^1 is the oscillatory term. This latter term may be expressed by*

$$f^1 = \frac{1}{(2\pi)^{1/2}}\int f_k \exp i(kx - \omega_k t)\mathrm{d}k. \tag{14.34}$$

The electric field E is also expressed by a Fourier expansion:

$$E = \frac{1}{(2\pi)^{1/2}}\int E_k \exp i(kx - \omega_k t)\mathrm{d}k. \tag{14.35}$$

From eq. (14.33), the kth component f_k is given by

$$f_k = \frac{e}{m}\frac{i}{\omega_k - kv}E_k\frac{\partial f^0}{\partial v}. \tag{14.36}$$

Since E_k is expressed by

* ω_k is the function of k satisfying the dispersion relation.

$$E_k = -\frac{e}{ik\varepsilon_0}\int f_k \mathrm{d}v, \tag{14.37}$$

one can write the dispersion equation for the electron plasma wave in the case $(\omega/k)^2 \gg v^2$ as

$$\left.\begin{aligned} \omega_r^{\,2} &= \Pi_e^{\,2} + 3\langle v^2\rangle k^2 + \cdots \\ \gamma_k &\approx \frac{\pi \Pi_e^{\,3}(\pm 1)}{2k^2 n_0}\frac{\partial f^0}{\partial v}\bigg|_{v=\omega/k}, \end{aligned}\right\} \tag{14.38}$$

where the $\pm$ signs correspond to $v = \omega/k \gtrless 0$. As E is real, it follow that $E_k{}^* = E_{-k}$, $\omega_k{}^* = -\omega_{-k}$, $\omega_k = \omega_{rk} + i\gamma_k$. When we take the statistical average of $f = f^0 + f^1$ which satisfies eq. (14.33), we find

$$\frac{\partial f^0}{\partial t} = \left(\frac{e}{m}\right)^2 \frac{\partial}{\partial v}\left\langle \frac{1}{2\pi}\int E_{k'} \exp i(k'x - \omega_{k'}t)\mathrm{d}k' \times \int \frac{i\partial f^0/\partial v}{\omega_k - kv} E_k \exp i(kx - \omega_k t)\mathrm{d}k \right\rangle$$

$$= \frac{\partial}{\partial v}\left(D_v \frac{\partial f^0}{\partial v}\right) \tag{14.39}$$

$$D_v = \frac{e^2}{m^2}\int \frac{\gamma_k |E_k|^2 \exp(2\gamma_k t)}{(\omega_{rk} - kv)^2 + \gamma_k{}^2}\mathrm{d}k. \tag{14.40}$$

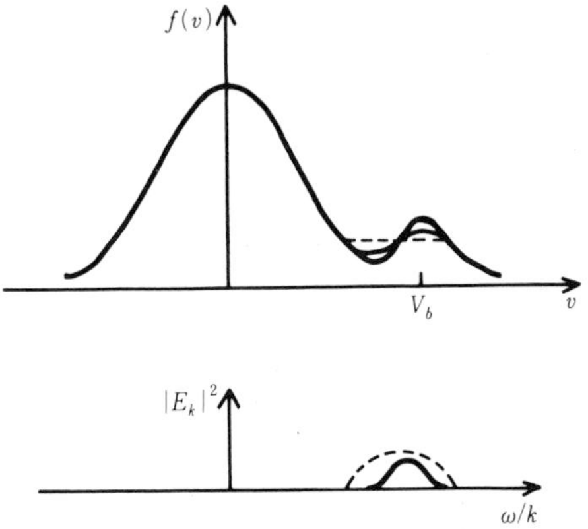

Fig. 14.10
Decay of a hump in the velocity-space distribution function, and the amplitude $|E_k|^2$ of the electric field.

When $|\gamma_k| \ll |\omega_k|$, the diffusion coefficient in velocity space is

$$D_v \approx \frac{e^2\pi}{m^2}\int |E_k|^2 \exp(2\gamma_k t)\delta(\omega_{rk} - kv)dk$$

$$\approx \frac{e^2\pi}{m^2|v|}|E_v|^2 \exp(2\gamma_k t), \tag{14.41}$$

where $\omega_{rk}/k = v$. From eqs. (14.38), (14.39), and (14.41), it follows that*

$$\frac{\partial}{\partial t}\left(f^0 - \frac{\partial}{\partial v}\left(\frac{e^2}{m^2}\frac{n_0}{\Pi_e v^3}|E_v|^2\right)\exp(2\gamma_k t)\right) = 0. \tag{14.42}$$

When $E_v = 0$ at $t = 0$ and $E(v_1, t) = 0$, $E_v(v, t)$ is given by

$$|E_v(v, t)|^2 = \left(\frac{m}{e}\right)^2 \frac{\Pi_e v^3}{n_0}\int_{v1}^{v}(f^0(v, t) - f^0(v, 0))dv$$

$$|E_v(v, \infty)|^2 \rightarrow \left(\frac{m}{e}\right)^2 \frac{\Pi_e v^3}{n_0}\int_{v1}^{v}(f^0(v, \infty) - f^0(v, 0))dv. \tag{14.43}$$

The amplitude of the electric field increases in the region $v\,\partial f^0/\partial v > 0$, so that the diffusion coefficient D_v in this velocity region becomes large. Accordingly, the region $v\,\partial f^0/\partial v > 0$ is gradually smoothed out and the distribution function becomes that of the steady state.[39] Figure 14.10 shows how the distribution function and the electric field change with time. This is one of the important results of quasi-linear theory. A generalization of quasi-linear theory is Dupree's theory.[40] Any wave in a plasma distorts the orbits of charged particles. This distortion in turn affects the behavior of other waves that are present. This interaction between waves via orbit distortion has not heretofore been taken into account. Dupree's theory is considered to be applicable to strongly turbulent plasmas.

*14.4b Wave-Wave Interaction, Manley-Rowe Relations, and Nonlinear Landau Damping

In this subsection, a general concept of nonlinear interaction between waves will be described. (The parametric instabilities related to plasma heating are discussed in ref. 41. General discussions of nonlinear phenomena are given in the texts 33 and 42–44 and in refs. 45 and 46.)

(i) Wave-Wave Interaction Let us consider, as a simple model, the interaction of three harmonic oscillators. The Hamiltonian is

* γ_k is a function of time because it depends on $\partial f^0/\partial v$, as is seen from eq. (14.38).

$$H = \sum_{i=1}^{3} \frac{p_i^{\,2}}{2} + \omega_i^{\,2}\frac{x_i^{\,2}}{2} + Vx_1x_2x_3. \tag{14.44}$$

The equations of motion of the three oscillators are

$$\begin{aligned}
\ddot{x}_1 + \omega_1^{\,2}x_1 &= -Vx_2x_3\\
\ddot{x}_2 + \omega_2^{\,2}x_2 &= -Vx_1x_3\\
\ddot{x}_3 + \omega_3^{\,2}x_3 &= -Vx_2x_1.
\end{aligned} \tag{14.45}$$

V is the coefficient of the weak-interaction term for the three oscillators. Let us examine how the energy is transferred to the oscillators x_2, x_3 when the oscillator x_1 is excited ($|x_1| \gg |x_2|, |x_3|$). The solutions of eq. (14.45) are assumed to be

$$x_j = C_j(t)\exp(i\omega_j t) + C_j^*(t)\exp(-i\omega_j t), \tag{14.46}$$

where $C_j(t)$ is a slowly changing function. The right-hand side of the equation for x_1 in eqs. (14.45) can be neglected, since $|x_2|, |x_3| \ll |x_1|$. Accordingly, C_1 is constant. The equations for x_2 and x_3 are

$$\begin{aligned}
\frac{\mathrm{d}^2C_2}{\mathrm{d}t^2} + 2i\omega_2\frac{\mathrm{d}C_2}{\mathrm{d}t} = {}&-\exp(-2i\omega_2 t)\left(\frac{\mathrm{d}^2C_2^*}{\mathrm{d}t^2} - 2i\omega_2\frac{\mathrm{d}C_2^*}{\mathrm{d}t}\right)\\
&- VC_1C_3\exp(i(\omega_1 - \omega_2 + \omega_3)t)\\
&- VC_1^*C_3\exp(i(-\omega_1 - \omega_2 + \omega_3)t)\\
&- VC_1C_3^*\exp(i(\omega_1 - \omega_2 - \omega_3)t)\\
&- VC_1^*C_3^*\exp(i(-\omega_1 - \omega_2 - \omega_3)t)
\end{aligned} \tag{14.47}$$

$$\begin{aligned}
\frac{\mathrm{d}^2C_3}{\mathrm{d}t^2} + 2i\omega_3\frac{\mathrm{d}C_3}{\mathrm{d}t} = {}&-\exp(-2i\omega_3 t)\left(\frac{\mathrm{d}^2C_3^*}{\mathrm{d}t^2} - 2i\omega_3\frac{\mathrm{d}C_3^*}{\mathrm{d}t}\right)\\
&- VC_1C_2\exp(i(\omega_1 + \omega_2 - \omega_3)t)\\
&- VC_1^*C_2\exp(i(-\omega_1 + \omega_2 - \omega_3)t)\\
&- VC_1C_2^*\exp(i(\omega_1 - \omega_2 - \omega_3)t)\\
&- VC_1^*C_2^*\exp(i(-\omega_1 - \omega_2 - \omega_3)t).
\end{aligned} \tag{14.48}$$

We have assumed that C_2, C_3 change slowly. When the averages of eqs. (14.47) and (14.48) are taken, the right-hand sides of these equations are usually set equal to zero, so that there are no interactions between the oscillators. However, where there is some resonance condition between ω_1, ω_2, and ω_3, i.e., if, for example,

$$\omega_2 = \omega_1 + \omega_3, \tag{14.49}$$

then eqs. (14.47) and (14.48) are reduced to

$$\left.\begin{aligned}\frac{d^2C_2}{dt^2} + 2i\omega_2\frac{dC_2}{dt} &= -VC_1C_3\\ \frac{d^2C_3}{dt^2} + 2i\omega_3\frac{dC_3}{dt} &= -VC_1{}^*C_2.\end{aligned}\right\} \tag{14.50}$$

C_2 and C_3 must now be taken to be proportional to $\exp i\nu t$, and ν is given by

$$\nu = \pm\frac{|C_1||V|}{2(\omega_2\omega_3)^{1/2}}. \tag{14.51}$$

When the product of ω_2 and ω_3 is negative, C_2 and C_3 grow. This condition means that $|\omega_1| > |\omega_2|, |\omega_3|$ since $\omega_2 = \omega_1 + \omega_3$. In other words, the energy can be transferred from the oscillator x_1 to the others when the frequency of the excited oscillator x_1 is larger than the others. It has been shown by Nishikawa[47] that there is a threshold value in the amplitude of the oscillator x_1 for excitation of the other waves when ω_1 and ω_3 are complex and x_1 and x_3 are damped. Waves in plasmas are characterized by ω as well as the propagation vector $\boldsymbol{k}$. In line with the foregoing discussion, we may take the resonance conditions for three plasma waves as

$$\left.\begin{aligned}\omega_0 &= \omega_1 + \omega_2\\ \boldsymbol{k}_0 &= \boldsymbol{k}_1 + \boldsymbol{k}_2.\end{aligned}\right\} \tag{14.52}$$

These equations, in fact, correspond to the conservation of energy and momentum in quantum-mechanical transition processes.

The parametric instabilities that arise when electron plasma, lower hybrid, and Bernstein waves coexist in a plasma are discussed in ref. 41.

(ii) Manley-Rowe Relations A pair of relations can be shown to hold in general for nonlinear phenomena that are independent of the details of the processes involved. These relations, first derived by J. M. Manley and H. E. Rowe,[48] relate the energy flowing in or out of several sources to the frequencies of these sources. Let us take the simple example of a nonlinear capacitor. The voltage v is a nonlinear function of the charge q,

$$v = f(q). \tag{14.53}$$

The function $f(q)$ is a single-valued function (there is no hysteresis). Let the capacitor be connected to a number of oscillators, each in series with

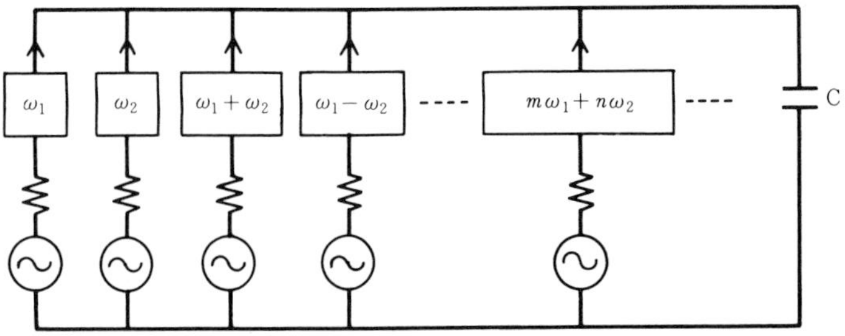

Fig. 14.11
Interaction between waves in a nonlinear system. C is nonlinear capacitor. The rectangles represent ideal filters which each pass only the waves of the frequencies shown. An oscillator is connected to each filter in series, and these oscillators are assumed to interact with each other only through the nonlinear capacitor.

a perfect filter, so that the outputs of the oscillator-filter combinations are ω_1, ω_2, $\omega_1+\omega_2$, $\omega_1-\omega_2$, ..., $m\omega_1+n\omega_2$. That is to say, ω_1 and ω_2 are the fundamental frequencies of the system, and the other terms of the sequence are the allowable higher harmonic frequencies (fig. 14.11). Using the notation

$$x = \omega_1 t$$
$$y = \omega_2 t,$$

we can express the charge q on the capacitor by

$$q = \sum_{m=-\infty}^{\infty} \sum_{n=-\infty}^{\infty} Q_{mn} \exp(-i(mx+ny)). \tag{14.54}$$

As q is real, it follows that

$$Q_{mn} = Q_{-m,-n}^*.$$

The current $I = \mathrm{d}q/\mathrm{d}t$ is

$$I = \sum_{m=-\infty}^{\infty} \sum_{n=-\infty}^{\infty} I_{mn} \exp(-i(mx+ny)), \tag{14.55}$$

where

$$I_{mn} = -i(m\omega_1 + n\omega_2)Q_{mn} \tag{14.56}$$

$$I_{mn} = I_{-m,-n}^*.$$

The voltage $v = f(q) = f(q(x, y)) \equiv F(x, y)$ is

$$v = \sum_{m=-\infty}^{\infty} \sum_{n=-\infty}^{\infty} V_{mn} \exp(-i(mx + ny)) \tag{14.57}$$

$V_{mn} = V_{-m,n}{}^*$.

The coefficients V_{mn} are given by

$$V_{mn} = \frac{1}{4\pi^2} \int_0^{2\pi} \mathrm{d}y \int_0^{2\pi} \mathrm{d}x F(x, y) \exp i(mx + ny). \tag{14.58}$$

Multiply eq. (14.58) by $-imQ_{mn}{}^*$ and take their summation with respect to m and n; then we find

$$\sum_{m=-\infty}^{\infty} \sum_{n=-\infty}^{\infty} -imQ_{mn}{}^* V_{mn}$$

$$= \frac{1}{4\pi^2} \sum_{m=-\infty}^{\infty} \sum_{n=-\infty}^{\infty} \int_0^{2\pi} \mathrm{d}y \int_0^{2\pi} \mathrm{d}x F(x, y)(-im) Q_{mn}{}^* \exp i(mx + ny). \tag{14.59}$$

The complex conjugate of the differentiation of eq. (14.54) is

$$\frac{\partial q}{\partial x} = \sum_{m=-\infty}^{\infty} \sum_{n=-\infty}^{\infty} imQ_{mn}{}^* \exp(i(mx + ny)). \tag{14.60}$$

The substitution of eq. (14.56) into the left-hand side of eq. (14.59) and the substitution of eq. (14.60) into the right-hand side of eq. (14.59) yield

$$\sum_{m=-\infty}^{\infty} \sum_{n=-\infty}^{\infty} \frac{mV_{mn}I_{mn}{}^*}{m\omega_1 + n\omega_2} = \frac{1}{2\pi} \int_0^{2\pi} \mathrm{d}y \int_0^{2\pi} \mathrm{d}x \frac{\partial q}{\partial x} F(x, y)$$

$$= \frac{1}{2\pi} \int_0^{2\pi} \mathrm{d}y \int_{q(0,y)}^{q(2\pi,y)} f(q) \mathrm{d}q. \tag{14.61}$$

A similar calculation with x and y interchanged gives

$$\sum_{m=-\infty}^{\infty} \sum_{n=-\infty}^{\infty} \frac{nV_{mn}I_{mn}{}^*}{m\omega_1 + n\omega_2} = \frac{1}{2\pi} \int_0^{2\pi} \mathrm{d}x \int_{q(x,0)}^{q(x,2\pi)} f(q) \mathrm{d}q. \tag{14.62}$$

Since $f(q)$ is a single-valued function, the right-hand sides of eqs. (14.61) and (14.62) are zero. The quantity

$$W_{mn} = V_{mn}I_{mn}{}^* + V_{mn}{}^* I_{mn} \tag{14.63}$$

expresses the average energy flow from the oscillators to the capacitor. Equations (14.61) and (14.62) are reduced to

$$\left.\begin{aligned}&\sum_{m=0}^{\infty}\sum_{n=-\infty}^{\infty}\frac{mW_{mn}}{m\omega_1+n\omega_2}=0\\&\sum_{n=0}^{\infty}\sum_{m=-\infty}^{\infty}\frac{nW_{mn}}{m\omega_1+n\omega_2}=0.\end{aligned}\right\}\tag{14.64}$$

These are the Manley-Rowe relations. When all terms except W_{10}, W_{01}, $W_{1,-1}$ are zero, the Manley-Rowe relations are

$$\left.\begin{aligned}&\frac{W_{10}}{\omega_1}+\frac{W_{11}}{\omega_1+\omega_2}=0\\&\frac{W_{01}}{\omega_2}+\frac{W_{11}}{\omega_1+\omega_2}=0.\end{aligned}\right\}\tag{14.65}$$

When the oscillator $\omega_0\equiv\omega_1+\omega_2$ is excited, the energy transferred to the oscillators ω_1 and ω_2 is

$$\left.\begin{aligned}&W_{10}=-\frac{\omega_1}{\omega_0}W_{11}\\&W_{01}=-\frac{\omega_2}{\omega_0}W_{11}.\end{aligned}\right\}\tag{14.66}$$

This relation holds in general. When a low-frequency wave is generated parametrically by the excitation of a high-frequency wave, the ratio of the energy flows is the same as the ratio of the frequencies. Accordingly, the Manley-Rowe relations indicate that the conversion efficiency from high-frequency to low-frequency waves is very low.

(iii) Nonlinear Landau Damping The ponderomotive force due to non-linear effects will be explained here. Let us consider the motion of electrons in an oscillatory electric field and a magnetic field. It is assumed that the zeroth-order electric and magnetic field are both zero. The equation of motion of an electron is

$$m\frac{\mathrm{d}\boldsymbol{v}}{\mathrm{d}t}=-e(\boldsymbol{E}+\boldsymbol{v}\times\boldsymbol{B})$$

$$\boldsymbol{E}(\boldsymbol{r},t)=\bar{\boldsymbol{E}}(\boldsymbol{r})\cos\omega t.$$

The 1st-order solution is

$$m\frac{d\boldsymbol{v}_1}{dt} = -e\bar{\boldsymbol{E}}(\boldsymbol{r}_0)\cos\omega t$$

$$\boldsymbol{v}_1 = -\frac{e}{m\omega}\bar{\boldsymbol{E}}\sin\omega t$$

$$\boldsymbol{r}_1 = \frac{e}{m\omega^2}\bar{\boldsymbol{E}}\cos\omega t.$$

In order to obtain the 2nd-order solution, we expand $\bar{\boldsymbol{E}}(\boldsymbol{r})$ around $\boldsymbol{r} = \boldsymbol{r}_0$:

$$\bar{\boldsymbol{E}}(\boldsymbol{r}) = \bar{\boldsymbol{E}}(\boldsymbol{r}_0) + (\boldsymbol{r}_1 \cdot \nabla)\bar{\boldsymbol{E}}(\boldsymbol{r}_0) + \cdots.$$

From the relation $\nabla \times \boldsymbol{E} = -\partial \boldsymbol{B}/\partial t$, $\boldsymbol{B}_1$ is given by

$$\boldsymbol{B}_1 = -\frac{1}{\omega}\nabla \times \bar{\boldsymbol{E}}\sin\omega t.$$

The 2nd-order solution is

$$m\frac{d\boldsymbol{v}_2}{dt} = -e((\boldsymbol{r}_1 \cdot \nabla)\boldsymbol{E} + \boldsymbol{v}_1 \times \boldsymbol{B}_1).$$

The average of this equation is

$$m\left\langle\frac{d\boldsymbol{v}_2}{dt}\right\rangle = -\frac{e^2}{m\omega^2}\frac{1}{2}((\bar{\boldsymbol{E}}\cdot\nabla)\bar{\boldsymbol{E}} + \bar{\boldsymbol{E}} \times (\nabla \times \bar{\boldsymbol{E}})) = -\frac{1}{4}\frac{e^2}{m\omega^2}\nabla\bar{\boldsymbol{E}}^2.$$

This is the force acting on an electron. The force acting on a unit volume of plasma is

$$\boldsymbol{F}_{\mathrm{p}} = -\frac{\Pi_{\mathrm{e}}^2}{\omega^2}\nabla\frac{\varepsilon_0\langle \boldsymbol{E}^2\rangle}{2}, \tag{14.67}$$

where $\bar{\boldsymbol{E}}^2 = 2\langle \boldsymbol{E}^2\rangle$. Equation 14.67 gives the *ponderomotive force* due to the electric and magnetic fields.[49]

Let us consider two high-frequency waves (ω_1, k_1) and (ω_2, k_2). The amplitude of the beat-frequency wave is modulated, and the envelope of the beat-frequency wave propagates with the velocity $(\omega_2 - \omega_1)/(k_2 - k_1) \approx d\omega/dk$. If this velocity is the same as the ion thermal velocity, energy is exchanged between the resonant ions and the two waves, because the ions feel the ponderomotive force of the electric field. The effective potential of the ponderomotive force has the same form as that of Landau damping. This phenomenon is called *nonlinear Landau damping*. This effect is important as a mechanism of ion heating by means of two high-frequency waves.

*14.4c Stochastic Heating

Let $f(v, t)$ be the probability that a particle has the velocity v at t. The probability that the velocity v at t changes to $v + \Delta v$ at $t + \Delta t$ is denoted by $P(v, \Delta v, t)$. Then $f(v, t + \Delta t)$ is expressed by

$$f(v, t + \Delta t) = \int_{-\infty}^{\infty} \mathrm{d}(\Delta v) P(v - \Delta v, \Delta v, t) f(v - \Delta v, t). \tag{14.68}$$

The probability $P(v, \Delta v, t)$ depends only on the state at time t and does not depend on the past history; i.e., we have assumed that the process is a Markov process. Expansion of the right-hand side of eq. (14.68) in terms of Δv gives

$$\frac{\partial}{\partial t} f(v, t) = -\frac{\partial}{\partial v}(A f(v, t)) + \frac{\partial^2}{\partial v^2}(D f(v, t)) \tag{14.69}$$

$$A(v, t) = \frac{1}{\Delta t} \int_{-\infty}^{\infty} \mathrm{d}(\Delta v) \Delta v P(v, \Delta v, t) \equiv \frac{\langle \Delta v \rangle}{\Delta t} \tag{14.70}$$

$$D(v, t) = \frac{1}{2\Delta t} \int_{-\infty}^{\infty} \mathrm{d}(\Delta v)(\Delta v)^2 P(v, \Delta v, t) \equiv \frac{\langle (\Delta v)^2 \rangle}{2\Delta t}. \tag{14.72}$$

The equation of motion of a particle of mass m and charge q in an electric field is

$$\frac{\mathrm{d}v}{\mathrm{d}t} = \frac{q}{m} E,$$

and

$$\Delta v = \frac{q}{m} \int_0^{\Delta t} E(t) \mathrm{d}t.$$

The statistical averages of Δv and $(\Delta v)^2$ are

$$\langle \Delta v \rangle = \frac{q}{m} \int_0^{\Delta t} \langle E(t) \rangle \mathrm{d}t = 0$$

$$\langle(\Delta v)^2\rangle = \frac{q^2}{m^2}\int_0^{\Delta t} \mathrm{d}t \int_0^{\Delta t} \langle E(t)E(t')\rangle \mathrm{d}t'$$
$$= \frac{q^2}{m^2}\int_0^{\Delta t} \mathrm{d}t \int_0^{\Delta t} C(t'-t)\mathrm{d}t'$$
$$= \frac{q^2}{m^2}\int_0^{\Delta t} \mathrm{d}t \int_{-t}^{-t+\Delta t} C(\tau)\mathrm{d}\tau,$$

where $C(t-t') = \langle E(t)E(t')\rangle$ is the autocorrelation function. When $C(\tau)$ is nonzero only near $\tau = 0$, i.e., when $E(t)$ is a random field, the statistical average $\langle(\Delta v)^2\rangle$ is

$$\langle(\Delta v)^2\rangle \approx \frac{q^2}{m^2}\Delta t \int_{-\infty}^{\infty} \mathrm{d}\tau C(\tau).$$

Accordingly one has

$$A \approx 0 \qquad (14.72)$$

$$D = \frac{1}{2}\frac{q^2}{m^2}\langle E^2\rangle \tau_{\mathrm{AC}}, \qquad (14.73)$$

where $\tau_{\mathrm{AC}} \equiv \int_{-\infty}^{\infty} \mathrm{d}t\, C(\tau)/\langle E^2\rangle$ is the autocorrelation time. The equation for the distribution function is reduced to

$$\frac{\partial}{\partial t} f(v,t) = D\frac{\partial^2}{\partial v^2} f(v,t), \qquad (14.74)$$

and the solution is

$$f(v,t) = \frac{1}{(4\pi Dt)^{1/2}}\exp\left(-\frac{v^2}{4Dt}\right). \qquad (14.75)$$

This is a Maxwell distribution at the temperature $T = 2mDt$, so that the temperature T increases as

$$T = \left(q^2\langle E^2\rangle \frac{\tau_{\mathrm{AC}}{}^2}{m}\right)\left(\frac{t}{\tau_{\mathrm{AC}}}\right) \qquad (14.76)$$

due to the stochastic process. Notice that this result does not depend on the form of $P(v, \Delta v, t)$ very much, if this function is even with respect to Δv. Therefore if particles are accelerated or decelerated many times during the confinement time, stochastic heating is effective. Stochastic processes are described in refs. 50–53 in detail.

*14.4d Heating Time and Confinement Time

In the previous subsection, stochastic heating of a plasma has been treated for the case without a magnetic field. In this subsection the more realistic case with a magnetic field is discussed. Since the electric field used to heat the plasma may enhance diffusion, the relation between heating and confinement times is also considered here.

Let the magnetic field $\boldsymbol{B}$ be directed along the z axis and let the random electric field $\boldsymbol{E}_\perp(\mathrm{t})$ be directed along the y axis, i.e., normal to the magnetic field. The electric field is a function of $\boldsymbol{r}$ and t, and we take $\boldsymbol{E}_\perp(\mathrm{t})$ to mean $\boldsymbol{E}_\perp(\boldsymbol{r}(t), t)$. The equation of motion of a charged particle of the mass m and charge q is

$$\left.\begin{aligned} \dot{v}_x(t) &= -\Omega v_y(t) \\ \dot{v}_y(t) &= \Omega v_x(t) + \frac{q}{m} E_\perp(t), \end{aligned}\right\} \tag{14.77}$$

where $\Omega = -qB_0/m$. It is assumed that $E_\perp(t)$ lasts from $t = 0$ to $t = \tau$ and $E_\perp(t) = 0$ outside the region $0 \le t \le \tau$ (τ can be arbitrarily large). When $E_\perp(t)$ is expressed by a Fourier integral, one has

$$\left.\begin{aligned} E_\perp(t) &= \frac{1}{2\pi}\int_{-\infty}^{\infty} F_\perp(\omega)\exp(-i\omega t)\mathrm{d}\omega \\ F_\perp(\omega) &= \int_0^\tau E_\perp(t)\exp(i\omega t)\mathrm{d}t. \end{aligned}\right\} \tag{14.78}$$

As $E_\perp(t)$ is real, then $F_\perp(-\omega) = F_\perp{}^*(\omega)$. The solution of the equation of motion is

$$v_x(t) = v_{x0}\cos\Omega t - v_{y0}\sin\Omega t + \frac{q}{m}\frac{1}{2\pi}\int_{-\infty}^{\infty}\frac{\Omega}{\Omega^2 - \omega^2}F_\perp(\omega)\exp(-i\omega t)\mathrm{d}\omega \tag{14.79}$$

$$v_y(t) = v_{x0}\sin\Omega t + v_{y0}\cos\Omega t + \frac{q}{m}\frac{1}{2\pi}\int_{-\infty}^{\infty}\frac{i\omega}{\Omega^2 - \omega^2}F_\perp(\omega)\exp(-i\omega t)\mathrm{d}\omega. \tag{14.80}$$

We have the relations

$$\frac{\Omega}{\Omega^2 - \omega^2} = \frac{1}{2}\left(\frac{1}{\omega + \Omega} - \frac{1}{\omega - \Omega}\right)$$

$$\frac{\omega}{\Omega^2 - \omega^2} = \frac{1}{2}\left(\frac{-1}{\omega + \Omega} - \frac{1}{\omega - \Omega}\right),$$

and $-i/(\omega \pm \Omega)$ is the Fourier transform of $\exp(\pm i\Omega t)$, so that the 3rd term of eqs. (14.79) and (14.80) can be expressed by the convolution of $E(t)$ and $\exp(\pm i\Omega t)$, i.e.,

$$\begin{aligned}\frac{1}{2\pi}\int_{-\infty}^{\infty}\frac{\Omega}{\Omega^2 - \omega^2}F_\perp(\omega)\exp(-i\omega t)\mathrm{d}\omega &= \int_0^{\tau}\sin\Omega(t-\tau')E_\perp(\tau')\mathrm{d}\tau' \\ &= \mathrm{Im}\int_{-\infty}^{\infty}\exp i\Omega(t-\tau')E_\perp(\tau')\mathrm{d}\tau' \\ &= \mathrm{Im}(\exp i\Omega t\, F_\perp(-\Omega)).\end{aligned}$$

Similarly, one has

$$\frac{1}{2\pi}\int_{-\infty}^{\infty}\frac{i\omega}{\Omega^2 - \omega^2}F_\perp(\omega)\exp(-i\omega t)\mathrm{d}\omega = -\mathrm{Re}(\exp i\Omega t\cdot F_\perp(-\Omega)).$$

The sum of squares of eqs. (14.79) and (14.80) is

$$\begin{aligned}&(v_x^{\,2}(t) + v_y^{\,2}(t)) \\ &= v_{x0}^{\,2} + v_{y0}^{\,2} + \left(\frac{q}{m}\right)^2(\mathrm{Im}[\exp i\Omega t\cdot F_\perp(-\Omega)])^2 \\ &+ \left(\frac{q}{m}\right)^2(\mathrm{Re}[\exp i\Omega t\cdot F_\perp(-\Omega)])^2 \\ &+ 2\frac{q}{m}v_{x0}[-\mathrm{Re}(\exp i\Omega t\cdot F_\perp(-\Omega))\sin\Omega t + \mathrm{Im}(\exp i\Omega t\cdot F_\perp(-\Omega))\cos\Omega t] \\ &- 2\frac{q}{m}v_{y0}[\mathrm{Re}(\exp i\Omega t\cdot F_\perp(-\Omega))\cos\Omega t + \mathrm{Im}(\exp i\Omega t\cdot F_\perp(-\Omega))\sin\Omega t].\end{aligned}$$

Taking the time average and the averages over the initial values of v_{x0} and v_{y0}, the increment of $v_\perp^{\,2}(t)$ is

$$\Delta\langle v_\perp^{\,2}(t)\rangle \equiv \langle v_x^{\,2} + v_y^{\,2}\rangle - \langle v_{x0}^{\,2} + v_{y0}^{\,2}\rangle = \left(\frac{q}{m}\right)^2\frac{1}{2}|F_\perp(\Omega)|^2. \tag{14.81}$$

The power spectrum $\Phi_\perp(\Omega)$ of the random electric field $E_\perp(t)$ is given by

$$\Phi_\perp(\Omega) \equiv \lim_{\tau\to\infty}\frac{1}{\tau}\left|\int_0^\tau E_\perp(t)\exp i\Omega t\,\mathrm{d}t\right|^2 = \lim_{\tau\to\infty}\frac{1}{\tau}|F_\perp(\Omega)|^2; \tag{14.82}$$

so we find

$$\frac{\mathrm{d}}{\mathrm{d}t}\left(\frac{m}{2}\langle v_\perp{}^2\rangle\right) = \frac{\mathrm{d}}{\mathrm{d}t}T_\perp = \frac{m}{4}\left(\frac{q}{m}\right)^2\Phi_\perp(\Omega), \tag{14.83}$$

where $T_\perp = m\langle v_\perp{}^2\rangle/2$.

When a parallel electric field $E_\parallel(t)$ exists, a similar relation $(\Omega\to 0)$ holds:

$$\frac{\mathrm{d}}{\mathrm{d}t}\left(\frac{m}{2}\langle v_\parallel{}^2\rangle\right) = \frac{\mathrm{d}}{\mathrm{d}t}T_\parallel = \frac{m}{2}\left(\frac{q}{m}\right)^2\Phi_\parallel(0). \tag{14.84}$$

Equations (14.83) and (14.84) give the heating rate of the plasma.

Next let us consider the diffusion of particles. The orbit of a particle is

$$\left.\begin{aligned}\ddot{x}(t) &= -\Omega\dot{y}(t)\\ \ddot{y}(t) &= \Omega\dot{x}(t) + \frac{q}{m}E_\perp(t).\end{aligned}\right\} \tag{14.85}$$

The solution of the equation of motion is

$$\begin{aligned}\Omega x(t) = \Omega x_0 &+ v_{x0}\sin\Omega t + v_{y0}\cos\Omega t\\ &+\left(\frac{q}{m}\right)\frac{1}{2\pi}\int_{-\infty}^{\infty}\frac{i\Omega^2F_\perp(\omega)}{\omega(\omega^2-\Omega^2)}\exp(-i\omega t)\mathrm{d}\omega\end{aligned} \tag{14.86}$$

$$\begin{aligned}\Omega y(t) = \Omega y_0 &- v_{x0}\cos\Omega t + v_{y0}\sin\Omega t\\ &+\left(\frac{q}{m}\right)\frac{1}{2\pi}\int_{-\infty}^{\infty}\frac{\Omega F_\perp(\omega)}{\omega^2-\Omega^2}\exp(-i\omega t)\mathrm{d}\omega.\end{aligned} \tag{14.87}$$

From the relations

$$\begin{aligned}&\frac{1}{2\pi}\int_{-\infty}^{\infty}\frac{i\Omega^2F_\perp(\omega)}{\omega(\omega^2-\Omega^2)}\exp(-i\omega t)\mathrm{d}\omega\\ &=\frac{i}{2\pi}\int_{-\infty}^{\infty}\left(\frac{1}{\omega}-\frac{1/2}{\omega+\Omega}-\frac{1/2}{\omega-\Omega}\right)F_\perp(\omega)\exp(-i\omega t)\mathrm{d}\omega\\ &= F_\perp(0) - \mathrm{Re}(\exp i\Omega t\cdot F(-\Omega))\end{aligned}$$

$$\frac{1}{2\pi}\int_{-\infty}^{\infty}\frac{\Omega F_{\perp}(\omega)}{\omega^2-\Omega^2}\exp(-i\omega t)\mathrm{d}\omega = -\mathrm{Im}(\exp i\Omega t\cdot F(-\Omega)),$$

$\langle(\Delta x)^2\rangle$ is found to be

$$\langle(\Delta x)^2\rangle = \langle(x(t)-x_0)^2\rangle = \frac{1}{\Omega^2}\left(\frac{q}{m}\right)^2 + \left(|F_{\perp}(0)|^2 + \frac{1}{2}|F_{\perp}(\Omega)|^2\right).$$

Then $\mathrm{d}\langle(\Delta x)^2\rangle/\mathrm{d}t$ is given by

$$\frac{\mathrm{d}}{\mathrm{d}t}\langle(\Delta x)^2\rangle = \frac{1}{\Omega^2}\left(\frac{q}{m}\right)^2\left(\Phi_{\perp}(0)+\frac{1}{2}\Phi_{\perp}(\Omega)\right) = \frac{1}{B_0{}^2}\left(\Phi_{\perp}(0)+\frac{1}{2}\Phi_{\perp}(\Omega)\right), \tag{14.88}$$

where $\Phi_{\perp}(\omega)$ is the power spectrum of $E_{\perp}(t)$. Similarly, $\mathrm{d}\langle(\Delta y)^2\rangle/\mathrm{d}t$ is given by

$$\frac{\mathrm{d}}{\mathrm{d}t}\langle(\Delta y)^2\rangle = \frac{1}{B_0{}^2}\frac{1}{2}\Phi_{\perp}(\Omega). \tag{14.89}$$

The 1st term of eq. (14.88) comes from the $E_y \times B_z$ drift. From eqs. (14.88) and (14.89), the diffusion coefficient is

$$D_{\perp} = \frac{\mathrm{d}}{\mathrm{d}t}\langle(\Delta x)^2 + (\Delta y)^2\rangle = \frac{1}{B_0{}^2}(\Phi_{\perp}(0)+\Phi_{\perp}(\Omega)). \tag{14.90}$$

Equations (14.83) and (14.84) give heating rate H. The heating time τ_{H} is defined by

$$H \equiv \frac{T_{\perp}+T_{\parallel}}{\tau_{\mathrm{H}}}. \tag{14.91}$$

The confinement time τ_{c} is given in terms of the diffusion coefficient $D_{\perp}$ by

$$\tau_{\mathrm{c}} = \frac{a^2}{5.8D_{\perp}}, \tag{14.92}$$

when a is the plasma radius and the radial density distribution is assumed to be given by Bessel functions. As the diffusion coefficient includes the effect of a random electric field only, the time thus defined is the upper limit of the confinement time. The ratio of τ_{H} and τ_{c} is

$$\frac{\tau_{\mathrm{H}}}{\tau_{\mathrm{c}}} = \frac{5.8\times 4(T_{\perp}+T_{\parallel})m}{a^2 B_0{}^2 q^2} \approx 23\left(\frac{\rho_B}{a}\right)^2, \tag{14.93}$$

where $\rho_B{}^2 = (T_\perp + T_\parallel)/(m\Omega^2)$. Accordingly the ratio of τ_H to τ_c is proportional to the square of the ratio of the Larmor radius to the plasma radius. The smaller the ratio, the more easily the particles are heated. This means that electrons are more likely to be heated than ions by stochastic processes.

References

1. T. H. Stix: Plasma Phys. **14**, 367(1972)

2. G. G. Kelley, O. B. Morgan, L. D. Stewart, and W. L. Stirling: Nucl. Fusion **12**, 169 (1972)
R. J. Colchin: Nucl. Fusion **11**, 329(1971)

3. D. R. Sweetman: Nucl. Fusion **13**, 157(1973)
D. R. Sweetman, A. C. Riviere, H. C. Cole, D. P. Hammond, M. F. A. Harrison, J. Hugill, F. A. Julian, F. B. Marcus, G. M. McCracken, E. Thompson, and C. J. H. Watson: CLM-R 112, Culham Laboratory, July 1971

4. L. A. Berry, C. E. Bush, J. L. Dunlap, P. H. Edmonds, T. C. Jernigan, J. F. Lyon, M. Murakami, and W. R. Wing: Plasma Phys. and Controlled Nucl. Fusion Research **1**, 113 (Conf. Proceedings, Tokyo 1974) (IAEA, Vienna, 1975)
K. Bol. J. L. Cecchi, C. C. Daughney, R. A. Ellis Jr., H. P. Eubank, H. P. Furth, R. J. Goldston, H. Hsuan, E. Mazzucat, R. R. Smith, and D. E. Stott: ibid., 77

5. J. M. Dawson, H. P. Furth, and F. H. Tenney: Phys. Rev. Lett. **26**, 1156(1971)
H. P. Furth and D. L. Jassby: ibid. **32**, 1176(1974)

6. H. P. Furth and S. Yoshikawa: Phys. of Fluids **13**, 2593(1970)

7. K. Bol, R. A. Ellis, H. Eubank, H. P. Furth, R. A. Jacobsen, and L. C. Johnson: Phys. Rev. Lett. **29**, 1495(1972)

8. F. H. Coensgen, W. F. Cummins, R. E. Ellis, and W. E. Nexsen, Jr.: Plasma Phys. and Controlled Nucl. Fusion Research **2**, 225 (Conf. Proceedings, Novosibirsk 1968) (IAEA, Vienna, 1969)

9. I. Alexeff, J. G. Harris, C. Murphy, Won Nam Kung, M. Saylors, M. Pace, H. Ikegami, S. Aihara, M. Hosokawa, H. Aikawa, and S. Okamura: Phys. Rev. Lett. **32**, 1035(1974)

10. J. M. Berger, W. A. Newcomb, J. M. Dawson, E. A. Frieman, R. M. Kulsrud, and A. Lenard: Phys. of Fluids **1**, 301 (1958)

11. A. Schlüter: Z. Naturforsch. **12a**, 822 (1957)

12. E. M. Little, W. E. Quinn, and G. A. Sawyer: Phys. of Fluids **8**, 1168(1965)
W. E. Quinn, E. M. Little, F. L. Ribe, and G. A. Sawyer: Plasma Phys. and Controlled Nucl. Fusion Research **1**, 237 (Conf. Proceedings, Culham 1965) (IAEA, Vienna, 1966)

13. K. Hain, G. Hain, K. V. Roberts, S. J. Roberts and W. Köppendörfer: Z. Naturforsch. **15a**, 1039(1960)
I. Kawakami: IPPJ-DT-23, Dec. 1970, Institute of Plasma Physics, Nagoya University (in Japanese)

14. L. A. Artsimovich: Controlled Thermonuclear Reactors, p. 124, Oliver & Boyd, London, 1964

15. T. S. Green, D. L. Fisher, A. H. Gabriel, F. J. Morgan, and A. A. Newton: Phys. of Fluids **10**, 1663(1967)

16. D. A. Tidman and N. A. Krall: Shock Waves in Collisionless Plasmas, Wiley Interscience, New York, 1970

17. R. Z. Sagdeev: Rev. of Plasma Phys. **4**, 23 (ed. by M. A. Leontovich), Consultants Bureau, New York, 1966

18. E. Hintz: Method of Experimental Physics vol. 9, pt. A, Plasma Physics, p. 213 (ed. by H. R. Griem and R. Lovberg), Academic Press, New York, 1970

19. T. Stix: Symposium on Plasma Heating and Injection, Varenna, Sept.–Oct. 1972, p. 8, CNR Laboratory of Plasma Physics and Quantum Eletronics Milan; MATT 929, Plasma Physics Laboratory, Princeton University (1972)

20. T. Stix: The Theory of Plasma Waves, McGraw-Hill, New York. 1962

21. M. Rothman, R. M. Sinclair, I. G. Brown, and J. C. Hosea: Phys. of Fluids **12**, 2211 (1969)
J. Adam, M. Chance, H. Eubank, G. Getty, E. Hinnov, W. Hooke, J. Hosea, F. Jobes, F. Perkins, R. Sinclair, J. Sperling, and H. Takahasi: Plasma Phys. and Controlled Nucl. Fusion Research **1**, 65 (Conf. Proceedings, Tokyo 1974) (IAEA, Vienna, 1975)

22. S. S. Ovchinnikov, S. S. Kalinichenko, O. M. Shvets, and V. T. Tolok: Sov. Phys. JETP **34**, 1246(1972)
V. T. Tolok, V. A. Suprunenko: Proc. of 3rd Intern. Symposium on Toroidal Plasma Confinement E 4-I, Garching, March 1973

23. P. Lallia: Proceedings of 2nd Topical Conf. on RF. Plasma Heating C3, June 1974, Texas Technical University, Lubbock

24. V. V. Alikaev, G. A. Bobrovskii, M. M. Ofitserov, V. I. Poznyak, and K. A. Razumova: JETP Lett. **15**, 27(1972)

25. H. Ikegami, H. Ikezi, T. Kawamura, H. Momota, K. Takayama, and Y. Terashima: Plasma Phys. and Controlled Nucl. Fusion Research **2**, 423 (Conf. Proceedings, Novosibirsk 1968) (IAEA, Vienna, 1969)
R. A. Dandl, J. L. Dunlap, H. O. Eason, P. H. Edmonds, A. C. England, W. J. Herrmann, and N. H. Lagar: ibid. **2**, 423

26. V. E. Golant: Sov. Phys. Tech. Phys. **16**, 1980(1972)
V. E. Golant and A. D. Piliya: Sov. Phys. Uspekhi **14**, 413(1972)

27. V. M. Glagolev: Plasma Phys. **14**, 301, 315(1972)

28. S. Puri and M. Tutter: Nucl. Fusion **14**, 93(1974)

29. T. Consoli et R. B. Hall: ibid. **3**, 237(1963)
R. Bardet, T. Consoli, and R. Geller: ibid. **5**, 7(1965)

30. E. Canobbio: Plasma Phys. and Controlled Nucl. Fusion Research **3**, 491 (Conf. Proceedings, Madison 1971) (IAEA, Vienna, 1971)

31. S. M. Hamberger and M. Frieman: Phys. Rev. Lett. **21**, 674(1968)

32. Turbulence (session E) Plasma Phys. and Controlled Nucl. Fusion Research **2**, 3–353 (Conf. Proceedings, Madison 1971) (IAEA, Vienna, 1971)

33. V. N. Tsytovich: An Introduction to the Theory of Plasma Turbulence, Pergamon, Oxford, 1972

34. S. Tanaka: 12th summer school of plasma physics for students, Aug. 1973, p. 12 (in Japanese)

35. F. Winterberg: Nucl. Fusion **12**, 353(1972)
G. Yonas, J. W. Poukey, K. R. Prestwich, J. R. Freeman, A. J. Toepfer, and M. J. Clauser: ibid. **14**, 731(1974)

36. M. V. Babykin, Ye. K. Zavoisky, A. A. Ivanov, and L. I. Rudakov: Plasma Phys. and Controlled Nucl. Fusion Research **1**, 635 (Conf. Proceedings, Madison 1971) (IAEA, Vienna, 1971)

37. F. Floux: Nucl. Fusion **11**, 635(1971)
G. Charatis et al: Plasma Phys. and Controlled Nucl. Fusion Research **2**, 317 (Conf. Proceedings, Tokyo 1974) (IAEA, Vienna, 1975)

38. Laser Interaction and Related Plasma Phenomena vol. 2, Proc. of the 2nd Workshop at Rensselaer Polytechnic Institute, Aug.–Sept. 1971 (ed. by H. J. Schwarz and H. Hora), Plenum, New York, 1972
Laser Interaction with Matter, Proceedings of Japan-US Seminar at Kyoto, Sept. 1972 (ed. by C. Yamanaka), Japan Society for the Promotion of Science

39. A. A. Vedenov, E. P. Velikhov, and R. Z. Sagdeev: Nucl. Fusion **1**, 82(1961), Nucl. Fusion, suppl., pt. 2, p. 465 (1962)
W. E. Drummond and D. Pines: ibid., suppl., pt. 3, p. 1049 (1962)

40. T. H. Dupree: Phys. of Fluids **9**, 1773(1966)

41. M. Porkolab: Symposium on Plasma Heating and Injection, Varenna, Sept.–Oct. 1972, p. 46, CNR Laboratory of Plasma Physics and Quantum Electronics, Milan; MATT 939, Plasma Physics Laboratory, Princeton University (1972)

42. R. Z. Sagdeev and A. A. Galeev: Non-linear Plasma Theory, Benjamin, New York, 1969

43. R. C. Davidson: Methods in Nonlinear Plasma Theory, Academic Press, New York, 1972

44. B. B. Kadomtsev: Plasma Turbulence, Academic Press, New York, 1965

45. A. A. Galeev and R. Z. Sagdeev: Nucl. Fusion **13**, 603(1973)

46. F. Einaudi and R. N. Sudan: Plasma Phys. **11**, 359(1969)

47. K. Nishikawa: J. Phys. Soc. Japan **24**, 916, 1152(1968)

48. J. M. Manley and H. E. Rowe: Proc. IRE **44**, 904(1956)

49. F. F. Chen: Introduction to Plasma Physics, Plenum, New York, 1974

50. S. Chadrasekhar: Rev. Modern Phys. **15**, 1(1943)

51. P. A. Sturrock: Phys. Rev. **141**, 186(1966)

52. S. Puri: Phys. of Fluids **9**, 2043(1966)

53. T. Kawamura, H. Momota, C. Namba, and Y. Terashima: Nucl. Fusion **11**, 339(1971)

15 Plasma Diagnostics

Since the parameters characterizing plasmas assume wide ranges of numerical values, the experimental techniques of plasma diagnostics include a wide variety of methods. For example, plasmas radiate electromagnetic waves of various wavelengths according to their temperatures. The wavelength may range from microwave, far infrared, visible, vacuum ultraviolet, and extreme ultraviolet up to soft X-ray and X-ray. The plasma electron temperature can be obtained by observing the radiations from the plasma as described in sec. 15.5. By observing the energy spectrum of the fast neutral particles emitted from a plasma of high ion temperature, the ion temperature can be estimated. There are other diagnostic methods, such as laser scattering. An intense laser beam is injected into a plasma, and observation of the scattered laser light supplies a great deal of information (secs. 15.4 and 15.5). It is important to understand the various elementary processes that can occur within a plasma in order to use the techniques of plasma diagnostics to best advantage. Atomic processes, bremsstrahlung, and synchrotron radiation are explained in secs. 15.1, 15.2, and 15.3, respectively.

In this chapter, an integrated description of plasma diagnostics will be given. Each diagnostic method has a specific range of applicability (see sec. 15.8); hence an adequate set of various diagnostic methods can often be used augment the reliability a particular observation. Interested readers may refer to refs. 1–3 for a more detailed introduction to plasma diagnostics. The cross sections of atomic processes are described in ref. 4.

15.1 Elementary Processes in Plasmas

The elementary processes in plasmas are:

1. elastic collisions between charged particles (Coulomb collision)
2. elastic collisions of electrons and ions with neutral particles
3. charge exchange between ions and neutral particles
4. inelastic collisions of electrons with ions or neutral particles
5. interactions between photons and particles.

Here "ions" means not only those of that make up the bulk of the plasma but also impurity ions.

Process (1) has already been described in detail in ch. 4.

Process (2) involves cross sections of the order of $\pi a_0{}^2 = 0.88\,\text{Å}^2$ (a_0 is the Bohr radius and $1\,\text{Å} = 10^{-10}\,\text{m}$). The energy loss ΔE of an electron

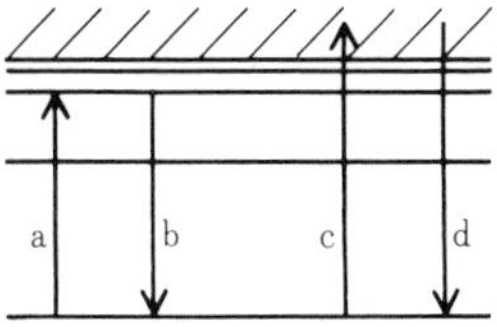

Fig. 15.1
Collisional excitation (a), de-excitation (b), ionization (c), and recombination (d).

due to a collision with a neutral particle (or an ion) is of the order of $\Delta E/E = m_e/M$, where m_e and M are the masses of the electron and the neutral particle or ion, respectively.[4] The cross sections for elastic collision of an electron with neutral argon (A), krypton (Kr), or xenon (Xe) become very small at an electron energy of about 1 eV due to the Ramsauer effect.[4a]

For process (3), when the ion is initially hot and the neutral particle is cold, charge exchange leaves the cold ions in the plasma while the fast neutrals escape across the magnetic field, so that the plasma ions are cooled. If the energy distribution of the escaping fast neutral particles is measured, the ion temperature can be estimated (see sec. 15.6).

The processes (4) include *collisional excitation* (a in fig. 15.1), in which an electron in an ion or neutral is excited to an upper energy level upon impact by a free electron. Its reverse is called *collisional de-excitation* (b). *Collisional ionization* or *collisional recombination* are inelastic collision processes in which the bound electron of an ion or neutral is excited to the free level by electron impact (c), or a free electron is de-excited to a bound level (d).

The processes (5) include *induced absorption* (a in fig. 15.2), *induced emission* (b), *spontaneous emission* (c), *photoionization* (d), and *radiative recombination* (e). Also, a free electron can be accelerated by collision with an ion, and will radiate *bremsstrahlung*. *Synchrotron radiation* arises from electrons gyrating in a magnetic field. These processes are indicated by f in fig. 15.2. Electromagnetic fields accelerate electrons, which themselve will radiate electromagnetic fields. This is the *scattering of an electromagnetic field* by electrons (g). Bremsstrahlung and the scattering of an electromagnetic field are very import processes for diagnostics, and they will be described in secs. 15.2 and 15.4. Synchrotron radiation is discussed in sec. 15.3.

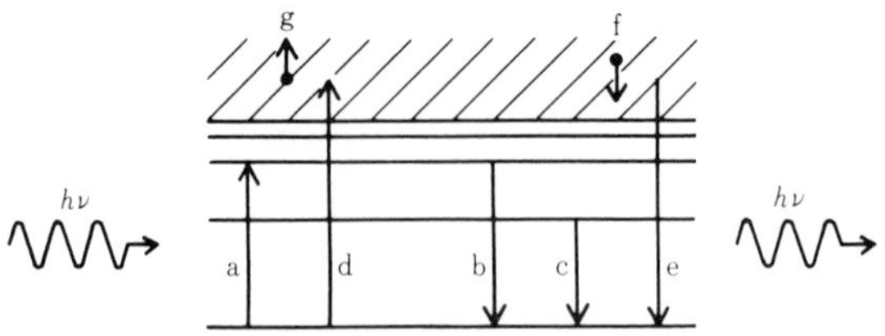

Fig. 15.2
Induced absorption (a), induced emission (b), spontaneous emission (c), photoionization (d), radiative recombination (e), bremsstrahlung and synchrotron radiation (f), and scattering of an electromagnetic field (g).

15.2 Bremsstrahlung

Plasma electrons are accelerated by collisions with ions, and bremsstrahlung is radiated as the electric fields around the electrons are spilled. Quantum mechanics must be used to calculate the bremsstrahlung accurately.[5,6] However, if the electron energy is not very large, a semiclassical analysis provides a good approximation.[7]

When a particle with the electric charge q moves with velocity $\boldsymbol{v}$ and acceleration $\dot{\boldsymbol{v}}$, the total radiated power is given by

$$\frac{\mathrm{d}U}{\mathrm{d}t} = \frac{q^2}{6\pi\varepsilon_0 c^3}\frac{\dot{v}^2 - (\boldsymbol{v}\times\dot{\boldsymbol{v}})^2/c^2}{(1 - v^2/c^2)^3}. \tag{15.1}$$

When $(v/c)^2 \ll 1$, the radiated power is

$$\frac{\mathrm{d}U}{\mathrm{d}t} = \frac{q^2\dot{v}^2}{6\pi\varepsilon_0 c^3}. \tag{15.2}$$

When a particle of charge q_e is Coulomb scattered by a particle of charge q_i, the relation between the scattering angle χ_0 and the impact parameter b (fig. 15.3) is

$$\cot\frac{\chi_0}{2} = \frac{4\pi\varepsilon_0 m b u^2}{q_e q_i}.$$

Here m is the reduced mass and u the relative velocity (see the beginning of ch. 4). The electrons that contribute the most to bremsstrahlung are those scattered at $\chi_0 \approx \pi$. The orbit of a scattered electron can be expressed in terms of a parameter α by

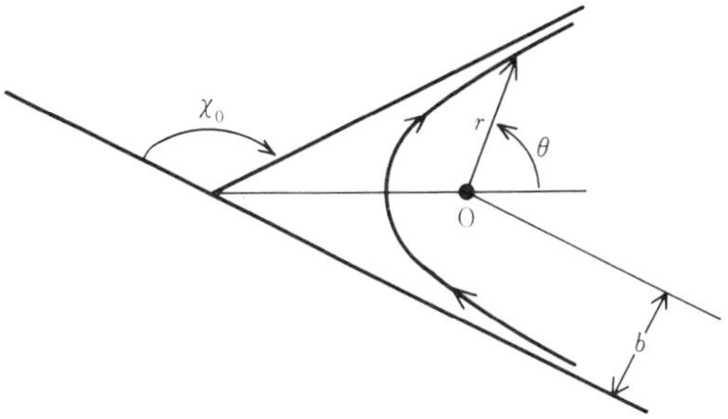

Fig. 15.3
Orbit of a charged particle scattered by a charged particle at O. The scattering angle is χ_0 and the impact parameter is b.

$$\frac{q_e{}^2 q_i{}^2}{(4\pi\varepsilon_0)^2 m^2 b^3 u^3} t = \frac{1}{2}\left(\alpha + \frac{\alpha^3}{3}\right)$$

$$\frac{|q_e q_i|}{(4\pi\varepsilon_0) m b^2 u^2} r = \frac{1+\alpha^2}{2}$$

$$\cos\theta = -\frac{1-\alpha^2}{1+\alpha^2} \qquad \sin\theta = \frac{2\alpha}{1+\alpha^2}.$$

The components of the acceleration are

$$a_\perp = \frac{\mathrm{d}^2(r\sin\theta)}{\mathrm{d}t^2} = +\left(\frac{q_e q_i}{4\pi\varepsilon_0 m b u^2}\right)^3 \frac{u^2}{b} \frac{8\alpha}{(1+\alpha^2)^3}$$

$$a_\parallel = \frac{\mathrm{d}^2(r\cos\theta)}{\mathrm{d}t^2} = -\left(\frac{q_e q_i}{4\pi\varepsilon_0 m b u^2}\right)^3 \frac{u^2}{b} \frac{4(1-\alpha^2)}{(1+\alpha^2)^3}.$$

Accordingly, the total radiated energy is given by

$$U = \frac{q_e{}^2}{6\pi\varepsilon_0 c^3}\int_{-\infty}^{\infty} (a_\perp{}^2 + a_\parallel{}^2)\mathrm{d}t = \frac{2\pi q_e{}^6 q_i{}^4}{(4\pi\varepsilon_0)^5 c^3 m^4 b^5 u^5}. \tag{15.3}$$

We introduce the dimensionless quantities

$$\tau \equiv \frac{q_e{}^2 q_i{}^2}{(4\pi\varepsilon_0)^2 m^2 b^3 u^3} t \qquad \alpha_\perp \equiv \left(\frac{4\pi\varepsilon_0 m b u^2}{q_e q_i}\right)^3 \frac{b}{u^2} a_\perp,$$

and take the Fourier expansions of $\alpha_\perp(\tau)$ and $\alpha_\parallel(\tau)$:

$$\left.\begin{aligned}\alpha_\perp(\tau) &= \int_0^\infty \phi(\gamma)\sin(\gamma\tau)\mathrm{d}\gamma \\ \alpha_\parallel(\tau) &= \int_0^\infty \psi(\gamma)\cos(\gamma\tau)\mathrm{d}\gamma\end{aligned}\right\} \tag{15.4}$$

and

$$\left.\begin{aligned}\phi(\gamma) &= \frac{1}{\pi}\int_{-\infty}^\infty \alpha_\perp(\tau)\sin(\gamma\tau)\mathrm{d}\tau \\ \psi(\gamma) &= \frac{1}{\pi}\int_{-\infty}^\infty \alpha_\parallel(\tau)\cos(\gamma\tau)\mathrm{d}\tau.\end{aligned}\right\} \tag{15.5}$$

The relation of the frequency v to γ is

$$\gamma\tau = 2\pi v t \qquad \gamma = 2\pi v \frac{m^2 b^3 u^3 (4\pi\varepsilon_0)^2}{q_e{}^2 q_i{}^2}.$$

From eqs. (15.3) and (15.4), one has

$$\int_{-\infty}^\infty (\alpha_\perp{}^2 + \alpha_\parallel{}^2)\mathrm{d}\tau = \pi \int_0^\infty (\phi^2(\gamma) + \psi^2(\gamma))\mathrm{d}\gamma = 3\pi,$$

so that $P(\gamma) = (\phi^2(\gamma) + \psi^2(\gamma))/3$ is normalized to $\int_0^\infty P(\gamma)\mathrm{d}\gamma = 1$. The energy radiated between v and $v + \mathrm{d}v$ is given by

$$UP(\gamma)\frac{\mathrm{d}\gamma}{\mathrm{d}v}\mathrm{d}v.$$

By taking the average of the radiated energy over the impact parameter b, the spectral distribution of the bremsstrahlung from an electron beam of unit intensity scattered by an ion of charge q_i is found:

$$\begin{aligned}B(v, u)\mathrm{d}v &= \mathrm{d}v \int_0^\infty UP(\gamma)\frac{\mathrm{d}\gamma}{\mathrm{d}v} 2\pi b\mathrm{d}b \\ &= \mathrm{d}v \frac{8\pi^3 q_i{}^2 q_e{}^4}{(4\pi\varepsilon_0)^3 3c^3 m^2 u^2}\int_0^\infty \frac{P(\gamma)}{\gamma}\mathrm{d}\gamma \\ &= \frac{32\pi^2}{3\cdot 3^{1/2}}\frac{q_i{}^2 q_e{}^4}{(4\pi\varepsilon_0)^3 c^3 m^2 u^2}\mathrm{d}v,\end{aligned} \tag{15.6}$$

where $\int_0^\infty P(\gamma)\gamma^{-1}\,d\gamma = 4/(\pi\cdot 3^{1/2})$. (This latter integral will be derived at the end of this section.) The spectral distribution (15.6) is valid only for

$$h\nu \leq \frac{1}{2}mu^2,$$

due to the quantum nature of light. When $h\nu > mu^2/2$, then $B(\nu, u) = 0$. For the semiclassical treatment to be valid, the minimum distance between the electron and the ion,

$$r_{\min} = \frac{1}{2}\frac{4\pi\varepsilon_0 mb^2u^2}{|q_e q_i|},$$

must be larger than the de Broglie wavelength $\lambda = h/mu$ corresponding to the electron relative velocity $u = 2|q_e q_i|/(4\pi\varepsilon_0 mbu)$, i.e., the semiclassical treatment is applicable only for

$$\frac{h}{mbu} \ll 1. \tag{15.7}$$

Since $\chi_0 \approx \pi$, then $\cot(\chi_0/2) = 4\pi\varepsilon_0 mbu^2/|q_e q_i| \ll 1$. From these conditions, the condition for validity of the semiclassical treatment is

$$\frac{u}{c} \ll \frac{|q_e q_i|2\pi}{4\pi\varepsilon_0 ch} = \frac{Ze^2 2\pi}{4\pi\varepsilon_0 ch} = \frac{Z}{137}. \tag{15.8}$$

Here we have set $q_e = -e$ and $q_i = Ze$.

When the ion number density is n_i and the velocity distribution function for the electrons is

$$n_e f(u, T_e) = n_e\left(\frac{m}{2\pi T_e}\right)^{3/2}\exp\left(-\frac{mu^2}{2T_e}\right)4\pi u^2,$$

the power $w(\nu)d\nu$ radiated per unit volume of plasma between the frequencies ν and $\nu + d\nu$ is

$$\begin{aligned} w(\nu)d\nu &= \left(\int_{(2h\nu/m)^{1/2}}^{\infty} B(\nu, u)n_i n_e u f(u, T_e)du\right)d\nu \\ &= \frac{32\pi}{3}\frac{Z^2e^6}{(4\pi\varepsilon_0)^3c^3m}\left(\frac{2\pi}{3mT_e}\right)^{1/2} n_e n_i \exp\left(-\frac{h\nu}{T_e}\right)d\nu \end{aligned}$$

$$= 6.3 \times 10^{-53} Z^2 \left(\frac{e}{T_e}\right)^{1/2} n_e n_i \exp\left(\frac{-h\nu}{T_e}\right) d\nu \qquad (\mathrm{W/m^3}). \qquad (15.9)$$

To derive the more rigorous quantum mechanical formula, eq. (15.9) must be multiplied by the so-called Gaunt factor[6] $g(\nu, T_e)$. However, $g(\nu, T_e)$ is of the order of 1 for $u/c \lesssim Z/137$. The total power w radiated per unit volume is

$$w \equiv \int_0^\infty w(\nu)d\nu = \frac{32\pi e^6 Z^2 n_e n_i}{3(4\pi\varepsilon_0)^3 c^3 mh}\left(\frac{2\pi T_e}{3m}\right)^{1/2}$$

$$\approx 1.5 \times 10^{-38} Z^2 n_e n_i \left(\frac{T_e}{e}\right)^{1/2} \qquad (\mathrm{W/m^3}) \qquad (15.10)$$

(n is in $\mathrm{m^{-3}}$, T_e/e is in eV). The bremsstrahlung spectral intensity is large in the range $h\nu \lesssim T_e$ and is proportional to $n_e{}^2$. The wavelength λ corresponding to $T_e = h\nu$ is

$$\lambda = \frac{12397}{(T_e/e)} \qquad (\text{Å}). \qquad (15.11)$$

As can be seen from eq. (15.10), the radiation intensity is proportional to Z^2. A plasma containing only 2% of stripped oxygen ($Z = 8$) radiates bremsstrahlung at 2.3 times the rate for a plasma without this impurity.

Proof of $\int_0^\infty P(\gamma)\gamma^{-1}d\gamma = 4(\pi \cdot 3^{1/2})^{-1}$: Using the integral variable α, we reduce the partial integrations of $\phi(\gamma)$ and $\psi(\gamma)$ to

$$\phi(\gamma) = \frac{2}{\pi}\gamma \int_0^\infty \cos\left(\frac{\gamma}{2}\left(\alpha + \frac{\alpha^3}{3}\right)\right) d\alpha$$

$$\psi(\gamma) = \frac{2}{\pi}\gamma \int_0^\infty \sin\left(\frac{\gamma}{2}\left(\alpha + \frac{\alpha^3}{3}\right)\right) \alpha d\alpha.$$

With the notation

$$\frac{\gamma}{6}\alpha^3 = s^3 \qquad \frac{\gamma}{2}\alpha = xs \qquad x = \left(\frac{3\gamma^2}{4}\right)^{1/3},$$

the integrals $\phi(\gamma)$ and $\psi(\gamma)$ may be written as

$$\phi(\gamma) = \frac{4}{\pi}x \int_0^\infty \cos(s^3 + xs)ds \equiv \frac{4x}{\pi}A(x)$$

$$\psi(\gamma) = \frac{4}{\pi}(3x)^{1/2} \int_0^\infty \sin(s^3 + xs)\mathrm{d}s = -\frac{4}{\pi}(3x)^{1/2}\frac{\mathrm{d}}{\mathrm{d}x}A(x),$$

where

$$A(x) = \int_0^\infty \cos(s^3 + xs)\mathrm{d}s.$$

It follows that

$$\int_0^\infty \frac{P(\gamma)}{\gamma}\mathrm{d}\gamma = \frac{1}{3}\int_0^\infty \frac{\phi^2(\gamma) + \psi^2(\gamma)}{\gamma}\mathrm{d}\gamma$$
$$= \frac{8}{\pi^2}\int_0^\infty \left(xA^2(x) + 3\left(\frac{\mathrm{d}A(x)}{\mathrm{d}x}\right)^2\right)\mathrm{d}x.$$

Using the relations

$$\int_0^\infty (A'(x))^2\,\mathrm{d}x = A(0)A'(0) - \int_0^\infty A(x)A''(x)\mathrm{d}x$$

$$\int_0^\infty (xA^2 - 3AA'')\mathrm{d}x = \int_0^\infty\int_0^\infty A(x)\cos(s^3 + xs)(3s^2 + x)\mathrm{d}s\mathrm{d}x = 0,$$

we find

$$\int_0^\infty \frac{\cos y}{y^{2/3}}\mathrm{d}y = \frac{\pi}{\Gamma(2/3)} \qquad \int_0^\infty \frac{\sin y}{y^{1/3}}\mathrm{d}y = \frac{\pi}{\Gamma(1/3)}.$$

Then the required formula is obtained:

$$\int_0^\infty \frac{P(\gamma)}{\gamma}\mathrm{d}\gamma = \frac{8}{3\Gamma(1/3)\Gamma(2/3)} = \frac{8}{3\pi}\sin\frac{\pi}{3} = \frac{4}{\pi 3^{1/2}}.$$

*15.3 Synchrotron Radiation

Plasma electrons in a magnetic field always gyrate, so that they always emit synchrotron radiation. Let us calculate the radiation from a single electron in Larmor motion.

The total radiated power is given by eq. (15.1). The angular frequency ω_0 is

$$\omega_0 = \frac{eB(1 - v^2/c^2)^{1/2}}{m} = \Omega_e\left(1 - \frac{v^2}{c^2}\right)^{1/2},$$

where Ω_e is the Larmor frequency and m the rest mass. When the acceleration $\dot{v}_\perp = \omega_0 v_\perp$ is substituted into eq. (15.1), the radiated power is

$$\frac{\mathrm{d}U}{\mathrm{d}t} = \frac{e^4 {v_\perp}^2 B^2}{6\pi\varepsilon_0 c^3 m^2 (1 - v^2/c^2)}. \qquad (15.12)$$

*15.3a Spectral Distribution of Synchrotron Radiation[8]

Let us consider the spectral distribution of synchrotron radiation. The electron moves in a circle of Larmor radius ρ_B in the x, y plane, at an angular frequency ω_0. The magnetic field $\boldsymbol{B}$ is in the z direction. Choose the y axis so that the observation point P is in the y, z plane. The vector from the guiding center O of the Larmor orbit to the observation point P is denoted by $\boldsymbol{R}$; the unit vector along $\boldsymbol{R}$ is denoted by $\boldsymbol{s}$. The vector from O to the electron is $\boldsymbol{r}$. Then the distance R' between the electron and P is

$$R' = R - (\boldsymbol{s} \cdot \boldsymbol{r}).$$

When $\phi = \omega_0 t$ and θ are as shown in fig. 15.4, the scalar product $\boldsymbol{s} \cdot \boldsymbol{r}$ is

$$\boldsymbol{s} \cdot \boldsymbol{r} = \rho_B \sin\theta \sin\varphi.$$

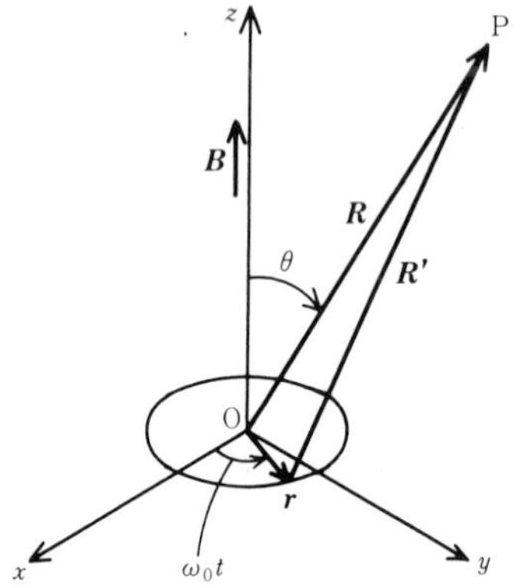

Fig. 15.4
An electron gyrating around the point O. The magnetic field is in the z direction and the observation point P is in the y, z plane.

The vector potential $\boldsymbol{A}$ at P and t is given by

$$\boldsymbol{A}(\boldsymbol{R}, t) = \frac{\mu_0}{4\pi}\int \frac{\boldsymbol{j}(\boldsymbol{r}, t - R'/c)}{R'}\mathrm{d}\boldsymbol{r}. \tag{15.13}$$

The current density $\boldsymbol{j}(t)$ at the Larmor orbit is a periodic function of period $\tau = 2\pi/\omega_0 = 2\pi\rho_B/v$:

$$\boldsymbol{I}(t) = -\hat{\boldsymbol{\phi}} e \sum_{g=0}^{\infty} \delta(t - q\tau), \tag{15.14}$$

where $\hat{\boldsymbol{\phi}}$ is the unit vector for the electron velocity. The average current is

$$\overline{I(t)} = \frac{-e}{\tau}\int_0^{\tau} \mathrm{d}t \sum_g \delta(t - g\tau) = \frac{-e}{\tau}.$$

The periodic function can be expressed as a Fourier series

$$\sum_{g=0}^{\infty} \delta(t - g\tau) = \frac{1}{\tau}\left(1 + 2\mathrm{Re}\sum_{g=1}^{\infty} \exp(ig\omega_0 t)\right), \tag{15.15}$$

where Re represents the real part. Substitution of eqs. (15.14) and (15.15) into eq. (15.13) yields the gth harmonic component $\boldsymbol{A}_g$:

$$\begin{aligned}\boldsymbol{A}_g &= \frac{-\mu_0 e}{2\pi\tau}\oint \mathrm{d}\boldsymbol{l}\frac{\exp ig\omega_0(t - R/c - (\boldsymbol{s}\cdot\boldsymbol{r})/c)}{R}\\ &= \frac{-\mu_0 e\exp(-ig\omega_0 R/c)}{2\pi\tau R}\oint \mathrm{d}\boldsymbol{l}\exp\left(ig\omega_0\left(t - \frac{(\boldsymbol{s}\cdot\boldsymbol{r})}{c}\right)\right).\end{aligned}$$

The x component A_{gx} of $\boldsymbol{A}_g$ is

$$A_{gx} = \frac{\mu_0 ev\exp(-ig\omega_0 R/c)}{4\pi^2 R}\int_0^{2\pi} \mathrm{d}\varphi\sin\varphi\exp\left(ig\left(\varphi - \frac{v}{c}\sin\theta\sin\varphi\right)\right).$$

As the Bessel function $J_g(z)$ is

$$J_g(z) = \frac{i^{-g}}{2\pi}\int_0^{2\pi} \mathrm{d}\psi\exp(i(z\cos\psi - g\psi)),$$

A_{gx} is

$$A_{gx} = \frac{i\mu_0 ev\exp(-ig\omega_0 R/c)}{2\pi R}J_g'\left(\frac{gv}{c}\sin\theta\right),$$

where $J_g'(z) \equiv \mathrm{d}J_g(z)/\mathrm{d}z$. The y component A_{gy} is

$$A_{gy} = \frac{-\mu_0 ev \exp(-ig\omega_0 R/c)}{4\pi^2 R} \int_0^{2\pi} \mathrm{d}\varphi \cos\varphi \exp\left(ig\left(\varphi - \frac{v}{c}\sin\theta\sin\varphi\right)\right)$$

$$= \frac{-\mu_0 ec \exp(-ig\omega_0 R/c)}{2\pi R \sin\theta} J_g\left(\frac{gv}{c}\sin\theta\right)$$

(the formula $J_g'(z) = (g/z) J_g(z) - J_{g+1}(z)$ has been used here). The z component is $A_{gz} = 0$.

The radiated power is given by the Poynting vector. At a distance R from the electron, the power radiated into a small solid angle $\mathrm{d}\Omega$ in the θ direction is

$$\dot{U}_g(\theta)\mathrm{d}\Omega = |\boldsymbol{E}_g \times \boldsymbol{H}_g^*| R^2 \mathrm{d}\Omega.$$

The notation * indicates the complex conjugate. Using the relations $\boldsymbol{B} = \mu_0 \boldsymbol{H}$, $\boldsymbol{B} = \nabla \times \boldsymbol{A}$, and $\boldsymbol{k} \equiv (g\omega_0/c)\boldsymbol{s}$, we may write

$$\dot{U}_g(\theta) = c|\boldsymbol{k} \times \boldsymbol{A}_g|^2 \frac{R^2}{\mu_0}$$

$$= ck^2(|A_{gx}|^2 + |A_{gy}|^2 \cos^2\theta)\frac{R^2}{\mu_0}$$

and

$$\dot{U}_g(\theta, v_\perp, \Omega_e) = \frac{g^2 e^2 \Omega_e^2 (1 - v_\perp^2/c^2)}{8\pi^2 \varepsilon_0 c} \left[\frac{v_\perp^2}{c^2}\left(J_g'\left(\frac{gv_\perp}{c}\sin\theta\right)\right)^2 + \cot^2\theta\left(J_g\left(\frac{gv_\perp}{c}\sin\theta\right)\right)^2\right]. \tag{15.16}$$

This last formula expresses the power radiated in the gth harmonic of the synchrotron radiation from a single electron for which the velocity $v_\parallel$ parallel to the magnetic field is zero. The synchrotron radiation from a group of electrons in a plasma may be considered as the sum of the radiations from the individual electrons, neglecting mutual correlations.[8] When the average of $\dot{U}$ is taken over the velocity distribution function, an expression for the overall synchrotron radiation from the plasma is deduced.

With the introduction of the momentum

$$\boldsymbol{p} = \frac{m\boldsymbol{v}}{(1 - (v/c)^2)^{1/2}}, \tag{15.17}$$

the relativistic distribution function $f(\boldsymbol{p})$ is given by

$$f(\boldsymbol{p}) = C_{\mathrm{p}} \exp\left(-\frac{U}{T}\right), \tag{15.18}$$

where $U = (p^2c^2 + m^2c^4)^{1/2} = mc^2(1 - (v/c)^2)^{-1/2}$. The normalizing factor C_{p} is

$$C_{\mathrm{p}} = \left(4\pi m^2 c T K_2\left(\frac{mc^2}{T}\right)\right)^{-1},$$

where $K_2(z)$ is the 2nd-order modified Bessel function. The power radiated by the gth harmonic (of frequency ω) into a unit solid angle in the θ direction, per unit volume of plasma, is given by

$$w_{\mathrm{c}}(\omega, \theta, \boldsymbol{r}, T, g) = \int_{\omega = \mathrm{const.}} \mathrm{d}\boldsymbol{v}\, \dot{U}_g(\theta, \boldsymbol{v}, \Omega_{\mathrm{e}}) f(\boldsymbol{r}, \boldsymbol{v}, T). \tag{15.19}$$

Here, as indicated, the integration must be carried out under the condition $\omega = \mathrm{const.}$ Equation (15.16) is the case for $v_{\parallel} = 0$. If $v_{\parallel} \neq 0$, then $\dot{U}_g(\theta, v, \Omega_{\mathrm{e}})$ is given by[8]

$$\begin{aligned} &\dot{U}_g(\theta, \boldsymbol{v}, \Omega_{\mathrm{e}}) \\ &= \frac{g^2 e^2 \Omega_{\mathrm{e}}{}^2}{8\pi\varepsilon_0 c} \frac{1 - v^2/c^2}{1 - (v_{\parallel}/c)\cos\theta} \\ &\quad \times \left[\left(\frac{v_{\perp}}{c}\right)^2 \left(J_g'\left(g\frac{v_{\perp}}{c}\sin\theta\right)\right)^2 + \frac{(\cos\theta - v_{\parallel}/c)^2}{\sin^2\theta} J_g{}^2\left(g\frac{v_{\perp}}{c}\sin\theta\right)\right], \end{aligned}$$

where $\omega = \dfrac{g\omega_0}{1 - (v_{\parallel}/c)^2\cos\theta} = \dfrac{g\Omega_{\mathrm{e}}(1 - v^2/c^2)}{1 - (v_{\parallel}/c)^2\cos\theta}$.

*15.3b Absorption and Emission of Radiation

The power radiated per unit area of surface of a black body is given by the Stefan-Boltzmann law,

$$S_{\mathrm{B}} = a(T')^4, \tag{15.20}$$

where $a = 5.67 \times 10^{-8}$ $Wm^{-2}(°K)^{-4}$, and T' is the absolute temperature. As will be made clear later, a plasma can be considered to be a black body only in the low-frequency region, and not in the high-frequency region (the region of high photon energy). The *spectral emissivity* $w_c(\omega, \Omega, \boldsymbol{r}, T)$ of a plasma is defined as the radiated power per unit frequency near ω emitted into a unit solid angle per unit volume of the plasma. The *spectral absorption coefficient* $\alpha(\omega, \Omega, \boldsymbol{r}, T)$ is the absorption coefficient of the plasma at $\boldsymbol{r}$ for radiation of frequency ω propagating in the direction of the solid angle Ω. The *spectral intensity of radiation of a black body* $S_B(\omega, \Omega, T)$ is the power emitted per unit area of the surface into a unit solid angle. A Kirchhoff's law relation holds between ω_c, α, and S_B when the plasma is in thermal equilibrium:

$$\frac{w_c(\omega, \Omega, \boldsymbol{r}, T)}{\alpha(\omega, \Omega, \boldsymbol{r}, T)} = S_B(\omega, \Omega, T) \tag{15.21}$$

$$S_B(\omega, \Omega, T)\,d\omega = \frac{2h}{c^2}\frac{v^3 dv}{\exp(hv/T) - 1}, \tag{15.22}$$

where $\omega = 2\pi v$. Let us assume that the radiation propagates as shown in fig. 15.5. When the radiation travels a distance dl in the direction of the solid angle Ω, the spectral intensity of radiation $S(\omega, \Omega, \boldsymbol{r}, T)$ satisfies

$$\frac{dS(\omega, \Omega, \boldsymbol{r}, T)}{dl} = -\alpha(\omega, \Omega, \boldsymbol{r}, T)S + w_c(\omega, \Omega, \boldsymbol{r}, T)$$

$$= -\alpha(S - S_B). \tag{15.23}$$

When S is zero at the point O and the temperature is uniform, the spectral intensity of radiation becomes

$$S(\omega, \Omega, P) = S_B(\omega, \Omega)\left(1 - \exp\left(-\int_0^P \alpha\, dl\right)\right). \tag{15.24}$$

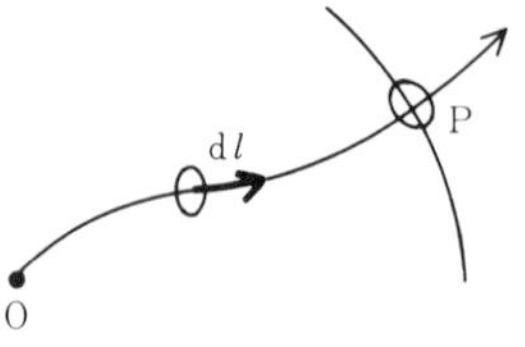

Fig. 15.5
Radiation path from a point in a plasma to a point P of the plasma surface.

Accordingly S is

$$S = S_B \quad \text{when} \quad \int \alpha \mathrm{d}l \gg 1$$

$$S = S_B \int \alpha \mathrm{d}l = \int w_c \mathrm{d}l \quad \text{when} \quad \int \alpha \mathrm{d}l \ll 1.$$

When the absorption is large, i.e., for $\int \alpha \mathrm{d}l \gg 1$, the plasma is called *optically thick* and the radiation becomes black body radiation. When the absorption is small, i.e., for $\int \alpha \mathrm{d}l \ll 1$, the plasma is called *optically thin*. In this latter case, integration of the spectral emissivity along the path gives the spectral radiation power from the plasma surface. This value thus obtained is always smaller than that for black body radiation.

*15.3c Loss due to Synchrotron Radiation

When a plasma is optically thin, the loss via synchrotron radiation can be estimated by means of

$$P_s^*(\omega, T) = \int_V \int_\Omega w_c(\omega, \Omega, \boldsymbol{r}, T)\mathrm{d}\Omega \mathrm{d}\boldsymbol{r} = \int_V w_c(\omega, \boldsymbol{r}, T)\mathrm{d}\boldsymbol{r}. \tag{15.25}$$

Usually plasmas are not optically thin and the radiation loss is given by

$$P_s(\omega, T) = \int_A \int_\Omega S(\omega, \Omega, \boldsymbol{r}, T)\mathrm{d}\Omega \mathrm{d}A = \int_A S(\omega, \boldsymbol{r}, T)\mathrm{d}A. \tag{15.26}$$

If the value of $\int \alpha \mathrm{d}l$,

$$\int \alpha \mathrm{d}l = \alpha L = \frac{w_c L}{S_B},$$

is known, $P_s(\omega, t)$ can be obtained from eq. (15.24). The normalized value of αL in terms of the dimensionless quantity Λ

$$\Lambda = \frac{\Pi_e^2 L}{\Omega_e c}, \tag{15.27}$$

i.e.,

$$\frac{\alpha L}{\Lambda} \equiv \frac{\Omega_e c}{\Pi_e^2} \frac{\int w_c d\Omega}{\int S_B \cos\varphi d\Omega},$$

is shown in fig. 15.6[8] for $v_{\parallel} \ll v_{\perp}$. As ω is of the order of $g\Omega$ ($g = 1, 2, 3, \cdots$), the photon energy is $h\nu \ll T_e$, so that the Rayleigh-Jeans approximation $S_B(\omega, \Omega, T) = T\omega^2/(4\pi^3c^2)$ can be used. In this case $\int S_B \cos\phi d\Omega$ is

$$\int S_B \cos\varphi d\Omega = \frac{T\omega^2}{4\pi^3c^2} \int \cos\varphi d\Omega = \frac{T\omega^2}{2\pi^2c^2}. \tag{15.28}$$

Next let us estimate the total loss of synchrotron radiation over the whole frequency spectrum. When the plasma is optically thin, the total loss is equal to the integration of $P_s^*(\omega, T)$ (eq. (15.25)) over ω. For this optically thin case, the total loss can be obtained by another simple method. The total radiation loss of a single electron is given by eq. (15.12), upon subsitution of eq. (15.17), and is

$$\frac{dU}{dt} = \frac{e^4B^2p_{\perp}^2}{6\pi\varepsilon_0 c^3 m^4}. \tag{15.29}$$

The integration of $(dU/dt)f(\boldsymbol{p})$ ($f(\boldsymbol{p})$ being the relativistic Maxwell distribution function) gives the total power radiated per unit volume of plasma:

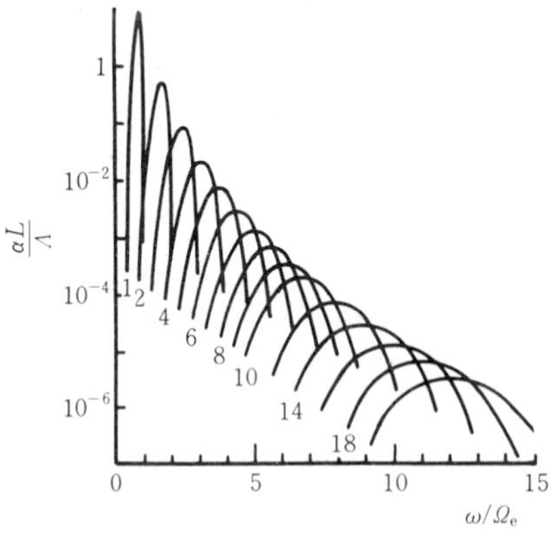

Fig. 15.6
The dependence of $\alpha L/\Lambda$ on ω/Ω_e for synchrotron radiation from a plasma of $T_e = 50$ keV.[8]

$$w_{\mathrm{c}}(T) = \frac{e^4 B^2 n_{\mathrm{e}}}{3\pi\varepsilon_0 m^2 c}\left(\frac{T_{\mathrm{e}}}{mc^2}\right)\frac{K_3(mc^2/T_{\mathrm{e}})}{K_2(mc^2/T_{\mathrm{e}})} = \frac{e^2 \Omega_{\mathrm{e}}{}^2 n_{\mathrm{e}}}{3\pi\varepsilon_0 c}\left(\frac{T_{\mathrm{e}}}{mc^2}\right)\left(1 + \frac{5}{2}\frac{T_{\mathrm{e}}}{mc^2} + \cdots\right)$$

$$= 6.2 \times 10^{-14} B^2 n_{\mathrm{e}}\left(\frac{T_{\mathrm{e}}}{e}\right)\left(1 + \left(\frac{T_{\mathrm{e}}}{e}\right)\frac{1}{2.04 \times 10^5} + \cdots\right) \quad (\mathrm{W/m^3}), \tag{15.30}$$

where the units of B, n_{e} and T_{e}/e are Wb/m^2, m^{-3}, and eV, respectively.

Usually the plasma is not optically thin in the low-frequency region of synchrotron radiation. In this case, the integration of $P_{\mathrm{s}}(\omega, T)$ over ω gives the total radiation loss

$$\int_A S(T)\mathrm{d}A, \tag{15.31}$$

where dA is an element of surface area. The ratio

$$K_L \equiv \frac{\displaystyle\int_A S(T)\mathrm{d}A}{\displaystyle\int_V w_{\mathrm{c}}(T)\mathrm{d}\boldsymbol{r}} \tag{15.32}$$

is a function of the electron temperature T_{e} and the parameter Λ given in eq. (15.27) (which is a parameter of plasma size). The ratio K_L has been calculated[9] for plasmas of uniform thickness L and homogeneous T_{e}; the result is shown in fig. 15.7. K_L is less than 1, which is a result of the effect of reabsorption of synchrotron radiation by the plasma.

The frequency of synchrotron radiation is of the order of harmonics of the electron cyclotron frequency, so that the radiation is of wavelength long enough to be reflected by the wall of the vacuum vessel. We can expect that the reflected radiation could be reabsorbed by the plasma, so that the radiation loss can be reduced. In fact, when the reflectivity of the wall is R, the radiation loss in the low-frequency region can be reduced to $1 - \mathrm{R}$, where the plasma, considered to be optically thick, reabsorbs the radiation. However the loss of the high-frequency components cannot be reduced, because in this region the plasma must be taken to be optically thin and does not reabsorb the radiation.

When the magnitude of the magnetic field changes with position, as in the case of a toroidal field, reabsorption of synchrotron radiation may be weak. This effect is analyzed in ref. 10.

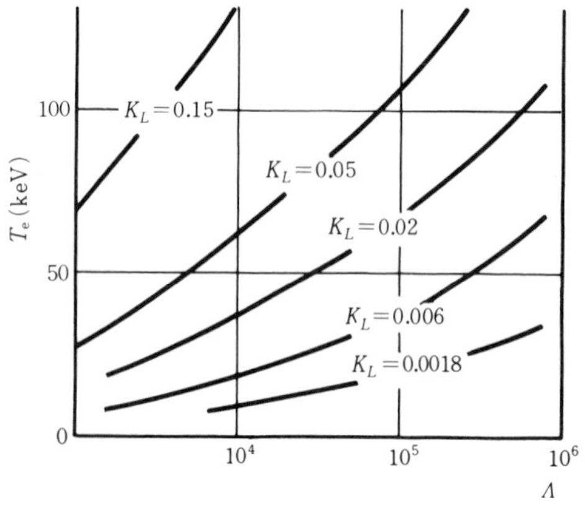

Fig. 15.7
Curves of $K_L(T_e, \Lambda)$ = const. on a T_e vs. Λ diagram. After B. A. Trubnikov et al.: Proc. 2nd UN Conf. on Peaceful Uses of Atomic Energy **31**, p. 93, Geneva, 1958.

15.4 Scattering of Electromagnetic Waves by Plasmas

The theory of scattering of electromagnetic waves by plasmas was first applied to studies of the interaction of the ionosphere and microwaves.[11–13] The cross section for scattering of electromagnetic waves by an electron is of the order of the electron size. Detection and measurement of electron-scattered light in the visible region was impossible until lasers began to be used as sufficiently intense monochromatic light sources. Several kinds of information can be extracted from observations of the light scattered by a plasma, and this technique plays an important role in plasma diagnostics.

15.4a Noninteracting Scattering

Let us assume a monochromatic, polarized electromagnetic field $\boldsymbol{E}_1$, of angular frequency ω_L and polarization $\boldsymbol{n}_1$. Then the field at the position $\boldsymbol{r}_1$ is

$$\boldsymbol{E}_1 = E_1 \boldsymbol{n}_1 \exp(-i(\omega_L t - \boldsymbol{k}_1 \cdot \boldsymbol{r}_1)), \tag{15.33}$$

where $\boldsymbol{k}_1$ is the propagation vector. If there is an electron at $\boldsymbol{r}_1$, the electron will be accelerated by the field ($\dot{\boldsymbol{v}} = -e\boldsymbol{E}_1/m$) and will emit radiation. This is the basis of the scattering of electromagnetic waves by electrons. Scattering by ions is usually negligible, due to the large mass. Let the scattering electron be at $Q(\boldsymbol{r}_1)$ and the observation point be at $P(\boldsymbol{r})$, and denote the scattered field by $\boldsymbol{E}_s(\boldsymbol{r}, t)$ and its propagation vector by $\boldsymbol{k}_s$. From the formula for dipole radiation, $\boldsymbol{E}_s(\boldsymbol{r}, t)$ is given by

$$\boldsymbol{E}_s(\boldsymbol{r}, t) = \frac{-\alpha_0}{R}\frac{(\boldsymbol{k}_s \times \boldsymbol{n}_1) \times \boldsymbol{k}_s}{|\boldsymbol{k}_s|^2} E_1 \exp(-i(\omega_L t - \boldsymbol{k}_s \cdot (\boldsymbol{r} - \boldsymbol{r}_1) - \boldsymbol{k}_1 \cdot \boldsymbol{r}_1))$$

$$= \frac{-\alpha_0}{R} \sin\phi \boldsymbol{n}_s E_1 \exp(-i((\omega_L t + \boldsymbol{k} \cdot \boldsymbol{r}_1) - \boldsymbol{k}_s \cdot \boldsymbol{r})) \tag{15.34}$$

where $\alpha_0 = e^2/(4\pi\varepsilon_0 mc^2)$ is the classical electron radius, $R \equiv |\boldsymbol{r} - \boldsymbol{r}_1|$, and

$$\boldsymbol{k} \equiv \boldsymbol{k}_s - \boldsymbol{k}_1. \tag{15.35}$$

ϕ is the angle between $\boldsymbol{n}_1$ and $\boldsymbol{k}_s$, and $\boldsymbol{n}_s$ is the unit vector in the direction $(\boldsymbol{k}_s \times \boldsymbol{n}_1) \times \boldsymbol{k}_s$ (fig. 15.8).

The electron is moving with the velocity $\boldsymbol{v}$, and $\boldsymbol{r}_1 = \boldsymbol{v}t + \boldsymbol{r}_0$. It is shown in eq. (15.34) that the frequency ω_s of the scattered light is Doppler shifted by the amount $\boldsymbol{k}_1 \cdot \boldsymbol{v}$, i.e.,

$$\Delta\omega = \omega_s - \omega_L = \boldsymbol{k} \cdot \boldsymbol{v} = v 2k_1 \sin\frac{\theta}{2} = \frac{4\pi}{\lambda}\sin\frac{\theta}{2} v, \tag{15.36}$$

where v is the component of the velocity in the $\boldsymbol{k}$ direction and θ is the angle between $\boldsymbol{k}_s$ and $\boldsymbol{k}_1$. If the distribution function $f(v)$ of the velocity component along $\boldsymbol{k}$ is

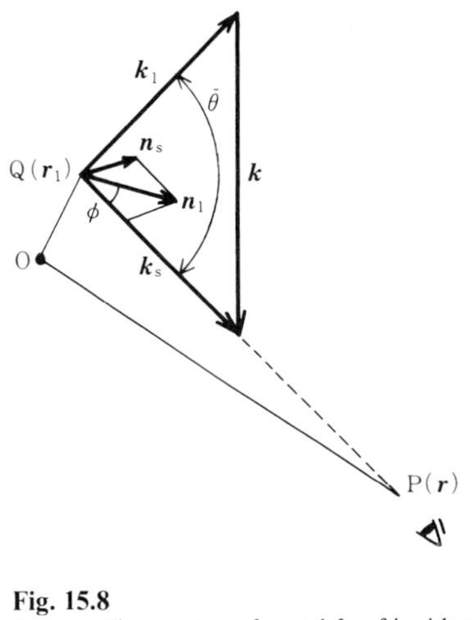

Fig. 15.8
Propagation vectors $\boldsymbol{k}_1$ and $\boldsymbol{k}_s$ of incident and scattered light.

$$f(v)\mathrm{d}v = \left(\frac{m}{2\pi T_\mathrm{e}}\right)^{1/2} \exp\left(-\frac{mv^2}{2T_\mathrm{e}}\right)\mathrm{d}v,$$

then the spectral distribution of the scattered light is given by

$$F(\Delta\omega)\mathrm{d}(\Delta\omega)$$
$$= \left(\frac{m}{2\pi T_\mathrm{e}}\right)^{1/2} \frac{\lambda}{4\pi\sin(\theta/2)} \exp\left(-\frac{m}{2T_\mathrm{e}}\left(\frac{\lambda\Delta\omega}{4\pi\sin(\theta/2)}\right)^2\right)\mathrm{d}(\Delta\omega). \tag{15.37}$$

This formula is based on the assumption that the individual electrons move and radiate independently. This is called *noninteracting scattering* or *Thompson scattering*. From eq. (15.34), the cross section $\mathrm{d}\sigma_1$ for scattering into a small solid angle $\mathrm{d}\Omega$ is given by

$$\frac{\mathrm{d}\sigma_1}{\mathrm{d}\Omega} = \frac{|E_\mathrm{s}|^2 R^2}{|E_1|^2} = \alpha_0{}^2 \sin^2\phi. \tag{15.38}$$

The total cross section for Thomson scattering by an electron is

$$\sigma_\mathrm{T} = \frac{8\pi\alpha_0{}^2}{3} = 0.665 \times 10^{-28} \quad (\mathrm{m}^2). \tag{15.39}$$

The half-width $\Delta\lambda_\mathrm{D}$ of the Doppler broadening of the scattered light is given by

$$\frac{\Delta\lambda_\mathrm{D}}{\lambda} = \frac{4\sin(\theta/2)}{c}\left(\frac{2T_\mathrm{e}}{m}\ln 2\right)^{1/2}. \tag{15.40a}$$

Thus, from the measurement of Doppler broadening of the scattered light, we can obtain the electron temperature. As will be explained later, the assumption of noninteraction between electrons is only valid when $1/k$ is smaller than the Debye radius $\lambda_\mathrm{d}{}^\mathrm{e}$, i.e., for

$$\alpha_\mathrm{e} \equiv \frac{1}{k\lambda_\mathrm{d}{}^\mathrm{e}} = \frac{1}{k}\left(\frac{ne^2}{\varepsilon_0 T_\mathrm{e}}\right)^{1/2} \ll 1 \tag{15.41a}$$

$$k = 2\frac{2\pi}{\lambda}\sin\frac{\theta}{2}. \tag{15.42}$$

When ruby laser light ($\lambda = 6943$ Å) is scattered at $\theta = 90^\circ$, Doppler broadening of the spectral distribution and the noninteraction condition become

$$\Delta\lambda_\mathrm{D} = 32.3\left(\frac{T_\mathrm{e}}{e}\right)^{1/2} \quad (\text{Å}) \tag{15.40b}$$

$$\alpha_e = 1.09\left(\frac{n}{10^{22}}\frac{e}{T_e}\right)^{1/2} \lesssim 0.2. \tag{15.41b}$$

*15.4b Noninteracting Scattering in a Magnetic Field[14–17]

When electrons gyrate in a magnetic field, the relation $\boldsymbol{k}\cdot\boldsymbol{r}_1 = (\boldsymbol{k}\cdot\boldsymbol{v})t + \phi$ cannot be applied. In terms of the unit vector $\boldsymbol{b}$ of the magnetic field, $\boldsymbol{k}\cdot\boldsymbol{r}_1$ is

$$\begin{aligned}\boldsymbol{k}\cdot\boldsymbol{r}_1 &= (\boldsymbol{k}\cdot\boldsymbol{b})v_{\parallel}t + k_{\perp}\rho_B\sin(\Omega_e t) + \phi \\ &= k_{\parallel}v_{\parallel}t + k_{\perp}(v_{\perp}/\Omega_e)\sin(\Omega_e t) + \phi,\end{aligned} \tag{15.43}$$

where the components of $\boldsymbol{k} = \boldsymbol{k}_s - \boldsymbol{k}_1$ parallel and perpendicular to the magnetic field are given by

$$k_{\parallel} = (\boldsymbol{k}\cdot\boldsymbol{b}) = k\sin\varepsilon = \frac{4\pi}{\lambda}\sin(\theta/2)\sin\varepsilon \tag{15.44}$$

$$k_{\perp} = |\boldsymbol{k}\times\boldsymbol{b}| = k\cos\varepsilon = \frac{4\pi}{\lambda}\sin(\theta/2)\cos\varepsilon. \tag{15.45}$$

Here ε is the complement of the angle between $\boldsymbol{k}$ and $\boldsymbol{b}$ (fig. 15.9).

Substituting eq. (15.43) into (15.34) and using the expansion formula

$$\exp(iz\sin\alpha) = \sum_{-\infty}^{\infty} J_m(z)\exp im\alpha,$$

we obtain the amplitudes of the harmonics of the electric field. The spectrum of the scattered light is given by the sum of the squares of the absolute values of these amplitudes. Multiplying the Maxwell distribution function

$$f(\boldsymbol{v})2\pi v_{\perp}dv_{\perp}dv_{\parallel} = \left(\frac{m}{2\pi T}\right)^{3/2}\exp\left(-\frac{m(v_{\parallel}^2 + v_{\perp}^2)}{2T}\right)2\pi v_{\perp}dv_{\perp}dv_{\parallel}$$

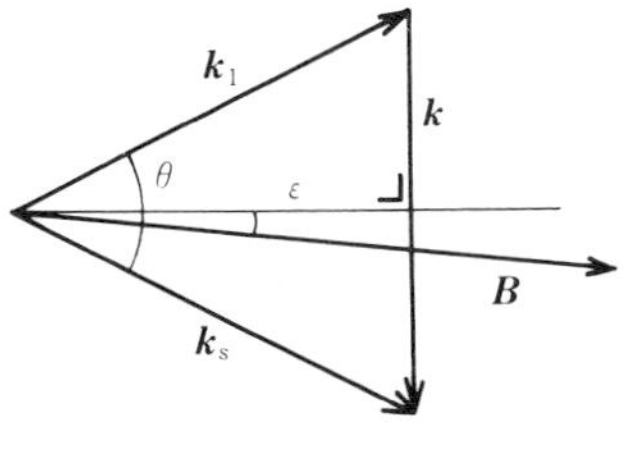

Fig. 15.9
Propagation vectors $\boldsymbol{k}_1$ and $\boldsymbol{k}_s$ of incident and scattered light ($\boldsymbol{k} = \boldsymbol{k}_s - \boldsymbol{k}_1$). $\boldsymbol{B}$ is the magnetic field.

by this spectrum and integrating over $v_\parallel$ and $v_\perp$, we have the spectral distribution of the scattered light. From the integral formula

$$\int_0^\infty \exp(-a^2x^2)\, J_\nu(px)\, J_\nu(qx) x \mathrm{d}x = \frac{1}{2a^2}\exp\left(-\left(\frac{p^2+q^2}{4a^2}\right)\right) I_\nu\left(\frac{pq}{2a^2}\right),$$

the spectral distribution is given by

$F(\Delta\omega)\mathrm{d}(\Delta\omega)$

$$= \frac{\mathrm{d}(\Delta\omega)}{2\pi}\exp\left(-\left(\frac{k_\perp v_\mathrm{T}}{\Omega_\mathrm{e}^2}\right)^2\right) 2\pi^{1/2} \sum_{l=-\infty}^{\infty} I_l\left(\frac{k_\perp{}^2 v_\mathrm{T}{}^2}{\Omega_\mathrm{e}{}^2}\right) \frac{\exp\left(-\frac{1}{2}\left(\frac{\Delta\omega + l\Omega_\mathrm{e}}{k_\parallel v_\perp}\right)^2\right)}{2^{1/2} k_\parallel v_\mathrm{T}} \tag{15.46}$$

where $v_\mathrm{T} = (T_\mathrm{e}/m)^{1/2}$. This equation has a maximum at

$$\Delta\omega + l\Omega_\mathrm{e} = 0. \tag{15.47}$$

The smaller the value of $k_\parallel$, the steeper the maximum. The envelope of the maximum is determined by $I_l(z)$ ($z \equiv (k_\perp v_\mathrm{T}/\Omega_\mathrm{e})^2 = (k_\perp \rho_B)^2$). Usually the Larmor radius is larger than the wavelength of light and $z \gg 1$. In this case, $I_l(z) = \exp(z - l^2/2z)/(2\pi z)^{1/2}$; so the spectral intensity is

$$F(\Delta\omega) = \frac{1}{\pi}\left(\frac{\Omega_\mathrm{e}}{k_\perp v_\mathrm{T}}\right) \sum \exp\left(-\left(\frac{l\Omega_\mathrm{e}}{k_\perp v_\mathrm{T}}\right)^2\right) \frac{\exp\left(-\left(\frac{\Delta\omega + l\Omega_\mathrm{e}}{2^{1/2} k_\parallel v_\mathrm{T}}\right)^2\right)}{2^{1/2} k_\parallel v_\mathrm{T}}. \tag{15.48}$$

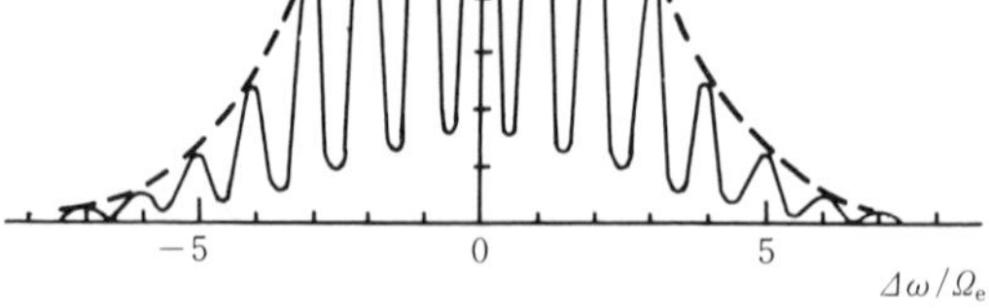

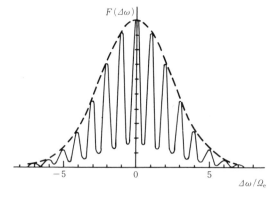

Fig. 15.10
Spectral distribution of the light scattered in the presence of a magnetic field.

A plot of this function is shown in fig. 15.10. The width of the maximum is

$$k_{\parallel} v_{\mathrm{T}} \approx 2k_1 v_{\mathrm{T}} \sin\frac{\theta}{2} \sin\varepsilon.$$

Unless this width is less than Ω_{e}, the fine structure of the distribution disappears and the function approaches that obtained in the absence of a magnetic field. Consequently, the complementary angle ε must satisfy

$$\varepsilon < \frac{\Omega_{\mathrm{e}}}{2k_1 v_{\mathrm{T}} \sin(\theta/2)} \tag{15.49}$$

in order that the fine structure be observable. This discussion is also based on the noninteracting assumption so that α_{e} must satisfy

$$\alpha_{\mathrm{e}} = \frac{1}{k\lambda_{\mathrm{d}}^{\mathrm{e}}} \ll 1.$$

*15.4c Collective Scattering[11-13,18]

The electric field of the light scattered by a single electron is given by eq. (15.34). The electric field $\boldsymbol{E}_{\mathrm{s}}(\boldsymbol{r}, t)$ of the light scattered by a group of electrons of number density $n(\boldsymbol{r}_1, t)$ is expressed by

$$\boldsymbol{E}_{\mathrm{s}}(\boldsymbol{r}, t) = \alpha_0 \sin\phi \boldsymbol{n}_{\mathrm{s}} E_1 \int_V \frac{n(\boldsymbol{r}_1, t')}{\boldsymbol{R}} \exp(-i(\omega_L t' - \boldsymbol{k}_1 \boldsymbol{r}_1)) \mathrm{d}\boldsymbol{r}_1, \tag{15.50}$$

where $t' = t - R/c = t - \boldsymbol{k}_{\mathrm{s}} \cdot (\boldsymbol{r} - \boldsymbol{r}_1)/\omega_L \approx t - |\boldsymbol{r}|/c$ (R is the distance between the electron and the observation point). If the density is expressed by a Fourier integral

$$n(\boldsymbol{x}, t) = \int_{-\infty}^{\infty} \frac{\mathrm{d}\boldsymbol{k}}{(2\pi)^3} \int_{-\infty}^{\infty} \frac{\mathrm{d}\omega}{2\pi} \exp(-i(\omega t - \boldsymbol{k} \cdot \boldsymbol{x})) N(\boldsymbol{k}, \omega) \tag{15.51}$$

$$N(\boldsymbol{k}, \omega) = \int_V \mathrm{d}\boldsymbol{x} \int_0^T \mathrm{d}t \exp(i(\omega t - \boldsymbol{k} \cdot \boldsymbol{x})) n(\boldsymbol{x}, t), \tag{15.52}$$

then $\boldsymbol{E}_{\mathrm{s}}(\boldsymbol{r}, t)$ may be reduced to

$$\boldsymbol{E}_{\mathrm{s}}(\boldsymbol{r}, t) = \frac{\alpha_0}{R_0} \sin\phi \boldsymbol{n}_{\mathrm{s}} E_1 \int_{-\infty}^{\infty} \frac{\mathrm{d}\boldsymbol{k}}{(2\pi)^3} \int_{-\infty}^{\infty} \frac{\mathrm{d}\omega}{2\pi} N(\boldsymbol{k}, \omega)$$

$$\times \int_V \mathrm{d}\boldsymbol{r}_1 \exp i[-\omega(t - |\boldsymbol{r}|/c) + \boldsymbol{k}\cdot\boldsymbol{r}_1 - \omega_L t + \boldsymbol{k}_s\cdot(\boldsymbol{r} - \boldsymbol{r}_1) + \boldsymbol{k}_1\cdot\boldsymbol{r}_1]$$

$$= \frac{\alpha_0}{R_0}\sin\phi \boldsymbol{n}_s E_1 \exp\left(i\omega\frac{R_0}{c} + i\boldsymbol{k}_s\cdot\boldsymbol{r}\right)\int_{-\infty}^{\infty}\frac{\mathrm{d}\omega}{2\pi}\int_{-\infty}^{\infty}\frac{\mathrm{d}\boldsymbol{k}}{(2\pi)^3}N(\boldsymbol{k}, \omega)$$

$$\times \int_V \mathrm{d}\boldsymbol{r}_1 \exp i(-(\omega + \omega_L)t + (\boldsymbol{k} - \boldsymbol{k}_s + \boldsymbol{k}_1)\cdot\boldsymbol{r}_1)$$

$$= \frac{\alpha_0}{R_0}\sin\phi \boldsymbol{n}_s E_1 \exp\left(i\omega\frac{R_0}{c} + i\boldsymbol{k}_s\cdot\boldsymbol{r}\right)$$

$$\times \int_{-\infty}^{\infty}\frac{\mathrm{d}\omega}{2\pi}N(\boldsymbol{k}_s - \boldsymbol{k}_1, \omega - \omega_L)\exp(-i\omega t), \tag{15.53}$$

where t' of $n(\boldsymbol{r}, t')$ is approximated by $t' = t - |\boldsymbol{r}|/c$, and $R_0 = |\boldsymbol{r}|$. This approximation is appropriate when $\omega - \omega_L \ll \omega_L$ and $|\boldsymbol{r}_1| \ll |\boldsymbol{r}|$.

When $\boldsymbol{E}_s(\boldsymbol{r}, t)$ is expressed by $\boldsymbol{E}_s(\boldsymbol{r}, t) = \boldsymbol{E}(\boldsymbol{k}, t)\exp(i\boldsymbol{k}_s\cdot\boldsymbol{r})$, the power spectrum $I(\boldsymbol{k}, \omega)$ of $\boldsymbol{E}_s(\boldsymbol{r}, t)$ is

$$I(\boldsymbol{k}, \omega) \equiv \lim_{T\to\infty}\frac{1}{T}\left|\int_0^T \mathrm{d}t\,\boldsymbol{E}(\boldsymbol{k}, t)\exp i\omega t\right|^2 \tag{15.54}$$

and the autocorrelation function $C_E(\boldsymbol{k}, \tau)$ is

$$C_E(\boldsymbol{k}, \tau) \equiv \lim_{T\to\infty}\frac{1}{T}\int_0^T \mathrm{d}t\,\boldsymbol{E}^*(\boldsymbol{k}, t)\cdot\boldsymbol{E}(\boldsymbol{k}, t + \tau). \tag{15.55}$$

From the formula for the Fourier integral we have the Wiener-Khintchine theorem:

$$I(\boldsymbol{k}, \omega) = \int_{-\infty}^{\infty}\mathrm{d}\tau\,C_E(\boldsymbol{k}, \tau)\exp(i\omega\tau). \tag{15.56}$$

The power spectrum $I(\boldsymbol{k}, \omega)$ is obtained from eqs. (15.53) and (15.54), and is

$$I(\boldsymbol{k}, \omega) = \frac{{\alpha_0}^2}{{R_0}^2}\sin^2\phi {E_1}^2 \lim_{T\to\infty}\frac{1}{T}|N(\boldsymbol{k}, \omega - \omega_L)|^2$$

$$= \frac{N{\alpha_0}^2}{R_0}\sin^2\phi {E_1}^2 S(\boldsymbol{k}, \omega - \omega_L) \tag{15.57}$$

$$S(\boldsymbol{k}_s - \boldsymbol{k}_1, \omega - \omega_L) \equiv \frac{1}{N}\lim_{T\to\infty}\frac{1}{T}|N(\boldsymbol{k}_s - \boldsymbol{k}_1, \omega - \omega_L)|^2. \tag{15.58}$$

N is the total number of scattering electrons in the volume V.

The power spectrum $S(\boldsymbol{k}, \omega)$ of the density $n(\boldsymbol{r}, t)$ can also be expressed in terms of the density autocorrelation function by using the Wiener-Khintchine theorem:

$$C_n(\boldsymbol{x}, \tau) \equiv \lim_{T \to \infty} \frac{1}{T} \int_0^T \mathrm{d}t \int_V \mathrm{d}\boldsymbol{r} n(\boldsymbol{r}, t) n(\boldsymbol{r} + \boldsymbol{x}, t + \tau) \tag{15.59}$$

and

$$S(\boldsymbol{k}, \omega) = \frac{1}{N} \int_{-\infty}^{\infty} \mathrm{d}\tau \int_V \mathrm{d}\boldsymbol{x} C_n(\boldsymbol{x}, \tau) \exp(i(\omega\tau - \boldsymbol{k} \cdot \boldsymbol{x})). \tag{15.60}$$

The cross section $\mathrm{d}\sigma$ for scattering into an element of solid angle $\mathrm{d}\Omega$ and the small frequency interval $\mathrm{d}\omega$ is given by

$$\frac{\mathrm{d}\sigma}{\mathrm{d}\Omega \mathrm{d}\omega / 2\pi} \equiv \frac{I(\boldsymbol{k}, \omega) R_0^2}{|E_1|^2}$$

$$= N \alpha_0^2 \sin^2 \phi S(\boldsymbol{k}_s - \boldsymbol{k}_1, \omega - \omega_L). \tag{15.61}$$

Therefore if the spectral distribution of the scattered light in the direction $\boldsymbol{k}_s$ is observed experimentally, the power spectrum of the density fluctuation corresponding to $\boldsymbol{k} = \boldsymbol{k}_s - \boldsymbol{k}_1$ and $\Delta\omega = \omega - \omega_L$ can be obtained.

Since $S(\boldsymbol{k}, \omega)$ can be expressed by the Fourier transform of $n(\boldsymbol{r}, t)$, $S(\boldsymbol{k}, \omega)$ is obtained by means of the Laplace transform of Vlasov's equation.[13] If the plasma is Maxwellian, with electron temperature T_e and ion temperature T_i, then $S(\boldsymbol{k}, \Delta\omega)$ is given by

$$S(\boldsymbol{k}, \Delta\omega) = \frac{2\pi}{k} \left(\frac{m}{2\pi T_e} \right)^{1/2} \Gamma_{\alpha_e}(x) + Z \left(\frac{\alpha_e^2}{1 + \alpha_e^2} \right)^2 \frac{2\pi}{k} \left(\frac{M}{2\pi T_i} \right)^{1/2} \Gamma_\beta(y) \tag{15.62}$$

$$\Gamma_\gamma(z) = \frac{\exp(-z^2)}{(1 + \gamma^2 - \gamma^2 g(z))^2 + \pi \gamma^4 z^2 \exp(-2z^2)},$$

where

$$g(z) = 2z \exp(-z^2) \int_0^z \exp(t^2) \mathrm{d}t$$

$$\beta^2 = \frac{\alpha_i^2}{1 + \alpha_e^2} \qquad x = \frac{\Delta\omega}{k} \left(\frac{m}{2T_e} \right)^{1/2} \qquad y = \frac{\Delta\omega}{k} \left(\frac{M}{2T_i} \right)^{1/2}$$

$$\alpha_e = \frac{1}{k\lambda_d^{\ e}} = \frac{1}{k}\left(\frac{ne^2}{\varepsilon_0 T_e}\right)^{1/2}$$

$$\alpha_i = \frac{1}{k\lambda_d^{\ i}} = \left(\frac{ZT_e}{T_i}\right)^{1/2}\alpha_e.$$

Here m and M are the electron and ion masses, respectively, $\lambda_d^{\ e}$ and $\lambda_d^{\ i}$ are the Debye lengths, and Ze is the ion charge.

If $\alpha_e \ll 1$ (say $\alpha_e < 0.2$), the 2nd term of $S(\boldsymbol{k}, \Delta\omega)$ is negligible and the 1st term approaches to the formula (15.37) for noninteracting scattering. For 90° scattering of ruby laser light, α_e is

$$\alpha_e = 1.09\left(\frac{n}{10^{22}}\frac{e}{T_e}\right)^{1/2} \tag{15.63}$$

(n in m^{-3}, T_e/e in eV). When $\alpha_e < 1$, $1/k < \lambda_D^{\ e}$, Debye shielding is not effective for electron motion and the total scattered light is the sum of the individual contributions of each electron.

However if $\alpha_e > 1.5$ and $1/k > \lambda_d^{\ e}$, the collective effect of ions becomes strong and the 2nd term of eq. (15.62) is important. The width of the spectral distribution of the scattered light is of the order of the Doppler broadening of the ion thermal velocity (fig. 15.11). Therefore the ion temperature can be estimated when $\alpha_e > 1.5$. This phenomenon is called *collective scattering.*

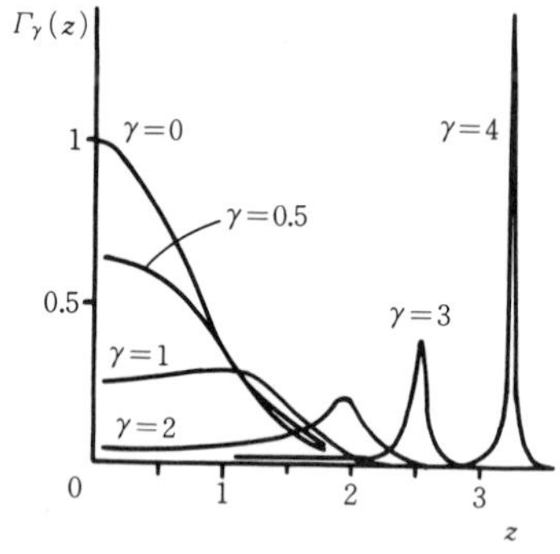

Fig. 15.11
The function $\Gamma_\gamma(z)$.

15.5 Diagnostics Using Electromagnetic Waves[19]

Diagnostics using electromagnetic waves can yield information without disturbing the plasma under study. The plasma electrons play the main role in the processes of absorption, propagation, and scattering of light, and there are several different methods whereby one can measure the electron temperature and electron density. For measurement of the ion temperature, Doppler broadening of a line spectrum or collective scattering of laser light can be used.

15.5a Measurement of Ion Temperature by Doppler Broadening

When the velocity distribution of the plasma ions is Maxwellian, the line spectrum $I(\Delta\lambda)$ is

$$I(\Delta\lambda) \propto \exp\left(-\frac{M}{2T_{\mathrm{i}}}\left(\frac{c}{\lambda}\Delta\lambda\right)^2\right), \tag{15.64}$$

where the Doppler shift is

$$\frac{\Delta\lambda}{\lambda} = \frac{v}{c}.$$

M is the ion mass, c is the velocity of light, and λ is the wavelength at the center of the observed line. Accordingly *the full half-width of Doppler broadening* $\Delta\lambda_{\mathrm{D}}$ is

$$\frac{\Delta\lambda_{\mathrm{D}}}{\lambda} = \frac{2(2\ln 2)^{1/2}}{c}\left(\frac{T_{\mathrm{i}}}{M}\right)^{1/2} = 7.70\times 10^{-5}\left(\frac{1}{A}\frac{T_{\mathrm{i}}}{e}\right)^{1/2}, \tag{15.65}$$

where A is the atomic weight and T_{i}/e is in eV. When $T_{\mathrm{i}}/(eA) = 10\,\mathrm{eV}$ and $\lambda = 5000$ Å, then $\Delta\lambda_{\mathrm{D}} = 1.2$ Å.

We must examine the effect of line broadening due to other mechanisms before we can estimate the ion temperature from the line width. Of course, hydrogen ions (protons) do not emit a line spectrum. Singly ionized helium ions emit an ion line spectrum, and it is possible to measure the helium ion temperature. The lines of impurity ions may also be used, but the relaxation of plasma ion and the impurity ion temperatures must be carefully examined.

Line broadening is caused by the *fine structure* of the line spectrum itself, the *Zeeman effect* due to the magnetic field, and *Stark broadening* due to the microscopic electric fields of the charged particles moving around the light-emitting ion. When the plasma is MHD unstable, random collective motions of plasma ions may also cause line broadening.

The Zeeman effect is explained in detail in ref. 20. When the change of energy level due to the Zeeman effect is small compared to the intervals between the different energy levels, Russell-Saunders coupling is applicable and the perturbed energy level is given by

$$\Delta E_z = \frac{eB}{2m}\frac{h}{2\pi} g M_z.$$

Lande's factor g is of the order of 1. M_z is the quantum number of the z component of the total angular momentum. The light corresponding to the transition $\Delta M_z = \pm 1$ is circularly polarized perpendicularly to the magnetic field (σ component). The light for $\Delta M_z = 0$ is linearly polarized parallel to the magnetic field (π component). Therefore one can select the components of Zeeman splitting by use of a polarizer or quarter-wave plate and reduce the effect of Zeeman broadening. The amount $\Delta\lambda_z$ of Zeeman splitting is

$$\frac{\Delta\lambda_z}{\lambda} \equiv \left(\frac{e}{4\pi mc}\right)\lambda B = 4.67 \times 10^{-9}\lambda B, \tag{15.66}$$

where λ is in Å and B is in $\mathrm{Wb/m^2}$.

Stark broadening is described in refs. 21–23. There are two ways to explain Stark broadening. One way is by means of an impact theory. The light-emitting ion collides with other particles and so the lifetime of light emission becomes shorter and the line width broadens. When the lifetime is T, the Fourier amplitude is

$$J(\omega, T) \propto \int_0^T \exp(-i\omega_0 t + i\omega t)\mathrm{d}t = \frac{\exp i(\omega - \omega_0)T - 1}{i(\omega - \omega_0)}.$$

The observed line spectrum is the average of $|J(\omega, T)|^2$. When the average of the collision times (i.e., the average of the emission lifetimes) is τ, the line profile is

$$I(\omega) = \tau^{-1}\int_0^\infty |J(\omega, T)|^2 \exp(-T/\tau)\mathrm{d}T \propto \frac{1}{(\omega - \omega_0)^2 + \omega_c{}^2},$$

where $\omega_c = 1/\tau$ is collision frequency and the half-width $\omega_{1/2}$ is $2\omega_c$. According to ref. 21, $\omega_{1/2}$ is

$$\omega_{1/2}{}^{c} = 2\pi^3 Z^2 (\Omega_s)^2 n \left(\frac{m}{3T}\right)^{1/2}, \tag{15.67}$$

where m, n, and Ze are the mass, the number density, and the electric charge of perturbing particles (electrons or ions). Ω_s is the coefficient of the 1st-order Stark effect of electrons on the light-emitting ion. Ω_s is the order of 10^{-3} m^2/s for hydrogen.

The other explanation of the Stark effect proceeds by statistical theories. In all of these, the light-emitting particle is assumed to be subjected to a perturbation during its emission lifetime. Holtzmark's theory is typical: Let us introduce the probability $P(r)$ that there is at least one charged particle inside the sphere of radius r centered on the light-emitting particle. $P_-(r) = 1 - P(r)$ is the probability that there is no particle inside the sphere. One can easily deduce the relation

$$P_-(r + \mathrm{d}r) = P_-(r)(1 - 4\pi n r^2 \mathrm{d}r),$$

where n is the number density of particles.

Solving this, one obtains $P(r) = 1 - \exp(-4nr^3/3)$. The probability that there is at least one particle inside the shell between r and $r + \mathrm{d}r$ is $\mathrm{d}P(r) = 4nr^2 \exp(-4nr^3/3)\mathrm{d}r$. When r_0, the average distance between particles, is defined by $4\pi n r_0{}^3 = 1$, $\mathrm{d}P(r)$ is expressed by

$$\mathrm{d}P(r) = \exp\left(-\left(\frac{r}{r_0}\right)^3\right)\mathrm{d}\left(\frac{r}{r_0}\right)^3.$$

The frequency shift due to the 1st-order Stark effect is

$$\Delta\omega = \frac{4\pi\varepsilon_0 \Omega_s}{e} E.$$

As the electric field is $E = Ze/(4\pi\varepsilon_0 r^2)$, the probability of the shift $\Delta\omega$ is

$$\mathrm{d}P(\Delta\omega) = \exp\left(-\left(\frac{\Delta\omega_0}{\Delta\omega}\right)^{3/2}\right)\mathrm{d}\left(\frac{\Delta\omega_0}{\Delta\omega}\right)^{3/2},$$

and the profile of the spectral line is

$$I(\Delta\omega)\mathrm{d}(\Delta\omega) = \frac{3}{2}\left(\frac{\Delta\omega_0}{\Delta\omega}\right)^{5/2} \exp\left(-\left(\frac{\Delta\omega_0}{\Delta\omega}\right)^{3/2}\right)\mathrm{d}\left(\frac{\Delta\omega}{\Delta\omega_0}\right)$$

$$\Delta\omega_0 = \frac{\Omega_s}{4\pi\varepsilon_0 e} E_0 = \frac{\Omega_s Z}{r_0{}^2}.$$

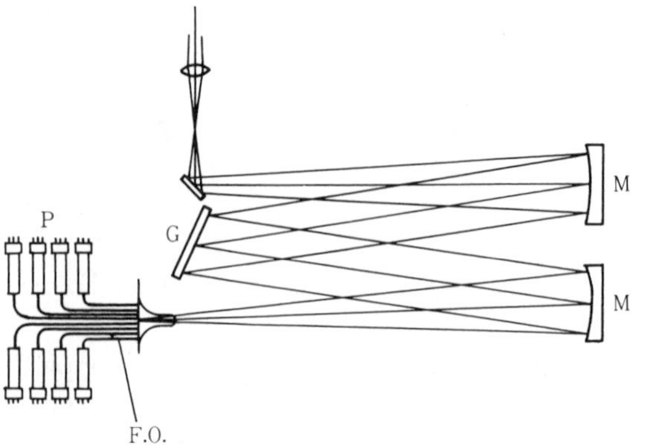

Fig. 15.12
Elements of a Czerney-Turner monochromator and multichannel photodetector system (P) using fiber optics (F. O.). G is the grating and M is the mirror.

From this result, one has the Holtzmark half-width (n in m^{-3})

$$\omega_{1/2}{}^{H} \approx 3\Delta\omega_0 = 7.8 Z \Omega_s n^{2/3}. \tag{15.68}$$

The 1st-order Stark effect occurs in for degenerate energy levels, such as occur in the hydrogen atom or helium ion. When the energy level is not degenerate, the 2nd-order effect may appear although the 2nd-order broadening is smaller. More detailed discussion will be found in refs. 23 and 24.

For sufficiently accurate measurement of line spectra dispersion systems of high resolving power are necessary. Monochrometers (Czerney-Turner type[25], Echelle type[26,27]) having gratings of large effective area, or Fabry-Perot interferometers[28] are often used (see figs. 15.12 and 15.13, respectively). Photomultipliers are used most frequently as photodetectors, because of their short resolving time. Fiber-optical systems are conveniently used with multichannel photodetector systems. Of course, it is necessary to calibrate and monitor the relative sensitivity of each channel.

15.5b Laser Scattering (Microwave Scattering)[18,29,30]

When an intense laser beam is directed into a plasma, the spectral distribution of the scattered light can be observed. This expresses the electron velocity distribution function of the velocity component parallel to $\boldsymbol{k}$ when $\alpha_e \ll 1$ (say $\alpha_e \lesssim 0.2$). α_e was given in sec. 15.4 as follows:

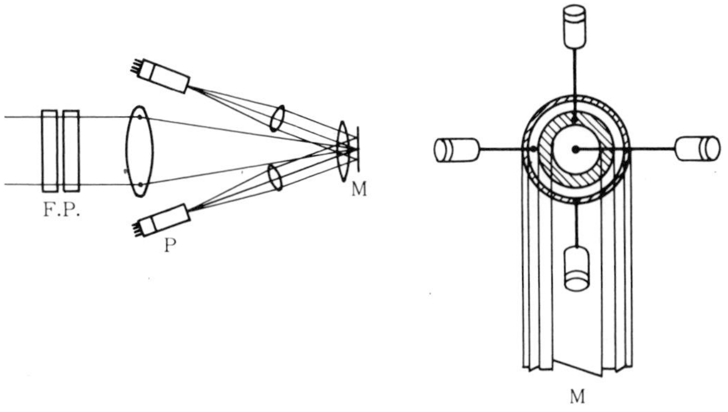

Fig. 15.13
Fabry-Perot interferometer (F. P.) and multichannel photodetector system (P) with concentric multimirrors (M) at different reflecting angles.

$$\alpha_e = \frac{1}{k\lambda_d{}^e} = \frac{1}{k}\left(\frac{ne^2}{\varepsilon_0 T_e}\right)^{1/2} \qquad k = \frac{4\pi}{\lambda}\sin\frac{\theta}{2}.$$

Accordingly, the electron temperature can be measured from the profile of the scattered light. When the intensity of the incident light of cross-sectional area S is I, the energy of the light scattered from the volume lS of a plasma of number electron density n_e into the small solid angle $d\Omega$ per unit time is

$$W = In_e lS\frac{d\sigma}{d\Omega}d\Omega = ISln_e\alpha_0{}^2 \sin^2\phi\, d\Omega. \tag{15.69}$$

When $n_e = 10^{20}\,\mathrm{m}^{-3} = 10^{14}\,\mathrm{cm}^{-3}$, $\phi = 90°$, $l = 0.01$ m, $d\Omega = 10^{-2}$, then $W/IS = 0.8 \times 10^{-3}$. Even if $IS = 100$ MW, the scattered power is only 8 μW. Accordingly, careful design is necessary in order to minimize instrumental scattering (e.g., scattering from the windows through which the laser light passes) and to shield the detector system from unavoidable instrumental scattering. So-called beam dumps and view dumps are necessary, as shown in fig. 15.14. As the spectrum of instrumentally scattered light does not broaden, it can be discriminated against spectroscopically.

When $\alpha_e > 1.5$, the collective effect of ions is dominant and the ion temperature can be estimated from the spectral distribution of the scattered light. In order to make α_e large enough, one must take advantage

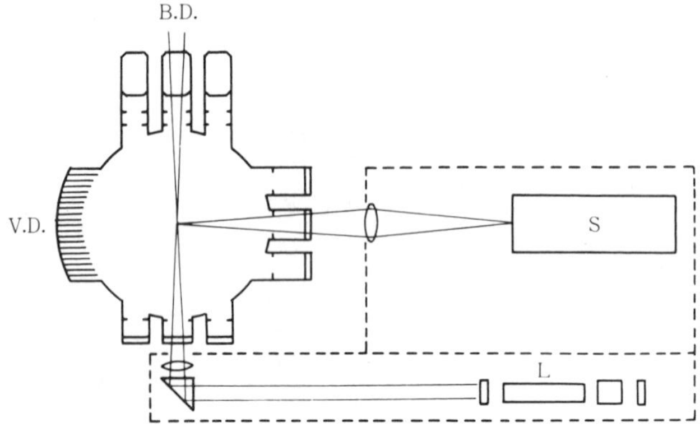

Fig. 15.14
Laser scattering device: laser L, spectrometer S, beam dump (B.D.) and view dump (V.D.).

of small-angle, or forward, scattering, or else work in the infrared or microwave region.

When the complement ε of the angle between the scattering propagation vector $\boldsymbol{k} = \boldsymbol{k}_s - \boldsymbol{k}_1$ and the magnetic field $\boldsymbol{B}$ is very small and when

$$\varepsilon < \frac{\lambda/\rho_B{}^{\mathrm{e}}}{4\pi \sin(\theta/2)},$$

then fine structure appears in the spectral profile of the scattered light as was described in sec. 15.4b. Usually $\lambda/\rho_B{}^{\mathrm{e}}$ is small, and one may easily find the direction in which the fine structure is most prominent. Accordingly the direction of the magnetic field $\boldsymbol{B}$ can be measured to high accuracy and the current distribution in the plasma may be estimated. It has been proposed that this method be applied to Tokamak experiments.[17]

As the intensity of the scattered light is proportional to the electron density and the scattering cross section is known, it is possible to measure the electron density. Calibration of the detector system can be carried out by observing Rayleigh scattering by a neutral gas at a specified pressure (the product of the number density and the cross section for Rayleigh scattering in dry air at 1 atmosphere is $n_{\mathrm{R}}\sigma_{\mathrm{R}} = 4.0 \times 10^{-6}\,\mathrm{m}^{-1}$).

There are many applications of laser-light scattering to plasma diagnostics. The advantage of the light-scattering method is the ability to make local measurements; and this method is the most suitable for measurement of the spatial distributions of the electron temperature and density.

15.5c Bremsstrahlung Measurement[31]

The power radiated as bremsstrahlung per unit volume of plasma is

$$W(\nu)\mathrm{d}\nu = 6.3 \times 10^{-53} Z^2 \left(\frac{e}{T_\mathrm{e}}\right)^{1/2} n_\mathrm{e} n_\mathrm{i} \exp\left(\frac{-h\nu}{T_\mathrm{e}}\right)\mathrm{d}\nu \qquad (\mathrm{W/m^3}).$$

Accordingly, the electron temperature can be obtained from the slope of the measured $\ln W(\nu)$ vs. ν. The relative electron density can also be estimated from the light intensity at $h\nu \ll T_\mathrm{e}$, since $n \propto T_\mathrm{e}^{1/4}(W(\nu))^{1/2}$.

For spectroscopic measurement of the electron temperature, the range of wavelengths around $h\nu \approx T_\mathrm{e}$ is most effective, i.e., $\lambda = 12397/(T_\mathrm{e}/e)$ Å (table 15.1). The range is in the vacuum ultraviolet, extreme ultraviolet, soft X-ray, or X-ray, according to the electron temperature. Normal incidence vacuum spectrometers ($\lambda > 400$ Å), grazing incidence vacuum spectrometers ($\lambda \gtrsim 10$–20 Å), and crystal spectrometers (the lattice distance $2d$ is 15.95 Å for beryl and 26.63 Å for KAP, potassium acid phthalate) ($\lambda \lesssim 10$–20 Å) may be used. As usual, calibration of the spectral sensitivity of the detector system is necessary.

Table 15.1
Wavelength λ vs. $h\nu = T_\mathrm{e}$.

T_e (eV)	10	100	1,000	10,000
λ (Å)	1,240	124	12.4	1.24

When the electron temperature is higher than 200 eV, the absorption method is applicable. The dependence of the absorption of bremsstrahlung on the thickness of the absorbers (Be, Al, Ni, Ti, and C are commonly used) can be estimated if the spectroscopic absorption coefficients are known. Absorption is different for different electron temperatures, as shown in fig. 15.15. The electron temperature can be estimated by comparison of the observed and calculated results.[32]

When the electron temperature is higher than, say, 1 keV, pulse height analysis by means of a scintillator/photomutiplier detector system is applicable. When X-rays pass through a plastic (polyvinyl, toluene, etc.) or, typically, a NaI crystal, fluorescent light is emitted. The intensity of the fluorescent light pulse is proportional to the energy of the incident X-ray photon. Accordingly when the distribution of the pulse heights from many pulse signals from the detector is obtained, the electron

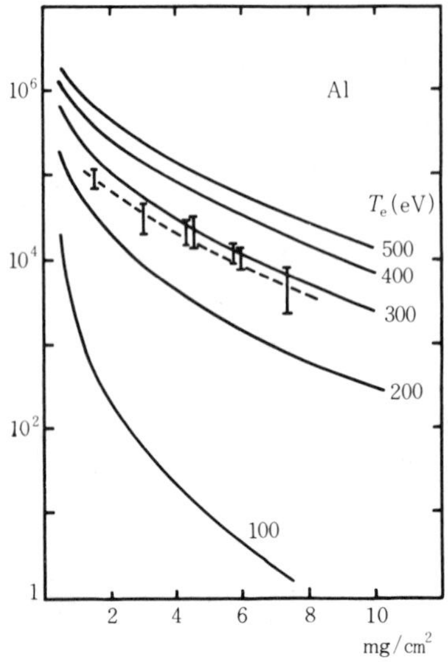

Fig. 15.15
The dependence of absorption of bremsstrahlung on the thickness, for an aluminum absorber. The parameter in the figure is the electron temperature. The ordinate is the relative intensity of absorbed bremsstrahlung and the abscissa is the thickness of the absorbing plate. The dotted curve is an experimental result. After F. C. Jahada et al: Phys. Rev. **119**, 843 (1960).

temperature can be estimated.[33] Sophisticated pulse-height analysers can be used to speed the data processing. So-called solid-state detectors are also used for soft X-ray and X-ray detection.

15.5d Density Measurement by Microwave and Laser Interferometers

When the electron density is denoted by n, the refractive index N for an electromagnetic wave of frequency f is

$$N = \left(1 - \left(\frac{f_{\rm p}}{f}\right)^2\right)^{1/2} \tag{15.70}$$

$$f_{\rm p} = \frac{1}{2\pi}\left(\frac{ne^2}{m\varepsilon_0}\right)^{1/2} = 8.98n^{1/2} \qquad ({\rm s}^{-1})$$

where $f_{\rm p}$ is the plasma frequency (n in ${\rm m}^{-3}$). When $f = f_{\rm p}$, N is zero and the electromagnetic wave is reflected (cut off). The relation of the cutoff density to the wavelength of the electromagnetic wave is

$$n_e = \frac{0.112}{\lambda^2} \times 10^{20}. \qquad (\lambda \text{ in cm}) \tag{15.71}$$

For 2 mm microwaves, the cutoff density is $n_c \approx 2.8 \times 10^{20}$ m^{-3}. When the length of plasma under investigation is d, the variation of the optical path length is $(N - 1)d$. A Mach-Zehnder or Michelson interferometer may be used for measurement of the fringe shift $(N - 1)d/\lambda$, as is shown in fig. 15.16. The fringe shift is given by

$$\frac{(N-1)\mathrm{d}}{\lambda} \approx \frac{1}{2}\left(\frac{f_p}{f}\right)^2 \frac{d}{\lambda} \approx 4.49 \times 10^{-16} d\lambda n\,,$$

where d, λ and n are in mks units. The product nd corresponding to 0.1 fringe is

$$(nd)_{0.1} = \frac{2.23 \times 10^{20}}{\lambda(\mu)} \qquad (\mathrm{m}^{-2}), \tag{15.72}$$

where λ is expressed in microns (table 15.2).

Table 15.2
The product $(nd)_{0.1}$ corresponding to a 0.1 fringe shift for a laser interferometer at the wavelength λ.

λ (μ)	0.63	10.6	340	2,000
$(nd)_{0.1}$ (m^{-2})	3.5×10^{20}	2.1×10^{19}	6.6×10^{17}	1.1×10^{17}

The wavelengths for a He-Ne laser are 0.63 μ and 3.39 μ. The wavelengths for CO_2 and HCN lasers are 10.6 μ and 340 μ, respectively.

Two-dimensional density distributions can be obtained by means of a Mach-Zehnder interferometer and a Q-switched ruby laser.[34] Interference fringes can be also obtained by double exposure interferometry or holography.[35]

15.5e Spectral Intensity and Abel Conversion

There are other optical diagnostic techniques in addition to those explained in the foregoing subsections. As long as the absolute intensity of a line, or the intensity ratio of different lines, can be measured, the electron temperature can be estimated.[36] In such measurements one must take account of many transition processes, and there are a number of approximate models that can be used to simplify the analysis. One is the local

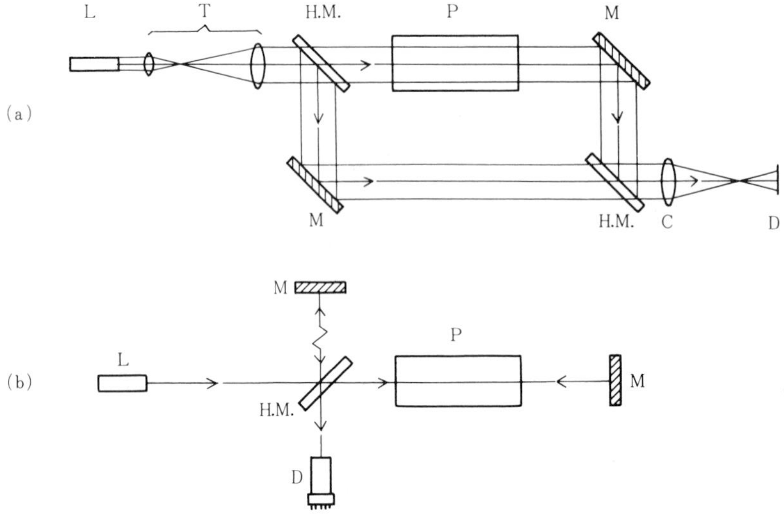

Fig. 15.16
Mach-Zehnder interferometer (a) and Michelson interferometer (b).

thermal equilibrium (LTE) model and another is the corona model. In the LTE model, it is assumed that the population density of electrons in each energy level is determined by collision processes. The velocity distribution of the free electron is taken to be Maxwellian and the population of the electrons in bound levels follows the Boltzmann distribution. These assumptions are satisfied when $n_e > 5 \times 10^{23}$ m^{-3}, so that this model is applicable to dense plasmas. In the case of low-density plasmas, the corona model is applicable. This model assumes that (i) the plasma electrons are Maxwellian; (ii) the number of electrons in excited levels is negligible compared to the number in the ground level; and (iii) the population of electrons in each excited level is determined by the balance of collisional excitation from the ground level to the excited level vs. spontaneous emission from the excited level. An appropriate pair of line spectra is taken and the dependence of their intensity ratio on the electron temperature is thus calculated, and the electron temperature is then obtained from the observed relative intensity ratio.[36] These methods are relatively easy to apply, although their accuracy depends on the validity of the assumptions made.

By measurement of the time histories of line intensities of impurity lines such as $O^{I}, O^{II}, \cdots, O^{VIII}$, it is possible to estimate the time variation of the electron temperature.[37]

The measurement of impurity-ion density is important for plasma transport or radiation loss. The identification of impurity-ion lines and the measurement of the intensities of these lines provide useful means to study the behavior of impurity ions.

Framing and streak photographs are used to observe plasma position and shape and the development of macroscopic instabilities, for which image-converter cameras are frequently used. Sometimes the schlieren method is employed.

So-called Abel conversion is often useful when techniques such as the foregoing are applied to cylindrical plasmas. When some physical quantity $\varepsilon(r)$ has rotational symmetry, and the observed quantity $I(y)$ is the average taken in the horizontal (x) direction (fig. 15.17), we have the relation

$$I(y) = 2\int_0^{(R^2-y^2)^{1/2}} \varepsilon(r)\mathrm{d}x = 2\int_y^R \frac{\varepsilon(r)r\mathrm{d}r}{(r^2-y^2)^{1/2}}. \tag{15.73}$$

The inverse statement is

$$\varepsilon(r) = -\frac{1}{\pi}\int_r^R \frac{\mathrm{d}I(y)}{\mathrm{d}y}\frac{\mathrm{d}y}{(y^2-r^2)^{1/2}}. \tag{15.74}$$

Accordingly $\varepsilon(r)$ can be estimated from the measured distribution $I(y)$. A numerical technique employing Abel conversion is described in ref. 38.

15.6 Particle Measurements

Charged particles or neutral particles escaping from a plasma and fusion-reaction neutrons can of course also provide information on plasma parameters. If a neutral-particle beam is injected and the beam-intensity damping is measured, the plasma ion density can be obtained.

15.6a Detection of Fast Ions[39]

Since ions can escape along lines of magnetic force, fast ions can be detected for open-ended systems. If the number of ions per unit energy leaving per unit area of plasma surface per unit solid angle during unit time is denoted by $J(W, t)$ and when the sensitivity of the detector is $\varepsilon(W)$, the output of the detector $I_{\mathrm{m}}(W, t)$ is

$$I_{\mathrm{m}}(W, t) = S\Omega\varepsilon(W)J(W, t)\Delta W. \tag{15.75}$$

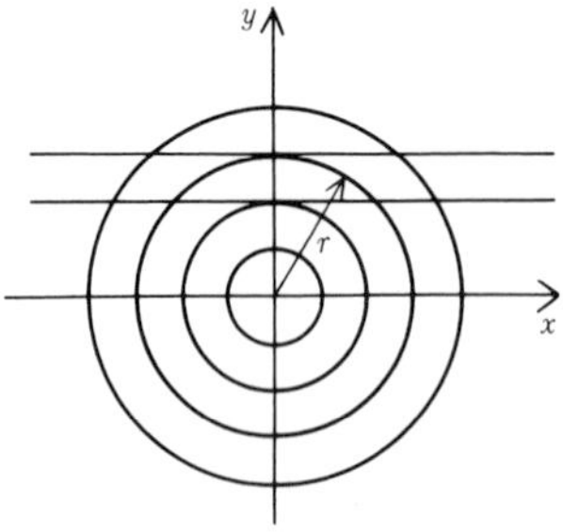

Fig. 15.17
Rotationally symmetric quantity $\varepsilon(r)$ and its average $I(y)$ in the horizontal (x) direction.

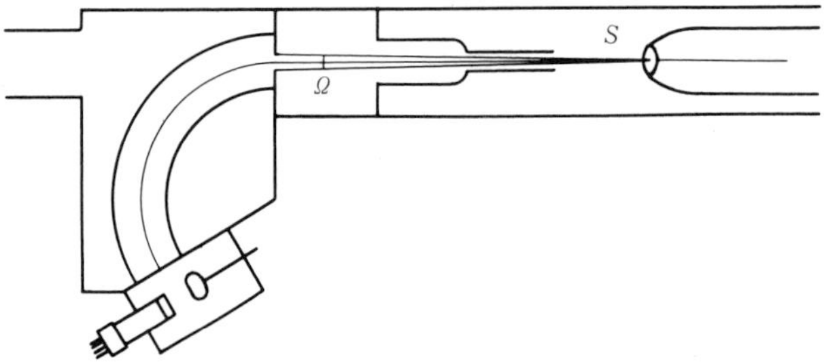

Fig. 15.18
Measurement of energy distribution of ions escaping from the surface of a plasma.

Ω is the effective solid angle (the aperture of the detector) and S is the surface area of the plasma (fig. 15.18). Accordingly,

$$J(W, t) = \frac{I_{\mathrm{m}}(W, t)}{S\Omega\varepsilon(W)\Delta W}.$$

If the normalized velocity distribution function of ions at the plasma surface is denoted by f_{s}, we have

$$J(W, t)\Omega\Delta W = \int\int n_{\mathrm{i}} v_z f_{\mathrm{s}}(v_x, v_y, v_z)\mathrm{d}v_x \mathrm{d}v_y \mathrm{d}v_z$$

$$= \Omega\frac{2n_{\mathrm{i}}}{M^2}\langle f_{\mathrm{s}}(W)\rangle W\Delta W, \tag{15.76}$$

where $W = (M/2)\cdot(v_x{}^2 + v_y{}^2 + v_z{}^2)$. Therefore the ion temperature can be obtained.

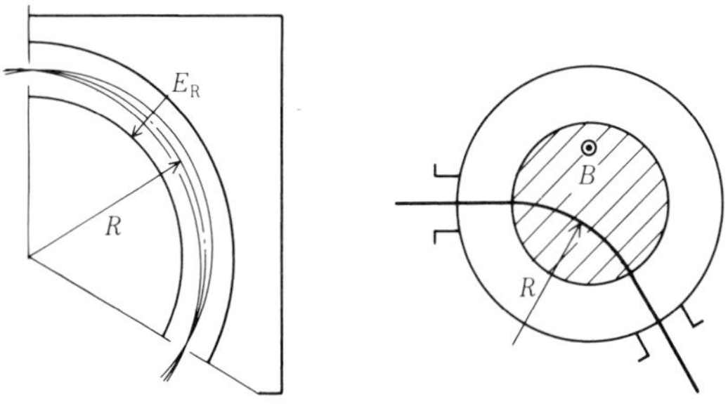

Fig. 15.19
Electrostatic analyzer (a) and magnetic analyzer (b).

Electrostatic analyzers and *magnetic analyzers* are used for analysis of charged particle (fig. 15.19). In electrostatic analyzers, the radial electric field E_R is applied as is shown in fig. 15.19a, so that the radius R of the ion orbit is given by $Mv^2/R = ZeE_R$, and ions of kinetic energy $Mv^2/2 = ZReE_R/2$ can pass through the analyzer. Accordingly this analyzer is also called an *energy analyzer*. In magnetic analyzers the magnetic field B is applied normal to the direction of the ion velocity. The radius of the ion orbit is $Mv^2/R = ZevB$ and ions of momentum $Mv = ZReB$ can pass through. Accordingly this is known as a *momentum analyzer*.

15.6b Energy Measurement of Fast Neutrals due to Charge Exchange

When charge is exchanged between a fast plasma ion P^+ and a slow neutral particle N ($P^+ + N \rightarrow P + N^+$), the slow ion N^+ is trapped while the fast neutral P escapes across the magnetic field. The cross section for charge exchange is denoted by $\sigma_{ex}(E)$. Let $I(E)$ be the number of fast neutrals of energy between E and $E + \Delta E$ escaping per unit volume of a plasma of ion density n_i, neutral density n_0, into the solid angle $d\Omega$ per unit time. When the energy distribution function of ions is $F(E)dE$, $I(E)$ is given by

$$I(E) = n_0 n_i F(E) dE \left\langle \frac{d\sigma_{ex}(E, \Omega)}{d\Omega} v \right\rangle d\Omega, \tag{15.77}$$

where v is the velocity of the fast plasma ion P^+ and $E = Mv^2/2$. Accordingly $F(E)$ can be obtained by measurement of $I(E)$, and the ion temperature can be estimated.

15.6c Density Measurement by Fast Neutral Beam

When a fast neutral beam is injected into a plasma, the fast neutrals in the beam are charge-exchanged with plasma ions or are ionized by the electrons and ions of the plasma. The fast neutrals are also scattered by plasma ions. The differential cross section of the jth process for ions of the ith kind is denoted by $\mathrm{d}\sigma_{ij}(\boldsymbol{v}_0 - \boldsymbol{v}_i)/\mathrm{d}\Omega$. Here $\boldsymbol{v}_0$ and $\boldsymbol{v}_i$ are the velocities of the neutral beam and the ions. The change Δn_0 in the beam density n_0 after penetration of a plasma to a depth l is given by

$$\Delta n_0 = -\int_0^l \int_\Omega \mathrm{d}\Omega n_0 \sum_{ij} \left\langle n_i(\boldsymbol{v}_i) \frac{\mathrm{d}\sigma_{ij}(\boldsymbol{v}_0 - \boldsymbol{v}_i)}{\mathrm{d}\Omega} |\boldsymbol{v}_0 - \boldsymbol{v}_i| \right\rangle \frac{\mathrm{d}l}{v_0}. \tag{15.78}$$

When $v_0 \gg v_i$, $\Delta n_0/n_0$ is

$$-\frac{\Delta n_0}{n_0} \approx \sum_{ij} \frac{\mathrm{d}\sigma_{ij}(v_0)}{\mathrm{d}\Omega} \Omega \int_0^l n_i \mathrm{d}l. \tag{15.79}$$

Accordingly, the densities measurable by means of this method lie near

$$\int n_i \mathrm{d}l \approx \Big(\sum_{ij} \sigma_{ij}(v_0)\Big)^{-1}.$$

When $v_i \gg v_0$, $\Delta n_0/n_0$ is

$$-\frac{\Delta n_0}{n_0} \approx \sum_{ij} \left\langle \frac{\mathrm{d}\sigma_{ij}(-\boldsymbol{v}_i)}{\mathrm{d}\Omega} v_i \right\rangle \frac{1}{v_0} \Omega \int_0^l n_i \mathrm{d}l. \tag{15.80}$$

The densities measurable here lie near $\int n_i \mathrm{d}l \approx \boldsymbol{v}_0(\langle \sigma(-\boldsymbol{v}_i) v_i \rangle)^{-1}$, and it is possible to change the region by choosing an appropriate value of v_0.

15.6d Heavy Ion Beam Probing

When a heavy ion of high energy is injected into a plasma across the magnetic field, the heavy ion can cross through the plasma if the Larmor radius is larger than the plasma size. When a singly ionized thallium ion Tl^+ of high energy (say 100 keV) is injected and the thallium ion is doubly ionized at a certain point in the plasma, one can deduce the electrostatic potential of the plasma at that point if the energy of Tl^{++} can be measured.[40] When the plasma is scanned by the ion beam, as is shown in fig. 15.20, one can observe the potential distribution across the plasma cross section.[41] The plasma density can also be measured from the intensity of the Tl^{++} beam.

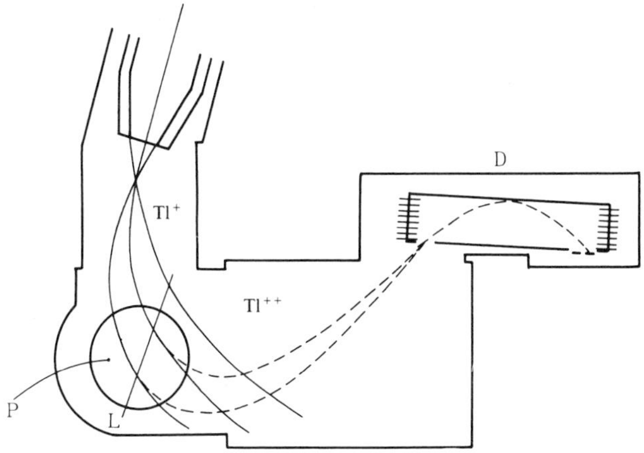

Fig. 15.20
Thallium ion beam probing. The Tl^{++} beam, which is produced at the line L in the plasma P, is detected at D.

The orbit of the Tl^{++} beam is curved by the magnetic field inside the plasma, so that one can estimate the spatial distribution of the magnetic field inside the plasma from the deflection of the orbit from that of the vacuum magnetic field. As the current density is given by $\mu_0 \boldsymbol{j} = \nabla \times \boldsymbol{B}$, some information on the current distribution can also be obtained.

15.6e Measurement of Ion Temperature by Neutron Detection[39]

The cross sections of nuclear fusion reactions are strongly energy dependent. The cross sections for D-D, D-T and D-He^3 reactions have been shown in fig. 1.1. When $W_D < 13$ keV, the cross section for the D-D reaction can be expressed by

$$\sigma_{DD} \approx \frac{182}{W_D} \exp\left(-\left(\frac{44.24}{W_D}\right)\right) \qquad \text{(barn)}.$$

W_D, the kinetic energy of the relative velocity of the deuteron, is expressed in keV, and 1 barn $= 10^{-28}$ m^2. The average $\langle \sigma v \rangle_{DD}$ for a Maxwellian deuterium plasma at ion temperature T_i^* (in keV) is given by

$$\langle \sigma v \rangle_{DD} = \frac{2.33 \times 10^{-20}}{(T_i^*)^{3/2}} \exp\left(-\frac{18.76}{(T_i^*)^{1/2}}\right) \qquad (\text{m}^3/\text{s}). \tag{15.81}$$

Since the number I_n of neutrons due to fusion per unit time per unit volume is given by

$$I_n = \frac{1}{4} n_D{}^2 \langle \sigma v \rangle_{DD}, \tag{15.82}$$

the ion temperature of a deuterium plasma can be obtained (n_D is the number density of the plasma deuterium).

15.7 Electrostatic Probe and Magnetic Probe

15.7a Electrostatic Probe[42,43]

The electrostatic probe is used for diagnostics of plasmas at low density and temperature. After its developer, it is also called the Langmuir probe. A tiny electrode is inserted into a plasma, and positive or negative bias voltage V_p is applied to the electrode. The probe current I_p is measured. From the characteristic curve (I_p vs. V_p), the local values of electron density, electron temperature, and plasma potential can be estimated. However, the disturbance of the plasma by the probe becomes serious when the plasma becomes dense and hot.

When the probe potential is equal to the space potential of the plasma, there is no net electric field around the probe. Then charged particles enter the probed region due to thermal motion. Since the velocity of electrons is much larger than that of ions, the probe current in this situation is mainly due to the electrons. As the probe potential is made more positive, electrons are accelerated to the probe, while ions are repelled. Accordingly there is built up a layer of negative charge around the probe; this layer is called the *sheath*. The thickness of the sheath is of the order of the Debye radius, and the electric field outside the sheath is zero, so that the disturbance due to the probe is limited to the region inside the sheath. Therefore, the probe current I_p is limited to the current of electrons which enter due to their thermal motion. This current is called the *electron saturation current*.

When the probe voltage V_p becomes negative with respect to the space potential V_s, electrons are repelled and ions are accelerated, so that the probe current decreases. The probe voltage at which the probe current is zero is called the *floating potential*. If the probe voltage is made more negative than this, most of the electrons are repelled and a sheath of positive charge is formed under certain conditions;[42] the probe current has then reached the *ion saturation current* (fig. 15.21).

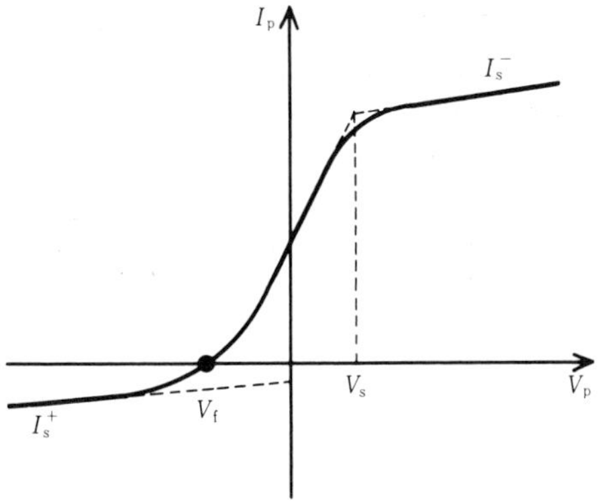

Fig. 15.21
The characteristic curve of probe current I_p vs. probe voltage V_p.

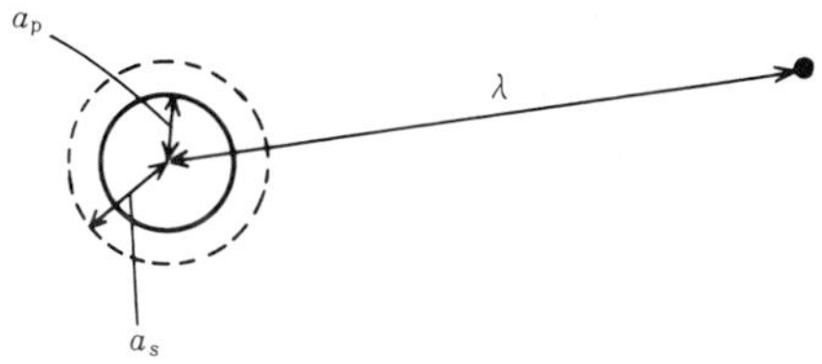

Fig. 15.22
Radius of the probe electrode a_p, sheath radius a_s, and mean free path length λ.

Let us treat the probe characteristic more analytically for the simple case in which the effect of collision on the charged particles reaching the probe from the outside of the sheath is negligible ($\lambda_d^e < \lambda$; λ_d^e is the Debye radius and λ is the mean free path), and where the thickness of the sheath is smaller than the probe radius $a_p(\lambda_d^e < a_p)$. The magnetic field is zero or weak so that the Larmor radius of the charged particles is larger than the probe radius ($a_p < \rho_B$) (fig. 15.22). It is also assumed that the velocity distribution function is Maxwellian. As the sheath is thin, all electrons which enter into the sheath can also reach the probe surface. Then the electron saturation current I_s^- is

$$I_s^- = -eJ_r A_s, \tag{15.83}$$

where A_s is the sheath surface area and J_r is given by

$$J_r = \frac{1}{4} n\bar{v} = \frac{n}{4}\left(\frac{8T_e}{\pi m}\right)^{1/2}. \tag{15.84}$$

The factor 1/4 comes from the fact that the density must be halved, since charged particles do not come from the probe; and the average of the direction cosine over the hemisphere is 1/2.

When the probe potential is below the space potential, only electrons of kinetic energy larger than $mv_0^2/2 = -e(V_p - V_s) \equiv -eV$ can reach the probe surface. If the distribution function is denoted by f, the electron flux J_r is given by

$$J_r = \int_{v_0}^{\infty} v_z f \, dv_z. \tag{15.85}$$

For a Maxwell distribution function, the random current I is

$$I = -eJ_r A_s = -e\frac{n_0}{4}\exp\left(\frac{eV}{T_e}\right)\left(\frac{8T_e}{\pi m}\right)^{1/2} A_s. \tag{15.86}$$

Deduction of the ion saturation current is not as simple as that for the electron saturation current; detailed discussions will be found in ref. 42. A useful approximate equation of the ion saturation current is

$$I_s^+ = \frac{e}{2} n_0 A_s \left(\frac{T_e}{M}\right)^{1/2}, \tag{15.87}$$

where M is the ion mass and I_s^+ does not depend on the ion temperature. These relations hold when λ_d^e, $a_p < \lambda$ and $a_p < \rho_B^i$. The slope of the characteristic curve ln I_p *vs.* V_p yields the electron temperature T_e/e and I_s^+ gives the electron density n_e (approximately). The relative density can be measured accurately, but it is advisable to calibrate for absolute sensitivity by other methods (e.g., by microwave interferometry).

When there is a magnetic field, the electron saturation current is much influenced because the electron Larmor radius ρ_B^e is usually smaller than a_p. However, if $\rho_B^i > a_p$, then eq. (15.87) for I_s^+ is still valid. The region near the ion saturation current on the characteristic curve is mainly due to the high-energy tail of the Maxwell distribution of electrons, and eqs. (15.85) and (15.86) are still valid in this region unless the magnetic field is very strong.

When a plasma is produced by gas discharge between electrodes, one electrode may be used as a reference for the plasma potential. When a plasma is produced without electrodes it is not possible to refer the voltage of a single probe to the plasma potential. In this case, two electrodes are

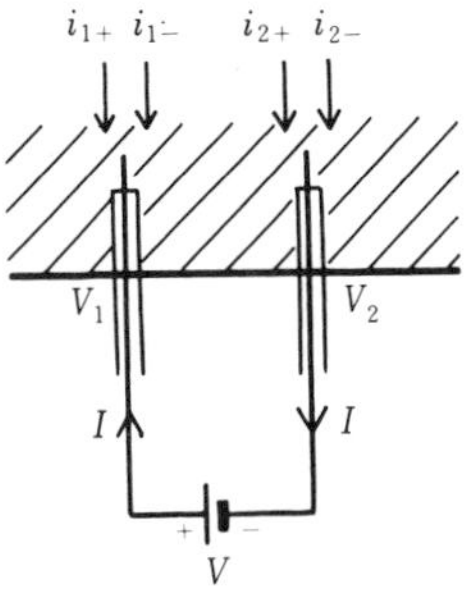

Fig. 15.23
Double probe.

inserted into the plasma and a voltage V is applied between them (fig. 15.23). Let the surface areas of probe 1 and probe 2 be A_1 and A_2 and the voltages be V_1 and V_2. Their difference is $V = V_1 - V_2$. If $V = 0$, then $V_1 = V_2 = V_f$, the floating potential. Let us assume that $V = V_1 - V_2 > 0$ and denote the ion currents and electron currents into probe 1 and probe 2 by i_{1+}, i_{1-} and i_{2+}, i_{2-}. As the *double probe* is floating as a whole, the total current entering is zero, i.e.,

$$i_{1+} + i_{1-} + i_{2+} + i_{2-} = 0. \tag{15.88}$$

The current I flowing from probe 2 to probe 1 is

$$I = i_{2+} + i_{2-} - (i_{1+} + i_{1-}). \tag{15.89}$$

Accordingly, the current I is expressed by

$$\frac{I}{2} = i_{2+} + i_{2-} = -(i_{1+} + i_{1-}).$$

As the electron currents i_{1-} and i_{2-} are given by

$$i_{1-} = -eA_1 J_r \exp\left(e\frac{(V_1 - V_s)}{T_e}\right) = -\frac{I}{2} - i_{1+}$$

$$i_{2-} = -eA_2 J_r \exp\left(e\frac{(V_2 - V_s)}{T_e}\right) = \frac{I}{2} - i_{2+},$$

it follows that

$$\frac{i_{1+} + I/2}{i_{2+} - I/2} = \frac{A_1}{A_2}\exp\left(\frac{eV}{T_e}\right).$$

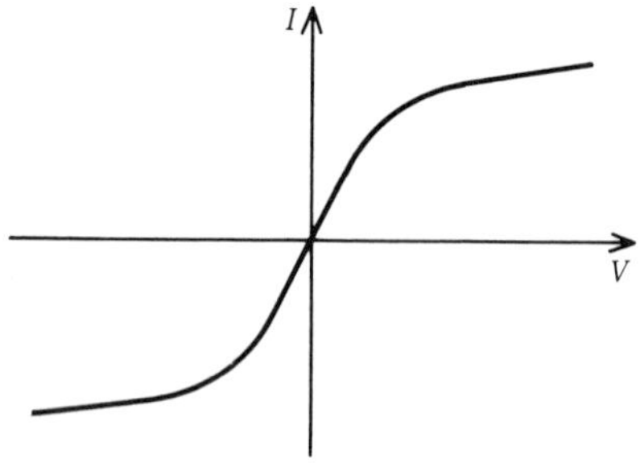

Fig. 15.24
Characteristic curve for a double probe.

If $A_1 = A_2$, then $i_{1+} = i_{2+} = i_+$ and I is reduced to (fig. 15.24)

$$I = 2i_+ \tanh\left(\frac{eV}{2T_e}\right). \tag{15.90}$$

Therefore, the electron temperature and the electron density can be estimated from the characteristic curve of the double probe. Since the electron current of the double probe samples the high-energy tail of electrons, the error may be large even if the velocity distribution function deviates only slightly from the Maxwell distribution.

It is always necessary in probe measurement to keep the probe surface clean. When the probe voltage is biased negatively, the effect of secondary electrons must be taken into account. When the ion saturation current is very small (in the very low density case) we must take into account the effect of a photoemission current due to the plasma light.

15.7b Magnetic Probe[44]

When the plasma is axially symmetric, the equation of pressure equilibrium is

$$\frac{\partial}{\partial r}\left(p + \frac{B_z{}^2 + B_\theta{}^2}{2\mu_0}\right) = -\frac{B_\theta{}^2}{r\mu_0} \tag{15.91}$$

(sec. 7.1), where $p(r)$ is the plasma pressure and $B_\theta(r)$, $B_z(r)$ are the θ and z components of the magnetic field. Accordingly if $B_\theta(r)$ and $B_z(r)$ are measured, $p(r)$ can be obtained.[44]

From the equation of pressure balance

$$p = \frac{1}{2\mu_0}(B_{out}{}^2 - B_{in}{}^2), \tag{15.92}$$

the integral of the pressure distribution over the plasma cross section is

$$\int p\mathrm{d}S = \frac{1}{2\mu_0}\int (B_{out}{}^2 - B_{in}{}^2)\mathrm{d}S \approx \frac{B_{out}}{\mu_0}\int (B_{out} - B_{in})\mathrm{d}S. \tag{15.93}$$

As $\int_S (B_{out} - B_{in})\mathrm{d}S$ is the change of magnetic flux over the cross section S, this quantity can be measured by the time integration of the voltage induced in the loop L shown in fig. 15.25a, so that the average of the pressure $\langle p \rangle = S^{-1}\int p\mathrm{d}S$ is estimated. The small magnetic probe M in fig. 15.25a is used to measure the local magnetic field.

When current flows in the plasma, then

$$\oint \boldsymbol{H}\mathrm{d}l = \int_S j\mathrm{d}S = \boldsymbol{I}_s.$$

A *Rogowsky coil* is shown in fig. 15.25b; the induced voltage is

$$V_R = \frac{\mathrm{d}}{\mathrm{d}t}\left(\mu_0 an \oint \boldsymbol{H}\mathrm{d}l\right) = na\mu_0 \frac{\mathrm{d}\boldsymbol{I}_s}{\mathrm{d}t}, \tag{15.94}$$

where a is the cross section of the small loops and n is the number of turns per unit length along the coil. Integration of the voltage induced in the Rogowsky coil gives the total current which passes the region encircled by the coil.

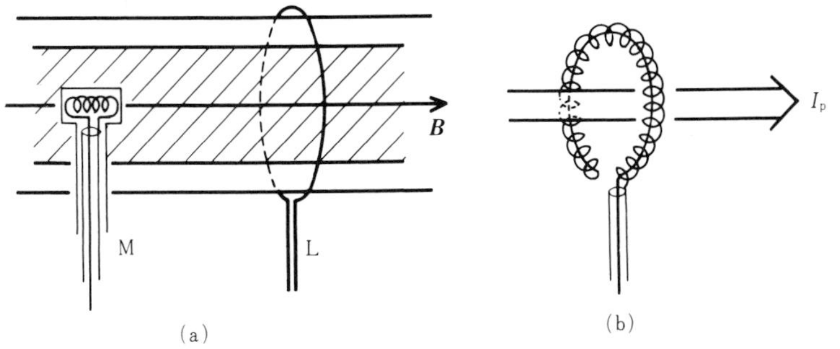

Fig. 15.25
(a) Magnetic probe M and magnetic loop L. (b) Rogowsky coil.

15.8 Applicability of Various Diagnostics

So far various methods of plasma diagnostics have been described. Each method has its own region of applicability, which will be briefly discussed in this section. For the measurement of ion temperature, Doppler broadening of the line spectrum, collective scattering of laser light, fast neutral measurement, and neutron detection can be used. The regions where these methods are applicable are shown on the n vs T_i diagram of fig. 15.26.

Diagnostics for electron temperature are based on laser scattering and bremsstrahlung. The regions of applicability are shown in fig. 15.27. The lower limit of measurable density for laser scattering depends on the laser output power (or energy).

The measurable region of electron densities by means of interferometry is very wide, as coherent sources of microwave, far infrared, infrared, and visible light are available (fig. 15.28). Density measurement by laser scattering is important, as local density measurement is possible with this technique (when $n_e > 10^{18}\,m^{-3}$). Methods involving damping of neutral beams form a distinct category, as ion density is directly obtained, instead of the electron density.

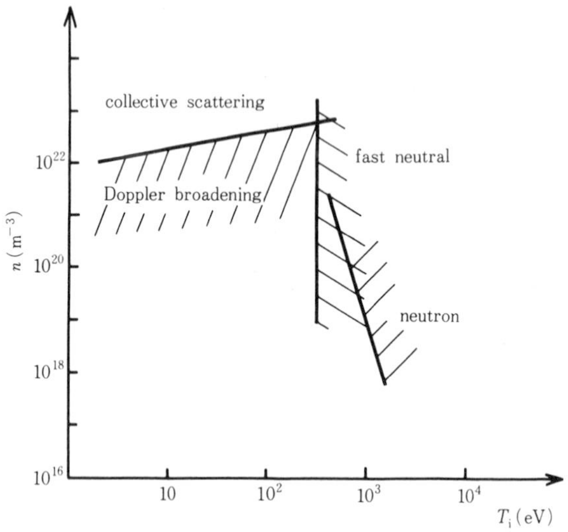

Fig. 15.26
Regions of applicability of methods for measurement of ion temperature. The boundaries are not as sharp as is shown in the figure.

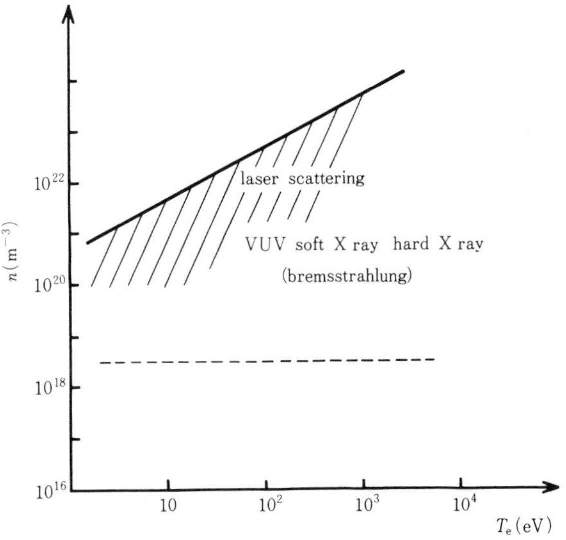

Fig. 15.27
Regions of applicability of methods for measurement of electron temperature. The boundaries are not as sharp as is shown in the figure.

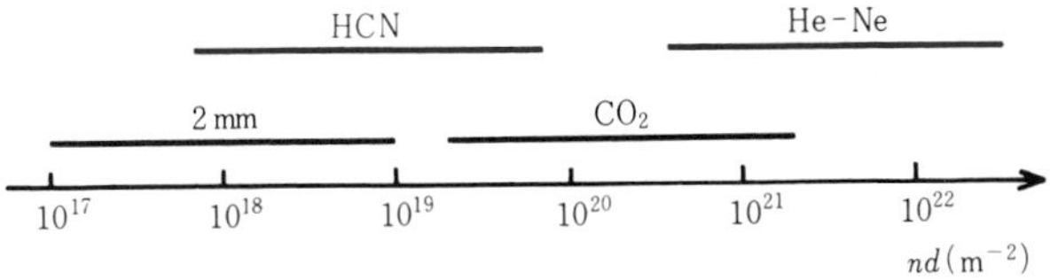

Fig. 15.28
The electron-density regions corresponding to 0.1–10 fringes for various microwave and laser interferometers.

References

1. R. H. Huddlestone and S. L. Leonard (es).: Plasma Diagnostic Techniques, Academic Press, New York, 1965

2. W. Lochte-Holtgreven (ed.): Plasma Diagnostics, North-Holland, Amsterdam, 1968

3. L. Marton (editor-in-chief): Methods of Experimental Physics vol. 9, pt. A, Plasma Physics A (ed. by H. R. Griem and R. Lovberg), Academic Press, New York, 1970

4. S. C. Brown: Basic Data of Plasma Physics, MIT Press, Cambridge, Mass., 1966

4a. See, for example, L. I. Schiff: Quantum Mechanics, p. 109, McGraw-Hill, New York, 1955

5. A. Sommerfeld: Atombau und Spektrallinien Band. II, Kap. 7, Vieweg, Braunschweig, 1960

6. W. J. Karzas and R. Latter: Astrophys. J., suppl. VI, 167 (1961)

7. T. Kihara: Kakuyogo Kenkyu **5**, 16(1960) (in Japanese)

8. J. L. Hirshfield, D. E. Baldwin, and S. C. Brown: Phys. of Fluids **4**, 198(1961)
G. Bekefi: Radiation Processes in Plasmas, Wiley, New York, 1966
D. J. Rose and M. Clark, Jr.: Plasmas and Controlled Fusion, MIT Press, Cambridge, Mass. 1961

9. B. A. Trubnikov and V. S. Kudryavtsev: Proc. 2nd UN Conf. on Peaceful Uses of Atomic Energy 31, p. 93, United Nation, Geneva (1958)

10. M. N. Rosenbluth: Nucl. Fusion **10**, 340(1970)

11. W. E. Gordon: Proc. JRE **46**, 1824(1958)

12. K. L. Bowles: Phys. Rev. Lett. **1**, 454(1958)

13. E. E. Salpeter: Phys. Rev. **120**, 1528(1960)

14. T. Laaspere: J. Geophys. Research **65**, 3955(1960)

15. J. A. Fejer: Canadian J. of Phys. **39**, 716(1961)

16. E. E. Salpeter: Phys. Rev. **122**, 1663(1961)

17. J. Sheffield: Plasma Physics **14**, 385(1972)

18. A. W. DeSilva and G. C. Goldenbaum: Chapter 3 of ref. 3

19. K. Miyamoto: Oyo-butsuri **35**, 772(1966) (in Japanese)

20. E. U. Condon and G. H. Shortley: The Theory of Atomic Spectra, Cambridge, 1959

21. H. Margenau and M. Lewis: Rev. Mod. Phys. **31**, 569(1959)

22. A. Unsöld: Physik der Sternatmosphären 2 Aufl. Springer, Berlin 1955
M. Baranger: Atomic and Molecular Processes (ed. by D. R. Bates), Academic Press, New York, 1962

23. H. R. Griem: Plasma Spectroscopy, McGraw-Hill, New York, 1964

24. H. R. Griem, A. C. Kolb and K. Y. Shen: Phys. Rev, **116**, 4(1959)

25. E. B. Turner: Chapter 7 of ref. 1

26. G. R. Harrison, J. E. Archer and J. Camus: J. Opt. Soc. Amer. **42**, 706(1952)

27. K. Miyamoto, K. Mori, K. Ando, M. Otsuka, T. Ishimura, J. Fujita, S. Kawasaki, N. Inoue, Y. Suzuki, and T. Uchida: Japan. J. Appl. Phys. **5**, 970(1966)

28. J. G. Hirschberg and P. Platz: Appl. Opt. **4**, 1375(1965)

29. H. J. Kunze: Chapter 9 of ref. 2

30. M. J. Forrest, N. J. Peacock, D. C. Robinson, V. V. Sannikov, and P. D. Wilcock: CLM-R 107, Culham Laboratory, July 1970

31. T. F. Stratton: Chapter 8 of ref. 1
R. Lincke: Chapters 6, 7 of ref. 2

32. G. A. Sawyer, F. C. Jahoda, F. L. Ribe, and T. F. Stratton: J. Quant. Spectrosc. Radiat. Transfer **2**, 467(1962)
F. C. Jahoda, E. M. Little, W. E. Quinn, G. A. Sawyer, and T. F. Stratton: Phys. Rev. **119**, 843(1960)

33. ed. by S. M. Shafroth, Scintillation Spectroscopy of Gamma Radiation vol. **1**, Gordon and Breach, London, 1966

34. W. E. Quinn, E. M. Little, F. L. Ribe, and G. A. Sawyer: Plasma Phys. and Controlled Nucl. Fusion Research **1**, 237 (Conf. Proceedings, Culham 1965) (IAEA, Vienna, 1966)

35. F. C. Jahoda, R. A. Jeffries, and G. A. Sawyer: Applied Optics **6**, 1407(1967)

36. R. W. P. McWhirter: Chapter 5 of ref. 1

37. L. M. Goldman and R. W. Kilb: Plasma Phys. **6**, 217(1964)

38. W. Lochte-Holtgreven: Chapter 3 of ref. 2
C. Fleurier and J. Chapelle: Computer Phys. Communications **7**, 200(1974)

39. J. E. Osher: Chapter 12 of ref. 1

40. R. L. Hickok, F. C. Jobes: AFOSR TR-72-0018, Mobil Research and Developement Corporation, Princeton, NJ, Feb. 1972

41. J. Hosea, F. C. Jobes, R. L. Hickok, and A. N. Dellis: Phys. Rev. Lett. **30**, 839(1973)

42. F. F. Chen: Chapter 4 of ref. 1

43. L. Schott: Chapter 11 of ref. 2

44. R. H. Lovberg: Chapter 3 of ref. 1

16 Confinement of High-Temperature Plasmas

The major research effort in the area of controlled nuclear fusion is focused on the confinement of hot plasmas by means of strong magnetic fields. The magnetic confinements most actively studied are toroidal and mirror configurations. Confinement in a linear mirror field may have advantages over toroidal confinement with respect to stability and anomalous diffusion across the magnetic field. However, the end loss due to particles leaving along lines of magnetic force is determined solely by diffusion in the velocity space; that is, the confinement time cannot be improved by increasing the intensity of the magnetic field or the plasma size. It is necessary to find ways to suppress the end loss.

Toroidal magnetic confinements have no open end. In the simple toroidal field, ions and electrons drift in opposite directions due to the gradient of the magnetic field. This gradient-B drift causes the charge separation that induces the electric field $\boldsymbol{E}$ directed parallel to the major axis of the torus. The subsequent $\boldsymbol{E} \times \boldsymbol{B}$ drift tends to carry the plasma ring outward, as described in sec. 3.4. In order to reduce the $\boldsymbol{E} \times \boldsymbol{B}$ drift, it is necessary to connect the upper and lower parts of the plasma by lines of magnetic force and to short-circuit the separated charges along these field lines. Accordingly, a poloidal component of the magnetic field is essential to the equilibrium of toroidal plasmas, and toroidal devices may be classified according to the method used to generate the poloidal field. The tokamak and the reversed field pinch devices use the plasma current along the toroid, whereas the toroidal stellarator has helical conductors outside the plasma that produce appropriate rotational transform angles.

Besides the study of magnetic confinement systems, inertial confinement approaches are being actively investigated. If a very dense and hot plasma could be produced within a very short time, it might be possible to complete the nuclear fusion reaction and to satisfy the Lawson criterion before the plasma starts to expand. An extreme example is a hydrogen bomb. This type of confinement is called *inertial confinement*. In laboratory experiments, high-power laser beams or particle beams are focused onto small solid deuterium and tritium targets, thereby producing very dense, hot plasma within a short time. Because of the development of the technologies of high-power energy drivers, the approaches along this line have some foundation in reality. At the end of this chapter inertial confinement will be discussed briefly.

The various kinds of approaches that are actively investigated in controlled thermonuclear fusion are classified as follows:

- Magnetic confinement
  - Toroidal system
    - Axially symmetric
      - Tokamak
      - Reversed field pinch
      - Spheromak
    - Axially asymmetric
      - Stellarator system
      - Heliac
      - Bumpy torus
  - Open system
    - Mirror, tandem mirror
    - Field reversed configuration
    - Cusp
- Inertial confinement
  - Laser
  - Ion beam, electron beam

16.1 History of Confinement Research

Basic research into controlled thermonuclear fusion probably began right after World War II in the United States, the Soviet Union, and the United Kingdom in strict secrecy. There are on record many speculations about research into controlled thermonuclear fusion even in the 1940s. The United States program, called Project Sherwood, has been described in detail by Bishop.[1] Bishop states that Z pinch experiments for linear and toroidal configurations at the Los Alamos Scientific Laboratory were carried out in an attempt to overcome sausage and kink instabilities. The astrophysicist L. Spitzer, Jr., started the figure-eight toroidal stellarator project at Princeton University in 1951. At the Lawrence Livermore National Laboratory, mirror confinement experiments were conducted. At the Atomic Energy Research Establishment in Harwell, United Kingdom, the Zeta experiment was started,[2] and at the I. V. Kurchatov Institute of Atomic Energy in the Soviet Union, experiments on a mirror called Ogra and on tokamaks were carried out.[3]

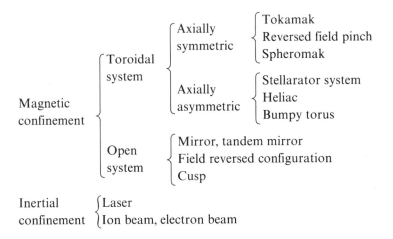

The first United Nations International Conference on the Peaceful Uses of Atomic Energy was held in Geneva in 1955. Although this conference was concerned with peaceful applications of nuclear fission, the chairman, H. J. Bhabha, hazarded the prediction that ways of controlling fusion energy that would render it industrially usable would be found in less than two decades. However, as we have seen, the research into controlled nuclear fusion encountered serious and unexpected difficulties. It was soon recognized that the realization of a practical fusion reactor was a long way off and that basic research on plasma physics and the international exchange of scientific information were absolutely necessary. From around that time articles on controlled nuclear fusion started appearing regularly in academic journals. Lawson's paper on the conditions for fusion was published in January 1957,[4] and several important theories on MHD instabilities had by that time begun to appear.[5,6] Experimental resuits of the Zeta[7] (Zero Energy Thermonuclear Assembly) and Stellarator[8] projects were made public in January 1958. In the fusion sessions of the second United Nations International Conference on the Peaceful Uses of Atomic Energy, held in Geneva, September 1–13, 1958,[9,10] many results of research that had proceeded in secrecy were revealed. Artsimovich[12] expressed his impression of this conference as "something that might be called a display of ideas." The second UN conference marks the start of open rather than secret international cooperation and competition in fusion research.

In Japan the Atomic Energy Committee set up a division of nuclear fusion research in 1958 under the chairmanship of H. Yukawa. The division began to inquire into the general direction that scientific activity might take. Thus the Institute of Plasma Physics was established at Nagoya University in 1961, with K. Husimi as director, as the center of cooperative research activities among the universities.[11]

The First International Conference on Plasma Physics and Controlled Nuclear Fusion Research was held in Salzburg in 1961 under the auspices of the International Atomic Energy Agency (IAEA). At the Salzburg conference[12] the big projects were fully discussed. Some of there were Zeta, Alpha, Stellarator C, Ogra, and DCX. Theta pinch experiments (Scylla, Thetatron, etc.) appeared to be more popular than linear pinches. The papers on the large-scale experimental projects such as Zeta or Stellarator C all reported struggles with various instabilities. L. A. Artsimovich said in the summary on the experimental results: "Our original beliefs that the doors into the desired regions of ultra-high temperature would open smoothly ... have proved as unfounded as the sinner's hope of entering Paradise without passing through Purgatory."

The importance of the PR-2 experiments of M. S. Ioffe and others was soon widely recognized (vol. 3, p. 1045). These experiments demonstrated that a plasma confined in a minimum-B configuration is MHD stable.

The Second International Conference on Plasma Physics and Controlled Nuclear Fusion Research was held at Culham in 1965.[13] The stabilizing effect of minimum-B configurations was confirmed by many experiments. An absolute minimum-B field cannot be realized in a toroidal configuration. Instead of this, the average minimum-B concept was introduced (vol. 1, pp. 103, 145). Ohkawa and others succeeded in confining plasmas for much longer than the Bohm time with toroidal multipole configurations (vol. 2, p. 531) and demonstrated the effectiveness of the average minimum-B configuration. Artsimovich and others reported on a series of tokamak experiments (T-5, vol. 2, p. 577; T-3, p. 595; T-2, p. 629; TM-2, p. 647; TM-1, p. 659). Further experiments with Zeta and Stellarator C were also reported. However, the confinement times for these big devices were only of the order of the Bohm time, and painful examinations of loss mechanisms had to be carried out. In the Wendelstein experiments WIa and Ib (vol. 2, p. 719), a steady-state cesium plasma of low temperature was produced, and it was shown that the diffusion in these cases could be explained in terms of classical diffusion. Theta pinch experiments were still the most actively pursued. The ion temperatures produced by means of theta pinches were several hundred eV to several keV, and confinement times were limited only by end losses. One of the important goals of the theta pinch experiments had thus been attained, and it was a turning point from linear theta pinch to toroidal pinch experiments.

In this conference, the effectiveness of minimum-B, average minimum-B, and shear configurations was thus confirmed. Many MHD instabilities were seen to be well understood experimentally as well as theoretically. Methods of stabilizing against MHD instabilities seemed to be becoming gradually clearer. The importance of velocity-space instabilities due to the non-Maxwellian distribution function of the confined plasma was recognized. There had been and were subsequently to be reports on loss-cone instabilities[17] (1965), Harris instability[18] (1959), drift instabilities[19] (1963, 1965), etc. The experiment by J. M. Malmberg and C. B. Wharton was the first experimental verification of Landau damping.

L. Spitzer, Jr., concluded in his summary talk at Culham that "most of the serious obstacles have been overcome, sometimes after years of effort by a great number of scientists. We can be sure that there will be many obstacles ahead but we have good reason to hope that these will be surmounted by the cooperative efforts of scientists in many nations."

The Third International Conference[14] was held in 1968 at Novosibirsk. The most remarkable topic in this conference was the report that Tokamak T-3 (vol. 1, p. 157) had confined a plasma up to 30 times the Bohm time (several milliseconds) at an electron temperature of 1 keV. In Zeta experiments a quiescent period was found during a discharge, and MHD stability of the magnetic field configuration of the quiescent period was discussed. This was the last report of Zeta, and HBTX succeeded this reversed field pinch experiment. Stellarator C (vol. 1, pp. 479, 495) was still confining plasmas only to several times the Bohm time at electron temperatures of only several tens to a hundred eV. This was the last report on Stellarator C; this mechine was converted into the ST tokamak before the next conference (Madison 1971).[15] However, various aspects of stellarator research were still pursued. The magnetic coil systems of Clasp (vol. 1, p. 465) were constructed accurately, and the confinement of high-energy electrons were examined using the β decay of tritium. It was demonstrated experimentally that the electrons ran around the torus more than 10^7 times and that the stellarator field had good charge-particle confinement properties. In WII the confinement of a barium plasma was tested, and resonant loss was observed when the magnetic surface was rational. Diffusion in a barium plasma in nonrational cases was classical. In 2X (vol. 2, p. 225) a deuterium plasma was confined up to an ion temperature of 6–8 keV at a density of $n = 5 \times 10^{13}\ \mathrm{cm}^{-3}$ for up to $\tau = 0.2$ ms. Laser plasmas appeared at this conference.

At the Novosibirsk conference toroidal confinement appeared to have the best overall prospects, and the mainstream of research shifted toward toroidal confinement. Artsimovich concluded this conference, saying: "We have rid ourselves of the gloomy specter of the enormous losses embodied in Bohm's formula and have opened the way for further increases in plasma temperature leading to the physical thermonuclear level."

The Tokamak results were seen to be epoch making if the estimates of the electron temperature were accurate. Pease, the director of the Culham Laboratory, sent a team of researchers to Kurchatov to measure the electron temperature of the T-3 plasma by laser scattering methods. The measurements supported the previous estimates by the tokamak group.[20] The experimental results of T-3 had a strong impact on the next phase of nuclear fusion research in various nations. At the Princeton Plasma Physics Laboratory, Stellarator C was converted to the ST tokamak device; newly built were ORMAK at Oak Ridge National Laboratory, TFR at the Center for Nuclear Research, Fontaney aux Rose, Cleo at the Culham Laboratory, Pulsator at the Max Planck Institute for Plasma Physics, and JFT-2 at the Japan Atomic Energy Research Institute.

The Fourth International Conference was held in Madison, Wisconsin, in 1971.[15] The main interest at Madison was naturally focused on the tokamak experiments. In T-4 (vol. 1, p. 443), the electron temperature approached 3 keV at a confinement time around 10 ms. The ions were heated to around 600 eV by collision with the electrons. ST (vol. 1, pp. 451, 465) produced similar results.

Since then the IAEA conference has been held every two years: Tokyo in 1974,[16] Berchtesgarden in 1976, Innsburg in 1978, Brussels in 1980, Baltimore in 1982, London in 1984, and Kyoto in 1986. Tokamak research has made steady progress as the mainstream of magnetic confinement. After the tokamaks of the first generation (T-4, T-6, ST, ORMAK, Alcator A, C, TFR, Pulsator DITE, FT, JFT-2, JFT-2a, JIPP T-II, etc.), second generation tokamaks (T-10, PLT, PDX, ISX-B, Doublet III, ASDEX, etc.) began appearing around 1976. The energy confinement time of ohmically heated plasmas was approximately described by the Alcator scaling law ($\tau_E \propto na^2$). The value of $\bar{n}\tau_E$ reached 2×10^{13} $cm^{-3}s$ in Alcator A in 1976. Heating experiments of neutral beam injection (NBI) in PLT achieved the ion temperature of 7 keV in 1978, and the effective wave heating in an ion cyclotron range of frequency was demonstrated in TFR and PLT around 1980. The average β value of 4.6% was realized in the Doublet III noncircular tokamak ($\kappa = 1.4$) in 1982 using 3.3 MW NBI.

Noninductive drives for plasma current have been pursued. Current drive by the tangential injection of a neutral beam was proposed by Ohkawa in 1970 and was demonstrated in DITE experimentally in 1980. Current drive by a lower hybrid wave was proposed by Fisch in 1978 and demonstrated in JFT-2 in 1980 and then in Versator 2, PLT, Alcator C, JIPP T-II, Wega, T-7, etc. Ramp-up experiments of plasma current from 0 were succeeded by WT-2 and then by PLT and JIPP T-II in 1984.

The suppression of impurity ions by a diverter was demonstrated in JFT-2a (DIVA) in 1978 and was investigated by ASDEX and Doublet III in detail (1982). At that time the energy confinement time had deteriorated compared with the ohmic heating case as the heating power of NBI was increased (according to the Kaye-Goldston scaling law). However, the improved mode (so-called H mode) of the confinement time—increased by 2 or 3 times compared with the ordinary mode (L mode)—was found in the diverter configuration of ASDEX in 1982. The H mode was also observed in Doublet III, PDX, JFT-2M, and DIII-D. Thus much progress had been made to solve many critical issues of tokamaks. Based on these achievements, experiments of third-generation large tokamaks started, with TFTR (United States) in the end of 1982, JET (European Community) in 1983, JT-60 (Japan) in 1985, and T-15 (Soviet Union) in 1988 (scheduled). TFTR achieved $n_e(0)\tau_E \sim 10^{13}$ cm^{-3}s, $T_i(0) = 20$ keV by supershot (H mode–like), and JET achieved $n_e(0)\tau_E \sim 2 \times 10^{13}$ cm^{-3}s, $T_i(0) = 10$ keV by diverter configuration (H mode) in 1986. JT-60 drove a plasma current of 1.7 MA ($\bar{n}_e = 0.3 \times 10^{13}$ cm^{-3}) by lower hybrid wave ($P_{\mathrm{R.F.}} = 1.2$ MW) in 1986 and achieved $\bar{n}_e\tau_E = 0.2 \times 10^{13}$ cm^{-3}, $T_i(0) = 11$ keV in 1987. Now these large tokamaks are aiming at the scientific demonstration of the critical condition of fusion reactors (fusion output power = heating input power).

Alternative methods have been investigated intensively to catch up with the achievements of tokamaks.

The stellarator program proceeded from small-scale experiments (Wendelstein IIb, Clasp, Uragan-1, L-1, JIPP-I, Heliotron D) to middle-scale experiments (Wendelstein VIIA, Cleo, Uragan-2, L-2, JIPP T-II, Heliotron E). The plasmas with $T_e \sim T_i \approx$ several hundred eV to 1 keV, $\bar{n}_e \sim$ several 10^{13} cm^{-3} were sustained by NBI heating without an ohmic heating current, and the possibility of steady-state operation of stellarators was demonstrated by WVIIA and then Heliotron E. The projects ATF and W VII AS are proceeding.

The reversed field pinch (RFP) configuration was found in the stable quiescent period of Zeta discharge just before the shutdown in 1968. J. B. Taylor pointed out that RFP configuration is the minimum energy state under the constraint of the conservation of magnetic helicity in 1974 (see sec. 16.3). RFP experiments have been conducted in HBTX 1B, ETA-BETA 2, TPE-IRM, TPE-IR15, ZT-40M, OHTE, REPUTE-1, STP-3M. An average β of 10–15% was realized. ZT-40M demonstrated that RFP configuration can be sustained by relaxation phenomena (the so-called dynamo effect) as long as the plasma current is sustained (1982). The projects FRX and ZT-H/CPRF are proceeding.

Spheromak configurations have been studied by Proto S-1, S-1, CTX, and CTCC-1, and field reversed configurations have been studied by FRX, TRX, OCT, and NUCTE.

In mirror research, 2XIIB confined a plasma with an ion temperature of 13 keV and $n\tau_E \sim 10^{11}$ cm^{-3}s in 1976. However, the suppression of the end loss is absolutely necessary. The concept of a tandem mirror, in which end losses are suppressed by electrostatic potential, was proposed in 1976–

1977 (sec. 16.5). TMX, TMX-U, PHAEDRUS, TARA, GAMMA 10, and AMBAL are the tandem mirror projects.

The bumpy torus is the toroidal linkage of many mirrors to avoid end loss, and this method was pursued in EBT and NBT.

Inertial confinement research has made great advances in the implosion experiment by using a Nd glass laser as the energy driver. Gekko XII (30 kJ, 1 ns, 12 beams), Nova (100 kJ, 1 ns, 10 beams), Omega X (4 kJ, 1 ns, 24 beams), and Octal (2 kJ, 1 ns, 8 beams) investigated implosion using laser light of $\lambda = 1.06$ μm and its higher harmonics $\lambda = 0.53$ μm and 0.35 μm. It was shown that a short wavelength is favorable because of the better absorption and less preheating of the core. A high-density plasma, 100 times as dense as the solid state, was produced by laser implosion. CO_2 lasers (Antares, Lekko VIII), iodine lasers (Asterix III), and KrF lasers have been developed as the energy driver. Light ion beam drivers have also been developed in PBFA I, PBFA II, and Reiden IV.

Nuclear fusion research has been making steady progress through international collaboration and competition.[21,22] R. S. Pease stated in his summary talk of the IAEA conference at Berchtesgarden in 1976 that "one can see the surprisingly steady progress that has been maintained. Furthermore, looked at logarithmically, we have now covered the greater part of the total distance. What remains is difficult, but the difficulties are finite and can be summed up by saying that we do not yet have an adequate understanding or control of cross-field electron thermal conduction."

A summary of the progress of magnetic confinement is given in the $\bar{n}_e \tau_E - T_i(0)$ diagram in fig. 16.1.

16.2 Tokamaks

The word "tokamak" is said to be a contraction of the Russian words for current (ТОК), vessel (КАМЕР), magnet (МАГНЕТ), and coil (КАТУЩКА). Tokamaks are axisymmetric, with the plasma current itself giving rise to the poloidal field essential to the equilibrium of toroidal plasmas. In a tokamak the toroidal field, used to stabilize against MHD instabilities, is strong enough to satisfy the Kruskal-Shafranov condition. This characteristic is quite different from that of reversed field pinch, with its relatively weak toroidal field. There are excellent reviews and textbooks of tokamak experiments,[23,24] equilibrium,[25] and diagnostics.[26,27]

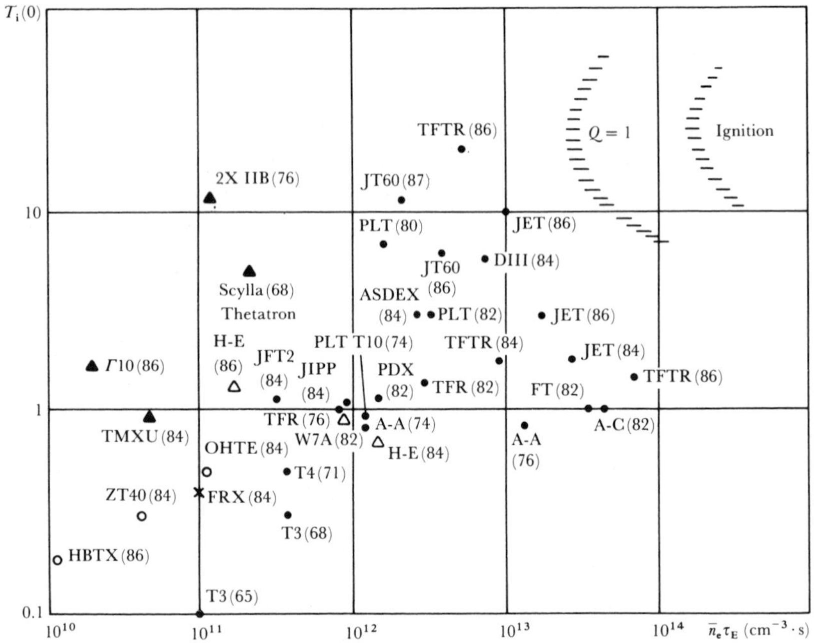

Fig. 16.1
Development of confinement experiments in $\bar{n}_e\tau_E - T_i(0)$ diagram ($\bar{n}_e$ is the line average electron density; τ_E is the energy confinement time; $T_i(0)$ is the ion temperature at the center). ●, tokamak. ○, RFP. △, stellarator. ▲, mirror, tandem mirror, theta pinch. ×, field reversal configuration. $Q = 1$ is the critical condition.

16.2a Tokamak Devices

The structure of the JET device is shown in fig. 16.2 as a typical example. The toroidal field coils, equilibrium field coils (also called the poloidal field coils, which produce the vertical field and shaping field), ohmic heating coils (the primary windings of the current transformer), and vacuum vessel can be seen in the figure. Sometimes "poloidal field coils" means both the equilibrium field coils and the ohmic heating coils. By raising the current of the primary windings of the current transformer (ohmic heating coils), a current is induced in the plasma, which acts as the secondary winding. In the JET device, the current transformer is of the iron core type. The air-core type of current transformer is also utilized in many tokamak devices. The vacuum vessel is usually made of thin stainless steel or inconel so that it has enough electric resistance in the toroidal direction. Therefore the

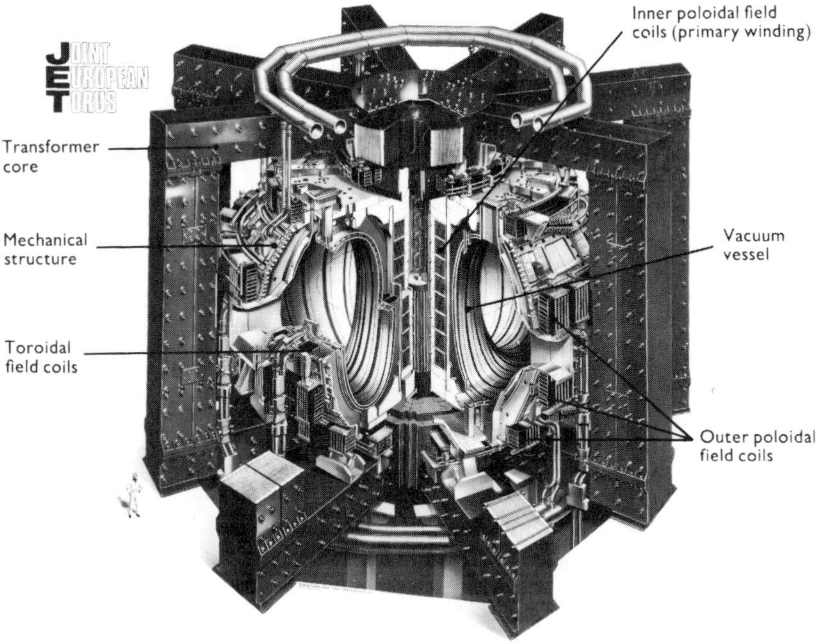

Fig. 16.2
Artist's drawing of JET (Joint European Torus). The toroidal field coils are arranged around the vacuum vessel. The outer poloidal field coils (equilibrium field coils) and inner poloidal field coils (ohmic heating coils) are wound in the toroidal direction outside the toroidal field coils. JET uses an ion-core current transformer. The mechanical structures support the toroidal field coils against the large amount of torque due to the equilibrium field. (JET Joint Undertaking, Abingdon, Oxfordshire, England. Reprinted with permission.)

Table 16.1
Parameters of tokamak devices.

	$R(m)$	$a(\times b)$	R/a	$B_\varphi(T)$	$I_p(MA)$	Remarks
T-4	1.0	0.17	5.9	5.0	0.3	
T-10	1.5	0.39	3.8	5.0	0.65	
PLT	1.32	0.4	3.3	3.2	0.5	
TFTR	2.48	0.85	2.9	6.2	2.5	high field
JET	2.96	1.25($\times$2.1)	2.4	3.45	5.1	noncircular
JT-60	3.03	0.95	3.2	4.5	2.7	diverter
T-15	2.4	0.75	3.2	3.5(4.5)	1.4(2.0)	Nb_3Sn TF coil

voltage induced by the primary windings can penetrate it. The thin vacuum vessel is called the liner. Before starting an experiment, the liner is outgassed by baking at a temperature of 150–400°C for a long time under high vacuum. Furthermore, before running an experiment, a plasma is run with a weak toroidal field in order to discharge-clean the wall of the liner. Inside the liner there is a diaphragm made of tungsten, molybdenum, or graphite that limits the plasma size and minimizes the interaction of the plasma with the wall. This diaphragm is called a limiter. Recently a diverter configuration was introduced instead of the limiter. In this case the magnetic surface, including the separatrix, determines the plasma boundary (see sec. 16.2e). A conducting shell surrounds the plasma outside the liner and is used to maintain the positional equilibrium or to stabilize MHD instabilities during the *skin time* scale. The magnitude of the vertical field is feedback controlled to keep the plasma at the center of the liner always. Many improvements have been made in tokamak devices over the years. Accuracy of the magnetic field is also important to improve the plasma performance in tokamak and other toroidal devices. The parameters of typical tokamak devices are listed in table 16.1.

Measurements by magnetic probes are a simple and useful way to monitor plasma behavior. Loop voltage V_L and the plasma current I_p can be measured by the magnetic loop and Rogowsky coil, respectively.[26] Then the electron temperature can be estimated by the Spitzer formula from the resistivity of the plasma, which can be evaluated using V_L and I_p. From eq. (7.206), the poloidal beta ratio β_p is given by

$$\beta_p = 1 + \frac{2B_\varphi}{B_\omega^2}\langle B_{\varphi v} - B_\varphi \rangle, \tag{16.1}$$

where $B_\omega = \mu_0 I_p / 2\pi a$. Since the diamagnetic flux $\delta\Phi$ is

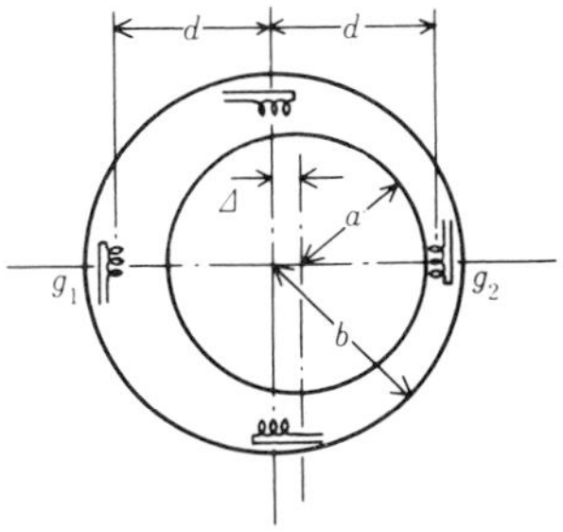

Fig. 16.3
Location of magnetic probes around a plasma (the sign of Δ shown in the figure is minus).

$$\delta\Phi = \pi a^2 \langle B_{\varphi\mathrm{v}} - B_\varphi \rangle, \tag{16.2}$$

we have

$$\beta_{\mathrm{p}} = \frac{p}{B_\omega{}^2/2\mu_0} = 1 + \frac{8\pi B_\varphi}{\mu_0{}^2 I_{\mathrm{p}}{}^2}\delta\Phi. \tag{16.3}$$

Therefore measurement of the diamagnetic flux $\delta\Phi$ yields β_{p} and the plasma pressure. Magnetic probes g_1, g_2 located around the plasma, as shown in fig. 16.3, can be used to determine the plasma position (see sec. 16.2b). Since the necessary magnitude of the vertical field for the equilibrium $B_\perp$ is related to the quantity $\Lambda = \beta_{\mathrm{p}} + l_{\mathrm{i}}/2 - 1$, $\beta_{\mathrm{p}} + l_{\mathrm{i}}/2$ can be estimated from $B_\perp$ (l_{i} is the normalized internal inductance).

The fluctuations in the soft X-ray (bremsstrahlung) signal follow the fluctuations in electron temperature. The fluctuations occur at the rational surfaces ($q(r_{\mathrm{s}}) = 1, 2, \ldots$). The mode number and the direction of the propagation can be estimated by arrays of solid-state detectors, as shown in fig. 16.4. When the positions of the rational surfaces can be measured, the radial current profile can be estimated for use in studies of MHD stability.

16.2b Equilibrium

The solution of the Grad-Shafranov equation (7.168) for the equilibrium gives the magnetic surface function $\psi = rA_\varphi$ (A_φ is the φ component of the vector potential). Then the magnetic field $\boldsymbol{B}$ is described by

$$rB_r = -\frac{\partial\psi}{\partial z}, \qquad rB_z = \frac{\partial\psi}{\partial r}, \tag{16.4}$$

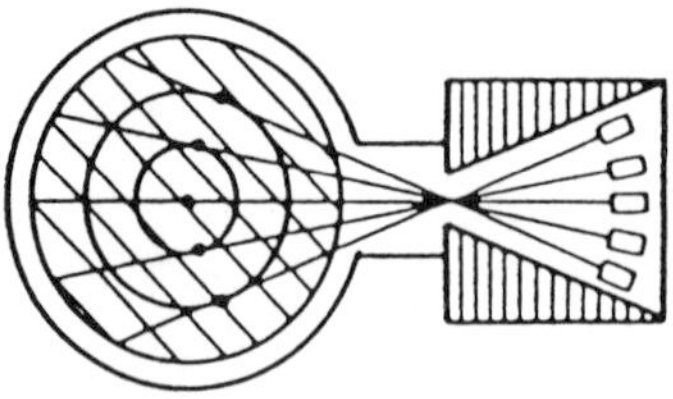

Fig. 16.4
An array of soft X-ray solid-state detectors. Each detector's main contribution to a signal comes from the emission at the peak temperature along the line of sight of the detector. The fluctuation of the electron temperature at this point can be detected.

and the current density $\boldsymbol{j}$ is

$$\boldsymbol{j} = \frac{I'}{2\pi}\boldsymbol{B} + rp'\boldsymbol{e}_{\varphi}. \tag{16.5}$$

When the plasma cross section is circular, the magnetic surface $\psi(\rho, \omega)$ outside the plasma is given by eq. (7.214) as follows:

$$\psi(\rho, \omega) = \frac{\mu_0 I_p R}{2\pi}\left(\ln\frac{8R}{\rho} - 2\right) - \frac{\mu_0 I_p}{4\pi}\left(\ln\frac{\rho}{a} + \left(\Lambda + \frac{1}{2}\right)\left(1 - \frac{a^2}{\rho^2}\right)\right)\rho\cos\omega. \tag{16.6}$$

The plasma boundary is given by $\rho = a$, that is, by

$$\psi(\rho, \omega) = \frac{\mu_0 I_p R}{2\pi}\left(\ln\frac{8R}{a} - 2\right). \tag{16.7}$$

(i) Case with Conducting Shell When a conducting shell of radius b surrounds the plasma, the magnetic surface ψ must be constant at the conducting shell unless an external stationary magnetic field is applied. Therefore the location of the shell is given by

$$\psi(\rho, \omega) = \frac{\mu_0 I_p R}{2\pi}\left(\ln\frac{8R}{b} - 2\right). \tag{16.8}$$

(In practice, the position of the shell is fixed, and the plasma settles down to the appropriate equilibrium position; the important point is their relative position.) Therefore the plasma center is displaced from the center of the shell by an amount Δ_0 given by* (see eq. (7.208))

$$\Delta_0(b) = -\frac{b^2}{2R}\left(\ln\frac{b}{a} + \left(\Lambda + \frac{1}{2}\right)\left(1 - \frac{a^2}{b^2}\right)\right) \tag{16.9}$$

($\Delta_0 < 0$ means that the plasma center is outside the center of the shell).

When the stationary vertical field $\boldsymbol{B}_\perp$ is applied, the normal component of the magnetic field at the shell must be equal to the normal component of $\boldsymbol{B}_\perp$. Therefore the overall displacement Δ of the shell center form the plasma center is

$$\Delta = \Delta_0 + \Delta_{\mathrm{H}},$$

$$\Delta_{\mathrm{H}} = b\frac{B_\perp}{B_\omega(b)}. \tag{16.10}$$

The magnetic field is obtained from the magnetic surface $\psi(\rho, \omega)$ by means of eq. (7.207):

$$B_\rho = \frac{-\mu_0 I_{\mathrm{p}}}{2\pi a}\frac{a}{2R}\left(\ln\frac{\rho}{a} + \left(\Lambda + \frac{1}{2}\right)\left(1 - \frac{a^2}{\rho^2}\right)\right)\sin\omega,$$

$$B_\omega = \frac{-\mu_0 I_{\mathrm{p}}}{2\pi a}\left[\frac{a}{\rho} + \frac{a}{2R}\left(\ln\frac{\rho}{a} + \left(\Lambda + \frac{1}{2}\right)\left(1 + \frac{a^2}{\rho^2}\right) + 1\right)\cos\omega\right]. \tag{16.11}$$

From these equations it is possible to obtain the ratio of the difference of $B_{\omega 1} = B_\omega\,(d + \Delta, 0)$ and $B_{\omega 2} = B(d - \Delta, \pi)$ to their sum, which can be measured by the magnetic probes g_1 and g_2 shown in fig. 16.3. The term $\Lambda + \frac{1}{2}$ in the equation of B_ω can be expressed by $\Delta_0 = \Delta - \Delta_{\mathrm{H}}$, and we find

$$\frac{B_{\omega 1} - B_{\omega 2}}{B_{\omega 1} + B_{\omega 2}} = -\frac{\left(1 + \frac{b^2}{d^2}\right)d}{b^2 - a^2}(\Delta - \Delta_{\mathrm{H}}) - \frac{\Delta_{\mathrm{H}}}{d} + \frac{d}{2R}\left(1 - \frac{2a^2\ln b/a}{b^2 - a^2}\right). \tag{16.12}$$

Here the approximation $d \approx b$ is used in the last term of the right-hand side. When $B_\perp$ is stationary, this field cannot be detected by the magnetic probes. Taking acount of this fact, the ratio of the difference u_- to the sum

* The image of the shell current flows at the distance b^2/Δ from the shell center, which provides the vertical field $B_\perp = -(\mu_0 I_{\mathrm{p}}\Delta/2\pi b^2) = -(\mu_0 I_{\mathrm{p}}/4\pi R)(\ln b/a + \Lambda - \frac{1}{2})$. This is the value necessary for equilibrium (see sec. 7.7).

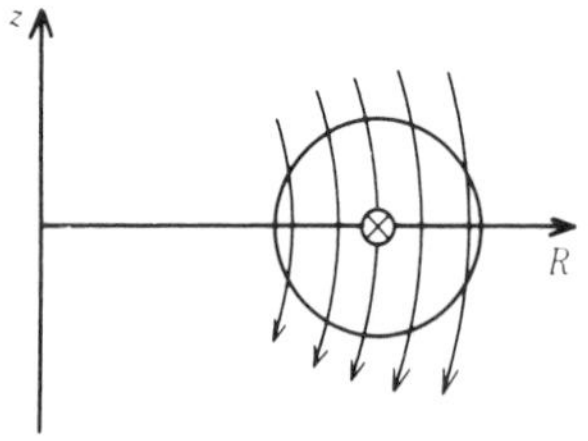

Fig. 16.5
Vertical field for plasma equilibrium.

u_+ of the probe signals is given by

$$\frac{u_-}{u_+} = -\frac{\left(1+\frac{b^2}{d^2}\right)d}{b^2-a^2}\left(\Delta - \frac{a^2}{b^2}\Delta_{\mathrm{H}}\right) + \frac{d}{2R}\left(1 - \frac{2a^2 \ln b/a}{b^2-a^2}\right). \tag{16.13}$$

(ii) Case without Conducting Shell If the vertical field $B_\perp$ is uniform in space, the equilibrium is neutral with regard to changes to plasma position. When the lines of the vertical field are curved, as shown in fig. 16.5, the plasma position is stable with regard to up and down motion. The z component F_z of the magnetic force applied to a plasma current ring of mass M is

$$F_z = -2\pi R I_{\mathrm{p}} B_R. \tag{16.14}$$

From the relation $\partial B_R/\partial z - \partial B_z/\partial R = 0$,

$$M\ddot{z} = -2\pi R I_{\mathrm{p}} \frac{\partial B_R}{\partial z} z = 2\pi I_{\mathrm{p}} B_z \left(-\frac{R}{B_z}\frac{\partial B_z}{\partial R}\right) z. \tag{16.15}$$

As $I_{\mathrm{p}} B_z < 0$, the stability condition for decay index is[25]

$$n \equiv -\frac{R}{B_z}\frac{\partial B_z}{\partial R} > 0. \tag{16.16}$$

The horizontal component F_R of the magnetic force is

$$M(\ddot{\Delta} R) = F_R = 2\pi R I_{\mathrm{p}}(B_z - B_\perp) = 2\pi \frac{\partial}{\partial R}(R I_{\mathrm{p}}(B_z - B_\perp))\Delta R. \tag{16.17}$$

The amount of $B_\perp$ necessary for plasma equilibrium (see eq. (7.218)) is

$$B_\perp = \frac{-\mu_0 I_p}{4\pi R}\left(\ln\frac{8R}{a} + \Lambda - \frac{1}{2}\right), \qquad \Lambda = \frac{l_i}{2} + \beta_p - 1.$$

When the plasma is ideally conductive, the magnetic flux is conserved and

$$\frac{\partial}{\partial R} L_p I_p + 2\pi R B_\perp = 0. \tag{16.18}$$

Here the self-inductance is $L_p = \mu_0 R(\ln(8R/a) + l_i/2 - 2)$. Therefore the equation of motion is

$$M(\ddot{\Delta R}) = 2\pi I_p B_\perp \left(\frac{3}{2} - n\right)\Delta R \tag{16.19}$$

under the assumption $\ln(8R/a) \gg 1$. Then the stability condition for horizontal movement is[25]

$$\frac{3}{2} > n. \tag{16.20}$$

(iii) Equilibrium Beta Limit of Tokamaks with Elongated Plasma Cross Sections The poloidal beta limit of a circular tokamak is given by $\beta_p = 0.5R/a$, as was shown in sec. 7.7. The same poloidal beta limit is derived by similar consideration for the elongated tokamak with horizontal radius a and vertical radius b. Since the length of circumference along the poloidal direction is $2\pi aK$ for the elongated plasma and the average of poloidal field is $\bar{B}_p = \mu_0 I_p/(2\pi aK)$, the ratio of the poloidal and toroidal field is $\bar{B}_p/B_t = aK/(Rq_I)$, where $K = [(1 + (b/a)^2)/2]^{1/2}$. Therefore the beta limit of an elongated tokamak is $\beta = \beta_p(aK/Rq_I)^2 = 0.5K^2a/(Rq_I^2)$ and is K^2 times as large as that of circular one. In order to make the plasma cross section elongated, the decay index n of the vertical field must be negative, and the elongated plasma is positionally unstable in the up-down motion. Therefore feedback control of the variable horizontal field is necessary to keep the plasma position stable.[28]

16.2c MHD Stability

A possible MHD instability in the low-beta tokamak is kink modes, which were treated in sec. 9.3d. Kink modes can be stabilized by tailoring the current profile and by appropriate choice of the safety factor q_a. The localized MHD modes around the rational surface (see sec. 9.4c) are stable when $q > 1$.

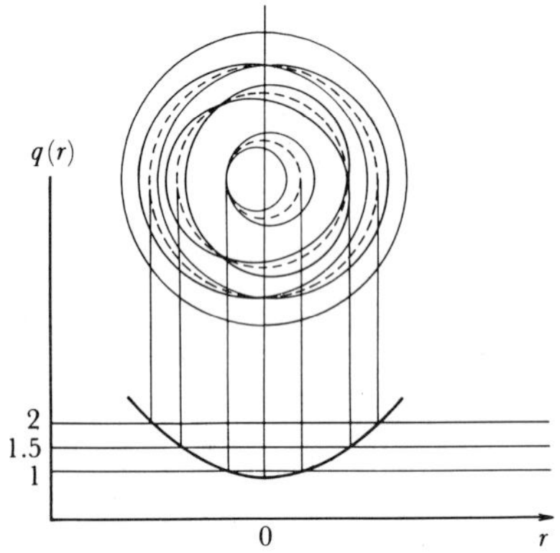

Fig. 16.6
Magnetic islands of $m = 1$, $m = 3/n = 2$, and $m = 2$ appear at $q(r) = 1, \frac{3}{2}$, and 2.

When the plasma pressure is increased, the beta value is limited by the ballooning modes (sec. 9.6a). This instability is an interchange mode localized in the bad curvature region driven by a pressure gradient. The beta limit β_c is given by $\beta_c \sim \alpha(r/Rq^2)$, $\alpha \sim O(1)$ by eq. (9.206a) with the connection length set to $L \approx 2\pi Rq$.

The β limit by kink and ballooning modes depends on the radial profile of the plasma current (shear) and the shape of the plasma cross section. The limit of the average beta, $\beta_c = \langle p \rangle/(B^2/2\mu_0)$, of the optimized condition is derived by MHD simulation codes to be $\beta_c(\%) = cI_p(MA)/a(m)B_t(T)$ $(c \sim 3.5)$.[29,30] When the relations $\bar{B}_p = \mu_0 I_p/(2\pi a K)$ and $q_I = K(a/R)(B_t/\bar{B}_p)$ are used, the critical β is reduced to $\beta_c \approx 0.18K^2 a/(Rq_I)$, where $K^2 = [1 + (b/a)^2]/2$, a is the horizonal radius, and b is the vertical radius.

Even if a plasma is ideal MHD stable, tearing modes can be unstable for a finite resistive plasma. When Δ' defined in eq. (9.177a) is positive at the rational surfaces in which the safety factor $q(r)$ is rational ($q(r) = 1, \frac{3}{2}, 2, \ldots$), tearing modes grow (see eq. (9.198)) and magnetic islands are formed, as shown in fig. 16.6. When the profile of the plasma current is peaked, then the safety factor at the center becomes $q(0) < 1$ and the tearing mode with $m = 1$, $n = 1$ grows at the rational surface $q(r) = 1$. The reconnection of the magnetic surface occurs at the rational surface, and the hot core of the plasma is pushed out and the current profile is flattened (fig. 16.7). The thermal energy in the central hot core is lost in this way.[24,31] Since the electron temperature in the central part is higher than in the outer region

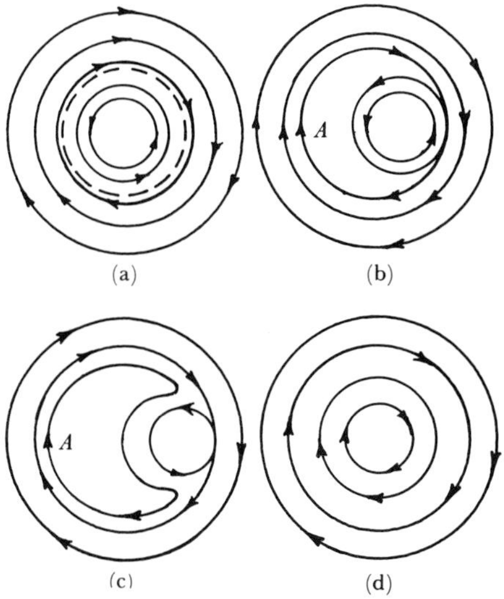

Fig. 16.7
The hot core in the center is expelled by the reconnection of magnetic surfaces.

and the resistance in the central part is smaller, the current profile is peaked again and the same process is repeated. This type of phenomenon is called *internal disruption* or *minor disruption.*

The stable operational region of a tokamak with plasma current I_p and density $\bar{n}_e$ is limited. With the normalized quantities $1/q = (\mu_0 I_p/2\pi a)(R/aB_t)$ and $n_M = \bar{n}_{13}/(B_t/R)$ (where $\bar{n}_{13}$ is the average line density in $10^{13}\ \mathrm{cm}^{-3}$, the units of B_t and R are the tesla and meter, respectively, and n_M is the Murakami parameter), the stable operational region is $1/q < 0.5$ and $n_M < 3$ for ohmically heated plasmas and $n_M < 10$ for NBI plasmas. Beyond the stable region, strong instability, called *disruptive instability*, occurs. Negative spikes appear in the loop voltage due to the rapid expansion of the current channel (flattened current profile), that is, the rapid reduction of the internal inductance. The thermal energy of the plasma is lost suddenly. The electron temperature drops rapidly, and the plasma resistance increases. A positive pulse appears in the loop voltage. Then the plasma discharge is terminated rapidly. Just before the disruptive instability, the magnetic perturbation due to the $m = 2$ tearing mode is observed. Sometimes $m = 3/n = 2$ or $m = 1/n = 1$ perturbations are observed. For possible mechanisms of the disruptive instability, overlapping of the magnetic islands of

$m = 2/n = 1$ ($q(r) = 2$) and $m = 3/n = 2$ ($q(r) = 1.5$) or the reconnection of $m = 2/n = 1$, $m = 1/n = 1$ magnetic islands are discussed. Reviews of the MHD instabilities of tokamak plasmas are given in refs. 32–34, and reviews of plasma transport and turbulence are given in ref. 35.

16.2d Beta Limit

The output power density of nuclear fusion is proportional to $n^2\langle\sigma v\rangle$. Since $\langle\sigma v\rangle$ is proportional to T_i^2 in the region near $T_i \sim 10$ keV, the fusion output power is proportional to the square of plasma pressure $p = nT$. Therefore the higher the beta ratio $\beta = p/(B/2\mu_0)$, the more economical the possible fusion reactor. The average beta of $\langle\beta\rangle \sim 3\%$ was realized by NBI experiments in ISX-B, JET-2, and PLT. All these tokamaks have a circular plasma cross section. The theoretical upper limit of the average beta β_c of an elongated tokamak plasma is

$$\beta_c(\%) \sim 3.5I_p(\text{MA})/a(\text{m})B_t(\text{T}) = 18((1 + \kappa^2)/2)a/(Rq_1), \tag{16.21}$$

where κ is the degree of noncircularity, that is, the ratio of the vertical radius to the horizontal radius (when $\kappa = 1$, the cross section is circular). The average beta $\langle\beta\rangle = 4.6\%$ was realized with 3.3 MW NBI in the noncircular tokamak Doublet III, in which $R = 1.43$ m, $a = 0.45$ m, $\kappa = 1.4$, $I_p = 0.33$ MA, $\bar{n}_e = 6.5 \times 10^{13}$ cm^{-3}, and $B_t = 0.62$ T. The value of $\langle\beta\rangle = 5.5\%$ was obtained in PBX, in which the plasma has a bean shape (in 1986).

16.2e Impurity Control

When the temperature of the plasma increases, the ions from the plasma hit the walls of the vacuum vessel and impurity ions are sputtered. When the sputtered impurities penetrate the plasma, the impurities are highly ionized and yield a large amount of radiation loss, which causes radiation cooling of the plasma. The light impurities, such as C and O, can be removed by baking and discharge-cleaning of the vacuum vessel. The sputtering of heavy atoms (Fe, etc.) of the wall material itself can be avoided by using a diverter, as shown in fig. 16.8.

Scrap-off plasmas flow at the velocity of sound along the lines of magnetic force on the separatrix S into the neutralized plates, where the plasmas are neutralized. Even if the material of the neutralized plates is sputtered, the atoms are ionized within the diverter regions near the neutralized plates. Since the thermal velocity of the heavy ions is much smaller than the flow velocity of the plasma (which is the same as the thermal velocity of hydro-

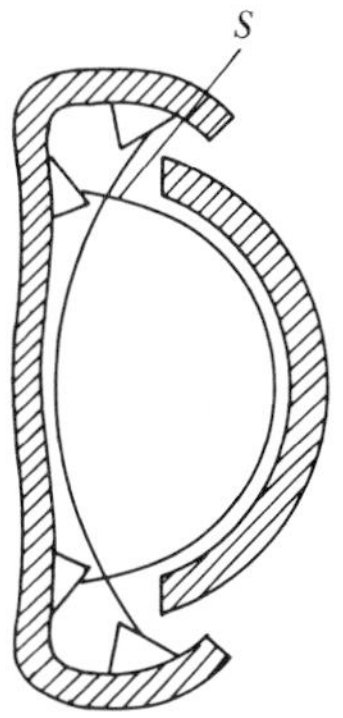

Fig. 16.8
Diverter configuration using separatrix S of the magnetic surface.

gen ions), they are unlikely to flow back into the main plasma. In the diverter region the electron temperature of the plasma becomes low because of impurity radiation cooling. Because of pressure equilibrium along the lines of magnetic force, the density in the diverter region near the neutralized plates becomes high. Therefore the velocity of ions from the plasma into the neutralized plates is collisionally damped and sputtering is suppressed. A decrease in the impurity radiation in the main plasma can be observed by using a diverter configuration.

16.2f Confinement Scaling

The energy flow of ions and electrons inside the plasma is schematically shown in fig. 16.9. Denote the heating power into the electrons per unit volume by P_{he} and the radiation loss and the energy relaxation of electrons with ions by R and P_{ei}, respectively; then the time derivative of the electron thermal energy per unit volume is given by

$$\frac{\mathrm{d}}{\mathrm{d}t}\left(\frac{3}{2}n_e T_e\right) = P_{he} - R - P_{ei} + \frac{1}{r}\frac{\partial}{\partial r}r\left(\chi_e \frac{\partial T_e}{\partial r} + D_e \frac{3}{2} T_e \frac{\partial n_e}{\partial r}\right), \tag{16.22}$$

where χ_e is the electron thermal conductivity and D_e is the electron diffusion coefficient. Concerning the ions, the same relation is derived, but instead of the radiation loss the charge exchange loss L_{ex} of ions with neutrals must be taken into account, and then

$$\frac{\mathrm{d}}{\mathrm{d}t}\left(\frac{3}{2}n_i T_i\right) = P_{hi} - L_{ex} + P_{ei} + \frac{1}{r}\frac{\partial}{\partial r}r\left(\chi_i \frac{\partial T_i}{\partial r} + D_i \frac{3}{2} T_i \frac{\partial n_i}{\partial r}\right). \tag{16.23}$$

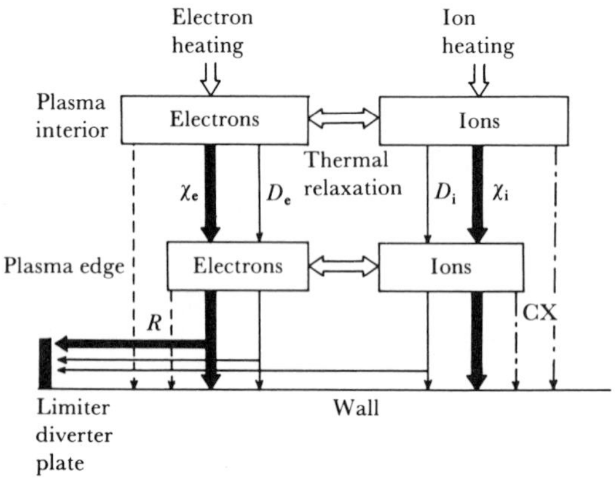

Fig. 16.9
Energy flow of ions and electrons in a plasma. Bold arrows, thermal conduction (χ). Light arrows, convective loss (D). Dashed arrows, radiation loss (R). Dot-dashed arrows, charge exchange loss (CX).

The experimental results of ohmic heating and neutral beam injection can be explained by classical processes. The efficiency of wave heating can be estimated fairly accurately by theoretical analysis. The radiation and the charge exchange loss are classical processes. In order to evaluate the energy balance of the plasma experimentally, it is necessary to measure the fundamental quantities $n_e(r, t)$, $T_i(r, t)$, $T_e(r, t)$, and others.[26] According to the many experimental results, the energy relaxation between ions and electrons is classical (sec. 4.3), and the observed ion thermal conductivity is around 2–3 times the neoclassical thermal conductivity:

$$\chi_{ic} = n_i f(q, \varepsilon) q^2 (\rho_B{}^i)^2 \nu_{ii} \tag{16.24}$$

($f = 1$ in the Pfirsch-Schlüter region and $f = \varepsilon_t{}^{-3/2}$ in the banana region). However, the electron thermal conduction estimated by the experimental results is anomalous and is much larger than the neoclassical one. Therefore the energy confinement time of the plasma is determined mostly by electron thermal conduction loss in many cases. The energy confinement time τ_{OH} of an ohmically heated plasma is well described by Alcator (neo-Alcator) scaling as follows:

$$\tau_{OH}(\mathrm{s}) = 7.1 \times 10^{-22} q^{0.5} \bar{n}_e(\mathrm{cm}^{-3}) a^{1.04}(\mathrm{cm}) R^{2.04}(\mathrm{cm}). \tag{16.25}$$

However, the linearity of τ_{OH} on the average electron density $\bar{n}_e$ deviates in the high-density region $\bar{n}_e > 2.5 \times 10^{14}\ \mathrm{cm}^{-3}$ and τ_{OH} tends to saturate. When the plasma is heated by high-power NBI or wave heating, the energy confinement time degrades as the heating power increases. Kaye and Goldstone examined the many experimental results of NBI heated plasma and derived the so-called Kaye-Goldstone scaling on the energy confinement time,[36] that is,

$$\tau_E = (1/\tau_{OH}{}^2 + 1/\tau_{AUX}{}^2)^{-1/2},$$

$$\tau_{AUX}(\mathrm{s}) = 6.4 \times 10^{-8}\kappa^{0.5}I_p(\mathrm{A})P_{tot}{}^{-0.5}(\mathrm{W})a^{-0.37}(\mathrm{cm})R^{1.75}(\mathrm{cm}), \qquad (16.26)$$

where κ is the degree of noncircularity and P_{tot} is the total heating power. Connor, Taylor, and Turner[37] derived the following CTT scaling:

$$\tau_E \propto I_p{}^{1.44}n_e{}^{0.28}P_{tot}{}^{-0.72}a^{-0.16}R^{1.44}, \qquad (16.27)$$

according to the assumption of profile consistency, in which the radial profiles of the plasma current density and the electron temperature are assumed to be self-adjusted so that the MHD stability condition can be satisfied.

More recently, an improved confinement state "H mode" was found in the ASDEX experiments with diverter configuration. When the NBI heating power is larger than a threshold value in the diverter configuration, the D_α line of deuterium in the edge region of the deuterium plasma decreases during discharge, and recycling of deuterium atoms near the boundary decreases. At the same time the electron density and the thermal energy density increase and the energy confinement time of NBI heated plasma is improved by a factor of 2 and becomes almost equal to that of ohmically heated plasma in some cases. (We call the confinement state following Kaye-Goldstone scaling the "L mode"). In the H mode the gradients of electron temperature and the electron density become steep at the plasma boundary determined by the separatrix, and the electron temperature at the separatrix increases from 130 eV to 250 eV in ASDEX. Several trials of the theoretical interpretation of the H mode have been carried out,[39] and the mechanism of thermal insulation (energy confinement) at the plasma boundary is a key issue. The H mode has also been observed in Doublet III, PDX, JFT-2M, DIII-D, and JET. In JFT-2M experiments (1986), the H mode was observed even in a limiter configuration without a diverter in which the limiter is located on the inner side of the vacuum torus. In the TFTR experiment, outgassing of deuterium from the wall and the carbon limiter located on the inner (high-field) side of the vacuum torus was extensively carried out before the experiments. Then balanced neutral beam injections of co-injection (beam direction parallel to the plasma current) and counterinjection (beam direction opposite to that of co-injection) were applied to the deuterium plasma, and an improved confinement "super-shot" was observed. The experimental results of JET and TFTR are described in sec. 16.2g.

16.2g Additional Heating

If a tokamak plasma is heated only by ohmic heating, the ohmic heating power density ηj^2 decreases and the bremsstrahlung loss P_b increases according to the increase of the electron temperature. Therefore the upper limit of the electron temperature can be estimated from $\eta j^2 = P_b$ as follows:

$$\frac{T_e}{e} = 4.1 \times 10^3 Z^{-1/2} \frac{B_t}{qR} \frac{10^{20}}{n_e} \tag{16.28}$$

(T_e/e in eV; toroidal field B_t, major radius R, and the electron density n_e are in MKS units). Because of the Kruskal-Shafranov limit ($q > 1$–2), the upper limit is less than several keV, so that additional heatings are necessary to approach the ignition condition.

Most of the NBI heating experimental results can be explained by the classical process of Coulomb collision, as described in sec. 14.1. In the NBI experiment of TFTR in 1986, the power of co-injection and counterinjection was balanced in "supershot" condition, as explained in sec. 16.2f. An ion temperature of $T_i(0) = 20$ keV at the center was obtained with density $n_e(0) = 7 \times 10^{13}\,\mathrm{cm}^{-3}$ ($\bar{n}_e \approx 3 \times 10^{13}\,\mathrm{cm}^{-3}$) for NBI power $P_{\mathrm{NBI}} = 12$ MW ($D^0 \rightarrow D^+$ plasma; beam energy ≈ 100 keV). The other main parameters were $\tau_E = 0.17$ s, $T_e(0) = 6.5$ keV, $Z_{\mathrm{eff}} = 3$, $I_p = 0.9$ MA, $B_t = 4.7$ T. The increase of the central ion temperature $\Delta T_i(0)$ scales $\bar{n}_{20} \Delta T_i(0)(\mathrm{keV}) = 0.38 P_{\mathrm{NBI}}$ (MW) ($\bar{n}_{20}$ is the electron density in units of $10^{20}\,\mathrm{cm}^{-3}$). In the NBI experiment of JET with a diverter configuration, the H mode was realized in 1986, and the results were $T_i(0) = 10$ keV, $T_e(0) = 5.4$ keV, $n_e(0) = 3 \times 10^{13}\,\mathrm{cm}^{-3}$ ($\bar{n}_e \approx 1.5 \times 10^{13}\,\mathrm{cm}^{-3}$), $\tau_E = 0.6$ s, $Z_{\mathrm{eff}} = 2$–3, $I_p = 3$ MA, $B_t = 3.3$ T.

In the present neutral beam source, the positive hydrogen ions are accelerated and then passed through the cell filled with neutral hydrogen gas, where they are converted to a fast neutral beam by charge exchange.

However, the conversion ratio of positive hydrogen ions to neutral becomes small when the ion energy is larger than 100 keV (2.5% at 200 keV of H^+). On the other hand, the conversion ratio of negative hydrogen ions (H^-) to neutral does not decrease in the high energy range; a neutral beam source with a negative ion source is being developed as a high-efficiency source.

Wave heating is another method of additional heating and will be described in detail in sec. 16.6. The similar heating efficiency of wave heating in ICRF (ion cyclotron range of frequency) to that of NBI was observed in PLT. In the ICRF experiments of JET, the parameters $T_i(0) \approx 5.4$ keV, $T_e(0) = 5.6$ keV, $n_e(0) = 3.7 \times 10^{13}$ cm^{-3}, $\tau_E \approx 0.3$ s were obtained by $P_{ICRF} = 7$ MW.

16.2h Noninductive Current Drive

As long as the plasma current is driven by electromagnetic induction of the current transformer in a tokamak device, the discharge is a necessarily pulsed operation with finite duration. If the plasma current can be driven in a noninductive way, a steady-state tokamak reactor is possible in principle. Current drive by neutral beam injection has been proposed by Ohkawa,[40] and current drive by traveling wave has been proposed by Wort.[41] The momenta of particles co-injected by NBI or of traveling waves are transferred to the charged particles of the plasma, and the resultant charged particle flow produces the plasma current. Current drive by NBI was demonstrated by DITE, TFTR, etc. Current drive by a lower hybrid wave (LHW), proposed by Fisch,[42] was demonstrated by JFT-2, JIPPT-II, WT-2, PLT, Alcator C, Versator 2, T-7, Wega, JT-60, and others.

(i) Lower Hybrid Wave The theory of current drive by waves is described here according to Fisch and Karney.[42] When a wave is traveling along the line of magnetic force, the velocity distribution function near the phase velocity of the wave is flattened by the diffusion in velocity space (see sec. 14.4a). Denote the diffusion coefficient in velocity space by the wave by D_{rf}; then the Fokker-Planck equation is given by

$$\frac{\partial f}{\partial t} + \boldsymbol{v}\cdot\nabla_r f + \left(\frac{\boldsymbol{F}}{m}\right)\cdot\nabla_v f = \frac{\partial}{\partial v_z}\left(D_{rf}\frac{\partial f}{\partial v_z}\right) + \left(\frac{\delta f}{\delta t}\right)_{F.P.}. \qquad (16.29)$$

Here $(\delta f/\delta t)_{F.P.}$ is the Fokker-Planck collision term (see sec. 5.4):

$$-\left(\frac{\delta f}{\delta t}\right)_{\text{F.P.}} = \sum_{\text{i,e}}\left[\frac{1}{v^2}\frac{\partial}{\partial v}(v^2 J_v) + \frac{1}{v \sin\theta}\frac{\partial}{\partial\theta}(\sin\theta J_\theta)\right],$$

$$J_v = -D_v\left(\frac{\partial f}{\partial v} + \frac{v}{v_{\text{T}}^2} f\right), \qquad J_\theta = -D_\perp \frac{1}{v}\frac{\partial f}{\partial\theta},$$

$$D_v = \frac{v_{\text{T}}^2 \nu_0}{2}\left(\frac{v_{\text{T}}}{v}\right)^3, \qquad D_\perp = \frac{v_{\text{T}}^2 \nu_0}{2}\frac{v_{\text{T}}}{2v},$$

$$v_{\text{T}}^2 = \frac{T^*}{m^*}, \qquad \nu_0 = \left(\frac{qq^*}{\varepsilon_0}\right)^2 \frac{n^* \ln \Lambda}{2\pi v_{\text{T}}^3 m^2}, \tag{16.30}$$

where (v, θ, φ) are spherical coordinates in velocity space. v_{T}, q^*, n^* are the thermal velocity, charge, and density of field particles, respectively, and v, q, n are quantities of test particles. Let us consider the electron distribution function in a homogeneous case in space without external force ($F = 0$). Collision terms of electron-electron and electron-ion (charge number $= Z$) are taken into account. When dimensionless quanties $\tau = \nu_{0\text{e}} t$, $u = v/v_{\text{Te}}$, $w = v_z/v_{\text{Te}}$, $D(w) = D_{\text{rf}}/v_{\text{Te}}^2 \nu_{0\text{e}}$ are introduced, the Fokker-Planck equation reduces to

$$\frac{\partial f}{\partial \tau} = \frac{\partial}{\partial w}\left(D(w)\frac{\partial f}{\partial w}\right) + \frac{1}{2u^2}\frac{\partial}{\partial u}\left(\frac{1}{u}\frac{\partial f}{\partial u} + f\right) + \frac{1+Z}{4u^3}\frac{1}{\sin\theta}\frac{\partial}{\partial\theta}\left(\sin\theta\frac{\partial f}{\partial\theta}\right). \tag{16.31}$$

When Cartesian coordinates in velocity space $(v_x, v_y, v_z) = (v_1, v_2, v_3)$ are used, the Fokker-Planck collision term in Cartesian coordinates is given by sec. 5.5 as follows:

$$A_i = -D_0 v_{\text{T}} \frac{m}{m^*}\frac{v_i}{v^3},$$

$$D_{ij} = \frac{D_0}{2}\frac{v_{\text{T}}}{v^3}\left[(v^2\delta_{ij} - v_i v_j) + \frac{v_{\text{T}}^2}{v^2}(3v_i v_j - v^2\delta_{ij})\right],$$

$$J_i = A_i f - \sum_j D_{ij}\frac{\partial f}{\partial v_j},$$

$$D_0 \equiv \frac{(qq^*)^2 n^* \ln\Lambda}{4\pi\varepsilon_0^2 m^2 v_{\text{T}}} \equiv \frac{v_{\text{T}}^2 \nu_0}{2},$$

$$\left(\frac{\delta f}{\delta t}\right)_{\text{F.P.}} = -\nabla_v \cdot \boldsymbol{J}. \tag{16.32}$$

Let us assume that the distribution function of the perpendicular velocities v_x, v_y to the line of magnetic force is Maxwellian. Then the one-dimensional Fokker-Planck equation on the distribution function $F(v_z) = \int f \, dv_x \, dv_y$ of parallel velocity v_z can be deduced by (v_x, v_y) integration:

$$\iint \left(\frac{\delta f}{\delta t}\right)_{\text{F.P.}} dv_x \, dv_y = \iint (-\nabla_v \cdot \boldsymbol{J}) \, dv_x \, dv_y$$

$$= \iint \frac{\partial}{\partial v_z}\left[-A_z f + \sum_j D_{zj} \frac{\partial f}{\partial v_j}\right] dv_x \, dv_y. \tag{16.33}$$

When $|v_z| \gg |v_x|, |v_y|$, the approximation $v \approx |v_z|$ can be used. The resultant one-dimensional Fokker-Planck equation on $F(w)$ $(w = v_z/v_{\text{Te}})$ is

$$\frac{\partial F}{\partial \tau} = \frac{\partial}{\partial w}\left(D(w)\frac{\partial F}{\partial w}\right) + \left(1 + \frac{Z}{2}\right)\frac{\partial}{\partial w}\left(\frac{1}{w^3}\frac{\partial}{\partial w} + \frac{1}{w^2}\right)F(w). \tag{16.34}$$

The steady-state solution is

$$F(w) = C \exp \int^w \frac{-w \, dw}{1 + w^3 D(w)/(1 + Z/2)}, \tag{16.35}$$

and $F(w)$ is shown in fig. 16.10 schematically (when $D(w) = 0$, this solution

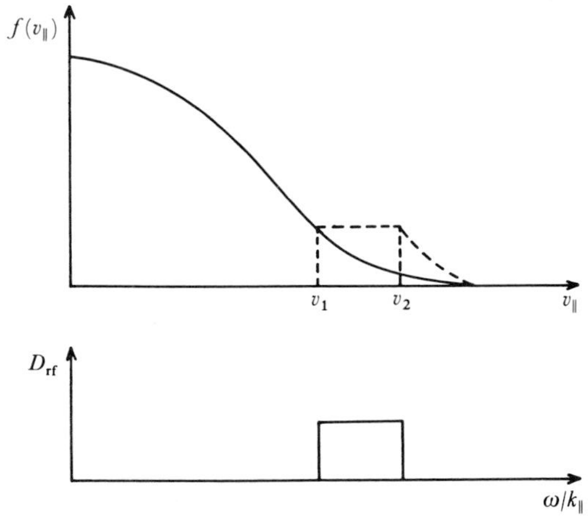

Fig. 16.10
The velocity distribution function $f(v_\parallel)$ of electrons is flattened in the region $v_1 = c/N_1$, $v_2 = c/n_2$ due to the interaction with the lower hybrid wave whose spectra of parallel index $N_\parallel$ ranges from N_1 to N_2.

is Maxwellian). $F(w)$ is asymmetric with respect to $w = 0$, so that the current is induced. The induced current density J is

$$J = env_{\mathrm{Te}}\, j, \tag{16.36}$$

where $j = \int wF(w)\, \mathrm{d}w$, that is,

$$j \approx \frac{w_1 + w_2}{2} F(w_1)(w_2 - w_1). \tag{16.37}$$

On the other hand, this current tends to dissipate by Coulomb collison. Dissipated energy must be supplied by the input energy from the wave in order to sustain the current. Necessary input power P_{d} is

$$\begin{aligned} P_{\mathrm{d}} &= \int \frac{nmv^2}{2} \left(\frac{\delta f}{\delta t}\right)_{\mathrm{F.P.}} \mathrm{d}v \\ &= \int \frac{nmv^2}{2} \frac{\partial}{\partial v_z} \left(D_{\mathrm{rf}} \frac{\partial f}{\partial v_z}\right) \mathrm{d}v \\ &= nmv_{\mathrm{Te}}{}^2 \nu_0 \int \frac{w^2}{2} \frac{\partial}{\partial w} \left(D(w) \frac{\partial F}{\partial w}\right) \mathrm{d}w \\ &= nmv_{\mathrm{Te}}{}^2 \nu_0 p_{\mathrm{d}}, \end{aligned} \tag{16.38}$$

where p_{d} is given by use of the steady-state solution of $F(w)$, as follows:

$$p_{\mathrm{d}} = \left(1 + \frac{Z}{2}\right) F(w_1) \ln\left(\frac{w_2}{w_1}\right) \approx \left(1 + \frac{Z}{2}\right) F(w_1) \frac{w_2 - w_1}{w_1}. \tag{16.39}$$

When $Z = 1$, $j_{\mathrm{d}}/p_{\mathrm{d}} \approx 0.7w^2$. More accurately, this ratio is $1.7\langle w^2 \rangle$. The ratio of the current density J and the necessary input power P_{d} per unit volume to sustain the current is given by

$$\frac{J}{P_{\mathrm{d}}} = \frac{env_{\mathrm{Te}}}{nT\nu_0} \frac{j}{p_{\mathrm{d}}} = 0.16 \frac{T_{10}}{n_{20}} \langle w^2 \rangle \left(\frac{\mathrm{A/m^2}}{\mathrm{W/m^3}}\right), \tag{16.40}$$

where T_{10} is the electron temperature in 10 keV units and n_{20} is the electron density in $10^{20}\ \mathrm{m^{-3}}$ units. When the specific resistivity $\eta = m_{\mathrm{e}}\nu_{\mathrm{ei}}/e^2 n_{\mathrm{e}}$ is used, P_{d} is described by

$$P_{\mathrm{d}} = \frac{v_{\mathrm{Te}}}{v_{\mathrm{d}} \langle w^2 \rangle} \frac{10(1 + Z/2)}{Z} \eta J^2 \qquad (v_{\mathrm{d}} \equiv J/en). \tag{16.41}$$

The ratio of total driven current $I \sim \pi a^2 J$ to the necessary input power

$W \sim \pi a^2 2\pi R P_d$ is $I/W = (J/P_d)(2\pi R)^{-1} \sim 0.025\alpha.\ T_{10}\langle w^2\rangle/(R_1 n_{20})$ (A/W), where α is a coefficient of order 1 that depends on the spatial distribution of current density, electron temperature, and the density (R_1 is the major radius in meters). The square average $\langle w^2\rangle$ of the ratio of the phase velocity (in the direction of magnetic field) of traveling waves to the electron thermal velocity is of the order of 20–50. In the experiment of PLT with $R = 1.32$ m, a plasma current of $I_p = 165$ kA was driven by a lower hybrid wave with $P_{rf} = 70$ kW when $T_e \approx 1$ keV, $n = 4 \times 10^{18}$ m^{-3}. These results are consistent with the theoretical results. A ramp-up experiment of the plasma current from zero was carried out first by WT-2 and then PLT and JIPP T-II by the application of a lower hybrid wave to the target plasma produced by electron cyclotron heating and other types of heating.

The efficiency of current drives by lower hybrid waves or NBI is inversely proportional to the density, and the current cannot be driven beyond a threshold density in the case of lower hybrid current drive. When the application of a current drive to the fusion grade plasma with $n \sim 10^{20}$ m^{-3} is considered, the necessary input power for the current drive occupies a considerable part ($\sim$0.1) of the fusion thermal output. Other possible methods, such as current drive by waves in the electron cyclotron range of frequencies,[43] are also being studied.

A few scenarios for the application of current drive by a lower hybrid wave to a fusion reactor have been proposed. When the plasma current is ramped up in the low-density plasma and the density is increased after the plasma current reaches a specified value, all the available magnetic flux of the current transformer can be used only for sustaining the plasma current, so that the discharge duration can be increased by one order of magnitude. Furthermore, if the plasma current can be driven by a lower hybrid wave efficiently, even when a reversed electric field is applied to the plasma current, the magnetic flux of the current transformer can be recovered. The efficiency of the magnetic flux recovery can be increased by lowering the density and the electron temperature. Therefore the plasma current can be sustained continuously. However, the fusion energy output power changes periodically (it decreases during the recovery of the magnetic flux), and the first wall suffers from the change of heat load and the thermal stress.

(ii) Fast Neutral Beam Injection When a fast neutral beam is injected into a plasma, it changes to a fast ion beam by charge exchange or ionization processes. When the fast ions have higher energy than $\varepsilon_{cr} = m_b v_{cr}^2/2$ given by eq. (4.42), they are decelerated mainly by electrons in the plasma and the fast ions with $\varepsilon < \varepsilon_{cr}$ are decelerated mainly by ions in the plasma. The distribution function of the ion beam can be obtained by solving the Fokker-Planck equations. The Fokker-Planck collision term, eq. (5.25), of the fast ions with $\varepsilon \gg \varepsilon_{cr}$ is dominated by the dynamic friction term due to electrons and is given by

$$\frac{\partial f_b}{\partial t} + \frac{\partial}{\partial v}\left(\frac{-vf_b}{2\tau_\varepsilon{}^{be}}\right) = \phi\delta(v - v_b), \tag{16.42}$$

where v_b is the initial injection velocity and $\tau_\varepsilon{}^{be}$ is the energy relaxation time of beam ions and electrons as described by eq. (4.49). The steady-state solution of the Fokker-Planck equation is $f_b \propto 1/v$; however, the dynamic friction term due to ions or the diffusion term dominates the collision term in the region of $v < v_{cr}$. Therefore the approximate distribution function of the ion beam is given by $f_b \propto v^2/(v^3 + v_{cr}{}^3)$, that is,

$$f_b(v) = \begin{cases} \dfrac{n_b}{\ln[1 + (v_b/v_{cr})^3]^{1/3}} \dfrac{v^2}{(v^3 + v_{cr}{}^3)} & (v \le v_b) \\ 0 & (v > v_b) \end{cases}. \tag{16.43}$$

The necessary ion injection rate ϕ per unit time per unit volume to keep the steady-state condition of the beam is derived from the Fokker-Planck equation (16.42),

$$\phi = \frac{n_b}{2\tau_\varepsilon{}^{be}} \frac{[1 + (v_{cr}/v_b)^3]^{-1}}{\ln[1 + (v_b/v_{cr})^3]^{1/3}}, \tag{16.44}$$

and the power per unit volume is

$$P_b = \frac{m_b v_b{}^2}{2}\phi \approx \frac{m_b v_b{}^2 n_b}{4\ln(v_b/v_{cr})\tau_\varepsilon{}^{be}}. \tag{16.45}$$

The average velocity of the decelerating ion beam is $\bar{v}_b = v_b[\ln(v_b/v_{cr})]^{-1}$. Then the current density J driven by the fast ions' beam consists of terms due to fast ions and bulk ions and electrons of the plasma:

$$J = Z_i e n_i \bar{v}_i + Z_b e \bar{v}_b - e n_e \bar{v}_e,$$

$$n_e = Z_i n_i + Z_b n_b, \tag{16.46}$$

where $\bar{v}_i$ and $\bar{v}_e$ are the average velocities of ions with density n_i and electrons with density n_e, respectively. The electrons of the plasma receive momentum by collision with fast ions and lose it by collision with plasma ions, that is,

$$m_e n_e \frac{d\bar{v}_e}{dt} = m_e n_e(\bar{v}_b - \bar{v}_e)\nu_{eb\|} + m_e n_e(\bar{v}_i - \bar{v}_e)\nu_{ei\|} = 0, \tag{16.47}$$

so that

$$(Z_i^2 n_i + Z_b^2 n_b)\bar{v}_e = Z_b^2 n_b \bar{v}_b + Z_i^2 n_i \bar{v}_i. \tag{16.48}$$

Since $n_b \ll n_i$,

$$n_e \bar{v}_e = \frac{Z_b^2}{Z_i} n_b \bar{v}_b + Z_i n_i \bar{v}_i, \tag{16.49}$$

so that we have

$$J = Z_b\left(1 - \frac{Z_b}{Z_i}\right) e n_b \bar{v}_b. \tag{16.50}$$

The driven current density consists of the fast ion beam term $Z_b e n_b v_b$ and the term of dragged electrons by the fast ion beam, $-Z_b^2 e n_b v_b / Z_i$. Since $\tau_\varepsilon^{be} \sim 0.36\,(Z_i m_b / Z_b^2 m_e)\nu_{ei\|}^{-1}$, the following relation holds:

$$\frac{Z_b e n_e \eta J v_b}{P_b} = \frac{Z_b \cdot 1.4 \cdot \ln(v_b/v_{cr})(Z_i m_b / Z_b^2 m_e)}{m_b v_b^2 n_b \nu_{ei\|}} \frac{m_e \nu_{ei\|}}{e} Z_b\left(1 - \frac{Z_b}{Z_i}\right) e n_b \bar{v}_b v_b$$

$$\approx 1.4(Z_i - Z_b). \tag{16.51}$$

When v_d is defined by $J \equiv e n_e v_d$, the beam power P_b per unit volume is described by

$$P_b \approx \frac{0.7}{(Z_i/Z_b - 1)} \frac{v_b}{v_d} \eta J^2. \tag{16.52}$$

The efficiency of current drive J/P_b is derived as follows:

$$\frac{J}{P_b} = \frac{1.4(Z_i/Z_b - 1)}{\eta e n v_b} \approx 64\left(\frac{Z_i - Z_b}{Z_b Z_i}\right) \frac{T_{10}^{3/2}}{n_{20}(\varepsilon_{10}/A_b)^{1/2}} \quad \left(\frac{\mathrm{A/m^2}}{\mathrm{W/m^3}}\right), \tag{16.53}$$

where T_{10} and ε_{10} are the electron temperature and the beam energy, respectively, in 10 keV units and n_{20} is the electron density in $10^{20}\ \mathrm{m}^{-3}$ units. When the charge number of beam ions is equal to that of the plasma ions, that is, when $Z_b = Z_i$, the current density becomes zero, $J = 0$, for linear (cylindrical) plasmas. For toroidal plasmas, the motion of circulating electrons is disturbed by collision with the trapped electrons (banana electrons), and the term of the dragged electrons is reduced. Thus we have $J > 0$ for toroidal plasmas.[44] The current drive by NBI is demonstrated by the experiments of DITE,[45] TFTR,[46] and others.

(iii) Bootstrap Current When a toroidal plasma has a radial pressure gradient in the banana region, it is theoretically predicted that a bootstrap current with a current density of $j_b = -\varepsilon^{1/2} B_p^{-1}\, dp/dr$ flows in the toroidal direction, as shown in eq. (8.112b).[47-50] When the pressure profile is $[1 - (r/a)^2]^n$ and the average poloidal beta is $\beta_p = \langle p \rangle / (B_p^2/2\mu_0)$, the ratio of the total bootstrap current I_b to the plasma current I_p to form B_p is given by $I_b/I_p \approx 0.3(n+1)(a/R)^{1/2}(\beta_p/q_a)$ (q_a is the safety factor). This value can be near 1 if β_p is high ($\beta_p \sim R/a$). It is predicted that the bootstrap current can be detected more easily in high-beta and high-temperature plasmas. Experiments on bootstrap current were carried out in TFTR (1986).

16.2i Tokamak Reactor Concept[127]

The conceptual design of tokamak reactors has been actively pursued according to the development of tokamak experimental research. INTOR[51] (International Tokamak Reactor) is representative of international activity in this field. The main parameters of INTOR are $R = 5.0$ m, $a = 1.2$ m ($R/a = 4.2$), noncircularity $\kappa = b/a = 1.6$. The safety factor is $q_I = 1.8$ (see sec. 16.2c) (the exact safety factor is $q_\psi > 2$); $B_t = 5.56$ T, $I_p = 8$ MA, $\langle T_e \rangle = \langle T_i \rangle = 10$ keV, $\langle n \rangle = 1.4 \times 10^{14}$ cm^{-3}, $\langle \beta \rangle = 4.9\%$.

16.3 Reversed Field Pinch

16.3a Reversed Field Pinch Configurations

Reversed field pinch (RFP) is an axisymmetric toroidal field used as a tokamak. The magnetic field configuration is composed of the poloidal field B_p produced by the toroidal component of the plasma current and the toroidal field B_t produced by the external toroidal field coil and the poloidal component of the plasma current. The particle orbit loss is as small as that of a tokamak. However, RFP and tokamaks have quite different characteristics. In RFP, the magnitudes of the poloidal field B_p and the toroidal field B_t are comparable and the safety factor

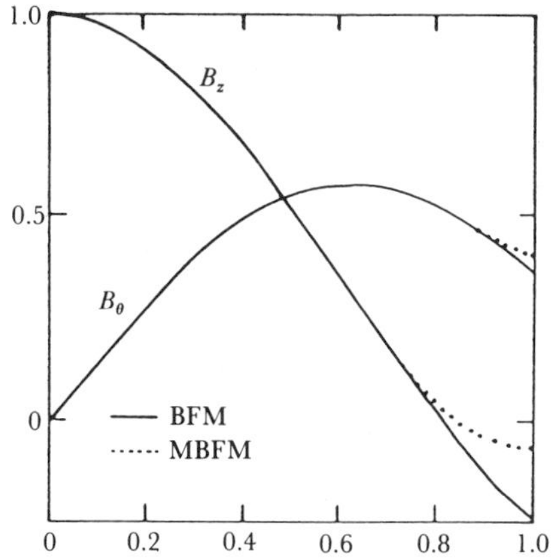

Fig. 16.11
Toroidal field $B_z(r)$ and poloidal field $B_\theta(r)$ of a reversed field pinch (RFP). The radial profiles of the Bessel function model (BFM) and the modified Bessel function model (MBFM) are shown.

$$q_s(r) = \frac{r}{R}\frac{B_t(r)}{B_p(r)} \tag{16.54}$$

is much less than 1 ($q_s(0) \sim a/(R/\Theta)$, $\Theta \sim 1.6$). The radial profile of the toroidal field is shown in fig. 16.11. The direction of the boundary toroidal field is reversed with respect to the direction of the on-axis field, and the magnetic shear is strong. Therefore high-beta ($\langle\beta\rangle \sim 10$–$20\%$) plasmas can be confined in an MHD stable way. Since the plasma current can be larger than the Kruskal-Shafranov limit ($q < 1$), there is a possibility of reaching the ignition condition by ohmic heating only (although it depends on the confinement scaling).

RFP started in an early phase of nuclear fusion research. A stable quiescent phase of discharge was found in Zeta in Harwell in 1968.[52] The configuration of the magnetic field in the quiescent phase was the reversed field pinch configuration, as shown in fig. 16.11. The electron temperature, the energy confinement time, and the average beta of Zeta were $T_e =$ 100–150 eV, $\tau_E = 2$ ms, $\langle\beta\rangle \sim 10\%$ at the time of IAEA conference at Novosibirsk. However, the epoch-making result of tokamak T-3 ($T_e = 1$ keV, $\tau_E =$ several ms, $\beta \sim 0.2\%$) was also presented in the same conference, and Zeta was shut down because of the better confinement characteristics in tokamaks. On the other hand, RFP can confine much higher beta plasmas and has been actively investigated to improve the confinement characteristics (ZT-40 M, OHTE, HBTX1B, TPE-IRM15, etc.).[53–55] The important issues of RFP are confinement scaling and impurity control in the high-temperature region.

16.3b Relaxation of Reversed Field Pinch

Even if the plasma is initially MHD unstable in the formation phase, it has been observed in RFP experiments that the plasma turns out to be a stable RFP configuration irrespective of the initial condition. J. B. Taylor pointed out in 1974 that RFP configuration is a minimum energy state by relaxation processes under certain constraints.[56] The change of $\boldsymbol{B}$ in an ideally conductive fluid is

$$\frac{\partial \boldsymbol{B}}{\partial t} - \nabla \times (\boldsymbol{v} \times \boldsymbol{B}) = 0 \tag{16.55}$$

due to $\boldsymbol{E} + \boldsymbol{v} \times \boldsymbol{B} = 0$, so that the vector potential satisfies $\partial \boldsymbol{A}/\partial t = \boldsymbol{v} \times \boldsymbol{B} + \nabla \chi$. We have the relation $\boldsymbol{B} \cdot \partial \boldsymbol{A}/\partial t = \nabla \cdot (\chi \boldsymbol{B})$, and the integration of $\boldsymbol{B} \cdot \partial \boldsymbol{A}/\partial t$ along the lines of magnetic force is zero. The volume integral defined by

$$K = \int_v \boldsymbol{A} \cdot \boldsymbol{B} \, \mathrm{d}\boldsymbol{r} \tag{16.56}$$

within the region surrounded by the lines of magnetic force (by the magnetic surfaces) is called the magnetic helicity. The time derivative is

$$\begin{aligned} \frac{\mathrm{d}K}{\mathrm{d}t} &= \int \frac{\partial}{\partial t}(\boldsymbol{A} \cdot \boldsymbol{B}) \, \mathrm{d}\boldsymbol{r} + \int (\boldsymbol{A} \cdot \boldsymbol{B})(\boldsymbol{v} \cdot \boldsymbol{n}) \, \mathrm{d}S \\ &= \int \left(\boldsymbol{A} \cdot \frac{\partial \boldsymbol{B}}{\partial t} \right) \mathrm{d}\boldsymbol{r} + \int (\boldsymbol{A} \cdot \boldsymbol{B})(\boldsymbol{v} \cdot \boldsymbol{n}) \, \mathrm{d}S. \end{aligned} \tag{16.57}$$

The substitution of $\partial \boldsymbol{B}/\partial t = \nabla \times (\boldsymbol{v} \times \boldsymbol{B})$ and the relation

$$\int \boldsymbol{A} \cdot \nabla \times \boldsymbol{C} \, \mathrm{d}\boldsymbol{r} = \int \nabla \times \boldsymbol{A} \cdot \boldsymbol{C} \, \mathrm{d}\boldsymbol{r} + \int (\boldsymbol{C} \times \boldsymbol{A}) \cdot \boldsymbol{n} \, \mathrm{d}S \tag{16.58}$$

give $\mathrm{d}K/\mathrm{d}t = 0$.* When the plasma is a perfect conductor, K is constant

* For finite resistive plasmas, $\eta \boldsymbol{j} = \boldsymbol{E} + \boldsymbol{v} \times \boldsymbol{B}$, we have

$$\frac{\mathrm{d}}{\mathrm{d}t} \int \boldsymbol{A} \cdot \boldsymbol{B} \, \mathrm{d}\boldsymbol{r} = -2 \int \eta \boldsymbol{j} \cdot \boldsymbol{B} \, \mathrm{d}\boldsymbol{r} \qquad \frac{\mathrm{d}}{\mathrm{d}t} \int \frac{\boldsymbol{B} \cdot \boldsymbol{B}}{2\mu_0} \, \mathrm{d}\boldsymbol{r} = -\int \eta \boldsymbol{j} \cdot \boldsymbol{j} \, \mathrm{d}\boldsymbol{r} \tag{16.59}$$

in a similar way.

for arbitrary integral regions surrounded by closed magnetic surfaces. However, if there is small resistivity in the plasma, the local reconnections of the lines of magnetic force are possible and the plasma can relax to a more stable state and the magnetic helicity K may change. But the global magnetic helicity K_{T} integrated over the whole region of the plasma changes much more slowly. It is assumed that K_{T} is constant within the time scale of relaxation processes. Under the constraint of K_{T} invariant,

$$\delta K_{\mathrm{T}} = \int \boldsymbol{B}\cdot\delta\boldsymbol{A}\,\mathrm{d}\boldsymbol{r} + \int \delta\boldsymbol{B}\cdot\boldsymbol{A}\,\mathrm{d}\boldsymbol{r} = 2\int \boldsymbol{B}\cdot\delta\boldsymbol{A}\,\mathrm{d}\boldsymbol{r} = 0, \tag{16.60}$$

the minimum energy state

$$(2\mu_0)^{-1}\delta\int(\boldsymbol{B}\cdot\boldsymbol{B})\,\mathrm{d}\boldsymbol{r} = {\mu_0}^{-1}\int \boldsymbol{B}\cdot\nabla\times\delta\boldsymbol{A}\,\mathrm{d}\boldsymbol{r} = {\mu_0}^{-1}\int(\nabla\times\boldsymbol{B})\cdot\delta\boldsymbol{A}\,\mathrm{d}\boldsymbol{r} \tag{16.61}$$

can be obtained by the method of undetermined multipliers, and we have

$$\nabla\times\boldsymbol{B} - \lambda\boldsymbol{B} = 0. \tag{16.62}$$

This solution is the force-free ($\boldsymbol{j}\times\boldsymbol{B} = \nabla p = 0, \boldsymbol{j}\|\boldsymbol{B}$) or pressureless plasma. The axisymmetric solution in cylindrical coordinates is

$$B_r = 0, \qquad B_\theta = B_0 J_1(\lambda r), \qquad B_z = B_0 J_0(\lambda r) \tag{16.63}$$

and is called a Bessel function model.

The profiles of $B_\theta(r)$ and $B_z(r)$ are shown in fig. 16.11. In the region $r > 2.405$, the toroidal field B_z is reversed. The pinch parameter Θ and the field reversal ratio F are used commonly to characterize the RFP magnetic field as follows:

$$\Theta = \frac{B_\theta(a)}{\langle B_z\rangle} = \frac{(\mu_0/2)I_{\mathrm{p}}a}{\int B_z 2\pi r\,\mathrm{d}r},$$

$$F = \frac{B_z(a)}{\langle B_z\rangle}, \tag{16.64}$$

where $\langle B_z\rangle$ is the volume average of the toroidal field. The values of F and Θ for the Bessel function model are

$$\Theta = \frac{\lambda a}{2}, \qquad F = \frac{\Theta J_0(2\theta)}{J_1(2\theta)}, \tag{16.65}$$

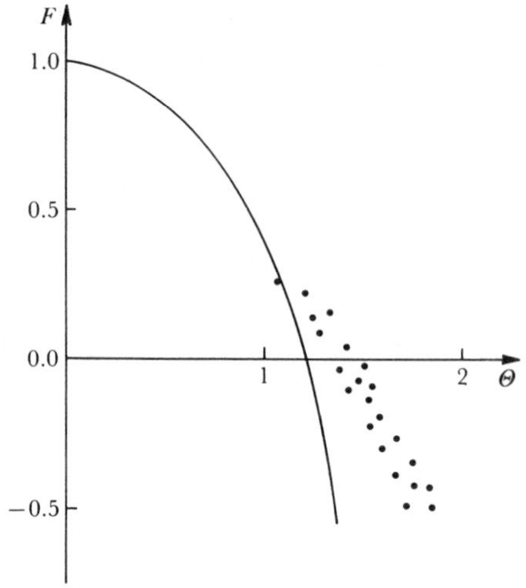

Fig. 16.12
F-θ curve. The solid line is the Bessel function model (BFM) and the dots are the experimental values.

and the F-Θ curve is plotted in fig. 16.12. The observed RFP fields in experiments deviate from the Bessel function model due to the finite beta effect and the imperfect relaxation state.

The stability condition of the local MHD mode (sec. 9.4c) is

$$\frac{1}{4}\left(\frac{q_s'}{q_s}\right)^2 + \frac{2\mu_0 p'}{rB_z^2}(1 - q_s^2) > 0. \tag{16.66}$$

This formula indicates that the strong shear can stabilize the RFP plasma in the $p'(r) < 0$ region but that the flat pressure profile, $p'(r) \sim 0$, is preferable in the central region of weak shear. The growth rate of ideal MHD instabilities can be obtained by solving the Hain-Lüst MHD equation.[57,58]

When the effect of finite resistivity of a plasma is taken into account, it is expected by the classical process of magnetic dissipation that the RFP configuration can be sustained only when $\tau_{cl} = \mu_0 \sigma a^2$, where σ is the specific conductivity. However, ZT-40M experiments[59] demonstrated that RFP discharge was sustained more than three times (~ 20 ms) as long as τ_{cl}. This is clear evidence that the regeneration process of the toroidal flux exists during the relaxation process, which is consumed by classical magnetic dissipation. This suggests that the RFP configuration can be sustained as long as the plasma current is sustained. This kind of relaxation phenomenon is also called the dynamo mechanism. Much research on relaxation phenomena has been done.[55,60,61]

So-called "F-Θ pumping" to drive the plasma current has been proposed,[62] and the experiments are being carried out in ZT-40M.

16.4 Stellarators

A stellarator field can provide a steady-state magnetohydrodynamic equilibrium configuration of plasma only by the external field produced by the coils outside the plasma. The rotational transform, which is necessary to confine the toroidal plasma, is formed by the external coils so that the stellarator has the merit of steady-state confinement. Although Stellarator C[63] was rebuilt as the ST tokamak in 1969, at the Princeton Plasma Physics Laboratory, confinement experiments by Wendelstein VIIA/AS, Heliotron-E, and ATF are being actively carried out, since there is a merit of steady-state confinement, possibly without current-driven instabilities.

16.4a Stellarator Devices

As described in sec. 2.3c, familiar stellarator fields are of pole number $l = 2$ or $l = 3$. The solid magnetic axis system of Heliac has $l = 1$. When the ratio of the minor radius a_{h} of a helical coil to the helical pitch length R/m (R is the major radius and m is the number of field periods) is much less than 1, that is, $ma_{\mathrm{h}}/R \ll 1$, the rotational transform angle is $\iota_2(r) = \text{const.}$ for $l = 2$ and $\iota_3(r) = \iota(r/a)^2$ for $l = 3$. In this case the shear is small for the $l = 2$ configuration, and $\iota_3(r)$ is very small in the central region for the $l = 3$ configuration. However, if $ma_{\mathrm{h}}/R \sim 1$, then $\iota_2(r) = \iota_0 + \iota_2(r/a)^2 + \ldots$, so that the shear can be large even when $l = 2$ (sec. 2.3c).

The arrangement of coils in the $l = 2$ case is shown in fig. 16.13. Figure 16.13a is the standard type of stellarator,[64,65] and fig. 16.13b is a torsatron/heliotron type.[66,67] In general, stellarator fields are produced by the toroidal field coils and the helical coils. However, it is possible to realize a stellarator field without helical coils. When elliptical coils are arranged as shown in fig. 16.14, an $l = 2$ stellarator field can be obtained.[68] The currents produced by the twisted toroidal coil system shown in fig. 16.15 can simulate the currents of toroidal field coils and the helical coils taken together.[69] A typical device of this type is Wendelstein VIIAS.

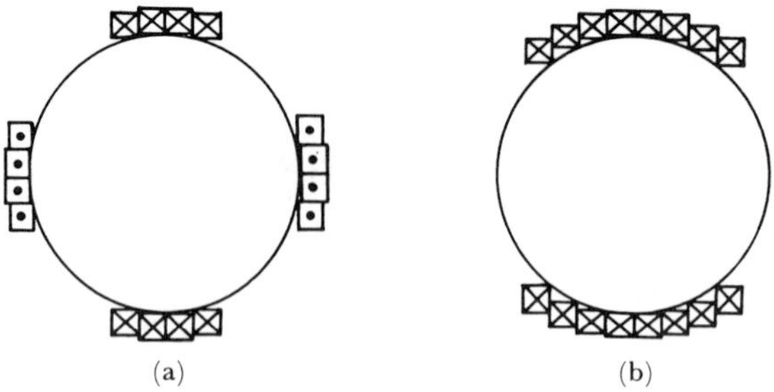

Fig. 16.13
Cross-sectional views of helical coils in the $l = 2$ case. (a) Standard stellarator. (b) Torsatron/heliotron.

Fig. 16.14
Arrangement of elliptical coils used to produce an $l = 2$ linear stellarator field.

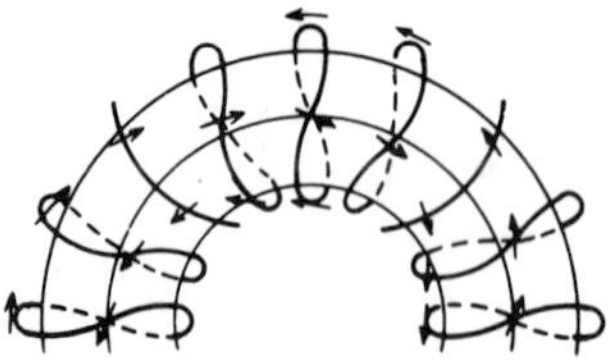

Fig. 16.15
Twisted toroidal field coils that produce the toroidal stellarator field ($l = 2$).

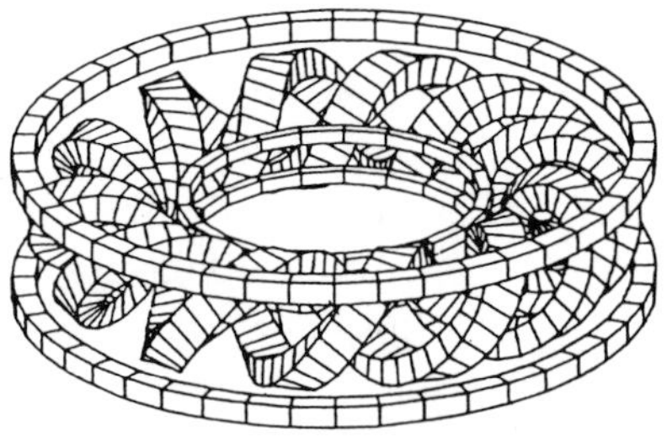

Fig. 16.16
Coils of the toroidal torsatron/heliotron. When the linear helical coils are changed to toroidal helical coils, an extra vertical field is produced. The vertical field coils are necessary to compensate for the unfavorable vertical field. After L. Garcia, B. A. Carreras, and J. H. Harris: Nucl. Fusion **24**, 115(1984).

In the torsatron/heliotron configuration the current directions of the helical coils are the same so that the toroidal field and the helical field can be produced by the helical coils alone.[70,71] So, if the pitch is properly chosen, closed magnetic surfaces can be formed even without toroidal field coils.[72,73] The typical devices of this type are Heliotron E and ATF-1. The coil systems are shown in fig. 16.16.

For linear stellarators the magnetic surface $\psi = rA_\theta$ exists due to its helical symmetry. However, the existence of magnetic surfaces in toroidal stellarators has not yet been proven in the strict mathematical sense. According to numerical calculations, the magnetic surfaces exist in the central region near the magnetic axis, but in the outer region the lines of magnetic force behave ergodically and the magnetic surfaces are destroyed. Although the helical coils have a relatively complicated structure, the lines of magnetic force can be traced by computer, and the design of stellarator field devices becomes less elaborate. The effect of the geometrical error to the stellarator field can be estimated, and accurate coil windings are possible with numerically controlled devices ($\Delta l/R < 0.1\%$).

The orbit of the charged particles in stellarators has been studied in detail. The orbit in linear stellarators is obtained analytically due to the helical symmetry. Banana particles trapped by the helical ripples move to the poloidal direction with the angular velocity ω_{h} given by eq. (3.125) in sec. 3.5b due to ∇B drift. For toroidal stellarators the toroidal term is added to the inhomogeneity of the magnetic field. The orbit of the (untrapped) banana deviates by Δ, which is given in eq. (3.132), due to toroidal drift.

When the degree of helical ripple ε_h is much larger than $\varepsilon_t = r/R$ (r and R are the minor and the major radius, respectively), the superbanana (trapped banana) is expected theoretically, as described in sec. 8.3. However, both terms are of the same order, $\varepsilon_h \sim \varepsilon_t$, in real devices, and the orbit trace by numerical calculations shows that the type of superbanana discussed in sec. 8.3 does not appear when $\varepsilon_h \sim \varepsilon_t$.[74] The orbit loss of particles due to the crossing of the orbit to the wall and the loss region in the velocity space due to the orbit loss have been evaluated.[75] When the plasma temperature becomes high and the collisional frequency becomes low, the neoclassical thermal diffusion coefficient due to trapped ions by helical ripple becomes large, and then the neoclassical thermal diffusion coefficient due to trapped electrons becomes larger, as follows:

$$D_h{}^i = \gamma_h{}^i \varepsilon_h{}^{1/2} \varepsilon_t{}^2 \left(\frac{\omega_h{}^i}{\nu_{ii}}\right) \frac{T_i}{ZeB} \propto \varepsilon_h{}^{1/2} \varepsilon_t{}^2 \frac{T_i^{3.5}}{nB^2 r^2},$$

$$D_h{}^e = \gamma_h{}^e \varepsilon_h{}^{1/2} \varepsilon_t{}^2 \left(\frac{\omega_h{}^e}{\nu_{ee}}\right) \frac{T_e}{eB} \propto \varepsilon_h{}^{1/2} \varepsilon_t{}^2 \frac{T_e^{3.5}}{nB^2 r^2}, \tag{16.67}$$

where $\gamma_h{}^{i,e} \sim 50$. Furthermore, when the collisional frequency becomes low, the effect of the radial electric field appears, and the dependence of the thermal diffusion coefficients on collisional frequencies changes (see sec. 8.3).[76–79]

16.4b Confinement Experiments in Stellarators

After Stellerator C, the basic experiments were carried out in small but accurate stellarator devices (clasp, Proto Cleo, Wendelstein IIb, JIPP I, Heliotron D, L1, Uragan 1). Alkali plasmas, or afterglow plasmas produced by wave heating or gun injection, were confined quiescently. The effect of shear on the stability and confinement scaling were investigated.[80]

The $l = 2$ stellarators with long helical pitch, such as Wendelstein IIa[81] or JIPP I-b,[82] have nearly constant rotational transform angles and the shears are small. When the transform angle is rational, $\iota/2\pi = n/m$, a line of magnetic force comes back to the initial position after m turns of the torus and is closed. If electric charges are localized in some place, they cannot be dispersed uniformly within the magnetic surface in the case of rational surfaces. A resistive drift wave or resistive MHD instabilities are likely to be excited, and convective loss is also possible. The enhanced loss is observed in the rational case (fig. 16.17). This is called resonant loss. Resonant loss can be reduced by the introduction of shear.

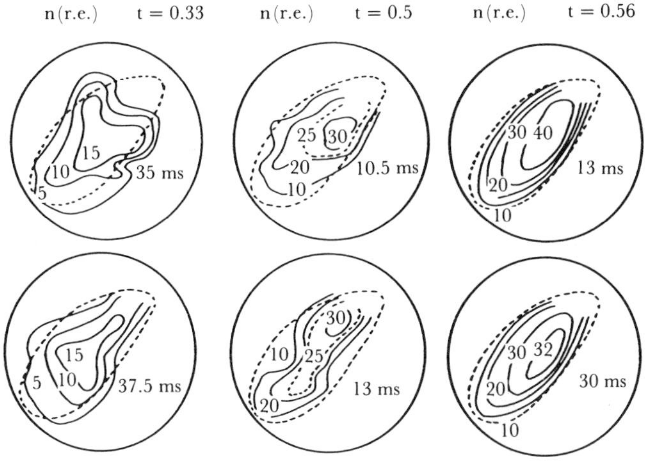

Fig. 16.17
Equidensity contours of plasmas confined in JIPP I-b stellarator ($l = 2$) with the rotational transform angles of $\iota/2\pi = \frac{1}{3}, \frac{1}{2}$, and 0.56.

Medium-scale stellarators (Wendelstein VIIA, Cleo, JIPP T-II, Heliotron-E, L2, Uragan 2, Uragan 3) have been constructed.[80] The confinement time of the ohmically heated plasmas is roughly equal to that of tokamaks with similar scale. When the rotational transform angle is larger than $\iota_h/2\pi > 0.14$, the major disruption observed in tokamaks is suppressed (W VIIA, JIPP T-II). NBI heating or wave heating, which was developed in tokamaks, has been applied to plasma production in stellarators. In Wendelstein VIIA, a target plasma was produced by ohmic heating; then the target plasma was sustained by NBI heating while the plasma current was gradually decreased, and finally a high-temperature plasma with $T_i \approx$ several hundred eV, $n_e \approx$ several 10^{13} cm^{-3} was confined without plasma current (1982).[64] In Heliotron-E, a target plasma was produced by electron cyclotron resonance heating with $T_e \approx 800$ eV, $n_e \sim 0.5 \times 10^{13}$ cm^{-3}, and the target plasma was heated by NBI heating with 1.8 MW to the plasma with $T_i \approx 1$ keV, $n_e = 2 \times 10^{13}$ cm^{-3} (1984). The average beta, $\langle\beta\rangle \approx 2\%$, was obtained in the case of $B = 0.94$ T and NBI power $P_{NB} \approx 1$ MW.[66] These experimental results are important because they demonstrate the possibility of steady-state confinement by stellarators. The important issues for stellarators are studies of beta limit, impurity control, and confinement scaling in the high-temperature region.

16.5 Open-End Systems

Open-end magnetic field systems are of a simpler configuration than toroidal systems. The attainment of absolute minimum-B configurations is possible with mirror systems, whereas only average minimum-B configurations can be realized in toroidal systems. Although absolute minimum-B configurations are MHD stable, the velocity distribution of the plasma becomes non-Maxwellian due to end losses, and the plasma will be prone to velocity-space instabilities.

The most critical issue of open-end systems is the suppression of end loss. The end plug of the mirror due to electrostatic potential has been studied by tandem mirrors. The bumpy torus, which is a toroidal linkage of many mirrors, was used in other trials to avoid end loss.

A cusp field is another type of open-end system. It is rotationally symmetric and absolute minimum-B. However, the magnetic field becomes zero at the center, so the magnetic momentum is no longer invariant (sec. 3.3c) and the end loss from the line and point cusps are severe.

16.5a Confinement Times in Mirrors and Cusps

Particles are trapped in a mirror field when the velocity components $v_\perp$ and $v_\parallel$, perpendicular and parallel to the magnetic field, satisfy the condition

$$\frac{v_\perp^{\ 2}}{v^2} > \frac{b_0}{b_{\mathrm{L}}} \qquad (v^2 = v_\perp^{\ 2} + v_\parallel^{\ 2}), \tag{16.68}$$

where b_0 and b_{L} are the magnitudes of the magnetic field at the center and at the end, respectively. Denoting the mirror ratio b_{L}/b_0 by R, so that

$$\frac{b_0}{b_{\mathrm{L}}} = \frac{1}{R} = \sin^2 \alpha_{\mathrm{L}}, \tag{16.69}$$

the trapping condition is reduced to

$$\frac{v_\perp}{v} > \sin \alpha_{\mathrm{L}}. \tag{16.70}$$

When the particles enter the loss cone, they escape immediately, so that the confinement time is determined by the velocity space diffusion to the loss cone. As was described in sec. 5.4 in detail, the confinement time of a mirror

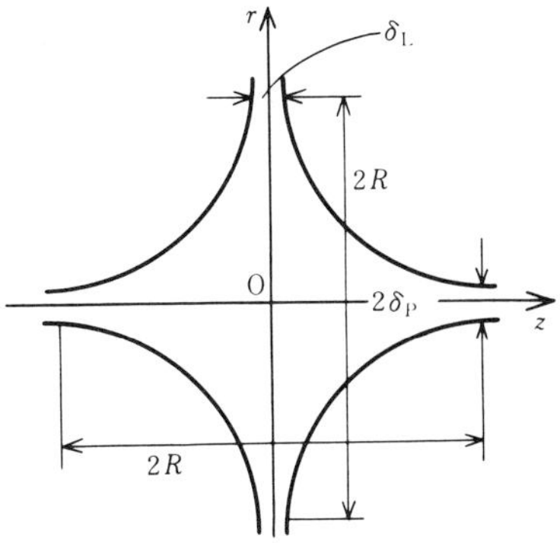

Fig. 16.18
Line cusp width δ_L and point cusp radius δ_p.

field is given by

$$\tau \approx \tau_i \ln R, \tag{16.71}$$

where τ_i is the ion-ion collision time. Even if the mirror ratio R is increased, the confinement time is increased only as fast as ln R. The time is independent of the magnitude of the magnetic field and the plasma size. If the density is $n \approx 10^{20}\ \mathrm{m}^{-3}$, the ion temperature $T_i \approx 100$ keV, the atomic mass number $A = 2$, and ion charge $Z = 1$, the ion-ion collision time is $\tau_i =$ 0.3 s, so that $n\tau_i = 0.3 \times 10^{20}\ \mathrm{m}^{-3}\mathrm{s}$. This value is quite marginal for a fusion reactor. There is only a narrow allowance for enhanced loss due to instability. It is also necessary to increase the efficiency of recovery of the kinetic energy of charged particles escaping through the ends.

Next let us consider cusp losses. Since the magnetic field is zero at the cusp center 0, the magnetic moment is not conserved. Therefore the end losses from the line cusp and the point cusp (fig. 16.18) are determined by the hole size and average ion velocity $\bar{v}_i$. The flux of particles from the line cusp of width δ_L and radius R is given by

$$F_L = \frac{1}{4} n_L \bar{v}_i 2\pi R \cdot \delta_L, \qquad \bar{v}_i = \left(\frac{8}{\pi}\frac{T_i}{M}\right)^{1/2}, \tag{16.72}$$

where n_L is the particle density at the line cusp and T_i and M are the temperature and mass of the ions. The flux of particles from two point cusps of radius δ_p is[83,84]

$$F_{\mathrm{p}} = \frac{1}{4} n_{\mathrm{p}} \bar{v}_{\mathrm{i}} \pi \delta_{\mathrm{p}}^{2} \times 2. \tag{16.73}$$

The observed value of δ_{L} is about that of the ion Larmor radius. As the lines of magnetic force are given by

$$rA_{\theta} = ar^2 z = \text{const.} \tag{16.74}$$

(sec. 2.3c), we find that

$$\delta_{\mathrm{p}}^{2} R \approx R^2 \frac{\delta_{\mathrm{L}}}{2},$$

$$\delta_{\mathrm{p}}^{2} = \frac{R\delta_{\mathrm{L}}}{2}, \tag{16.75}$$

so that $F_{\mathrm{p}} \approx F_{\mathrm{L}}/2$. The total end-loss flux is

$$F \approx \frac{1}{4} \bar{v}_{\mathrm{i}} 2\pi R \delta_{\mathrm{L}} \left(n_{\mathrm{L}} + \frac{n_{\mathrm{p}}}{2} \right), \tag{16.76}$$

and the volume V of the cusp field is (fig. 16.18)

$$V = \pi R^2 \delta_{\mathrm{L}} \left(\ln \frac{2R}{\delta_{\mathrm{L}}} + 1 \right). \tag{16.77}$$

The confinement time is now given by

$$\tau = \frac{nV}{F} \approx \frac{R}{\bar{v}_{\mathrm{i}}} \left(\ln \frac{2R}{\delta_{\mathrm{L}}} + 1 \right) \frac{4}{3} \frac{n}{n_{\mathrm{L}} + n_{\mathrm{P}}/2}, \tag{16.78}$$

where n is the average density. When $R = 10$ m, $B = 10$ Wb/m^2, $T_{\mathrm{i}} =$ 20 keV, $A = 2$, and $\delta_{\mathrm{L}} = \rho_B{}^{\mathrm{i}}$, then $\tau \approx 0.1$ ms, which is 10–20 times the transit time $R/\bar{v}_{\mathrm{i}}$. Even if the density is $n \approx 10^{22}$ m^{-3}, the product of $n\tau$ is of the order of 10^{18} m^{-3}s, or 10^{-2} times smaller than Lawson's condition.

16.5b Confinement Experiments with Mirrors

The most important mirror confinement experiment was done by Ioffe and his colleagues, who demonstrated that the minimum-B mirror configuration is quite stable.[85] Addition of stabilizing coils as shown in fig. 16.19a results in a mirror field with the minimum-B property (sec. 2.3c).

Define the wall mirror ratio as the ratio of the magnitude of the magnetic field at the radial boundary to its magnitude at the center. When the wall

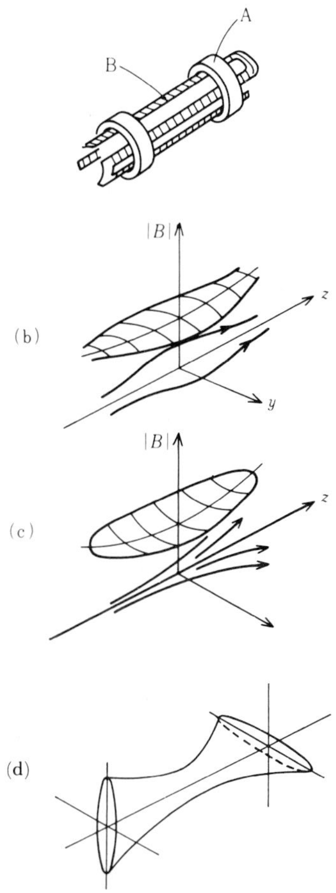

Fig. 16.19
Minimum-B mirror field. (a) Coil system with mirror coil (A) and stabilizing coils (Ioffe bars) (B). (b) Magnitude of the magnetic field in a simple mirror. (c) Magnitude of the magnetic field in the minimum-B mirror field. (d) The plasma shape of a quadrupole minimum-B (fishtail).

mirror ratio is increased, flute instability disappears and the confinement time becomes long. In the PR-5 device, a plasma of density 10^9–10^{10} cm^{-3} and ion temperature 3–4 keV has been confined for 0.1 s.[86] However, when the density is larger than 10^{10} cm^{-3} ($\Pi_i > |\Omega_i|$), the plasma suffers from loss-cone instability (sec. 16.6c).

One of the most promising mirror confinement results was brought about in the 2X experiments.[87] Here a plasmoid of mean energy of 2.5 keV produced by a plasma gun is injected into a quadrupole mirror and then compressed. The initial density of the trapped plasma is $n \approx 3 \times 10^{13}$ cm^{-3}, and $n\tau \approx 10^{10}$ cm^{-3}s. The magnitude of the magnetic field is 13 kG and $\beta \approx 5\%$. The average energy of the ions is 6–8 keV, and the electron temperature is about 200 eV. Since $n\tau$ for an ideal case (classical Coulomb collision) is $n\tau \approx 3 \times 10^{10}$ cm^{-3}s, the results obtained are 1/3–1/15 that of the ideal.

In the 2XIIB experiment,[88] microscopic instability is suppressed by adding a small stream of warm plasma to smooth out the loss cone. A 15–19 keV, 260 A neutral beam is injected, and the resulting plasma, of 13 keV ion temperature, is confined to $n\tau \approx 10^{11}$ cm^{-3}s ($n \approx 4 \times 10^{13}$ cm^{-3}).

16.5c Instabilities in Mirror Systems

Instabilities of mirror systems are reviewed in refs. 89 and 90. MHD instability can be suppressed by the minimum-B configuration, and that obstacle can be considered to have been surmounted. However, the particles in the loss-cone region are not confined, and the non-Maxwellian distribution gives rise to electrostatic perturbations at the ion cyclotron frequency and its harmonics, which scatter the particles into the loss cone and so enhance the end loss. It can be shown that instabilities are induced when the cyclotron wave couples with other modes, such as the electron plasma wave or the drift wave.

(i) Instability in the Low-Density Region ($\Pi_e \approx l|\Omega_i|$) Let us consider the low-density case. When the plasma electron frequency Π_e reaches the ion cyclotron frequency $|\Omega_i|$, there is an interaction between the ion Larmor motion and the electron Langmuir oscillation, and Harris instability occurs[91] (sec. 12.3). When the density increases further, the oblique Langmuir wave with $\omega = (k_\parallel/k)\Pi_e$ couples with the harmonics $l|\Omega_i|$ of the ion cyclotron wave. The condition $\omega \approx k_\perp v_{\perp i}$ is necessary for the ions to excite the instability effectively; and, if $\omega > 3k_\parallel v_{\parallel e}$, then the Landau damping due to electrons is ineffective. Thus excitation will occur when

$$\omega \approx l|\Omega_{\mathrm{i}}| \approx \Pi_{\mathrm{e}} k_{\parallel}/k \approx k_{\perp} v_{\perp \mathrm{i}} > 3k_{\parallel} v_{\parallel \mathrm{e}}, \tag{16.79}$$

where $k_{\parallel}, k_{\perp}$ are the parallel and perpendicular components of the propagation vector and $v_{\parallel}$, $v_{\perp}$ are the parallel and perpendicular components of the velocity. Therefore the instability condition is

$$\frac{\Pi_{\mathrm{e}}}{l|\Omega_{\mathrm{i}}|} > (1 + 9v_{\parallel \mathrm{e}}{}^2/v_{\perp \mathrm{i}}{}^2)^{1/2} \approx (1 + 9MT_{\mathrm{e}}/mT_{\mathrm{i}})^{1/2}. \tag{16.80}$$

Let L be the length of the device. As the relation $k_{\parallel} > 2\pi/L$ holds, eq. (16.79) is reduced to[92,93]

$$\frac{L}{\rho_B} > \frac{6\pi}{l} \frac{v_{\parallel \mathrm{e}}}{v_{\perp \mathrm{i}}}. \tag{16.81}$$

Harris instability has been studied in detail experimentally.[94]

(ii) Instability in the High-Density Region ($\Pi_{\mathrm{i}} > |\Omega_{\mathrm{i}}|$) When the density increases further, so that Π_{i} is larger than $|\Omega_{\mathrm{i}}|$ (while $\Omega_{\mathrm{e}} > \Pi_{\mathrm{i}}$ still holds), loss-cone instability with $\omega_{\mathrm{r}} \approx \omega_{\mathrm{i}} \approx \Pi_{\mathrm{i}}$ will occur[95] (sec. 12.3). This is a convective mode and the length of the device must be less than a critical length, given by

$$L_{\mathrm{crit}} = 20A\rho_B{}^{\mathrm{i}} \left(\frac{\Omega_{\mathrm{e}}{}^2}{\Pi_{\mathrm{e}}{}^2} + 1 \right)^{1/2}, \tag{16.82}$$

for stability. Here A is of the order of 5 ($A \approx 1$ for complete reflection and $A \approx 10$ for no reflection at the open end). Therefore the instability condition is $L > 100\rho_B{}^{\mathrm{i}}$. Loss-cone instability can occur for a homogeneous plasma. When there is a density gradient, this type of instability couples with the drift wave, and drift cyclotron loss-cone instability may occur.[95] When the characteristic length of the density gradient is comparable to the radial dimension R_{P} of the plasma, the instability condition of this mode is

$$R_{\mathrm{P}} < \rho_{\mathrm{i}} \left(\frac{\Pi_{\mathrm{i}}}{|\Omega_{\mathrm{i}}|} \left(1 + \frac{\Pi_{\mathrm{e}}{}^2}{\Omega_{\mathrm{e}}{}^2} \right)^{-1/2} \right)^{4/3}. \tag{16.83}$$

(iii) Mirror Instability When the beta ratio becomes large, the anisotropy of plasma pressure induces electromagnetic-mode mirror instabilities (sec. 13.3). (Note that the Harris and loss-cone instabilities are electrostatic.) The instability condition is

$$\left(\frac{T_{\perp i}}{T_{\parallel i}} - 1\right) 2\beta > 1. \tag{16.84}$$

To avoid the instabilities described in (i)–(iii), the following stability conditions must be met:

$$L < 100\text{–}200\rho_B{}^{i}, \tag{16.85}$$

$$R_p > 200\rho_B{}^{i}, \tag{16.86}$$

$$\beta < \frac{1}{2}\left(\frac{T_{\perp i}}{T_{\parallel i}} - 1\right)^{-1} \approx 0.3\text{–}0.5. \tag{16.87}$$

(iv) Negative Mass Instability Let us assume that charged particles in Larmor motion are uniformly distributed at first and then a small perturbation occurs, so that positive charge (for example) accumulates at the region A shown in fig. 16.20. The electric field decelerates ions in the region to the right of A, and their kinetic energy $\varepsilon_\perp$ corresponding to the velocity component perpendicular to the magnetic field is decreased. Ions to the left of A are accelerated, and their $\varepsilon_\perp$ is increased. When the rotational frequency ω depends on $\varepsilon_\perp$ through the relation $d\omega/d\varepsilon_\perp < 0$, the frequency ω of the ions in the region to the right of A is increased and the ions approach A despite the deceleration. These ions behave as if they had negative mass. This situation is exactly the same for the ions in the left-hand region. Therefore the charge accumulates at A, and the perturbation is unstable. This type of instability is called *negative mass instability*. The condition $d\omega/d\varepsilon_\perp < 0$ is satisfied when the magnitude of the magnetic field decreases radially. Thus, as expected, simple mirrors, where B decreases radially and the Larmor radius is large, have been reported to exhibit negative mass instability.[96]

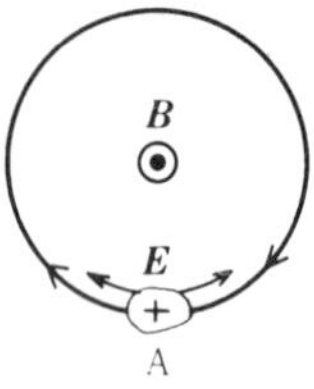

Fig. 16.20
Negative mass instability.

The PR-5 device is of minimum-B configuration, so that the magnitude of the magnetic field increases radially. However, here another type of negative mass instability is observed.[97] When the perpendicular energy $\varepsilon_\perp$ decreases, ions can enter the mirror region more deeply, so that the ion cyclotron frequency is increased. Thus the condition $\mathrm{d}\omega/\mathrm{d}\varepsilon_\perp < 0$ is satisfied even in PR-5.

(v) Instability in Hot Electron Plasmas A hot electron plasma can be produced by electron cyclotron resonant heating in mirror fields. The temperature of the hot component is raised up to the range of several keV to several hundred keV, and the density range is 10^{10}–10^{11} cm^{-3}. Along with the electrostatic Harris instability, the electromagnetic whistler instability[98,99] is excited by anistropy of the velocity distribution function (sec. 13.3). This whistler instability of hot electron plasmas has been observed experimentally.[100]

16.5d Tandem Mirrors

As described in sec. 16.5a, the input and output energy balance of a classical mirror reactor is quite marginal even in the ideal confinement condition. Therefore the suppression of the end loss is the critical issue for realistic mirror reactors. This section describes the research on the end plug by the use of electrostatic potential. In a simple mirror case, the ion confinement time is of the order of ion-ion collision time, and the electron confinement time is of the order of electron-electron or electron-ion collision times, $\tau_{ee} \sim \tau_{ei}$. Since $\tau_{ee} \ll \tau_{ii}$, the plasma is likely to be ion rich, and the plasma potential in the mirror tends to be positive. When the two mirrors are arranged in the ends of the central mirror in tandem, it is expected that the plasma potentials in the plug mirrors (plug cell) at both ends become positive with respect to the potential of the central mirror (central cell). This configuration is called a *tandem mirror*.[101,102] The loss-cone region in velocity space of the tandem mirror, as shown in fig. 16.21, is given by

$$\left(\frac{v_\perp}{v}\right)^2 < \frac{1}{R_\mathrm{m}}\left(1 - \frac{q\phi_\mathrm{c}}{mv^2/2}\right), \tag{16.88}$$

where q is the charge of the particle and R_m is the mirror ratio. The loss-cone regions of electrons and ions are shown in fig. 16.22 in the positive case of the potential ϕ_c. Solving the Fokker-Planck equation, Pastukhov derived the ion confinement time of a tandem mirror with positive potential as follows:[103]

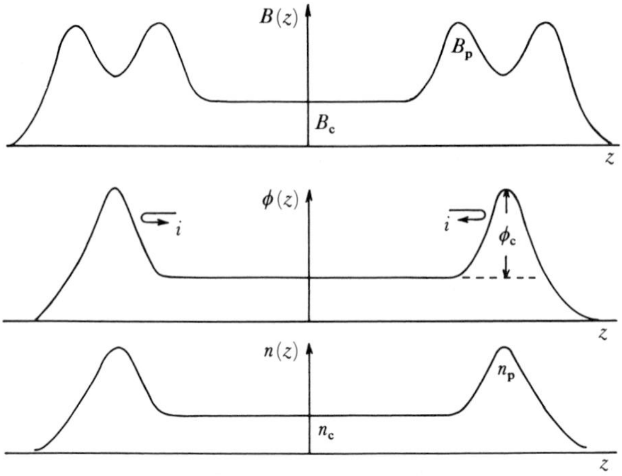

Fig. 16.21
The magnitudes of magnetic field $B(z)$, electrostatic potential $\phi(z)$, and density $n(z)$ along the z axis (mirror axis) of a tandem mirror.

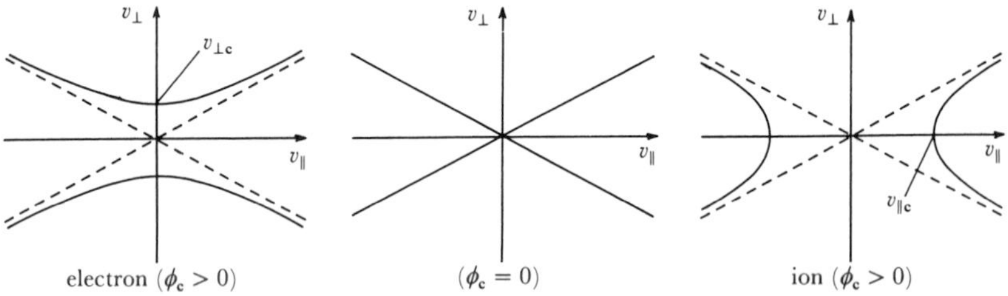

Fig. 16.22
Loss-cone regions of an ion ($q = Ze$) and electron ($-e$) for positive electrostatic potential $\phi_c > 0$ and the mirror ratio R_m. $(v_{\perp c}/v)^2 = [e\phi_c/(R_m - 1)]/(m_e v^2/2)$ in the electron case. $(v_{\perp c}/v)^2 = Ze\phi_c/(m_i v^2/2)$ in the ion case.

$$\tau_{\text{PAST}} = \tau_{\text{ii}} g(R_{\text{m}}) \left(\frac{e\phi_{\text{c}}}{T_{\text{ic}}}\right) \exp\left(\frac{e\phi_{\text{c}}}{T_{\text{ic}}}\right),$$

$$g(R_{\text{m}}) = \pi^{1/2}(2R_{\text{m}} + 1)(4R_{\text{m}})^{-1} \ln(4R_{\text{m}} + 2), \tag{16.89}$$

where T_{ic} is the ion temperature of the central cell and ϕ_{c} is the potential difference of the plug cells in both ends with respect to the central cell. Denote the electron densities of the central cell and the plug cell by n_{c} and n_{p}, respectively; then the Boltzmann relation $n_{\text{p}} = n_{\text{c}} \exp(e\phi_{\text{c}}/T_{\text{e}})$ gives

$$\phi_{\text{c}} = \frac{T_{\text{e}}}{e} \ln \frac{n_{\text{p}}}{n_{\text{c}}}. \tag{16.90}$$

By application of neutral beam injection into the plug cell, it is possible to increase the density in the plug cell to larger than that of the central cell. When $R_{\text{m}} \sim 10$ and $e\phi_{\text{c}}/T_{\text{ic}} \sim 2.5$, $\tau_{\text{PAST}} \sim 100\ \tau_{\text{ii}}$, so that the theoretical confinement time of the tandem mirror is longer than that of the simple mirror.

In this type of tandem mirror it is necessary to increase the density n_{p} in the plug cell in order to increase the plug potential ϕ_{c}, and the necessary power of the neutral beam injection becomes large. Since the plug potential ϕ_{c} is proportional to the electron temperature T_{e}, an increase in T_{e} also increases ϕ_{c}. If the electrons in the central cell and the electrons in the plug cells can be thermally isolated, the electrons in the plug cells only can be heated, so that efficient potential plugging may be expected. For this purpose a thermal barrier[104] is introduced between the central cell and the plug cell, as shown in fig. 16.23. When a potential dip is formed in the thermal barrier in an appropriate way, the electrons in the plug cell and the central cell are thermally isolated. Since the electrons in the central cell are considered to be Maxwellian with temperature T_{ec}, we have the relation

$$n_{\text{c}} = n_{\text{b}} \exp\frac{e\phi_{\text{b}}}{T_{\text{ec}}}. \tag{16.91}$$

The electrons in the plug cell are modified Maxwellian,[105] and the following relation holds:

$$n_{\text{p}} = n_{\text{b}} \exp\frac{e(\phi_{\text{b}} + \phi_{\text{c}})}{T_{\text{ep}}} \times \left(\frac{T_{\text{ep}}}{T_{\text{ec}}}\right)^{\nu}, \tag{16.92}$$

where $\nu \sim 0.5$. These relations reduce to

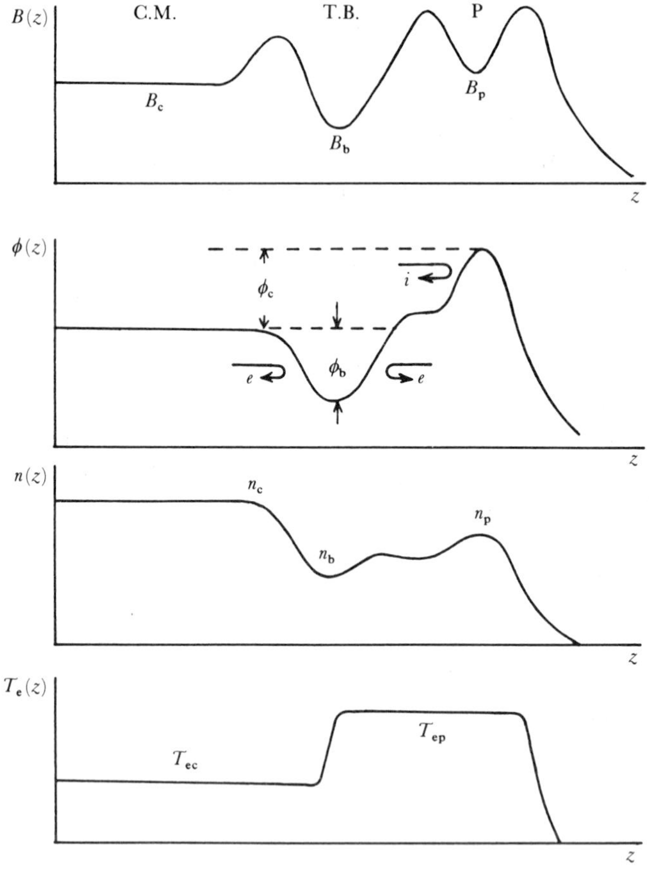

Fig. 16.23
The magnitudes of magnetic field $B(z)$, electrostatic potential $\phi(z)$, density $n(z)$, and the electron temperature $T_e(z)$ at the thermal barrier. C.M., central mirror. T.B., thermal barrier. P, plug mirror.

$$\phi_c = \frac{T_{ep}}{e}\ln\left[\frac{n_p}{n_b}\left(\frac{T_{ec}}{T_{ep}}\right)^{\nu}\right] - \frac{T_{ec}}{e}\ln\left(\frac{n_c}{n_b}\right)$$

$$= \frac{T_{ep}}{e}\ln\left[\frac{n_p}{n_c}\left(\frac{T_{ec}}{T_{ep}}\right)^{\nu}\right] + \left(\frac{T_{ep}}{T_{ec}} - 1\right)\frac{T_{ec}}{e}\ln\left(\frac{n_c}{n_b}\right),$$

$$\phi_b = \frac{T_{ec}}{e}\ln\left(\frac{n_c}{n_b}\right). \tag{16.93}$$

Therefore, if the electron temperature T_{ep} in the plug cell is increased, ϕ_c can be increased without the condition $n_p > n_c$; thus efficient potential plugging may be expected. Experiments with tandem mirrors have been carried out in TMX-U, GAMMA-10, PHAEDRUS, TARA, and AMBAL.

16.6 Wave Heating

16.6a Propagation and Absorption of Waves

Wave heating in ion cyclotron range of frequency (ICRF), lower hybrid wave heating (LHH), electron cyclotron heating (ECH), and other heating processes are being studied actively. The power sources of high-frequency waves are generally easier to construct than the beam source of neutral beam injection (NBI), with the exception of the microwave power source for ECH. Although the heating mechanism of NBI can be well explained by the classical process of Coulomb collision, the physical processes of wave heating are complicated and the interactions of waves and plasmas have a lot of variety, so that various applications are possible depending on the development of wave heating.

Waves are excited in the plasma by antennae or waveguides located outside the plasma. Excited waves may propagate and pass through the plasma center without damping (heating) in some cases and may refract and turn back to the external region without passing the plasma center or may be reflected by the cutoff layer.

The waves propagating in the plasma are absorbed and damped at the locations where Landau damping and cyclotron damping occur and heat the plasma. Therefore it is necessary for heating the plasma center that the waves be able to propagate into the plasma center without absorption and that they be absorbed when they reach the plasma center.

The experimental results of wave propagation can be well explained by the linear theory. Although nonlinear or stochastic processes accompany wave heating in many cases, the experimental results of wave heating or absorption can usually be described by linear or quasi-linear theories. The basis of the linear theory is the dispersion relation with the dielectric tensor (eq. (13.17b)) of a finite-temperature plasma.

In this section the propagation vector is expressed in (x, y, z) coordinates by $\boldsymbol{k} = k_\perp \hat{\boldsymbol{x}} + k_z \hat{\boldsymbol{z}}$ (although $\boldsymbol{k} = k_\perp \hat{\boldsymbol{y}} + k_z \hat{\boldsymbol{z}}$ was used in chapter 13). When the plasma is Maxwellian, the dielectric tensor $\boldsymbol{K}$ is given by

$$\boldsymbol{K} = \boldsymbol{I} + \sum_j \frac{\Pi^2}{\omega^2} \sum_{n=-\infty}^{\infty} \zeta_0 Z(\zeta_n) e^{-b} \boldsymbol{X}_n + \sum_j \frac{\Pi^2}{\omega^2} 2\zeta_0{}^2 \boldsymbol{L}, \tag{16.94}$$

$$X_n = \begin{bmatrix} n^2 I_n/b & in(I_n' - I_n) & -2^{1/2}\zeta_n (n/\alpha) I_n \\ -in(I_n' - I_n) & (n^2/b + 2b)I_n - 2bI_n' & i2^{1/2}\zeta_n \alpha (I_n' - I_n) \\ -2^{1/2}\zeta_n (n/\alpha) I_n & -i2^{1/2}\zeta_n \alpha (I_n' - I_n) & 2\zeta_n{}^2 I_n \end{bmatrix}, \tag{16.95}$$

$$\zeta_n \equiv \frac{\omega + n\Omega}{2^{1/2} k_z v_\perp}, \qquad \alpha \equiv \frac{k_\perp v_T}{\Omega}, \qquad b \equiv \alpha^2, \qquad I_{-n} = I_n, \tag{16.96}$$

and the components of the $\boldsymbol{L}$ matrix are all zero except $L_{zz} = 1$.

Usually the square of the product of the normal component $k_\perp$ of the propagation vector and the Larmor radius ρ, $b = (k_\perp \rho)^2 = (k_\perp v_\perp/\Omega)^2$, is less than 1 ($b < 1$). The expansion in b and the inclusion of terms up to the second harmonic in $\boldsymbol{K}$ give

$$K_{xx} = 1 + \sum_j \left(\frac{\Pi_j}{\omega^2}\right)^2 \zeta_0 \Bigg[(Z_1 + Z_{-1})\left(\frac{1}{2} - \frac{b}{2} + \cdots\right) + (Z_2 + Z_{-2})\left(\frac{b}{2} - \frac{b^2}{2} + \cdots\right) + \cdots \Bigg]_j,$$

$$K_{yy} = 1 + \sum_j \left(\frac{\Pi_j}{\omega}\right)^2 \zeta_0 \Bigg[Z_0(2b + \cdots) + (Z_1 + Z_{-1})\left(\frac{1}{2} - \frac{3b}{2} - \cdots\right) + (Z_2 + Z_{-2})\left(\frac{b}{2} - b^2 + \cdots\right) - \cdots \Bigg]_j,$$

$$K_{zz} = 1 - \sum_j \left(\frac{\Pi_j}{\omega}\right)^2 \zeta_0 \Bigg[2\zeta_0 W_0 (1 - b + \cdots) + (\zeta_1 W_1 + \zeta_{-1} W_{-1})(b + \cdots) + (\zeta_2 W_2 + \zeta_{-2} W_{-2})\left(\frac{b^2}{4} + \cdots\right) \Bigg]_j,$$

$$K_{xy} = i\sum_j \left(\frac{\Pi_j}{\omega}\right)^2 \zeta_0\left[(Z_1 - Z_{-1})\left(\frac{1}{2} - b + \cdots\right)\right.$$

$$\left. + (Z_2 - Z_{-2})\left(\frac{b}{2} + \cdots\right) + \cdots\right]_j,$$

$$K_{xz} = 2^{1/2}\sum_j \left(\frac{\Pi_j}{\omega}\right)^2 b^{1/2}\zeta_0\left[(W_1 - W_{-1})\left(\frac{1}{2} + \cdots\right)\right.$$

$$\left. + (W_2 - W_{-2})\left(\frac{b}{4} + \cdots\right) + \cdots\right]_j,$$

$$K_{yz} = -2^{1/2} i\sum_j \left(\frac{\Pi_j}{\omega}\right)^2 b^{1/2}\zeta_0\left[W_0\left(-1 + \frac{3}{2}b + \cdots\right)\right.$$

$$\left. + (W_1 + W_{-1})\left(\frac{1}{2} + \cdots\right) + (W_2 + W_{-2})\left(\frac{b}{4} + \cdots\right) + \cdots\right]_j, \quad (16.97)$$

where $Z_{\pm n} = Z(\zeta_{\pm n})$, $W_n = (1 + \zeta_n Z(\zeta_n))$, and $\zeta_n = (\omega - n\Omega)/[2^{1/2} \cdot k_z(T_z/m)^{1/2}]$. When $\zeta \gg 1$ (cold plasma), then $Z(\zeta) = -\zeta^{-1}(1 + (1/2)\zeta^{-2} + (3/4)\zeta^{-4} + \cdots)$ and $W(\zeta) = (1/2)\zeta^{-2}(1 + (3/2)\zeta^{-2} + \cdots)$. The Maxwell equation is

$$(K_{xx} - N_\parallel{}^2)E_x + K_{xy}E + (K_{xz} + N_\perp N_\parallel)E_z = 0,$$

$$-K_{xy}E_x + (K_{yy} - N_\parallel{}^2 - N_\perp{}^2)E_y + K_{yz}E_z = 0,$$

$$(K_{xz} + N_\perp N_\parallel)E_x - K_{yz}E_y + (K_{zz} - N_\perp{}^2)E_z = 0. \quad (16.98)$$

The dispersion relation is given by making the determinant of the coefficient matrix of eq. (16.98) zero. The absorption power P per unit volume is given by $(\boldsymbol{j}\cdot\boldsymbol{E})$. When $\boldsymbol{j}$ and $\boldsymbol{E}$ are the complex amplitudes, the average absorption power during one period is $P = (\boldsymbol{j}^*\cdot\boldsymbol{E} + \boldsymbol{j}\cdot\boldsymbol{E}^*)/4$. Since $\boldsymbol{j}$ is given by $\boldsymbol{j} = -i\varepsilon_0\omega(\boldsymbol{K} - 1)\boldsymbol{E}$ from eq. (10.4), the absorption power P is described by

$$P = -\frac{i\varepsilon_0\omega}{4}\boldsymbol{E}^*(\boldsymbol{K} - 1)\boldsymbol{E} + \text{c.c.}$$

$$= \omega\frac{\varepsilon_0}{2}[|E_x|^2 \operatorname{Im} K_{xx} + |E_y|^2 \operatorname{Im} K_{yy} + |E_z|^2 \operatorname{Im} K_{zz}$$

$$+ 2\operatorname{Im}(E_x{}^*E_y)\operatorname{Re} K_{xy} + 2\operatorname{Im}(E_y{}^*E_z)\operatorname{Re} K_{yz}$$

$$+ 2\operatorname{Re}(E_x{}^*E_z)\operatorname{Im} K_{xz}], \quad (16.99)$$

where c.c. is the complex conjugate.

The absorption power P_0 due to Landau damping (including transit time damping) can be obtained by the contribution from the terms Im $\zeta_0 Z(\zeta_0) = \pi^{1/2}\zeta_0 \exp(-\zeta_0{}^2) = G_0$ in K_{ij}, that is,

$$(\mathrm{Im}\ K_{yy})_0 = \left(\frac{\Pi_j}{\omega}\right)^2 2bG_0, \tag{16.100a}$$

$$(\mathrm{Im}\ K_{zz})_0 = \left(\frac{\Pi_j}{\omega}\right)^2 2\zeta_0{}^2 G_0, \tag{16.100b}$$

$$(\mathrm{Re}\ K_{yz})_0 = \left(\frac{\Pi_j}{\omega}\right)^2 2^{1/2} b^{1/2} \zeta_0 G_0, \tag{16.100c}$$

and

$$P_0 = 2\omega\left(\frac{\Pi_j}{\omega}\right)^2 G_0 \frac{\varepsilon_0}{2}[|E_y|^2 b + |E_z|^2\zeta_0{}^2 + \mathrm{Im}\,|E_y{}^*E_z|(2b)^{1/2}\zeta_0]. \tag{16.101}$$

Equation (16.100a) is the term of the transit time damping (14.27), eq. (16.100b) is the term of the Landau damping (14.25), and the eq. (16.100c) is the term of the interference of both.

The absorption power P_n due to cyclotron damping can be obtained by the contribution from the terms Im $\zeta_0 Z_{\pm n} = \pi^{1/2}\zeta_0 \exp(-\zeta_{\pm n}{}^2) = G_{\pm n}$, that is,

$$(\mathrm{Im}\ K_{xx})_{\pm n} = (\mathrm{Im}\ K_{yy})_{\pm n} = \left(\frac{\Pi_j}{\omega}\right)^2 G_{\pm n}\alpha_n,$$

$$(\mathrm{Im}\ K_{zz})_{\pm n} = \left(\frac{\Pi_j}{\omega}\right)^2 2\zeta_{\pm n}{}^2 G_{\pm n} b\alpha_n n^{-2},$$

$$(\mathrm{Re}\ K_{xy})_{\pm n} = -\left(\frac{\Pi_j}{\omega}\right)^2 G_{\pm n}(\pm\alpha_n),$$

$$(\mathrm{Re}\ K_{yz})_{\pm n} = -\left(\frac{\Pi_j}{\omega}\right)^2 (2b)^{1/2}\zeta_{\pm n} G_{\pm n}\alpha_n n^{-1},$$

$$(\mathrm{Im}\ K_{xz})_{\pm n} = -\left(\frac{\Pi_j}{\omega}\right)^2 (2b)^{1/2}\zeta_{\pm n} G_{\pm n}(\pm\alpha_n) n^{-1},$$

$$\alpha_n = n^2(2\cdot n!)^{-1}\left(\frac{b}{2}\right)^{n-1}, \tag{16.102}$$

and

$$P_{\pm n} = \omega\left(\frac{\Pi_j}{\omega}\right)^2 G_n\left(\frac{\varepsilon_0}{2}\right)\alpha_n|E_x \pm iE_y|^2. \tag{16.103}$$

The term of the ion cyclotron damping comes from the + sign of eq. (16.103) since $\zeta_n = (\omega + n\Omega_{\mathrm{i}})/(2^{1/2}k_z v_{\mathrm{Ti}}) = (\omega - n|\Omega_{\mathrm{i}}|)/(2^{1/2}k_z v_{\mathrm{Ti}})$. The term of the electron cyclotron damping comes from the − sign of eq. (16.103).

The relative ratio of $\boldsymbol{E}$ components can be estimated from eq. (16.98). For cold plasmas, $K_{xx}, K_{yy} \to K_\perp$, $K_{zz} \to K_\parallel$, $K_{xy} \to -iK_\times$, $K_{xz}, K_{yz} \to 0$ can be substituted in eq. (16.98), and the relative ratio is $E_x{:}E_y{:}E_z = (K_\perp - N^2)(K_\parallel - N_\perp{}^2){:}-iK_\times(K_\parallel - N_\perp{}^2){:}-N_\parallel N_\perp(K_\perp - N^2)$. In order to obtain the magnitude of the electric field, it is necessary to solve the Maxwell equation with the dielectric tensor of eq. (16.94). In this case the density, the temperature, and the magnetic field are functions of the coordinates. Therefore the simplified model must be used for analytical solutions; otherwise numerical calculations are necessary to derive the wave field.

When the wavelength of waves in the plasma is much less than the characteristic length (typically the minor radius a), the WKB approximation (geometrical optical approximation) can be applied. Let the dispersion relation be $D(\boldsymbol{k}, \omega, \boldsymbol{r}, t) = 0$. The direction of wave propagation is given by the group velocity $\boldsymbol{v}_g = \partial\omega/\partial\boldsymbol{k} \equiv (\partial\omega/\partial k_x, \partial\omega/\partial k_y, \partial\omega/\partial k_z)$, so that the optical ray can be given by $\mathrm{d}\boldsymbol{r}/\mathrm{d}t = \boldsymbol{v}_g$. Although the quantities $(\boldsymbol{k}, \omega)$ change according to the change of $(\boldsymbol{r}, t)$, they always satisfy $D = 0$. Then the optical ray can be obtained by[106]

$$\frac{\mathrm{d}\boldsymbol{r}}{\mathrm{d}s} = \frac{\partial D}{\partial \boldsymbol{k}}, \qquad \frac{\mathrm{d}\boldsymbol{k}}{\mathrm{d}s} = -\frac{\partial D}{\partial \boldsymbol{r}}, \tag{16.104a}$$

$$\frac{\mathrm{d}t}{\mathrm{d}s} = -\frac{\partial D}{\partial \omega}, \qquad \frac{\mathrm{d}\omega}{\mathrm{d}s} = \frac{\partial D}{\partial t}. \tag{16.104b}$$

Here s is a measure of the length along the optical ray. Along the optical ray the variation δD becomes zero,

$$\delta D = \frac{\partial D}{\partial \boldsymbol{k}}\cdot\delta\boldsymbol{k} + \frac{\partial D}{\partial \omega}\delta\omega + \frac{\partial D}{\partial \boldsymbol{r}}\cdot\partial\boldsymbol{r} + \frac{\partial D}{\partial t}\delta t = 0, \tag{16.105}$$

and $D(\boldsymbol{k}. \omega, \boldsymbol{r}, t) = 0$ is satisfied. Equation (16.104) reduces to $\mathrm{d}\boldsymbol{r}/\mathrm{d}t = (\mathrm{d}\boldsymbol{r}/\mathrm{d}s)/(\mathrm{d}t/\mathrm{d}s) = (\partial D/\partial\boldsymbol{k})/(\partial D/\partial\omega) = \partial\omega/\partial\boldsymbol{k} = \boldsymbol{v}_g$. Equation (16.104a) has the same formula as the equation of motion with Hamiltonian D. $D =$ const. $= 0$ corresponds to the energy conservation law. If the plasma medium does not depend on z, $k_z = \text{const.}$ corresponds to the momentum conservation law and is the same as the Snell law, $N_\parallel = \text{const.}$

When $\boldsymbol{k} = \boldsymbol{k}_r + i\boldsymbol{k}_i$ is a solution of $D = 0$ for a given ω and $|\boldsymbol{k}_i| \ll |\boldsymbol{k}_r|$ is satisfied, we have

$$D(\boldsymbol{k}_r + i\boldsymbol{k}_i, \omega) = \text{Re}\, D(\boldsymbol{k}_r, \omega) + \frac{\partial\, \text{Re}\, D(\boldsymbol{k}_r, \omega)}{\partial \boldsymbol{k}_r} \cdot i\boldsymbol{k}_i + i\, \text{Im}\, D(\boldsymbol{k}_r, \omega) = 0. \tag{16.106}$$

Since $\text{Re}\, D(\boldsymbol{k}_r, \omega) = 0$, this reduces to

$$\boldsymbol{k}_i \cdot \frac{\partial\, \text{Re}\, D(\boldsymbol{k}_r, \omega)}{\partial \boldsymbol{k}_r} = -\text{Im}\, D(\boldsymbol{k}_r, \omega). \tag{16.107}$$

Then the wave intensity $I(\boldsymbol{r})$ becomes

$$I(\boldsymbol{r}) = I(\boldsymbol{r}_0) \exp\left(-2 \int_{\boldsymbol{r}_0}^{\boldsymbol{r}} \boldsymbol{k}_i\, d\boldsymbol{r}\right), \tag{16.108}$$

$$\begin{aligned} \int \boldsymbol{k}_i\, d\boldsymbol{r} &= \int \boldsymbol{k}_i \cdot \frac{\partial D}{\partial \boldsymbol{k}}\, ds = -\int \text{Im}\, D(\boldsymbol{k}_r, \omega)\, ds \\ &= -\int \frac{\text{Im}\, D(\boldsymbol{k}_r, \omega)}{|\partial D / \partial \boldsymbol{k}|}\, dl, \end{aligned} \tag{16.109}$$

where dl is the length along the optical ray. Therefore the wave absorption can be estimated from the eqs. (16.108) and (16.109) by tracing many optical rays. The geometrical optical approximation can provide the wave intensity with a space resolution of, say, two or three times the wavelength.

16.6b Wave Heating in Ion Cyclotron Range of Frequency

The dispersion relation of waves in the ion cyclotron range of frequency (ICRF) is given by eqs. (10.64) or (10.98):

$$\begin{aligned} N_\parallel^2 = \frac{N_\perp^2}{2[1 - (\omega/\Omega_i)^2]} \Bigg\{ &-\left(1 - \left(\frac{\omega}{\Omega_i}\right)^2\right) + \frac{2\omega^2}{k_\perp^2 v_A^2} \\ &\pm \left[\left(1 - \left(\frac{\omega}{\Omega_i}\right)^2\right)^2 + 4\left(\frac{\omega}{\Omega_i}\right)^2 \left(\frac{\omega}{k_\perp v_A}\right)^4\right]^{1/2} \Bigg\}. \end{aligned} \tag{16.110}$$

The plus sign corresponds to the slow wave (L wave, ion cyclotron wave), and the minus sign corresponds to the fast wave (R wave, extraordinary wave). When $1 - \omega^2/\Omega_i^2 \ll 2(\omega/k_\perp v_A)^2$, the dispersion relation becomes

$$k_z^2 = \frac{c^2}{v_A^2} 2\left(1 - \frac{\omega^2}{\Omega_i^2}\right)^{-1} \quad \text{(for the slow wave)},$$

$$k_z^2 = -\frac{k_\perp^2}{2} + \frac{\omega^2}{2v_A^2} \quad \text{(for the fast wave)}. \qquad (16.111)$$

Since the externally excited waves have propagation vectors with $0 < k_z^2 < (1/a)^2$, $k_\perp^2 > (\pi/a)^2$ usually, there are constraints[107]

$$\frac{c^2}{v_A^2} \frac{2}{(1 - \omega^2/\Omega_i^2)} \lesssim \left(\frac{\pi}{a}\right)^2,$$

$$n_{20} a^2 \lesssim 2.6 \times 10^{-4} \frac{B^2(1 - (\omega/\Omega_i)^2)}{A} \qquad (16.112)$$

for the slow wave and

$$\frac{\omega^2}{2v_A^2} \gtrsim \left(\frac{\pi}{a}\right)^2,$$

$$n_{20} a^2 \gtrsim 0.5 \times 10^{-2} \frac{(\Omega_i/\omega)^2}{A/Z^2} \qquad (16.113)$$

for the fast wave, where n_{20} is the ion density in 10^{20} m^{-3}, a is the plasma radius in meters, and A is the atomic number.

An ion cyclotron wave (slow wave) can be excited by a Stix coil[106] and can propagate and heat ions in a low-density plasma. But it cannot propagate in a high-density plasma like that of a tokamak.

The fast wave is an extraordinary wave in this frequency range and can be excited by a loop antenna, which generates a high-frequency electric field perpendicular to the magnetic field (see sec. 10.2). The fast wave can propagate in a high-density plasma. The fast wave in a plasma with a single ion species has $E_x + iE_y = 0$ at $\omega = |\Omega_i|$ in cold plasma approximation, so that it is not absorbed by the ion cyclotron damping. However, the electric field of the fast wave in a plasma with two ion species is $E_x + iE_y \neq 0$, so that the fast wave can be absorbed; that is, the fast wave can heat the ions in this case.

Let us consider the heating of a plasma with two ion species, M and m, by a fast wave. The masses, charge numbers, and densities of the M ion and m ion are denoted by m_M, Z_M, n_M and m_m, Z_m, n_m, respectively. When we use

$$\eta_{\mathrm{M}} \equiv \frac{Z_{\mathrm{M}}{}^2 n_{\mathrm{M}}}{n_{\mathrm{e}}}, \qquad \eta_{\mathrm{m}} \equiv \frac{Z_{\mathrm{m}}{}^2 n_{\mathrm{m}}}{n_{\mathrm{e}}}, \tag{16.114}$$

we have $\eta_{\mathrm{M}}/Z_{\mathrm{M}} + \eta_{\mathrm{m}}/Z_{\mathrm{m}} = 1$ since $n_{\mathrm{e}} = Z_{\mathrm{M}}n_{\mathrm{M}} + Z_{\mathrm{m}}n_{\mathrm{m}}$. Since $(\Pi_{\mathrm{e}}/\omega)^2 \gg 1$ in ICRF wave, the dispersion relation in the cold plasma approximation is given by eq. (10.20) as follows:

$$N_{\perp}{}^2 = \frac{(R - N_{\parallel}{}^2)(L - N_{\parallel}{}^2)}{K_{\perp} - N_{\parallel}{}^2},$$

$$R = -\frac{\Pi_{\mathrm{i}}{}^2}{\omega^2}\left(\frac{(m_{\mathrm{M}}/m_{\mathrm{m}})\eta_{\mathrm{m}}\omega}{\omega + |\Omega_{\mathrm{m}}|} + \frac{\eta_{\mathrm{M}}\omega}{\omega + |\Omega_{\mathrm{M}}|} - \frac{\omega}{|\Omega_{\mathrm{M}}|/Z_{\mathrm{M}}}\right),$$

$$L = -\frac{\Pi_{\mathrm{i}}{}^2}{\omega^2}\left(\frac{(m_{\mathrm{M}}/m_{\mathrm{m}})\eta_{\mathrm{m}}\omega}{\omega - |\Omega_{\mathrm{m}}|} + \frac{\eta_{\mathrm{M}}\omega}{\omega - |\Omega_{\mathrm{M}}|} + \frac{\omega}{|\Omega_{\mathrm{M}}|/Z_{\mathrm{M}}}\right),$$

$$K_{\perp} = -\frac{\Pi_{\mathrm{i}}{}^2}{\omega^2}\left(\frac{(m_{\mathrm{M}}/m_{\mathrm{m}})\eta_{\mathrm{m}}\omega^2}{\omega^2 - \Omega_{\mathrm{m}}{}^2} + \frac{\eta_{\mathrm{M}}\omega^2}{\omega^2 - \Omega_{\mathrm{M}}{}^2}\right),$$

$$\Pi_{\mathrm{i}}{}^2 \equiv \frac{n_{\mathrm{e}}e^2}{\varepsilon_0 m_{\mathrm{M}}}. \tag{16.115}$$

Therefore ion-ion hybrid resonance occurs at $K_{\perp} - N_{\parallel}{}^2 = 0$, that is,

$$\frac{\eta_{\mathrm{m}}(m_{\mathrm{M}}/m_{\mathrm{m}})\omega^2}{\omega^2 - \Omega_{\mathrm{m}}{}^2} + \frac{\eta_{\mathrm{M}}\omega^2}{\omega^2 - \Omega_{\mathrm{M}}{}^2} \approx \frac{-\omega^2}{\Pi_{\mathrm{i}}{}^2}N_{\parallel}{}^2 \approx 0, \tag{16.116}$$

where

$$\omega^2 \approx \omega_{\mathrm{IH}}{}^2 \equiv \frac{\eta_{\mathrm{M}} + \eta_{\mathrm{m}}(\mu^2/\mu')}{\eta_{\mathrm{M}} + \eta_{\mathrm{m}}/\mu'}\Omega_{\mathrm{m}}{}^2, \tag{16.117}$$

$$\mu' \equiv \frac{m_{\mathrm{m}}}{m_{\mathrm{M}}}, \qquad \mu \equiv \frac{\Omega_{\mathrm{M}}}{\Omega_{\mathrm{m}}} = \frac{m_{\mathrm{m}}Z_{\mathrm{M}}}{m_{\mathrm{M}}Z_{\mathrm{m}}}. \tag{16.118}$$

Figure 16.24 shows the ion-ion hybrid resonance layer ($K_{\perp} - N_{\parallel}{}^2 = 0$), the L cutoff layer ($L - N_{\parallel}{}^2 = 0$), and the R cutoff layer ($R - N_{\parallel}{}^2 = 0$) of a tokamak plasma with the two ion species D^+ (M ion) and H^+ (m ion).

Since the K_{zz} component of the dielectric tensor is much larger than the other components, even in a hot plasma, the dispersion relation of a hot plasma is[108]

$$\begin{vmatrix} K_{xx} - N_{\parallel}{}^2 & K_{xy} \\ -K_{xy} & K_{yy} - N_{\parallel}{}^2 - N_{\perp}{}^2 \end{vmatrix} = 0. \tag{16.119}$$

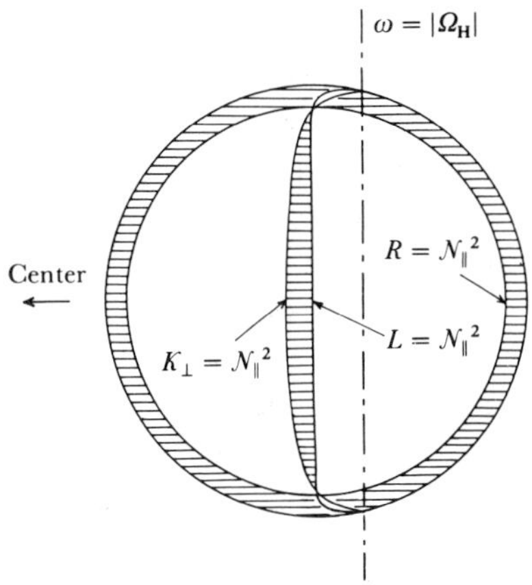

Fig. 16.24
L cutoff layer, R cutoff layer, and the ion-ion hybrid resonance layer of ICRF wave in a tokamak plasma with two ion components D^+ and H^+. The shaded area is the region of $N_\perp^2 < 0$.

When we use the relation $K_{yy} = K_{xx} + \Delta K_{yy}$, $|\Delta K_{yy}| \ll K_{xx}$,

$$N_\perp^2 = \frac{(K_{xx} - N_\parallel^2)(K_{xx} + \Delta K_{yy} - N_\parallel^2) + K_{xy}^2}{K_{xx} - N_\parallel^2}$$

$$\approx \frac{(K_{xx} + iK_{xy} - N_\parallel^2)(K_{xx} - iK_{xy} - N_\parallel^2)}{K_{xx} - N_\parallel^2}. \tag{16.120}$$

When ω^2 is near $\omega_{\rm IH}^2$, K_{xx} is given by

$$K_{xx} = -\frac{\Pi_{\rm i}^2}{\omega^2}\left(\frac{m_{\rm M}}{2m_{\rm m}}\eta_{\rm m}\zeta_0 Z(\zeta_1) + \frac{\zeta_{\rm M}\omega^2}{\omega^2 - \Omega_{\rm M}^2}\right). \tag{16.121}$$

The resonance condition is $K_{xx} = N_\parallel^2$. The value of $Z(\zeta_1)$ that appears in the dispersion relation is finite and $0 > Z(\zeta_1) > -1.08$. The condition

$$\eta_{\rm m} \geq \eta_{\rm er} \equiv \frac{2}{1.08}\frac{m_{\rm m}}{m_{\rm M}} 2^{1/2} N_\parallel \frac{v_{\rm Ti}}{c}\left(\frac{\eta_{\rm M}\omega^2}{\omega^2 - \Omega_{\rm M}^2} + N_\parallel^2 \frac{\omega^2}{\Pi_{\rm i}^2}\right) \tag{16.122}$$

is necessary to satisfy the resonance condition. This point is different from the cold plasma dispersion relation (note the difference between K_{xx} and $K_\perp$).

It is deduced from the dispersion relation (eq. (16.119)) that the mode conversion from the fast wave to the ion Bernstein wave occurs at the resonance layer when $\eta_m > \eta_{cr}$.[110] When the L cutoff layer and the ion-ion hybrid resonance layer are close to each other, as shown in fig. 16.24, the fast wave propagating from the outside torus penetrates the L cutoff layer partly by the tunneling effect and is converted to the ion Bernstein wave. The mode converted wave is absorbed by ion cyclotron damping or electron Landau damping. The theory of mode conversion is described in chapter 10 of ref. 106. ICRF experiments related to this topic were carried out in TFR.

When $\eta_m < \eta_{cr}$, $K = N_\parallel^2$ cannot be satisfied and the ion-ion hybrid resonance layer disappears. In this case a fast wave excited by the loop antenna outside the torus can pass through the R cutoff region (because the width is small) and is reflected by the L cutoff layer and bounced back and forth in the region surrounded by $R = N_\parallel^2$ and $L = N_\parallel^2$. In this region, there is a layer satisfying $\omega = |\Omega_m|$, and the minority m ions (since $\eta_m < \eta_{cr}$) are heated by the fundamental ion cyclotron damping. The majority M ions are heated by the Coulomb collisions with m ions. If the mass of M ions is l times as large as the mass of m ions, the M ions are also heated by the lth harmonic ion cyclotron damping. This type of ICRF heating is called minority ICRF heating. This type of experiment was carried out in PLT with good heating efficiency. The absorption power P_{e0} due to electron Landau damping per unit volume is given by eq. (16.101), and it is important only in the case $\zeta_0 < 1$. In this case we have $E_y/E_z \approx K_{zz}/K_{yz} \approx 2\zeta_0/(2^{1/2}b^{1/2}\zeta_0(-i))$ and P_{e0} is[108]

$$P_{eo} = \frac{\omega\varepsilon_0}{4}|E_y|^2\left(\frac{\Pi_e}{\omega}\right)^2\left(\frac{k_\perp v_{Te}}{\Omega_e}\right)^2 2\zeta_{oe}\pi^{1/2}\exp(-\zeta_{oe}{}^2). \tag{16.123}$$

The absorption power P_{in} by the nth harmonic ion cyclotron damping is given by eq. (16.103) as follows:

$$P_{in} = \frac{\omega\varepsilon_0}{2}|E_x + iE_y|^2\left(\frac{\Pi_i}{\omega}\right)^2\left(\frac{n^2}{2\times n!}\right)\left(\frac{b}{2}\right)^{n-1}\frac{\omega}{2^{1/2}k_z v_{Ti}}\pi^{1/2}$$
$$\times\exp\left(-\frac{(\omega - n|\Omega_i|)^2}{2(k_z v_{Ti})^2}\right). \tag{16.124}$$

The average absorption power $\langle P_{in}\rangle$ within the region $r_1 < r < r_2$ is

$$2\pi r(r_2 - r_1)\langle P_{in}\rangle = \iint_{r_1<r<r_2} P_{in}\,dx\,dy. \tag{16.125}$$

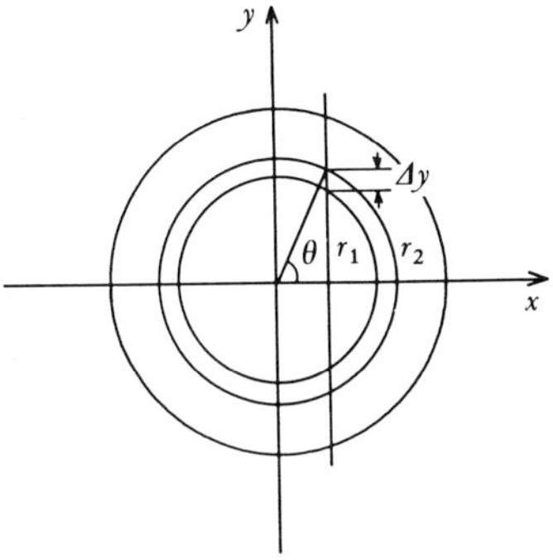

Fig. 16.25
The coordinates used for calculation of absorption power $\langle P_{in}\rangle$ of ICRF wave per unit volume.

Changing variables from x to $\xi = (k_z v_{Ti})^{-1}(\omega - n|\Omega_{i0}|(1 - x/R))$ and using the relation $y = (r_2 - r_1)/\sin\theta$ reduce this to (see fig. 16.25)[108,109]

$$\langle P_{in}\rangle = \frac{\varepsilon_0 |E_x + iE_y|^2}{2} \frac{\Pi_i^2}{|\Omega_i|} \frac{R}{r\sin\theta}\left(\frac{n}{2\times n!}\right)\left(\frac{b}{2}\right)^{n-1}. \tag{16.126}$$

For fundamental ion cyclotron damping, $\Lambda_n \equiv (n/(2\times n!))(b/2)^{n-1}$ is $\Lambda_1 = 1/2$ and $\Lambda_2 = 4^{-1}k_\perp^2(T_i/m_i)\Omega_i^2$ for the second harmonic.

The absorption power due to the second cyclotron damping is proportional to the beta value of the plasma. In order to evaluate the absorption power by eqs. (16.123) and (16.124), we need the spatial distributions of E_x and E_y. They can be calculated easily in the simple case of a sheet model.[111]

In the range of the higher harmonic ion cyclotron frequencies ($\omega \sim 2\Omega_i$, $3\Omega_i$), the direct excitation of the ion Bernstein wave has been studied[112] by an external antenna or waveguide, which generates a high-frequency electric field parallel to the magnetic field. Since $k_\perp \rho_i \sim O(1)$ in the ion Bernstein wave, the bulk ions can be heated effectively.

16.6c Lower Hybrid Wave Heating

Since $|\Omega|_i \ll \Pi_i$ in a tokamak plasma ($n_e > 10^{13}\ \mathrm{cm}^{-3}$), the lower hybrid resonance frequency ω_{LH} becomes

$$\omega_{LH}^2 = \frac{\Pi_i^2 + \Omega_i^2}{1 + \Pi_e^2/\Omega_e^2 + Zm_e/m_i} \approx \frac{\Pi_i^2}{1 + \Pi_e^2/\Omega_e^2}. \tag{16.127}$$

There is the relation $\Omega_e \gg \omega_{LH} \gg |\Omega_i|$. For a given frequency ω, lower hybrid resonance occurs at the position where the electron density satisfies the following condition:

$$\frac{\Pi_e^2(x)}{\Omega_e^2} = \frac{\Pi_{res}^2}{\Omega_e^2} \equiv p, \qquad p = \frac{\omega^2}{\Omega_e|\Omega_i| - \omega^2}. \tag{16.128}$$

When the dispersion relation eq. (10.20) of a cold plasma is solved about $N_\perp^2$ using $N^2 = N_\perp^2 + N_\parallel$, we have

$$N_\perp^2 = \frac{K_\perp \tilde{K}_\perp - K_\times^2 + K_\parallel \tilde{K}_\perp}{2K_\perp} \pm \left[\left(\frac{K_\perp \tilde{K}_\perp - K_\times^2 + K_\parallel \tilde{K}_\perp}{2K_\perp}\right)^2 + \frac{K_\parallel}{K_\perp}(K_\times^2 - \tilde{K}_\perp^2)\right]^{1/2}, \tag{16.129}$$

where $\tilde{K}_\perp = K_\perp - N_\parallel^2$. The relations $h(x) \equiv \Pi_e^2(x)/\Pi_{res}^2$, $K_\perp = 1 - h(x)$, $K_\times = -ph(x)\Omega_e/\omega$, $K_\parallel = 1 - \beta h(x)$, $\beta \equiv \Pi_{res}^2/\omega^2 \sim O(m_i/m_e)$, $\alpha \equiv \Pi_{res}^2/(\omega\Omega_e) \sim O(m_i/m_e)^{1/2}$, and $\beta h \gg 1$ reduce this to

$$N_\perp^2(x) = \frac{\beta h}{2(1-h)}\{N_\parallel^2 - (1 - h + ph) \pm [(N_\parallel^2 - (1 - h + ph))^2 - 4(1-h)ph]^{1/2}\}. \tag{16.130}$$

The slow wave corresponds to the case of the plus sign in eq. (16.130). In order for the slow wave to propagate from the plasma edge with low density ($h \ll 1$) to the plasma center with high density ($\Pi_e^2 = \Pi_{res}^2$, $h = 1$), $N_\perp(x)$ must be real. Therefore the condition $N_\parallel > (1-h)^{1/2} + (ph)^{1/2}$ is necessary. The right-hand side of the inequality has the maximum value $(1+p)^{1/2}$ in the range $0 < h < 1$, so that the accessibility condition of the resonant region to the lower hybrid wave becomes

$$N_\parallel^2 > N_{\parallel\,cr}^2 = 1 + p = 1 + \Pi_{res}^2/\Omega_e^2. \tag{16.131}$$

If this condition is not satisfied, the externally excited slow wave propagates into the position where the square root term in eq. (16.130) becomes zero and transforms to the fast wave there. Then the fast wave returns to the low-density region. The slow wave that satisfies the accessibility condition can approach the resonance region and $N_\perp$ can become large, so that the dispersion relation of hot plasma must be used to examine the behavior of this wave. Near the lower hybrid resonance region, the approximation of the electrostatic wave, eq. (11.43), is applicable. Since $|\Omega_i| \ll \omega \ll \Omega_e$, the terms of ion contribution and electron contribution are given by eqs. (11.44) and (11.45) respectively, that is,

$$1 + \frac{\Pi_e^2 m_e}{k^2} \frac{m_e}{T_e}(1 + I_0 e^{-b}\zeta_0 Z(\zeta_0)) + \frac{\Pi_i^2}{k^2} \frac{m_i}{T_i}(1 + \zeta Z(\zeta)) = 0, \tag{16.132}$$

where $\zeta_0 = \omega/(2^{1/2}k_z v_{Te})$ and $\zeta = \omega/(2^{1/2}kv_{Ti}) \approx \omega/(2^{1/2}k_\perp v_{Ti})$. Since $I_0 e^{-b} = 1 - b + (3/4)b^2$, $\zeta_0 \gg 1$, $\zeta \gg 1$, $1 + \zeta Z(\zeta) \approx -(1/2)\zeta^{-2} - (3/4)\zeta^{-4}$, we have

$$\left(\frac{3\Pi_i^2}{\omega^4}\frac{T_i}{m_i} + \frac{3}{4}\frac{\Pi_e^2}{\Omega_e^2}\frac{T_e}{m_e}\right)k_\perp^4 - \left(1 + \frac{\Pi_e^2}{\Omega_e^2} - \frac{\Pi_i^2}{\omega^2}\right)k_\perp^2 - \left(1 - \frac{\Pi_e^2}{\omega^2}\right)k_z^2 = 0. \tag{16.133}$$

The dimensionless form is

$$(k_\perp \rho_i)^4 - \frac{1-h}{h}\frac{m_i}{m_e}\frac{1}{(1+p)s^2}(k_\perp \rho_i)^2 + \left(\frac{m_i}{m_e}\right)^2 \frac{1}{s^2}(k_z \rho_i)^2 = 0. \tag{16.134}$$

Here $\rho_i = v_{Ti}/|\Omega_i|$ and

$$s^2 \equiv 3\left(\frac{1+p}{p} + \frac{1}{4}\frac{T_e}{T_i}\frac{p}{1+p}\right) = 3\left(\frac{|\Omega_i \Omega_e|}{\omega^2} + \frac{1}{4}\frac{T_e}{T_i}\frac{\omega^2}{|\Omega_i \Omega_e|}\right). \tag{16.135}$$

This dispersion equation has two solutions. One corresponds to the slow wave in a cold plasma and the other to the plasma wave in a hot plasma. The slow wave transforms to the plasma wave at the location where eq. (16.133) or (16.134) has equal roots (see fig. 14.9).[113–115] The condition of zero discriminant is $1/h = 1 + 2k_z\rho_i(1+p)s$ and

$$\frac{\Pi_e^2(x)}{\Omega_e^2} = \frac{\Pi_{\mathrm{M.C.}}^2}{\Omega_e^2} \equiv \frac{p}{1 + 2k_z \rho_i (1+p)s}. \tag{16.136}$$

Accordingly, the mode conversion occurs at the position satisfying

$$\frac{\omega^2}{\Pi_i^2} = \left(1 - \frac{\omega^2}{|\Omega_i|\Omega_e}\right) + \frac{N_\parallel v_{Te} 2\sqrt{3}}{c}\left(\frac{T_i}{T_e} + \frac{1}{4}\left(\frac{\omega^2}{\Omega_i \Omega_e}\right)^2\right)^{1/2}, \tag{16.137}$$

and the value of $k^2 \rho_i^2$ at this position becomes

$$k_\perp^2 \rho_i^2|_{\mathrm{M.C.}} = \frac{m_i}{m_e}\frac{k_z \rho_i}{s}. \tag{16.138}$$

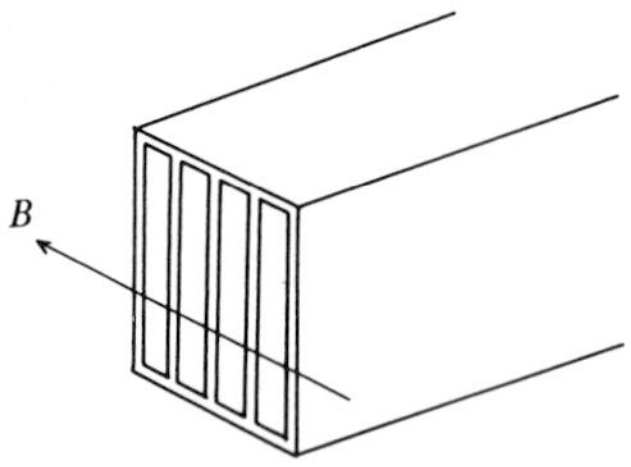

Fig. 16.26
Array of waveguides to excite a lower hybrid wave (slow wave).

If the electron temperature is high enough at the plasma center to satisfy $v_{\mathrm{Te}} > (1/3)c/N_{\parallel}$, the wave is absorbed by electrons due to electron Landau damping.

After the mode conversion, the value $N_{\perp}$ becomes large so that $c/N_{\perp}$ becomes comparable to the ion thermal velocity ($c/N \sim v_{\mathrm{Ti}}$). Since $\omega \gg |\Omega_{\mathrm{i}}|$, the ion motion is not affected by the magnetic field within the time scale of ω^{-1}. Therefore the wave with phase velocity c/N is absorbed by ions due to ion Landau damping. When ions have velocity v_{i} larger than $c/N_{\perp}$ ($v_{\mathrm{i}} > c/N$), the ions are accelerated or decelerated at each time satisfying $v_{\mathrm{i}} \cos(\Omega_{\mathrm{i}} t) = c/N_{\perp}$ and are subjected to stochastic heating.

The wave is excited by the array of waveguides, as shown in fig. 16.26, with an appropriate phase difference to provide the necessary parallel index $N_{\parallel} = k_z c/\omega = 2\pi c/(\lambda_z \omega)$. In the low-density region at the plasma boundary, the component of the electric field parallel to the magnetic field is larger for the slow wave than for the fast wave. Therefore the direction of waveguides is arranged to excite the electric field parallel to the line of magnetic force. Reference 118 discusses the coupling of waves to plasmas in detail.

For the current drive by lower hybrid wave, the accessibility condition (16.131) and $c/N \gg v_{\mathrm{Te}}$ are necessary. If the electron temperature is high and $T_{\mathrm{e}} \sim 10$ keV, then v_{Te}/c is already $\sim 1/7$. Even if $N_{\parallel}$ is chosen to be small under the accessibility condition, eq. (16.131), the wave is subjected to absorbtion by electron damping in the outer part of the plasma, and it can be expected that the wave cannot propagate into the central part of the plasma.

When the value of $N_{\parallel}$ is chosen to be $N_{\parallel} \sim (1/3)(c/v_{\mathrm{Te}})$, electron heating can be expected and has been observed experimentally. Under the condition that the mode conversion can occur, ion heating can be expected.

However, the experimental results are less clear than those for electron heating.

16.6d Electron Cyclotron Heating

The dispersion relation of waves in the electron cyclotron range of frequency in a cold plasma is given by eq. (10.80). The plus and minus signs in eq. (10.80) correspond to ordinary and extraordinary waves, respectively. The ordinary wave can propagate only when $\omega^2 > \Pi_e^2$ (in the case of $\theta = \pi/2$). This wave can be excited by an array of waveguides, like that used for lower hybrid waves (fig. 16.26), which emit an electric field parallel to the magnetic field. The phase of each waveguide is selected to provide the appropriate value of the parallel index $N_\parallel = k_z c/\omega = 2\pi c/\omega\lambda_z$.

The dispersion relation of the extraordinary wave is given by eq. (10.188). When $\theta = \pi/2$, it is given by eq. (10.52). It is necessary to satisfy $\omega^2 > \omega_L^2$ ($\omega_L < \Pi_e$). As is seen from the CMA diagram of fig. 10.6, the extraordinary wave can access the plasma center from the high magnetic field side (region 6a of fig. 10.6). There is an R cutoff $\omega = \omega_R$ from the low field side. The extraordinary wave can be excited by the waveguide, which emits an electric field perpendicular to the magnetic field (see sec. 10.2a).

The ion's contribution to the dielectric tensor is negligible, and the relations $b \ll 1$, $\zeta_0 \gg 1$ are satisfied so that the dielectric tensor of a hot plasma is

$$K_{xx} = K_{yy} = 1 + X\zeta_0 Z_{-1}/2, \qquad K_{zz} = 1 - X + N_\perp^2 \chi_{zz},$$

$$K_{xy} = -iX\zeta_0 Z_{-1}/2, \qquad K_{xz} = N_\perp \chi_{xz}, \qquad K_{yz} = iN_\perp \chi_{yz},$$

$$\chi_{xz} \approx \chi_{yz} \approx 2^{-1/2} X Y^{-1} \frac{v_T}{c} \zeta_0 (1 + \zeta_{-1} Z_{-1}),$$

$$\chi_{zz} \approx X Y^{-2} \left(\frac{v_T}{c}\right)^2 \zeta_0 \zeta_{-1} (1 + \zeta_{-1} Z_{-1}),$$

$$X \equiv \frac{\Pi_e^2}{\omega^2}, \qquad Y = \frac{\Omega_e}{\omega}, \qquad \zeta_{-1} = \frac{\omega - \Omega_e}{2^{1/2} k_z v_T}, \qquad N_\perp = \frac{k_\perp c}{\omega}. \tag{16.139}$$

The Maxwell equation is

$$(K_{xx} - N_\parallel^2) E_x + K_{xy} E_y + N_\perp (N_\parallel + \chi_{xz}) E_z = 0,$$

$$-K_{xy} E_x + (K_{yy} - N_\parallel^2 - N_\perp^2) E_y + iN_\perp \chi_{yz} E_z = 0,$$

$$N_\perp (N_\parallel + \chi_{xz}) E_x - iN_\perp \chi_{yz} E_y + (1 - X - N_\perp^2 (1 - \chi_{zz})) E_z = 0. \tag{16.140}$$

The solution is

$$\frac{E_x}{E_z} = -\frac{iN_\perp{}^2\chi_{xz}(N_\parallel + \chi_{xz}) + K_{xy}(1 - X - N_\perp{}^2(1 - \chi_{zz}))}{N_\perp[i\chi_{xz}(K_{xx} - N_\parallel{}^2) + K_{xy}(N_\parallel + \chi_{xz})]},$$

$$\frac{E_y}{E_z} = -\frac{N_\perp{}^2(N_\parallel + \chi_{xz})^2 - (K_{xx} - N_\parallel{}^2)(1 - X - N_\perp{}^2(1 - \chi_{zz}))}{N_\perp[i\chi_{xz}(K_{xx} - N_\parallel{}^2) + K_{xy}(N_\parallel + \chi_{xz})]}. \quad (16.141)$$

The absorption power P_{-1} per unit volume is given by eq. (16.103) as follows:

$$P_{-1} = \omega X\zeta_0 \frac{\pi^{1/2}}{2} \exp\left(-\frac{(\omega - \Omega_e)^2}{2k_z{}^2 v_{Te}{}^2}\right)\frac{\varepsilon_0}{2}|E_x - iE_y|^2. \quad (16.142)$$

When $\omega = \Omega_e$, then $\zeta_{-1} = 0, z_{-1} = i\pi^{1/2}, K_{xx} = 1 + ih, K_{xy} = h, \chi_{yz} = \chi_{xz} = 2^{-1/2}X(v_{Te}/c)\zeta_0 = X/(2N_\parallel), \chi_{zz} = 0, h \equiv \pi^{1/2}\zeta_0 X/2$. Therefore the dielectric tensor becomes

$$\boldsymbol{K} = \begin{bmatrix} 1 + ih & h & N_\perp\chi_{xz} \\ -h & 1 + ih & iN_\perp\chi_{xz} \\ N_\perp\chi_{xz} & -iN_\perp\chi_{xz} & 1 - X \end{bmatrix}. \quad (16.143)$$

For an ordinary wave (O wave), we have

$$\frac{E_x - iE_y}{E_z} = \frac{iN_\perp{}^2(\mathrm{O})N_\parallel(N_\parallel + \chi_{xz}) - i(1 - N_\parallel{}^2)(1 - X - N_\perp{}^2(\mathrm{O}))}{N_\perp(\mathrm{O})(N_\parallel h + i\chi_{xz}(1 - N_\parallel{}^2))}. \quad (16.144)$$

When $N_\parallel \ll 1$ and the incident angle θ is nearly perpendicular, eq. (10.83) gives $1 - X - N_\perp{}^2(\mathrm{O}) = (1 - X)N_\parallel{}^2$. Since $\chi_{xz} = X/2N_\parallel$, $\chi_{xz} \gg N$. Therefore the foregoing equation reduces to

$$\frac{E_x - iE_y}{E_z} = \frac{iN_\perp(\mathrm{O})N_\parallel\chi_{xz}}{N_\parallel h + i\chi_{xz}}. \quad (16.145)$$

For an extraordinary wave (X wave), we have

$$\frac{E_x - iE_y}{E_y} = -\frac{iN_\perp{}^2(\mathrm{X})N_\parallel(N_\parallel + \chi_{xz}) - i(1 - N_\parallel{}^2)(1 - X - N_\perp{}^2(\mathrm{X}))}{N_\perp{}^2(\mathrm{X})(N_\parallel + \chi_{xz})^2 - (K_{xx} - N_\parallel{}^2)(1 - X - N_\perp{}^2(\mathrm{X}))}. \quad (16.146)$$

When $N_\parallel \ll 1$ and $\omega = \Omega_e$, eq. (10.84) gives $1 - X - N_\perp{}^2(\mathrm{X}) \approx -1 + N_\parallel{}^2$. Since $\chi_{xz}{}^2 = (2\pi)^{-1/2}(v_{Te}/cN_\parallel)Xh \ll h$, the foregoing equation reduces to

$$\frac{E_x - iE_y}{E_y} = \frac{-(1 + N_\perp{}^2(\mathrm{X})N_\parallel(N_\parallel + \chi_{xz}))}{h - i(1 + N_\perp{}^2(\mathrm{X})(N_\parallel + \chi_{xz})^2)} \sim \frac{-1}{h}. \tag{16.147}$$

The absorbed power per unit volume at $\omega = \Omega_e$ is

$$P_{-1}(\mathrm{O}) \approx \frac{\omega\varepsilon_0}{2}|E_z|^2 \frac{hN_\perp{}^2(\mathrm{O})N_\parallel{}^2\chi_{xz}{}^2}{(N_\parallel h)^2 + \chi_{xz}{}^2}\exp(-\zeta_{-1}{}^2) \tag{16.148}$$

$$\approx \frac{\omega\varepsilon_0}{2}|E_z|^2 \frac{1}{(2\pi)^{1/2}}\left(\frac{\Pi_e}{\omega}\right)^2\left(\frac{v_{Te}}{cN_\parallel}\right)\frac{N_\perp{}^2(\mathrm{O})N_\parallel{}^2}{N_\parallel{}^2 + (v_{Te}/c)^2(2\pi)} \tag{16.149}$$

for the ordinary wave and

$$P_{-1}(\mathrm{X}) \sim \frac{\omega\varepsilon_0}{2}|E_y|^2\frac{2}{h} = \frac{\omega\varepsilon_0}{2}|E_y|^2 2\left(\frac{2}{\pi}\right)^{1/2}\left(\frac{\Pi_e}{\omega}\right)^{-2}\left(\frac{N_\parallel v_{Te}}{c}\right) \tag{16.150}$$

for the extraordinary wave.[118,119]

Since $P(\mathrm{O}) \propto n_e T_e^{1/2}/N_\parallel$, $P(\mathrm{X}) \propto N_\parallel T_e^{1/2}/n_e$, the ordinary wave is absorbed more in the case of higher density and perpendicular incidence, but the extraordinary wave has the opposite tendency.

The experiments of electron cyclotron heating have been carried out by T-10, ISX-B, JFT-2, and so on, and the good heating efficiency of ECH has been demonstrated.

16.7 Inertial Confinement

The characteristic of inertial confinement is that the extremely high-density plasma is produced by means of an intense energy driver, such as a laser or particle beam, within a short period so that the fusion reactions can occur before the plasma starts to expand. Magnetic confinement plays no part in this process, which has come to be called *inertial confinement.* For fusion conditions to be reached by inertial confinement, a small solid deuterium-tritium pellet must be compressed to a particle density 10^3–10^4 times that of the solid pellet particle density $n_S = 4.5 \times 10^{22}\ \mathrm{cm}^{-3}$. One cannot expect the laser light pressure or the momentum carried by the particle beam to compress the solid pellet: they are too small. A more feasible method of compression involves irradiating the pellet from all sides, as shown in fig. 16.27. The plasma is produced on the surface of the pellet and is heated instantaneously. The plasma expands immediately. The reaction of the outward plasma jet accelerates and compresses the inner

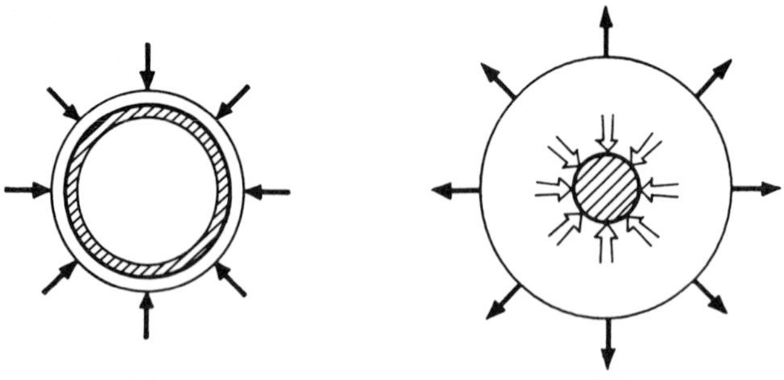

Fig. 16.27
Conceptual drawing of implosion. (a) Irradiation by laser or particle beam from all sides. (b) Expansion of plasma from the pellet surface and the implosion due to the reaction of the outward plasma jet.

pellet inward like a spherical rocket. This process is called *implosion.* The study of implosion processes is one of the most important current issues, and theoretical and experimental research is being carried out intensively.

16.7a Pellet Gain[120,121]

Let the pellet gain η_G be the ratio of the output nuclear fusion energy E_F to the input driver energy E_D delivered to the pellet. The heating efficiency η_h of the incident driver is defined by the conversion ratio of the driver energy E_D to the thermal energy of the compressed pellet core. Denote the density and the volume of the compressed core plasma by n and V, respectively, and assume that $T_e = T_i$; the following relations can then be obtained:

$$E_F = \frac{n^2}{4}\langle\sigma v\rangle Q_{NF}\alpha V\tau \equiv \eta_G E_D, \tag{16.151}$$

$$3nTV \equiv \eta_h E_D. \tag{16.152}$$

α is an enhancement factor due to the fusion reaction occurring in the surrounding plasma of the core, which is heated by α particles released from the hot compressed core. The confinement time τ is the characteristic expansion time and is expressed by

$$\tau = \frac{r_c}{v_T}, \qquad V = \frac{4\pi}{3}r_c{}^3, \tag{16.153}$$

where v_T is the ion thermal velocity and r_c is the radius of the compressed plasma core. $Q_{NF} = 17.6$ MeV is the sum of the energies of the released neutron $Q_n = 14.1$ MeV and of the α particle $Q_\alpha = 3.5$ MeV. Equation (16.151) reduces to

$$E_F = nV\alpha \frac{Q_{NF}}{2} \frac{n}{2} \langle \sigma v \rangle \tau = nV\alpha \frac{Q_{NF}}{2} \eta_b, \tag{16.154}$$

$$\eta_b \equiv \frac{n}{2} \langle \sigma v \rangle \tau. \tag{16.155}$$

Since $Q_{NF}/2$ is the fusion energy output per ion, η_b defined by eq. (16.155) is the fuel-burn ratio. The ratio

$$\eta_{NF} \equiv \frac{Q_{NF}/2}{3T} \alpha \tag{16.156}$$

is the thermonuclear gain, since one ion and electron are of energy $3T$ before fusion. Equations (16.151), (16.152), (16.155), and (16.156) yield the pellet gain

$$\eta_G = \eta_h \eta_b \eta_{NF} \tag{16.157}$$

and

$$\left.\begin{aligned} \tau &= \frac{2\eta_b}{n\langle \sigma v \rangle}, \\ r_c &= v_T \tau = \frac{2\eta_b v_T}{n\langle \sigma v \rangle}, \\ E_D &= \frac{1}{\eta_h} 3nT \frac{4\pi}{3} \left(\frac{2\eta_b v_T}{n\langle \sigma v \rangle} \right)^3. \end{aligned}\right\} \tag{16.158}$$

When $T = T_i = T_e = 10$ keV and the plasma density is expressed in units of solid density $n_S = 4.5 \times 10^{28}$ m^{-3}, one has

$$\left.\begin{aligned} \tau &= 0.3 \times 10^{-6} \eta_b \left(\frac{n_s}{n} \right) && \text{(s)}, \\ r_c &= 0.2 \eta_b \left(\frac{n_s}{n} \right) && \text{(m)}, \\ E_D &= 8 \times 10^{12} \frac{\eta_b^3}{\eta_h} \left(\frac{n_s}{n} \right)^2 && \text{(J)}. \end{aligned}\right\} \tag{16.159}$$

Here we use $v_T = 7 \times 10^5$ ms^{-1} and $\langle \sigma v \rangle = 1.5 \times 10^{-22}$ m^{3}s^{-1}. The thermonuclear gain is $\eta_T = 293\alpha$. If we assume that the fuel burn ratio is $\eta_b = 0.3$ and the heating efficiency is $\eta_h = 0.05$, then the pellet gain is $\eta_G = 4.4\alpha$, and the necessary input power E_D, the compressed radius r_c, and the confinement time τ are given by

for $n/n_s = 10^3$:	for $n/n_s = 10^4$:
$E_D = 4.4$ MJ,	$E_D = 44$ KJ,
$\tau = 90$ ps,	$\tau = 9$ ps,
$r_c = 63$ μm,	$r_c = 6.3$ μm.

r_c in eq. (16.159) is equivalent to $\rho_c r_c = 3.75\eta_b$ g/cm^2 when the mass density of the compressed core is $\rho_c = 2.5 m_p n$ (m_p is the proton mass).

Consider the energy balance of a possible inertial fusion reactor. The conversion efficiency of the thermal-to-electric energy is about $\eta_E \sim 0.4$, and the conversion efficiency of the electric energy to the output energy E_D of the driver is denoted by η_D. Then the condition $\eta_E \eta_D \eta_G > 2$ is at least necessary to obtain usable net energy from the reactor (fig. 16.28). When the expected value of η_D of a laser driver is assumed to be $\eta_D \sim 0.05$ ($\eta_D \sim 0.3$ in the case of a beam driver), the necessary pellet gain must be $\eta_G > 100$ ($\eta_G > 17$ in the case of a beam driver). The pellet gain is the product of three terms, $\eta_G = \eta_h \eta_b \eta_{NF}$. The heating efficiency η_h is considered to be ~ 0.05, and the burn ratio and the thermonuclear fusion gain are $\eta_b \sim 0.3$ and $\eta_{NF} \sim 293\alpha$, so that we have $\eta_G \sim 4.4\alpha$. The necessary enhancement factor is around $\alpha > 23$ in the case of a laser driver ($\alpha > 4$ for a beam driver).

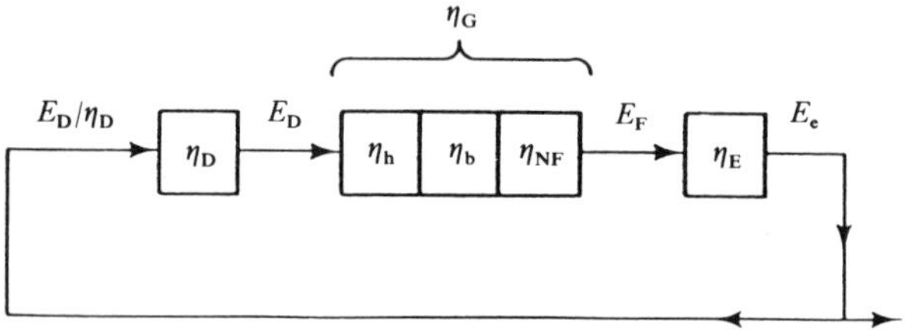

Fig. 16.28
Energy flow diagram of an inertial confinement reactor.

The critical issues for an inertial fusion reactor are how extremely high-density plasmas can be produced by implosion and how the enhancement factor of fusion output in the surrounding plasma heated by α particles released from the compressed core can be analyzed. Optimum design of fuel pellet structures and materials are also important.

Resolution of technological issues of energy drivers should increase the efficiency of laser drivers and improve the focusing of electrons, light ions, and heavy ion beams.[122]

16.7b Implosion

A typical structure of a pellet is shown in fig. 16.29. Outside the spherical shell of deuterium-tritium fuel, there is a pusher cell, which plays the role of a piston during the compression; an ablator cell with low-Z material surrounds the pusher cell and the fuel. The heating efficiency η_h is the conversion ratio of the driver energy to the thermal energy of the compressed core fuel. The heating efficiency depends on the interaction of the driver energy with the ablator, the transport process of the particles, and the energy and motion of the plasma fluid. The driver energy is absorbed on the surface of the ablator, and the plasma is produced and heated. Then the plasma expands and the inner deuterium-tritium fuel shell is accelerated inward by the reaction of the outward plasma jet. The implosion takes place at the center. Therefore the heating efficiency η_h is the product of three terms, that is, the absorption ratio η_{ab} of the driver energy by the ablator, the conversion ratio η_{hydro} of the absorbed driver energy to the kinetic energy of the hydrodynamic fluid motion, and the conversion ratio η_T of

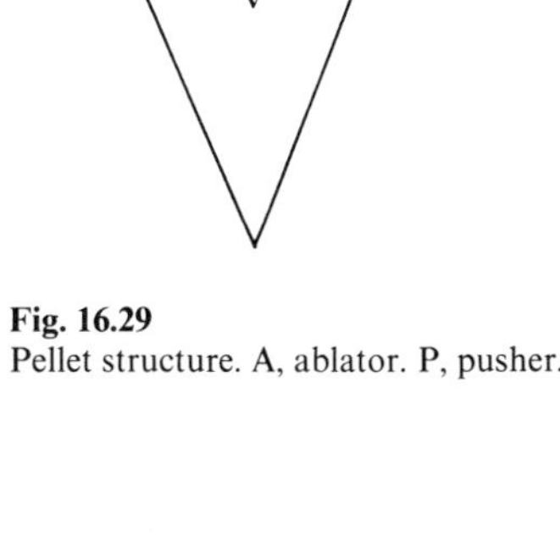

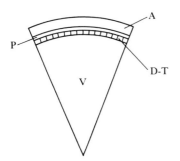

Fig. 16.29
Pellet structure. A, ablator. P, pusher. D-T, solid deuterium-tritium fuel. V, vacuum.

the kinetic energy of the hydrodynamic motion to the thermal energy of compressed core energy:

$$\eta_h = \eta_{ab}\eta_{hydro}\eta_T. \tag{16.160}$$

The internal energy of the solid deuterium-tritium fuel per unit volume is given by the product of Fermi energy $\varepsilon_F = (\hbar^2/2m_e)(2\pi^2 n)^{2/3}$ and the density n with a factor 3/5 ($\hbar = h/2\pi$ is Planck's constant, m_e is the electron mass). The internal energy of solid deuterium-tritium per unit mass w_0 can be estimated to be $w_0 = 1.0 \times 10^8$ J/kg. If the preheating occurs before the compression starts, the initial internal energy is increased to αw_0, and then the solid deuterium-tritium fuel is compressed adiabatically. By use of the equation of state for an ideal gas, the internal energy w after the compression is

$$w = \alpha w_0 \left(\frac{\rho}{\rho_0}\right)^{2/3}, \tag{16.161}$$

where ρ_0 and ρ are the mass densities before and after the compression. If the preheating is well suppressed and α is of the order of 3, the internal energy per unit mass after 1000 × compression is $w \sim 3 \times 10^{10}$ J/kg. This value w corresponds to the kinetic energy of unit mass with velocity $v \sim 2.5 \times 10^5$ m/s ($w = v^2/2$). Therefore, if the spherical fuel shell is accelerated to this velocity and if the kinetic energy is converted with good efficiency η_T into the thermal energy of the fuel core at the center, then compression with 1000 times mass density of the solid deuterium-tritium is possible.

When the pellet is irradiated from all sides by the energy driver, the plasma expands with velocity u from the surface of the ablator. Then the spherical shell with mass M is accelerated inward by the reaction with the ablation pressure P_a. The inward velocity v of the spherical shell can be analyzed by the rocket model with an outward plasma jet,[123,124] that is,

$$\frac{d(Mv)}{dt} = -\frac{dM}{dt} \cdot u = SP_a, \tag{16.162}$$

where S is the surface area of the shell. When the average mass density and the thickness of the spherical shell are denoted by ρ and Δ, respectively, the mass M is $M = \rho S \Delta$. Usually the outward velocity of the expanding plasma u is much larger than the inward velocity of the spherical shell v, and u is almost constant. The change of sum of the kinetic energies of the plasma jet and the spherical shell is equal to the absorbed power of the energy driver as follows:

$$\eta_{ab} I_D S = \frac{d}{dt}\left(\frac{1}{2} M v^2\right) + \frac{1}{2}\left(-\frac{dM}{dt}\right) u^2, \tag{16.163}$$

where I_D is the input power per unit area of the energy driver. From eqs. (16.162) and (16.163) the absorbed energy E_a is reduced to

$$E_a = \int \eta_{ab} I_D S \, dt = \frac{1}{2}(\Delta M) u^2, \tag{16.164}$$

where the approximations $u \gg v$ and $u =$ const. are used. The quantity ΔM is the absolute value of the change of mass of the spherical shell. The pressure P_a is estimated from eqs. (16.162) and (16.163) as follows:

$$P_a = \frac{u - v}{S}\left(-\frac{dM}{dt}\right) = \frac{2\eta_{ab} I_D}{u}\left(1 - \frac{v}{u}\right). \tag{16.165}$$

Then the conversion ratio η_{hydro} of the absorbed energy to the kinetic energy of the spherical shell is

$$\eta_{hydro} = \frac{1}{2E_a}(M_0 - \Delta M) v^2 = \frac{M_0 - \Delta M}{\Delta M}\left(\frac{v}{u}\right)^2. \tag{16.166}$$

Since the rocket equation (16.162) implies $v/u = -\ln[(M_0 - \Delta M)/M_0]$, the conversion ratio η_{hydro} is

$$\eta_{hydro} = \left(\frac{M_0}{\Delta M} - 1\right)\left[\ln\left(1 - \frac{\Delta M}{M_0}\right)\right]^2 \approx \frac{\Delta M}{M_0}. \tag{16.167}$$

Here $\Delta M/M_0 \ll 1$ is assumed.

The final inward velocity of the accelerated spherical shell must still be larger than $v \sim 3 \times 10^5$ m/s. The necessary ablation pressure P_a can be obtained from eq. (16.162) with relation $S = 4\pi r^2$ and the approximation $M \approx M_0$, $P_a =$ const. as follows:

$$\frac{dv}{dt} = \frac{4\pi P_a}{M_0} r^2 = \frac{P_a}{\rho_0 r_0{}^2 \Delta_0} r^2, \qquad v = -\frac{dr}{dt}. \tag{16.168}$$

Integration of $v \, dv/dt$ gives

$$P_a = \frac{3}{2}\rho_0 v^2 \frac{\Delta_0}{r_0}, \tag{16.169}$$

where ρ_0, r_0, and Δ_0 are the mass density, the radius, and the thickness of the spherical shell at the initial condition, respectively.

When $r_0/\Delta_0 = 30$ and $\rho_0 = 1$ g/cm^3, the necessary ablation pressure is $P_a = 4.5 \times 10^{12}$ N/m^2 = 45 Mbar (1 atm = 1.013 bar). Therefore the necessary energy flux intensity of driver I_D is

$$\eta_{ab} I_D \approx \frac{P_a u}{2}. \tag{16.170}$$

For the evaluation of the velocity u of the expanding plasma, the interaction of the driver energy and the ablator cell must be taken into account. In this section the case of the laser driver is described. Let the sound velocity of the plasma at the ablator surface be c_s and the mass density be ρ_c. The energy extracted by the plasma jet from the ablator surface per unit time is $4\rho_c c_s^3$ and this must be equal to the absorbed energy power $\eta_{ab} I_D$. The plasma density is around the cutoff density corresponding to the laser light frequency (wavelength), that is,

$$\eta_{ab} I_D \sim 4 m_{DT} n_c c_s^3, \tag{16.171}$$

where $m_{DT} = 2.5 \times 1.67 \times 10^{-27}$ kg is the average mass of deuterium and tritium and the cutoff density is $n_c = 1.1 \times 10^{27}/\lambda^2$ m^{-3} (λ in units of μm). From eq. (16.165) we have $\eta_{ab} I_D \sim 2c_s P_a$ and

$$P_a = 13\left[\frac{(\eta_{ab} I_D)_{14}}{\lambda(\mu\text{m})}\right]^{2/3} \text{Mbar}, \tag{16.172}$$

where $(\eta_{ab} I_D)_{14}$ is the value in 10^{14} W/cm^2. This scaling is consistent with the experimental results in the range $1 < (\eta_{ab} I_D)_{14} < 10$.

Most implosion research is carried out by the laser driver. The observed absorption rate η_{ab} tends to decrease according to the increase of laser light intensity I_D.

The absorption rate is measured for a Nd glass laser with wavelength 1.06 μm (red), second harmonic 0.53 μm (green), and third harmonic 0.35 μm (blue). The absorption is better for shorter wavelengths, and it is $\eta_{ab} = 0.8$–0.9 in the case of $x = 0.35$ μm in the range of $I_D = 10^{14}$–10^{15} W/cm^2. The conversion ratio η_{hydro} determined by the experiments is $\eta_{hydro} = 0.1$–0.15. The conversion ratio η_T is expected to be $\eta_T \sim 0.5$. The

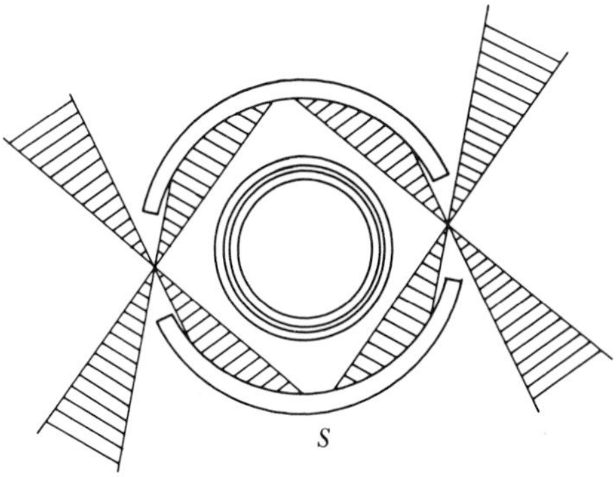

Fig. 16.30
Structure of cannonball/hohlraum target. S, outside shell.

necessary ablation pressure of $P_a = 45$ Mbar has been realized. In order to compress the fuel to extremely high density, it is necessary to avoid the preheating of the inner pellet during the implosion process, since the pressure of the inner part of pellet must be kept as low as possible before the compression. When laser light with a long wavelength (CO_2 laser $\lambda = 10.6\ \mu m$) is used, the high-energy electrons are produced by the laser-plasma interaction, and the high-energy electrons penetrate into the inner part of the pellet and preheat it. However, the production of high-energy electrons is much lower in short-wavelength experiments.

The implosion process just described is for directly irradiated pellets. The other case is indirectly irradiated pellets. The outer spherical shell surrounds the fuel pellet, as shown in fig. 16.30. The inner surface of the outer spherical shell is irradiated by the laser light, and the laser energy is converted to X-ray energy and plasma energy. The converted X-rays and the plasma particles irradiate the inner fuel pellet and implosion occurs. The X-ray and plasma energy are confined between the outer spherical shell and the inner fuel pellet and are used for the implosion effectively. This type of pellet is called a *cannonball target*[125] or *hohlraum target*.[126]

References

1. A. S. Bishop: Project Sherwood, Addison Wesley, Reading, Mass. 1958
2. R. S. Pease: Plasma Physics and Controlled Fusion **28**, 397(1986)
3. L. A. Artsimovich: Sov. Phys. Uspekhi **91**, 117(1967)
4. J. D. Lawson: Proc. Phys. Soc. **70B**, 6(1957)

5. M. D. Kruskal and M. Schwarzschild: Proc. Roy. Soc. A **223**, 348(1954)

6. M. N. Rosenbluth and C. L. Longmire: Ann. Phys. **1**, 120(1957)
I. B. Bernstein, E. A. Frieman, M. D. Kruskal and R. M. Kulsruid: Proc. Roy. Soc. A **244**, 17(1958)

7. Nature **181** No. 4604 p. 217 Jan. 25(1958)

8. L. Spitzer, Jr.: Phys. Fluids **1**, 253(1958)

9. Proceedings of the Second United Nations International Conference on the Peaceful Uses of Atomic Energy in Geneva Sep. 1–13(1958), Vol. 31 Theoretical and Experimental Aspects of Controlled Nuclear Fusion, Vol. 32 Controlled Fusion Devices, United Nations Publication, Geneva (1958)

10. The Second Geneva Series on the Peaceful Uses of Atomic Energy (series editor, J. G. Beckerley; Nuclear Fusion ed. by W. P. Allis) D. Van Nostrand Co. Inc., New York (1960)

11. S. Hayakawa and K. Kimura: Kakuyugo Kenkyu **57**, 201(1987) (in Japanese)

12. Plasma Physics and Controlled Nuclear Fusion Research (Conference Proceeding, Salzburg Sep. 4–9 1961) Nucl. Fusion Suppl. (1962)
(Translation of Russian Papers: U.S. Atomic Energy Commission, Division of Technical Information Office of Tech. Service, Depart. of Commerce, Washington 25 D.C. (1963))

13. ibid: (Conference Proceedings, Culham Sep. 6–10 1965)
International Atomic Energy Agency Vienna (1966)
(Translation of Russian Papers: U.S. Atomic Energy Commission, Division of Technical Information Oak Ridge Tenn. (1966))

14. ibid.: (Conference Proceedings, Novosibirsk Aug. 1–7 1968)
International Atomic Energy Agency Vienna (1969)
(Translation of Russian Papers: Nucl. Fusion Suppl. (1969))

15. ibid.: (Conference Proceedings, Madison, June 17–23 1971)
International Atomic Energy Agency, Vienna (1971)
(Translation of Russian Papers: Nucl. Fusion Suppl. (1972))

16. ibid.: (Conference Proceedings, Tokyo, Nov. 11–15 1974)
International Atomic Energy Agency, Vienna (1975)

17. M. N. Rosenbluth and R. F. Post: Phys. Fluids **8**, 547(1965)

18. E. G. Harris: Phys. Rev. Lett. **2**, 34(1959)

19. A. B. Mikhailovskii and L. I. Rudakov: Sov. Phys. JEPT **17**, 621(1963)
N. A. Krall and M. N. Rosenbluth: Phys. Fluids **8**, 1488(1965)

20. M. J. Forrest, N. J. Peacock, D. C. Robinson, V. V. Sannikov, and P. D. Wilcock: Culham Report CLM-R 107 July (1970)

21. The 20th Anniversary issue: Nuclear Fusion **25**, No. 9(1985)

22. World Survey of Activities in Controlled Fusion Research, International Atomic Energy Agency, Vienna (1986)

23. L. A. Artsimovich: Nucl. Fusion **12**, 215(1972)

24. H. P. Furth: Nucl. Fusion **15**, 487(1975)
H. P. Furth: Fusion (ed. by E. Teller) vol. 1, Academic Press, New York, 1981

25. V. S. Mukhovatov, V. D. Shafranov: Nucl. Fusion **11**, 605(1971)

26. Equipe TFR: Nucl. Fusion **18**, 647(1978)

27. J. Sheffield: Plasma Scattering of Electromagnetic Radiation, Academic Press, New York (1975)

28. For example, Y. Nagayama, Y. Ohki, K. Miyamoto: Nucl. Fusion **23**, 1447(1983)

29. F. Troyon, R. Gruber, H. Saurenmann, S. Semenzato, S. Succi: Plasma Physics and Controlled Fusion **26**, 209(1984)

30. A. Sykes, M. F. Turner, S. Patel: Proc. 11th European Conference on Controlled Fusion and Plasma Physics, Aachen Part II 363(1983)
T. Tuda, M. Azumi, K. Itoh, G. Kurita, T. Takeda et al.: Plasma Physics and Controlled Nuclear Fusion Research **2**, 173(1985) (Conference Proceedings, London 1984) IAEA Vienna

31. B. B. Kadomtsev: Sov. J. Plasma Phys. **1**, 389(1975)

32. J. A. Wesson: Nucl. Fusion **18**, 87(1978)

33. B. B. Kadomtsev and O. P. Pogutse: Nucl. Fusion **11**, 67(1971)

34. M. N. Rosenbluth and P. H. Rutherford: Fusion (ed. by E. Teller) vol. 1, Academic Press, New York (1981)

35. K. Itoh and S. Itoh: Kakuyuko Kenkyu **57**, 280, 379(1987) (in Japanese)
P. C. Liewer: Nucl. Fusion **25**, 543(1985).

36. S. M. Kaye: Phys. Fluids **28**, 2327(1985); R. J. Goldstone: Plasma Physics and Controlled Fusion **26**, 87 (1984)

37. J. W. Connor, J. B. Taylor, M. F. Turner: Nucl. Fusion **24**, 642(1984)

38. F. Wagner, G. Becker, K. Behringer, D. Campbell, M. Keilhacker et al.: Plasma Physics and Controlled Nuclear Fusion Research **1**, 43(1983) (Conference Proceedings, Baltimore 1982) IAEA Vienna
M. Keilhacker et al.: Plasma Phys. and Controlled Fusion **28**, 29(1986)

39. T. Ohkawa and F. L. Hinton: Plasma Phys. and Controlled Nuclear Fusion Research **2**, 221(1987) (Conference Proceedings, Kyoto 1986) IAEA, Vienna
H. P. Furth: Plasma Physics and Controlled Fusion **28**, 1305(1986)

40. T. Ohkawa: Nucl. Fusion **10**, 185(1970)

41. D. J. H. Wort: Plasma Phys. **13**, 258(1971)

42. N. J. Fisch: Phys. Rev. Lett. **41**, 873 (1978); C. F. F. Karney and N. J. Fisch: Phys. Fluids **22**, 1817(1979)

43. N. J. Fisch and A. H. Boozer: Phys. Rev. Lett. **45**, 720(1980)

44. D. F. H. Start, J. G. Cordey, and E. M. Jones: Plasma Phys. **22**, 303(1980)

45. W. H. M. Clark, J. G. Cordey, M. Cox, R. D. Gill, J. Hugill, J. W. M. Paul, and D. F. H. Start: Phys. Rev. Lett. **45**, 1101(1980)

46. R. J. Hawryluk, V. Arunasalam, M. G. Bell, M. Bitter, W. R. Blanchard et al.: Plasma Physics and Controlled Nuclear Fusion Research **1**, 51(1987) (Conference Proceedings, Kyoto 1986) IAEA, Vienna

47. R. J. Bickerton, J. W. Connor, J. B. Taylor: Nature Physical Science **229**, 110(1971)

48. A. A. Galeev: Sov. Phys. JETP **32**, 752(1971)

49. M. N. Rosenbluth, R. D. Hazeltine, F. L. Hinton: Phys. Fluids **15**, 116(1972)

50. D. J. Sigmar: Nucl. Fusion **13**, 17(1973)

51. International Tokamak Reactor, Zero Phase 1980, Phase One 1982, Phase Two A Part 1, 1984, Phase Two A Part 2, 1986, International Atomic Energy Agency, Vienna.

52. D. C. Robinson and R. E. King: Plasma Physics and Controlled Nuclear Fusion Research **1**, 263(1969) (Conference Proceedings, Novosibirsk 1968) IAEA Vienna

53. H. A. B. Bodin, A. A. Newton: Nucl. Fusion **20**, 1255(1980)

54. H. A. B. Bodin: Plasma Physics and Controlled Fusion **29**, 1297(1987)

55. K. Miyamoto: Plasma Physics and Controlled Fusion **30** (1988)

56. J. B. Taylor: Phys. Rev. Lett. **33**, 1139(1974)
Proc. 3rd Topical Conf. on Pulsed High Beta Plasma, p. 59(1975)

57. K. Hain and R. Lüst: Z. Naturforsch. **13a**, 936(1958)

58. K. Matsuoka and K. Miyamoto: Jpn. J. Appl. Phys. **18**, 817(1979)

59. D. A. Baker, M. D. Bausman, C. J. Buchenauer, L. C. Burkhardt, G. Chandler, J. N. Dimarco et al.: Plasma Physics and Controlled Nuclear Fusion Research **1**, 587(1983) (Conference Proceedings, Baltimore 1982) IAEA Vienna

60. D. D. Schnack, E. J. Caramana and R. A. Nebel: Phys. Fluids **28**, 321(1985)

61. K. Kusano, T. Sato: Nucl. Fusion **26**, 1051(1986), ibid **27**, 821(1987)

62. M. K. Bevir and J. W. Gray: Proc. of Reversed Field Pinch Theory Workshop (LANL Los Alamos, 1981) Report No-8944-C p. 176, M. K. Bevir and C. G. Gimblett: Phys. Fluids **28**, 1826(1985)

63. L. Spitzer, Jr.: Phys. Fluids **1**, 253(1958)

64. W VII-A Team: Plasma Physics and Controlled Nuclear Fusion Research **2**, 241(1983) (Conference Proceedings, Baltimore 1982) IAEA Vienna

65. E. D. Andryukhina, G. M. Batanov, M. S. Berezhetshij, M. A. Blokh, G. S. Voronov et al.: Plasma Physics and Controlled Nuclear Fusion Research **2**, 409(1985) (Conference Proceedings, London 1984) IAEA Vienna

66. K. Uo, A. Iiyoshi, T. Obiki, O. Motojima, S. Morimoto, N. Wakatani et al.: Plasma Physics and Controlled Nuclear Fusion Research **2**, 383(1985) (Conference Proceedings, London 1984) IAEA Vienna

67. L. Garcia, B. A. Carreras, J. H. Harris: Nucl. Fusion **24**, 115(1984)

68. Yu. N. Petrenlco and A. P. Popyyadukhin: The 3rd International Symposium on Toroidal Plasma Confinements, D 8 March 1973 Garching

69. H. Wobig and S. Rehker: Proceedings of the 7th Symposium on Fusion Technology, Grenoble 345(1972)
S. Rehker, H. Wobig: Proceedings of the 6th European Conference on Controlled Fusion and Plasma Physics p. 117 in Moscow (1973), IPP 2/215 Max Planck Inst. of Plasma Phys. (1973)

70. C. Gourdon, D. Marty, E. K. Maschke, J. P. Dumont: Plasma Physics and Controlled Nuclear Fusion Research **1**, 847(1969) (Conference Proceedings, Novosibirsk 1968) IAEA Vienna

71. K. Uo: Plasma Phys. **18**, 243(1971)

72. A. Mohri: J. Phys. Soc. Japan **28**, 1549(1970)

73. C. Gourdon, D. Marty, E. K. Maschke, J. Touche: Nucl. Fusion **11**, 161(1971)

74. J. A. Derr and J. L. Shohet: Phys. Rev. Lett. **43**, 1730(1979)

75. M. Wakatani, S. Kodama, M. Nakasuga, K. Hanatani: Nucl. Fusion **21**, 175(1981)

76. B. B. Kadomtsev and O. P. Pogutse: Nucl. Fusion **11**, 67(1971)

77. J. W. Connor and R. J. Hastie: Phys. Fluids **17**, 114(1974)

78. L. M. Kovrizhnykh: Nucl. Fusion **24**, 851(1984)

79. D. E. Hastings, W. A. Houlberg, and K. C. Shaing: Nucl. Fusion **25**, 445(1985)

80. K. Miyamoto: Nucl. Fusion **18**, 243(1978)
H. Wobig: Plasma Physics and Controlled Fusion **29**, 1389(1987)

81. G. Grieger, W. Ohlendorf, H. D. Pacher, H. Wobig, G. H. Wolf: Plasma Physics and Controlled Nuclear Fusion Research **3**, 37(1971) (Conference Proceedings, Madison 1971) IAEA Vienna

82. K. Miyamoto, M. Fujiwara, K. Kawahata, Y. Terashima, T. Dodo, K. Yatsu: Plasma Physics and Controlled Nuclear Fusion Research **2**, 115(1975) (Conference Proceedings, Tokyo 1974) IAEA Vienna

83. I. J. Spalding: Nucl. Fusion **8**, 161(1968)

84. I. J. Spalding: Advances in Plasma Phys. **4**, 79, ed. by A. Simon and W. B. Thompson, Interscience, New York 1971

85. Yu. V. Gott, M. S. Ioffe, V. G. Tel'kovskii: Nucl. Fusion Suppl. Pt. 3, 1045(1962) (Conference Proceedings, Salzburg 1961) IAEA, Vienna
Yu. T. Baiborodov, M. S. Ioffe, V. M. Petrov and R. I. Sobolev: Sov. Atomic Energy **14**, 459(1963)

86. Yu. T. Baiborodov, M. S. Ioffe, B. I. Kanaev, R. I. Sobolev, E. E. Yushmanov: Plasma Physics and Controlled Nuclear Fusion Research **2**, 697(1971) (Conference Proceedings, Madison 1971) IAEA, Vienna

87. F. H. Coensgen, W. F. Cummins. V. A. Finlayson, W. E. Nexsen, Jr., T. C. Simonen: ibid. **2**, 721(1971)

88. F. H. Coensgen, W. F. Cummins, B. G. Logan, A. W. Halvik, W. E. Nexsen, T. C. Simonen, B. W. Stallard, and W. C. Turne: Phys. Rev. Lett **35**, 1501(1975)

89. T. K. Fowler, ed.: Nucl. Fusion **9**, 3(1969)
R. F. Post: Nucl. Fusion **27**, 1579(1987)

90. M. S. Ioffe and B. B. Kadomtsev: Soviet Physics Uspekhi **13**, 225(1970)

91. E. G. Harris: Phys. Rev. Lett. **2**, 34(1959)

92. R. A. Dory, G. E. Guest, E. G. Harris: Phys. Rev. Lett. **14**, 131(1965)

93. G. E. Guest, R. A. Dory: Phys. Fluids **8**, 1853(1965)

94. J. Gordey, G. Kuo-Petravic, E. Murphy, M. Petravie, D. Sweetman, and E. Thompson: Plasma Physics and Controlled Nuclear Fusion Research **2**, 267(1969) (Conference Proceedings, Novosibirsk 1968) IAEA, Vienna

95. R. F. Post and M. N. Rosenbluth: Phys. Fluids **9**, 730(1966)

96. H. Postma, H. Dunlap, R. Dory, G. Haste, and R. Young: Phys. Rev. Lett. **16**, 265(1966)

97. B. B. Kadomtsev and O. P. Pogutse: Plasma Physics and Controlled Nuclear Fusion Research **2**, 125(1969) (Conference Proceedings, Novosibirsk 1968) IAEA, Vienna

98. R. Z. Sagdeev and V. D. Shafranov: Sov. Phys. JETP **12**, 130(1961)

99. J. Sharer and A. Trivelpiece: Phys. Fluids **10**, 591(1967)
J. Sharer: Phys. Fluids **10**, 652(1967)

100. H. Ikegami, H. Ikezi, T. Kawamura, H. Momota, K. Takayama, and Y. Terashima: Plasma Physics and Controlled Nuclear Fusion Research **2**, 423(1969) (Conference Proceedings, Novosivirsk 1968) IAEA, Vienna

101. G. I. Dimov, V. V. Zakaidakov, and M. E. Kishinevskii: Sov. J. Plasma Phys. **2**, 326(1976)

102. T. K. Fowler and B. G. Logan: Comments Plasma Phys. Controlled Fusion Res. **2**, 167(1977)

103. V. P. Pastukhov: Nucl. Fusion **14**, 3(1974)

104. D. E. Baldwin and B. G. Logan: Phys. Rev. Lett. **43**, 1318(1979)

105. R. H. Cohen, I. B. Bernstein, J. J. Dorning, G. Rowland: Nucl. Fusion **20**, 1421(1980)

106. T. H. Stix: The Theory of Plasma Waves, McGraw Hill, New York (1962)

107. M. Porkolab: Fusion vol. 1, Part B, p. 151 (ed. by E. Teller) Academic Press (1981)

108. T. H. Stix: Nucl. Fusion **15**, 737(1975)

109. J. Adam and A. Samain: Report EUR-CEA-FC 579(1971) Association Euratom-CEA, Fontenay-aux-Roses

110. J. E. Scharer, B. D. McVey, T. K. Mau: Nucl. Fusion **17**, 297(1977)

111. Y. Lapierre: 2nd Joint Grenoble-Varenna International Symposium on Heating in Toroidal Plasmas, Como (1980)
A. Fukuyama, S. Nishiyama, K. Itoh, S. I. Itoh: Nucl. Fusion **23**, 1005(1983)

112. M. Ono, T. Watari, R. Ando, J. Fujita et al.: Phys. Rev. Lett. **54**, 2339(1985)

113. T. H. Stix: Phys. Rev. Lett. **15**, 878(1965)

114. V. M. Glagolev: Plasma Phys. **14**, 301, 315(1972)

115. M. Brambilla: Plasma Phys. **18**, 669(1976)

116. S. Bernabei, M. A. Heald, W. M. Hooke, R. W. Motley, F. J. Paoloni, M. Brambilla, W. D. Getty: Nucl. Fusion **17**, 929(1977)

117. H. Takamura: Fundamentals of Plasma Heating, Nagoya Univ. Press, 1986 (in Japanese)

118. I. Fidone, G. Granata, and G. Ramponi: Phys. Fluids **21**, 645(1978)

119. A. G. Litvak, G. V. Permitin, E. V. Suvorov, A. A. Frajman: Nucl. Fusion **17**, 659 (1977)

120. K. A. Brueckner and S. Jorna: Rev. of Modern Phys. **46**, 325(1974)

121. J. L. Emmett, J. Nuckolls, and L. Wood: Scientific American **230**, No. 6 p. 2 (1974)

122. C. Yamanaka: Plasma Physics and Controlled Nuclear Fusion Research **3**, 473(1983) (Conference Proceedings, Baltimore 1982) IAEA Vienna

123. R. Decoste, S. E. Bodner, B. H. Ripin, E. A. McLean, S. P. Obenshain, and C. M. Armstrong: Phys. Rev. Lett. **42**, 1673(1979)

124. K. Mima: Kakuyuko Kenkyu **51**, 400(1984) (in Japanese)

125. C. Yamanaka, H. Azechi, E. Fujiwara, S. Ido, Y. Izawa et al.: Plasma Physics and Controlled Nuclear Fusion Research **3**, 3(1985) (Conference Proceedings, London 1984) IAEA Vienna

126. J. H. Nuckolls: Physics Today **35**, Sept. 25(1982)

127. (added in proof). In 1988 INTOR was succeeded by ITER (International Thermonuclear Experimental Reactor) on the basis of recent experimental results, especially of confinement scaling. One reference parameter set under consideration is: $R = 5.5$ m, $a = 1.8$ m ($R/a = 3.1$), $\kappa = 2.0$, $q_\psi = 3.1$, $B_t = 5.3$ T, $I_p = 18$ MA, $\langle T_e \rangle \approx \langle T_i \rangle \approx 10$ keV, $\langle n \rangle \approx 1.2 \times 10^{14}$ cm^{-3}, $\langle \beta \rangle \approx 4\%$, $\tau_E \approx 2.5$ s. See Plasma Physics and Controlled Nuclear Fusion Research IAEA-CN-50/F-I-4, F-II-1,2,3,4 (Conference Proceedings, Nice 1988) IAEA Vienna.

Appendix 1 Vector Relations

$$\boldsymbol{a}\cdot(\boldsymbol{b}\times\boldsymbol{c}) = \boldsymbol{b}\cdot(\boldsymbol{c}\times\boldsymbol{a}) = \boldsymbol{c}\cdot(\boldsymbol{a}\times\boldsymbol{b})$$

$$\boldsymbol{a}\times(\boldsymbol{b}\times\boldsymbol{c}) = (\boldsymbol{a}\cdot\boldsymbol{c})\boldsymbol{b} - (\boldsymbol{a}\cdot\boldsymbol{b})\boldsymbol{c}$$

$$\begin{aligned}(\boldsymbol{a}\times\boldsymbol{b})\cdot(\boldsymbol{c}\times\boldsymbol{d}) &= \boldsymbol{a}\cdot\boldsymbol{b}\times(\boldsymbol{c}\times\boldsymbol{d}) = \boldsymbol{a}\cdot[(\boldsymbol{b}\cdot\boldsymbol{d})\boldsymbol{c} - (\boldsymbol{b}\cdot\boldsymbol{c})\boldsymbol{d}] \\ &= (\boldsymbol{a}\cdot\boldsymbol{c})(\boldsymbol{b}\cdot\boldsymbol{d}) - (\boldsymbol{a}\cdot\boldsymbol{d})(\boldsymbol{b}\cdot\boldsymbol{c})\end{aligned}$$

$$(\boldsymbol{a}\times\boldsymbol{b})\times(\boldsymbol{c}\times\boldsymbol{d}) = (\boldsymbol{a}\times\boldsymbol{b}\cdot\boldsymbol{d})\boldsymbol{c} - (\boldsymbol{a}\times\boldsymbol{b}\cdot\boldsymbol{c})\boldsymbol{d}$$

$$\nabla\cdot(\phi\boldsymbol{a}) = \phi\nabla\cdot\boldsymbol{a} + (\boldsymbol{a}\cdot\nabla)\phi$$

$$\nabla\times(\phi\boldsymbol{a}) = \nabla\phi\times\boldsymbol{a} + \phi\nabla\times\boldsymbol{a}$$

$$\nabla(\boldsymbol{a}\cdot\boldsymbol{b}) = (\boldsymbol{a}\cdot\nabla)\boldsymbol{b} + (\boldsymbol{b}\cdot\nabla)\boldsymbol{a} + \boldsymbol{a}\times(\nabla\times\boldsymbol{b}) + \boldsymbol{b}\times(\nabla\times\boldsymbol{a})$$

$$\nabla\cdot(\boldsymbol{a}\times\boldsymbol{b}) = \boldsymbol{b}\cdot\nabla\times\boldsymbol{a} - \boldsymbol{a}\cdot\nabla\times\boldsymbol{b}$$

$$\nabla\times(\boldsymbol{a}\times\boldsymbol{b}) = \boldsymbol{a}(\nabla\cdot\boldsymbol{b}) - \boldsymbol{b}(\nabla\cdot\boldsymbol{a}) + (\boldsymbol{b}\cdot\nabla)\boldsymbol{a} - (\boldsymbol{a}\cdot\nabla)\boldsymbol{b}$$

$$\nabla\times\nabla\times\boldsymbol{a} = \nabla(\nabla\cdot\boldsymbol{a}) - \nabla^2\boldsymbol{a}$$

$$\nabla\times\nabla\phi = 0$$

$$\nabla\cdot(\nabla\times\boldsymbol{a}) = 0$$

$$\boldsymbol{r} = x\boldsymbol{i} + y\boldsymbol{j} + z\boldsymbol{k} \quad \nabla\cdot\boldsymbol{r} = 3 \quad \nabla\times\boldsymbol{r} = 0$$

Volume integral V enclosed by the closed surface S

$$\int_V \nabla\phi\cdot\mathrm{d}V = \int_S \phi\boldsymbol{n}\mathrm{d}a$$

$$\int_V \nabla\cdot\boldsymbol{a}\mathrm{d}V = \int_S \boldsymbol{a}\cdot\boldsymbol{n}\mathrm{d}a$$

$$\int_V \nabla\times\boldsymbol{a}\mathrm{d}V = \int_S \boldsymbol{n}\times\boldsymbol{a}\mathrm{d}a$$

Surface integral S enclosed by the closed curve C

$$\int_S \boldsymbol{n}\times\nabla\phi\mathrm{d}a = \oint_C \phi\mathrm{d}\boldsymbol{s}$$

$$\int_S \nabla\times\boldsymbol{a}\cdot\boldsymbol{n}\mathrm{d}a = \oint_C \boldsymbol{a}\mathrm{d}\boldsymbol{s}$$

Appendix 2 Differential Operators

Curvilinear coordinates $x = \varphi_1(u^1, u^2, u^3)$, $y = \varphi_2(u^1, u^2, u^3)$, $z = \varphi_3(u^1, u^2, u^3)$

$$\mathrm{d}\boldsymbol{r} = \sum_j \frac{\partial \boldsymbol{r}}{\partial u^j} \mathrm{d}u^j$$

$$\boldsymbol{a}_j \equiv \frac{\partial \boldsymbol{r}}{\partial u^j}$$

$$\mathrm{d}\boldsymbol{r} = \sum_j \mathrm{d}u^j \boldsymbol{a}_j$$

$$V \equiv \boldsymbol{a}_1 \cdot (\boldsymbol{a}_2 \times \boldsymbol{a}_3) = \boldsymbol{a}_2 \cdot (\boldsymbol{a}_3 \times \boldsymbol{a}_1) = \boldsymbol{a}_3 \cdot (\boldsymbol{a}_1 \times \boldsymbol{a}_2)$$

$$\boldsymbol{a}^1 \equiv \frac{1}{V}(\boldsymbol{a}_2 \times \boldsymbol{a}_3) \quad \boldsymbol{a}^2 \equiv \frac{1}{V}(\boldsymbol{a}_3 \times \boldsymbol{a}_1) \quad \boldsymbol{a}^3 \equiv \frac{1}{V}(\boldsymbol{a}_1 \times \boldsymbol{a}_2)$$

$$V^{-1} = \boldsymbol{a}^1 \cdot (\boldsymbol{a}^2 \times \boldsymbol{a}^3) = \boldsymbol{a}^2 \cdot (\boldsymbol{a}^3 \times \boldsymbol{a}^1) = \boldsymbol{a}^3 \cdot (\boldsymbol{a}^1 \times \boldsymbol{a}^2)$$

$$\boldsymbol{a}_1 = V(\boldsymbol{a}^2 \times \boldsymbol{a}^3) \quad \boldsymbol{a}_2 = V(\boldsymbol{a}^3 \times \boldsymbol{a}^1) \quad \boldsymbol{a}_3 = V(\boldsymbol{a}^1 \times \boldsymbol{a}^2)$$

$$\boldsymbol{a}^i \cdot \boldsymbol{a}_j = \delta_{ij}$$

$$g_{ij} \equiv \boldsymbol{a}_i \cdot \boldsymbol{a}_j = g_{ji} \quad g^{ij} \equiv \boldsymbol{a}^i \cdot \boldsymbol{a}^j = g^{ji}$$

$$\boldsymbol{F} = \sum f^i \boldsymbol{a}_i \quad f^i = \boldsymbol{F} \cdot \boldsymbol{a}^i \quad \text{Contravariant}$$

$$\boldsymbol{F} = \sum f_j \boldsymbol{a}^j \quad f_j = \boldsymbol{F} \cdot \boldsymbol{a}_j \quad \text{Covariant}$$

$$f_j = \sum_i g_{ji} f^i \quad f^i = \sum_j g^{ij} f_j$$

$$\mathrm{d}s^2 = (\mathrm{d}\boldsymbol{r})^2 = \sum_{ij} g_{ij} \mathrm{d}u^i \mathrm{d}u^j = \sum_{ij} g^{ij} \mathrm{d}u_i \mathrm{d}u_j$$

$$g \equiv |g_{ik}| = V^2, \quad \boldsymbol{a}^i = \nabla u^i$$

$$(\boldsymbol{a} \times \boldsymbol{b})^1 = g^{-1/2}(a_2 b_3 - a_3 b_2)$$

$$(\boldsymbol{a} \times \boldsymbol{b})_1 = g^{1/2}(a^2 b^3 - a^3 b^2)$$

$$\therefore \nabla\phi = \sum_i \frac{\partial \phi}{\partial u^i} \boldsymbol{a}^i$$

$$\nabla \cdot \boldsymbol{F} = \frac{1}{g^{1/2}} \sum \frac{\partial}{\partial u^i}(g^{1/2} f^i)$$

$$\nabla \times \boldsymbol{F} = \frac{1}{g^{1/2}}\left(\left(\frac{\partial f_3}{\partial u^2} - \frac{\partial f_2}{\partial u^3}\right)\boldsymbol{a}_1 + \left(\frac{\partial f_1}{\partial u^3} - \frac{\partial f_3}{\partial u^1}\right)\boldsymbol{a}_2 + \left(\frac{\partial f_2}{\partial u^1} - \frac{\partial f_1}{\partial u^2}\right)\boldsymbol{a}_3\right)$$

$$\nabla^2\phi = \frac{1}{g^{1/2}}\sum_{ij}\frac{\partial}{\partial u^i}\left(g^{ij}g^{1/2}\frac{\partial\phi}{\partial u^j}\right)$$

Orthogonal coordinates u^1, u^2, u^3

$$ds^2 = h_1{}^2(du^1)^2 + h_2{}^2(du^2)^2 + h_3{}^2(du^3)^2,\ g^{1/2} = h_1h_2h_3$$

$$\nabla\phi = \sum_j \frac{1}{h_j}\frac{\partial\phi}{\partial u^j}\boldsymbol{i}_j$$

$$\nabla\cdot\boldsymbol{F} = \frac{1}{h_1h_2h_3}\left(\frac{\partial}{\partial u^1}(h_2h_3F_1) + \frac{\partial}{\partial u^2}(h_3h_1F_2) + \frac{\partial}{\partial u^3}(h_1h_2F_3)\right)$$

$$\nabla\times\boldsymbol{F} = \frac{1}{h_2h_3}\left(\frac{\partial}{\partial u^2}(h_3F_3) - \frac{\partial}{\partial u^3}(h_2F_2)\right)\boldsymbol{i}_1 + \frac{1}{h_3h_1}\left(\frac{\partial}{\partial u^3}(h_1F_1) - \frac{\partial}{\partial u^1}(h_3F_3)\right)\boldsymbol{i}_2 + \frac{1}{h_1h_2}\left(\frac{\partial}{\partial u^1}(h_2F_2) - \frac{\partial}{\partial u^2}(h_1F_1)\right)\boldsymbol{i}_3$$

$$= \frac{1}{h_1h_2h_3}\begin{vmatrix} h_1\boldsymbol{i}_1 & h_2\boldsymbol{i}_2 & h_3\boldsymbol{i}_3 \\ \dfrac{\partial}{\partial u^1} & \dfrac{\partial}{\partial u^2} & \dfrac{\partial}{\partial u^3} \\ h_1F_1 & h_2F_2 & h_3F_3 \end{vmatrix}$$

$$\nabla^2\phi = \frac{1}{h_1h_2h_3}\left(\frac{\partial}{\partial u^1}\left(\frac{h_2h_3}{h_1}\frac{\partial\phi}{\partial u^1}\right) + \frac{\partial}{\partial u^2}\left(\frac{h_3h_1}{h_2}\frac{\partial\phi}{\partial u^2}\right) + \frac{\partial}{\partial u^3}\left(\frac{h_1h_2}{h_3}\frac{\partial\phi}{\partial u^3}\right)\right)$$

Cylindrical coordinates r, θ, z

$$ds^2 = dr^2 + r^2d\theta^2 + dz^2$$

$$\nabla\psi = \frac{\partial\psi}{\partial r}\boldsymbol{i}_1 + \frac{1}{r}\frac{\partial\psi}{\partial\theta}\boldsymbol{i}_2 + \frac{\partial\psi}{\partial z}\boldsymbol{i}_3$$

$$\nabla\cdot\boldsymbol{F} = \frac{1}{r}\frac{\partial}{\partial r}(rF_1) + \frac{1}{r}\frac{\partial F_2}{\partial\theta} + \frac{\partial F_3}{\partial z}$$

$$\nabla\times\boldsymbol{F} = \left(\frac{1}{r}\frac{\partial F_3}{\partial\theta} - \frac{\partial F_2}{\partial z}\right)\boldsymbol{i}_1 + \left(\frac{\partial F_1}{\partial z} - \frac{\partial F_3}{\partial r}\right)\boldsymbol{i}_2 + \left(\frac{1}{r}\frac{\partial}{\partial r}(rF_2) - \frac{1}{r}\frac{\partial F_1}{\partial\theta}\right)\boldsymbol{i}_3$$

$$\nabla^2\psi = \frac{1}{r}\frac{\partial}{\partial r}\left(r\frac{\partial\psi}{\partial r}\right) + \frac{1}{r^2}\frac{\partial^2\psi}{\partial\theta^2} + \frac{\partial^2\psi}{\partial z^2}$$

Spherical coordinates r, θ, ϕ

$$ds^2 = dr^2 + r^2 d\theta^2 + r^2 \sin^2\theta \, d\phi^2$$

$$\nabla\psi = \frac{\partial\psi}{\partial r}\boldsymbol{i}_1 + \frac{1}{r}\frac{\partial\psi}{\partial\theta}\boldsymbol{i}_2 + \frac{1}{r\sin\theta}\frac{\partial\psi}{\partial\phi}\boldsymbol{i}_3$$

$$\nabla\cdot\boldsymbol{F} = \frac{1}{r^2}\frac{\partial}{\partial r}(r^2F_1) + \frac{1}{r\sin\theta}\frac{\partial}{\partial\theta}(\sin\theta F_2) + \frac{1}{r\sin\theta}\frac{\partial F_3}{\partial\phi}$$

$$\nabla\times\boldsymbol{F} = \frac{1}{r\sin\theta}\left(\frac{\partial}{\partial\theta}(\sin\theta F_3) - \frac{\partial F_2}{\partial\phi}\right)\boldsymbol{i}_1 + \frac{1}{r}\left(\frac{1}{\sin\theta}\frac{\partial F_1}{\partial\phi} - \frac{\partial}{\partial r}(rF_3)\right)\boldsymbol{i}_2 + \frac{1}{r}\left(\frac{\partial}{\partial r}(rF_2) - \frac{\partial F_1}{\partial\theta}\right)\boldsymbol{i}_3$$

$$\nabla^2\psi = \frac{1}{r^2}\frac{\partial}{\partial r}\left(r^2\frac{\partial\psi}{\partial r}\right) + \frac{1}{r^2\sin\theta}\frac{\partial}{\partial\theta}\left(\sin\theta\frac{\partial\psi}{\partial\theta}\right) + \frac{1}{r^2\sin^2\theta}\frac{\partial^2\psi}{\partial\phi^2}$$

Toroidal coordinates ρ, ω, φ

$$r = \frac{R_0\sinh\rho}{\cosh\rho - \cos\omega} \qquad z = \frac{R_0\sin\omega}{\cosh\rho - \cos\omega} \qquad \varphi = \varphi$$

(r, φ, z are cylindrical coordinates)

$$ds^2 = \frac{R_0^{\,2}}{(\cosh\rho - \cos\omega)^2}(d\rho^2 + d\omega^2 + \sinh^2\rho(d\varphi)^2)$$

$$\nabla\phi = \frac{(\cosh\rho - \cos\omega)}{R_0}\left(\frac{\partial\phi}{\partial\rho}\boldsymbol{i}_\rho + \frac{\partial\phi}{\partial\omega}\boldsymbol{i}_\omega + \frac{1}{\sinh\rho}\frac{\partial\phi}{\partial\varphi}\boldsymbol{i}_\phi\right)$$

$$\nabla\cdot\boldsymbol{F} = \frac{(\cosh\rho - \cos\omega)^3}{R_0\sinh\rho}\left[\frac{\partial}{\partial\rho}\left(\frac{\sinh\rho F_1}{(\cosh\rho - \cos\omega)^2}\right) + \sinh\rho\frac{\partial}{\partial\omega}\left(\frac{F_2}{(\cosh\rho - \cos\omega)^2}\right) + \frac{1}{(\cosh\rho - \cos\omega)^2}\frac{\partial F_3}{\partial\varphi}\right]$$

$$\nabla \times \boldsymbol{F} = \frac{(\cosh\rho - \cos\omega)^2}{R_0 \sinh\rho} \times \begin{vmatrix} \boldsymbol{i}_\rho & \boldsymbol{i}_\omega & \sinh\rho \boldsymbol{i}_\varphi \\ \dfrac{\partial}{\partial\rho} & \dfrac{\partial}{\partial\omega} & \dfrac{\partial}{\partial\varphi} \\ \dfrac{F_1}{(\cosh\rho - \cos\omega)} & \dfrac{F_2}{(\cosh\rho - \cos\omega)} & \dfrac{\sinh\rho F_3}{(\cosh\rho - \cos\omega)} \end{vmatrix}$$

$$\nabla^2\phi = \frac{(\cosh\rho - \cos\omega)^3}{R_0 \sinh\rho}\left[\frac{\partial}{\partial\rho}\left(\frac{\sinh\rho}{(\cosh\rho - \cos\omega)}\frac{\partial\phi}{\partial\rho}\right) + \frac{\partial}{\partial\omega}\left(\frac{\sinh\rho}{(\cosh\rho - \cos\omega)}\frac{\partial\phi}{\partial\omega}\right) + \frac{\partial^2\phi/\partial^2\varphi}{(\cosh\rho - \cos\omega)\sinh\rho}\right]$$

Appendix 3 Physical Constants

Speed of light in vacuum	2.99793×10^{8} m/s
Dielectric constant of vacuum	8.854×10^{-12} F/m$(=(4\pi \times 10^{-7} c^{2})^{-1})$
Permeability of vacuum	1.257×10^{-6} H/m$(= 4\pi \times 10^{-7})$
Planck's constant	6.625×10^{-34} J-s
Boltzmann's constant	1.3804×10^{-23} J/K
Avogadro's number	6.025×10^{23}/mol (760 torr, 0°C, 22.41)
Gas density at 0°C and 1 torr	3.53×10^{22} m^{-3}
Charge of the electron	1.6021×10^{-19} C
1 electron volt	1.6021×10^{-19} J
e/κ	11600 K/V
m_p/m_e	1836
$(m_p/m_e)^{1/2}$	42.9

Particle	Mass (kg $\times 10^{-27}$)
e	9.108×10^{-4} ($m_e c^2 = 0.511$ MeV)
n	1.674
H	1.673
D	3.343
T	5.006
He^{+}	6.643

Appendix 4 Formulas for Plasma Parameters

Units are mks, T_e/e in eV; $\ln \Lambda = 20$, $10^{1/2} = 3.16$

Collision time

$$\tau_{ei} = \frac{25.8\pi^{1/2}\varepsilon_0{}^2 m_e{}^{1/2} T_e{}^{3/2}}{n_e Z e^4 \ln \Lambda} = 1.6 \times 10^{-10} \frac{1}{Z}\left(\frac{T_e}{e}\right)^{3/2}\left(\frac{n_e}{10^{20}}\right)^{-1}$$

$$\tau_{ii} = \frac{25.8\pi^{1/2}\varepsilon_0{}^2 m_i{}^{1/2} T_i^{3/2}}{n_i Z^4 e^4 \ln \Lambda} = 7.0 \times 10^{-9} \frac{A^{1/2}}{Z^4}\left(\frac{T_i}{e}\right)^{3/2}\left(\frac{n_i}{10^{20}}\right)^{-1}$$

$$\tau_{ie} = \frac{(2\pi)^{1/2} 3\pi\varepsilon_0{}^2 m_i T_e{}^{3/2}}{n_e Z^2 e^4 \ln \Lambda m_e{}^{1/2}} = 1.5 \times 10^{-7} \frac{A}{Z^2}\left(\frac{T_e}{e}\right)^{3/2}\left(\frac{n_e}{10^{20}}\right)^{-1}$$

Collision frequency

$$\nu_{ei} = \tau_{ei}{}^{-1} = 6.3 \times 10^9 Z\left(\frac{T_e}{e}\right)^{-3/2}\left(\frac{n_e}{10^{20}}\right)$$

$$\nu_{ii} = \tau_{ii}{}^{-1} = 1.4 \times 10^8 \frac{Z^4}{A^{1/2}}\left(\frac{T_i}{e}\right)^{-3/2}\left(\frac{n_i}{10^{20}}\right)$$

$$\nu_{ie} = \tau_{ie}{}^{-1} = 6.7 \times 10^6 \frac{Z^2}{A}\left(\frac{T_e}{e}\right)^{-3/2}\left(\frac{n_e}{10^{20}}\right)$$

Cyclotron frequency

$$\Omega_e = \frac{eB}{m_e} = 1.76 \times 10^{11} B \qquad \frac{\Omega_e}{2\pi} = 2.80 \times 10^{10} B$$

$$-\Omega_i = \frac{ZeB}{m_i} = 9.58 \times 10^7 \frac{Z}{A} B \qquad \frac{-\Omega_i}{2\pi} = 1.52 \times 10^7 \frac{Z}{A} B$$

Plasma frequency

$$\Pi_e = \left(\frac{n_e e^2}{m_e \varepsilon_0}\right)^{1/2} = 5.64 \times 10^{11}\left(\frac{n_e}{10^{20}}\right)^{1/2} \qquad \frac{\Pi_e}{2\pi} = 8.98 \times 10^{10}\left(\frac{n_e}{10^{20}}\right)^{1/2}$$

$$\Pi_i = \left(\frac{Z^2 n_i e^2}{m_i \varepsilon_0}\right)^{1/2} = 1.32 \times 10^{10} Z\left(\frac{n_i}{A 10^{20}}\right)^{1/2}$$

$$\frac{\Pi_i}{2\pi} = 2.10 \times 10^9 Z\left(\frac{n_i}{A 10^{20}}\right)^{1/2}$$

Mean free path

$$\lambda_{ei} = \left(\frac{3T_e}{m_e}\right)^{1/2} \tau_{ei} = 1.2 \times 10^{-4} \frac{1}{Z}\left(\frac{T_e}{e}\right)^2\left(\frac{n_e}{10^{20}}\right)^{-1}$$

$$\lambda_{ii} = \left(\frac{3T_i}{m_i}\right)^{1/2} \tau_{ii} = 1.2 \times 10^{-4} \frac{1}{Z^4}\left(\frac{T_i}{e}\right)^2 \left(\frac{n_i}{10^{20}}\right)^{-1}$$

Debye length

$$\lambda_d^{\,e} = \left(\frac{\varepsilon_0 T_e}{n_e e^2}\right)^{1/2} = 7.45 \times 10^{-7}\left(\frac{T_e}{e}\right)^{1/2}\left(\frac{n_e}{10^{20}}\right)^{-1/2}$$

Larmor radius

$$\rho_B^{\,e} = \left(\frac{T_e}{m_e}\right)^{1/2}\frac{1}{\Omega_e} = \frac{(m_e T_e)^{1/2}}{eB} = 2.38 \times 10^{-6}\left(\frac{T_e}{e}\right)^{1/2}\frac{1}{B}$$

$$\rho_B^{\,i} = \left(\frac{T_i}{m_i}\right)^{1/2}\frac{1}{\Omega_i} = \frac{(m_i T_i)^{1/2}}{ZeB} = 1.02 \times 10^{-4}\frac{1}{Z}\left(\frac{AT_i}{e}\right)^{1/2}\frac{1}{B}$$

Thermal velocity

$$v_T^{\,e} = \left(\frac{T_e}{m_e}\right)^{1/2} = 4.19 \times 10^5\left(\frac{T_e}{e}\right)^{1/2}$$

$$v_T^{\,i} = \left(\frac{T_i}{m_i}\right)^{1/2} = 9.79 \times 10^3\left(\frac{T_i}{Ae}\right)^{1/2}$$

Alfvén velocity

$$v_A = \left(\frac{B^2/\mu_0}{n_i m_i}\right)^{1/2} = 2.18 \times 10^6 \frac{B}{(An_i/10^{20})^{1/2}}$$

Dimensionless parameters

$$\frac{\Omega_e^2}{\Pi_e^2} = \frac{\varepsilon_0 B^2}{(m_e n_e)} = 0.097 B^2\left(\frac{n_e}{10^{20}}\right)^{-1}$$

$$S \equiv \frac{\tau_R}{\tau_H} = \frac{\mu_0 a^2}{\eta}\cdot\frac{B}{a(\mu_0\rho)^{1/2}} = 2.6 \times 10^3 \frac{aB(T_e/e)^{3/2}}{ZA^{1/2}(n/10^{20})^{1/2}}$$

$$\beta = \frac{nT}{(B^2/2\mu_0)} = 4.03 \times 10^{-5}\frac{1}{B^2}\left(\frac{T}{e}\right)\left(\frac{n}{10^{20}}\right)$$

Specific resistivity

$$\eta_{\parallel} = \frac{Ze^2 m_e^{1/2}\ln\Lambda}{51.6\pi^{1/2}\varepsilon_0^2 T_e^{3/2}} = 5.23 \times 10^{-5} Z\ln\Lambda\left(\frac{T_e}{e}\right)^{-3/2}$$

Classical diffusion coefficient

$$D_{\text{cl}} = \frac{m_{\text{e}} T_{\text{e}}}{e^2 B^2} \nu_{\text{ei}} = 3.55 \times 10^{-2} \frac{Z}{B^2} \left(\frac{n}{10^{20}} \right) \left(\frac{T_{\text{e}}}{e} \right)^{-1/2}$$

Bohm diffusion coefficient

$$D_{\text{B}} = \frac{1}{16} \frac{T_{\text{e}}}{eB}$$

Relations between parameters

$$\eta_{\parallel} = m_{\text{e}} (\nu_{\text{ei}})_{\parallel} / (n_{\text{e}} e^2) = m_{\text{e}} \nu_{\text{ei}} / (2 n_{\text{e}} e^2)$$

$$D_{\text{cl}} = (\rho_B{}^{\text{e}})^2 \nu_{\text{ei}} = \frac{\beta_{\text{e}} \eta_{\parallel}}{\mu_0}$$

$$\frac{D_{\text{B}}}{D_{\text{cl}}} = \frac{1}{16.3^{1/2}} \frac{\lambda_{\text{ei}}}{\rho_B{}^{\text{e}}} = \frac{1}{16} \frac{\tau_{\text{ei}}}{\Omega_{\text{e}}}$$

$$\tau_{\text{ei}} = \frac{25.8 \pi^{1/2}}{\ln \Lambda \cdot Z} \frac{n_{\text{e}} (\lambda_{\text{d}}{}^{\text{e}})^3}{\Pi_{\text{e}}}$$

$$(\lambda_{\text{d}}{}^{\text{e}} \Pi_{\text{e}})^2 = \frac{T_{\text{e}}}{m_{\text{e}}} = (v_{\text{T}}{}^{\text{e}})^2$$

$$\left(\frac{v_{\text{A}}}{v_{\text{T}}{}^{\text{e}}} \right)^2 = \frac{2 m_{\text{e}}}{m_{\text{i}}} \frac{1}{\beta_{\text{e}}}$$

$$\left(\frac{v_{\text{A}}}{v_{\text{T}}{}^{\text{i}}} \right)^2 = \frac{2}{\beta_{\text{i}}}$$

$$\left(\frac{v_{\text{A}}}{c} \right)^2 = \left(\frac{\lambda_{\text{d}}{}^{\text{i}}}{\rho_B{}^{\text{i}}} \right)^2$$

$$\left(\frac{\Omega_{\text{e}}}{\Pi_{\text{e}}} \right)^2 = \left(\frac{v_{\text{A}}}{c} \right)^2 \frac{m_{\text{i}}}{m_{\text{e}}} = \left(\frac{T_{\text{e}}}{m_{\text{e}} c^2} \right) \left(\frac{2}{\beta_{\text{e}}} \right)$$

$$\left(\frac{\Omega_{\text{i}}}{\Pi_{\text{i}}} \right)^2 = \left(\frac{v_{\text{A}}}{c} \right)^2 \frac{1}{Z^2}$$

Index